Bayesian Methods

A Social and Behavioral Sciences Approach

Third Edition

Chapman & Hall/CRC
Statistics in the Social and Behavioral Sciences Series

Series Editors

Jeff Gill
Washington University, USA

Steven Heeringa
University of Michigan, USA

Wim van der Linden
CTB/McGraw-Hill, USA

J. Scott Long
Indiana University, USA

Tom Snijders
Oxford University, UK
University of Groningen, NL

Aims and scope

Large and complex datasets are becoming prevalent in the social and behavioral sciences and statistical methods are crucial for the analysis and interpretation of such data. This series aims to capture new developments in statistical methodology with particular relevance to applications in the social and behavioral sciences. It seeks to promote appropriate use of statistical, econometric and psychometric methods in these applied sciences by publishing a broad range of reference works, textbooks and handbooks.

The scope of the series is wide, including applications of statistical methodology in sociology, psychology, economics, education, marketing research, political science, criminology, public policy, demography, survey methodology and official statistics. The titles included in the series are designed to appeal to applied statisticians, as well as students, researchers and practitioners from the above disciplines. The inclusion of real examples and case studies is therefore essential.

Published Titles

Analyzing Spatial Models of Choice and Judgment with R
David A. Armstrong II, Ryan Bakker, Royce Carroll, Christopher Hare, Keith T. Poole, and Howard Rosenthal

Analysis of Multivariate Social Science Data, Second Edition
David J. Bartholomew, Fiona Steele, Irini Moustaki, and Jane I. Galbraith

Latent Markov Models for Longitudinal Data
Francesco Bartolucci, Alessio Farcomeni, and Fulvia Pennoni

Statistical Test Theory for the Behavioral Sciences
Dato N. M. de Gruijter and Leo J. Th. van der Kamp

Multivariable Modeling and Multivariate Analysis for the Behavioral Sciences
Brian S. Everitt

Multilevel Modeling Using R
W. Holmes Finch, Jocelyn E. Bolin, and Ken Kelley

Bayesian Methods: A Social and Behavioral Sciences Approach, Third Edition
Jeff Gill

Multiple Correspondence Analysis and Related Methods
Michael Greenacre and Jorg Blasius

Applied Survey Data Analysis
Steven G. Heeringa, Brady T. West, and Patricia A. Berglund

Informative Hypotheses: Theory and Practice for Behavioral and Social Scientists
Herbert Hoijtink

Generalized Structured Component Analysis: A Component-Based Approach to Structural Equation Modeling
Heungsun Hwang and Yoshio Takane

Statistical Studies of Income, Poverty and Inequality in Europe: Computing and Graphics in R Using EU-SILC
Nicholas T. Longford

Foundations of Factor Analysis, Second Edition
Stanley A. Mulaik

Linear Causal Modeling with Structural Equations
Stanley A. Mulaik

Age–Period–Cohort Models: Approaches and Analyses with Aggregate Data
Robert M. O'Brien

Handbook of International Large-Scale Assessment: Background, Technical Issues, and Methods of Data Analysis
Leslie Rutkowski, Matthias von Davier, and David Rutkowski

Generalized Linear Models for Categorical and Continuous Limited Dependent Variables
Michael Smithson and Edgar C. Merkle

Incomplete Categorical Data Design: Non-Randomized Response Techniques for Sensitive Questions in Surveys
Guo-Liang Tian and Man-Lai Tang

Computerized Multistage Testing: Theory and Applications
Duanli Yan, Alina A. von Davier, and Charles Lewis

Chapman & Hall/CRC
Statistics in the Social and Behavioral Sciences Series

Bayesian Methods

A Social and Behavioral Sciences Approach

Third Edition

Jeff Gill

Washington University
St. Louis, Missouri, USA

CRC Press
Taylor & Francis Group
Boca Raton   London   New York

CRC Press is an imprint of the
Taylor & Francis Group, an **informa** business

A CHAPMAN & HALL BOOK

CRC Press
Taylor & Francis Group
6000 Broken Sound Parkway NW, Suite 300
Boca Raton, FL 33487-2742

© 2015 by Taylor & Francis Group, LLC
CRC Press is an imprint of Taylor & Francis Group, an Informa business

No claim to original U.S. Government works

Printed on acid-free paper
Version Date: 20141105

International Standard Book Number-13: 978-1-4398-6248-3 (Hardback)

Visit the Taylor & Francis Web site at
http://www.taylorandfrancis.com

and the CRC Press Web site at
http://www.crcpress.com

This book is again dedicated to Jack Gill, who is still there when I need him.

Contents

List of Figures

List of Tables

Preface to the Third Edition

General Comments

Welcome to the third edition of BMSBSA. When the first edition appeared in 2002 Bayesian methods were still considered a bit exotic in the social sciences. Many distrusted the use of prior distributions and some mysterious simulation process that involved non-iid sampling and uncertainty about convergence. The world is completely different now, and Bayesian modeling has become standard and MCMC is well-understood and trusted. Of course it helps that Moore's Law (doubling of computing power every two years, presumably until we reach the 7 nanometer threshold) continues without pause making our computers notably faster and allowing longer sampling procedures and parallel process without an agonizingly long wait. In this context the third edition spends less time justifying procedures and more time providing implementation details on these procedures. This is also an opportunity to expand the set of examples.

Changes from the Second Edition

As expected there are a number of additions in the new edition. First, and most laboriously, the number of exercises has been doubled such that there are now twenty in each chapter. The former exercises are now the odd-numbered exercises with the answer key being fully publicly distributed (not just to instructors). The new exercises emphasize recent developments in Bayesian inference and Bayesian computing as a way to include more modern material. All of the chapters have been refreshed, although some more than others. The basic material has not changed in over a century, so there is no need to dramatically alter basic material. Conversely, Bayesian stochastic simulation (MCMC) has undergone dramatic developments in the last decade, including having become routine in applied settings. New MCMC material includes Hamiltonian Monte Carlo and expanded model implementation. Second, there are two new chapters. A chapter on Bayesian decision theory is long overdue, and this is now Chapter 8. It includes discussion of both Bayesian and frequentist decision theory since this is where the two paradigms are most intertwined. Included topics are: loss functions, risk, decision rules, and regression-model applications. The chapter finishes with two important topics not featured in previous editions: James-Stein estimation and empirical Bayes. While empirical Bayes was briefly covered in the second edition, its connection with James-Stein estimation was ignored. This section now covers the important topics in this area and provides Bayesian context. Also new is a chapter on practical implementation of MCMC methods (Chapter 11). This covers mechanical issues with the BUGS language, including calling the software from R. The goal is to provide a detailed introduction to the essential software for running Bayesian hierarchical regression models. Relatedly the chapter on hierarchical models is greatly expanded. This is an area of great applied interest right now and provides a strong motivation for the Bayesian paradigm. Finally, on a more practical side, there is a wealth of new examples and applications. These are chosen from a variety of social science disciplines and are intended to illustrate the key principles of the relevant chapter. In addition, the BaM package in R that accompanies this manuscript has been greatly expanded with new datasets and new code. This includes new procedures for calling BUGS packages from R.

Course Plans

The recommended course plans remain essentially the same as outlined in the preface to the second edition. The one critical difference is adding Chapter 11 (Implementing Bayesian Models with Markov Chain Monte Carlo) to a basic course or comprehensive course. The longer length of the text means that not all chapters are practical in a one-semester course. For a standard introductory Bayesian social science graduate course, the most succinct set of chapters are:

▷ Chapter 1: Background and Introduction

▷ Chapter 2: Specifying Bayesian Models

▷ Chapter 3: The Normal and Students'-t Models

▷ Chapter 4: The Bayesian Prior

▷ Chapter 5: The Bayesian Linear Model

▷ Chapter 10: Basics of Markov Chain Monte Carlo

▷ Chapter 11: Implementing Bayesian Models with Markov Chain Monte Carlo

▷ Chapter 12: Bayesian Hierarchical Models

▷ Chapter 14: Utilitarian Markov Chain Monte Carlo.

This assumes some knowledge of basic Monte Carlo methods of the students. Chapter 11 and Chapter 14 could also be assigned reading rather than part of lectures since the focus is on very practical concerns.

Support

As done in the last two editions of this text, there is a dedicated website provided to support readers: `http://stats.wustl.edu/BMSBSA3`. This site has software, errata, comments, and the answer key for odd-numbered exercises. All of the code is also provided in the associated R package, `BaM`, which has been substantially updated to include new code and data. Where possible `BUGS` code is included in this package. Note that in many cases the code relies on multiple R packages, as well as stand-alone software such as `JAGS` and `WinBUGS`, so changes over time may introduce incompatibilities that need to be worked out. In many cases this amounts to downloading the most recent version of some software. Relevant updates will be posted at the dedicated webpage when they come to my attention.

Acknowledgments

Since this edition has been seven years in the making, there are many people to thank. First, I appreciate the support from the Department of Political Science, the Division of Biostatistics, and the Department of Surgery (Public Health Sciences) at Washington University for being supportive home environments. This includes support from the Transdisciplinary Research on Energetics and Cancer (TREC) grant and Graham Colditz and Sarah Gehlert. I also thank the Summer School in Social Science Data Analysis at The University of Essex leadership, Thomas Plümper and Vera Troeger, since some of the datasets and software were developed in the process of teaching there. In particular, new ideas for the chapter on hierarchical models were created in the welcoming community that is Essex. The manuscript was finished while on sabbatical Spring 2014 at the University of Minnesota Division of Biostatistics (with additional financial support from the Departments of Political Science, Psychology, Sociology, and Statistics). I thank Brad Carlin, Sudipto Banerjee, John Freeman, and Niels Waller for making this happen and providing a pleasant (but really cold!) environment to wrap up this project. Finally, I acknowledge the support of NSF grants: DMS-0631632 and SES-0631588, which are my last in cooperation with George Casella. Numerous people have sent useful comments, notably Gary McDonald, Kentaro Fukumoto, Ben Begozzi, Patrick Brandt, Yuta Kamahara, Bruce Desmarais, Skyler Cranmer, Ryan Bakker, and Gary in particular was relentless in his comments (in a good way!). In the Spring of 2011 twelve graduate students at Washington University participated in a weekly reading group around the manuscript giving useful and insightful comments on methods, computing, and even the writing. I thank Peter Casey, Adriana Crespo-Tenorio, Constanza Figueroa Schibber, Morgan Hazelton, Rachael Hinkley, Chia-yi Lee, Michael Nelson, Christopher Pope, Elizabeth Rose, and Alicia Uribe. All of these students have finished their studies and gone on to successful careers. This is the first edition of this book after George Casella left us. He made a big impact on the first two issues as my neighbor, close friend, colleague, and coauthor. As Christian Robert states, George simply made everyone's work better for those fortunate to be around him. This work is certainly better because of George's influence.

Preface to the Second Edition

Starters

Wow, over five years have elapsed since the first edition appeared. Bayesian methods in the social sciences have grown and changed dramatically. This is a positive and encouraging development. When I was writing the first version I would often get questions from social science colleagues about why I would write on a seemingly obscure branch of statistics. This is clearly no longer an issue and Bayesian approaches appear to have a prominent role in social science methodology. I hope that the first edition contributed to this development.

Bayesian methods continue to become more important and central to statistical analysis, broadly speaking. Seemingly, no issue of the *Journal of the American Statistical Association* arrives without at least one Bayesian application or theoretical development. While this upward trend started in the 1990s after we discovered Markov chain Monte Carlo hiding in statistical physics, the trend accelerates in the 21st century. A nice foretelling is found in the 1999 *Science* article by David Malakoff, "Bayesian Boom," complete with anecdotes about popular uses in biology and computing as well as quotes from John Geweke. Back in 1995, the Bayesian luminary Bruno de Finetti predicted that by the year 2020 we would see a paradigm shift to Bayesian thinking (quoted in Smith [1995]). I believe we are fully on track to meet this schedule.

Bayesian computing is broader and more varied than it was at the writing of the first edition. In addition to BUGS and WinBUGS, we now routinely use MCMCpack, JAGS, openbugs, bayesm, and even the new SAS MCMC procedure. The diagnostic routines in R, BOA, and CODA continue to be useful and are more stable than they were. Of the course the lingua franca of R is critical, and many researchers use C or C++ for efficiency. Issues of statistical computing remain an important component of the book. It is also necessary to download and use the R packages CODA and BOA for MCMC diagnostics.

Bayesian approaches are also increasingly popular in related fields not directly addressed in this text. There is now an interesting literature in archaeology that is enjoyable to read (Reese 1994, Freeman 1976, Laxton *et al.* 1994), and the best starting point is the seminal paper by Litton and Buck (1995) that sets the agenda for Bayesian archaeometrics. Researchers in this area have also become frustrated with the pathetic state of the null hypothesis significance test in the social and behavioral sciences (Cowgill 1977). One area where

Bayesian modeling is particularly useful is in archaeological forensics, where researchers make adult-age estimates of early humans (Lucy *et al.* 1996, Aykroyd *et al.* 1999).

Changes from the First Edition

A reader of the first edition will notice many changes in this revision. Hopefully these constitute improvements (they certainly constituted a lot of work). First, the coverage of Markov chain Monte Carlo is greatly expanded. The reason for this is obvious, but bears mentioning. Modern applied Bayesian work is integrally tied to stochastic simulation and there are now several high-quality software alternatives for implementation. Unfortunately these solutions can be complex and the theoretical issues are often demanding. Coupling this with easy-to-use software, such as `WinBUGS` and `MCMCpack`, means that there are users who are unaware of the dangers inherent in MCMC work. I get a fair number of journal and book press manuscripts to review supporting this point. There is now a dedicated chapter on MCMC theory covering issues like ergodicity, convergence, and mixing. The last chapter is an extension of sections from the first edition that now covers in greater detail tools like: simulated annealing (including its many variants), reversible jump MCMC, and coupling from the past. Markov chain Monte Carlo research is an incredibly dynamic and fast growing literature and the need to get some of these ideas before a social science audience was strong. The reader will also note a substantial increase on MCMC examples and practical guidance. The objective is to provide detailed advice on day-to-day issues of implementation. Markov chain Monte Carlo is now discussed in detail in the first chapter, giving it the prominent position that it deserves. It is my belief that Gibbs sampling is as fundamental to estimation as maximum likelihood, but we (collectively) just do not realize it yet. Recall that there was about 40 years between Fisher's important papers and the publication of Birnbaum's Likelihood Principle. This second edition now provides a separate chapter on Bayesian linear models. Regression remains the favorite tool of quantitative social scientists, and it makes sense to focus on the associated Bayesian issues in a full chapter. Most of the questions I get by email and at conferences are about priors, reflecting sensitivity about how priors may affect final inferences. Hence, the chapter on forms of prior distributions is longer and more detailed. I have found that some forms are particularly well-suited to the type of work that social and behavioral researchers do. One of the strengths of Bayesian methods is the ease with which hierarchical models can be specified to recognize different levels and sources in the data. So there is now an expanded chapter on this topic alone, and while Chapter 12 focuses exclusively on hierarchical model specifications, these models appear throughout the text reflecting their importance in Bayesian statistics.

Additional topics have crept into this edition, and these are covered at varied levels from a basic introduction to detailed discussions. Some of these topics are older and well-known, such as Bayesian time-series, empirical Bayes, Bayesian decision theory, additional prior specifications, model checking with posterior data prediction, the deviance information criterion (DIC), methods for computing highest posterior density (HPD) intervals, convergence theory, metropolis-coupling, tempering, reversible jump MCMC, perfect sampling, software packages related to BUGS, and additional models based on normal and Student's-t assumptions.

Some new features are more structural. There is now a dedicated R package to accompany this book, BaM (for "**Bayesian Methods**"). This package includes data and code for the examples as well as a set of functions for practical purposes like calculated HPD intervals. These materials and more associated with the book are available at the dedicated Washington University website: http://stats.wustl.edu/BMSBSA. The second edition includes three appendices covering basic maximum likelihood theory, distributions, and BUGS software. These were moved to separate sections to make referencing easier and to preserve the flow of theoretical discussions. References are now contained in a single bibliography at the end for similar reasons. Some changes are more subtle. I've changed all instances of "noninformative" to "uninformative" since the first term does not really describe prior distributions. Markov chain Monte Carlo techniques are infused throughout, befitting their central role in Bayesian work. Experience has been that social science graduate students remain fairly tepid about empirical examples that focus on rats, lizards, beetles, and nuclear pumps. Furthermore, as of this writing there is no other comprehensive Bayesian text in the social sciences, outside of economics (except the out-of-print text by Phillips [1973]).

Road Map

To begin, the prerequisites remain the same. Readers will need to have a basic working knowledge of linear algebra and calculus to follow many of the sections. My math text, *Essential Mathematics for Political and Social Research* (2006), provides an overview of such material. Chapter 1 gives a brief review of the probability basics required here, but it is certainly helpful to have studied this material before. Finally, one cannot understand Bayesian modeling without knowledge of maximum likelihood theory. I recognize graduate programs differ in their emphasis on this core material, so Appendix A covers these essential ideas.

The second edition is constructed in a somewhat different fashion than the first. The most obvious difference is that the chapter on generalized linear models has been recast as an appendix, as mentioned. Now the introductory material flows directly into the construction of basic Bayesian statistical models and the procession of core ideas is not interrupted by a non-Bayesian discussion of standard models. Nonetheless, this material is important to have close at hand and hopefully the appendix approach is convenient. Another notable change is the "promotion" of linear models to their own chapter. This material is important enough to stand on its own despite the overlap with Bayesian normal and Student's-t models. Other organization changes are found in the computational section where considerable extra material has been added, both in terms of theory and practice. Markov chain Monte Carlo set the Bayesians free, and remains an extremely active research field. Keeping up with this literature is a time-consuming, but enjoyable, avocation.

There are a number of ways that a graduate course could be structured around this text. For a basic-level introductory course that emphasizes theoretical ideas, the first seven chapters provide a detailed overview without considering many computational challenges. Some of the latter chapters are directed squarely at sophisticated social scientists who have not yet explored some of the subtle theory of Markov chains. Among the possible structures, consider the following curricula.

Basic Introductory Course

▷ Chapter 1: **Background and Introduction**

▷ Chapter 2: **Specifying Bayesian Models**

▷ Chapter 3: **The Normal and Student's-t Models**

▷ Chapter 5: **The Bayesian Linear Model**

▷ Chapter 12: **Bayesian Hierarchical Models**

Thorough Course without an Emphasis on Computing

▷ Chapter 1: **Background and Introduction**

▷ Chapter 2: **Specifying Bayesian Models**

▷ Chapter 3: **The Normal and Student's-t Models**

▷ Chapter 5: **The Bayesian Linear Model**

▷ Chapter 4: **The Bayesian Prior**

▷ Chapter 6: **Assessing Model Quality**

▷ Chapter 7: **Bayesian Hypothesis Testing and the Bayes Factor**

▷ Chapter 12: **Bayesian Hierarchical Models**

A Component of a Statistical Computing Course

▷ Chapter 2: **Specifying Bayesian Models**

▷ Chapter 9: **Monte Carlo and Related Iterative Methods**

▷ Chapter 10: **Basics of Markov Chain Monte Carlo**

▷ Chapter 13: **Some Markov Chain Monte Carlo Theory**

▷ Chapter 14: **Utilitarian Markov Chain Monte Carlo**

▷ Chapter 15: **Markov Chain Monte Carlo Extensions**

A Component of an Estimation Course

▷ A: **Generalized Linear Model Review**

▷ Chapter 1: **Background and Introduction**

▷ Chapter 2: **Specifying Bayesian Models**

▷ Chapter 5: **The Bayesian Linear Model**

▷ Chapter 7: **Bayesian Hypothesis Testing and the Bayes Factor**

Of course I am eager to learn about how instructors use these chapters independent of any advice here.

Acknowledgments

So many people have commented on this edition, the previous edition, related papers, associated conference presentations, and classes taught from the book that I am unlikely to remember them all. Apologies to anyone left out from this list. This edition received three detailed, formal reviews from Patrick Brandt, Andrew Gelman, and Andrew Martin. Their comments were invaluable and dramatically improved the quality of this work.

A substantial part of the writing of this book was done during the 2006-2007 academic year while I was Visiting Professor at Harvard University's Department of Government and Fellow at the Institute for Quantitative Social Science. I thank Gary King for inviting me into that dynamic and rewarding intellectual environment. The Fall semester of that year I taught a graduate seminar titled *Bayesian Hierarchical Modeling*, which enabled me to produce and distribute chapter material on a weekly basis. Participants in the seminar provided excellent critiques of the principles and exposition. These students and guests included: Justin Grimmer, Jonathan Harris, Jennifer Katkin, Elena Llaudet, Clayton Nall, Emre Ozaltin, Lindsay Page, Omar Wasow, Lefteris Anastasopoulos, Shelden Bond, Janet Lewis, Serban Tanasa, and Lynda Zhang. The Teaching Fellows for this seminar were Skyler Cranmer and Andrew Thomas, who were instrumental in improving various technical sections. I also thank Jens Hainmueller, Dominik Hangartner, and Holger Lutz Kern for productive discussions of statistical computing during the Spring semester of that year.

Since 2000 I have taught a course based on this book at the Inter-University Consortium for Political and Social Research (ICPSR) Summer Program at the University of Michigan. Many of the highly motivated students in this program had constructive comments on the material. I also benefited immensely from interactions with colleagues and administrators in the program, including Bob Andersen, David Armstrong, Ken Bollen, Dieter Burrell, John Fox, Charles Franklin, Hank Heitowit, Bill Jacoby, Jim Johnson, Dean Lacy, Scott Long, Jim Lynch, Tim McDaniel, Sandy Schneider, Bob Stine, and Lee Walker. The three teaching assistants over this period were incredibly helpful in developing homework and computer lab assignments: Ryan Bakker, Leslie Johns, and Yu-Sung Su. Overall, ICPSR is unmatched as a high-intensity teaching and learning Summer experience.

I have tried to record those making comments on the first edition or the manuscript version of the second edition. In addition to those already mentioned, useful critiques came from: Attic Access, Larry Bartels, Neal Beck, Jack Buckley, Sid Chib, Boyd Collier, Skyler J. Cranmer, Chris Dawes, Daniel J. Denis, Persi Diaconis, Alexis Dinno, Hanni Doss, George Duncan, James Fowler, Justin Gross, Josue Guzman, Michael Herrmann, Jim Hobert, Kosuke Imai, Brad Jones, Lucas Leemann, Jane Li, Rod Little, John Londregan, Enrico Luparini, Jonathan Nagler, Shunsuke Narita, Keith T. Poole, Kevin Quinn, Rafal Raciborski, Michael Smithson, John Sprague, Rick Waterman, Bruce Western, and Chris Zorn. I also want to give a special thank you to my friend and coauthor George Casella who has provided irreplaceable brotherly guidance over the last eight years.

The research on Dirichlet process priors (Chapter 15) was supported by National Science Foundation grants DMS-0631632 and SES-0631588. My work on elicited priors (Chapter 4) was helped by research with Lee Walker that appeared in the *Journal of Politics*. The discussion of dynamic tempered transitions (Chapter 15) draws from an article written with George Casella that appeared in *Political Analysis*. Comments from the editors, Bill Jacoby and Bob Erikson, made these works better and more focused. The education policy example (Chapter 5) using a Bayesian linear model is drawn from an article I wrote with my former graduate student Kevin Wagner. The discussion of convergence in Chapter 13 benefited from my recent article on this topic in *Political Analysis*. Finally, while there is no direct overlap, my understanding of detailed statistical computing principles benefited from the book project with Micah Altman and Michael McDonald.

References

Aykroyd, Robert G., Lucy, David, Pollard, Mark A., and Roberts, Charlotte A. (1999). Nasty, Brutish, But Not Necessarily Short: A Reconsideration of the Statistical Methods Used to Calculate Age At Death From Adult Human Skeletal and Dental Age Indicators. *American Antiquity* **64**, 55-70.

Cowgill, G. L. (1977). The Trouble With Significance Tests and What We Can Do About It. *Philosophical Transactions of the Royal Society* **327**, 331-338.

Freeman, P. R. (1976). A Bayesian Approach to the Megalithic Yard. *Journal of the Royal Statistical Association, Series A* **139**, 279-295.

Gill, Jeff. (2006). *Essential Mathematics for Political and Social Research.* Cambridge, England: Cambridge University Press.

Laxton, R. R., Cavanaugh, W. G., Litton, C. D., Buck, C. E., and Blair, R. (1994).   The Bayesian Approach to Archaeological Data Analysis: An Application of Change-Point Analysis to Prehistoric Domes. *Archeologia e Calcolatori* **5**, 53-68.

Litton, C. D. and Buck, C. E. (1995).   The Bayesian Approach to the Interpretation of Archaeological Data. *Archaeometry* **37**, 1-24.

Lucy, D., Aykroyd, R. G., Pollard, A. M., and Solheim, T. (1996).   A Bayesian Approach to Adult Human Age Estimation from Dental Observations by Johanson's Age Changes. *Journal of Forensic Sciences* **41**, 189-194.

Malakoff, David. (1999). Bayes Offers a 'New' Way to Make Sense of Numbers. *Science* **286**, 1460-1464.

Phillips, L. D. (1973). *Bayesian Statistics for Social Scientists.* Thomas Nelson and Sons, London.

Reese, R. (1994). Are Bayesian Statistics Useful to Archaeological Reasoning? *Antiquity* **68**, 848-850.

Smith, A. F. M. (1995). A Conversation with Dennis Lindley. *Statistical Science* **10**, 305-319.

Preface to the First Edition

Contextual Comments

This book is intended to fill a void. There is a reasonably wide gap between the background of the median empirically trained social or behavioral scientist and the full weight of Bayesian statistical inference. This is unfortunate because, as we will see in the forthcoming chapters, there is much about the Bayesian paradigm that suits the type of data and data analysis performed in the social and behavioral sciences. Consequently, the goal herein is to bridge this gap by connecting standard maximum likelihood inference to Bayesian methods by emphasizing linkages between the standard or classical approaches and full probability modeling via Bayesian methods.

This is far from being an exclusively theoretical book. I strongly agree that "theoretical satisfaction and practical implementation are the twin ideals of coherent statistics" (Lindley 1980), and substantial attention is paid to the mechanics of putting the ideas into practice. Hopefully the extensive attention to calculation and computation basics will enable the interested readers to immediately try these procedures on their own data. Coverage of various numerical techniques from detailed posterior calculations to computational-numerical integration is extensive because these are often the topics that separate theory and realistic practice.

The treatment of theoretical topics in this work is best described as "gentle but rigorous": more mathematical derivation details than related books, but with more explanation as well. This is not an attempt to create some sort of "Bayes-Lite" or "Bayes for Dummies" (to paraphrase the popular self-help works). Instead, the objective is to provide a Bayesian methods book tailored to the interests of the social and behavioral sciences. It therefore features data that these scholars care about, focuses more on the tools that they are likely to require, and speaks in a language that is more compatible with typical prerequisites in associated departments.

There is also a substantial effort to put the development of Bayesian methods in a historical context. To many, the principles of Bayesian inference appear to be something that "came out of left field," and it is important to show that not only are the fundamentals of Bayesian statistics older than the current dominant paradigms, but that their history and development are actually closely intertwined.

Outline of the Book

This book is laid out as follows. Chapter 1 gives a high-level, brief introduction to the basic philosophy of Bayesian inference. I provide some motivations to justify the time and effort required to learn a new way of thinking about statistics through Bayesian inference. Chapter 2 (*now Appendix A*) provides the necessary technical background for going on: basic likelihood theory, the generalized linear model, and numerical estimation algorithms.

Chapter 1 describes the core idea behind Bayesian thinking: updating prior knowledge with new data to give the *posterior distribution*. Examples are used to illustrate this process and some historical notes are included. The normal model and its relatives are no less important in Bayesian statistics than in non-Bayesian statistics, and Chapter 3 outlines the key basic normal models along with extensions.

Specifying prior distributions is a key component of the Bayesian inference process and Chapter 4 goes through the typology of priors. The Bayesian paradigm has a cleaner and more introspective approach to assessing the quality of fit and robustness of researcher-specified models, and Chapter 6 outlines procedures so that one can test the performance of various models. Chapter 7 is a bit more formal about this process; it outlines a number of ways to explicitly test models against each other and to make decisions about unknown parameters.

The most modern component of this book begins with Chapter 9, which is an introduction to Monte Carlo and related methods. These topics include the many varieties of numerical integration and importance sampling, and culminating with the EM algorithm. While none of these tools are exclusively Bayesian in nature, Bayesians generally make more use of them than others. Chapter 10 formally introduces Markov chain Monte Carlo (MCMC). These are the tools that revolutionized Bayesian statistics and led to the current renaissance. This chapter includes both theoretical background on Markov chains as well as practical algorithmic details. Chapter 12 discusses hierarchical models that give the Bayesian researcher great flexibility in specifying models through a general framework. These models often lead to the requirement of MCMC techniques and the examples in this chapter are illustrated with practical computing advice. Finally, Chapter 11 (*in the first edition*) discusses necessary details about the mechanics of running and testing MCMC inferences.

The structure of each chapter is reasonably uniform. The basic ideas are enumerated early in the chapter and several of the chapters include an advanced topics section to further explore ideas that are unlikely to be of interest to every reader or to the first-time reader. All chapters have exercises that are intended to give practice developing the central ideas of each topic, including computer-based assignments.

There are unfortunately several topics that I have not had the space to cover here. Foremost is Bayesian decision theory. Many social and behavioral scientists do not operate in a data-analytic environment where the explicit cost of making a wrong decision can be quantified and incorporated into the model. This may be changing and there are a number of areas that are currently oriented toward identifying loss and risk, such as applied public policy. In the meantime, readers who are focused accordingly are directed to the books by Berger (1985), Winkler (1972), Robert (2001), and the foundational work of Wald (1950). The second major topic that is mentioned only in passing is the growing area of empirical Bayes. The best introduction is the previously noted text of Carlin and Louis (2001). See also the extensive empirical Bayes reference list in Section 12.4. I would very much have liked to cover the early, but exciting developments in perfect sampling (coupling from the past). See the original work by Propp and Wilson (1996).

Bayesian game theory is an important topic that has been omitted. Some of the better known citations are Raiffa (1982), Blackwell and Girshick (1954), Savage (1954), and Bayarri and DeGroot (1991). The Bayesian analysis of survival data as a distinct subspecialty is somewhat understudied. The recent book by Ibrahim, Chen, and Sinha (2001) goes a long way toward changing that. Chapter 10 provides the essentials for understanding Markov chains in general. The study of Markov chains extends well beyond basic MCMC and the mathematical references that I often find myself reaching for are Meyn and Tweedie (1993), Norris (1997), and Nummelin (1984). The Bayesian hierarchical models covered in Chapter 12 naturally and easily extend into meta-analysis, a subject well-covered in the social sciences by Cooper and Hedges (1994), Hunter and Schmidt (1990), and Lipsey and Wilson (2001).

Background and Prerequisites

This is not a book for a first-semester course in social and behavioral statistics. Instead, it is intended to extend the training of graduate students and researchers who have already experienced a one-year (roughly) sequence in social statistics. Therefore good prerequisites include intermediate-level, regression-oriented texts such as Fox (1997), Gujarati (1995), Hanushek and Jackson (1977), Harrell (2001), Neter *et al.* (1996), and Montgomery *et al.* (2001). Essentially it is assumed that the reader is familiar with the basics of the linear model, simple inference, multivariate specifications, and some nonlinear specifications.

A rudimentary understanding of matrix algebra is required, but this does not need to go beyond the level of Chapter 1 in Greene (2000), or any basic undergraduate text. The essential manipulations that we will use are matrix multiplication, inversion, transposition, and segmentation. The calculus operations done here are more conceptual than mechanical; that is, it is more important to understand the meaning of differentiation and integration operations rather than to be an expert on the technicalities. A knowledge at the level of Kleppner and Ramsey's (1985) self-teaching primer is sufficient to follow the calculations.

The core philosophical approach taken with regard to model specification comes from the generalized linear model construct of Nelder and Wedderburn (1972), elaborated in McCullagh and Nelder (1989). This is an integrated theoretical framework that unifies disparate model specifications by re-expressing models based on making the appropriate choice of model configuration based on the structure of the outcome variable and the nature of the dispersion. This fundamental way of thinking is independent of whether the model is Bayesian (see Dey, Ghosh, and Mallick 2000) or classical (see Fahrmeir and Tutz 2001).

Software

The concepts and procedures in this book would be of little practical value without a means of directly applying them. Consequently, there is an emphasis here on demonstrating ideas with statistical software. *All* code in R and BUGS and *all* data are posted at the dedicated webpage:

http://web.clas.ufl.edu/~jgill/BMSBSA.

A great deal of the material in this book focuses on, developing examples using the R and BUGS statistical packages. Not only are these extremely high-quality analytical tools, they are also widely distributed free of charge.

It is hard to overstate the value of the R statistical environment. R is the Open Source implementation of the S statistical language (from AT&T-Bell Labs), which has become the *de facto* standard computing language in academic statistics because of its power, flexibility, and sense of community. R was initially written by Robert Gentleman and Ross Ihak at the University of Auckland, but is now supported by a growing group of dedicated scholars. An important aspect of R is the user community itself, and the user-written packages have been shown to be an effective way for scholars to share and improve new methods.

The homesite for R (see the details in Chapter 2, *now Appendix A*), contains documentation on installation and learning the language. In addition, because R is "non-unlike" S, any published book on S-Plus will be useful. The standard text for statistical modeling in S is the work of Venables and Ripley (1999). The forthcoming book by Fox (2002) is a particularly helpful and well-written introduction to doing applied work in S. In addition, an increasing number of applied methodology books that feature the S language have appeared, and I try to keep up with these on a webpage:

http://web.clas.ufl.edu/~jgill/s-language.help.html.

Any applied Bayesian today that wants to feel good about the state of the world with regard to software need only look at Press' 1980 summary of available Bayesian analysis programs. This is a disparate, even tortured, list of mainframe-based programs that generally only implement one or two procedures each and require such pleasantries as "Raw data on

Paper tape." In contrast, the BUGS package makes Bayesian analysis using MCMC pleasant and engaging by taking the odious mechanical aspects away from the user, allowing one to focus on the more interesting aspects of model specification and testing. This unbelievable gift to the Bayesian statistical community was developed at the MRC Biostatistics Unit in Cambridge:

http://www.mrc-bsu.cam.ac.uk/bugs/.

Acknowledgments

I am indebted to many people for criticisms, suggestions, formal reviews, and influential conversations. These include Alan Agresti, Micah Altman, Attic Access, Sammy Barkin, Neal Beck, Jim Booth, Brad Carlin, George Casella, Lauren Cowles, John Fox, Wayne Francis, Charles Franklin, Malay Ghosh, Hank Heitowit, Jim Hobert, Bill Jacoby, Renee Johnson, Gary King, Andrew Martin, Michael Martinez, Mike McDonald, Ken Meier, Elias Moreno, Brad Palmquist, Kevin Quinn, Christian Robert, Stephen Sen, Jason Wittenberg, Sam Wu, Chris Zorn, and anonymous reviewers. I am especially grateful to George Casella for his continued advice and wisdom; I've learned as much sitting around George's kitchen table or trying to keep up with him running on the beach as I have in a number of past seminars. Andrew Martin also stands out for having written detailed and indispensable reviews (formally and informally) of various drafts and for passing along insightful critiques from his students.

A small part of the material in Chapter 7 was presented as a conference paper at the Fifth International Conference on Social Science Methodology in Cologne, Germany, October 2000. Also, some pieces of Chapter 10 were taken from another conference paper given at the Midwestern Political Science Association Annual Meeting, Chicago, April 2001.

This book is a whole lot better due to input from students at the University of Florida and the University of Michigan Summer Program, including: Javier Aprricio-Castello, Jorge Aragon, Jessica Archer, Sam Austin, Ryan Bakker, David Conklin, Jason Gainous, Dukhong Kim, Eduardo Leoni, Carmela Lutmar, Abdel-hameed Hamdy Nawar, Per Simonnson, Jay Tate, Tim McGraw, Nathaniel Seavy, Lee Walker, and Natasha Zharinova. For research assistance, I would also like to thank Ryan Bakker, Kelly Billingsley, Simon Robertshaw, and Nick Theobald.

I would like to also thank my editor, Bob Stern, at Chapman & Hall for making this process much more pleasant than it would have been without his continual help. This volume was appreciatively produced and delivered "camera-ready" with LATEX using the $\mathcal{AMS}$ packages, pstricks, and other cool typesetting macros and tools from the TEX world. I cannot imagine the bleakness of a life restricted to the low-technology world of word processors.

References

Bayarri, M. J. and DeGroot, M. H. (1991). What Bayesians Expect of Each Other. *Journal of the American Statistical Association* **86**, 924-932.

Berger, J. O. (1985). *Statistical Decision Theory and Bayesian Analysis.* Second Edition. New York: Springer-Verlag.

Blackwell, D. and Girshick, M. A. (1954). *Theory of Games and Statistical Decisions.* New York: Wiley.

Carlin, B. P. and Louis, T. A. (2001). *Bayes and Empirical Bayes Methods for Data Analysis.* Second Edition. New York: Chapman & Hall.

Cooper, H. and Hedges, L. (eds.) (1994). *The Handbook of Research Synthesis.* New York: Russell Sage Foundation.

Dey, D. K., Ghosh, S. K., and Mallick, B. K. (2000). *Generalized Linear Models: A Bayesian Perspective.* New York: Marcel Dekker.

Fahrmeir, L. and Tutz, G. (2001). *Multivariate Statistical Modelling Based on Generalized Linear Models.* Second Edition. New York: Springer.

Fox, J. (1997). *Applied Regression Analysis, Linear Models, and Related Methods.* Thousand Oaks, CA: Sage.

Fox, J. (2002). *An R and S-Plus Companion to Applied Regression.* Thousand Oaks, CA: Sage.

Greene, W. (2000). *Econometric Analysis.* Fourth Edition. Upper Saddle River, NJ: Prentice Hall.

Gujarati, D. N. (1995). *Basic Econometrics.* New York: McGraw-Hill.

Hanushek, E. A. and Jackson, J. E. (1977). *Statistical Methods for Social Scientists.* San Diego: Academic Press.

Harrell, F. E. (2001). *Regression Modeling Strategies: With Applications to Linear Models, Logistic Regression, and Survival Analysis.* New York: Springer-Verlag.

Hunter, J. E. and Schmidt, F. L. (1990). *Methods of Meta-Analysis: Correcting Error and Bias in Research Findings.* Thousand Oaks, CA: Sage.

Ibrahim, J. G., Chen, M-H., and Sinha, D. (2001). *Bayesian Survival Analysis.* New York: Springer-Verlag.

Kleppner, D. and Ramsey, N. (1985). *Quick Calculus: A Self-Teaching Guide.* New York: Wiley Self Teaching Guides.

Lindley, D. V. (1980). Jeffreys's Contribution to Modern Statistical Thought. In *Bayesian Analysis in Econometrics and Statistics: Essays in Honor of Harold Jeffreys.* Arnold Zellner (ed.). Amsterdam: North Holland.

Lipsey, M. W. and Wilson, D. B. (2001). *Practical Meta-Analysis.* Thousand Oaks, CA: Sage.

McCullagh, P. and Nelder, J. A. (1989). *Generalized Linear Models.* Second Edition. New York: Chapman & Hall.

Meyn, S. P. and Tweedie, R. L. (1993). *Markov Chains and Stochastic Stability.* New York: Springer-Verlag.

Montgomery, D. C. C., Peck, E. A., and Vining, G. G. (2001). *Introduction to Linear Regression Analysis.* Third Edition. New York: John Wiley & Sons.

Nelder, J. A. and Wedderburn, R. W. M. (1972). "Generalized Linear Models." *Journal of the Royal Statistical Society, Series A* **135**, 370-85.

Neter, J., Kutner, M. H., Nachtsheim, C., and Wasserman, W. (1996). *Applied Linear Regression Models.* Chicago: Irwin.

Norris, J. R. (1997). *Markov Chains.* Cambridge: Cambridge University Press.

Nummelin, E. (1984). *General Irreducible Markov Chains and Non-negative Operators.* Cambridge: Cambridge University Press.

Press, S. J. (1980). Bayesian Computer Programs. In *Studies in Bayesian Econometrics and Statistics*, S. E. Fienberg and A. Zellner, eds., Amsterdam: North Holland, pp. 429-442..

Propp, J. G. and Wilson, D. B. (1996). Exact Sampling with Coupled Markov Chains and Applications to Statistical Mechanics. *Random Structures and Algorithms* **9**, 223-252.

Raiffa, H. (1982). *The Art and Science of Negotiation.* Cambridge: Cambridge University Press.

Robert, C. P. (2001). *The Bayesian Choice: A Decision Theoretic Motivation.* Second Edition. New York: Springer-Verlag.

Savage, L. J. (1954). *The Foundations of Statistics.* New York: Wiley.

Venables, W. N. and Ripley, B. D. (1999). *Modern Applied Statistics with S-Plus*, Third Edition. New York: Springer-Verlag.

Wald, A. (1950). *Statistical Decision Functions.* New York: Wiley.

Winkler, R. L. (1972). *Introduction to Bayesian Inference and Decision.* New York: Holt, Rinehart, and Winston.

Chapter 1

Background and Introduction

1.1   Introduction

Vitriolic arguments about the merits of Bayesian versus classical approaches seem to have faded into a quaint past of which current researchers in the social sciences are, for the most part, blissfully unaware. In fact, it almost seems odd well into the 21st century that deep philosophical conflicts dominated the last century on this issue. What happened? Bayesian methods always had a natural underlying advantage because all unknown quantities are treated probabilistically, and this is the way that statisticians and applied statisticians really prefer to think. However, without the computational mechanisms that entered into the field we were stuck with models that couldn't be estimated, prior distributions (distributions that describe what we know before the data analysis) that incorporated uncomfortable assumptions, and an adherence to some bankrupt testing notions. Not surprisingly, what changed all this was a dramatic increase in computational power and major advances in the algorithms used on these machines. We now live in a world where there are very few model limitations, other than perhaps our imaginations. We therefore live in world now where researchers are for the most part comfortable specifying Bayesian and classical models as it suits their purposes.

It is no secret that Bayesian methods require a knowledge of classical methods as well as some additional material. Most of this additional material is either applied calculus or statistical computing. That is where this book comes in. The material here is intended to provide an introduction to Bayesian methods all the way from basic concepts through advanced computational material. Some readers will therefore be primarily interested in different sections. Also it means that this book is not strictly a textbook, a polemic, nor a research monograph. It is intended to be all three.

Bayesian applications in medicine, the natural sciences, engineering, and the social sciences have been increasing at a dramatic rate since the middle of the early 1990s. Interestingly, the Mars Rovers are programmed to think Bayesianly while they traverse that planet. Currently seismologists perform Bayesian updates of aftershocks based on the mainshock and previous patterns of aftershocks in the region. Bayesian networks are built in computational biology, and the forefront of quantitative research in genomics is now firmly Bayesian.

So why has there been a noticeable increase in interest in Bayesian statistics? There are actually several visible reasons. *First*, and perhaps most critically, society has radically increased its demand for statistical analysis of all kinds. A combined increase in clinical trials, statistical genetics, survey research, general political studies, economic analysis, government policy work, Internet data distribution, and marketing research have led to golden times for applied statisticians. *Second*, many introspective scholars who seriously evaluate available paradigms find that alternatives to Bayesian approaches are fraught with logical inconsistencies and shortcomings. *Third*, until recent breakthroughs in statistical computing, it was easy to specify realistic Bayesian statistical models that could not provide analytically tractable summary results.

There is therefore ample motivation to understand the basics of Bayesian statistical methodology, and not just because it is increasingly important in mainstream analysis of data. The Bayesian paradigm rests on a superior set of underlying assumptions and includes procedures that allow researchers to include reliable information in addition to the sample, to talk about findings in intuitive ways, and to set up future research in a coherent manner. At the core of the data-analytic enterprise, these are key criteria to producing useful statistical results.

Statistical analysis is the process of "data reduction" with the goal of separating out underlying systematic effects from the noise inherent in all sets of observations. Obviously there is a lot more to it than that, but the essence of what we do is using models to distill findings out of murky data. There are actually three general steps in this process: collection, analysis, and assessment. For most people, data collection is not difficult in that we live in an age where data are omnipresent. More commonly, researchers possess an abundance of data and seek meaningful patterns lurking among the various dead-ends and distractions. Armed with a substantive theory, many are asking: what should I do now? Furthermore, these same people are often frustrated when receiving multiple, possibly conflicting, answers to that question.

Suppose that there exists a model-building and data analysis process with the following desirable characteristics:

▷ overt and clear model assumptions,

▷ a principled way to make probability statements about the real quantities of theoretical interest,

▷ an ability to update these statements (i.e., learn) as new information is received,

▷ systematic incorporation of previous knowledge on the subject,

▷ missing information handled seamlessly as part of the estimation process,

▷ recognition that population quantities can be changing over time rather than forever fixed,

▷ the means to model all data types including hierarchical forms,

▷ straightforward assessment of both model quality and sensitivity to assumptions.

As the title of this book suggests, the argument presented here is that the practice of Bayesian statistics possesses all of these qualities. Press (1989) adds the following practical advantages to this list:

▷ it often results in shorter confidence/credible intervals,

▷ it often gives smaller model variance,

▷ predictions are usually better,

▷ "proper" prior distributions (Chapter 4) give models with good frequentist properties,

▷ reasonably "objective" assumptions are available,

▷ hypotheses can be tested without pre-determination of testing quality measures.

This text will argue much beyond these points and assert that the type of data social and behavioral scientists routinely encounter makes the Bayesian approach ideal in ways that traditional statistical analysis cannot match. These natural advantages include avoiding the assumption of infinite amounts of forthcoming data, recognition that fixed-point assumptions about human behavior are dubious, and a direct way to include existing expertise in a scientific field.

What reasons are there for *not* worrying about Bayesian approaches and sticking with the, perhaps more comfortable, traditional mindset? There are several reasons why a reader may not want to worry about the principles in this text for use in their research, including:

▷ their population parameters of interest are truly fixed and unchanging under all realistic circumstances,

▷ they do *not* have any prior information to add to the model specification process,

▷ it is necessary for them to provide statistical results as if data were from a *controlled experiment*,

▷ they care more about "significance" than effect size,

▷ computers are slow or relatively unavailable for them,

▷ they prefer very automated, "cookbook" type procedures.

So why do so-called classical approaches dominate Bayesian usage in the social and behavioral sciences? There are several reasons for this phenomenon. *First*, key figures in the development of modern statistics had strong prejudices against aspects of Bayesian inference for narrow and subjective reasons. *Second*, the cost of admission is higher in the form of additional mathematical formalism. *Third*, until recently realistic model specifications sometimes led to unobtainable Bayesian solutions. *Finally*, there has been a lack of methodological introspection in a number of disciplines. The primary mission of this text is to make the second and third reasons less of a barrier through accessible explanation, detailed examples, and specific guidance on calculation and computing.

It is important to understand that the Bayesian way does not mean throwing away one's comfortable tools, and it is not itself just another tool. Instead it is a way of *organizing* one's toolbox and is also a way of doing statistical work that has sharply different philosophical underpinnings. So adopting Bayesian methods means keeping the usual set of methods, such as linear regression, ANOVA, generalized linear models, tabular analysis, and so on. In fact, many researchers applying statistics in the social sciences are not actually frequentists since they cannot assume an *infinite* stream of iid (independent and identically distributed) data coming from a controlled experimental setup. Instead, most of these analysts can be described as "likelihoodists," since they obtain one sample of observational data that is contextual and will not be repeated, then perform standard likelihood-based (Fisherian) inference to get coefficient estimates.

Aside from underlying philosophical differences, many readers will be comforted in finding that Bayesian and non-Bayesian analyses often agree. There are two important instances where this is *always* true. *First*, when the included prior information is very uninformative (there are several ways of providing this), summary statements from Bayesian inference will match likelihood point estimates. Therefore a great many researchers are Bayesians who do not know it yet. *Second*, when the data size is very large, the form of the prior information used does not matter and there is agreement again. Other circumstances also exist in which Bayesian and non-Bayesian statistical inferences lead to the same results, but these are less general than the two mentioned. In addition to these two important observations, all hierarchical models are overtly Bayesian since they define distributional assumptions at levels. These are popular models due to their flexibility with regard to the prevalence of different levels of observed aggregation in the same dataset. We will investigate Bayesian hierarchical models in Chapter 12.

We will now proceed to a detailed justification for the use of modern Bayesian methods.

1.2   General Motivation and Justification

With Bayesian analysis, assertions about unknown model parameters are not expressed in the conventional way as single point estimates along with associated reliability assessed through the standard null hypothesis significance test. Instead the emphasis is on making probabilistic statements using distributions. Since Bayesians make no fundamental distinction between unobserved data and unknown parameters, the world is divided into: immediately available quantities, and those that need to be described probabilistically. Before observing some data, these descriptions are called *prior distributions*, and after observing the data these descriptions are called *posterior distributions*. The quality of the modeling process is the degree to which a posterior distribution is more informed than a prior distribution for some unknown quantity of interest. Common descriptions of posterior distributions include standard quantile levels, the probability of occupying some affected region of the sample space, the predictive quantities from the posterior, and Bayesian forms of confidence intervals called credible intervals.

It is important to note here that the pseudo-frequentist null hypothesis significance test (NHST) is not just sub-optimal, it is *wrong*. This is the dominant hypothesis testing paradigm as practiced in the social sciences. Serious problems include: a logical inconsistency coming from probabilistic modus tollens, confusion over the order of the conditional probability, chasing significance but ignoring effect size, adherence to completely arbitrary significance thresholds, and confusion about the probability of rejection. There is a general consensus amongst those that have paid attention to this issue that the social sciences have been seriously harmed by the NHST since it has led to fixations with counting stars on tables rather than looking for effect sizes and general statistical reliability. See the recent discussions in Gill (1999) and Denis (2005) in particular for detailed descriptions of these problems and how they damage statistical inferences in the social sciences. Serious criticism of the NHST began shortly after its creation in the early 20th century by textbook writers who blended Fisherian likelihoodism with Neyman and Pearson frequentism in an effort to offend neither warring and evangelical camp. An early critic of this unholy union was Rozeboom (1960) who noticed its "strangle-hold" on social science inference. In 1962 Arthur Melton wrote a parting editorial in the *Journal of Experimental Psychology* revealing that he had held authors to a 0.01 p-value standard: "In editing the *Journal* there has been a strong reluctance to accept and publish results related to the principal concern of the research when those results were significant at the .05 level, whether by one- or two-tailed test!" This had the effect of accelerating the criticism and led to many analytical and soul-searching articles discussing the negative consequences of this procedure in the social sciences, including: Barnett (1973), Berger (2003), Berger, Boukai, and Wang (1997), Berger and Sellke (1987), Berkhardt and Schoenfeld (2003), Bernardo (1984), Brandstätter (1999), Carver (1978, 1993), Cohen (1962, 1977, 1988, 1992, 1994), Dar, Serlin and Omer (1994), Falk and Greenbaum (1995), Gigerenzer (1987, 1998a, 1998b, 2004), Gigerenzer and Murray (1987), Gliner, Leech and Morgan (2002), Greenwald (1975), Greenwald, *et al.* (1996), Goodman (1993, 1999), Haller and Krauss (2002), Howson and Urbach (1993), Hunter (1997), Hunter and Schmidt (1990), Kirk (1996), Krueger (2001), Lindsay (1995), Loftus (1991, 1993), Macdonald (1997), McCloskey and Ziliak (1996), Meehl (1978, 1990, 1997), Moran and Soloman (2004), Morrison and Henkel (1969, 1970), Nickerson (2000), Oakes (1986), Pollard (1993), Pollard and Richardson (1987), Robinson and Levin (1997), Rosnow and Rosenthal (1989), Schmidt (1996), Schmidt and Hunter (1977), Schervish (1996), Sedlmeier and Gigerenzer (1989), Thompson (1996, 1997, 2002a, 2002b, 2004), Wilkinson (1977), Ziliak and McCloskey (2007). And this is only a small sample of the vast literature describing the NHST as bankrupt. Conveniently some of the more influential articles listed above are reprinted in Harlow *et al.* (1997). We will return to this point in Chapter 7 (starting on page 209) in the discussion of Bayesian hypothesis testing and model comparison.

This focus on distributional inference leads to two key assumptions for Bayesian work.

First, a specific parametric form is assumed to describe the distribution of the data given parameter values. Practically, this is used to construct a *likelihood function* (A) to incorporate the contribution of the full sample of data. Note that this is an inherently parametric setup, and although nonparametric Bayesian modeling is a large and growing field, it exists beyond the scope of the basic setup. *Second*, since unknown parameters are treated as having distributional qualities rather than being fixed, an assumed prior distribution on the parameters of interest unconditional on the data is given. This reflects either uncertainty about a truly fixed parameter or recognition that the quantity of interest actually behaves in some stochastic fashion.

With those assumptions in hand, the essentials of Bayesian thinking can be stated in three general steps:

1. Specify a probability model that includes some prior knowledge about the parameters for unknown parameter values.

2. Update knowledge about the unknown parameters by conditioning this probability model on observed data.

3. Evaluate the fit of the model to the data and the sensitivity of the conclusions to the assumptions.

Notice that this process does not include an unrealistic and artificial step of making a contrived decision based on some arbitrary quality threshold. The value of a given Bayesian model is instead found in the description of the distribution of some parameter of interest in probabilistic terms. Also, there is nothing about the process contained in the three steps above that cannot be repeated as new data are observed. It is often convenient to use the conventional significance thresholds that come from Fisher, but Bayesians typically do not ascribe any major importance to being barely on one side or the other. That is, Bayesian inference often prescribes something like a 0.05 threshold, but it is rare to see work where a 0.06 finding is not taken seriously as a likely effect.

Another core principle of the Bayesian paradigm is the idea that the data are fixed once observed. Typically (but not always) these data values are assumed to be *exchangeable*; the model results are not changed by reordering the data values. This property is more general than, and implied by, the standard assumption that the data are *independent and identically distributed* (iid): independent draws from the same distribution, and also implies a common mean and variance for the data values (Leonard and Hsu 1999, p.1). Exchangeability allows us to say that the data generation process is conditional on the unknown model parameters in the same way for every data value (de Finetti 1974, Draper *et al.* 1993, Lindley and Novick 1981). Essentially this is a less restrictive version of the standard iid assumption. Details about the exchangeability assumption are given in Chapter 12. We now turn to a discussion of probability basics as a precursor to Bayesian mechanics.

1.3 Why Are We Uncertain about Uncertainty?

The fundamental principles of probability are well known, but worth repeating here. Actually, it is relatively easy to *intuitively* define the properties of a probability function: *(1)* its range is bounded by zero and one for all the values in the applicable domain, *(2)* it sums or integrates to one over this domain, and *(3)* the sum or integral of the functional outcome (probability) of disjoint events is equal to the functional outcome of the union of these events. These are the Kolmogorov axioms (1933), and are given in greater detail in Gill (2006), Chapter 7. The real problem lies in describing the actual meaning of probability statements. This difficulty is, in fact, at the heart of traditional disagreements between Bayesians and non-Bayesians.

The frequentist statistical interpretation of probability is that it is a limiting relative frequency: the long-run behavior of a nondeterministic outcome or just an observed proportion in a population. This idea can be traced back to Laplace (1814), who defined probability as the number of successful events out of trials observed. Thus if we could simply repeat the experiment or observe the phenomenon enough times, it would become apparent what the future probability of reoccurrence will be. This is an enormously useful way to think about probability but the drawback is that frequently it is not possible to obtain a large number of outcomes from exactly the same event-generating system (Kendall 1949, Placket 1966).

A competing view of probability is called "subjective" and is often associated with the phrase "degree of belief." Early proponents included Keynes (1921) and Jeffreys (1961), who observed that two people could look at the same situation and assign different probability statements about future occurrences. This perspective is that probability is *personally* defined by the conditions under which a person would make a bet or assume a risk in pursuit of some reward. Subjective probability is closely linked with the idea of decision-making as a field of study (see, for instance, Bernardo and Smith [1994, Chapter 2]) and the principle of selecting choices that maximize personal utility (Berger 1985).

These two characterizations are necessarily simplifications of the perspectives and de Finetti (1974, 1975) provides a much deeper and more detailed categorization, which we will return to in Chapter 12. To de Finetti, the ultimate arbiter of subjective probability assignment is the conditions under which individuals will wager their own money. In other words, a person will not violate a personal probability assessment if it has financial consequences. Good (1950) makes this idea more axiomatic by observing that people have personal probability assessments about many things around them rather than just one, and in order for these disparate comparative statements to form a *body of beliefs* they need to be free of contradictions. For example, if a person thinks that A is more likely to occur than B, and B is more likely to occur than C, then this person cannot coherently believe that C is more likely than A (transitivity). Furthermore, Good adds the explicitly Bayesian idea that people are constantly updating these personal probabilities as new information is observed, although there is evidence that people have subadditive notions of probability when making calculations (the probability of some event plus the probability of its complement do not add to certainty).

The position underlying nearly all Bayesian work is the subjective probability characterization, although there have been many attempts to "objectify" Bayesian analysis (see Chapter 4). Prior information is formalized in the Bayesian framework and this prior information can be subjective in the sense that the researcher's experience, intuition, and theoretical ideas are included. It is also common to base the prior information on previous studies, experiments, or just personal observations and this process is necessarily subject to a limited (although possibly large) number of observations rather than the infinite number assumed under the frequentist view. We will return to the theme of subjectivity contained in prior information in Chapter 4 and elsewhere, but the principal point is that *all* statistical models include subjective decisions, and therefore we should *ceteris paribus* prefer one that is the most explicit about assumptions. This is exactly the sense that the Bayesian prior provides readers with a specific, formalized statement of currently assumed knowledge in probabilistic terms.

1.3.1   Required Probability Principles

There are some simple but important principles and notational conventions that must be understood before proceeding. We will not worry too much about measure theory until Chapter 13, and the concerned reader is directed to the first chapter of any mathematical statistics text or the standard reference works of Billingsley (1995), Chung (1974), and Feller (1990, Volumes 1 and 2). Abstract events are indicated by capital Latin letters: A, B, C, etc. A probability function corresponding to some event A is always indicated by $p(A)$. The complement of the event A is denoted A^c, and it is a consequence of Kolmogorov's axioms listed above that $p(A^c) = 1 - p(A)$. The union of two events is indicated by $A \cup B$ and the intersection by $A \cap B$. For any two events: $p(A \cup B) = p(A) + p(B) - p(A \cap B)$. If two events are independent, then $p(A \cap B) = p(A)p(B)$, but not necessarily the converse (the product relationship does not imply independence).

Central to Bayesian thinking is the idea of conditionality. If an event B is material to another event A in the sense that the occurrence or non-occurrence of B affects the probability of A occurring, then we say that A is conditional on B. It is a basic tenet of Bayesian statistics that we update our probabilities as new relevant information is observed. This is done with the definition of conditional probability given by: $p(A|B) = p(A \cap B)/p(B)$, which is read as "the probability of A given B is equal to the probability of A and B divided by the probability of B."

In general the quantities of interest here are random variables rather than the simple discrete events above. A random variable X is defined as a measurable function from a probability space to a state space. This can be defined very technically (Shao 2005, p.7), but for our purposes it is enough to understand that the random variable connects possible occurrences of some data value with a probability structure that reflects the relative "frequency" of these occurrences. The function is thus defined over a specific state space of all possible realizations of the random variable, called support. Random variables can be discrete or continuous. For background details see Casella and Berger (2002), Shao (2005), or the essay by Doob (1996). The expression $p(X \cap Y)$ is usually denoted as $p(X, Y)$, and is referred to as the *joint distribution* of random variables X and Y. Marginal distributions are then simply $p(X)$ and $p(Y)$. Restating the principle above in this context, for two independent random variables the joint distribution is just the product of the marginals, $p(X, Y) = p(X)p(Y)$. Typically we will need to integrate expressions like $p(X, Y)$ to get marginal distributions of interest. Sometimes this is done analytically, but more commonly we will rely on computational techniques.

We will make extensive use of expected value calculations here. Recall that if a random variable X is distributed $p(X)$, the expected value of some function of the random variable, $h(X)$, is

$$E[h(X)] = \begin{cases} \sum_{i=1}^{k} h(x_i)p(x_i) & k\text{-category discrete case} \\ \int_{\mathcal{X}} h(X)p(X)dx & \text{continuous case.} \end{cases} \tag{1.1}$$

Commonly $h(X) = X$, and we are simply concerned with the expectation of the random variable itself. In the discrete case this is a very intuitive idea as the expected value can be thought of as a probability-weighted average over possible events. For the continuous case there are generally limits on the integral that are dictated by the support of the random variable, and sometimes these are just given by $[-\infty, \infty]$ with the idea that the PDF (probability density function) indicates zero and non-zero regions of density. Also $p(X)$ is typically a conditional statement: $p(X|\theta)$. For the $k \times 1$ vector $\mathbf{X}$ of discrete random variables, the expected value is: $E[\mathbf{X}] = \sum \mathbf{X}p(\mathbf{X})$. With the expected value of a function of the continuous random vector, it is common to use the *Riemann-Stieltjes integral* form (found in any basic calculus text): $E[f(\mathbf{X})] = \int_{\mathcal{X}} f(\mathbf{X})dF(\mathbf{X})$, where $F(\mathbf{X})$ denotes the joint distribution of the random variable vector $\mathbf{X}$. The principles now let us look at Bayes' Law in detail.

1.4 Bayes' Law

The Bayesian statistical approach is based on updating information using what is called Bayes' Law (and synonymously Bayes' Theorem) from his famous 1763 essay. The Reverend Thomas Bayes was an amateur mathematician whose major contribution (the others remain rather obscure and do not address the same topic) was an essay found and published two years after his death by his friend Richard Price. The enduring association of an important branch of statistics with his name actually is somewhat of an exaggeration of the generalizeability of this work (Stigler 1982). Bayes was the first to explicitly develop this famous law, but it was Laplace (1774, 1781) who (apparently independently) provided a more detailed analysis that is perhaps more relevant to the practice of Bayesian statistics today. See Stigler (1986) for an interesting historical discussion and Sheynin (1977) for a detailed technical analysis. Like Bayes, Laplace assumed a uniform distribution for the unknown parameter, but he worried much less than Bayes about the consequences of this assumption. Uniform prior distributions are simply "flat" distributions that assign equal probability for every possible outcome.

Suppose there are two events of interest A and B, which are not independent. We know from basic axioms of probability that the conditional probability of A given that B has occurred is given by:

$$p(A|B) = \frac{p(A, B)}{p(B)}, \tag{1.2}$$

where $p(A|B)$ is read as "the probability of A given that B has occurred, $p(A, B)$ is the "the probability that both A and B occur" (i.e., the joint probability) and $p(B)$ is just the unconditional probability that B occurs. Expression (1.2) gives the probability of A after some event B occurs. If A and B are independent here then $p(A, B) = p(A)p(B)$ and (1.2) becomes uninteresting.

We can also define a different conditional probability in which A occurs first:

$$p(B|A) = \frac{p(B, A)}{p(A)}. \tag{1.3}$$

Since the probability that A and B occur is the same as the probability that B and A occur ($p(A, B) = p(B, A)$), then we can rearrange (1.2) and (1.3) together in the following way:

$$p(A, B) = p(A|B)p(B)$$

$$p(B, A) = p(B|A)p(A)$$

$$p(A|B)p(B) = p(B|A)p(A)$$

$$p(A|B) = \frac{p(A)}{p(B)}p(B|A). \tag{1.4}$$

The last line is the famous Bayes' Law. This is really a device for "inverting" conditional probabilities. Notice that we could just as easily produce $p(B|A)$ in the last line above by moving the unconditional probabilities to the left-hand side in the last equality.

We can also use Bayes' Law with the use of odds, which is a common way to talk about uncertainty related to probability. The odds of an event is the ratio of the probability of an event happening to the probability of the event not happening. So for the event A, the odds of this event is simply:

$$Odds = \frac{p(A)}{1 - p(A)} = \frac{p(A)}{p(\neg A)}, \tag{1.5}$$

which is this ratio expressed in two different ways. Note the use of "$\neg A$" for "not A," which is better notation when the complement of A isn't specifically defined and we care only that event A did not happen. Since this statement is not conditional on any other quantity, we can call it a "prior odds." If we make it conditional on B, then it is called a "posterior odds," which is produced by multiplying the prior odds by the reverse conditional with regard to B:

$$\frac{p(A|B)}{p(\neg A|B)} = \frac{p(A)}{p(\neg A)}\frac{p(B|A)}{p(B|\neg A)}. \tag{1.6}$$

The last ratio, $p(B|A)/p(B|\neg A)$ is the "likelihood ratio" for B under the two conditions for A. This is actually the ratio of two expressions of Bayes' Law in the sense of (1.4), which we can see with the introduction of the ratio $p(B)/p(B)$:

$$\frac{p(A|B)}{p(\neg A|B)} = \frac{p(A)/p(B)}{p(\neg A)/p(B)}\frac{p(B|A)}{p(B|\neg A)}. \tag{1.7}$$

This ratio turns out to be very useful in Bayesian model consideration since it implies a test between the two states of nature, A and $\neg A$, given the observation of some pertinent information B.

■ **Example 1.1:** **Testing with Bayes' Law.** How is this useful? As an example, hypothetically assume that 2% of the population of the United States are members of some extremist Militia group ($p(M) = 0.02$), a fact that some members might attempt to hide and therefore not readily admit to an interviewer. A survey is 95% accurate on positive Classification, $p(C|M) = 0.95$, ("sensitivity") and the unconditional probability of classification (i.e., regardless of actual militia status) is given by $p(C) = 0.05$. To illustrate how $p(C)$ is really the normalizing constant obtained by accumulating over all possible events, we will stipulate the additional knowledge that the survey is 97% accurate on negative classification, $p(C^c|M^c) = 0.97$ ("specificity"). The unconditional probability of classifying a respondent as a militia member results from accumulation of the probability across the sample space of survey events using the Total Probability Law: $p(C) = p(C \cap M) + p(C \cap M^c) = p(C|M)p(M) + [1 - p(C^c|M^c)]p(M^c) = (0.95)(0.02) + (0.03)(0.98) \cong 0.05$.

Using Bayes' Law, we can now derive the probability that someone positively classified by the survey as being a militia member really *is* a militia member:

$$p(M|C) = \frac{p(M)}{p(C)}p(C|M) = \frac{0.02}{0.05}(0.95) = 0.38. \tag{1.8}$$

The startling result is that although the probability of correctly classifying an individual as a militia member given they really are a militia member is 0.95, the probability that an individual really is a militia member given that they are positively classified is only 0.38.

The highlighted difference here between the order of conditional probability is often substantively important in a policy or business context. Consider the problem of designing a home pregnancy test. Given that there exists a fundamental business trade-off between the reliability of the test and the cost to consumers, no commercially viable product will have perfect or near-perfect test results. In designing the chemistry and packaging of the test, designers will necessarily have to compromise between the probability of **PR**egnancy given positive **T**est results, $p(\mathbf{PR}|\mathbf{T})$, and the probability of positive test results given pregnancy, $p(\mathbf{T}|\mathbf{PR})$. Which one is more important? Clearly, it is better to maximize $p(\mathbf{T}|\mathbf{PR})$ at the expense of $p(\mathbf{PR}|\mathbf{T})$, as long as the reduction in the latter is reasonable: it is preferable to give a higher number of false positives, sending women to consult their physician to take a more sensitive test, than to fail to notify many pregnant women. This reduces the possibility that a woman who does not realize that she is pregnant might continue unhealthy practices such as smoking, drinking, or maintaining a poor diet. Similarly, from the perspective of general public health, it is better to have preliminary tests for deadly contagious diseases designed to be similarly conservative with respect to false positives.

1.4.1   Bayes' Law for Multiple Events

It would be extremely limiting if Bayes' Law only applied to two alternative events. Fortunately the extension to multiple events is quite easy. Suppose we observe some data $\mathbf{D}$ and are interested in the relative probabilities of three events A, B, and C conditional on these data. These might be rival hypotheses about some social phenomenon for which the data are possibly revealing. Thinking just about event A, although any of the three could be selected, we know from Bayes' Law that:

$$p(A|\mathbf{D}) = \frac{p(\mathbf{D}|A)p(A)}{p(\mathbf{D})}. \tag{1.9}$$

We also know from the Total Probability Law and the definition of conditional probability that:

$$p(\mathbf{D}) = p(A \cap \mathbf{D}) + p(B \cap \mathbf{D}) + p(C \cap \mathbf{D})$$

$$= p(\mathbf{D}|A)p(A) + p(\mathbf{D}|B)p(B) + p(\mathbf{D}|C)p(C). \tag{1.10}$$

This means that if we substitute the last line into the expression for Bayes' Law, we get:

$$p(A|\mathbf{D}) = \frac{p(\mathbf{D}|A)p(A)}{p(\mathbf{D}|A)p(A) + p(\mathbf{D}|B)p(B) + p(\mathbf{D}|C)p(C)}, \tag{1.11}$$

which demonstrates that the conditional distribution for any of the rival hypotheses can be produced as long as there exist unconditional distributions for the three rival hypotheses, $p(A)$, $p(B)$, and $p(C)$, and three statements about the probability of the data given these three hypotheses, $p(\mathbf{D}|A)$, $p(\mathbf{D}|B)$, $p(\mathbf{D}|C)$. The first three probability statements are called prior distributions because they are unconditional from the data and therefore presumably determined before observing the data. The second three probability statements are merely

PDF (probability density function) or PMF (probability mass function) statements in the conventional sense. All this means that a posterior distribution, $p(A|\mathbf{D})$, can be determined through Bayes' Law to look at the weight of evidence for any one of several rival hypotheses or claims.

There is a more efficient method for making statements like (1.11) when the number of outcomes increases. Rather than label the three hypotheses as we have done above, let us instead use θ as an unknown parameter whereby different regions of its support define alternative hypotheses. So statements may take the form of "Hypothesis A: $\theta < 0$," or any other desired statement. To keep track of the extra outcome, denote the three hypotheses as θ_i, $i = 1, 2, 3$. Now (1.11) is given more generally for $i = 1, 2, 3$ as:

$$p(\theta_i|\mathbf{D}) = \frac{p(\mathbf{D}|\theta_i)p(\theta_i)}{\sum_{j=1}^{3} p(\mathbf{D}|\theta_j)p(\theta_j)} \tag{1.12}$$

for the posterior distribution of θ_i. This is much more useful and much more in line with standard Bayesian models in the social and behavioral sciences because it allows us to compactly state Bayes' Law for any number of discrete outcomes/hypotheses, say k for instance:

$$p(\theta_i|\mathbf{D}) = \frac{p(\theta_i)p(\mathbf{D}|\theta_i)}{\sum_{j=1}^{k} p(\theta_j)p(\mathbf{D}|\theta_j)}. \tag{1.13}$$

Consider also that the denominator of this expression averages over the θ variables and therefore just produces the *marginal distribution of the sample data*, which we could overtly label as $p(\mathbf{D})$. Doing this provides a form that very clearly looks like the most basic form of Bayes' Law: $p(\theta_i|\mathbf{D}) = p(\theta_i)p(\mathbf{D}|\theta_i)/p(\mathbf{D})$. We can contrast this with the standard likelihood approach in the social sciences (King 1989, p.22), which overtly ignores information available through a prior and has no use for the denominator above: $L(\hat{\theta}|y) \propto p(y|\hat{\theta})$, in King's notation using proportionality since the objective is simply to find the mode and curvature around this mode, thus making constants unimportant. Furthermore, in the continuous case, where the support of θ is over some portion of the real line, and possibly all of it, the summation in (1.13) is replaced with an integral. The continuous case is covered in the next chapter.

■ **Example 1.2: Monty Hall.** The well-known Monty Hall problem (Selvin 1975) can be analyzed using Bayes' Law. Suppose that you are on the classic game show *Let's Make a Deal* with its personable host Monty Hall, and you are to choose one of three doors, A, B, or C. Behind two of the doors are goats and behind the third door is a new car, and each door is equally likely to award the car. Thus, the probabilities of selecting the car for each door at the beginning of the game are simply:

$$p(A) = \frac{1}{3}, \qquad p(B) = \frac{1}{3}, \qquad p(C) = \frac{1}{3}.$$

After you have picked a door, say A, before showing you what is behind that door Monty opens another door, say B, revealing a goat. At this point, Monty gives you the opportunity to switch doors from A to C if you want to. What should you do? The psychology of this approach is to suggest the idea to contestants that they must have picked the correct door and Monty is now trying to induce a change. A naïve interpretation is that you should be indifferent to switching due to a perceived probability of 0.5 of getting the car with either door since there are two doors left. To see that this is false, recall that Monty is not a benign player in this game. He is deliberately trying to deny you the car. Therefore consider his probability of opening door B. Once you have picked door A, success is clearly conditional on what door of the three possibilities actually provides the car since Monty has this knowledge and the contestant does not. After the first door selection, we can define the three conditional probabilities as follows:

The probability that Monty opens door B,
given the car is behind A: $p(B_{\text{Monty}}|A) = \frac{1}{2}$

The probability that Monty opens door B,
given the car is behind B: $p(B_{\text{Monty}}|B) = 0$

The probability that Monty opens door B,
given the car is behind C: $p(B_{\text{Monty}}|C) = 1.$

Using the definition of conditional probability, we can derive the following three joint probabilities:

$$p(B_{\text{Monty}}, A) = p(B_{\text{Monty}}|A)p(A) = \frac{1}{2} \times \frac{1}{3} = \frac{1}{6}$$

$$p(B_{\text{Monty}}, B) = p(B_{\text{Monty}}|B)p(B) = 0 \times \frac{1}{3} = 0$$

$$p(B_{\text{Monty}}, C) = p(B_{\text{Monty}}|C)p(C) = 1 \times \frac{1}{3} = \frac{1}{3}.$$

Because there are only three possible events that cover the complete sample space, and these events are non-overlapping (mutually exclusive), they form a partition of the sample space. Therefore the sum of these three events is the unconditional probability of Monty opening door B, which we obtain with the Total Probability Law:

$$p(B_{\text{Monty}}) = p(B_{\text{Monty}}, A) + p(B_{\text{Monty}}, B) + p(B_{\text{Monty}}, C)$$

$$= \frac{1}{6} + 0 + \frac{1}{3} = \frac{1}{2}.$$

Now we can apply Bayes' Law to obtain the two probabilities of interest:

$$p(A|B_{\text{Monty}}) = \frac{p(A)}{p(B_{\text{Monty}})} p(B_{\text{Monty}}|A) = \frac{\frac{1}{3}}{\frac{1}{2}} \times \frac{1}{2} = \frac{1}{3}$$

$$p(C|B_{\text{Monty}}) = \frac{p(C)}{p(B_{\text{Monty}})} p(B_{\text{Monty}}|C) = \frac{\frac{1}{3}}{\frac{1}{2}} \times 1 = \frac{2}{3}.$$

Therefore you are twice as likely to win the car if you switch to door C! This example demonstrates that Bayes' Law is a fundamental component of probability calculations, and the principle will be shown to be the basis for an inferential system of statistical analysis. For a nice generalization to N doors, see McDonald (1999).

1.5   Conditional Inference with Bayes' Law

To make the discussion more concrete and pertinent, consider a simple problem in sociology and crime studies. One quantity of interest to policy-makers is the recidivism rate of prisoners released after serving their sentence. The quantity of interest is the probability of committing an additional crime and returning to prison. Notice that this is a very elusive phenomenon. Not only are there regional, demographic, and individualistic differences, but the aggregate probability is also constantly in flux, given entries and exits from the population as well as exogenous factors (such as the changing condition of the economy). Typically, we would observe a change in law or policy at the state or federal level, and calculate a point estimate from observed recidivism that follows.

Perhaps we should not assume that there is some fixed value of the recidivism probability, A, and that it should be estimated with a single point, say $\bar{A}$. Instead, consider this unknown quantity in probabilistic terms as the random variable A, which means conceptualizing a distribution for the probability of recidivism. Looking at data from previous periods, we might have some reasonable guess about the distribution of this probability parameter, $p(A)$, which is of course the prior distribution since it is not conditional on the information at hand, B.

In all parametric statistical inference, a model is proposed and tested in which an event has some probability of occurring given a specific value of the parameter. This is the case for both Bayesian and traditional approaches, and is just a recognition that the researcher must specify a data generation model. Let us call this quantity $p(B|A)$, indicating that for posited values of recidivism, we would expect to see a particular pattern of events. For instance, if recidivism suddenly became much higher in a particular state, then there might be pressure on the legislature to toughen sentencing and parole laws. This is a probability model and we do not need to have a specific value of A to specify a parametric form (i.e., PMF or PDF). Of course what we are really interested in is $p(A|B)$, the (posterior) distribution of A after having observed an event, which we obtain using Bayes' Law: $p(A|B) = \frac{p(A)}{p(B)}p(B|A)$. From a public policy perspective, this is equivalent to asking how do recidivism rates change for given statutes.

We are still missing one component of the right-hand-side of Bayes' Law here, the *unconditional* probability of generating the legal or policy event, $p(B)$. This is interpretable as the denominator of (1.13), but to a Bayesian this is an unimportant *probability* statement since B has already been observed and therefore has probability one of occurring. Recall that for Bayesians, observed quantities are fixed and unobserved quantities are assigned probability statements. So there is no point in treating B probabilistically if the actual facts are sitting on our desk right now. This does not mean that everything is known about all possible events, missing events, or events occurring in the future. It just means that everything is known about *this* event. So the only purpose for $p(B)$ in this context is to make sure that $p(A|B)$ sums or integrates to one.

This last discussion suggests simply treating $p(B)$ as a normalizing constant since it does not change the *relative* probabilities for A. Maybe this is a big conceptual leap, but if we could recover unconditional $p(B)$ later, it is convenient to just use it then to make the conditional statement, $p(A|B)$, a properly scaled probability statement. So if $p(A|B)$ summed or integrated to five instead of one, we would simply divide everywhere by five and lose nothing but the agony of carrying $p(B)$ through the calculations. If we temporarily ignore $p(B)$, then:

$$p(A|B) \propto p(A)p(B|A), \tag{1.14}$$

where "$\propto$" means "proportional to" (i.e., the *relative* probabilities are preserved). So the final estimated probability of recidivism (in our example problem) given some observed behavior, is proportional to prior notions about the distribution of the probability times the parametric model assumed to be generating the new observed event. The conditional probability of interest on the left-hand side of (1.14) is a balance between things we have already seen or believe about recidivism, $p(A)$, and the contribution from the new observation, $p(B|A)$. It is important to remember that there are occasions where the data are more influential than the prior and vice-versa. This is comforting since if the data are poor in size or information we want to rely more on prior knowledge, prior research, researcher or practitioner information and so on. Conversely, if the data are plentiful and highly informed, then we should not care much about the form of the prior information. Remarkably, the Bayesian updating process in (1.14) has this trade-off automatically built-in to the process.

As described, this is an ideal paradigm for inference in the social and behavioral sciences, since it is consentaneously desirable to build models that test theories with newly observed events or data, but also based on previous research and knowledge. We never start a data analysis project with absolutely no *a priori* notions whatsoever about the state of nature (or at least we should not!). This story actually gets better. As the number of events increases, $p(B|A)$ becomes progressively more influential in determining $p(A|B)$. That is, the greater the number of our new observations, the less important are our previous convictions: $p(A)$. Also, if either of the two distributions, $p(A)$ and $p(B|A)$, are widely dispersed relative to the other, then this distribution will have less of an impact on the final probability statement. We will see this principle detailed-out in Chapter 2. The natural weighting of these two distributions suitably reflects relative levels of uncertainty in the two quantities.

1.5.1 Statistical Models with Bayes' Law

The statistical role of the quantities in (1.14) has not yet been identified since we have been talking abstractly about "events" rather than conventional data. The goal of inference is to make claims about unknown quantities using data currently in hand. Suppose that we designate a generic Greek character to denote an unobserved parameter that is the objective of our analysis. As is typical in these endeavors, we will use θ for this purpose. What we usually have available to us is generically (and perhaps a little vaguely) labeled $\mathbf{D}$ for data. Therefore, the objective is to obtain a probabilistic statement about θ given $\mathbf{D}$: $p(\theta|\mathbf{D})$.

Inferences in this book, and in the majority of Bayesian and non-Bayesian statistics, are made by first specifying a parametric model for the data generating process. This defines what the data should be expected to look like given a specific probabilistic function conditional on unknown variable values. These are the common probability density functions (continuous data) and probability mass functions (discrete data) that we already know, such as normal, binomial, chi-square, etc., denoted by $p(\mathbf{D}|\theta)$.

Now we can relate these two conditional probabilities using (1.14):

$$\pi(\theta|\mathbf{D}) \propto p(\theta)p(\mathbf{D}|\theta), \tag{1.15}$$

where $p(\theta)$ is a formalized statement of the prior knowledge about θ before observing the data. If we know little, then this prior distribution should be a vague probabilistic statement and if we know a lot then this should be a very narrow and specific claim. The right-hand side of (1.15) implies that the *post*-data inference for θ is a compromise between prior information and the information provided by the new data, and the left-hand side of (1.15) is the posterior distribution of θ since it provides the updated distribution for θ after conditioning on the data.

Bayesians describe $\pi(\theta|\mathbf{D})$ to readers via distributional summaries such as means, modes, quantiles, probabilities over regions, traditional-level probability intervals, and graphical displays. Once the posterior distribution has been calculated via (1.15), everything about it is known and it is entirely up to the researcher to highlight features of interest. Often it is convenient to report the posterior mean and variance in papers and reports since this is what non-Bayesians do by default. We can calculate the posterior mean using an expected value calculation, confining ourselves here to the continuous case:

$$E[\theta|\mathbf{D}] = \int_{-\infty}^{\infty} \theta\pi(\theta|\mathbf{D})d\theta \tag{1.16}$$

and the posterior variance via a similar process:

$$\text{Var}[\theta|\mathbf{D}] = E\left[(\theta - E[\theta|\mathbf{D}])^2|\mathbf{D}\right]$$

$$= \int_{-\infty}^{\infty} (\theta - E[\theta|\mathbf{D}])^2 \pi(\theta|\mathbf{D})d\theta$$

$$= \int_{-\infty}^{\infty} \left(\theta^2 - 2\theta E[\theta|\mathbf{D}]\right) + E[\theta|\mathbf{D}]^2\right) \pi(\theta|\mathbf{D})d\theta$$

$$= \int_{-\infty}^{\infty} \theta^2 \pi(\theta|\mathbf{D})d\theta - 2E[\theta|\mathbf{D}] \int_{-\infty}^{\infty} \theta\pi(\theta|\mathbf{D})d\theta + \left(\int_{-\infty}^{\infty} \theta\pi(\theta|\mathbf{D})d\theta\right)^2$$

$$= E[\theta^2|\mathbf{D}] - E[\theta|\mathbf{D}]^2 \tag{1.17}$$

given some minor regularity conditions about switching the order of integration (see Casella and Berger 2002, Chapter 1). An obvious summary of this posterior would then be the vector $(E[\theta|\mathbf{D}], \sqrt{\text{Var}[\theta|\mathbf{D}]})$, although practicing Bayesians tend to prefer reporting more information.

Researchers sometimes summarize the Bayesian posterior distribution in a deliberately traditional, non-Bayesian way in an effort to communicate with some readers. The posterior mode corresponds to the maximum likelihood point estimate and is calculated by:

$$M(\theta) = \underset{\theta}{\text{argmax}}\,\pi(\theta|\mathbf{D}), \tag{1.18}$$

where argmax function specifies the value of θ that maximizes $\pi(\theta|\mathbf{D})$. Note that the denominator of Bayes' Law is unnecessary here since the function has the same mode with or with including it. The accompanying measure of curvature (e.g., Fisher Information, defined in Appendix A) can be calculated with standard analytical tools or more conveniently from MCMC output with methods introduced in Chapter 9. The posterior median is a slightly less popular choice for a Bayesian point estimate, even though its calculation from MCMC output is trivial from just sorting empirical draws and determining the mid-point.

■ **Example 1.3:   Summarizing a Posterior Distribution from Exponential Data.**
Suppose we had generic data, $\mathbf{D}$, distributed $p(\mathbf{D}|\theta) = \theta e^{-\theta\mathbf{D}}$, which can be either a single scalar or a vector for our purposes. Thus $\mathbf{D}$ is exponentially distributed with the support $[0{:}\infty)$; see Appendix A for details on this probability density function. We also need to specify a prior distribution for θ: $p(\theta) = 1$, where $\theta \in [0{:}\infty)$. Obviously this prior distribution does not constitute a "proper" distribution in the Kolmogorov sense since it does not integrate to one (infinity, in fact). We should not let this bother us since this effect is canceled out due to its presence in both the numerator and denominator of Bayes' Law (a principle revisited in Chapter 4 in greater detail).

This type of prior is often used to represent high levels of prior uncertainty, although it is not completely *uninformative*. Using (1.15) now, we get:

$$\pi(\theta|\mathbf{D}) \propto p(\theta)p(\mathbf{D}|\theta) = (1)\theta e^{-\theta\mathbf{D}} = \theta e^{-\theta\mathbf{D}}. \tag{1.19}$$

This posterior distribution has mean:

$$E[\theta|\mathbf{D}] = \int_0^\infty (\theta)\left(\theta e^{-\theta\mathbf{D}}\right) d\theta = \frac{2}{\mathbf{D}^3}, \tag{1.20}$$

which is found easily using two iterations of integration-by-parts. Also, the expectation of $\theta^2|\mathbf{D}$ is:

$$E[\theta^2|\mathbf{D}] = \int_0^\infty (\theta^2)\left(\theta e^{-\theta\mathbf{D}}\right) d\theta = \frac{6}{\mathbf{D}^4}, \tag{1.21}$$

which is found using three iterations of integration-by-parts now. So the posterior variance is:

$$\text{Var}[\theta|\mathbf{D}] = E[\theta^2|\mathbf{D}] - E[\theta|\mathbf{D}]^2 = 6\mathbf{D}^{-4} - 4\mathbf{D}^{-6}. \tag{1.22}$$

The notation would be slightly different if $\mathbf{D}$ were a vector.

Using these quantities we can perform an intuitive Bayesian hypothesis test, such as asking what is the posterior probability that θ is positive $p(\theta|\mathbf{D}) > 0$. In the context of a regression coefficient, this would be the probability that increases in the corresponding X explanatory variable have a positive affect on the Y outcome variable. Testing will be discussed in detail in Chapter 7. We can also use these derived quantities to create a Bayesian version of a confidence interval, the credible interval for some chosen α level:

$$\left[E[\theta|\mathbf{D}] - \sqrt{\text{Var}[\theta|\mathbf{D}]}f_{\alpha/2} : E[\theta|\mathbf{D}] + \sqrt{\text{Var}[\theta|\mathbf{D}]}f_{1-\alpha/2}\right], \tag{1.23}$$

where $f_{\alpha/2}$ and $f_{1-\alpha/2}$ are lower and upper tail values for some assumed or empirically observed distribution for θ (Chapter 2).

The purpose of this brief discussion is to highlight the fact that conditional probability underlies the ability to update previous knowledge about the distribution of some unknown quantity. This is precisely in line with the iterative scientific method, which postulates theory improvement through repeated specification and testing with data. The Bayesian approach combines a formal structure of rules with the mathematical convenience of probability theory to develop a process that "learns" from the data. The result is a powerful and elegant tool for scientific progress in many disciplines.

1.6   Science and Inference

This is a book about the scientific process of discovery in the social and behavioral sciences. Data analysis is best practiced as a theory-driven exploration of collected observations with the goal of uncovering important and unknown effects. This is true regardless of academic discipline. Yet some fields of study are considered more rigorously analytical in this pursuit than others.

The process described herein is that of *inference*: making probabilistic assertions about unknown quantities. It is important to remember that "in the case of uncertain inference, however, the very uncertainty of uncertain predictions renders question of their proof or disproof almost meaningless" (Wilkinson 1977). Thus, confusion sometimes arises in the interpretation of the inferential process as a scientific, investigative endeavor.

1.6.1   The Scientific Process in Our Social Sciences

Are the social and behavioral sciences truly "scientific"? This is a question asked about fields such as sociology, political science, economics, anthropology, and others. It is not a question about whether serious, rigorous, and important work has been done in these endeavors; it is a question about the research process and whether it conforms to the empirico-deductive model that is historically associated with the natural sciences. From a simplistic view, this is an issue of the conformance of research in the social and behavioral sciences to the so-called scientific method. Briefly summarized, the scientific method is characterized by the following steps:

▷ Observe or consider some phenomenon.

▷ Develop a theory about the cause(s) of this phenomenon and articulate it in a specific hypothesis.

▷ Test this hypothesis by developing a model to fit experimentally generated or collected observational data.

▷ Assess the quality of the fit to the model and modify the theory if necessary, repeating the process.

This is sometimes phrased in terms of "prediction" instead of theory development, but we will use the more general term. If the scientific method as a process were the defining criterion for determining what is scientific and what is not, then it would be easy to classify a large proportion of the research activities in the social and behavioral sciences as scientific. However useful this typology is in teaching children about empirical investigation, it is a poor standard for judging academic work.

Many authors have posited more serviceable definitions. Braithwaite (1953, p.1) notes:

The function of a science, in this sense of the word, is to establish general laws covering the behavior of the empirical events or objects with which the science in question is concerned, and thereby to enable us to connect together our knowledge of the separately known events, and to make reliable predictions of events as yet unknown.

The core of this description is the centrality of empirical observation and subsequent accumulation of knowledge. Actually, "science" is the Latin word for knowledge. Legendary psychologist B. F. Skinner (1953, p.11) once observed that "science is unique in showing a cumulative process." It is clear from the volume and preservation of published research that social and behavioral scientists *are* actively engaged in empirical research and knowledge accumulation (although the quality and permanence of this foundational knowledge might be judged to differ widely by field). So what is it about these academic pursuits that makes them only suspiciously scientific to some? The three defining characteristics about the *process* of scientific investigation are empiricism, objectivity, and control (Singleton and Straight 2004). This is where there is lingering and sometimes legitimate criticism of the social and behavioral sciences as being "unscientific."

The social and behavioral sciences are partially empirical (data-oriented) and partially normative (value-oriented), the latter because societies develop norms about human behavior, and these norms permeate academic thought prior to the research process. For instance, researchers investigating the onset and development of AIDS initially missed the effects of interrelated social factors such as changes in behavioral risk factors, personal denial, and reluctance to seek early medical care on the progress of the disease as a sociological phenomenon (Kaplan *et al.* 1987). This is partially because academic investigators as well as health professionals made normative assumptions about individual responses to sociological effects. Specifically, researchers investigating human behavior, whether political, economic, sociological, psychological, or otherwise, cannot completely divorce their prior attitudes about some phenomenon of interest the way a physicist or chemist can approach the study of the properties of thorium: atomic number 90, atomic symbol Th, atomic weight 232.0381, electron configuration $[Rn]7^s26d^2$. This criticism is distinct from the question of objectivity; it is a statement that students of human behavior are themselves human.

We are also to some extent driven by the quality and applicability of our tools. Many fields have radically progressed after the introduction of new analytical devices. Therefore, some researchers may have a temporary advantage over others, and may be able to answer more complex questions: "It comes as no particular surprise to discover that a scientist formulates problems in a way which requires for their solution just those techniques in which he himself is especially skilled" (Kaplan 1964). The objective of this book is to "level the pitch" by making an especially useful tool more accessible to those who have thus far been accordingly disadvantaged.

1.6.2   Bayesian Statistics as a Scientific Approach to Social and Behavioral Data Analysis

The standard frequentist interpretation of probability and inference assumes an infinite series of trials, replications, or experiments using the same research design. The "objectivist" paradigm is typically explained and justified through examples like multiple tosses of a coin, repeated measurements of some physical quantity, or samples from some ongoing process like a factory output. This perspective, which comes directly from Neyman and Pearson (1928a, 1928b, 1933a, 1933b, 1936a, 1936b), and was formalized by Von Mises (1957) among others, is combined with an added Fisherian fixation with p-values in typical inference in the social and behavioral sciences (Gill 1999). Efron (1986), perhaps overly kindly, calls this a "rather uneasy alliance."

Very few, if any, social scientists would be willing to seriously argue that human behavior fits this objectivist long-run probability model. Ideas like "personal utility," "legislative ideal points," "cultural influence," "mental states," "personality types," and "principal-agent goal discrepancy" do not exist as parametrically uniform phenomena in some physically tangible manner. In direct contrast, the Bayesian or "subjective" conceptualization of probability is the degree of belief that the individual researcher is willing to personally assign and defend. This is the idea that an individual *personally* assigns a probability measure to some event as an expression of uncertainty about some event that may only be relevant to one observational situation or experiment.

The central idea behind subjective probability is the assignment of a prior probability based on what information one currently possesses and under what circumstances one would be willing to place an even wager. Naturally, this probability is updated as new events occur, therefore incorporating serial events in a systematic manner. The core disagreement between the frequentist notion of objective probability and the Bayesian idea of subjective probability is that frequentists see probability measure as a property of the outside world and Bayesians view probability as a personal internalization of observed uncertainty. The key defense of the latter view is the inarguable point that all statistical models are subjective: decisions about variable specifications, significance thresholds, functional forms, and error distributions are completely nonobjective.[1] In fact, there are instances when Bayesian subjectivism is more "objective" than frequentist objectivism with regard to the impact of irrelevant information and arbitrary decision rules (e.g., Edwards, Lindman, and Savage 1963, p.239).

---

[1]As a brief example, consider common discussions of reported analyses in social science journals and books that talk about reported model parameters being "of the wrong sign." What does this statement mean? The author is asserting that the statistical model has produced a regression coefficient that is positive when it was *a priori* expected to be negative or vice versa. What is this statement in effect? It is a prior statement about knowledge that existed before the model was constructed. Obviously this is a form of the Bayesian prior without being specifically articulated as such.

Given the existence of subjectivity in all scientific data analysis endeavors,[2] one should prefer the inferential paradigm that gives the most *overt* presentation of model assumptions. This is clearly the Bayesian subjective approach since both prior information and posterior uncertainty are given with specific, clearly stated model assumptions. Conversely, frequentist models are rarely presented with caveats such as "Caution: the scientific conclusions presented here depend on repeated trials that were never performed," or "Warning: prior assumptions made in this model are not discussed or clarified." If there is a single fundamental scientific tenet that underlies the practice and reporting of empirical evidence, it is the idea that all important model characteristics should be provided to the reader. It is clear then which of the two approaches is more "scientific" by this criterion. While this discussion specifically contrasts Bayesian and frequentist approaches, likelihood inference is equally subjective in every way, and as already explained, ignores available information.

These ideas of what sort of inferences social scientists make are certainly not new or novel. There is a rich literature to support the notion that the Bayesian approach is more in conformance with widely accepted scientific norms and practices. Poirer (1988, p.130) stridently makes this point in the case of prior specifications:

> I believe that subjective prior beliefs should play a *formal* role so that it is easier to investigate their impact on the results of the analysis. Bayesians must live with such honesty whereas those who introduce such beliefs informally need not.

The core of this argument is the idea that if the prior contains information that pertains to the estimation problem, then we are foolish to ignore it simply because it does not neatly fit into some familiar statistical process. For instance, Theil and Goldberger (1961) suggested "mixed" estimation some time ago, which is a way to incorporate prior knowledge about coefficients in a standard linear regression model by mixing earlier estimates into the estimation process and under very general assumptions is found to be simultaneously best linear unbiased with respect to both sample and prior information (see also Theil [1963]). This notion of combining information from multiple sources is not particularly controversial among statisticians, as observed by Samaniego and Reneau (1994, p.957):

> If a prior distribution contains "useful" information about an unknown parameter, then the Bayes estimator with respect to that prior will outperform the best frequentist rule. Otherwise, it will not.

A more fundamental advantage to Bayesian statistics is that both prior and posterior parameter estimates are assumed to have a distribution and therefore give a more realistic picture of uncertainty that is also more useful in applied work:

> With conventional statistics, the only uncertainty admitted to the analysis is

---

[2]See Press and Tanur (2001) for a fascinating account of the role of researcher-introduced subjectivity in a number of specific famous scientific breakthroughs, including discoveries by Galileo, Newton, Darwin, Freud, and Einstein.

sampling uncertainty. The Bayesian approach offers guidance for dealing with the myriad sources of uncertainty faced by applied researchers in real analyses.

Western (1999, p.20). Lindley (1986, p.7) expresses a more biting statement of preference:

> Every statistician would be a Bayesian if he took the trouble to read the literature thoroughly and was honest enough to admit he might have been wrong.

This book rests on the perspective, sampled above, that the Bayesian approach is not only useful for social and behavioral scientists, but it also provides a more compatible methodology for analyzing data in the manner and form in which it arrives in these disciplines. As we describe in subsequent chapters, Bayesian statistics establishes a rigorous analytical platform with clear assumptions, straightforward interpretations, and sophisticated extensions. For more extended discussions of the advantages of Bayesian analysis over alternatives, see Berger (1986b), Dawid (1982), Efron (1986), Good (1976), Jaynes (1976), and Zellner (1985). We now look at how the Bayesian paradigm emerged over the last 250 years.

1.7   Introducing Markov Chain Monte Carlo Techniques

In this section we briefly discuss Bayesian computation and give a preview of later chapters. The core message is that these algorithms are relatively simple to understand in the abstract.

Markov chain Monte Carlo (MCMC) set the Bayesians free. Prior to 1990, it was relatively easy to specify an interesting and realistic model with actual data whereby standard results were unobtainable. Specifically, faced with a high dimension posterior resulting from a regression-style model, it was often very difficult or even impossible to perform multiple integration across the parameter space to produce a regression table of marginal summaries. The purpose of MCMC techniques is to replace this difficult analytical integration process with iterative work by the computer. When calculations similar to (1.16) are multidimensional, there is a need to summarize each marginal distribution to provide useful results to readers in a table or other format for journal submission. The basic principle behind MCMC techniques is that if an iterative chain of computer-generated values can be set up carefully enough, and run long enough, then *empirical* estimates of integral quantities of interest can be obtained from summarizing the observed output. If each visited multidimensional location is recorded as a row vector in a matrix, then the marginalization for some parameter of interest is obtained simply by summarizing the individual dimension down the corresponding column. So we replace an analytical problem with a sampling problem, where the sampling process has the computer perform the difficult and repetitive processes. This is an enormously important idea to Bayesians and to others since it frees researchers from having to make artificial simplifications to their model specifications just to obtain describable results.

These Markov chains are successive quantities that depend probabilistically only on the value of their immediate predecessor: the *Markovian property*. In general, it is possible to set up a chain to estimate multidimensional probability structures (i.e., desired probability distributions), by starting a Markov chain in the appropriate sample space and letting it run until it settles into the target distribution. Then when it runs for some time confined to this particular distribution, we can collect summary statistics such as means, variances, and quantiles from the simulated values. This idea has revolutionized Bayesian statistics by allowing the empirical estimation of probability distributions that could not be analytically calculated.

1.7.1   Simple Gibbs Sampling

As a means of continuing the discussion about conditional probability and covering some basic principles of the R language, this section introduces an important, and frequently used Markov chain Monte Carlo tool, the Gibbs sampler. The idea behind a Gibbs sampler is to get a marginal distribution for each variable by iteratively conditioning on interim values of the others in a continuing cycle until samples from this process empirically approximate the desired marginal distribution. Standard regression tables that appear in journals are simply marginal descriptions. There will be much more on this topic in Chapter 10 and elsewhere, but here we will implement a simple but instructive example.

As outlined by Example 2 in Casella and George (1992), suppose that we have two conditional distributions, where they are conditional on each other such that the parameter of one is the variable of interest in the other:

$$f(x|y) \propto y \exp[-yx], \quad f(y|x) \propto x \exp[-xy], \quad 0 < x, y < B < \infty. \tag{1.24}$$

These conditional distributions are both exponential probability density functions (see Appendix B for details). The upper bound, B, is important since without it there is no finite joint density and the Gibbs sampler will not work. It is possible, but not particularly pleasant, to perform the correct integration steps to obtain the desired marginal distributions: $f(x)$ and $f(y)$. Instead we will let the Gibbs sampler do the work computationally rather than us do it analytically.

The Gibbs sampler is defined by first identifying conditional distributions for each parameter in the model. These are conditional in the sense that they have dependencies on other parameters, and of course the data, which emerge from the model specification. The "transition kernel" for the Markov chain is created by iteratively cycling through these distributions, drawing values that are conditional on the latest draws of the dependencies. It is proven that this allows us to run a Markov chain that eventually settles into the desired limiting distribution that characterizes the marginals. In other language, it is an iterative process that cycles through conditional distributions until it reaches a stable status whereby future samples characterize the desired distributions. The important theorem here assures us that when we reach this stable distribution, the autocorrelated sequence of values can be treated as an iid sample from the marginal distributions of interest. The amazing part is that this is accomplished simply by ignoring the time index, i.e., putting the values in a "bag" and just "shaking it up" to lose track of the order of occurrence. Gibbs sampling is actually even more general than this. Chib (1995) showed how Gibbs sampling can be used to compute the marginal distribution of the sample data, i.e., the denominator of (1.13), by using the individual parameter draws. This quantity is especially useful in Bayesian hypothesis testing and model comparison, as we shall see in Chapter 7. The second half of this text applies this tool and similar methods of estimation.

For two parameters, x and y, this process involves a starting point, $[x_0, y_0]$, and the cycles are defined by drawing random values from the conditionals according to:

$$x_1 \sim f(x|y_0), \qquad\qquad y_1 \sim f(y|x_1)$$
$$x_2 \sim f(x|y_1), \qquad\qquad y_2 \sim f(y|x_2)$$
$$x_3 \sim f(x|y_2), \qquad\qquad y_3 \sim f(y|x_3)$$
$$\vdots \qquad\qquad\qquad\qquad \vdots$$
$$\vdots \qquad\qquad\qquad\qquad \vdots$$
$$x_m \sim f(x|y_{m-1}), \qquad\qquad y_m \sim f(y|x_m).$$

If we are successful, then after some reasonable period the values x_j, y_j are safely assumed to be empirical samples from the correct marginal distribution. There are many theoretical and practical concerns that we are ignoring here, and the immediate objective here is to give a rough overview.

The following steps indicate how the Gibbs sampler is set up and run:

▷ Set the initial values: $B = 10$, and $m = 50,000$. B is the parameter that ensures that the joint distribution is finite, and m is the desired number of generated values for x and y.

▷ Create x and y vectors of length m where the first value of each is a starting point uniformly distributed over the support of x and y, and all other vector values are filled in with unacceptable entries greater than B.

▷ Run the chain for $m = 50,000 - 1$ iterations beginning at the starting points. At each iteration, fill-in and save only sampled exponential values that are less than B, repeating this sampling procedure until an acceptable value is drawn to replace the unacceptable $B + 1$ in that position.

▷ Throw away some early part of the chain where it has not yet converged.

▷ Describe the marginal distributions of x and y with the remaining empirical values.

This leads to the following R code, which can be retyped verbatim, obtained from the book's webpage, or the book's R package BaM to replicate this example:

```
B <- 10; m <- 50000
gibbs.expo <- function(B,m) {
    x <- c(runif(1,0,B),rep((B+1),length=(m-1)))
    y <- c(runif(1,0,B),rep((B+1),length=(m-1)))
    for (i in 2:m)  {
        while(x[i] > B) x[i] <- rexp(1,y[i-1])
        while(y[i] > B) y[i] <- rexp(1,x[i])
    }
    return(cbind(x,y))
}

gibbs.expo(B=5, m=500)
```

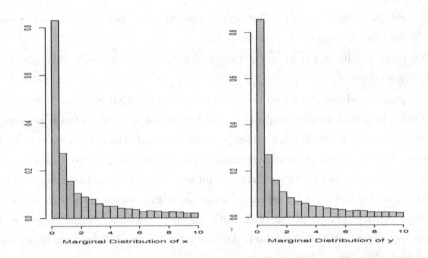

FIGURE 1.1: GIBBS SAMPLING, MARGINAL EXPONENTIALS

These samples are summarized by histograms of the empirical results for x and y in Figure 1.1, where $m = 50,000$ samples are drawn and the first $40,000$ are discarded (these are called "burn-in" values). It is clear from the figure that the marginal distributions are exponentially distributed. We can recover parameters by using the empirical draws to calculate sample statistics. This part of the MCMC process is actually quite trivial once we are convinced that there has been convergence of the Markov chain. In later chapters we will see this process in a more realistic, and therefore detailed, setting. This example is intended to give an indication of activities to come and to reinforce the linkage between Bayesian inference and modern statistical computing.

1.7.2 Simple Metropolis Sampling

Another Markov chain Monte Carlo tool with wide use is the Metropolis algorithm from statistical physics (Metropolis *et al.* 1953). The Metropolis algorithm is more flexible than

Gibbs sampling because it works with the joint distribution rather than a full listing of conditional distributions for the parameters in the model. As a result, many variations have been developed to satisfy particular sampling challenges posed by complicated models and ill-behaved target functions. Later chapters will cover these extensions in detail.

The essential idea behind the Metropolis algorithm is that, while we cannot easily generate values from the joint (posterior) distribution of interest, we can often find a "similar" distribution that is easy to sample from. Obviously we need to make sure that this alternative distribution is defined over the same support as the target distribution and that it does not radically favor areas of low density of this target. Once a candidate point in multivariate space has been produced by this candidate-generating distribution we will accept or reject it based upon characteristics of the target distribution. The algorithm is characterized by the following steps.

1. The candidate-generating distribution proposes that we move to some other point by drawing a point from *its* generating mechanism.

2. If this point produces a step on the target distribution that is of *higher* density, then we will always go there.

3. If this point produces a step on the target distribution that is of *lower* density, then we will go there probabilistically proportional to how much lower the step is in density.

Thus it is easy to see that the Markov chain "wanders around" the target density describing it as it goes and favoring higher density regions. The nice part is that the Markov chain will also explore other lower density regions as well, but with lower probability as we would want. Analogously, consider locking a house cat in large room with features that are attractive to cats (the high density regions of the posterior), and features that are unattractive to cats (the low density regions of the posterior). Anyone who has spent time with house cats can see at least some Markovian feature to their nature, as well as an innate curiosity. As our feline Markov chain wanders the room in a memory-less state, we record the coordinates of their travel. Over time we will find that the cat spends more time in the attractive areas, but still occasionally investigates the unattractive areas. If this attractiveness is proportional to density we want to describe, then the cat eventually produces description of the posterior distribution.

We can more precisely describe this algorithm. Suppose we have a two-dimensional target distribution, $p(x, y)$, which can be a posterior distribution from a Bayesian model, or any other form that is hard to marginalize, i.e., produce individual distributions $p(x)$ and $p(y)$. A single Metropolis step is produced by:

1. Sample (x', y') from the candidate-generating distribution, $q(x', y')$.

2. Sample a value u from $u[0:1]$.

3. If

$$a(x', y'|x, y) = \frac{p(x', y')}{p(x, y)} > u$$

then accept (x', y') as the new destination.

4. Otherwise keep (x, y) as the new destination.

The result is a chain of values, $[(x_0, y_0), (x_1, y_1), (x_2, y_2), \ldots]$. Note that unlike the Gibbs sampler, the Metropolis algorithm does not necessarily have to move to a new position at each iteration, and the decision to stay put is considered a Markovian step to the current position (time is consumed by the step).

There are a few technical details that we will worry about in much more detail later beginning in Chapter 10. Often the candidate-generating distribution produces values conditional on the current position, $q(x', y'|x, y)$, but this is not strictly necessary. The basic version described here requires that the candidate-generating distribution be symmetrical in its arguments, $q(x', y'|x, y) = q(x, y|x', y')$. Also, the choice of candidate-generating distribution can be complicated by the need to match irregularities in the target distribution. Finally, it is important in real applications to run the Markov chain for some initial period to let it settle into the distribution of interest before recording values.

Consider a problem similar to that above, but where we have a joint distribution for the parameters and not the desired marginals (or conditionals as used in the Gibbs sampler). The bivariate exponential distribution for $x, y \in [0{:}\infty]$ is given by the function:

$$p(x, y) = \exp[-(\lambda_1 + \lambda)x - (\lambda_2 + \lambda)y - \lambda \max(x, y)], \tag{1.25}$$

with non-negative parameters λ_1, λ_2, and λ. This model is common in reliability analysis (Marshall and Olkin 1967), where the interpretation is that the first two parameters are the event intensities for systems 1 and 2, and the non-subscripted parameter is the shared intensity between systems. In this literature events are usually machine failures, but for our purposes they can be death, graduation, cabinet dissolutions, divorce, cessation of war, and so on. In this example we have the parameters:

$$\lambda_1 = 0.5, \quad \lambda_2 = 0.5, \quad \lambda = 0.01, \quad B = \max(x) = \max(y) = 8,$$

which produces the bivariate distribution shown in the first panel of Figure 1.2. The maximum in the function makes it a little harder to analytically calculate marginal distributions with integration, so we might want to apply MCMC to save trouble. This is exactly analogous to the process where complicated Bayesian model specifications sometimes make it difficult to describe marginal posteriors for parameters of interest.

To implement the Metropolis algorithm we need a candidate-generating distribution from which to draw potential destinations for the Markov chain. Typically researchers look for some convenient distribution from the commonly used form since software such as R makes drawing values trivial. Here we will exploit the stipulated bounds on the problem and note that the bivariate exponential is enclosed in a big box with length and width equal to $B = 8$ and maximum height equal to one from the form of (1.25). The process is further covered in Chapter 9, but note here that it is easy to draw points inside this box from scaled uniforms. Nicely, we do not have to rescale the distribution of $q(x', y')$ because the values are drawn from this distribution but inserted into $p()$. It is important to note, without getting too far ahead of ourselves, that a better fitting candidate-generating distribution could be found and that drawing from uniform boxes is not particularly efficient.

To begin we define our function in R according to:

```
biv.exp <- function(x,y,L1,L2,L)
          exp( -(L1+L)*x - (L2+L)*y -L*max(x,y) )
```

So it will return density values for given (x, y) pairs and specific parameters. The candidate-generating function is:

```
cand.gen <- function(max.x,max.y)
            c(runif(1,0,max.x),runif(1,0,max.y))
```

where we could have stipulated the B value but left the function slightly more general. Markov chains require starting positions and we arbitrarily select $(x = 0.5, y = 0.5$ here. The algorithm is now given to be the following R code, which (again) can be retyped verbatim to replicate the example:

```
m <-5000; x<-0.5; y<-0.5; L1<-0.5; L2<-0.5; L<-0.01; B<-8
for (i in 1:m)  {
    cand.val <- cand.gen(B,B)
    a <- biv.exp(cand.val[1],cand.val[2],L1,L2,L)
            / biv.exp(x[i],y[i],L1,L2,L)
    if (a > runif(1)) {
        x <- c(x,cand.val[1])
        y <- c(y,cand.val[2])
    }
    else  {
        x <- c(x,x[i])
        y <- c(y,y[i])
    }
}
```

The resulting values are shown by the histograms in the second and third panels of Figure 1.2, where the algorithm has been run for $m = 5,000$ iterations but the first $3,000$ are discarded. We could also simply summarize the resulting marginals for x and y empirically with means, quantiles, or other simple statistics. The Metropolis algorithm shown here will be expanded and generalized in Chapter 10 by loosening restrictions on the candidate-generating distribution and allowing for hybrid processes that accommodate difficult features in the target distribution. The two MCMC algorithms described here form the basis for all practical work needed to estimate complex Bayesian models in the social sciences.

1.8   Historical Comments

Statistics is a relatively new field of scientific endeavor. In fact, for much of its history it was subsumed to various natural sciences as a combination of foster-child and household

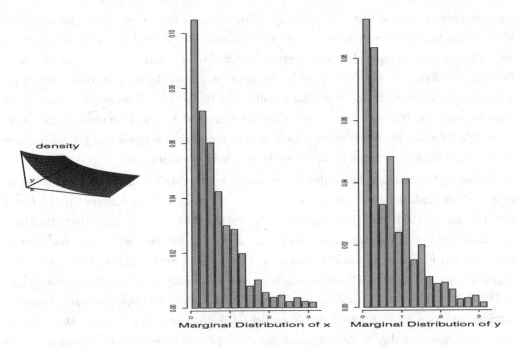

FIGURE 1.2: METROPOLIS SAMPLING, BIVARIATE EXPONENTIAL

maid: unwanted by its natural parents (mathematics and philosophy), yet necessary to clean things up. Beginning with the work of Laplace (1774, 1781, 1811), Gauss (1809, 1823, 1855), Legendre (1805), and de Morgan (1837, 1838, 1847), statistics began to emerge as a discipline worthy of study on its own merits. The first renaissance occurred around the turn of the last century due to the monumental efforts of Galton (1869, 1875, 1886, 1892), Fisher (1922, 1925a, 1925b, 1934), Neyman and (Egon) Pearson (1928a, 1928b, 1933a, 1933b, 1936a, 1936b), Gossett (as Student, 1908a, 1908b), Edgeworth (1892a, 1892b, 1893a, 1893b), (Karl) Pearson (1892, 1900, 1907, 1920), and Venn (1866). Left out of the twin intellectual developments of frequentist inference from Neyman and Pearson and likelihood inference from Fisher (see Chapter 2, Section 2.6 for details), was the Bayesian paradigm. Sir Thomas Bayes' famous (and only) essay was published in 1763, two years after his death (Bayes chose to perish before publishing), suggesting to some that he was ambivalent about the approach of applying a uniform prior to a binomial probability parameter. This ingenious work unintendedly precipitated a philosophy about how researcher-specified theories are fit to empirical observations. Interestingly, it was not until the early 1950s that Bayesian statistics became a self-aware branch (Fienberg 2006).

Fisher in particular was hostile to the Bayesian approach and was often highly critical, though not always with substantiated claims: Bayesianism "which like an impenetrable jungle arrests progress towards precision of statistical concepts" (1922, p.311). Fisher also worked·to discredit Bayesianism and inverse probability (Bayesianism with an assumed uniform prior) by pressuring peers and even misquoting other scholars (Zabell 1989). Yet Fisher (1935) develops *fiducial inference*, which is an attempt to apply inverse probability without uniform priors, but this approach fails; Efron (1998, p.105) calls this "Fisher's biggest blunder." In fact, Lindley (1958) later proved that fiducial inference is consistent *only* when it is made equivalent to Bayesian inference with a uniform prior. The Neyman-Pearson paradigm was equally unkind to the development of Bayesian statistics, albeit on a less vindictive level. If one is willing to subscribe to the idea of an infinite sequence of samples, then the Bayesian prior is unimportant since the data will overwhelm this prior. Although there are scenarios where this is a very reasonable supposition, generally these are far more difficult to come by in the social and behavioral sciences.

Although Bayesianism had suffered "a nearly lethal blow" from Fisher and Neyman by the 1930s (Zabell 1989), it was far from dead. Scholars such as Jeffreys (1961), Good (1950), Savage (1954, 1962), de Finetti (1972, 1974, 1975), and Lindley (1961, 1965) reactivated interest in Bayesian methods in the middle of the last century in response to observed deficiencies in classical techniques. Lindley and Novick (1978, 1981) published important applied work in education psychology that carefully studied exchangeability and utility from a Bayesian perspective, and Novick *et al.* (1976) developed an early Bayesian software program for estimating simple models: CADA, Computer Assisted Data Analysis. Unfortunately many of the specifications developed by these modern Bayesians, while superior in theoretical foundation, led to mathematical forms that were intractable.[3] Fortunately, this problem has been largely resolved in recent years by a revolution in statistical computing techniques, and this has led to a second renaissance for the Bayesian paradigm (Berger 2001).

Markov chain Monte Carlo (MCMC) techniques solve a lingering problem in Bayesian analysis, and thus earn a special place in this work. Often Bayesian model specifications considered either interesting or realistic produced inference problems that were analytically intractable because they led to high-dimension integral calculations that were impossible to solve analytically. Previous numerical techniques for performing these integrations were often difficult and highly specialized tasks (e.g., Shaw 1988, Stewart and Davis 1986, van Dijk and Kloek 1982, Tierney and Kadane 1986). Beginning with the foundational work of Metropolis *et al.* (1953), Hastings (1970), Peskun (1973), Geman and Geman (1984), and the critical synthesizing essay of Gelfand and Smith (1990), there is now a voluminous literature on Markov chain Monte Carlo. In fact, modern Bayesian statistical practice is intimately and intrinsically tied to stochastic simulation techniques and as a result, these tools are an integral part of this book. We introduce these tools in this chapter in Section 1.7 and in much greater detail in Chapter 10.

Currently the most popular method for generating samples from posterior distributions using Markov chains is the WinBUGS program and its Unix-based precursor BUGS and the more recent functional equivalent JAGS. The name BUGS is a pseudo-acronym for *Bayesian inference Using Gibbs Sampling*, referring to the most frequently used method for producing Markov chains. In what constitutes a notable and noble contribution to the Bayesian

---

[3]This led one observer (Evans 1994) to compare Bayesians to "unmarried marriage guidance counselors."

statistical world, the Community Statistical Research Project at the MRC Biostatistics Unit and the Imperial College School of Medicine at St. Mary's, London provide this high-quality software to users free of charge, and it can be downloaded from their web page: `http://www.mrc-bsu.cam.ac.uk/software/bugs/`. These authors have even made available extensive documentation at the same site by Spiegelhalter *et al.* (1996a, 1996b, 2000, 2012). Alternative ways to use `WinBUGS` with R as the interface are: `BRugs`, `rbugs`, and `R2WinBUGS`. There are also facilities for calling `WinBUGS` from `SAS`, `stata`, and `excel`. The `JAGS` program (Just Another Gibbs Sampler) is an engine for the `BUGS` language that has nearly the same structure as `WinBUGS`, with only a few syntactical differences. Authored by Martyn Plummer, it is extremely well-developed software that runs on non-windows platforms and is command-line driven rather than point-and-click. It can be downloaded at `http://www-ice.iarc.fr/~martyn/software/jags/`. There are also facilities for calling `JAGS` from R: `R2jags`, `Rjags`, and `runjags`. Most of the `BUGS` code in this text are run with `JAGS` from the command window. Other high-quality R packages using or providing MCMC computing include: `BMS`, `dclone`, `eco`, `glmdm`, `HI`, `lmm`, `MasterBayes`, `mcmc`, `MCMCglmm`, `MCMCpack`, `MNP`, `pscl`, `spBayes`, `tgp`, and `zic`. Of these `MCMCpack` is the most general, whereas most of the others are MCMC implementations to solve a specific problem. Given the rapid pace of R package development, this list is growing rapidly.

1.9 Exercises

1.1   Restate the three general steps of Bayesian inference from page 6 in your own words.

1.2   Given k possible disjoint (non-overlapping) events labeled: $E_1, \ldots, E_k$ where k could even be infinity, denote $p(E_i)$ as the mapping from events E_i to $[0:1]$ space. Write the Kolmogorov axioms of probability in technical detail.

1.3   Rewrite Bayes' Law when the two events are independent. How do you interpret this?

1.4   Equation (1.11) on page 12 showed that $p(A|\mathbf{D}) = p(\mathbf{D}|A)p(A)/(p(\mathbf{D}|A)p(A) + p(\mathbf{D}|B)p(B) + p(\mathbf{D}|C)p(C))$. Rewrite this expression for $p(A|\mathbf{D})$ when there are arbitrary $k \in \mathcal{I}^+$ events including A.

1.5   Suppose $f(\theta|\mathbf{X})$ is the posterior distribution of θ given the data $\mathbf{X}$. Describe the shape of this distribution when the mode, $\underset{\theta}{\mathrm{argmax}} f(\theta|\mathbf{X})$, is equal to the mean, $\int_\theta \theta f(\theta|\mathbf{X}) d\theta$.

1.6   The Rényi countable additivity axiom is defined by: (1) for any events E_1 and E_2, $p(E_1|E_2) \geq 0$ (and reversed), $p(E_i|E_i) = 1$, (2) for disjoint sets $E_1, \ldots$ and another arbitrary event D, $p(\cup_{i=1}^{\infty}E_i|D) = \sum_{i=1}^{\infty}p(E_i|D)$, and (3) for every sub-set of events, E_i, E_j, E_k, with $E_j \subseteq E_k$ and $p(E_j|E_k) > 0$, we get $p(E_i|E_j) = p(E_i, E_j|E_k)/p(E_j|E_k)$. Show that the Kolmogorov axioms are a special case.

1.7   Using R run the Gibbs sampling function given on page 26. What effect do you see in varying the B parameter? What is the effect of producing 200 sampled values instead of 50,000?

1.8   Some authors have objected to the uniform prior, $p(\theta) = 1, \theta \in [0:1]$ to describe unknown probabilities in a binomial model and suggested instead the Haldane prior: $p(\theta) \propto [\theta^{-1}(1-\theta)]^{-1}$ (Haldane [1938], Novick and Hall [1965], Villegas [1977],). Plot this prior and the uniform prior over $[0:1]$ in the same graph.

1.9   Rerun the Metropolis algorithm on page 30 in R but replacing the uniform gen-eration of candidate values in cand.gen with a normal truncated to fit in the appropriate range. What differences do you observe?

1.10   The Gibbs sampler described in Section 1.7.1 from Casella and George (1992) was originally done as follows: (1) set initial values for $B = 5$, $k = 15$, $m = 5,000$, and the set of accepted values (x, y) as an empty object, (2) run m chains of length $k+1$ where the first value is the uniformly distributed starting point $[0:B]$) and the rest are sampled conditional exponential values that are less than B, (3) save only the last value from the x and y series, x_{16} and y_{16} to the stored Markov chain until 5,000 of each are obtained. Implement this alternative algorithm in R and compare it to the output shown in Figure 1.1 on page 27.

1.11   If $p(\mathbf{D}|\theta) = 0.5$, and $p(\mathbf{D}) = 1$, calculate the value of $p(\theta|\mathbf{D})$ for priors $p(\theta)$, $[0.001, 0.01, 0.1, 0.9]$.

1.12   Buck, Cavanaugh, and Litton (1996) demonstrate the use of Bayesian statistics for radiocarbon dating of Early Bronze Age archaeological samples (seeds and bones) from St. Veit-Klinglberg, Austria. These ten age data points are produced by the Oxford accelerator dating facilities:

Context #	μ_i	σ_i
758	3275	75
814	3270	80
1235	3400	75
493	3190	75
925	3420	65
923	3435	60
1168	3160	70
358	3340	80
813	3270	75
1210	3200	70

Given the model $X_i | \mu_i, \sigma_i \sim \mathcal{N}(\mu_i, \sigma_i^2)$, (1) calculate the probability that sample 358 originates between 3300 to 3400 years ago, (2) generate 10,000 samples from the distribution for sample 493 and sample 923 and plot a histogram of these in the same figure (side-by-side), (3) give the proportion of values that overlap, and (4) how do you interpret this overlap probabilistically with regard to the age of the samples?

1.13 Sometimes Bayesian results are given as *posterior odds ratios*, which for two possible alternative hypotheses is expressed as:

$$\text{odds}(\theta_1, \theta_2) = \frac{p(\theta_1 | \mathbf{D})}{p(\theta_2 | \mathbf{D})}.$$

If the prior probabilities for θ_1 and θ_2 are identical, how can this be re-expressed using Bayes' Law?

1.14 Using the posterior distribution in (1.19) on page 19, produce the posterior mean for θ in (1.20) and the posterior variance for θ in (1.21).

1.15 Suppose we had data, $\mathbf{D}$, distributed $p(\mathbf{D}|\theta) = \theta e^{-\theta \mathbf{D}}$ as in Section 1.5.1 starting on page 18, but now $p(\theta) = 1/\theta$, for $\theta \in (0{:}\infty)$. Calculate the posterior mean.

1.16 Modify the Gibbs sampler in Section 1.7.1 starting on page 25 to sample from two mutually conditional gamma distributions instead of exponential distributions. The exponential distribution is a simplified form of the rate parameter gamma distribution where the first (shape) parameter is 1 (Appendix B, page 580). Set the two relevant shape parameters to values of your choosing $\alpha > 1$. Produce a graphs of the marginal draws.

1.17 Since the posterior distribution is a compromise between prior information and the information provided by the new data, then it is interesting to compare relative strengths. Perform an experiment where you flip a coin 10 times, recording the data as zeros and ones. Produce the posterior expected value (mean) for two priors on p (the probability of a heads): a uniform distribution between zero and one, and a beta distribution (Appendix B) with parameters $[10, 1]$. Which prior appears to influence the posterior mean more than the other?

1.18 Fisher (1930) defined fiducial inference for a parameter θ with maximum likelihood estimate $\hat{\theta}$ by making the CDF (cumulative distribution function) of $\hat{\theta}|\theta$ uniform over $[0:1]$. Taking a derivative of this CDF with respect to $\hat{\theta}$ then gives the associated PDF:

$$f(\hat{\theta}|\theta) = \left| \frac{d}{d\hat{\theta}} F(\hat{\theta}|\theta) \right|,$$

and taking a derivative of the CDF with respect to θ produced what Fisher called the fiducial distribution of the parameter θ given the statistic $\hat{\theta}$. Fisher (1956) later claimed that this approach "uses the observations only to change the logical status of the parameter from one in which nothing is known of it, and no probability statement about it can be made, to the status of a random variable having a well defined distribution." Show that this cannot be true and that this is not Bayesian inference.

1.19   The *Stopping Rule Principle* states that if a sequence of experiments (trials) is governed by a stopping rule, η, that dictates cessation, then inference about the unknown parameters of interest must depend on η only through the collected sample. Obvious stopping rules include setting the number of trials in advance (i.e., reaching n is the stopping rule), and stopping once a certain number of successes are achieved. Consider the following experiment and stopping rule. Standard normal distributed data are collected until the absolute value of the mean of the data exceeds $1.96/\sqrt{n}$. Explain why this fails as a non-Bayesian stopping rule for testing the hypothesis that $\mu = 0$ (the underlying population is zero), but is perfectly acceptable for Bayesian inference.

1.20   The PDF of a C-finite mixture of normal distributions is given by:

$$f(y_i|\boldsymbol{\mu}, \boldsymbol{\Sigma}) = \sum_{c=1}^{C} \omega_c \mathcal{N}(y_i|\mu_c, \sigma_c^2),$$

where ω_c is the weight placed on distribution c with $\sum \omega_c = 1$, and $\mathcal{N}()$ denotes a normal distribution. For a mixture with means $\boldsymbol{\mu} = [-3, 1, 4]$, variances $\boldsymbol{\Sigma} = [0.3, 1, 0.2]$, and weights $\Omega = [0.2, 0.5, 0.3]$ develop a simple Metropolis (symmetric) algorithm to get draws from this mixture using a Student's-t candidate distribution where you determine a good choice for the degrees of freedom parameter, ν. Describe the mixture distribution empirically with summary statistics and a graph from producing 10,000 draws and disposing of the first 5,000.

Chapter 2

Specifying Bayesian Models

2.1 Purpose

This chapter changes the discussion from the basic workings of Bayes' Law in a probability context to a focus on the use of Bayes' Law for realistic statistical models. Consequently, the first order of business is to go from our previous vague definition of data, $\mathbf{D}$, to a rectangular $n \times k$ matrix of data, $\mathbf{X}$. In this chapter we also make the move from the unspecific $p()$ for posterior distributions to the more clear $\pi()$ notation in order to distinguish them from priors, likelihoods, and other functions. Also, from now on we use the vector form of theta, $\boldsymbol{\theta}$, since nearly all interesting social science models are multidimensional.

In the immediately forthcoming material, we cover the core idea of Bayesian statistics: updating prior distributions by conditioning on data through the likelihood function. We will also look at repeating this updating process as new information becomes available. There is an additional historical discussion placing this modeling approach into context.

2.2 Likelihood Theory and Estimation

In order to make inferences about unknown model parameters in generalized linear models, Bayesian probability models, or any other parametric specification, we would like to have a description of parameter values that are more or less probable given the observed data and the parametric form of the model. In other words, some values of an unknown parameter are certainly more likely to have generated the data than others, and if there is one value that is more likely than all others, we would typically prefer to report that one.

For instance, suppose we wanted to know the probability of getting a heads with a possibly unfair coin. Flipping it ten times, we observe 5 heads. It seems logical to infer that $p = 0.5$ is more likely than $p = 0.4$, or $p = 0.6$, or any other value for that matter. In this case, $p = 0.5$ is the value that maximizes the likelihood function given the observed series of flips. Maximizing a likelihood function with regard to coefficient values is without question the most frequently used estimation technique in applied statistics.

Stipulate now that we are interested in analyzing a model for a k-dimensional unknown $\boldsymbol{\theta}$ vector, $k - 1$ explanatory variables, a constant, and n data points. Asymptotic theory assures us that for sufficiently large samples the likelihood surface is unimodal in k dimensions for the commonly used forms (Lehmann 1999). Denote this likelihood function as $L(\boldsymbol{\theta}|\mathbf{X})$ even though it is constructed as the joint distribution of the iid outcomes: $p(\mathbf{X}|\boldsymbol{\theta}) = f(x_1|\boldsymbol{\theta})f(x_2|\boldsymbol{\theta})\cdots f(x_n|\boldsymbol{\theta})$.

The likelihood function differs from the inverse probability, $p(\boldsymbol{\theta}|\mathbf{X})$, in that it is necessarily a *relative* function since it is not a normalized probability measure bounded by zero and one. From a frequentist standpoint, the probabilistic uncertainty is a characteristic of the random variable $\mathbf{X}$, not the unknown but fixed $\boldsymbol{\theta}$. Barnett (1973, p.131) clarifies this distinction: "Probability remains attached to X, not θ; it simply reflects inferentially on θ." Thus maximum likelihood estimation substitutes the unbounded notion of likelihood for the bounded definition of probability (Casella and Berger 2002, p.316; Fisher 1922, p.327; King 1989, p.23). This is an important theoretical distinction, but of little significance in applied practice. If we regard $p(\mathbf{X}|\boldsymbol{\theta})$ as a function of $\boldsymbol{\theta}$ for some given observed data $\mathbf{X}$, then $L(\boldsymbol{\theta}|\mathbf{X}) = \prod_{i=1}^{n} p(\mathbf{X}|\boldsymbol{\theta})$ (DeGroot 1986, p.339).

Typically it is mathematically more convenient to work with the natural log of the likelihood function. This does not change any of the resulting parameter estimates because the likelihood function and the log likelihood function have identical modal points for commonly used forms. Using a PDF for a single parameter of interest, the basic *log* likelihood function is very simple:

$$\ell(\boldsymbol{\theta}|\mathbf{X}) = \log(L(\boldsymbol{\theta}|\mathbf{X})), \tag{2.1}$$

where we use $\ell(\boldsymbol{\theta}|\mathbf{X})$ as shorthand to distinguish the log likelihood function from the likelihood function, $L(\boldsymbol{\theta}|\mathbf{X})$.

The score function is the first derivative of the log likelihood function with respect to the parameters of interest:

$$\dot{\ell}(\boldsymbol{\theta}|\mathbf{X}) = \frac{\partial}{\partial\boldsymbol{\theta}}\ell(\boldsymbol{\theta}|\mathbf{X}). \tag{2.2}$$

Setting $\dot{\ell}(\boldsymbol{\theta}|\mathbf{X})$ equal to zero and solving gives the maximum likelihood estimate, $\hat{\boldsymbol{\theta}}$. This is now the "most likely" value of $\boldsymbol{\theta}$ from the parameter space Θ treating the observed data as given: $\hat{\boldsymbol{\theta}}$ maximizes the likelihood function at the observed values. The *Likelihood Principle* (Birnbaum 1962) states that once the data are observed, and therefore treated as given, all of the available evidence for estimating $\boldsymbol{\theta}$ is contained in the (log) likelihood function, $\ell(\boldsymbol{\theta}|\mathbf{X})$. This is a handy data reduction tool because it tells us exactly what treatment of the data is important to us and allows us to ignore an infinite number of alternates (Poirer 1988, p.127). The key difference between the classic likelihood approach and the Bayesian inference is that more information is used in the analysis and more information is provided through descriptions of the posterior beyond modal summaries. Thus the likelihood principle only has relevance here for part of the Bayesian model.

The maximum likelihood doctrine states that an admissible $\boldsymbol{\theta}$ that maximizes the likelihood function probability (discrete case) or density (continuous case), relative to alternative values of $\boldsymbol{\theta}$, provides the $\boldsymbol{\theta}$ that is "most likely" to have generated the observed data, $\mathbf{X}$, given the assumed parametric form. Restated, if $\hat{\boldsymbol{\theta}}$ is the maximum likelihood estimator for the unknown parameter vector, then it is necessarily true that $L(\hat{\boldsymbol{\theta}}|\mathbf{X}) \geq L(\boldsymbol{\theta}|\mathbf{X}) \ \forall \ \boldsymbol{\theta} \in \Theta$, where Θ is the admissible set of $\boldsymbol{\theta}$. Admissible here means values of $\boldsymbol{\theta}$ are taken from the valid parameter space (Θ): values of $\boldsymbol{\theta}$ that are unreasonable according to the form of the sampling distribution of $\boldsymbol{\theta}$ are not considered (integrated over).

Setting the score function from the joint PDF or PMF equal to zero and rearranging gives the likelihood equation:

$$\sum t(X_i) = n \frac{\partial}{\partial \boldsymbol{\theta}} E[\mathbf{X}] \tag{2.3}$$

where $\sum t(X_i)$ is the remaining function of the data, depending on the form of the probability density function (PDF) or probability mass function (PMF), and $E[\mathbf{X}]$ is the expectation over the kernel of the density function for $\mathbf{X}$. The kernel of a PDF or PMF is the component of the parametric expression that directly depends on the form of the random variable, i.e., what is left when normalizing constants are omitted. We can often work with kernels of distributions for convenience and recover all probabilistic information at the last stage of analysis by renormalizing (ensuring summation or integration to one). The kernel is the component of the distribution that assigns *relative* probabilities to levels of the random variable (see Gill 2000, Chapter 2). For example the kernel of a gamma distribution is just the part $x^{\alpha-1} \exp[-x\beta]$, without the normalizing constant $\beta^\alpha / \Gamma(\alpha)$.

The underlying theory here is remarkably strong. Solving (2.3) for the unknown coefficient produces an estimator that is unique (due to a unimodal posterior distribution), consistent (converges in probability to the population value), and asymptotically unbiased, but not necessarily unbiased in finite sample situations. On the latter point, the maximum likelihood estimate for the variance of a normal model, $\hat{\sigma}^2 = \frac{1}{n} \sum (x_i - \bar{x})^2$ is biased by $n/(n-1)$. This difference is rarely of significance and clearly the bias disappears in the limit, but it does illustrate that unbiasedness of the maximum likelihood estimate is guaranteed only in asymptotic circumstances. It is also asymptotically efficient (the variance of the estimator achieves the lowest possible value as the sample size becomes adequately large: the Cramér-Rao lower bound, see Shao 2005). This result combined with the central limit theorem gives the asymptotic normal form for the estimator: $\sqrt{n}(\hat{\boldsymbol{\theta}} - \boldsymbol{\theta}) \xrightarrow{\mathcal{P}} \mathcal{N}(\mathbf{0}, \Sigma_{\boldsymbol{\theta}})$. This means that as the sample size gets large, the difference between the estimated value of $\boldsymbol{\theta}$ and the true value of $\boldsymbol{\theta}$ gets progressively close to zero, with a variance governed by $\frac{1}{\sqrt{n}} \Sigma_{\boldsymbol{\theta}}$, where $\Sigma_{\boldsymbol{\theta}}$ is the $k \times k$ variance-covariance matrix for $\boldsymbol{\theta}$. Furthermore, $\sum t(x_i)$ is a sufficient statistic for $\boldsymbol{\theta}$, meaning that all of the relevant information about $\boldsymbol{\theta}$ in the data is contained in $\sum t(x_i)$. For example, the normal log likelihood expressed as a joint exponential family form is $\ell(\boldsymbol{\theta}|\mathbf{X}) = \left(\mu \sum x_i - \frac{n\mu^2}{2}\right) / \sigma^2 - \frac{1}{2\sigma^2} \sum x_i^2 - \frac{n}{2} \log(2\pi\sigma^2)$. So $t(\mathbf{X}) = \sum X_i$, $\frac{d}{d\mu} \frac{n\mu^2}{2} = n\mu$, and equating gives the maximum likelihood estimate of μ to be the sample average that we know from basic texts: $\frac{1}{n} \sum x_i$. Bayesian inference builds upon this strong foundation by combining likelihood information, as just described, with prior information in a way describes all unknown quantities distributionally.

2.3   The Basic Bayesian Framework

Our real interest lies in obtaining the distribution of the unknown k-dimensional $\boldsymbol{\theta}$ coefficient vector, given an observed $\mathbf{X}$ matrix of data values: $p(\boldsymbol{\theta}|\mathbf{X})$. If we choose to here, we can still determine the "most likely" values of the $\boldsymbol{\theta}$ vector using the k-dimensional posterior mode or mean, but it is better to more fully describe the shape of the posterior distribution, given by Bayes' Law:

$$p(\boldsymbol{\theta}|\mathbf{X}) = p(\mathbf{X}|\boldsymbol{\theta})\frac{p(\boldsymbol{\theta})}{p(\mathbf{X})} \tag{2.4}$$

where $p(\mathbf{X}|\boldsymbol{\theta})$ is the joint probability function for data (the *probability of the sample* for a fixed $\boldsymbol{\theta}$) under the assumption that the data are independent and identically distributed according to $p(X_i|\boldsymbol{\theta}) \ \forall \ i = 1, \ldots, n$, and $p(\boldsymbol{\theta})$, $p(\mathbf{X})$ are the corresponding unconditional probabilities. This is mechanically correct but it does not fully represent Bayesian thinking or notation about the inference process.

2.3.1   Developing the Bayesian Inference Engine

From the Bayesian perspective, there are only two fundamental types of quantities: known and unknown. The goal is to use the known quantities along with a specified parametric expression to make inferential statements about the unknown quantities. The definition of such unknown quantities is very general; they can be any missing data or unknown parameters. When quantities are observed, they are considered fixed and conditioned upon. Suppose we fully observe the data $\mathbf{X}$. This is now a fixed and given quantity in the inferential process. The first implication is that $p(\mathbf{X}|\boldsymbol{\theta})$ in (2.4) does not make notational sense since the known quantity is conditional on the unknown quantity. Instead label this quantity as $L(\boldsymbol{\theta}|\mathbf{X})$ and treat it as a likelihood function. It is a likelihood function of course, but note that the justification is inherently Bayesian (i.e., probabilistic). Also, since the $\mathbf{X}$ are treated as fixed, $p(\mathbf{X})$ is not especially useful here. However, this quantity performs an important role in model comparison as we shall see in Chapter 7.

The prior distribution, $p(\boldsymbol{\theta})$, must be specified, but need not be highly influential. This is simply a distributional statement about the unknown parameter vector $\boldsymbol{\theta}$, *before* observing or conditioning on the data. Much controversy has developed about the nature of prior distributions and we will look at alternative forms in detail in Chapter 4. It is *essential* to supply a prior distribution in Bayesian models and well over 100 years of futile searching for a way to avoid doing so have clearly demonstrated this. Currently an approach called *objective Bayes* (O-Bayes) seeks to mathematically minimize the effect of prior specifications.

Start with the form of Bayes' Law defined with conditional probability, giving the posterior of interest:

$$\pi(\boldsymbol{\theta}|\mathbf{X}) = \frac{p(\boldsymbol{\theta})L(\boldsymbol{\theta}|\mathbf{X})}{p(\mathbf{X})}, \tag{2.5}$$

which is an update of (2.4) that gives the desired probability statement on the left-hand side now using the $\pi()$ notation as a reminder. This states that the distribution of the unknown parameter conditioned on the observed data is equal to the product of the prior distribution assigned to the parameter and the likelihood function, divided by the unconditional probability of the data. The form of (2.5) can also be expressed as:

$$\pi(\boldsymbol{\theta}|\mathbf{X}) = \frac{p(\boldsymbol{\theta})L(\boldsymbol{\theta}|\mathbf{X})}{\int_{\Theta} p(\boldsymbol{\theta})L(\boldsymbol{\theta}|\mathbf{X})d\boldsymbol{\theta}}, \tag{2.6}$$

where $\int_{\Theta} p(\boldsymbol{\theta})L(\boldsymbol{\theta}|\mathbf{X})d\boldsymbol{\theta}$ is an expression for $p(\mathbf{X})$ explicitly integrating the numerator over the support of $\boldsymbol{\theta}$. This term has several names in the literature: the *normalizing constant*, the *normalizing factor*, the *marginal likelihood*, and the *prior predictive distribution*, although it is actually the marginal distribution of the data, and it ensures that $\pi(\boldsymbol{\theta}|\mathbf{X})$ integrates to one as required by the definition of a probability function. A more compact and succinct form of (2.6) is developed by dropping the denominator and using proportional notation since $p(\mathbf{X})$ does not depend on θ and therefore provides no relative inferential information about more or less likely values of θ:

$$\pi(\theta|\mathbf{X}) \propto p(\theta)L(\theta|\mathbf{X}), \tag{2.7}$$

meaning that the unnormalized *posterior* (sampling) distribution of the parameter of interest is proportional to the prior distribution times the likelihood function:

Posterior Probability $\propto$ Prior Probability $\times$ Likelihood Function.

It is typically (but not always, see later chapters) easy to renormalize the posterior distribution as the last stage of the analysis to return to (2.6).

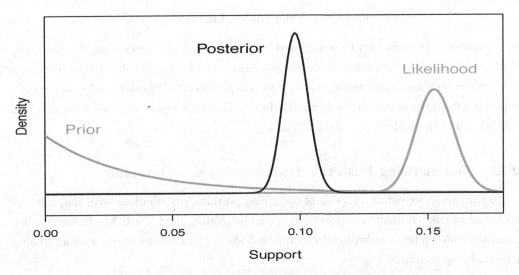

FIGURE 2.1: POSTERIOR $\propto$ PRIOR $\times$ LIKELIHOOD

As an illustration, suppose we have data that are iid exponentially distributed $f(X|\theta) = \theta e^{-\theta X}$, $X, \theta \in (0, \infty)$, and an exponential prior distribution for the unknown parameter $p(\theta) = \beta e^{-\theta\beta}$, $\beta \in (0, \infty)$, where $\beta = 30$ here as an arbitrary modeling choice. These data are actually taken from Example 2.3.2.1 on page 44 below but the specific data context is not yet important here. The exponential assumption for the data means that the likelihood function is a gamma distribution with parameters $n + 1$ and $\sum X_i$: $L(\theta) \propto \theta^n e^{-\theta \sum X_i}$. Multiplying the likelihood and the prior produces a another gamma distribution with new parameters $n + 1$ and $\sum X_i + \beta$ (Exercise 2 in this chapter). This is illustrated in Figure 2.1 where we see that the prior distribution on the left pulls the likelihood function towards it in the creation of the posterior distribution. This is called *shrinkage* in the Bayesian literature and it means that the posterior mean "shrinks" towards the prior mean. Figure 2.1 is purposefully over-dramatic in showing this effect, but movement such as this is a characteristic of all Bayesian models: the posterior distribution is always a compromise between the prior distribution and the likelihood function. The question is how influential is the prior distribution in this calculation.

We can also state Bayes' Law in odds form as done in (1.6) on page 11. Suppose we have two competing models expressed by θ_1 and θ_2, which are considered to exhaust the possible states of nature. This latter assumption may be unrealistic, but it is often the case that a researcher will consider only two alternatives at a time. If we now observe the data, $\mathbf{X}$, then Bayes' Law in odds form is:

$$\frac{\pi(\theta_1|\mathbf{X})}{\pi(\theta_2|\mathbf{X})} = \frac{\frac{p(\theta_1)}{p(\mathbf{X})} L(\theta_1|\mathbf{X})}{\frac{p(\theta_2)}{p(\mathbf{X})} L(\theta_2|\mathbf{X})}$$

$$= \frac{p(\theta_1)}{p(\theta_2)} \frac{L(\theta_1|\mathbf{X})}{L(\theta_2|\mathbf{X})}$$

Posterior Odds = Prior Odds × Likelihood Ratio.

This likelihood ratio will later be generalized in Chapter 7 as the Bayes Factor. Furthermore, if we assume equal prior probabilities, the posterior odds is simply equal to the likelihood ratio. Since likelihood ratio testing is a very popular tool in non-Bayesian model comparison, this is a nice linkage: under basic circumstances Bayesian posterior odds comparison is equivalent to simple likelihood ratio testing.

2.3.2 Summarizing Posterior Distributions with Intervals

In Chapter 1, we noted the value of describing posterior distributions with simple quantiles, and calculated analytical posterior moments: $E[\theta|\mathbf{X}]$, and $\text{Var}[\theta|\mathbf{X}]$. However, such summaries may miss distributional features and should be complemented with additional interval-based measures.

The first descriptive improvement here is found by moving from confidence intervals to Bayesian credible intervals. Recall that confidence intervals are intimately tied with frequentist (not likelihoodist!) theory since a $100(1-\alpha)\%$ confidence interval covers the *true* underlying parameter value across $1-\alpha$ proportion of the replications in the experiment, *on average*. So confidence is a property of frequentist replication from a large number of repeated iid samples and underlying parameters that are fixed immemorial. In fact, the confidence interval may be considered *the most* frequentist summary possible since it does not have an interpretation without multiple replications of the exact same experiment. One major problem with the confidence interval lies in its interpretation. Most consumers of statistics *want* confidence intervals to be probabilistic statements about some region of the parameter space, but careful writers discourage this by explaining the actual nature of confidence intervals: "a 95% confidence interval covers the true value of the parameter in nineteen out of twenty trials on average." In most social science settings with observational data it is not practical to repeat some experiment nineteen more times with an assumed iid data-generating source.

2.3.2.1 Bayesian Credible Intervals

The Bayesian analogue to the confidence interval is the credible interval and more generally the credible set, which does not have to be contiguous. Most of the time in practice it is calculated in *exactly the same way* as the confidence interval for unimodal symmetric forms. For instance calculating a 95% credible interval under the Gaussian normal assumption means marching out 1.96 standard errors from the mean in either direction, just like the analogous confidence interval is created. However, for asymmetric distributions this algorithm would produce a credible interval with unequal tails and incorrect coverage.

The difference between confidence intervals and Bayesian credible intervals is in the interpretation of what the interval means. A $100(1-\alpha)\%$ credible interval gives the region of the parameter space where the probability of covering $\boldsymbol{\theta}$ is equal to $1-\alpha$ (it may actually be a little more than $1-\alpha$ for discrete parameter spaces in order to guarantee at least this level of coverage). In contrast, applying this new definition to the *confidence interval* means that the probability of coverage is either zero or one, since it either covers the true $\boldsymbol{\theta}$ or it doesn't.

Formally, an equal tail credible set for the posterior distribution is defined as follows. Define C as a subset of the parameter space, $\boldsymbol{\Theta}$, such that a $100(1-\alpha)\%$ credible interval meets the condition:

$$1 - \alpha = \int_C \pi(\boldsymbol{\theta}|\mathbf{X})d\theta \qquad (2.8)$$

(this is summation instead of an integration for discrete parameter spaces, but we will discuss mostly continuous parameter spaces here). It is important to note that credible intervals are not unique. That is, we can easily define C in different ways to cover varying parts of $\boldsymbol{\Theta}$ and still meet the probabilistic condition in (2.8). It is not necessary that we center these intervals at a mean or mode. Important differences arise in asymmetric and multimodal distributions, and the convention is to create *equal tail intervals*: no matter what the shape of the posterior distribution. This means that the $100(1-\alpha)\%$ credible interval is created such that $\alpha/2$ of the density is put in both the left and right tails outside of the designated credible interval.

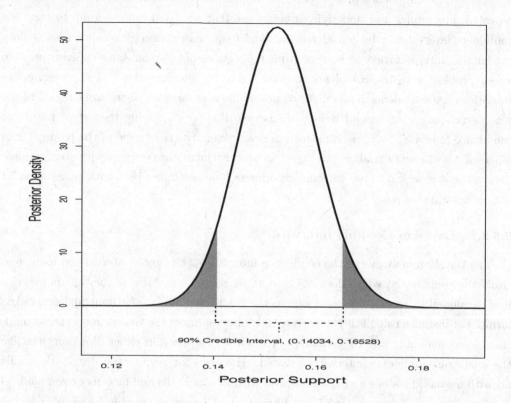

FIGURE 2.2: CI for State Duration Time to Adaptation, 1998-2005

■ **Example 2.1: Credible Interval, Fifty U.S. States Time to Adoption for Health Bills.** Boehmke (2009) counts bills passed in the fifty states between 1998 and 2005 that contain policy implications for the increasing obesity rates in the U.S. These include limits on sugary drinks at schools, requiring insurers to cover particular medical procedures, as well as limitations on lawsuits from consumer groups on the fast food industry. We define duration data, $\mathbf{X}$, to be the time in years through this period for a bill to be passed. Assume that $\mathbf{X}$ is exponentially distributed $p(X|\theta) = \theta e^{-\theta X}$ defined over $[0, \infty)$, where interest is in the posterior distribution of the unknown parameter θ. Specify the prior distribution as $p(\theta) = 1/\theta$, for $\theta \in [0:\infty)$. The resulting posterior is given by:

$$\pi(\theta|\mathbf{X}) \propto p(\theta)L(\theta|\mathbf{X}) = \left(\frac{1}{\theta}\right)\theta^n \exp\left[-\theta\sum_{i=1}^{n}x_i\right] = \theta^{n-1}\exp\left[-\theta\sum_{i=1}^{n}x_i\right] \quad (2.9)$$

(note that we are using the proportionality shortcut from (2.7)). If we stare at this for a few moments we can see that $\theta|\mathbf{X} \sim \mathcal{G}(\theta|n, \sum x_i)$ with the "rate" specification for the second parameter. Putting the constants back in front to recover the full form of this gamma posterior distribution produces:

$$\pi(\theta|\mathbf{X}) = \frac{\left(\sum x_i\right)^n}{\Gamma(n)} \theta^{n-1} \exp\left[-\theta \sum x_i\right]$$

(details in Appendix B). As stated, once we produce the posterior distribution, we know everything about the distribution of θ and can convey to our readers any summary we would like.

TABLE 2.1: STATE DURATION TIME TO ADAPTATION, 1998-2005

State	N	Mean Duration	State	N	Mean Duration	State	N	Mean Duration
AL	2	7.500	LA	14	5.571	OK	12	6.583
AK	12	6.667	ME	2	5.500	OH	0	NaN
AZ	12	6.250	MD	11	6.455	OR	1	8.000
AR	6	6.167	MA	7	7.143	PA	12	7.083
CA	46	6.000	MI	4	7.000	RI	7	7.000
CO	11	6.636	MN	2	7.000	SC	6	6.333
CT	2	7.000	MS	7	7.143	SD	1	7.000
DE	4	7.000	MO	18	5.556	TN	17	7.235
FL	11	6.364	MT	2	7.000	TX	16	6.250
GA	7	5.857	NE	5	7.400	UT	3	7.667
HI	8	6.375	NV	4	8.000	VT	8	6.625
ID	6	6.000	NH	1	5.000	VA	15	6.533
IL	4	6.750	NJ	6	7.333	WA	12	6.083
IN	31	7.065	NM	6	6.500	WV	2	7.500
IA	3	5.000	NY	9	6.556	WI	4	7.750
KS	4	8.000	NC	8	7.250	WY	1	8.000
KY	4	7.500	ND	9	6.111			

The complete data are given in Table 2.1 for annualized periods, as well as in the R package BaM. Note the "NaN" value for the Ohio mean duration given by R since there is nothing to average. We will leave this case out of the subsequent analysis since the time to adoption is infinity, or more realistically, censored from us. The state averages from the third column of the table are weighted by N in the second column to reflect the number of such events: $\mathbf{X}_i N_i$. Since the sufficient statistic in the posterior distribution is a sum, there is no loss of information from not having the full original data from the authors (sums of means times n equal the total sum). The end-points of the equal tail credible interval are created by solving for the limits (L and H) in the two integrals:

$$\frac{\alpha}{2} = \int_0^L \pi(\boldsymbol{\theta}|\mathbf{X})d\theta \qquad \frac{\alpha}{2} = \int_H^\infty \pi(\boldsymbol{\theta}|\mathbf{X})d\theta \qquad (2.10)$$

or, more simply, we could use basic R functions to manipulate the `state.df` dataframe containing the data in the table above:

```
state.df <- state.df[-35,] # REMOVES OHIO
qgamma(0.05,shape=sum(state.df$N),rate=sum(state.df$N*state.df$dur))
```

```
[1] 0.14034
qgamma(0.95,shape=sum(state.df$N),rate=sum(state.df$N*state.df$dur))
[1] 0.16528
```

for a 90% credible interval. These points and a plot of the posterior distribution are given in Figure 2.2. The slight asymmetry of this gamma distribution means that the left tail region needs to reach higher (moving the boundary to the right) in order to equal the right tail in total posterior density. Therefore the density values (y-axis) at endpoints differ: 14.384 versus 12.898. To contrast with these results, the maximum likelihood value is $\hat{\theta} = 0.018$, (the inverse of the weighted mean of the data, Casella and Berger [2002]), whereas the posterior mean $E_\pi[\theta|\mathbf{X}] = 0.153$, showing that the prior $p(\theta) = 1/\theta$ has some influence.

2.3.2.2 Bayesian Highest Posterior Density Intervals

Credible intervals are common and useful, but a theoretically more defensible interval can be produced by incorporating some additional flexibility. When looking at posterior distributions, we really care where the highest density exists on the support of the posterior density, regardless of whether it is contiguous or not. So the big idea behind highest posterior density (HPD) regions is that no region outside of the interval will have higher posterior density than any region inside the HPD region. Hence for multimodal distributions the HPD region may actually be a set of individually non-contiguous intervals. See Hyndman (1996) for interesting forms as well as a general introduction to HPD regions. The use of the word "intervals" is common instead of "regions," but HPD regions possess an automatic ability to be non-contiguous so the latter is

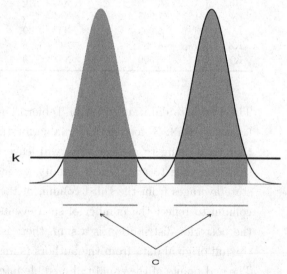

FIGURE 2.3: Bimodal Distribution Highest Posterior Density Interval

more correct. For symmetric unimodal forms the HPD interval will be contiguous and identical to an equal tail credible interval.

More specifically, a $100(1 - \alpha)\%$ highest posterior density region is the subset of the support of the posterior distribution for some parameter, θ, that meets the criteria:

$$C = \{\theta : \pi(\theta|\mathbf{x}) \geq k\},$$

where k is the largest number such that:

$$1 - \alpha = \int_{\theta:\pi(\theta|\mathbf{x})>k} \pi(\theta|\mathbf{x})d\theta \qquad (2.11)$$

(Casella and Berger 2002, p.448). This is the $1 - \alpha$ proportion subset of the sample space of θ where the posterior density of θ is maximized. So k is a horizontal line that slices across the density producing inside HPD areas and outside HPD areas. This will be a regular interval if the posterior distribution is unimodal, and it may be a discontiguous region if the posterior distribution is multimodal. This is shown in Figure 2.3 where it is clear that a bimodal distribution having a deep trough in the middle produces a non-contiguous HPD region emphasizing the undesirability of incorporating the middle region. Multimodal forms appear in Bayesian mixture models, and an example appears in Chapter 3 starting on page 87.

Chen and Shao (1999) provide another way to conceptualize the HPD region. For a unimodal posterior form, given by $\pi(\theta|\mathbf{X})$, our objective is to find the values $[\theta_L, \theta_U]$ that define a $(1 - \alpha)$ HPD region. It turns out that the answer to this question is given by also using cumulative (Π) differences:

$$\min_{\theta_L < \theta_U} \left[\underbrace{|\pi(\theta_U) - \pi(\theta_L)|}_{\text{difference in "height"}} + \underbrace{|\Pi(\theta_U) - \Pi(\theta_L) - (1 - \alpha)|}_{\text{difference in "width"}} \right]. \qquad (2.12)$$

So the first difference lines up k across the two HPD region endpoints and the second difference gives the coverage probability. In many circumstances this minimization gives zero, but for posteriors with flat regions it would need to be modified with some additional criteria to provide a unique interval such as picking the one with smallest θ_L value.

■ **Example 2.2:** **HPD Region, Fifty U.S. States Time to Adoption for Health Bills.** Returning to the example from time to adopt health laws, Figure 2.4, shows the HPD region for this posterior along with the determining line at $k = 16.873$. Notice in comparing the HPD region to the credible interval for this model, that the HPD region has equal height at the end-points at $k = 13.602$, and that the endpoints of the interval differ slightly from the corresponding equal-tail credible interval. The HPD region is constructed in a very simple way by starting at the posterior mode, then incrementing a horizontal line down vertically until the separation between the higher density and lower density regions reflects the desired coverage. So for each value of k, the level on the y-axis, we separately sum the area inside and outside the coverage area, regardless of contiguity. The **Computational Addendum** at the end of this chapter provides the R code used here for a posterior gamma distributed form. This process was quite easy to implement in R since we know the exact form of the gamma distribution from the model. Later when estimating marginal posterior forms with Bayesian stochastic simulation (MCMC), we will see that there are similarly easy ways to make this calculation even when we do not have an exact parametric description of

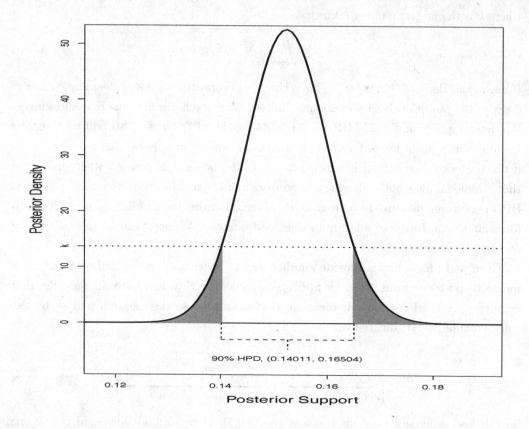

FIGURE 2.4: HPD REGION, STATE DURATION TIME TO ADAPTATION, 1998-2005

the posterior distribution. There is also the `HPDinterval` function in the `CODA` library for this, but it does not illustrate the underlying theory as directly as the exposition here.

2.3.3   Quantile Posterior Summaries

Often interval results are given basic quantile summaries without explicitly labeling these as credible intervals or HPD regions. Of course quantile summaries for unimodal forms are credible interval definitions so this is an analogous procedure. However, for highly multimodal forms basic quantile summaries can be misleading in the same way that boxplots hide such characteristics. Frequently summary tables for regression models contain quantile summaries since they are efficient with printed space and reveal characteristics of the distribution of interest.

■ **Example 2.3:   Quantiles, Fifty U.S. States Time to Adoption for Health Bills.** Using the data in Example 2.3.2.1 (starting on page 44), we can also calculate quantiles in addition to the credible intervals done before. We know that the time to adopt health laws example has a unimodal posterior, so simple quantiles are applicable. Consider the simple R commands:

```
q.vals <- c(0.025,0.05,0.25,0.5,0.75,0.95,0.975)
rbind( quant.vals, "quantiles"=
   qgamma(quant.vals,shape=sum(state.df$N),
        rate=sum(state.df$N*state.df$dur)) )
            [,1]     [,2]     [,3]     [,4]     [,5]     [,6]     [,7]
quant.vals 0.0250 0.05000 0.25000 0.50000 0.75000 0.95000 0.97500
quantiles  0.1381 0.14034 0.14742 0.15247 0.15764 0.16528 0.16781
```

using the `state.df` dataframe described above. This gives quantiles for common interval definitions ($\alpha = 0.05$ and $\alpha = 0.10$), the interquartile range (IQR), and the median). Note that the values $(0.14034, 0.16528)$ were given before with the 90% credible interval.

2.3.4   Beta-Binomial Model

This model illustrates the development of a posterior distribution for an interesting implication and shows how the properties of this posterior distribution can be described in conventional terms. Let $X_1, X_2, \ldots, X_n$ be independent random variables, all produced by the same probability mass function (iid): $\mathcal{BR}(p)$, and place a $\mathcal{BE}(A, B)$ prior distribution on the unknown population probability, p (see Appendix B for details on these forms). The goal is to get a posterior distribution for p, and this process is greatly simplified by noting that the sum of n Bernoulli(p) random variables is distributed binomial(n, p). Define a new variable: $Y = \sum_{i=1}^{n} X_i$. The joint distribution of Y and p is the product of the conditional distribution of Y and the marginal (prior) distribution of p:

$$f(y, p) = f(y|p)f(p)$$

$$= \left[\binom{n}{y} p^y (1-p)^{n-y} \right] \times \left[\frac{\Gamma(A+B)}{\Gamma(A)\Gamma(B)} p^{A-1}(1-p)^{B-1} \right]$$

$$= \frac{\Gamma(n+1)\Gamma(A+B)}{\Gamma(y+1)\Gamma(n-y+1)\Gamma(A)\Gamma(B)} p^{y+A-1}(1-p)^{n-y+B-1}. \tag{2.13}$$

The marginal distribution of Y is easy to calculate by integrating (2.13) with respect to p using a standard trick:

$$f(y) = \int_0^1 \frac{\Gamma(n+1)\Gamma(A+B)}{\Gamma(y+1)\Gamma(n-y+1)\Gamma(A)\Gamma(B)} p^{y+A-1}(1-p)^{n-y+B-1} dp$$

$$= \int_0^1 \frac{\Gamma(n+1)\Gamma(A+B)}{\Gamma(y+1)\Gamma(n-y+1)\Gamma(A)\Gamma(B)} \frac{\Gamma(y+A)\Gamma(n-y+B)}{\Gamma(n+A+B)}$$

$$\times \frac{\Gamma(n+A+B)}{\Gamma(y+A)\Gamma(n-y+B)} p^{y+A-1}(1-p)^{n-y+B-1} dp$$

$$= \frac{\Gamma(n+1)\Gamma(A+B)}{\Gamma(y+1)\Gamma(n-y+1)\Gamma(A)\Gamma(B)} \frac{\Gamma(y+A)\Gamma(n-y+B)}{\Gamma(n+A+B)}$$

$$\times \underbrace{\int_0^1 \frac{\Gamma(n+A+B)}{\Gamma(y+A)\Gamma(n-y+B)} p^{y+A-1}(1-p)^{n-y+B-1} dp}_{\text{equal to one}}$$

$$= \frac{\Gamma(n+1)\Gamma(A+B)}{\Gamma(y+1)\Gamma(n-y+1)\Gamma(A)\Gamma(B)} \frac{\Gamma(y+A)\Gamma(n-y+B)}{\Gamma(n+A+B)}. \tag{2.14}$$

The trick here is rearranging the terms such that a complete beta distribution with differing parameters is integrated across the support of the unknown random variable. The probability density function given in the last line of (2.14) is called (not surprisingly) the beta-binomial. Obtaining the posterior distribution of p is now a simple application of the definition of conditional probability:

$$f(p|y) = \frac{f(y,p)}{f(y)}$$

$$= \left(\frac{\Gamma(n+1)\Gamma(A+B)}{\Gamma(y+1)\Gamma(n-y+1)\Gamma(A)\Gamma(B)} p^{y+A-1}(1-p)^{n-y+B-1} \right) \Bigg/$$

$$\left(\frac{\Gamma(n+1)\Gamma(A+B)}{\Gamma(y+1)\Gamma(n-y+1)\Gamma(A)\Gamma(B)} \frac{\Gamma(y+A)\Gamma(n-y+B)}{\Gamma(n+A+B)} \right)$$

$$= \frac{\Gamma(n+A+B)}{\Gamma(y+A)\Gamma(n-y+B)} p^{(y+A)-1}(1-p)^{(n-y+B)-1}. \tag{2.15}$$

This can easily be seen as a new beta distribution with parameters $A' = y + A$ and $B' = n - y + B$. While it would be more typical of a Bayesian to describe this posterior distribution with quantiles, credible sets or highest posterior density intervals, we can get a point estimate of p here by taking the mean of this beta distribution:

$$\hat{p} = \frac{y+A}{A+B+n}. \tag{2.16}$$

Rearrange (2.16) algebraically to produce:

$$\hat{p} = \left[\frac{n}{A+B+n} \right] \left(\frac{y}{n} \right) + \left[\frac{A+B}{A+B+n} \right] \left(\frac{A}{A+B} \right), \tag{2.17}$$

which is the weighted combination of the sample mean from the binomial and the mean of the prior beta distribution where the weights are determined by the beta parameters, A and B, along with the sample size n. Holding A and B constant at reasonable values and increasing the sample size places more weight on the y/n term since the weight for the beta mean, $\frac{A}{A+B}$, has n only in the denominator and the weights necessarily add to one. This turns out to be theoretically more interesting than it would first appear and it highlights an important and desirable property of Bayesian data analysis: as the sample size increases, the likelihood function, $f(y|p)$, is iteratively updated to incorporate this new information and eventually subsumes the choice of prior, $f(p)$, because of sample size. Conversely, when the sample size is very small it makes sense to rely upon reliable prior information if it exists.

This hierarchical parameterization of the binomial with a *random effects component* (the name commonly used in non-Bayesian settings) is often done when there is evidence of overdispersion in the data: the variance exceeds that of the binomial: $np(1-p)$ (see Agresti 2002, p.151, Lehmann and Casella 1998, p.230, Carlin and Louis 2001, p.44, McCullagh and Nelder 1989, p.140). When the posterior has the same distributional family as the prior, as in this case, we say that the prior and the likelihood distributions are *conjugate*. This is an attractive property since it not only assures that there is a closed form for the prior, it means that it is also easy to calculate. Conjugate priors are discussed in detail in Chapter 4, Section 4.3, and Appendix B lists conjugate relationships, if they exist for commonly used distributions.

■ **Example 2.4:** **A Cultural Consensus Model in Anthropology.** Romney (1999) looks at the level of consensus among 24 Guatemalan women on whether or not 27 diseases known to all respondents are contagious. The premise is that a high level of consensus about something as important as the spread of diseases indicates to what extent knowledge is a component of culture in this setting. The survey data for polio are given by a vector, x, containing:

1	1	1	1	0	1	1	0	1	0	1	1	1	0	1	1	1	1	1	1	0	0	0	1

where 1 indicates that the respondent believes polio to be noncontagious and 0 indicates that the respondent believes polio to be contagious (Romney ranks it 13 out of 27 on a contagious scale). Here we apply the beta-binomial model with two different beta priors exactly in the manner discussed above. The first prior is a $\mathcal{BE}(p|15, 2)$ prior that imposes a great deal of prior knowledge about the unknown p parameter. The second prior is a $\mathcal{BE}(p|1, 1)$ prior that is actually a uniform prior over $[0\!:\!1]$ indicating a great deal of uncertainty about the location of p. Because the beta distribution is conjugate to the binomial the resulting posterior distributions are also beta and we obtain $\mathcal{BE}(\sum x_i + 15, n - \sum x_i + 2) = \mathcal{BE}(32, 9)$, and $\mathcal{BE}(\sum x_i + 1, n - \sum x_i + 1) = \mathcal{BE}(18, 8)$, respectively. Notice that there is almost no work to be done here (e.g. just plugging-in the specified values) since we have already worked out the analytical form of the posterior distribution. These posteriors along with the specified priors are shown in Figure 2.5 where the posterior is illustrated with the darker line and 95% HPD intervals are shown.

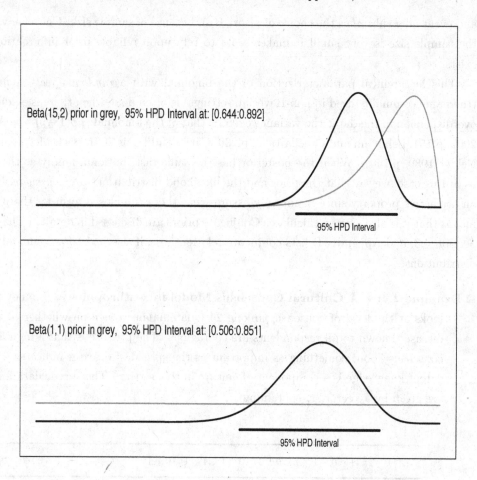

Beta(15,2) prior in grey, 95% HPD Interval at: [0.644:0.892]

95% HPD Interval

Beta(1,1) prior in grey, 95% HPD Interval at: [0.506:0.851]

95% HPD Interval

FIGURE 2.5: PRIOR AND POSTERIOR DISTRIBUTIONS IN THE BETA-BINOMIAL MODEL

It is clear from this figure that even though the prior specifications are very distinct, the resulting posteriors differ only modestly as evidenced by the highest posterior density regions indicated below the distributions. The uniform prior clearly pulls the posterior density to the left in contrast to the $\mathcal{BE}(15, 2)$, but it is apparent that the likelihood has a substantial effect *even though there are only 24 data points*. Substantively, we would have to say that this analysis only provides modest support for Romney's theory that believing specific diseases are contagious is a learned cultural response (although he looks at 26 additional diseases as well).

This relatively simple inference engine, as described, belies an immensely powerful procedure for developing and testing parametric models of inference. The central philosophical point is that the posterior distribution, which summarizes everything we know about the unknown parameter, is a weighted average of knowledge we possess before observing the data and the most likely value given the data. Furthermore, as the size of the data increases the likelihood becomes increasingly influential relative to the prior where in the limit the prior is immaterial. This is powerful because it explicitly incorporates knowledge about the parameter that researchers possess before developing the empirical model and collecting the data.

2.4 Bayesian "Learning"

There is actually no restriction on what constitutes "prior information," provided it can be expressed in a distributional form, and as new data are observed a new posterior can be created by treating the old posterior as a prior and updating with the new data through the new likelihood function. This is a rigorous formulation for the way people think: we have opinions about the manner in which something works and this opinion is updated or altered as new behavior is observed. Suppose there exist three serial events, C, B, A, that are nonindependent and we wanted to update the joint probability distribution as each event occurs. The first is just $p(C)$, and no updating is needed. The second event occurs conditionally on the first and we get the joint probability distribution by serially updating: $p(B, C) = p(B|C)p(C)$. Now the third event occurs conditional on the first two, and the resulting joint distribution is a new update of the previous update: $p(C, B, A) = p(A|B, C)p(B|C)p(C)$. It is easy to see that this can continue as long as we want, and that as long as the last conditional is multiplied by the string of previous conditionals, we always obtain the complete joint. In this simple example, we could think of $p(C)$ as the first prior since it is not conditional on any other event.

To be a bit more statistically concrete about this point, start with a univariate prior distribution, $p(\theta)$, on an unknown variable θ. We observe the first set of iid data, $\mathbf{x}_1$, and calculate the posterior from the likelihood function along with the specified prior:

$$\pi_1(\theta|\mathbf{x}_1) \propto p(\theta)L(\theta|\mathbf{x}_1). \tag{2.18}$$

Subsequent to calculating this posterior, we observe a second set of iid data, $\mathbf{x}_2$, independent of the first set but from the same data-generating process. To *update* our posterior and therefore improve our state of knowledge, we simply treat the previous posterior as a prior and proceed to calculate using a likelihood function from the new data, and this is exactly the same result we would have obtained if all the data had arrived at once:

$$\begin{aligned} \pi_2(\theta|\mathbf{x}_1, \mathbf{x}_2) &\propto \pi_1(\theta|\mathbf{x}_1)L(\theta|\mathbf{x}_2) \\ &= p(\theta)L(\theta|\mathbf{x}_1)L(\theta|\mathbf{x}_2) \\ &= p(\theta)L(\theta|\mathbf{x}_1, \mathbf{x}_2). \end{aligned} \tag{2.19}$$

Needless to say this process can be repeated *ad infinitum* and the model will continue to update the posterior conclusions as new information arrives. This cycle of prior to posterior is actually a very principled way of conceptualizing the scientific process: we take what knowledge we have in hand and update it with new information when such results become available.

■ **Example 2.5:   Timing of Campaign Polls.** The updating characteristic of the Bayesian framework is ideal for analyzing time-series, either statically (after all the data have been collected) or dynamically (as it comes in). In many commercial and economic settings, data analysis is often performed continually as the daily, monthly, or yearly figures arrive. In modern elections for political office, campaigns will do multiple polls of the relevant electorate in order to update their strategies. Suppose a campaign observes the candidate support figures for period t, $support_t$, after the conclusion of the period. The question is: given the known information, including previous polling data prior to period t (D_t), what does this tell us about support for the next period? Formalized, this means that we have a posterior at the end of time period t that reflects the current understanding of the underlying nature of the support pattern for the candidate:

$$\pi(support_t | D_t). \tag{2.20}$$

Now before viewing the next period's support figures, we treat this exact same distribution as a new prior:

$$p(support_{t+1}) = \pi(support_t | D_t). \tag{2.21}$$

After period $t + 1$, we have new support data and create a new posterior:

$$\pi(support_{t+1} | D_{t+1}) = p(support_{t+1})p(D_{t+1}). \tag{2.22}$$

Suppose that after period t, we were informed that a competing candidate suffered a public scandal. This is also new information (I_t) and would be incorporated in our prior for the next period's support:

$$p(support_{t+1}) = \pi(support_t | D_t, I_t). \tag{2.23}$$

What this means is that we should adjust our electoral expectations based on past support *and* this new information. The determination is still probabilistic because we cannot know for certain whether support will increase due to the competitors' misfortune: it could be that many of their supporters will move their intended vote to a third candidate, or they could be sufficiently loyal that the scandal is disregarded.

In a campaign environment, we might also want to predict support for future periods beyond a single period into the future. This is done by applying the sequential property of Bayes' Law:

$$p(support_{t+2}) = p(support_{t+1} | D_t)\pi(support_t | D_t). \tag{2.24}$$

This process can also be extended further into the future: $p(support_{t+k})$, albeit at the cost of progressively increasing uncertainty.

■ **Example 2.6:   Example: A Bayesian Meta-Estimate of Deaths in Stalin's Gulags.** It is very difficult to obtain a reliable estimate of the number of people that perished in the Soviet Gulags (forced labor camps) during Stalin's era as dictator (1924-1953). While there will apparently never be a *definitive* answer (Solzhenitsyn 1997), Blyth (1995) uses Bayesian conditional inference to provide a meta-estimate based on the best guesses of multiple historical researchers. The basic idea is to take summary notions of the number of deaths by these experts and translate them into workable probability functions using Lindley's (1983) location-scale translation (and adjusting them by subjective assessments of possible prejudices). Building a likelihood function based on these estimates would generally be straightforward multiplicatively with independent guesses, except that the experts have seen and are influenced by each others' work. Blyth's solution to this nonindependence is to explicitly recognize the chronology, and to build the likelihood function by conditional updating.

Denote all widely available knowledge on the scale of Gulag deaths from demographics, journalistic descriptions, and published personal accounts as **X**. The opinions of four experts and their associated estimates are considered by translating each best guess and level of uncertainty into a normal specification, or in the case where no uncertainty is given a uniform specification. For instance, in the case of one expert who estimates 10 to 20 million deaths, this is treated as a 95% credible interval. The normal distribution then is obtained by backing out the resulting coverage probability. The result is the following list of chronologically conditional statements:

Wiles, 1965:	$p(\theta_1|\mathbf{X}) \sim \mathcal{U}(\theta)$
Kurganov, 1973:	$p(\theta_2|\theta_1,\mathbf{X}) \sim \mathcal{U}(\theta)$
Conquest, 1978:	$p(\theta_3|\theta_2,\theta_1,\mathbf{X}) \sim \mathcal{N}(18.2, 8.5)$
Medvedev, 1989:	$p(\theta_4|\theta_3,\theta_2,\theta_1,\mathbf{X}) \sim \mathcal{N}(12, 9).$

Therefore the likelihood function from these "data" is:

$$L(\theta|\theta_4,\theta_3,\theta_2,\theta_1,\mathbf{X}) = p(\theta_4|\theta_3,\theta_2,\theta_1,\mathbf{X})p(\theta_3|\theta_2,\theta_1,\mathbf{X})p(\theta_2|\theta_1,\mathbf{X})p(\theta_1|\mathbf{X}). \quad (2.25)$$

The "supra-Bayesian" posterior developed here is modeled as a normal weighted by the precisions with the assumption that the intermediate conditionals are normal, $\mathcal{N}(\mu_i,\sigma_i^2)$, and this posterior form is therefore given by $\mathcal{N}(\mu_\pi,\sigma_\pi^2)$ where $\sigma_\pi^2 = \left(\sum_{i=1}^4 \sigma_i^{-2}\right)^{-1}$ and $\mu_\pi = \sigma_\pi^2 \sum_{i=1}^4 (\mu_i/\sigma_i^2)$. Blyth assigns the relatively "diffuse" normal prior (i.e., widely spread out by specifying a large variance parameter) $\mathcal{N}(8,12)$ and produces the posterior:

$$\pi(\theta|\theta_4,\theta_3,\theta_2,\theta_1,\mathbf{X}) \propto p(\theta)L(\theta|\theta_4,\theta_3,\theta_2,\theta_1,\mathbf{X}) \sim \mathcal{N}(13.2, 3.2).$$

This translates to a 95% credible interval of $[9.7:16.7]$ million deaths, which is a compromise between the four experts and the author of the meta-analysis.

2.5   Comments on Prior Distributions

The most controversial aspect of Bayesian statistics is the necessary assignment of a prior distribution. The primary criticism here is that this is a subjective process or worse yet, that it is a tool that allows researchers to manipulate the probability calculations to obtain a desired result. In truth, there exist subjective aspects of *every* statistical model, including: the experimental design or observational setting that produces the data, the parametric form of the model, the specification of explanatory variables, the choice of hypotheses to be tested, the selected significance level, and the determination of an adequate sample size (Barnett 1973, p.160; Howson and Urbach 1993, p.12). Obviously we should add the choice of prior distribution to this list, but note that Bayesians spend considerably more time and energy defending a prior distribution than non-Bayesians do justifying other subjective decisions.

Prior distributions can be categorized as either proper or improper. *Proper priors* meet the Kolmogorov axioms, most specifically that they integrate or sum to a finite value. A non-normalized prior that integrates or sums to some positive value other than one can always be renormalized, and this distinction is immaterial with Bayes' Law expressed proportionally anyway. *Improper priors* are those that sum or integrate to infinity, and yet they are useful and play an important role in Bayesian inference.

Importantly, a standard maximum likelihood inferential model is identical to a Bayesian model in which the prior probability distribution is an appropriate (correctly bounded for the parameter at hand) uniform distribution function, and the two models are asymptotically identical for *any* proper prior distribution. Specifically, if $\hat{\theta}$ is the MLE and $\tilde{\theta}$ is the posterior mean from a Bayesian model using the *same* likelihood, but any proper prior (and most improper priors), then:

$$\sqrt{n}(\tilde{\theta} - \hat{\theta}) \underset{n \to \infty}{\longrightarrow} 0 \tag{2.26}$$

almost assuredly for reasonable starting values of θ (Chao 1970). This is not to say that prior distributions are actually *unimportant*, but rather that in the presence of overwhelming data size we should not care about whether the inferential model puts non-zero mass on the prior or not. More practically, there are plenty of instances where we cannot rely on data size alone to drive the quality of statistical inference. Is it unreasonable to study 25 European Union countries, 7 Central American countries, a set of small group experiments, presidential nominees, 15 CIS countries, classroom level education, or other comparable problems?

The strongest substantive argument for inclusion of priors is that there often exists scientific evidence at hand before the statistical model is developed and it would be foolish to ignore such previous knowledge (Tiao and Zellner 1964a; Press 1989, Section 2.7.1). Furthermore, a formal statement of the prior distribution is an overt, nonambiguous assertion within the model specification that the reader can accept or dismiss (Box and Tiao 1973, p.9; Gelman *et al.* 2003, p.14). Also, imprecise or vague knowledge often justifies a diffuse (very large variance) or even uniform (flat) prior if bounded (Jeffreys 1961, Chapter III; Zellner 1971, p.41ff), and certain probability models logically lead to particular forms of the prior for mathematical reasons (Good 1950; Press 1989).

An immediate payoff for applying this Bayesian framework is that it facilitates the explicit comparison of rival models about the system under study: H_1 and H_2 (even if these are not nested models). In a preview of Chapter 6, suppose Γ_1 and Γ_2 represent two competing hypotheses about the location of some unknown parameter, γ, which together form a partition of the sample space: $\Gamma = \Gamma_1 \cup \Gamma_2$. Initially prior probabilities are assigned to each of the two outcomes:

$$p_1 = p(\gamma \in \Gamma_1) \quad \text{and} \quad p_2 = p(\gamma \in \Gamma_2). \tag{2.27}$$

This allows us to calculate the posterior probabilities from the two alternative priors and the likelihood function:

$$\pi_1 = p(\gamma \in \Gamma_1 | D, H_1) \quad \text{and} \quad \pi_2 = p(\gamma \in \Gamma_2 | D, H_2). \tag{2.28}$$

The Bayes Factor combines the prior odds, p_1/p_2, and the posterior odds, π_1/π_2, as evidence for H_1 versus H_2 by calculating the ratio:

$$B = \frac{(\pi_1/\pi_2)}{(p_1/p_2)} \tag{2.29}$$

(Berger 1985; Kass and Raftery 1995; Lee 2004). Thus the Bayes Factor is the odds favoring H_1 versus H_2, given the observed data incorporating both prior and posterior information. As we will see in Chapter 7, this Bayes Factor model testing framework is even more flexible than this discussion implies.

2.6 Bayesian versus Non-Bayesian Approaches

There is a long history of antagonism between Bayesians and those adhering to strongly classical approaches: frequentist methods from Neyman and Pearson, and likelihood based methods from Fisher. However, this disagreement has greatly diminished over the last three decades. The core frequentist paradigm bases a sampling model on an imagined infinite series of replications of the same analysis where the reliability of the calculated statistics is derived from their asymptotic properties. The likelihood approach is different in that only the currently observed sample is considered and statistics are produced from these data to estimate unknown population parameters by determining the value that is most likely, given that observed sample.

The likelihood theorem (Birnbaum 1962) states that all information that can be used for estimating some unknown parameter is contained in the likelihood function: a statement about the most likely value of the parameter given the observed data, and the Bayesian approach uses this same likelihood function (albeit with the addition of a prior distribution on the unknown parameters). Thus for likelihood inference, all information needed from the sample comes from the likelihood function. This does not mean that there is no additional information in the sample for other forms of inference. We know that every likelihood model is actually a Bayesian model with the appropriately bounded uniform prior and every Bayesian model is asymptotically equivalent to a corresponding likelihood model for any given prior. Therefore likelihoodists are simply Bayesians that do not know it or do not care to worry about the convenience of describing unknown quantities probabilistically. This means that the real differences are with classical frequentists.

There is a second distinction that causes more disagreement than it should. In classical inference, one assumes that the population parameters are fixed and unknown and therefore estimated with sample quantities. Conversely in Bayesian inference unknown inference parameters are treated as random quantities as a consequence of the application of Bayes' Law to invert the conditional probability statement. Actually this distinction is not very important in practice as the frequentist "sampling distribution" is exactly the same principle as the Bayesian posterior distribution, except that the imagined asymptote is unavailable. This explains why one hears social science researchers applying traditional inference procedures still using the word "posterior." Perhaps they have become frustrated with the difficulty in teaching the distinction between a *sample* distribution and a *sampling* distribution. Furthermore, Lewis (1994) points out the easily observed, but often forgotten fact that "Most applied statisticians have little interest in confrontation between rival philosophies but have a keen interest in pragmatic solutions to real problems . . . ." This is true of quantitative social scientists in particular.

A substantial amount of frequentist theory is built on the asymptotic normality of the sampling distribution of calculated statistics, and the associated calculation of such properties (Barndorff-Nielson and Cox 1989). Associated with this is the assumption of an unending stream of iid data. While Bayesian inference does not assume infinite replications to define sampling distributions, the posterior, being a compromise between the prior and the likelihood, will be affected by the same asymptotic properties as the amount of the data increases. Laplace was the first to note the near-normality of posterior distributions, as long ago as 1811! This property was later fully explored around the 1960s (Chao 1963, 1965), and Diaconis and Freedman (1986) subsequently gave mathematically rigorous conditions for the consistency of these Bayesian estimates, thus subjecting frequentist and Bayesian procedures to the same quality standard. 1970; Fabius 1964; Freedman

So it is important to understand where Bayesian inference stands relative to the Neyman-Pearson frequentist paradigm. We can now tabulate core differences between Bayesian and frequentist approaches. Most of these contrasts have been noted already in this chapter, and simply summarized here. This is done in the context of the following categories:

Interpretation of Probability

Frequentist: Observed result from infinite series of trials performed or imagined under identical conditions.
Probabilistic quantity of interest is $p(\text{data}|H_0)$.

Bayesian: Probability is the researcher/observer "degree of belief" before or after the data are observed.
Probabilistic quantity of interest is $p(\theta|\text{data})$.

What Is Fixed and Variable

Frequentist: Data are an iid random sample from continuous stream.
Parameters are fixed by nature.

Bayesian: Data observed and so fixed by the sample generated.
Parameters are unknown and described distributionally.

How Results Are Summarized

Frequentist: Point estimates and standard errors.
95% confidence intervals indicating that 19/20 times the interval covers the true parameter value, on average.

Bayesian: Descriptions of posteriors such as means and quantiles.
Highest posterior density intervals indicating region of highest posterior probability.

Inference Engine

Frequentist: Deduction from $p(\text{data}|H_0)$, by setting α in advance.
Accept H_1 if $p(\text{data}|H_0) < \alpha$.
Accept H_0 if $p(\text{data}|H_0) \geq \alpha$.

Bayesian: Induction from $p(\theta|data)$, starting with $p(\theta)$.
$100(1 - \alpha)\%$ of highest probability levels in $1 - \alpha$ HPD region.

Quality Checks

Frequentist: Calculation of Type I and Type II errors.
Sometimes: effect size and/or power.
Usually: attention to small differences in p-values.

Bayesian: Posterior predictive checks.
Sensitivity of the posterior to forms of the prior.
Bayes Factors, BIC, DIC (see Chapter 7).

In some ways, the seemingly wide gap between frequentist/likelihoodist and Bayesian thinking outlined above is an artificial and superficial divide. The maximum likelihood estimate is equal to the Bayesian posterior mode with the appropriate (correctly bounded) uniform prior, and they are asymptotically equal and normal given *any* proper prior (i.e., meeting the Kolmogorov axioms). Both approaches make extensive use of the central limit theorem and normal theory in general. However, differences are seen particularly in small sample problems where the asymptotic equivalence is obviously not applicable. A common frequentist criticism of the Bayesian approach is that "subjective" priors have great impact on the posterior distribution for problems with small sample sizes. There is a developing literature on robust Bayesian analysis that seeks to mitigate this problem by developing estimators that are relatively insensitive to a wide range of prior distributions (Berger 1984).

■ **Example 2.7: The Timing of Polls.** Bernardo (1984) developed a precinct-level Bayesian hierarchical model of vote choice for the Spanish election of 1982 in which the Socialist party obtained control of the government for the first time since the Civil War. Bayesian hierarchical models recognize and organize differing levels of data and prior information (see Chapter 12 for more details). The author defines n_{ij} as the number of voters in the ith precinct voting for the jth party. The data from the m precincts surveyed ($\{n_{1,j}, n_{2,j}, \ldots, n_{m,j}\}$, $j = 1$ to 5 major political parties) are assumed to be from an underlying multinomial distribution with unknown parameters θ_{ij} representing the *probability* of a vote for the jth party in the ith precinct with the constraints that these values are nonnegative and sum to one. Bernardo specifies a uniform prior distribution on the θ_{ij} values and this leads naturally to a Dirichlet form (a multivariate generalization of the beta) of the posterior.

TABLE 2.2: HPD REGIONS: PREDICTED 1982 VOTE PERCENTAGES FOR FIVE MAJOR PARTIES

Valencia	Party				
	Socialist	Conservative	Center	Center-Left	Communist
Four weeks before election	[39.0:48.9]	[12.6:19.4]	[7.9:13.2]	[4.0: 7.8]	[5.1:9.2]
One week before election	[47.3:54.2]	[13.3:24.9]	[5.0:11.8]	[7.0:11.9]	[4.0:6.7]
First 100 votes from 20 polls	[49.0:57.7]	[23.5:31.2]	[2.3: 4.6]	[1.1: 2.9]	[4.0:6.2]
Total vote from 20 polls	[50.1:56.8]	[26.6:32.6]	[3.3: 4.6]	[1.8: 2.7]	[3.7:5.6]
Actual results	53.5	29.4	4.4	2.3	5.3

A substantively interesting aspect of the methodology is the scheduling of data collection and analysis. Data are collected in the province of Valencia at four points in time: by a survey four weeks before the election ($n = 1,000$), by a survey one week before the election ($n = 1,000$), using the first 100 valid votes from 20 representative polling stations, and all valid votes from these same polling stations after the polls are closed. Data collection was performed with the full cooperation of the Spanish government and the results were immediately provided to the national media.

Bernardo presents the Bayesian estimates of predicted vote proportion by party as 0.90 highest posterior density regions. These results are summarized in Table 2.2.

One interesting result from this analysis is that the 0.90 HPD regions shrink as the final tally nears and better polling data are received. This reflects the growing certainty about the estimates as data quality improves. In addition, the estimates from actual polling data are remarkably accurate for the two parties receiving the largest vote share. Note that the uniform prior does not constrain the final, highly nonuniform, posterior distribution.

The last example demonstrates that Bayesian data analysis is essentially free from the well-known problems with the null hypothesis significance test. Inferences are communicated to the reader without artificial decisions, p-values, and confused conditional probability statements. The Bayesian approach also interprets sample size increases in a more desirable manner: larger sample sizes reduce the importance of prior information rather than guarantee a low but meaningless p-value.

2.7 Exercises

2.1   Suppose that 25 out of 30 firms develop new marketing plans during the next year. Using the beta-binomial model from Section 2.3.4 starting on page 49, apply a $\mathcal{BE}(0.5, 0.5)$ (Jeffreys prior) and then specify a normal prior centered at zero and truncated to fit on $[0:1]$ as prior distributions and plot the respective posterior densities. What differences do you observe?

2.2   Derive the posterior distribution for a sample of size n of iid data distributed $f(X|\theta) = \theta e^{-\theta X}$, $X, \theta \in (0, \infty)$), with a prior for θ, $p(\theta) = \beta e^{-\theta\beta}$, $\beta \in (0, \infty)$. What is common to all three distributions?

2.3   Prove that the gamma distribution,

$$f(\mu|\alpha, \beta) = \frac{1}{\Gamma(\alpha)}\beta^\alpha \mu^{\alpha-1} e^{-\beta\mu}, \qquad \mu, \alpha, \beta > 0,$$

is the conjugate prior distribution for μ in a Poisson likelihood function,

$$f(\mathbf{y}|\mu) = \left(\prod_{i=1}^{n} y_i!\right)^{-1} \exp\left[\log(\mu) \sum_{i=1}^{n} y_i\right] \exp[-n\mu],$$

that is, calculate a form for the posterior distribution of μ and show that it is also gamma distributed.

2.4   One requirement for specifying prior distributions is that the support of the assigned prior must match the allowable range of the parameter being modeled. What common distributions, without modification, can be used as priors on model variance components?

2.5   Use the gamma-Poisson conjugate specification developed in Exercise 2.3 to analyze the following count data on worker strikes in Argentina over the period 1984 to 1993, from McGuire (1996). Assign your best guess as to reasonable values for the two parameters of the gamma distribution: α and β. Produce the posterior distribution for μ and describe it with quantiles and graphs using empirically simulated values according to the following procedure:

▷ The posterior distribution for μ is gamma(δ_1, δ_2) according to some parameters δ_1 and δ_2 that you derived above, which of course depends on your choice of the gamma parameters.

▷ Generate a large number of values from this distribution in R, say 10,000 or so, using the command:
```
posterior.sample <- rgamma(10000,d1,d2)
```

▷ Produce posterior quantiles, such as the interquartile range, according to:
```
iqr.posterior <- c(sort(posterior.sample)[2500],
                   sort(posterior.sample)[7500])
```
Note: the IQR function in R gives a single value for the difference, which is not as useful.

▷ Graph the posterior in different ways, such as with a smoother like lowess (a local-neighborhood smoother, see Cleveland [1979, 1981]):
```
post.hist <- hist(posterior.sample,plot=F,breaks=100)
plot(lowess(post.hist$mids,post.hist$intensities),
     type="l")
```

Economic Sector	Number of Strikes		
Public Administrators	496	Meat Packers	56
Teachers	421	Paper Industry Workers	55
Metalworkers	199	Sugar Industry Workers	50
Municipal Workers	186	Public Services	47
Private Hospital Workers	181	University Staff Employees	43
Bank Employees	133	Telephone Workers	39
Court Clerks	128	Textile Workers	37
Bus Drivers	113	State Petroleum Workers	32
Construction Workers	92	Food Industry Workers	28
Doctors	83	Post Office Workers	26
Nationalized Industries	77	Locomotive Drivers	25
Railway Workers	76	Light and Power Workers	21
Maritime Workers	57	TOTAL	2701

2.6   For $\theta \sim$ binomial$(10, 0.5)$ construct an even tail credible interval that has *at least* 0.90 coverage. Is it possible to get exact coverage?

2.7   In his original essay (1763, p.376) Bayes offers the following question:

> *Given* the number of times in which an unknown event has happened and failed: *Required* the chance that the probability of its happening in a single trial lies somewhere between any two degrees of probability that can be named.

Provide an analytical expression for this quantity using an appropriate uniform prior (Bayes argued reluctantly for the use of the uniform as a "no information" prior: *Bayes postulate*).

2.8   Given a proper prior distribution, $p(\theta)$, and a likelihood function, $L(\theta|\mathbf{X})$, demonstrate that the only way that the prior distribution and the resulting posterior distribution, $\pi(\theta|\mathbf{X})$ can be identical is when the likelihood function does not contain θ.

2.9   Suppose we have two urns containing marbles; the first contains 6 red marbles and 4 green marbles, and the second contains 9 red marbles and 1 green marble. Now we take one marble from the first urn (without looking at it) and put it in the second urn. Subsequently, we take one marble from the second urn (again without looking at it) and put it in the first urn. Give the full probabilistic statement of the probability of now drawing a red marble from the first urn, and calculate its value.

2.10   In an experimental context Gill and Freeman (2013) ask participants to answer a wide range of background questions prior to eliciting prior distributions from watching video clips. One of these,

> "What proportion (percent) of undergraduate students at the University of Minnesota are women?"

generates the following response times in seconds to this question:

7	7	11	7	7	10	7	5	8	7	5	7	12	6	8	7
8	7	28	13	6	4	10	6	13	11	6	14	4	7	12	16
8	9	8	9	4	5	8	4	5	15	9	7	7	8	4	9
7	9	19	19	9	7	5	6	6	17	7	6	10	7	15	

Assume that the distribution of these times is $\mathcal{G}(4.5, 2)$ (shape and scale). Find and graph a 95% HPD region for an additional sample point drawn from the same population. Now suppose we do not know the distribution with certainly but impose a prior distribution that is $\mathcal{G}(3, 3)$. Find and graph the resulting 95% HPD regions for the posterior distribution.

2.11   This is the famous envelope problem. You and another contestant are each given one sealed envelope containing some quantity of money with equal probability of receiving either envelope. All you know at the moment is that one envelope contains twice the cash as the other. So if you open your envelope and observe $10, then the other envelope contains either $5 or $20 with equal probability. You are now given the opportunity to trade with the other contestant. Should you? The expected value of the unseen envelope is $E[other] = 0.5(5) + 0.5(20) = 12.50$, meaning that you have a higher expected value by trading. Interestingly, so does the other player for analogous reasons. Now suppose you are offered the opportunity to trade again before you open the newly traded envelope. Should you? What is the expected value of doing so? Explain how this game leads to infinite cycling. There is a Bayesian solution. Define M as the known maximum value in either envelope, stipulate a probability distribution, and identify a suitable prior.

2.12   Radiocarbon dating of the famous *Shroud of Turin* cloth that some believe was used to wrap Jesus Christ's body (since it has front and rear impressions of a bearded male with whipping and crucifixion injuries) was done by the "Arizona Group" (Linick *et al.* 1986) using accelerator mass spectrometry. Their serial process in 1988 produced the following estimated ages in years and associated standard errors:

Iteration	Mean	SE
1	606	51
2	574	52
3	753	51
4	632	49
5	676	59
6	540	57
7	701	47
8	701	47

As done in Example 2.4 starting on page 55, treat these as consecutive updates on the posterior distribution and produce the set of posterior distributions under normally distributed assumptions for each. Stipulate a reasonable prior to begin the process.

2.13   If the posterior distribution of θ is $\mathcal{N}(1,3)$, then calculate a 99% HPD region for θ.

2.14   Given a posterior distribution for θ that is $\mathcal{BE}(0.5, 0.5)$, calculate the 95% HPD region for θ.

2.15   Assume that the data $[1, 1, 1, 1, 1, 1, 0, 0, 0, 1, 1, 1, 0, 1, 0, 0, 1, 1, 1, 1]$ are produced from iid Bernoulli trials. Produce a $1 - \alpha$ credible set for the unknown value of p using a uniform prior distribution.

2.16 Browne, Frendreis, and Gleiber (1986) tabulate complete cabinet duration (constitutional inter-election period) for eleven Western European countries from 1945 to 1980 for annualized periods:

Country	N	Average Duration
Italy	38	0.833
Finland	28	1.070
Belgium	27	1.234
Denmark	20	1.671
Norway	17	2.065
Iceland	15	2.080
Austria	15	2.114
West Germany	15	2.168
Sweden	15	2.274
Ireland	14	2.629
Netherlands	12	2.637

The country averages from the third column of the table are weighted by N in the second column to reflect the number of such events: $\mathbf{X}_i N_i$. Assume that the durations, $\mathbf{X}$, are exponentially distributed $p(X|\theta) = \theta e^{-\theta X}$ defined over $(0, \infty)$, and like Example 2.3.2.1 on page 44 specify the prior distribution of $p(\theta) = 1/\theta$, for $\theta \in (0{:}\infty)$. Calculate an equal tail credible interval and an HPD region for the resulting posterior distribution of θ. Plot the posterior density and indicate the location of these intervals.

2.17 The beta distribution, $f(x|\alpha, \beta) = \frac{\Gamma(\alpha+\beta)}{\Gamma(\alpha)\Gamma(\beta)} x^{\alpha-1}(1-x)^{\beta-1}, 0 < x < 1, \alpha > 0, \beta > 0$, is often used to model the probability parameter in a binomial setup. If you were very unsure about the prior distribution of p, what values would you assign to α and β to make it relatively "flat"?

2.18 An *improper prior distribution* is a function that does not sum or integrate to a finite constant. Show that it is possible to still get a proper posterior distribution through (2.6). A possible prior for μ in a Poisson likelihood function is $p(\mu) = 1/\mu$. Show that this is improper.

2.19 Laplace (1774, p.28) derives Bayes' Law for uniform priors. His claim is

> ... je me propose de déterminer la probabilité des causes par les événements matière neuve à bien des égards et qui mérite d'autant plus d'être cultivée que c'est principalement sous ce point de vue que la science des hasards peut être utile dans la vie civile.

He starts with two events: E_1 and E_2 and n causes: $A_1, A_2, \ldots, A_n$. The assumptions are: *(1)* E_i are *conditionally independent* given A_i, and *(2)* A_i are equally probable. Derive Laplace's inverse probability relation:

$$p(A_i|E) = \frac{p(E|A_i)}{\sum_j p(E|A_j)}.$$

2.20   Martins (2009) is concerned with Bayesian updating by interacting actors who pay attention to each others' choices. Actor i has a prior distribution $f_i(\theta)$, and $E[\theta] = x_i$. This actor's posterior for θ is affected by the average estimates of others x_j, giving $f_i(\theta|x_j)$. Show that the mixture likelihood:

$$f(x_j|\theta) = \omega\mathcal{N}(\theta, \sigma_j) + (1-\omega)\mathcal{U}(0,1),$$

(for mixture parameter ω), leads to the posterior:

$$f(\theta|x_j) \propto \omega\exp\left[-\frac{1}{2\sigma_i^2}((\theta-x_i)^2 + (x_j-\theta)^2)\right](1-\omega)\exp\left[-\frac{1}{2\sigma_i^2}(x_i-x_j)^2\right].$$

2.8   Computational Addendum: R for Basic Analysis

This code gives the analysis and graphing of the cultural anthropology example in Section 2.3.4 starting on page 51.

```
par(oma=c(1,1,1,1),mar=c(0,0,0,0),mfrow=c(2,1))
x <- c(1,1,1,1,0,1,1,0,1,0,1,1,1,0,1,1,1,1,1,1,0,0,0,1)
ruler <- seq(0,1,length=300)


A <- 15; B <- 2
beta.prior <- dbeta(ruler,A,B)
beta.posterior <- dbeta(ruler,sum(x)+A,length(x)-sum(x)+B)
plot(ruler,beta.prior, ylim=c(-0.7,9.5),
        xaxt="n", yaxt="n", xlab="", ylab="", pch=".")
lines(ruler,beta.posterior)
hpd.95 <- qbeta(c(0.025,0.975),sum(x)+A,length(x)-sum(x)+B)
segments(hpd.95[1],0,hpd.95[2],0,lwd=4)
text(mean(hpd.95),-0.4,"95% HPD Region",cex=0.6)
text(0.25,5,paste("Beta(",A,",",B,
    ") prior,  95% HPD Regionat: [",round(hpd.95[1],3),
    ":",round(hpd.95[2],3),"]",sep=""),cex=1.1)


A <- 1; B <- 1
beta.prior <- dbeta(ruler,A,B)
beta.posterior <- dbeta(ruler,sum(x)+A,length(x)-sum(x)+B)
plot(ruler,beta.prior, ylim=c(-0.7,9.5),
        xaxt="n", yaxt="n", xlab="", ylab="", pch=".")
lines(ruler,beta.posterior)
```

```
hpd.95 <- qbeta(c(0.025,0.975),sum(x)+A,length(x)-sum(x)+B)
segments(hpd.95[1],0,hpd.95[2],0,lwd=4)
text(mean(hpd.95),-0.4,"95% HPD Region",cex=0.6)
text(0.25,5,paste("Beta(",A,",",B,
    ") prior,  95% HPD Region at: [",round(hpd.95[1],3),
    ":",round(hpd.95[2],3),"]",sep=""),cex=1.1)
```

The following is the simple HPD region calculation used in Example 2.3.2.2.

```
hpd.gamma <- function(g.shape,g.rate,target=0.90,steps=300,tol=0.01)  {
    if (steps %% 2 == 1) steps <- steps + 1
    g.mode  <- sum(state.df$N)/sum(state.df$N*state.df$dur)
    g.range <- seq(qgamma(0.001,g.shape,g.rate), qgamma(0.999,g.shape,
                g.rate),length=steps)
    g.range <- c(g.range[1:(steps/2)],g.mode,g.range[(steps/2+1):steps])
    g.dens  <- dgamma(g.range,g.shape,g.rate)
    g.probs <- pgamma(g.range,g.shape,g.rate)
    for (i in 1:(steps/2)) {
        k.dir <- which(c(g.dens[(steps/2-i)],g.dens[(steps/2+i)]) ==
                    max(g.dens[(steps/2-i)],g.dens[(steps/2+i)]))
        k <- c(g.dens[(steps/2-i)],g.dens[(steps/2+i)])[k.dir]
        k.loc <- c((steps/2-i),(steps/2+i))[k.dir]
        if (k.dir == 2) k2.range <- c(1:(steps/2))
else k2.range <- c((steps/2 + 1):steps)
        k2.min <- which(abs(k-g.dens[k2.range])==min(abs(k-g.dens[k2.range])))
        if (k.dir == 1) k2.min <- k2.min + steps/2
        if (g.probs[k.loc] + (1-g.probs[k2.min]) < 1-target)  break
        bounds <- c(g.range[k.loc],g.range[k2.min])
    }
    return(list("cdf.vals"=c(g.probs[k.loc],g.probs[k2.min]),
                "bounds"=bounds,"k"=k))
}


state.hpd <- hpd.gamma(g.shape=sum(state.df$N),
                    g.rate=sum(state.df$N*state.df$dur))
```

Chapter 3

The Normal and Student's-*t* Models

Statistical models built on the normal distribution are as common in Bayesian statistics as they are in non-Bayesian approaches. There are several reasons for this. *First*, nature seems to have an affinity for this form as evidenced through empirical observation as well as from the central limit theorem. The weakest form of the central limit theorem essentially says that an interval measured statistic with bounded variance will eventually be normally distributed provided sufficient sample size. Therefore it is quite common to see situations where quantities behave approximately normally. *Second*, a huge class of posterior distributions can be modeled by combining an assumed normal likelihood function with differing priors. Finally, when Bayesian models were more difficult to estimate numerically, the normal distribution sometimes provided an analytically tractable posterior when other forms were less compliant.

3.1 Why Be Normal?

Often when the posterior is known to be unimodal and symmetric, we can effectively model it with a normal distribution even if we know that the form is only nearly normal. In cases where the researcher has a rough idea of where an unknown parameter is centered, the normal provides a useful way of modeling this guess that allows the level of uncertainty to be described by the normal variance term. This convenience can provide good *approximations* to the desired posterior density with the knowledge that as more data are observed, such assumptions decrease in importance.

As demonstrated below, the Bayesian normal model has desirable frequentist properties. While the emphasis in Bayesian analysis is not on point estimates, it can be shown that with increasingly large samples the mean of the Bayesian posterior approaches the maximum likelihood estimate. This property exists because the posterior is a weighted compromise between the user specified prior distribution, normal in this chapter, and the data-driven likelihood function, also normal in this chapter. As the data size increases, the likelihood becomes increasingly dominant in this weighting.

In the case of a normal mean, illustrated here, the variance of the frequentist sampling distribution decreases with increases in sample size. In the Bayesian context, the shrinking mean variance from the likelihood function eventually overwhelms even a deliberately large prior variance. Therefore if the expected size of the dataset is large, researchers can afford to be liberal in specifying the prior variance.

3.2 The Normal Model with Variance Known

We begin with a very simple case. Suppose the data are assumed to follow a normal distribution with known variance (σ_0^2), and the unknown parameter of interest is the mean of this distribution, which is itself given a normal prior with assigned parameters m and s^2. These assumptions are summarized as follows:

$$X|\mu, \sigma_0^2 \sim \mathcal{N}(\mu, \sigma_0^2) = (2\pi\sigma_0^2)^{-\frac{1}{2}} \exp\left[-\frac{1}{2\sigma_0^2}(X - \mu)^2\right]$$

$$-\infty < \mu < \infty, \sigma_0^2 \quad \text{known}$$

$$\mu|m, s^2 \sim \mathcal{N}(m, s^2) = (2\pi s^2)^{-\frac{1}{2}} \exp\left[-\frac{1}{2s^2}(\mu - m)^2\right] \qquad m, s^2 \text{ known.} \qquad (3.1)$$

In this setup the parameters on the prior distribution (m,s^2) are either known or assigned for substantive reasons. After the data, $\mathbf{X}$, are observed, the posterior distribution of μ is produced by the product of the prior and the likelihood function:

$$\pi(\mu|\mathbf{x}) \propto p(\mathbf{x}|\mu)p(\mu)$$

$$\propto \prod_{i=1}^{n} \exp\left[-\frac{1}{2\sigma_0^2}(x_i - \mu)^2\right] \exp\left[-\frac{1}{2s^2}(\mu - m)^2\right]$$

$$= \exp\left[-\frac{1}{2}\left(\frac{1}{\sigma_0^2}\sum_{i=1}^{n}(x_i - \mu)^2 + \frac{1}{s^2}(\mu - m)^2\right)\right].$$

Now expand the squares.

$$= \exp\left[-\frac{1}{2}\left(\frac{1}{\sigma_0^2}\sum_{i=1}^{n}(x_i^2 - 2x_i\mu + \mu^2) + \frac{1}{s^2}(\mu^2 - 2\mu m + m^2)\right)\right]$$

$$= \exp\left[-\frac{1}{2}\frac{1}{\sigma_0^2 s^2}\left(s^2\sum_{i=1}^{n}x_i^2 - 2s^2\mu n\bar{x} + n\mu^2 s^2 + \sigma_0^2\mu^2 - 2\sigma_0^2\mu m + \sigma_0^2 m^2\right)\right]$$

$$= \exp\left[-\frac{1}{2}\frac{1}{\sigma_0^2 s^2}\left(\mu^2(\sigma_0^2 + ns^2) - 2\mu(m\sigma_0^2 + s^2 n\bar{x}) + (m^2\sigma_0^2 + s^2\sum_{i=1}^{n}x_i^2)\right)\right].$$

The last term in the expansion can be treated as part of the normalizing constant, k.

$$\propto \exp\left[-\frac{1}{2}\left(\mu^2\left(\frac{1}{s^2}+\frac{n}{\sigma_0^2}\right)-2\mu\left(\frac{m}{s^2}+\frac{n\bar{x}}{\sigma_0^2}\right)+k\right)\right]$$

$$=\exp\left[-\frac{1}{2}\left(\frac{1}{s^2}+\frac{n}{\sigma_0^2}\right)\left(\mu^2\frac{\left(\frac{1}{s^2}+\frac{n}{\sigma_0^2}\right)}{\left(\frac{1}{s^2}+\frac{n}{\sigma_0^2}\right)}-2\mu\frac{\left(\frac{m}{s^2}+\frac{n\bar{x}}{\sigma_0^2}\right)}{\left(\frac{1}{s^2}+\frac{n}{\sigma_0^2}\right)}+k\right)\right]$$

$$\propto \exp\left[-\frac{1}{2}\left(\frac{1}{s^2}+\frac{n}{\sigma_0^2}\right)\left(\mu-\frac{\left(\frac{m}{s^2}+\frac{n\bar{x}}{\sigma_0^2}\right)}{\left(\frac{1}{s^2}+\frac{n}{\sigma_0^2}\right)}\right)^2\right]. \tag{3.2}$$

Therefore the posterior for μ is a normal distribution with mean:

$$\hat{\mu}=\left(\frac{m}{s^2}+\frac{n\bar{x}}{\sigma_0^2}\right)\bigg/\left(\frac{1}{s^2}+\frac{n}{\sigma_0^2}\right), \tag{3.3}$$

and variance:

$$\hat{\sigma}_\mu^2=\left(\frac{1}{s^2}+\frac{n}{\sigma_0^2}\right)^{-1}=\frac{s^2\sigma_0^2}{\sigma_0^2+ns^2}. \tag{3.4}$$

Several points bear mentioning here. *First*, notice that the posterior for μ depends on the data only through $\bar{x}$. Thus $\bar{x}$ is a *sufficient statistic* in this context, meaning that this data summary is all that is required from the data to estimate the unknown mean.[1] *Second*, in several instances we dropped constant terms out of the derivation above, because as long as these are not changing relative values of the unknown parameter it is only necessary to normalize at the end.

The term $\frac{1}{s^2}$ is called the *prior precision*, and the term $\frac{n}{\sigma_0^2}$ is called the *data precision*. These terms play an important role in Bayesian statistics because they link uncertainty from the prior assumptions and uncertainty from the observed data. The *posterior precision*, easily obtained in this example, is just the sum of the prior precision and the data precision: $\frac{1}{\hat{\sigma}_\mu^2}=\frac{1}{s^2}+\frac{n}{\sigma_0^2}$. This is true in all normal-normal cases, not just the single example given here.

As the data size increases, the value of the posterior mean, $\hat{\mu}$, is increasingly determined by the data mean, $\bar{x}$, making prior assumptions less important. Consider a fixed value for the data variance as we have done above, σ_0^2, and let n go to infinity.

The asymptotic posterior mean is given by:

$$\lim_{n\to\infty}\hat{\mu}=\lim_{n\to\infty}\frac{\frac{m}{s^2}+\frac{n\bar{x}}{\sigma_0^2}}{\frac{1}{s^2}+\frac{n}{\sigma_0^2}}=\lim_{n\to\infty}\frac{\frac{m\sigma_0^2}{ns^2}+\bar{x}}{\frac{\sigma_0^2}{ns^2}+1}=\bar{x}, \tag{3.5}$$

---

[1] More technically, given the sample $\mathbf{X}$ whose distribution is conditional on a (possibly vector) parameter τ, a (possibly vector) $t(\mathbf{X})$ is jointly sufficient for τ if the likelihood, $\ell(\tau|\mathbf{X})$ is proportional to some function $g(\tau|t(\mathbf{X}))$ (Box and Tiao 1973). The sufficiency principle states that the distribution of $\mathbf{X}$ depends on τ only through $t(\mathbf{X})$ (Casella and Berger 2002, p.272; Hogg and Craig 1978, p.343). Sufficiency is one of the central contributions of Fisher (1922, 1925a, 1925b).

and the asymptotic posterior variance is given by:

$$\lim_{n\to\infty} \hat{\sigma}^2_\mu = \lim_{n\to\infty} \frac{1}{\frac{1}{s^2} + \frac{n}{\sigma_0^2}} = \lim_{n\to\infty} \frac{\sigma_0^2}{\frac{\sigma_0^2}{s^2} + n} = \frac{\sigma_0^2}{n}. \tag{3.6}$$

The asymptotic *distribution* of this Bayesian posterior mean is $\mathcal{N}(\bar{x}, \frac{\sigma_0^2}{n})$, which is exactly the result from frequentist theory (Berger 1985, p.224). In fact this result is sometimes called the "Bayesian Central Limit Theorem" (Carlin *et al.* 1993), and the importance of this is that Bayesian and frequentist results are identical in the limit. Furthermore, if we choose a huge value for the prior variance of μ, including the assignment $s^2 = \infty$, then we also get the frequentist result. This indicates that total or near total prior ignorance about the variance of μ translates to the same results as the standard likelihood model. Conversely, the frequentist approach unrealistically *assumes* the presence of asymptotic properties even in small sample analyses.

3.3   The Normal Model with Mean Known

Suppose, instead of knowing the variance parameter for normal data and obtaining a posterior distribution for the mean, that we know the mean parameter and estimate the variance. Specifically:

$$p(X|\mu_0, \sigma^2) = (2\pi\sigma^2)^{-\frac{1}{2}} \exp\left[-\frac{1}{2\sigma^2}(X - \mu_0)^2\right].$$

Here the notation μ_0 indicates a known value for μ and the absence of a subscript on σ^2 reminds us that this is the target of our investigation. Given an iid sample of size n, we have the likelihood function:

$$L(\sigma^2|\mathbf{x}) \propto (\sigma^2)^{-\frac{n}{2}} \exp\left[-\frac{n}{2\sigma^2}\underbrace{\left(\frac{1}{n}\sum_{i=1}^{n}(x_i - \mu_0)^2\right)}_{\text{sufficient statistic}}\right]. \tag{3.7}$$

Here the expression $\frac{1}{n}\sum_{i=1}^{n}(x_i - \mu_0)^2$ is a sufficient statistic for σ^2, and we need no other information from the data. In fact, we can label this sufficient statistic as $\tilde{\mathbf{x}}$ to make the notation easier from this point on.

The conjugate prior for the variance parameter in the normal likelihood model is the inverse gamma distribution, meaning that an inverse gamma form for the prior on σ^2 gives an inverse gamma posterior for σ^2 where parameters values change. The inverse gamma, for this reason, is a common choice for Bayesian specifications, and is less well-known in non-Bayesian settings. If a random variable X is distributed gamma (shape and rate parameters), then the random variable $1/X$ is distributed inverse gamma (Berger 1985, p.561, Shao 2005). The probability density function of the inverse gamma is given in Appendix B by:

$$\mathcal{IG}(\sigma^2|\alpha,\beta) = \frac{\beta^\alpha}{\Gamma(\alpha)}(\sigma^2)^{-(\alpha+1)}\exp[-\beta/\sigma^2]$$

$$\text{where:} \quad \sigma^2 > 0, \ \alpha > 0, \ \beta > 0. \tag{3.8}$$

The inverse gamma PDF is not as well behaved as one would expect for a distribution linked theoretically to the normal. For instance the first two moments need to be defined with the parameter restrictions:

$$E[\sigma^2] = \frac{\beta}{\alpha - 1}, \quad \alpha > 1$$

$$\text{Var}[\sigma^2] = \frac{\beta^2}{(\alpha-1)^2(\alpha-2)}, \quad \alpha > 2. \tag{3.9}$$

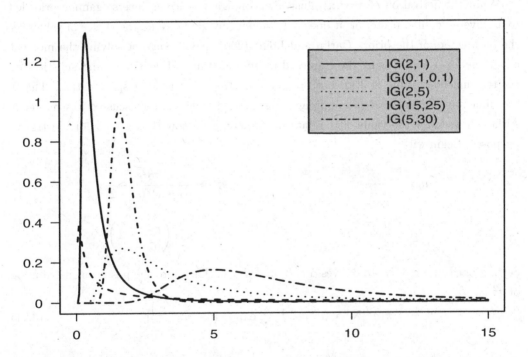

FIGURE 3.1: A MENAGERIE OF INVERSE GAMMA FORMS

The inverse gamma distribution is also slightly less intuitive to specify a desired shape than the gamma distribution. Figure 3.1 shows five differently parameterized forms from using the `dinvgamma(x,shape=,scale=)` command in R (package `MCMCpack`). Notice that the $\mathcal{IG}(0.1, 0.1)$ form rapidly flattens out as it asymptotes to zero on the y-axis. Inverse gamma priors with small and equal parameters are popular with `BUGS` users and appear frequently in the example models provided with the software. Yet this form would hardly be called a low information prior based on its form. An even more pronounced effect comes from using $\mathcal{IG}(1, \epsilon)$, where ϵ is a very small number. This is also a popular variance component prior specification in the BUGS world, and it very rapidly declines and then near flattens out moving to the right.

We now have enough tools to develop the posterior distribution of σ^2:

$$\pi(\sigma^2|\mathbf{x}) \propto L(\sigma^2|\mathbf{x})p(\sigma^2|\alpha, \beta)$$

$$= (\sigma^2)^{-\frac{n}{2}} \exp\left[-\frac{n}{2\sigma^2}\tilde{\mathbf{x}}\right] \frac{\beta^\alpha}{\Gamma(\alpha)}(\sigma^2)^{-(\alpha+1)} \exp\left[-\beta/\sigma^2\right]$$

$$\propto (\sigma^2)^{-((\alpha+\frac{n}{2})+1)} \exp\left[-\left(\beta + \frac{n}{2}\tilde{\mathbf{x}}\right)/\sigma^2\right]. \tag{3.10}$$

It is easy to see the last line as the kernel (core component of the PDF or PMF stripped of normalizing factors) of an $\mathcal{IG}(\alpha + \frac{n}{2}, \beta + \frac{n}{2}\tilde{\mathbf{x}})$ density function. This is a nice result since we now have the exact form of the prior for σ^2 without a lot of involved calculations.

While the derivation of $\pi(\sigma^2|\mathbf{x})$ illustrates conjugacy, since an inverse gamma prior led to an inverse gamma posterior, it does not provide any guidance on the choice of values for the parameters of the prior. Carlin and Louis (2001, p.326) suggest solving the moment equations for α and β using the empirical mean and standard deviation, and substituting the recommended value of 3 for both to get a relatively vague prior specification. This is a rudimentary *empirical Bayes* approach, and more details will be discussed in Chapter 8. Although we know the value of the data mean here, we follow their advice for illustrative purposes. Begin with:

$$\mu_{\sigma^2} = \frac{\beta}{\alpha - 1} \qquad\qquad s_{\sigma^2}^2 = \frac{\beta^2}{(\alpha-1)^2(\alpha-2)}$$

$$\alpha = \frac{\mu_{\sigma^2}^2}{s_{\sigma^2}^2} + 2 \qquad\qquad \beta = \mu_{\sigma^2}\left(\frac{\mu_{\sigma^2}^2}{s_{\sigma^2}^2} + 1\right).$$

Setting both μ_{σ^2} and s_{σ^2} to 3 gives $\alpha = 3$, and $\beta = 6$. Therefore the posterior distribution of σ^2 is:

$$\pi(\sigma^2|\mathbf{X}) \propto (\sigma^2)^{-((\frac{n}{2})+4)} \exp\left[-(6 + n\tilde{\mathbf{x}})/\sigma^2\right]. \tag{3.11}$$

3.4   The Normal Model with Both Mean and Variance Unknown

We can now develop the model with both μ and σ^2 unknown, with conjugate priors for both. The new wrinkle is that the conjugate prior for the mean is conditional on the variance. This is a mathematical necessity to preserve the "pass-through" nature of the normal distributional form for conjugacy. Start again with normally distributed data with mean μ and variance σ^2. We will use the inverse gamma and normal conjugate priors:

$$p(\sigma^2|\alpha,\beta) \propto (\sigma^2)^{-(\alpha+1)} \exp\left[-\beta/\sigma^2\right] \qquad (3.12)$$

$$p(\mu|m,\sigma^2/s_0) \propto (\sigma^2)^{-\frac{1}{2}} \exp\left[-\frac{1}{2\sigma^2/s_0}(\mu-m)^2\right], \qquad (3.13)$$

where the prior for μ is explicitly conditional on σ^2. The parameter s_0 is a so-called "confidence parameter" that measures the researcher's strength of belief on σ^2 on μ in the prior specification. This is intended to be a convenient feature of this model, but if one does not want to be bothered by an additional parameter specification, it can be set to 1 and dismissed (although this too is a parameter specification). Some authors choose to specify more intricate parameterizations for the prior on σ^2 in order to produce simplified marginal posterior expressions, but we will stay with the same parameterizations as given above for continuity and comparison. The resulting posterior is then:

$$\pi(\mu,\sigma^2|\mathbf{x}) = p(\mathbf{x}|\mu,\sigma^2)p(\sigma^2|\alpha,\beta)p(\mu|\sigma^2/s_0,m)$$

$$\propto (\sigma^2)^{-\alpha-\frac{n}{2}-\frac{3}{2}} \exp\left[-\frac{1}{\sigma^2}\beta - \frac{1}{2\sigma^2}\sum_{i=1}^{n}(x_i-\mu)^2 - \frac{1}{2\sigma^2/s_0}(\mu-m)^2\right]$$

$$= (\sigma^2)^{-\alpha-\frac{n}{2}-\frac{3}{2}} \exp\left[-\frac{1}{\sigma^2}\beta - \frac{1}{2\sigma^2}\left(\sum_{i=1}^{n}x_i^2 - 2n\bar{x}\mu + n\mu^2\right)\right.$$

$$\left. - \frac{1}{2\sigma^2/s_0}(\mu^2 - 2\mu m + m^2)\right]$$

$$= (\sigma^2)^{-\alpha-\frac{n}{2}-1} \exp\left[-\frac{1}{\sigma^2}\beta - \frac{1}{2\sigma^2}\left(\sum_{i=1}^{n}x_i^2 - n\bar{x}^2\right)\right]$$

$$\times (\sigma^2)^{-\frac{1}{2}} \exp\left[-\frac{1}{2\sigma^2}\left((n+s_0)\mu^2 - 2(n\bar{x}+ms_0)\mu + (n\bar{x}^2 + s_0m^2)\right)\right]$$

$$(3.14)$$

The last form of the posterior obviously sets up easy integration across the two parameters to get the marginal posterior distributions since the second term (line 2) is a normal kernel for μ and the first term (line 1) does not contain μ. First we perform this operation over μ to get the posterior for σ^2:

$$\pi(\sigma^2|\mathbf{x}) = \int_{-\infty}^{\infty} \pi(\mu,\sigma^2|\mathbf{x})d\mu$$

$$\propto (\sigma^2)^{-\alpha-\frac{n}{2}-1} \exp\left[-\frac{1}{\sigma^2}\left(\beta + \frac{1}{2}\left(\sum_{i=1}^{n}x_i^2 - n\bar{x}^2\right)\right)\right]. \qquad (3.15)$$

Therefore, observing the trivial fact that $-\alpha - \frac{n}{2} - 1 = -(\alpha + n/2) - 1$, the posterior distribution of σ^2 is another inverse gamma according to:

$$\sigma^2|\mathbf{x} \sim \mathcal{IG}\left(\alpha + \frac{n}{2}, \beta + \frac{1}{2}\sum_{i=1}^{n}(x_i - \bar{x})^2\right). \tag{3.16}$$

Since the prior (and posterior) distribution of μ is conditional on σ^2, we cannot integrate as done above. Fortunately, there is a simple trick based on an obvious cancellation:

$$\pi(\mu|\sigma^2\mathbf{x}) = \frac{\pi(\mu,\sigma^2|\mathbf{x})}{\pi(\sigma^2|\mathbf{x})}$$

$$\propto \sigma^{-1}\exp\left[-\frac{1}{2\sigma^2}\left((n+s_0)\mu^2 - 2(n\bar{x} + ms_0)\mu + (n\bar{x}^2 + s_0m^2)\right)\right]$$

$$= \sigma^{-1}\exp\left[-\frac{1}{2\sigma^2/(n+s_0)}\left(\mu^2 - 2\frac{n\bar{x} + ms_0}{n+s_0}\mu + \frac{n\bar{x}^2 + s_0m^2}{n+s_0}\right)\right]. \tag{3.17}$$

We can see now that the posterior distribution of μ is the normal:

$$\mu|\sigma^2,\mathbf{x} \sim \mathcal{N}\left[\frac{n\bar{x} + ms_0}{n+s_0}, \frac{\sigma^2}{n+s_0}\right]. \tag{3.18}$$

Note that the prior dependence on σ^2 flows through to the posterior for μ.

3.5   Multivariate Normal Model, μ and Σ Both Unknown

We will conclude our discussion of normal-conjugate models with the most intricate case: both parameters unknown, and multivariate data. While this is the most complex specification in this family, it is also the most realistic and therefore worthy of considerable attention here. The conjugate prior specification has an added complexity because it is only possible to specify it with a dependency of the mean on the variance: $p(\mu|\sigma^2)$, and $p(\sigma^2)$. If this is unrealistic, then a nonconjugate prior should be specified. The posterior is then determined by:

$$\pi(\mu,\sigma^2,\mathbf{x}) \propto L(\mu,\sigma^2|\mathbf{x})p(\mu|\sigma^2)p(\sigma^2) = L(\mu,\sigma^2|\mathbf{x})p(\mu,\sigma^2). \tag{3.19}$$

This is given here for a univariate normal case only. For the multivariate normal, we assume that each of the n rows of $\mathbf{X}$ is a k-dimensional vector representing a single case, so now μ is a k-length vector and Σ is a $k \times k$ matrix, both to be estimated. From the PDF of the multivariate normal (Appendix B), the likelihood function can be expressed and manipulated as follows:

$$L(\boldsymbol{\mu}, \boldsymbol{\Sigma}|\mathbf{X}) = \prod_{i=1}^{n} \left((2\pi)^{-k/2} |\boldsymbol{\Sigma}|^{-1/2} \exp\left[-\frac{1}{2}(\mathbf{x}_i - \boldsymbol{\mu})' \boldsymbol{\Sigma}^{-1} (\mathbf{x}_i - \boldsymbol{\mu}) \right] \right)$$

$$\propto |\boldsymbol{\Sigma}|^{-n/2} \exp\left[-\frac{1}{2} \sum_{i=1}^{n} (\mathbf{x}_i - \boldsymbol{\mu})' \boldsymbol{\Sigma}^{-1} (\mathbf{x}_i - \boldsymbol{\mu}) \right]$$

$$= |\boldsymbol{\Sigma}|^{-n/2} \exp\left[-\frac{1}{2} \sum_{i=1}^{n} \left(\mathbf{x}_i' \boldsymbol{\Sigma}^{-1} \mathbf{x}_i - 2\mathbf{x}_i' \boldsymbol{\Sigma}^{-1} \boldsymbol{\mu} + \boldsymbol{\mu}' \boldsymbol{\Sigma}^{-1} \boldsymbol{\mu} \right) \right]$$

$$= |\boldsymbol{\Sigma}|^{-n/2} \exp\left[-\frac{1}{2} \left(\sum_{i=1}^{n} \mathbf{x}_i' \boldsymbol{\Sigma}^{-1} \mathbf{x}_i - n\boldsymbol{\Sigma}^{-1} \bar{\mathbf{x}}' \bar{\mathbf{x}} + n\boldsymbol{\Sigma}^{-1} \bar{\mathbf{x}}' \bar{\mathbf{x}} \right. \right.$$

$$\left. \left. - 2n\bar{\mathbf{x}}' \boldsymbol{\Sigma}^{-1} \boldsymbol{\mu} + n\boldsymbol{\mu}' \boldsymbol{\Sigma}^{-1} \boldsymbol{\mu} \right) \right]$$

$$= |\boldsymbol{\Sigma}|^{-n/2} \exp\left[-\frac{1}{2} \left(\mathrm{tr}(\boldsymbol{\Sigma}^{-1}) \left(\sum_{i=1}^{n} \mathbf{x}_i' \mathbf{x}_i - n\bar{\mathbf{x}}' \bar{\mathbf{x}} \right) + n(\bar{\mathbf{x}} - \boldsymbol{\mu})' \boldsymbol{\Sigma}^{-1} (\bar{\mathbf{x}} - \boldsymbol{\mu}) \right) \right].$$

$$(3.20)$$

Since $\sum_{i=1}^{n}(\mathbf{x}_i'\mathbf{x}_i) - n\bar{\mathbf{x}}'\bar{\mathbf{x}} = \sum_{i=1}^{n}(\mathbf{x}_i - \bar{\mathbf{x}})'(\mathbf{x}_i - \bar{\mathbf{x}}) \equiv S^2$, then $L(\boldsymbol{\mu}, \boldsymbol{\Sigma}|\mathbf{X})$ is a function of the data only through the two-component sufficient statistic: $[\bar{\mathbf{x}}, S^2]$.

Robert (2001, p.189) suggests the following form of the conjugate priors:

$$\boldsymbol{\mu}|\boldsymbol{\Sigma} \sim \mathcal{N}_k \left(\mathbf{m}, \frac{1}{n_0} \boldsymbol{\Sigma} \right), \qquad \boldsymbol{\Sigma}^{-1} \sim \mathcal{W}(\alpha, \boldsymbol{\beta}), \qquad (3.21)$$

where $\mathcal{W}()$ denotes the Wishart distribution, which is a multivariate generalization of the gamma PDF (an obvious choice for modeling multivariate variances):[2]

$$\mathcal{W}(\boldsymbol{\Sigma}^{-1}|\alpha, \boldsymbol{\beta}) = \frac{|\boldsymbol{\Sigma}^{-1}|^{(\alpha - (k+1))/2}}{\Gamma_k(\alpha)|\boldsymbol{\beta}|^{\alpha/2}} \exp[-\mathrm{tr}(\boldsymbol{\beta}^{-1}\boldsymbol{\Sigma}^{-1})/2]$$

$$\text{where: } \Gamma_k(\alpha) = 2^{\alpha k/2} \pi^{k(k-1)/4} \prod_{i=1}^{k} \Gamma\left(\frac{\alpha + 1 - i}{2} \right),$$

$$2\alpha > k - 1, \quad \text{and } \boldsymbol{\beta} \text{ nonsingular.} \qquad (3.22)$$

The $\boldsymbol{\beta}$ matrix in the conjugate prior for $\boldsymbol{\Sigma}^{-1}$ need only be non-singular and symmetric (the latter is an assumption of the application not the PDF), but in practice it is generally given a diagonal form unless there are strong *a priori* reasons for assuming covariance terms. The term $\Gamma_k(\alpha)$ is the k-dimensional generalized gamma function, and is ignorable except for normalizing considerations. The parameter n_0 in Robert's prior parameterization is not a

---

[2]For mathematical and statistical properties of the Wishart distribution see Appendix B and the references: Krzanowski (1988, pp.209-210), Tong (1990, pp.51-55), and Stuart and Ord (1994), p.573-579). For generating procedures see Gentle (1998, p.107), Kleibergen and van Dijk (1993), and Smith and Hocking (1972).

prior sample size; it is intended to be a reflection of prior precision relative to the sample size that is tunable by the researcher to reflect prior confidence as specified before in (3.13). The smaller the ratio n_0/n, the less weight on the prior, and therefore the closer the results will be to classical results.

This setup leads to an articulation of the joint posterior:

$$\pi(\boldsymbol{\mu}, \boldsymbol{\Sigma}|\mathbf{x}) \propto |\boldsymbol{\Sigma}|^{-n/2} \exp\left[-\frac{1}{2}\left(\text{tr}(\boldsymbol{\Sigma}^{-1}S^2) + n(\bar{\mathbf{x}} - \boldsymbol{\mu})'\boldsymbol{\Sigma}^{-1}(\bar{\mathbf{x}} - \boldsymbol{\mu})\right)\right]$$

$$\times \left|\frac{\boldsymbol{\Sigma}}{n_0}\right|^{-1/2} \exp\left[-\frac{1}{2}(\boldsymbol{\mu} - \mathbf{m})'\left(\frac{1}{n_0}\boldsymbol{\Sigma}\right)^{-1}(\boldsymbol{\mu} - \mathbf{m})\right]$$

$$\times |\boldsymbol{\beta}|^{-\alpha/2}|\boldsymbol{\Sigma}^{-1}|^{(\alpha - (k+1))/2} \exp[-\text{tr}(\boldsymbol{\beta}^{-1}\boldsymbol{\Sigma}^{-1})/2]. \tag{3.23}$$

We can then proceed (with steps analogous to the production of marginals (3.16) and (3.18), only more involved) to the posterior distributions of interest:

$$\boldsymbol{\mu}|\boldsymbol{\Sigma} \sim \mathcal{N}_k\left(\frac{n_0\mathbf{m} + n\bar{\mathbf{x}}}{n_0 + n}, \frac{1}{n_0 + n}\boldsymbol{\Sigma}\right)$$

$$\boldsymbol{\Sigma}^{-1} \sim \mathcal{W}_k\left(\alpha + n, \left[\boldsymbol{\beta}^{-1} + S^2 + \frac{n_0 n}{n_0 + n}(\bar{\mathbf{x}} - \mathbf{m})(\bar{\mathbf{x}} - \mathbf{m})'\right]^{-1}\right). \tag{3.24}$$

Note that the dependency exists here in the multivariate case as well. We will see a more elegant application of a Bayesian normal model in the context of linear regression in Chapter 6. The distributional assignment for the variance described here is sometimes called the inverse-Wishart since it is inversely assigned to the matrix. Gelman and Hill (2007, p.286) note that the choice of degrees of freedom parameter (α) can add undesirable constraints to estimation of $\boldsymbol{\Sigma}$ components, effectively imposing a computational trade-off between diagonal and off-diagonal elements. Their solution, the *scaled inverse-Wishart distribution*, pre- and post-multiplies the inverse-Wishart with a vector of scale parameters.

■ **Example 3.1:   Variance Estimation with Public Health Data.**   Consider data from the 2000 U.S. census and North Carolina public records (North Carolina Division of Public Health, Women's and Children's Health Section in Conjunction with State Center for Health Statistics), which is available in the BaM package. Each case is one of 100 North Carolina counties, and we will use only the following subset of the variables.

▷ Substantiated.Abuse: within family documented abuse for the county.

▷ Percent.Poverty: percent within the county living in poverty, U.S. definition (see http://www.census.gov/hhes/www/poverty /threshld/thresh98.html).

▷ Total.Population: county population/1000.

So each $\mathbf{X}$ row is a 3-dimensional vector representing a single case, distributed $\mathcal{N}(\boldsymbol{\mu}, \boldsymbol{\Sigma})$ as in the model given above where both the mean and variance are unknown.

First we specify the relatively uninformed prior parameters $\alpha = 3$, $\mathbf{m} = (250, 16, 88)$, $n_0 = 0.01$, along with a $\boldsymbol{\beta}$, an identity matrix divided by 100 in this case (a selection not related to n here), to define the priors from (3.21). This produces marginal posteriors according to (3.24), which we describe with posterior quantiles in the case of the mean $\boldsymbol{\mu}$ in the left-side of Table 3.1.

TABLE 3.1: POSTERIOR SUMMARIES, NORMAL MODELS

μ Quantile	Uninformed Prior			Semi-Informed Prior		
	Abuse	%Poverty	Population	Abuse	%Poverty	Population
0.01	195.8976	14.2399	77.9827	138.4181	9.19079	82.3306
0.25	199.6618	14.3123	79.7873	147.1816	9.90282	83.6443
0.50	201.2110	14.3409	80.5230	150.7495	10.18752	84.1989
0.75	202.7294	14.3698	81.2590	154.3365	10.47835	84.7918
0.99	206.4080	14.4400	83.0124	163.0994	11.15920	86.2538

Since $\boldsymbol{\Sigma}$ is a 3×3 matrix, each cell possessing a distinct posterior, we need to describe nine posteriors. This description can be in tabular form like the posterior for $\boldsymbol{\mu}$, or we can abbreviate it and just give nine posterior means arranged in the same tabular positions:

$$\bar{\boldsymbol{\Sigma}} = \begin{bmatrix} 531.5540 & -3.2724 & 200.2079 \\ -3.2724 & 0.1871 & -1.6727 \\ 200.2079 & -1.6727 & 117.9017 \end{bmatrix}$$

To provide a contrast, specify strong priors, $\alpha = 3$, $\mathbf{m} = (100, 6, 88)$, $n_0 = 99$, and $\boldsymbol{\beta}$ an identity matrix divided by 10. This now gives posterior values summarized in the right-side of Table 3.1 and with the following matrix means for the posterior of $\boldsymbol{\Sigma}$:

$$\bar{\boldsymbol{\Sigma}} = \begin{bmatrix} 5678.6595 & 421.2349 & -181.0511 \\ 421.2349 & 35.2098 & -33.1597 \\ -181.0511 & -33.1597 & 146.3097 \end{bmatrix}$$

As a way of seeing the effect of the priors and the differences provided by the model, we now compare the marginal posterior distribution of the first cell value of the variance matrix from the diffuse model above having mean $\bar{\boldsymbol{\Sigma}}[1,1] = 531.5540$ with the likelihood function from a conventional non-Bayesian model. Figure 3.2 shows the two forms along with the associated 95% confidence and 95% credible interval endpoints. Note that there are differences in the form of the distributions, as well as the fact that interval inference would produce notably different conclusions. Such differences are observable even though the priors were set up to be relatively uninformed.

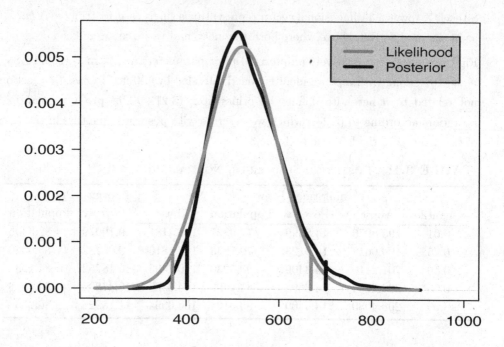

FIGURE 3.2: Posterior Distribution versus Likelihood Function

---

3.6   Simulated Effects of Differing Priors

To illustrate the diminishing effect of the prior as the size of the data increases, we first generate progressively larger contrived data sets distributed bivariate normal with mean vector $\begin{bmatrix} 0 \\ 0 \end{bmatrix}$, and variance matrix $\begin{bmatrix} 1 & 0 \\ 0 & 1 \end{bmatrix}$. By giving the data this simple distribution, the relative effects of differing and intentionally wrong priors will be more apparent. We calculate and summarize the posterior distributions for three different sample sizes, given in Table 3.2.

In each case, 1,000 simulated values from the posterior distribution are generated and the reported posterior statistics are the mean ($\hat{\mu}$), the standard error ($\hat{\sigma}_\mu$), and the 95% HPD region. Note that this generation of simulated values is a very different aspect of this demonstration than the generation of the original data. Here the size of the original data set is the quantity being varied. Once we have a parametric form of the prior, we can be as arbitrarily accurate as our patience permits (some computers being faster than others), since this is just a function of how long we want to run the simulation. The posterior simulation size of 1,000 was chosen for convenience.

The prior given here is a reasonably misguided specification: the prior normal mean vector for μ is $\mathbf{m} = \begin{bmatrix} 5 \\ 5 \end{bmatrix}$, $\alpha = 3 > (k - 1)/2$, and the prior Wishart parameters for Σ are $\beta = \begin{bmatrix} 5 & 0 \\ 0 & 5 \end{bmatrix}$. Table 3.2 shows the posterior effect for all estimated parameters as sample size goes from 10 to 100 to 1000.

TABLE 3.2: A BIVARIATE NORMAL, 1000 SIMULATIONS

Specification		Mean	Standard Error	95% HPD Interval
$n = 10,$	μ_1	1.736	0.726	[0.276: 3.113]
$p \sim \mathcal{N}\left[\begin{smallmatrix} 5 \\ 5 \end{smallmatrix} \middle\| \begin{smallmatrix} 5 & 0 \\ 0 & 5 \end{smallmatrix}\right]$	μ_2	1.399	0.893	[-0.559: 3.090]
	σ_1	8.175	3.618	[3.598:17.429]
	σ_2	12.076	5.762	[5.251:27.849]
	ρ	0.836	0.088	[0.617: 0.952]
$n = 100,$	μ_1	0.174	0.137	[-0.100: 0.435]
$p \sim \mathcal{N}\left[\begin{smallmatrix} 5 \\ 5 \end{smallmatrix} \middle\| \begin{smallmatrix} 5 & 0 \\ 0 & 5 \end{smallmatrix}\right]$	μ_2	0.279	0.146	[0.001: 0.565]
	σ_1	2.106	0.286	[1.614: 2.718]
	σ_2	2.158	0.310	[1.639: 2.848]
	ρ	0.563	0.066	[0.425: 0.679]
$n = 1000,$	μ_1	0.023	0.034	[-0.051: 0.087]
$p \sim \mathcal{N}\left[\begin{smallmatrix} 5 \\ 5 \end{smallmatrix} \middle\| \begin{smallmatrix} 5 & 0 \\ 0 & 5 \end{smallmatrix}\right]$	μ_2	-0.017	0.033	[-0.081: 0.043]
	σ_1	1.083	0.049	[0.991: 1.189]
	σ_2	1.164	0.051	[1.074: 1.272]
	ρ	0.059	0.032	[-0.007: 0.124]

Original data generated from the $\mathcal{N}\left[\begin{smallmatrix} 0 \\ 0 \end{smallmatrix} \middle\| \begin{smallmatrix} 1 & 0 \\ 0 & 1 \end{smallmatrix}\right]$ distribution.

As the sample size increases by orders of magnitude, the posterior variance for each estimated coefficient decreases as indicated in (3.2). This example shows that the likelihood function eventually overwhelms the prior specification. Were we to carry this example further and produce larger initial sample sizes for the bivariate normal data, we could come arbitrarily close to the standard maximum likelihood estimates.

Now we will replicate this simulation analysis with a slightly more interesting data specification. The three sample sizes will be generated as before, but now according to mean $\left[\begin{smallmatrix} 1 \\ 3 \end{smallmatrix}\right]$, and variance matrix $\left[\begin{smallmatrix} 1.0 & 0.7 \\ 0.7 & 3.0 \end{smallmatrix}\right]$, the main point being the introduction of fairly high data correlation. This time the prior specification will again be deliberately incorrect, but far more dispersed. This means that the increase in sample size according to the previous scheme should take longer to overwhelm the prior. This is exactly what we observe in Table 3.3.

Although the examples developed in these simulations are contrived, in that we assumed knowledge of the true form of the population data, the principles observed are exactly those that apply in more complex specifications. The point is to illustrate that the estimation technique is reasonably easy to implement (the R code is contained in this chapter's **Computational Addendum**).

TABLE 3.3:   ANOTHER BIVARIATE NORMAL, 1000 SIMULATIONS

Specification		Mean	Standard Error	95% HPD Interval
$n = 10,$	μ_1	1.220	0.303	[0.567: 1.820]
$p \sim \mathcal{N}\left[\begin{smallmatrix} 2 \\ 2 \end{smallmatrix} \middle\| \begin{smallmatrix} 10 & 5 \\ 5 & 10 \end{smallmatrix}\right]$	μ_2	2.676	0.409	[1.863: 3.498]
	σ_1	1.304	0.644	[0.546: 3.029]
	σ_2	2.490	1.286	[1.025: 5.836]
	ρ	0.313	0.264	[-0.295: 0.749]
$n = 100,$	μ_1	0.923	0.109	[0.707: 1.126]
$p \sim \mathcal{N}\left[\begin{smallmatrix} 2 \\ 2 \end{smallmatrix} \middle\| \begin{smallmatrix} 10 & 5 \\ 5 & 10 \end{smallmatrix}\right]$	μ_2	2.909	0.186	[2.542: 3.289]
	σ_1	1.208	0.176	[0.925: 1.616]
	σ_2	3.507	0.476	[2.634: 4.532]
	ρ	0.427	0.079	[0.260: 0.567]
$n = 1000,$	μ_1	1.059	0.031	[0.996: 1.119]
$p \sim \mathcal{N}\left[\begin{smallmatrix} 2 \\ 2 \end{smallmatrix} \middle\| \begin{smallmatrix} 10 & 5 \\ 5 & 10 \end{smallmatrix}\right]$	μ_2	2.994	0.055	[2.885: 3.105]
	σ_1	0.996	0.044	[0.919: 1.091]
	σ_2	2.914	0.135	[2.649: 3.176]
	ρ	0.376	0.028	[0.320: 0.430]

Original data generated from the $\mathcal{N}\left[\begin{smallmatrix} 1 \\ 3 \end{smallmatrix} \middle\| \begin{smallmatrix} 1.0 & 0.7 \\ 0.7 & 3.0 \end{smallmatrix}\right]$ distribution.

3.7   Some Normal Comments

The primary purpose of this chapter is to illustrate the Bayesian normal model in detail. While it is often not reasonable to make normal assumptions about the distribution of data or parameters, such approaches are effective surprisingly often. Given the knowledge that normality is an asymptotic property of many of the estimators studied, the normal model is well worth careful inspection. Excellent general expositions on further consequences of the Bayesian linear model specification can be found in Box and Tiao (1973, Chapter 2) and Zellner (1971, Chapters 3 and 8). Hartigan (1983, Chapter 9) gives a detailed mathematical overview of normality in the Bayesian linear model, and Tanner (1996, Chapter 1) discusses normal approximations in a Bayesian context. The role of normal assumptions in various Bayesian time-series models is covered by Bauwens, Lubrano, and Richard (1999), Broemeling (1985), Pole, West, and Harrison (1994), Sims (1988), and West and Harrison (1997). In addition, the normal specification is at the heart of the Bayesian linear model specification and leads nicely to the full Bayesian generalized linear model (Dey, Ghosh, and Mallick 2000).

In Chapter 5, we will use normal and t-distribution assumptions in the development of the Bayesian linear regression model by treating the standard linear model, $\mathbf{y} = \mathbf{X}\boldsymbol{\beta} + \boldsymbol{\epsilon}$, with priors on $\boldsymbol{\beta}$ and $\boldsymbol{\epsilon}$. Chapter 12 develops the linear *hierarchical* Bayesian model in which structured levels of terms and data are specified. A substantial amount of work in this area is based on normal and t-distribution assumptions for coefficient or error terms. Also, Lee (2004) gives the Bayesian treatment of classical two-sample problems by Student (Gossett actually, 1908a, 1908b) in the original derivation of the t-distribution theory.

3.8 The Student's-t Model

In this section we will see how the t-distribution naturally arises in the case where we specify vague prior information for the normal parameters μ and σ^2 in the normal model. The t-distribution resembles the normal except that it has heavier tails, thus expressing greater posterior uncertainty. Furthermore, the Student's-t is a natural choice for small sample applications because as the sample size increases, the distribution converges to a normal.

Assume that we have normally distributed data with both parameters unknown and the goal is to estimate these. We start with the most basic likelihood function for assumed normally distributed data, and rearrange it slightly:

$$L(\mu, \sigma^2 | \mathbf{x}) = (2\pi\sigma^2)^{-\frac{n}{2}} \exp\left[-\frac{1}{2\sigma^2} \sum_{i=1}^{n} (\mathbf{x}_i - \mu)^2\right]$$

$$= (2\pi\sigma^2)^{-\frac{n}{2}} \exp\left[-\frac{1}{2\sigma^2} \sum_{i=1}^{n} [(\mathbf{x}_i - \bar{\mathbf{x}}) - (\mu - \bar{\mathbf{x}})]^2\right]$$

$$= (2\pi\sigma^2)^{-\frac{n}{2}} \exp\left[-\frac{1}{2\sigma^2} \left(\sum_{i=1}^{n} (x_i - \bar{\mathbf{x}})^2 - 2\sum_{i=1}^{n} (x_i \mu - x_i \bar{\mathbf{x}} - \bar{\mathbf{x}}\mu + \bar{\mathbf{x}}^2) + n(\bar{\mathbf{x}} - \mu)^2\right)\right]$$

$$\propto \sigma^{-n} \exp\left[-\frac{1}{2\sigma^2} \left((n-1)s^2 + n(\bar{\mathbf{x}} - \mu)^2\right)\right]. \tag{3.25}$$

The purpose here is to re-express the likelihood function strictly in terms of two sufficient statistics: $s^2 = \frac{1}{n-1} \sum (\mathbf{x}_i - \bar{\mathbf{x}})^2$ and $\bar{\mathbf{x}} = \frac{1}{n} \sum (\mathbf{x}_i)$, which are distributed $(\sigma^2/(n-1))\chi^2_{n-1}$ and $\mathcal{N}(\mu, \sigma^2/n)$, respectively, under classical results.

Sometimes it is desirable to insert relatively little subjective information into the analysis. One way to do this is to stipulate vague or diffuse prior distributions for the unknown parameters. A common assignment of vague priors for the normal model is given by the pair:

$$p(\mu) \propto c, \quad -\infty < \mu < \infty$$

$$p(\sigma^2) \propto \sigma^{-2}, \quad 0 < \sigma < \infty, \tag{3.26}$$

where c is an arbitrary constant. These are examples of "improper" prior distributions in that they do not integrate to a finite quantity and therefore do not meet the technical requirements for density functions. This turns out to be a minor issue and we will discuss the characteristics of improper priors extensively in Chapter 4. Observe also that we are assuming independent priors for the normal parameters, unlike the conjugate model in Section 3.4.

The resulting posterior is created by $\pi(\mu, \sigma^2|\mathbf{x}) \propto L(\mu, \sigma|\mathbf{x})p(\mu)p(\sigma)$ according to:

$$\pi(\mu, \sigma^2|\mathbf{x}) \propto \sigma^{-(n+2)} \exp\left[-\frac{1}{2\sigma^2}\left((n-1)s^2 + n(\mu - \bar{\mathbf{x}})^2\right)\right]. \tag{3.27}$$

The real objective here is to get the marginal posterior distribution for both parameters. We could obtain the marginal posterior for μ by integrating out σ: $\int \pi(\mu, \sigma|\mathbf{x})d\sigma$, except that this is a difficult integration to do by brute force. Instead consider the handy integral formula based on the inverse gamma PDF:

$$\int_0^\infty x^{-b-1} \exp[-a/x^2]dx = \frac{1}{2}a^{-\frac{b}{2}}\Gamma\left(\frac{b}{2}\right), \tag{3.28}$$

which is used by Box and Tiao (1973, p.145) as well as by many calculus texts. Setting $\sigma^2 = x^2$, $n+1 = b$, and $\frac{1}{2}\left((n-1)s^2 + n(\mu - \bar{\mathbf{x}})^2\right) = a$ means that integrating (3.27) with respect to σ can be done by:

$$\pi(\mu|\mathbf{x}) = \int_0^\infty \pi(\mu, \sigma^2|\mathbf{x})d\sigma^2$$

$$= \frac{1}{2}\left[\frac{1}{2}\left((n-1)s^2 + n(\mu - \bar{\mathbf{x}})^2\right)\right]^{-\frac{(n+1)}{2}}\Gamma\left(\frac{n+1}{2}\right)$$

$$= \left(\frac{1}{2}\right)s^{-(n+1)}\left(\frac{1}{2}(n-1)\right)^{-\frac{(n+1)}{2}}\Gamma\left(\frac{n+1}{2}\right)\left[1 + \frac{1}{n-1}\left(\frac{\mu - \bar{\mathbf{x}}}{s/\sqrt{n}}\right)^2\right]^{-\frac{(n+1)}{2}}.$$

The structure here is much more revealing if we make the transformation $t = \frac{\mu - \bar{\mathbf{x}}}{s/\sqrt{n}}$ with the Jacobian $J = \frac{d}{dt}\mu = \frac{s}{\sqrt{n}}$:

$$\pi(t|\mathbf{x}) = \left(\frac{1}{2}\right)s^{-(n+1)}\left(\frac{1}{2}(n-1)\right)^{-\frac{(n+1)}{2}}\Gamma\left(\frac{n+1}{2}\right)\left[1 + \frac{1}{n-1}t^2\right]^{-\frac{(n+1)}{2}}\left[\frac{s}{\sqrt{n}}\right]$$

$$\propto \frac{\Gamma\left(\frac{n+1}{2}\right)}{\Gamma\left(\frac{n}{2}\right)}\frac{1}{(n\pi)^{\frac{1}{2}}}\left(1 + \frac{1}{n-1}t^2\right)^{-\frac{(n+1)}{2}}. \tag{3.29}$$

This is exactly the form of the Student's-t given in Appendix B. Therefore the marginal posterior of $\frac{\mu - \bar{\mathbf{x}}}{s/\sqrt{n}}$ is Student's-t with $\theta = n$ degrees of freedom, and the marginal posterior of μ itself is:

$$\pi(\mu|\mathbf{x}) = \frac{\Gamma\left(\frac{\theta+1}{2}\right)}{\Gamma\left(\frac{\theta}{2}\right)}\frac{s/\sqrt{n}}{(\theta\pi)^{\frac{1}{2}}(1 + \mu^2/\theta)^{(\theta+1)/2}} + \bar{\mathbf{x}}, \tag{3.30}$$

where the role of the two sufficient statistics, $s^2 = \frac{1}{n-1}\sum(\mathbf{x}_i - \bar{\mathbf{x}})^2$ and $\bar{\mathbf{x}} = \frac{1}{n}\sum(\mathbf{x}_i)$, is clear.

The calculation of the marginal posterior for σ^2 is considerably less involved because we can once again use the conditional probability property:

$$\pi(\sigma|\mathbf{x}) = \frac{\pi(\mu, \sigma|\mathbf{x})}{\pi(\mu|\mathbf{x})}. \tag{3.31}$$

This means that we can obtain the marginal posterior of σ by dividing the joint posterior by the marginal distribution of μ assuming that σ is independent. This works out very nicely, provided sufficient sample size:

$$\pi(\sigma|\mathbf{x}) = \frac{\left(\frac{n}{2\pi}\right)^{\frac{1}{2}} \frac{\left(\frac{(n-1)s^2}{2}\right)^{\frac{n-1}{2}}}{\frac{1}{2}\Gamma\left(\frac{n-1}{2}\right)} \sigma^{-(n+2)} \exp\left[-\frac{1}{2\sigma^2}\left((n-1)s^2 + n(\mu - \bar{\mathbf{x}})^2\right)\right]}{\frac{\Gamma\left(\frac{\theta+1}{2}\right)}{\Gamma\left(\frac{\theta}{2}\right)} \frac{s/\sqrt{n}}{(\theta\pi)^{\frac{1}{2}}(1+\mu^2/\theta)^{(\theta+1)/2}} + \bar{\mathbf{x}}}$$

$$\propto \sigma^{-((n+1)+1)} \exp\left[-\frac{1}{2}(n-1)s^2/\sigma^2\right]. \tag{3.32}$$

The last line shows that the marginal posterior of σ^2 (note the change) is distributed inverse gamma since this line is the kernel from $\mathcal{IG}(\frac{n+1}{2}, \frac{n-1}{2}s^2)$ (see Appendix B).

From this specification, we see that the resulting marginal posteriors for both μ and σ are more diffuse than those we produced by using conjugate priors. This makes a lot of sense since we used vague priors and it is logical that this decision should have some influence on the resulting posterior. However, consistent with all Bayesian specifications, if the data size becomes very large, then it does not matter whether a conjugate prior, giving a high level of *a priori* information, or a diffuse improper prior, giving a very low level of *a priori* information, is used.

■ **Example 3.2: National IQ Scores.** Standard IQ tests are designed to measure intelligence and reasoning with a mean of 100 and a standard deviation of 15. However, these tests are also said to have economic and cultural biases that favor some groups over others. An additional complication is added when IQ scores are aggregated at the national level because within-country features are masked. This example analyzes internationally collected IQ data (Lynn and Vanhanen 2001) for 80 countries from published national sources. The key idea in describing the posterior distribution is whether national effects alter the intended parameterization.

Average IQ Score By Country

Argentina	96	Australia	98	Austria	102	Barbados	78
Belgium	100	Brazil	87	Bulgaria	93	Canada	97
China	100	Congo (Br.)	73	Congo (Zr.)	65	Croatia	90
Cuba	85	Czech Repub.	97	Denmark	98	Ecuador	80
Egypt	83	Eq. Guinea	59	Ethiopia	63	Fiji	84
Finland	97	France	98	Germany	102	Ghana	71
Greece	92	Guatemala	79	Guinea	66	Hong Kong	107
Hungary	99	India	81	Indonesia	89	Iran	84
Iraq	87	Ireland	93	Israel	94	Italy	102
Jamaica	72	Japan	105	Kenya	72	Korea (S.)	106
Lebanon	86	Malaysia	92	Marshall I.	84	Mexico	87
Morocco	85	Nepal	78	Netherlands	102	New Zealand	100
Nigeria	67	Norway	98	Peru	90	Philippines	86
Poland	99	Portugal	95	Puerto Rico	84	Qatar	78
Romania	94	Russia	96	Samoa	87	Sierra Leone	64
Singapore	103	Slovakia	96	Slovenia	95	South Africa	72
Spain	97	Sudan	72	Suriname	89	Sweden	101
Switzerland	101	Taiwan	104	Tanzania	72	Thailand	91
Tonga	87	Turkey	90	Uganda	73	U.K.	100
U.S.	98	Uruguay	96	Zambia	77	Zimbabwe	66

The IQ testing instrument is designed to have a mean response of 100 with a standard deviation of 15 (the Stanford-Binet version has a standard deviation of 16). The data provide the mean IQ score for 80 countries. Industrialized Asian countries seem to do relatively quite well, and poorer developing countries generally fair less well.

TABLE 3.4: POSTERIOR SUMMARY: NATIONAL IQ SCORES

Quantile:	0.01	0.10	0.25	0.50	0.75	0.90	0.99
μ	85.081	86.489	87.311	88.213	89.106	89.920	91.379
σ	10.006	10.784	11.284	11.882	12.539	13.191	14.427

From (3.29), (3.32), and using the improper priors $p(\mu) \propto c$, $p(\sigma^2) \propto \sigma^{-2}$, we get the posterior quantiles in Table 3.4. This result demonstrates that the national level summary through the posterior no longer resembles the original distributional goal of the test. Noticeably, the distribution of μ is centered at about 88 rather than 100, and the median of the posterior variance implies a standard error of roughly 12 (recall that we are analyzing unweighted *means* by country, which will lower variance). Part of the results can be attributed to the aggregation effects that show up in the posterior mean. This may be due to some cultural impact, as the instrument was designed originally for use in an English-speaking industrialized country and then subsequently used around the world. The R code is provided below:

```
data(iq)
n <- length(iq)
t.iq <- (iq-mean(iq))/(sd(iq)/sqrt(n))
r.t <- (rt(100000, n-1)*(sd(iq)/sqrt(n))) + mean(iq)
quantile(r.t,c(0.01,0.10,0.25,0.5,0.75,0.90,0.99))
r.sigma.sq <- 1/rgamma(100000,shape=(n+1)/2,
                       rate=var(iq)*(n-1)/2)
quantile(sqrt(r.sigma.sq),
         c(0.01,0.10,0.25,0.5,0.75,0.90,0.99))
```

Some brief caveats are warranted. There is nothing implied here about innate natural intelligence of nationalities since test results can be a function of health, sociological, and political factors. Furthermore, there are differences by country on who takes (or is allowed to take) the test. Finally, since these data are unweighted means, population size is not taken into account.

3.9 Normal Mixture Models

Mixture models allow the parametric description of distributions that cannot be described with conventional PDFs and PMFs. This approach is a compromise between fully nonparametric models that completely avoid imposing an underlying probability generating form, and the standard assumption that all the observed data result from a single identifiable process. The basic idea is to incorporate more than one simple distributional form into a single model in a way that recognizes heterogeneity.

For example, suppose that a sample is observed in which every data point, $x_1, \ldots, x_n$ is generated by the same density function but with a different indexing parameter: $\theta_1, \ldots, \theta_n$. Then the "true" density of the sample can be approximated by:

$$g(x_i|\boldsymbol{\theta}) = \frac{1}{n}f(x_i|\theta_1) + \frac{1}{n}f(x_i|\theta_2) + \ldots + \frac{1}{n}f(x_i|\theta_n), \tag{3.33}$$

which for data value x_i substitutes a mean value, $g(x_i|\boldsymbol{\theta})$, for the exact value, $f(x_i|\theta_i)$.

The specification in (3.33) is a somewhat pessimistic view of the world in that it is unlikely that *every* single data point is produced by a unique generating process, given that we have collected these cases together for some reason. Instead it is far more likely that there is some number, $m \ll n$, of these functions that are required. So now we can condense (3.33) by collecting terms using a simple weighting scheme that indicates the proportion of the x_i with identical θ parameters:

$$g(x_i|\boldsymbol{\omega}, \boldsymbol{\theta}) = \omega_1 f(x_i|\theta_1) + \omega_2 f(x_i|\theta_2) + \ldots + \omega_m f(x_i|\theta_m)$$

$$= \sum_{j=1}^{m} \omega_j f(x_i|\theta_j), \tag{3.34}$$

where $\sum \omega_j \leq 1$. Thus the vector $\boldsymbol{\omega}$ describes variation in the θ_i and therefore variation in the $f(\mathbf{x}|\boldsymbol{\theta})$. This is very straightforward but requires the unrealistic assumption that we have perfect knowledge of the weights. It is far more likely that we have to make this assignment through an unobserved *component indicator vector*:

$$z_{ij} = \begin{cases} 1, & \text{iff } x_i \sim f(x_i|\theta_j) \\ 0, & \text{otherwise,} \end{cases} \tag{3.35}$$

where the $n \times m$ matrix $\mathbf{z}$, containing a single 1 in each row, is the stacking of these vectors. Although unobserved, each row of $\mathbf{z}$ can be modeled with the multinomial, a multicategory generalization of the binomial ($\mathcal{MN}(\mathbf{z}_i|m, \boldsymbol{\omega})$, see Appendix B). Now the joint distribution of the observed data, $\mathbf{x}$, and the unobserved latent data, $\mathbf{z}$ is just:

$$f(\mathbf{x}, \mathbf{z}) = \prod_{i=1}^{n} \prod_{j=1}^{m} \left[(\omega_j f(x_i|\theta_j))^{z_{ij}} \right]. \tag{3.36}$$

This model is much more explicitly Bayesian when we consider $\boldsymbol{\omega}$ to be the parameters of the multinomial and then assign them a prior distribution.

To fully elaborate this model, consider now the x_i to be normally distributed so that $f(x_i|\theta_j)$ in (3.36) is $\mathcal{N}(\mu_j, \sigma_j)$ provided that $z_{ij} = 1$, where θ_j in (3.34) is now a vector: $\theta_j = \left[\begin{smallmatrix} \mu_j \\ \sigma_j \end{smallmatrix} \right]$. Consider the following priors to complete the model:

$$p_j(\sigma_j^2) \sim \mathcal{IG}\left(\frac{\tau_j}{2}, \frac{\rho_j^2}{2} \right)$$

$$p_j(\mu_j|\sigma_j^2) \sim \mathcal{N}\left(\nu_j, \frac{\sigma_j^2}{n_j} \right).$$

$$p(\omega) \sim \mathcal{D}(\boldsymbol{\omega}|\alpha_1, \dots, \alpha_m), \tag{3.37}$$

and the τ_j, ρ_j, ν_j, and $\alpha_1, \dots, \alpha_m$ are all assigned specific parameter values. Here n_j is the number of cases within the jth category. The last assignment from (3.37) is a k-category Dirichlet prior, the multicategory generalization of the beta PDF (B). Accordingly, this model is now a generalization of the beta-binomial model from Chapter 2. This is a completely conjugate specification and the product of all of these priors and the likelihood function leads to the following posterior distributions:

$$\mu_j|\sigma_j \sim \mathcal{N}\left(\frac{n_j \nu_j + \bar{z}_j \bar{x}_j(\mathbf{z})}{n_j + \bar{z}_j}, \frac{\sigma_j^2}{\sigma_j^2 + n} \right)$$

$$\sigma_j^2 \sim \mathcal{IG}\left(\frac{\tau_j + n_j}{2}, \frac{1}{2} \left[\rho_j^2 + \sum_{i=1}^{n} z_{ij}(x_j - \bar{x}_j(\mathbf{z}))^2 + \frac{n_j \bar{z}_j(\bar{x}_j(\mathbf{z}) - \nu_j)^2}{n_j + \bar{z}_j} \right] \right)$$

$$\omega \sim \mathcal{D}(\alpha_1 + \bar{z}_1, \alpha_2 + \bar{z}_2, \dots, \alpha_k + \bar{z}_m), \tag{3.38}$$

where:

$$\bar{z}_j = \sum_{i=1}^{n} z_{ij} \qquad \bar{x}_j(\mathbf{z}) = \frac{1}{\bar{z}_j} \sum_{i=1}^{n} z_{ij} x_i.$$

If we knew the **z** values, this would be a very easy estimation problem. Unfortunately this is often not the case. However, from (3.38) and the multinomial specification for **z**, we have a full set of full conditional distributions in order to use Gibbs sampling. Another method that can be used here is to treat **z** as missing information and apply the EM algorithm described in Chapter 9. Both Gelman *et al.* (2003) and Robert (1996) provide extensive implementation details on both of these estimation approaches. The original EM algorithm article of Dempster, Laird, and Rubin (1977, Section 4.3) discusses a similar model. Carlin and Louis (2009, p.184) show how the scale mixture of normals developed here can be extended to models with non-normal (heavier tail) errors. Also, the classic reference on mixture models belongs to Titterington, Smith, and Makov (1985), and contains both Bayesian and non-Bayesian methods.

3.10   Exercises

3.1   The most important case of a two-parameter exponential family is when the second parameter is a scale parameter. Designate ψ as such a scale parameter, then the *exponential family form* expression of a PDF or PMF is rewritten:

$$f(y|\theta) = \exp\left[\frac{y\theta - b(\theta)}{a(\psi)} + c(y,\psi)\right].$$

Rewrite the normal PDF in exponential family form.

3.2   Show that the Student's-t PDF cannot be put into the exponential family form.

3.3   Suppose the random variable X is distributed $\mathcal{N}(\mu,\sigma^2)$. Prove that the random variable $Y = (X - \mu)/\sigma$ is $\mathcal{N}(0,1)$: standard normal.

3.4   Suppose that $X_1,\ldots,X_n$ and $Y_1,\ldots,Y_n$ are independent normally distributed samples with means μ_X, μ_Y, and finite variances σ_X^2, σ_Y^2. Let r be the standard (Pearson Product Moment) correlation coefficient between these two samples. Show that $\sqrt{n-2}r/\sqrt{1-r^2}$ has a Student's-t distribution with $\nu = n-2$ degrees of freedom, but $(n-2)r^2/(1-r^2)$ has an F distribution with 1 and $n-2$ degrees of freedom.

3.5   Missing Data. Suppose we have an iid sample of collected data: $X_1, X_2, \ldots, X_k,$ $Y_{k+1}, \ldots, Y_n \sim \mathcal{N}(\mu,1)$, where the Y_i values represent data that has gone missing. Specify a joint posterior for μ and the missing data with a $\mathcal{N}(0,1/n)$ prior.

3.6   Obesity is an increasing public health and policy problem in the United States, but rates differ by the individual states. Consider the following average body mass index $(BMI = [kilograms]/[meters]^2 = [4.88 \times pounds]/[feet]^2)$ data by state from 2009:

Alabama	31.0	Alaska	24.8	Arizona	25.5
Arkansas	30.5	California	24.8	Colorado	18.6
Connecticut	20.6	Delaware	27.0	Florida	25.2
Georgia	27.2	Hawaii	22.3	Idaho	24.5
Illinois	26.5	Indiana	29.5	Iowa	27.9
Kansas	28.1	Kentucky	31.5	Louisiana	33.0
Maine	25.8	Maryland	26.2	Massachusetts	21.4
Michigan	29.6	Minnesota	24.6	Mississippi	34.4
Missouri	30.0	Montana	23.2	Nebraska	27.2
Nevada	25.8	New Hampshire	25.7	New Jersey	23.3
New Mexico	25.1	New York	24.2	North Carolina	29.3
North Dakota	27.9	Ohio	28.8	Oklahoma	31.4
Oregon	23.0	Pennsylvania	27.4	Rhode Island	24.6
South Carolina	29.4	South Dakota	29.6	Tennessee	32.3
Texas	28.7	Utah	23.5	Vermont	22.8
Virginia	25.0	Washington	26.4	Washington DC	19.7
West Virginia	31.1	Wisconsin	28.7	Wyoming	24.6

(also available as `bmi.2009` in the `BaM` package in `R`). Plot these data with a histogram and overlay a normal PDF with the mean and variance from the data. Can you assert that these data are normally distributed? If you need to use these data as a prior distribution for the analysis of future BMI data, how would you construct this prior?

3.7   Suppose $X \sim \mathcal{N}(\mu, \mu)$, $\mu > 0$. Give an expression for the conjugate prior for μ.

3.8   Suppose that $X_1, \ldots, X_n$ and $Y_1, \ldots, Y_n$ are independent normally distributed samples with means μ_X, μ_Y, variances $\sigma_X^2 = \sigma_Y^2 = 1$. Calculate a 95% credible interval for $\mu_X - \mu_Y$, using normal priors on μ_X, μ_Y. Compare this to the frequentist 95% confidence interval for μ_X, μ_Y.

3.9   Returning to the example where the normal mean is known and the posterior distribution for the variance parameter is developed (equation (3.10), on page 74), plot a figure that illustrates how the likelihood increases in relative importance to the prior by performing the following steps:

(a) Plot the $\mathcal{IG}(5, 5)$ density over its support.

(b) Specify a posterior distribution for σ^2 using the $\mathcal{IG}(5, 5)$ prior.

(c) Generate four random vectors of size: 10, 100, 200, and 500 distributed standard normal. Using these create four different posterior distributions of σ^2, and add the new density curves to the existing plot.

(d) Label all distributions, axes, and legends.

Hint: modify the R code in the **Computational Addendum** to provide a posterior for the variance parameter instead of the mean parameter.

3.10 An 1854 study on mental health in the fourteen counties of Massachusetts yields data on 14 cases. This study was performed by Edward Jarvis (then president of the American Statistical Association), and has variables for: the number of "lunatics" per county (NBR), distance to the nearest mental healthcare center (DIST), population in the county by thousands (POP), population per square county mile (PDEN), and the percent of "lunatics" cared for in the home (PHOME). Use graphical tools to make a claim about whether any of these variables can be treated as coming from an underlying normal data generation process. Do you suspect that a larger dataset would alter your conclusions? Use `data(lunatics)` from the BaM package in R to start this exercise.

3.11 Rejection Method. Like the normal, the Cauchy distribution is a unimodal, symmetric density of the location-scale family: $C(X|\theta,\sigma) = \frac{1}{\pi\sigma}\frac{1}{1+\left(\frac{x-\theta}{\sigma}\right)^2}$, where $-\infty < X, \theta < \infty, 0 < \sigma < \infty$. Unlike the normal, the Cauchy distribution has very heavy tails, heavy enough so that the Cauchy distribution has no moments and is therefore less easy to work with. Given a $C(0,3)$ distribution, find the probability of observing a point between 3 and 7. To do this, simulate 100,000 values of $C(0,3)$ (in R this is done by `rcauchy(100000,0,3)`), and count the number of points in the desired range. Graph your results.

3.12 The bivariate Cauchy distribution (Student's-t with degrees of freedom $\nu = 1$) is given by $f(X_1, X_2) = k(1+X_1^2+X_2^2)$ for $X_1, X_2 \in \Re$, with k a normalizing constant. Is this form log-concave to the x-axis? Can this bivariate form be used as a joint prior for the mean vector?

3.13 The expressions for the mean and variance of the inverse gamma were supplied in (3.9). Derive these from the inverse gamma PDF. Show all steps.

3.14 Replicate the simulated results in Table 3.2 and Table 3.3 (Section 3.6, page 80). Why are your results very slightly different (besides rounding)?

3.15 Modify the function `biv.norm.post` given in the **Computational Addendum** so that it provides posterior samples for multivariate normals, given specified priors and data. Also modify the function `normal.posterior.summary` so that the user can specify any level of density coverage.

3.16 Consider the replication below of Figure 6.2 from Meier and Gill (2000). The left-hand side is a histogram of equity data and the right-hand side is a qqnorm plot of the same data (quantiles of the data plotted against quantiles of a standard normal distribution). Explain the deviance from normality indicated by the bracket in the second panel. Why is this not as apparent in the histogram?

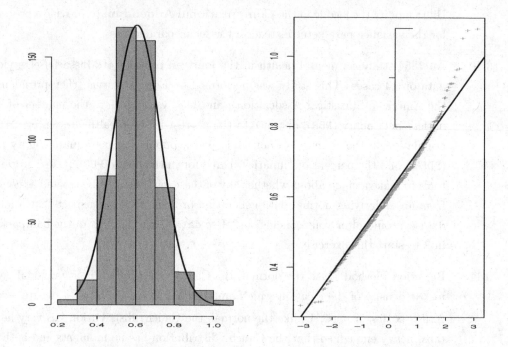

3.17   Specify a normal mixture model for the following data from Brooks, Dellaportas, and Roberts (1997), and summarize the posterior distributions of the model parameters with tabulated quantiles: $\mathbf{X} = \{2.3, 3.7, 4.1, 10.9, 11.6, 13.8, 20.1, 21.4, 22.3\}$.

3.18   Using Bettina Gruen's `bayesmix` package in R, run the following Bayesian mixture model (which calls JAGS) using the Swiss suicide data in BaM.

```
lapply(c("bayesmix","BaM"),library, character.only=TRUE)
data(suicide)
model1 <- BMMmodel(suicide$age, k = 3, initialValues = list(S0 = 2),
        priors = list(kind = "independence", parameter = "priorsFish",
        hierarchical = "tau"))
j.control <- JAGScontrol(variables = c("mu", "tau", "eta", "S"),
            burn.in = 1000, n.iter = 5000, seed = 10)
jags.out <- JAGSrun(suicide$age, model = model1, control = j.control)
( jags.sort <- Sort(jags.out, by = "mu") )
```

Interpret the model results including the fit of the selected number of mixtures.

3.19   Use the following data on the race of convicted murderers in the United States in 1999 to specify a bivariate model. Specify reasonable priors and produce posterior descriptions of the unknown parameters according to Section 3.5 starting on page 76. (Source: FBI, Crime in the United States 1999, http://www.fbi.gov/about-us/cjis/ucr/crime-in-the-u.s/1999.)

Age Group	White	Black
9 to 12	7	11
13 to 16	218	247
17 to 19	672	976
20 to 24	987	1,285
25 to 29	619	660
30 to 34	493	429
35 to 39	444	303
40 to 44	334	228
45 to 49	236	134
50 to 54	153	73
55 to 59	89	60
60 to 64	55	24
65 to 69	47	17
70 to 74	23	10
75 and up	57	14

3.20   For a bivariate normal distribution for x_1 and x_2, show that the conditional distributions are $\mathcal{N}(\mu_1+\rho(\sigma_1/\sigma_2)(x_2-\mu_2), (1-\rho^2)\sigma_1^2)$ and $\mathcal{N}(\mu_2+\rho(\sigma_2/\sigma_1)(x_1-\mu_1), (1-\rho^2)\sigma_2^2)$, where ρ is the correlation coefficient. Design and code a Gibbs sampler using these conditional distributions and run it with the White and Black variables as x_1 and x_2 from the data on the race of convicted murderers in exercise 19. Interpret your results.

3.11   Computational Addendum: Normal Examples

3.11.1   Normal Example with Variance Known

This R language function produces the posterior mean, variance, and 95% credible interval for user-specified prior, in the normal example with known population variance.

```
n.post1 <- function(data.vec,pop.var,prior.mean,prior.var) {
  if(length(data.vec) <= 1)
    stop("n.post1: input data must be a vector")
  mu.hat <-    (prior.mean/prior.var +
            length(data.vec)*mean(data.vec)/pop.var)/
            (1/prior.var + length(data.vec)/pop.var)
  sigma.hat <- 1/(1/prior.var + length(data.vec)/pop.var)
  credible.int<-c(mu.hat-1.96*sqrt(sigma.hat),mu.hat+1.96*sqrt(sigma.hat))
  list(mu.hat=mu.hat,sigma.hat=sigma.hat,
      credible.interval=credible.int)
}
```

3.11.2   Bivariate Normal Simulation Example

This code runs the bivariate normal simulation example by first producing a set of bivariate normal data according to the user specification, and then running the estimation procedure described in Section 3.5 on page 76. The `rwishart` function for generating random Wishart matrices (Appendix B) is found in multiple R packages.

```
# FUNCTION FOR GENERATING POSTERIOR QUANTITIES, BVT. NORMAL
biv.norm.post <- function(data.mat,alpha,beta,m,n0=5) {
  n <- nrow(data.mat)
  xbar <- apply(data.mat,2,mean)
  S2 <- (n-1)*var(data.mat)
  Wp.inv <-
    solve(beta)+S2+((n0*n)/(n0+n))*(xbar-m)%*%t(xbar-m)
  Sigma<-solve(rwishart(alpha+n,solve(Wp.inv))$W)
  mu <- rmultinorm(1, (n0*m + n*xbar)/(n0+n), Sigma/(n0+n))
  return(c(mu1=mu[1],mu2=mu[2],sig1=Sigma[1,1],
          sig2=Sigma[2,2],rho=Sigma[2,1]))
}

# A SIMPLE POSTERIOR SUMMARIZING FUNCTION
normal.posterior.summary <- function(reps)  {
  reps[,5] <- reps[,5]/sqrt(reps[,3]*reps[,4])
  reps <- apply(reps,2,sort)
  out.mat <- cbind("mean"=apply(reps,2,mean),
                  "std.err"=apply(reps,2,sd),
                  "95% HPD Lower"=reps[25,],
                  "95% HPD Upper"=reps[975,])
  return(out.mat)
}

# GENERATE THE CONTRIVED DATA, OBTAIN POSTERIOR SAMPLES
data.n10 <- rmultinorm(10, c(1,3),
                       matrix(c(1.0,0.7,0.7,3.0),2,2))
rep.mat <- NULL; reps <- 1000
for (i in 1:reps)
  rep.mat <- rbind(rep.mat,
    biv.norm.post(data.n10,3,
                matrix(c(10,5,5,10),2,2),c(2,2)))
round(normal.posterior.summary(rep.mat),3)
```

3.11.3 Multivariate Normal Example, Health Data

This last set of example code runs the multivariate normal model where the mean and variance are both unknown. The data used are the North Carolina county health data. The rwishart function here is from the bayesm package in R by Peter Rossi and Rob McCulloch.

```
data(nc.sub.dat)
library(bayesm)


# FIRST SPECIFICATION        # SECOND SPECIFICATION
Alpha <- 3 + nrow(nc.sub.dat)  # Alpha <- 3 + nrow(nc.sub.dat)
Beta.inv <- solve(diag(3)*100) # Beta.inv <- solve(diag(3)*10)
m <- c(250,16,88)              # m <- c(100,6,88)
n0 <- 0.01                     # n0 <- 99


x.bar <- apply(nc.sub.dat,2,mean)
S.sq <- var(nc.sub.dat)
k <- (n0 * nrow(nc.sub.dat))/(n0 + nrow(nc.sub.dat))
p.Beta <- solve( Beta.inv + S.sq
            + k * round((x.bar-m) %*% t(x.bar-m),2) )
Sigma <- array(NA,dim=c(3,3,1))
for (i in 1:10000)  Sigma <- array(c(Sigma,
                  rwishart(Alpha,p.Beta)$IW),dim=c(3,3,(i+1)))
Sigma <- Sigma[,,-1]


Sigma.Mean <- apply(Sigma,c(1,2),mean)
Sigma.SD <- apply(Sigma,c(1,2),sd)


# ANALYTICAL MEAN OF THE INVERSE WISHART:
( (Alpha-ncol(nc.sub.dat)-1)^(-1) )*solve(p.Beta)


# MEAN BY SIMULATION
Mu <- rmultinorm(5000,(n0*m + nrow(nc.sub.dat)*x.bar)/
      (n0 + nrow(nc.sub.dat)),Sigma.Mean/(n0+nrow(nc.sub.dat)))
apply(Mu,2,quantile, probs = c(0.01,0.25,0.50,0.75,0.99))
```

Chapter 4

The Bayesian Prior

4.1 A Prior Discussion of Priors

Specifying Bayesian models necessarily means providing prior distributions for unknown parameters. The prior plays a critical role in Bayesian inference through the updating statement: $\pi(\theta) \propto p(\theta)L(\theta|\mathbf{X})$. Central to the Bayesian philosophy is that all unknown quantities are described probabilistically, even before the data has been observed. Generally these unknown quantities are the model parameters, θ in the general notational sense, but missing data are handled in the same fashion. This is very much at odds with the frequentist notion that unknown parameters are fixed, unyielding quantities that can be estimated with procedures that are either repeated many times or imagined to be repeated many times. The immobile parameter perspective, although widespread, is contradictory to the way that most social and behavioral scientists conduct research. It simply is not possible to rerun elections, repeat surveys under exactly the same conditions, replay the stock market with exactly matching economic forces, fight the same war, or re-expose clinical subjects to identical stimuli.

The core of the disagreement between frequentists and Bayesians is the differing fundamental interpretations of probability. Frequentists believe that probabilities are long-run tendencies of events that eventually converge on some true population proportion that can be interpreted as a probability of such events even in the short term. Bayesians generally interpret probability as "degree of belief," meaning that prior distributions are descriptions of relative likelihoods of events based on the researcher's past experience, personal intuition, or expert opinion, and posterior distributions are these prior distributions updated by conditioning on new observed data. Some authors also focus on the distinction between priors that are explicitly based on previous empirical work versus priors that represent general knowledge possessed by the researcher (Zellner 1971, p.18). Mathematically this does not change the mechanics of prior inclusion, however.

It is important to understand that priors are not merely annoyances that must be dealt with before moving on to more interesting parts of the specification process. They are actually an opportunity to systematically include qualitative, narrative, and intuitive knowledge into statistical models. Because there is a lot of historical controversy concerning the impact of prior distributions, many applied Bayesians in the social and behavioral sciences seek to use only highly diffuse forms such as the uniform distribution. Some of these researchers merely want the probabilistic inferential process and the powerful machinery of MCMC without being "fully Bayesian." These "Bayesians of convenience" miss the point that informed prior distributions are incredibly useful for integrating non-quantitative information into the statistical model. Specifically, see Gill and Walker (2005) in political science, Vanpaemel (2010) in psychology, Raftery (1999) in sociology, Gill and Meier (2000) in public administration, Wolfson, Kadane and Small (1996) in policy studies, and Martin and Quinn (2007) in legal studies.

4.2   A Plethora of Priors

We have already seen several types of prior distributions in previous chapters. This chapter lays out the three general categories of prior distributions used in Bayesian models: conjugate, uninformed, and informative. The categorization is artificial (but convenient) since the boundaries are blurred: conjugate forms are sometimes highly informed, the distinction between uninformed and informed is across a relative spectrum, and there does not exist a prior distribution with absolutely no information (hence the use of "uninformative" rather than "noninformative" here). This distinction was made clear in a historical context by Diaconis and Ylvisaker (1985). In their words, the classical Bayesian views the prior distribution as a necessary inconvenience and typically attempts to specify a "flat" prior so as to interject the least amount of prior knowledge as possible. These priors are correctly called *uninformative* and are sometimes difficult to find. Note that these are frequently called *non*informative, but this is actually an inaccurate term for reasons that we will see shortly. Modern parametric Bayesians specify priors possessing deliberate characteristics, such as conjugacy, and assign prior parameter values according to specific criteria unrelated to substantive knowledge. Subjective Bayesians elicit prior distributions according to pre-existing scientific knowledge in a substantive field. This can come from previous empirical work in the field or from expert opinions by non-statisticians. In practice these categories are far from mutually exclusive, and it is more common to see a mixed approach that combines aspects of previous knowledge, mathematical convenience, and a desire not to overly affect the final conclusions with a strongly influential prior.

Prior distributions remain the most controversial aspect of Bayesian inference. Critics have focused on the supposedly personal-subjective nature of priors, generally neglecting to notice that *all* statistical models involve subjective choices. One ecumenical position taken by this book is that it is rare to approach a problem with absolutely no prior information about parameters of interest, and that even in these situations there will exist a class of priors designed to supply relatively little prior information. In fact, most readers would be reluctant to accept model specifications by authors who knew *nothing* before analyzing the data. While it is coy to say "everyone is a Bayesian, some of us know it," most researchers tell us about their prior knowledge even if it is not put directly in the form of a prior distribution. Consider the following quotation from Canes-Wrone, Brady, and Cogan (2002) in political science.

> The effects of campaign spending and district ideology are *consistently in the expected direction* and statistically significant. Those on challenger quality also have the *correct sign* in each regime and sample and are significant with the exception of the marginal regime of the 1980-1996 test. In addition, the coefficients for the remaining variables that are not included as a main effect typically have the *predicted sign* and they are significant *only with the expected sign*.

(Italics added.) Obviously there is a lot of prior information revealed in that passage. Not only are they indicating the direction of effects, they tell us which ones were expected to be statistically reliable. The Bayesian prior provides a way for researchers to be more overt about their knowledge, attitudes, and opinions on studied social phenomena.

Leamer (1983) specifies a hierarchy of priors based on the level of confidence in some effect of interest: truths (axioms) > facts (proven from axioms/observations) > opinions > conventions. So modeling decisions should be based on the most supportable level of this hierarchy. For instance, if we design an experiment in which one of two possible events occurs and the number of trials is established beforehand such that each trial is produced identically from the same distribution, then the *truth* is that a binomial distribution should be used to model the likelihood for the event of primary interest. The weakest form of prior evidence comes from conventions. These include not only defensible model choices, but also things like linearity assumptions, normal specifications, and pre-set α levels.[1] Leamer's point is that "...the choice of a particular sampling distribution, or a particular prior distribution, is inherently whimsical," so we should not pretend to find "objective" priors, and should not bury assumptions, but instead should seek to overtly obtain and describe the highest current level in the hierarchy. Of course we don't always have truths or facts and therefore need to rely on opinions: "As I see it, the fundamental problem facing econometrics is how adequately to control the whimsical character of inference, how sensibly to base inferences on opinions when facts are unavailable" (Leamer 1983, p.38).

What may have seemed like a weakness to 20th century critics of Bayesian inference is actually a core strength. Priors are a means of systematically incorporating existing human knowledge, quantitative or qualitative, into the statistical specification. For excellent discussions of the nuances see Berger (1985, p.74-82), Savage (1972, Chapters 3 and 4), Barnett (1973, p.80-88), and the thoughtful essays contained in Wright and Ayton (1994).

This chapter covers the previously noted trilogy of prior distributions commonly applied in Bayesian work: conjugate, uninformed, and informative. All priors are subjective in the

[1]The common α levels of 0.1, 0.05, and 0.01 come from Fisher's tables and the reluctance of mid-20th century scientists to challenge or recalculate normal tail values. Fisher's justification actually rests on no scientific principle other than the assertion that these levels represented some standard convention in human thought: "It is usual and convenient for experimenters to take 5 per cent as a standard level of significance..." (1934, p.15).

sense that the decision to use any prior is completely that of the researcher, subject to mathematical constraints that may apply. It is important to remember that the choice of priors is no more subjective than the choice of likelihood, the selection or collection of a given sample, the algorithm for estimation, or the statistic used for data reduction. The set of priors described herein are by no means the complete set of available forms. For a review of additional prior specifications see Kass and Wasserman (1996).

4.3　Conjugate Prior Forms

One difficult aspect of Bayesian inference is that the posterior distribution of the $\boldsymbol{\theta}$ vector might not have an analytically tractable form (*the* primary motivation for the simulation techniques discussed in later chapters), particularly in higher dimensions. Specifically, producing marginals from high-dimension $\pi(\boldsymbol{\theta})$ by repeated analytical integration may be difficult or even impossible mathematically. One way to guarantee that the posterior has an easily calculable form is to specify a *conjugate prior*. As briefly described in Chapter 2 in Section 2.3 and developed in some detail for the normal model in Chapter 3 in Section 3.3, conjugacy is a joint property of the prior and the likelihood function that provides a posterior from the same distributional family as the prior. In other words, the mathematical form of the prior distribution "passes through" the data-conditioning phase and endures in the posterior: closure under sampling. Thus the general form for the distribution of the effect of interest is invariant to Bayesian inference. We have already seen two important cases in Chapter 3 where a normal prior along with a normal joint likelihood function produced a normal posterior for the mean parameter, and an inverse gamma prior with the same normal likelihood produced an inverse gamma posterior for the variance parameter. It is important to keep in mind that inverse gamma distributions with small parameter values are not necessarily low information forms since they place most of the density near one, implying very small precision (see Figure 3.1). In variance terms these forms then specify a great amount of the density at large values that may be unrealistic in modeling terms. (Gelman 2006, Hodges and Sargent 2001, Natarajan and Kass 2000).

4.3.1　Example: Conjugacy in Exponential Specifications

A very simple way to model the time that something endures (wars, lifetimes, regimes, bull markets, marriages, etc.) is to use the exponential PDF, which is a special case of the gamma distribution where the shape parameter is fixed at one (Appendix B). This has the form:

$$\mathcal{E}(X|\theta) = \theta \exp[-\theta X], \quad 0 \le X, 0 < \theta \tag{4.1}$$

(note the use of the "rate" form of the gamma distribution). An attractive candidate prior for θ in the exponential PDF is the gamma distribution because not only does a random variable distributed gamma have the same support as that of an exponential,[2] but also because the gamma is an extremely flexible parametric form. The gamma PDF from Appendix B is:

$$f(\theta|\alpha, \beta) = \frac{1}{\Gamma(\alpha)} \beta^\alpha \theta^{\alpha-1} \exp[-\beta\theta], \qquad \theta, \alpha, \beta > 0.$$

Suppose we now observe $x_1, x_2, \ldots, x_n \sim$ iid $\mathcal{E}(X|\theta)$ and produce the likelihood function:

$$L(\theta|\mathbf{x}) = \prod_{i=1}^{n} \theta e^{-\theta x_i} = \theta^n \exp\left[-\theta \sum_{i=1}^{n} x_i\right].$$

Note that $\sum_{i=1}^{n} x_i$ is a sufficient statistic for θ. The posterior distribution is produced as follows:

$$\pi(\theta|\mathbf{x}) \propto L(\theta|\mathbf{x})p(\theta)$$

$$= \theta^n \exp\left[-\theta \sum_{i=1}^{n} x_i\right] \frac{1}{\Gamma(\alpha)} \beta^\alpha \theta^{\alpha-1} \exp[-\beta\theta]$$

$$\propto \theta^{(\alpha+n)-1} \exp\left[-\theta \left(\sum_{i=1}^{n} x_i + \beta\right)\right]. \tag{4.2}$$

It is easy to see that this is the kernel of a $\mathcal{G}(\alpha + n, \sum x_i + \beta)$ PDF, and therefore the gamma distribution is shown to be conjugate to the exponential likelihood function.

4.3.2 The Exponential Family Form

Recall that a family of PDFs or PMFs qualify as an exponential family form if they can be rearranged in a specific manner that recharacterizes familiar functions into a formula that is useful theoretically and demonstrates similarity between seemingly disparate mathematical forms. Suppose we temporarily consider only a one-parameter PDF or PMF, $f(x|\theta)$, with one datum. This function is classified as an exponential family only if it can be rearranged to be of the form: $f(x|\theta) = \exp\left[t(x)u(\theta)\right]r(x)s(\theta)$, where r and t are real-valued functions of x that do not depend on θ, and s and u are real-valued functions of θ that do not depend on x, and $r(x) > 0, s(\theta) > 0 \; \forall x, \theta$. Also, many forms, such as the normal, include a scale parameter making the form of the exponential family: $f(x|\theta) = \exp\left[t(x)u(\theta)/\phi\right]r(x)s(\theta)$, where ϕ (sometimes denoted $a(\phi)$) is a scale parameter (σ^2 for the normal), and sometimes explicitly weighted as in ϕ/ω_i. Because of the properties of logs and exponentiation, this form can always be rewritten as:

$$f(x|\theta) = \exp\Big[\underbrace{t(x)u(\theta)}_{\substack{\text{interaction} \\ \text{component}}} + \underbrace{\log(r(x)) + \log(s(\theta))}_{\text{additive component}} \Big], \tag{4.3}$$

---

[2] Actually the exponential PDF is a special case of the gamma PDF where the first (shape) parameter is fixed at one. See Exercise 4.1.

(Gill 2000). The labeled interaction component in (4.3), $t(x)u(\theta)$, reflects the product-indistinguishable relationship between x and θ. It should be noted that this component must specify $t(x)u(\theta)$ in a strictly multiplicative manner.[3] The structure of (4.3) is preserved under sampling so the joint density function of iid random variables $\mathbf{X} = \{X_1, X_2, \ldots, X_n\}$ is simply:

$$f(\mathbf{x}|\theta) = \exp\left[u(\theta)\sum_{i=1}^{n}t(x_i) + \sum_{i=1}^{n}\log(r(x_i)) + n\log(s(\theta))\right]. \tag{4.4}$$

Fisher (1934) originated this idea to show that many common PDFs and PMFs are actually all special cases of the more general classification labeled *exponential family* because subfunctions are contained within the exponent component. This contrasts sharply with the econometric approach of seeing these distributions as leading to separate modeling constructs. Fisher also showed that these isolated subfunctions quite naturally produce a small number of sufficient statistics that compactly summarize even large data sets without any loss of information. Barndorff-Nielsen (1978, p.114) demonstrated that exponential family probability functions possess all of their moments (defined in Appendix A), and are therefore easier to characterize in a Bayesian framework. Morris (1982, 1983a), building upon the results of Diaconis and Ylvisaker (1979) showed that the most commonly used exponential families of probability functions when specified as likelihood functions possess conjugate priors. Consonni and Veronese (1992) proved that this is true for any mean parameter in an exponential family form with quadratic variance (the mean parameter is contained in an algebraic term with an exponent equal to 2), and Gutiérrez-Peña and Smith (1995) extend this result with transformations of the mean prior.

We can develop a general form of the posterior that corresponds to the exponential family expression of the joint likelihood function in (4.4). Start with a corresponding conjugate prior form in the same generalized notation:

$$p(\theta|k, \gamma) = c(k, \gamma)\exp[ku(\theta)\gamma + k\log(s(\theta))]. \tag{4.5}$$

Now calculate the posterior from (4.5) and (4.4):

$$\pi(\theta|\mathbf{x}, k, \gamma) \propto f(\mathbf{x}|\theta)p(\theta|k, \gamma)$$

$$\propto \exp\left[u(\theta)\left(\sum_{i=1}^{n}t(x_i) + k\gamma\right) + (n+k)\log(s(\theta))\right]$$

$$= \exp\left[u(\theta)(n+k)\left(\frac{\sum_{i=1}^{n}t(x_i) + k\gamma}{n+k}\right) + (n+k)\log(s(\theta))\right]. \tag{4.6}$$

So the posterior distribution is an exponential family form like the prior with parameters: $k' = (n+k), \gamma' = \frac{\sum_{i=1}^{n}t(x_i)+k\gamma}{n+k}$. The connection between the exponential family form and

---

[3]For instance, consider the Weibull PDF (useful for modeling general failure times): $f(x|\gamma, \beta) = \frac{\gamma}{\beta}x^{\gamma-1}\exp(-x^{\gamma}/\beta)$ for $x \geq 0, \gamma, \beta > 0$. The term $-\frac{1}{\beta}x^{\gamma}$ in the exponent disqualifies this PDF from the exponential family classification since it cannot be expressed in the additive or multiplicative form of (4.3). However, if γ is known (or we are willing to assign an estimate), then the Weibull PDF reduces to an exponential family form.

conjugacy is a very useful result because if we know that the form of our likelihood function is an exponential family form, then it is almost certain to have a conjugate prior. Table 4.1 provides a list of common PDFs and PMFs with their associated conjugate prior.

TABLE 4.1: SOME EXPONENTIAL FAMILY FORMS AND THEIR CONJUGATE PRIORS

Likelihood Form	Conjugate Prior Distribution	Hyperparameters
Bernoulli	Beta	$\alpha > 0, \beta > 0$
Binomial	Beta	$\alpha > 0, \beta > 0$
Multinomial	Dirichlet	$\theta_j > 0, \Sigma\theta_j = \theta_0$
Negative Binomial	Beta	$\alpha > 0, \beta > 0$
Poisson	Gamma	$\alpha > 0, \beta > 0$
Exponential	Gamma	$\alpha > 0, \beta > 0$
Gamma (incl. χ^2)	Gamma	$\alpha > 0, \beta > 0$
Normal for μ	Normal	$\mu \in \mathbb{R}, \sigma^2 > 0$
Normal for σ^2	Inverse Gamma	$\alpha > 0, \beta > 0$
Pareto for α	Gamma	$\alpha > 0, \beta > 0$
Pareto for β	Pareto	$\alpha > 0, \beta > 0$
Uniform	Pareto	$\alpha > 0, \beta > 0$

Two important classes of probability density functions are not members of the exponential family. The Student's-t and the uniform distribution cannot be put into the form of (4.3) above. In general, a probability function in which the parameterization is dependent on the bounds, such as the uniform distribution, are not members of the exponential family. Even if a probability function is not an exponential family member, it can sometimes qualify under particular circumstances.

In fact, the exponential family form provides more practical Bayesian help than it appears at first. Since the exponential family form guarantees a finite-dimension sufficient statistic, there are no difficulties in finding maximum likelihood estimates for unknown parameters. This is formalized in the Darmois-Pitman-Koopman Theorem (independently and almost concurrently proven by Darmois [1935], Pitman [1936], and Koopman [1936]), which states that a distribution function possesses an associated fixed dimensional sufficient statistic for its parameter (Pitman) or parameter vector (Koopman) if and only if the distribution can be expressed in exponential family form (Anderson 1970, p.1248; Barankin and Maitra 1963, p.217; Hipp 1974, p.1283; Jeffreys 1961, p.168). This well-known property just assures us that with the exponential family form the right thing happens and that with other forms we may or may not encounter additional difficulties; see Bickel and Doksum (1977) or DeGroot (1986) for basic discussions. Furthermore, Crain and Morgan

(1975) show that Bayesian models based on exponential family form produce asymptotically normal posteriors.

The theoretical linkage between sufficient statistics and the likelihood function is important for our purposes. The minimum dimension sufficient statistic is actually the smallest collection of statistics that completely summarizes the likelihood function (Brown 1964, p.1458; Fraser 1963, p.117). Therefore under nonrestrictive regularity conditions, the likelihood function from an exponential family distribution can be characterized by a finite, usually small dimension, sufficient statistic *regardless of the sample size underlying it.* Therefore, we gain an immense amount of inferential leverage by restricting ourselves to exponential family forms and summarizing the sample information through likelihood functions. In subsequent sections, we will take advantage of small-dimensional sufficient statistics to simplify the analyses.

4.3.3 Limitations of Conjugacy

The primary advantage in specifying conjugate priors is their mathematical convenience in producing posterior inferences. This is not, however, identical to uncovering some absolute truth in the data-generation process. A conjugate prior, no matter how mathematically convenient or easily interpretable, should not be construed as *the right answer* in a data-analytic exercise and in fact may seriously misrepresent the actual truth (Barnett 1973, p.188). It is also not the case that conjugate priors are no-information default alternatives. One should be particularly cautious about using conjugate priors as ignorance priors, such as normal distributions centered at zero with small precision, in that they still indicate specific parametric prior knowledge. If the form of the conjugate prior through its alternative parametric values (some are more flexible than others) fits prior knowledge about the distribution of the parameter, then this is a fortunate and desirable outcome rather than an inevitability.

There was a time when conjugate specifications were very important in Bayesian statistics. Before the advent of MCMC techniques (Chapter 10), many proposed models were simply too hard to estimate without some trick like conjugacy. This is no longer the case, but conjugate priors can still be useful in practice and they are an excellent expository tool.

4.4 Uninformative Prior Distributions

An uninformative prior is one in which little new explanatory power about the unknown parameter is provided by intention. Despite the unfortunate name, uninformative priors are very useful from the perspective of traditional Bayesianism that sought to mitigate frequentist criticisms of intentional subjectivity. Consider a situation in which absolutely no previous subjective information is known about the phenomenon of interest. Does this preclude a Bayesian analysis? If we could make a probabilistic statement that did not favor any outcome over another, then it would be a simple matter to use this as a prior distribution allowing us to continue the analysis. Unfortunately this is often more difficult than one would expect.

4.4.1 Uniform Priors

An obvious candidate for the uninformative prior is the uniform distribution. Uniform priors are particularly easy to specify in the case of a parameter with bounded support. For instance, a uniform prior for the probability parameter in a Bernoulli, binomial, or negative binomial model can be specified by: $p(\theta) = 1$, $0 \leq \theta \leq 1$, or if there is some reason to specify a non-normalized uniform: $p(\theta) = 1$, $0 \leq \theta \leq k$. The second is non-normalized because it does not integrate to one, but as we demonstrated in Chapter 2, this provides no problem whatsoever in Bayesian analysis. Both of these forms are referred to as *proper* since they integrate to a finite quantity. Proper uniform priors can be specified for parameters defined over unbounded space if we are willing to impose prior restrictions. Thus if it is reasonable to restrict the range of values for a variance parameter in a normal model, instead of specifying it over $[0{:}\infty]$, we restrict it to $[0{:}\nu]$ and can now articulate it as $p(\sigma) = 1/\nu$, $0 \leq \theta \leq \nu$.

It is also possible to specify *improper* uniform priors that do not possess bounded integrals and surprisingly, these result in fully proper posteriors under most circumstances (although this is far from guaranteed, see Section 4.4.4). Consider the common case of an uninformative uniform prior for the mean of a normal distribution. It would necessarily have uniform mass over the interval: $p(\theta) = c$, $[-\infty \leq \theta \leq \infty]$. Therefore to give *any* nonzero probability to values on this support, $p(\theta) = \epsilon > 0$, would lead to a prior with infinite density: $\int_{-\infty}^{\infty} p(\theta) d\theta = \infty$.

The uniform prior is not invariant under transformation: simple transformations of the uniform prior produce a re-expression that is not uniform and loses whatever sense of uninformedness that the equiprobability characteristic of the uniform gives. For example, suppose again that we are interested in developing an uninformative prior for a normal model variance term and specify the improper uniform prior: $p(\sigma) = c$, $0 \leq \sigma < \infty$. A simple transformation that provides a parameter space over the entire real line is given by: $\tau = \log(\sigma)$, and the new PDF is given by applying the transformation with Jacobian $(J = |\frac{d}{d\tau} g^{-1}(\tau)|)$, to account for the rate of change difference, to the original PDF:

$$\tau = g(\sigma) = \log(\sigma) \longrightarrow g^{-1}(\tau) = \sigma = e^{\tau}$$

$$p(\tau) = p(g^{-1}(\tau)) \left| \frac{d}{d\tau} g^{-1}(\tau) \right| = (c) \left| \frac{d}{d\tau} e^{\tau} \right| \propto e^{\tau}.$$

This resulting prior clearly violates even the vaguest sense of uninformedness or "flatness," and makes a strong statement about values that are *a priori* more likely than others. Lest one think that this problem is restricted to the class of improper uniform priors, consider a proper uniform prior on a Bernoulli probability parameter: $f(p) = 1$, $0 \leq p \leq 1$. If we change from the probability metric to the odds ratio metric (fairly common), then we impose the transformation: $q = \frac{p}{1-p}$ and the new distribution is given by:

$$q = g(p) = \frac{p}{1-p} \longrightarrow g^{-1}(q) = \frac{q}{1+q}$$

$$f(q) = f(g^{-1}(q)) \left| \frac{d}{dq} g^{-1}(q) \right| = (1) \left| \frac{d}{dq} \frac{q}{1+q} \right| = (1+q)^{-2}.$$

Once again, a straightforward change imposes a serious departure from the uniform characteristic: no prior information about p does not imply no prior information about a simple transformation of p.

This is shown in the adjacent figure (with the right-hand-tail truncated for convenience). Notice that it is asymmetrical and places more prior density on the region $[0:1]$ than on the region $[1:2]$, which may violate some people's notion of symmetry around the odds ratio of one (equally probable success and failure). Nevertheless, it is interesting to note that it has a symmetric property around one by area:

$$\int_0^1 (1+q)^{-2} dq = \int_1^\infty (1+q)^{-2} dq = \frac{1}{2}.$$

(4.7)

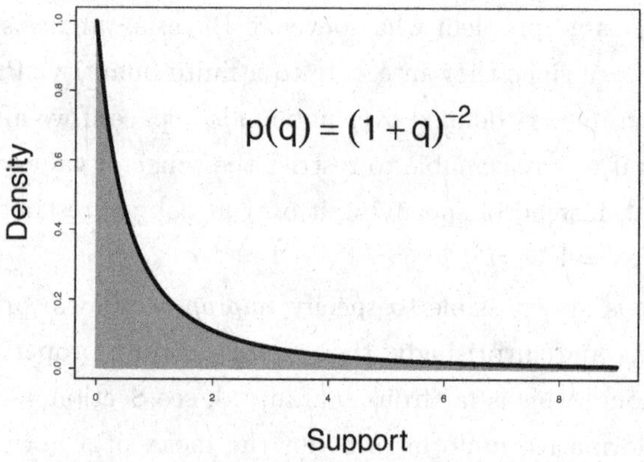

FIGURE 4.1: PRIOR FROM TRANSFORMATION

In the early 20th century, there was a great amount of controversy focused around the use of uniform priors and their role in calculating "inverse probability" (i.e., the early application of Bayes' Law). Fisher (1930, p.531) was characteristically negative on the subject: "...how are we to avoid the staggering falsity of saying that however extensive our knowledge of the values of x may be, yet we know nothing and can know nothing about the values of θ?" Other predecessors and contemporaries had notable criticisms of the default uniform prior specified in Bayes (1763) and Laplace (1774). Boole (1854, p.370) objected to its "arbitrary nature," and Venn (1866, p.182) called it "completely arbitrary." de Morgan (1847, p.188) was uncomfortable with the implication that events that were observed to occur have the same probability as events that were observed not to occur. Edgeworth (1921, pp.82-83 footnote) in comments directed at Pearson pointed out that there are many continuous prior specifications that can cancel out in the calculation of inverse probabilities so fixating on the uniform is unnecessary, and in 1884 he may have been the first scholar to point out the invariance property of the uniform distribution discussed above.

What is really clear from reading these early authors is that they rarely decoupled the use of uniform priors from Bayesian inference in general. Therefore, for a long period of time, producing posterior probabilities with Bayes' Law was synonymous with flat prior specifications and the associated philosophical and mathematical problems that ensued. Because of this inflexibility the uniform prior has long been a primary means of critiquing the Bayes-Laplace construct, and too little thought was given to broadening the scope of priors. Certainly no thought was given to finding other "no information" priors.

Others have pointed out that the uniform priors have an inherent bias against the endpoints of the specified interval (Novick and Hall 1965, Villegas 1977), and therefore do not necessarily provide the coverage that the researcher desires (particularly if these endpoints are of theoretical importance as might be the case for the $\mathcal{U}(0, 1)$ specification). Nonetheless, uniform priors remain very popular in applied work and there are certainly situations where the uniformness is a desired property for subjective reasons rather than as an uninformative choice.

Bauwens, Lubrano, and Richard (1999) give some justifications for using ignorance priors like the uniform: *(1)* as the sample size increases any effect from the uniform diminishes, *(2)* they are a suitable and convenient choice for so-called nuisance parameters that are going to be integrated out of the posterior anyway, and (3) the uniform distribution is a limit of some conjugate prior distributions (for example, a bounded normal with increasing scale parameter).

4.4.2 Jeffreys Prior

Jeffreys (1961, p.181) addresses the problems associated with uniform priors by suggesting a prior that is invariant under transformation:

> ...if we took the prior probability density for the parameters to be proportional to $||g_{ik}||$, it could be stated for any law that is differentiable with respect to all parameters in it, and would have the property that the total probability in any region of the α_i would be equal to the total probability in the corresponding region of α_i'; in other words, it satisfies the rule that equivalent propositions have the same probability.

He means that the transformation from α to α' is invariant with respect to probabilities:[4]

$$p_J(\alpha) = p_J(\alpha') \left| \frac{d\alpha'}{d\alpha} \right|. \tag{4.8}$$

The Jeffreys prior for a single parameter, θ, is produced by the square root of negative expected value of the second derivative of $f(\mathbf{x}|\theta)$ (and more generally the likelihood function):

$$p(\theta) \propto \left[-E_{\mathbf{X}|\theta} \left(\frac{d^2}{d\theta^2} \log f(\mathbf{x}|\theta) \right) \right]^{\frac{1}{2}}. \tag{4.9}$$

---

[4]See Dawid (1983) and Hartigan (1964) for broader definitions of invariance.

Note that the expectation here is taken over $f(\mathbf{X}|\theta)$. This is also the square root of the determinant of the familiar negative expected Fisher Information matrix. Expression 4.9 is given in the single-dimension case, for a parameter vector, θ, the Jeffreys prior is:

$$p(\theta) \propto \left[-E_{\mathbf{X}|\theta} \left(\left[\frac{\partial}{\partial \theta} \log f(\mathbf{x}|\theta) \right]' \left[\frac{\partial}{\partial \theta} \log f(\mathbf{x}|\theta) \right] \right) \right]^{1/2}. \tag{4.10}$$

The Jeffreys prior is straightforward to calculate and use for many parametric forms. Ibrahim and Laud (1991) give a general theoretical justification for using the Jeffreys prior with exponential family forms and therefore generalized linear models by showing that proper posteriors are produced. Hartigan (1983, p.74) tabulates a number of exponential family distributions with their associated Jeffreys prior. Gutiérrez-Peña and Smith (1995) relate the Jeffreys prior to standard conjugate priors, Poirer (1994) gives the Jeffreys prior for the logit model, and Kass (1989) carefully describes the full properties of the Jeffreys prior, including a geometric interpretation.

4.4.2.1   Bernoulli Trials and Jeffreys Prior

Consider a repeated Bernoulli trial with x successes out of n attempts in which we are interested in obtaining a posterior distribution for the unknown probability of success p. The binomial PMF, for the sum of the trials, is given by:

$$\mathcal{BN}(x|n,p) = \binom{n}{x} p^x (1-p)^{n-x}, \quad x = 0, 1, \ldots, n, \quad 0 \leq p \leq 1,$$

and the log likelihood is given by:

$$\ell(p|n,x) = \log \binom{n}{x} + x \log(p) + (n-x) \log(1-p).$$

The first and second derivatives are given by:

$$\frac{d}{dp} \ell(p|n,x) = \frac{x}{p} + \frac{n-x}{1-p}(-1)$$

$$= xp^{-1} - (n-x)(1-p)^{-1},$$

$$\frac{d^2}{dp^2} \ell(p|n,x) = -xp^{-2} - (n-x)(1-p)^{-2}(-1)(-1)$$

$$= -\frac{x}{p^2} - \frac{n-x}{(1-p)^2}.$$

Since $E[x] = np$, the last stage is trivial:

$$J = \left(E[-\frac{d^2}{dp^2}\ell(p|n,x)] \right)^{\frac{1}{2}} = \left(\frac{np}{p^2} + \frac{n-np}{(1-p)^2} \right)^{\frac{1}{2}} = \left(\frac{n}{p(1-p)} \right)^{\frac{1}{2}},$$

which suggests the prior $p(p) = p^{-\frac{1}{2}}(1-p)^{-\frac{1}{2}}$, a $\mathcal{BE}\left(\frac{1}{2}, \frac{1}{2}\right)$ distribution. This is shown in Figure 4.2. Notice that this form is far from uniform in shape. The Jeffreys prior is frequently used, meaning that researchers either highly value the invariance property or they want a form, in this case, that is bimodal with most of the density at the extremes. Clearly this is not an "uninformed" choice and should not be treated as a default prior distribution for general purposes. This highlights an important point. So-called "default" priors are common forms, perhaps even conjugate, that some researchers gravitate to out of convenience. This is a dangerous practice and all prior specifications should be carefully considered in the context of the data and the model. Since the advent of modern Bayesian computing (MCMC), analytical limitations on prior specifications are rare, meaning that a wealth of alternatives exists for any given problem.

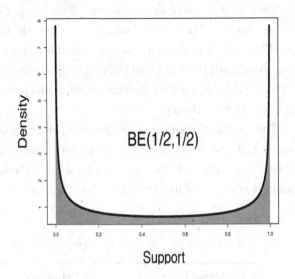

FIGURE 4.2: BINOMIAL: JEFFREYS PRIOR

4.4.2.2   Other Forms of Jeffreys Priors

In the binomial case the Jeffreys prior is quite easy to calculate, but this is not always true in other cases. Another simple form is the Jeffreys prior for a Poisson log-likelihood function. Starting with:

$$\ell(\lambda|\mathbf{x}) = -n\lambda + \left(\sum \mathbf{x}_i\right)\log\lambda - \log\sum(\mathbf{x}_i!),$$

taking derivatives produces:

$$\frac{d}{d\lambda}\ell(\lambda|\mathbf{x}) = -n + \lambda^{-1}\sum \mathbf{x}_i$$

$$\frac{d^2}{d\lambda^2}\ell(\lambda|\mathbf{x}) = -\lambda^{-2}\sum \mathbf{x}_i,$$

followed by the negative expectation over $\mathbf{x}$:

$$E_{\mathbf{x}|\lambda}\left(-\frac{d^2}{d\lambda^2}\ell(\lambda|\mathbf{x})\right) = E_{\mathbf{x}|\lambda}\left(\lambda^{-2}\sum \mathbf{x}_i\right) = (n\lambda)\lambda^{-2} \propto \lambda^{-1}.$$

Finally, the square root produces $p_J(\lambda) = \lambda^{-\frac{1}{2}}$.

Invariance to transformation is not the only useful characteristic of the Jeffreys prior. *Jeffreys' Rule* (Box and Tiao 1973, p.42) states that any prior that is proportional to the Jeffreys prior is uninformative in the sense that it interjects as little subjective information into the posterior as possible in terms of distributional invariance. Furthermore, the Jeffreys prior in the single parameter case, such as the binomial above, can be thought of as specifying a uniform distribution to the parameterization of the random variable where the Fisher Information is constant (Perks 1947). Hartigan (1964) extends Jeffreys' idea to relative parameter invariance priors and asymptotically locally invariant priors based on inverses or ratios of log-forms, but these can get quite complicated without substantial added benefit. These forms and a more general definition of invariance are covered by Dawid (1983). In addition, there is a strong debate on priors for time-series models with unit roots, with Phillips (1991) advocating a Jeffreys prior and others (Sims 1988, Sims and Uhlig 1991) believing that uniform distributions provide better comparisons to classical unit root distribution theory.

The most useful aspect of the Jeffreys prior is that it comes from a very mechanical process, yet almost always produces an uninformative form, and with the invariance property. Thus it can serve as a starting point for a more thoughtful determination of an uninformed prior. It is therefore handy to have basic forms tabulated for reference, see Table 4.2.

TABLE 4.2: SOME JEFFREYS PRIORS FOR COMMON FORMS

Likelihood Form	Parameter	Jeffreys Proportional
Bernoulli	$p \in [0{:}1]$	$(p(1-p))^{-\frac{1}{2}}$
Binomial	$p \in [0{:}1]$	$(p(1-p))^{-\frac{1}{2}}$
Multinomial(k)	$p_i \in [0{:}1], \sum p_i = 1$	$(p_1 p_2 \cdots p_k)^{-\frac{1}{2}}$
Negative Binomial	$p \in [0{:}1]$	$(1-p)^{-1/2} p^{-1}$
Poisson	$\lambda \geq 0$	$\lambda^{-1/2}$
Normal for $\sigma^2 = 1$	$-\infty < \mu < \infty$	1
Normal for $\mu = 0$	$\sigma^2 > 0$	σ^{-1}
Normal μ, σ^2 independent	$\sigma^2 > 0$	σ^{-1}
Normal μ, σ^2 nonindependent	$-\infty < \mu < \infty, \sigma^2 > 0$	σ^{-2}

4.4.2.3 Jeffreys Prior in the Multiparameter Case

While the Jeffreys prior is straightforward in one dimension, unfortunately it can be quite difficult in multiparameter models, except the normal, and the researcher may be required to seek approximate solutions. In fact it is possible to construct multidimensional priors that violate the likelihood principle (Birnbaum 1962, Evans, Fraser, and Monette 1986), and can lead to poor performing estimators even with the normal model (Dawid, Stone, and Zidek 1973). In some cases, multivariate Jeffreys priors can be constructed from independent univariate forms if we are willing to live up to a strict prior independence assumption.

There are situations where Jeffreys priors can be constructed quite easily even in high dimension. A common application is the multivariate generalized linear model (A). In this case common regression models can be easily specified with Jeffreys priors, producing a closed-form posterior expression that can often be integrated to obtain marginals. Following the setup of Ibrahim and Laud (1991), establish $\mathbf{y}$ as $n \times 1$ outcome vector, $\mathbf{X}$ as an

$n \times k$ matrix of explanatory variables including a constant, and $\boldsymbol{\beta}$ as the k coefficient whose posterior we wish to describe. The likelihood of interest is $L(\boldsymbol{\beta}|\mathbf{X}, \mathbf{y}, \phi) = \prod_{i=1}^{n} f(y_i|\theta_i, \phi)$. Using standard exponential family language, $b(\theta)$ from the canonical form, scale parameter ϕ, and $\theta_i = g^{-1}(\mathbf{X}'_i\boldsymbol{\beta})$, the Fisher Information matrix for this GLM is given by the determinant operation:

$$\mathbf{I}_{\boldsymbol{\beta}} \propto |\mathbf{X}'\Omega v(\boldsymbol{\beta})\delta(\boldsymbol{\beta})\mathbf{X}|/\phi, \tag{4.11}$$

with:

term	structure	elements $(i = 1 : n)$
Ω	$n \times n$ (optional) diagonal weight matrix	ω_i
$v(\boldsymbol{\beta})$	$n \times n$ diagonal matrix	$d^2b(\theta_i)/d\theta_i^2$
$\delta(\boldsymbol{\beta})$	$n \times n$ diagonal matrix	$d\theta_i/d\mathbf{X}'_i\boldsymbol{\beta}$

Notice that $\delta(\boldsymbol{\beta})$ expresses the link function here and would be just an identity matrix for the standard linear model. The general form of the Jeffrey's prior for this GLM is:

$$p_J(\boldsymbol{\beta}) \propto E[\mathbf{I}_{\boldsymbol{\beta}}]^{\frac{1}{2}} = |\mathbf{X}'\Omega v(\boldsymbol{\beta})\delta(\boldsymbol{\beta})\mathbf{X}|^{\frac{1}{2}}, \tag{4.12}$$

producing the joint posterior distribution of the $\boldsymbol{\beta}$:

$$\pi(\boldsymbol{\beta}|\mathbf{X}, \mathbf{y}, \phi) = \exp\left[\sum_{i=1}^{n} \omega_i(y_i\theta_i - b(\theta_i)/\phi)\right] |\mathbf{X}'\Omega v(\boldsymbol{\beta})\delta(\boldsymbol{\beta})\mathbf{X}|^{\frac{1}{2}}, \tag{4.13}$$

which varies in complexity depending on the particular link function. In the standard linear model with no weighting, the Jeffreys prior above simplifies down to $p_J(\boldsymbol{\beta}) \propto |\mathbf{X}'\mathbf{X}|^{\frac{1}{2}} = c$, for some constant c, over the support $[-\infty:\infty]$.

4.4.3 Reference Priors

A reference prior is a, not necessarily flat, prior distribution such that for the given problem (only), the likelihood is data translated (imposed on the posterior). The distinction between a reference prior and an uninformative prior is murky and author-dependent. Also, a number of authors refer to reference priors as "automatic priors" (Jeffreys 1961, Zellner and Siow 1980). Box and Tiao (1973, p.23) define a reference prior as "a prior which it is convenient to use as a standard" and is "dominated by the likelihood." Thus it need not be uninformative although some authors see it that way (e.g., Bernardo 1979). Informally Bernardo states that "Reference analysis may be described as a method to derive model-based, nonsubjective posteriors, based on information-theoretical ideas, and intended to describe the inferential content of the data for scientific communication" (quoted in Irony and Singpurwalla [1996]). Reference priors can be enormously helpful in dealing with nuisance parameters (Robert 2001, p.136), and have been shown to possess desirable asymptotic qualities under certain circumstances (Ghosh and Mukerjee 1992). Kass and

Wasserman (1995) identify two distinct interpretations of reference priors: as an expression of ignorance or as a socially agreed upon standard (model specific) alternative to subjective priors, and they proceed to identify a litany of associated problems with the use of reference priors.

The original idea for the modern definition of a reference prior is from Bernardo (1979), who introduces the useful notion that a difficult parameter vector (perhaps difficult in the sense that the Jeffreys prior does not work well), can be segmented into two components, parameters of high interest and nuisance parameters. This often leads to more complex derivations of the prior even before the data are considered (Berger and Bernardo 1989), but does have the advantage that higher order problems can be dealt with systematically (Berger and Bernardo 1992).

So a reference prior for the parameters of high interest is found by minimizing the distance between the chosen likelihood and the resulting posterior according to some criteria like the Kullback-Leibler distance (see Section 7.6 for details). This is also called a *dominant likelihood prior*, since it is a prior that is dominated by the likelihood function over the region of interest. One example is a very diffuse normal distribution centered somewhere near the expected mode of the posterior. The Zellner-Siow (1980) prior is similar but more dispersed: a Cauchy distribution restricted to the range of interest. These priors can be nearly uniform through the region of interest, but have the advantage of being mathematically easy to deal with.

Another reference prior of note is the previously mentioned locally uniform prior (see Section 4.4.1). This is just a normalized or non-normalized uniform prior over a bounded region of the support of the unknown parameter. While these can be tremendously helpful as a reference (hence the name), they suffer from the invariance problem discussed previously, the frequent difficulty in determining the bounds required to make the prior proper, the unrealistic notion that *nothing* is known in advance, and the classical arguments against the imposition of equiprobability through uniform priors. Where such a prior can be useful as a reference benchmark is in comparison with some subjective prior as a means of demonstrating the posterior effect of the prior deviance from uniformness. However, not a lot of theoretical importance should be attached to this difference.

There is a philosophical problem with such reference priors in the social sciences. To what degree is the model actually Bayesian if the goal is to reduce the influence of the prior to zero or near zero? Such a model still retains the Bayesian interpretation of a posterior, but it removes the ability to include a wealth of information that is usually available in the social sciences. In fact, suppose that one simply ran a maximum likelihood estimation on some non-Bayesian specification and declared it equivalent to a Bayesian model with the "best possible" reference prior. This would allow our hypothetical researcher to discuss results probabilistically (HPD regions, quantiles, etc.) without the trouble of being specific about prior specification. For most researchers in this area, the idea runs counter to the core tenets of Bayesian inferences and would create an artificial class of "lazy Bayesians," which is surely not in the interest of scientific progress in any field.

4.4.4 Improper Priors

As discussed previously, it is possible to specify so-called improper priors. These are prior distributions that do not add (PMF case) or integrate (PDF case) to a finite value. These improper forms typically arise in the search for an uninformative prior specification. An interesting and common case is the normal distribution with the improper prior: $p(\mu) = c, \ -\infty < \mu < \infty$. Obviously the equality in this specification can be replaced with proportionality as the integral is infinity in any case. We can now produce the posterior distribution for μ under the assumption that σ^2 is known (assumed simply for the clarity of exposition). We will also not use the standard Bayesian convenience of reducing notational complexity by using proportionality until the last step. This is done in order to make the point about the form of the posterior more emphatic. The posterior for μ is produced by:

$$
\begin{aligned}
p(\mu|\mathbf{x}, \sigma_0) &= \frac{p(\mu)L(\mu|\mathbf{x}, \sigma_0)}{\int p(\mu)L(\mu|\mathbf{x}, \sigma_0)d\mu} \\[2ex]
&= \frac{c(2\pi\sigma_0^2)^{-n/2}\exp\left[-\frac{1}{2\sigma_0^2}\sum_{i=1}^n(x_i - \mu)^2\right]}{\int c(2\pi\sigma_0^2)^{-n/2}\exp\left[-\frac{1}{2\sigma_0^2}\sum_{i=1}^n(x_i - \mu)^2\right]d\mu} \\[2ex]
&= \frac{c(2\pi\sigma_0^2)^{-n/2}\exp\left[-\frac{1}{2\sigma_0^2}(\sum_{i=1}^n x_i^2 - n\bar{x}^2) - \frac{1}{2\sigma_0^2}(n\mu^2 - 2n\bar{x}\mu + n\bar{x}^2)\right]}{\int c(2\pi\sigma_0^2)^{-n/2}\exp\left[-\frac{1}{2\sigma_0^2}(\sum_{i=1}^n x_i^2 - n\bar{x}^2) - \frac{1}{2\sigma_0^2}(n\mu^2 - 2n\bar{x}\mu + n\bar{x}^2)\right]d\mu} \\[2ex]
&= \frac{c(2\pi\sigma_0^2)^{-n/2}\exp\left[-\frac{1}{2\sigma_0^2}(\sum_{i=1}^n x_i^2 - n\bar{x}^2)\right]\exp\left[-\frac{n}{2\sigma_0^2}(\mu - \bar{x})^2\right]}{c(2\pi\sigma_0^2)^{-n/2}\exp\left[-\frac{1}{2\sigma_0^2}(\sum_{i=1}^n x_i^2 - n\bar{x}^2)\right]\int \exp\left[-\frac{n}{2\sigma_0^2}(\mu - \bar{x})^2\right]d\mu} \\[2ex]
&= \frac{\exp\left[-\frac{n}{2\sigma_0^2}(\mu - \bar{x})^2\right]}{\int \exp\left[-\frac{n}{2\sigma_0^2}(\mu - \bar{x})^2\right]d\mu} \\[2ex]
&\propto \exp\left[-\frac{n}{2\sigma_0^2}(\mu - \bar{x})^2\right].
\end{aligned}
\tag{4.14}
$$

So it is clear that the posterior for μ is a non-normalized, but proper $\mathcal{N}(\bar{x}, \sigma_0^2/n)$ distribution. It may be surprising that an improper prior distribution can still lead to a proper posterior distribution, and conversely that data-dependent proper priors can become improper priors in the limit (Akaike 1980). Also, improper priors do not necessarily imply a lack of attention to prior specifications and some authors prefer these forms as diffuse alternatives without heavy parametric decisions (Taralsden and Lindqvist 2010).

This example suggests a compromise between the artificiality of conjugate priors and the frequent difficulties of uninformed priors. Recall that the posterior distribution of the mean parameter in the conjugate normal model with the variance known is:

$$\mu \sim \mathcal{N}\left[\left(\frac{m}{s^2} + \frac{n\bar{x}}{\sigma_0^2}\right) \Big/ \left(\frac{1}{s^2} + \frac{n}{\sigma_0^2}\right), \left(\frac{1}{s^2} + \frac{n}{\sigma_0^2}\right)^{-1}\right]. \tag{4.15}$$

If an improper prior for μ is specified as $\mathcal{N}(m = 0, s^2 = \infty)$, then the resulting posterior is clearly proper $\mathcal{N}(\bar{x}, \sigma_0^2/n)$. The compromise suggested is to make the prior variance (s^2) very large so that the prior is normal and proper but very spread out. This is often considered a "conservative" choice of prior since the relative probability structure is quite flat. These *diffuse priors* or *vague priors* are quite popular in hierarchical models where there are many regression-style parameters of only moderate interest.

The best rationale for improper priors, provided that there is appreciable substantive motivation for assigning them, is that if the model is set up so that the likelihood dominates the prior to such an extent that the posterior is still proper, then their use is not harmful. However, the only way to get an improper posterior is to specify an improper prior, and this decision rests entirely with the researcher.

4.5   Informative Prior Distributions

This section introduces several forms of informed (or informative) priors. In this chapter we looked closely at conjugate prior specifications, which are generally informed (although specifying ∞ for various parameters can dilute such qualities). Informative priors are those that deliberately insert information that researchers have at hand. On one level this seems like a reasonable and reasoned approach since previous scientific knowledge should play a role in statistical inference. The key concern that some readers, reviewers, and editors harbor is that the author is deliberately manipulating prior information to obtain a desired posterior result. Therefore there are two important requirements to any written research using informative priors: overt declaration of prior specifications, and detailed sensitivity analysis to show the effect of these priors relative to uninformed types. The latter requirement is the subject of Chapter 6, and the former requirement is discussed periodically in this section.

So where do informative priors come from? Generally there is an abundance of previous work in the social and behavioral sciences that can guide the researcher, including her own. So generally, these priors are derived from:

▷ previous studies, published work,

▷ researcher intuition,

▷ interviewing substantive experts,

▷ convenience through conjugacy,

▷ nonparametrics and other data derived sources,

which can obviously be overlapping definitions. Prior information from previous studies need not be in agreement. One fruitful strategy is to construct prior specifications from competing intellectual strains in order to contrast the resulting posteriors and say something informed about the relative strength of each. The last item on this list can be productive if the data used are distinct from that at hand to be used to construct the likelihood functions. There is considerable controversy, otherwise, about "double-use" of the data.

4.5.1 Power Priors

Ibrahim and Chen (2000a) introduce an informed prior that explicitly uses data from previous studies (also discussed by Ibrahim, Chen, and Sinha [2003] as well as Chen and Ibrahim [2006]). Their idea is to weight data from earlier work as input for the prior used in the current model. Define $\mathbf{x}_0$ as these older data and $\mathbf{x}$ as the current data. Their primary application is to clinical trials for AIDS drugs where a considerable amount of previous data exist. In a social science context, there are many settings where previous research informs extant model specifications. Our interest centers on the unknown parameter θ, which is studied in both periods. Specify a regular prior for θ, $p(\theta)$ that would have been used un-modified if the previous data were not included. This can be a diffuse prior if desired, although it will become informed through this process.

An elementary power prior is created by updating the regular prior with a likelihood function from the previous data in a very simple manner, which is scaled by a value $a_0 \in [0:1]$:

$$p(\theta|\mathbf{x}_0, a_0) \propto p(\theta)[L(\theta|\mathbf{x}_0)]^{a_0}. \tag{4.16}$$

It is important to remember that this is still a *prior* form and the regular process follows wherein the posterior is obtained by conditioning this distribution on the data through the likelihood function based on the current data:

$$\pi(\theta|\mathbf{x}, \mathbf{x}_0, a_0) \propto p(\theta|\mathbf{x}_0, a_0)L(\theta|\mathbf{x}). \tag{4.17}$$

The parameter a_0 scales our confidence in the similarity or applicability of the previous data for current inferences. If it is close to zero then we do not particularly value the older observations or studies, and if it is close to one then we believe strongly in the ties to the current data. So lower values favor the regular prior specification, $p(\theta)$, and the choice of this parameter can be very influential. Note that we would not want this value to exceed one, since that would be equivalent to valuing older over newer data.

To reduce the influence of a single choice for a_0, we specify a mixture of these priors using a specified distribution for this parameter, $p(a_0|\cdot)$. Thus (4.16) becomes

$$p(\theta|\mathbf{x}_0) = \int_0^1 p(\theta|\mathbf{x}_0, a_0)p(a_0|\cdot)da_0$$

$$= \int_0^1 p(\theta)[L(\theta|\mathbf{x}_0)]^{a_0}p(a_0|\cdot)da_0. \tag{4.18}$$

Here the parameterization for $p(a_0|\cdot)$ is left vague since its parametric form remains undefined. Chen and Ibrahim recommend a beta distribution as the "natural" choice, but point out that truncated forms of the normal or gamma work as well. The mixture specification has the effect of inducing heavier tails in the marginal distribution of θ and thus represents a more conservative choice of prior.

4.5.2   Elicited Priors

A completely different class of priors is derived not from real or desired mathematical properties, but from previous human knowledge on the subject of investigation. These elicited priors are discussed in detail in Gill and Walker (2005), with a detailed application to attitudes towards the judicial system in Nicaragua. Typically the source for elicited priors is from subject area experts with little or no concern for the statistical aspects of the project. These include physicians, policy-makers, theoretical economists, and qualitative researchers in various fields. However, there is no reason that politicians, study participants, outside experts, or opinion leaders in general could not be used as a source for informative priors as well.

The bulk of the published work on elicited priors is on the Bayesian analysis of clinical trials. In these settings, it is typical to elicit qualitative priors from the clinicians as a means of incorporating local expertise into the calculations of posteriors and trial stopping points (Freedman and Spiegelhalter 1983, Kadane 1986, Spiegelhalter, Abrams, and Myles 2004, Chapter 5). There is also a small literature on elicitation of priors for variable selection (Garthwaite and Dickey 1988, 1992; Ibrahim and Chen 2000b). Here we will concentrate on the more basic task of using elicitation to specify a particular parametric form for the prior. It is relatively common to use conjugate priors or mixtures of conjugate priors for this task so as to remove additional complications. However, this is certainly not a mathematical or theoretical restriction.

Although an overwhelming proportion of the studies employing elicited priors are in the medical and biological sciences, the methodology is ideal for a wide range of social science applications. In virtually every field and subfield of the various disciplines there are practicing "experts" whose opinions can be directly or indirectly elicited. Furthermore, the fact that the social and behavioral sciences are focused on varying aspects of human behavior means that describing current knowledge and thinking about some specific behavioral phenomenon probabilistically is a more realistic way to incorporate disparate judgments. Restated, uncertain and divided opinion is better summarized in probabilistic language than with deterministic alternatives.

The central challenge here is how to translate expert knowledge into a specific probability statement. This process ranges from informal assignments to detailed elicitation plans and even regression analysis across multiple experts (Johnson and Albert 1999, Chapter 5). Spetzler and Staël von Holstein (1975) outline three general steps in the process:

1. **Deterministic Phase.** The problem is codified and operationalized into specific variables and definitions.

2. **Probabilistic Phase.** Experts are interviewed and tested in order to assign probabilistic values to specific outcomes.

3. **Informational Phase.** The assigned probabilities are tested for inconsistencies and completeness is verified.

The deterministic phase includes specifying the explanatory variables and possibly their assumed parametric role in the model (Steffey 1992), determining data sources and data collection processes fitted to this methodology (Garthwaite and Dickey 1992), determining how many experts to query and where to find these experts (Carlin, *et al.* 1993), and finally judging their contributions (Hogarth 1975). Some of this work is far from trivial: experts might need to be trained prior to elicitation (Winkler 1967), variable selection can be influenced by the difficulty of elicitation (Garthwaite and Dickey 1992), and cost projections can be difficult.

The informational phase is somewhat mechanical and it includes testing elicitation responses for consistency, calibrating responses with known data, and perhaps weighting expert opinions. Determining consistency is an important requirement and experts differ in their familiarity with the details of the project at hand. Less experienced respondents tend to show more inconsistencies (especially with continuous rather than discrete choices), and more experienced respondents as well as normative experts show high levels of consistency (Hogarth 1975, Winkler 1967). By consistency it is meant that answers do not contradict each other, for instance, the subset of an event having a higher probability than the event itself. Calibration generally involves comparison of results after the rest of the analysis and can be a safety check for future work as well as a confirmation of the reliability of the experts (Seidenfeld 1985). Sometimes these checks are further complicated when the subject is a rapidly changing area and the experts' earlier statements can quickly become outdated (Carlin, *et al.* 1995). Leamer (1992) also gives a diagnostic approach that helps categorize elicitations into blunt responses removing the necessity of further inquiry.

By far the most challenging is the probabilistic phase and this has consumed the bulk of the literature. For instance, the experts can be asked fixed value ("P-methods") and/or fixed probability ("V-methods") questions where specific estimates of the probability or relative likelihood of events are queried (Spetzler and Staël von Holstein 1975, p.347). In addition, the experts can be asked these questions directly with regard to a cumulative density function (CDF), or indirectly by way of physical devices or hypothetical constructions. A more challenging, but perhaps informative, approach is to ask open-ended questions and code the response. In all of these cases, it is important to clarify to experts that they are giving probability estimates rather than utility assessments (Kadane and Winkler 1988). The concern is that these experts will otherwise express their normative ideals about outcomes, and preferred outcomes will be given unrealistic probabilities.

In general it is not feasible to ask subject-matter experts to make determinations about coefficient estimates or about moments for specified PDFs and PMFs convenient to the statistician. So the most common strategy, dating back to the seminal paper of Kadane *et al.* (1980), is to query these experts about outcome variable quantiles for given (hypothetical) levels of explanatory variables. For example, in one study an emergency room physician is asked about survival probabilities of patients with specified injury type, injury severity score, trauma score, age, and type of injury (Bedrick, Christensen, and Johnson 1997). The idea is then to take these quantiles and solve for the parameters of an assumed distributional form for the (often conjugate) prior.

One particularly simple application is in the case of a binomial outcome. For psychological reasons, it appears to be easier to elicit hypothetical binary outcomes. Using the beta conjugate prior several authors have suggested algorithms for elicitation (Chaloner and Duncan 1983, 1987; Gavasakar 1988). The basic process is to hypothesize a fixed set of Bernoulli trials, ask the expert to give a most likely number of successes given the particular scenario and reasonable bounds on the uncertainty, work these values backward into the beta-binomial PMF to get the beta parameters, and finally show the expert the posterior implications of these values. If they are found to be unreasonable, then adjustments are made and the process repeats itself.

4.5.2.1   The Community of Elicited Priors

The priors that are elicited from experts can have a variety of characterizations. Kass and Greenhouse (1989) coined the phrase "community of priors" to describe the range of attitudes that equally qualified experts may have about the same phenomenon. These can be categorized as well:

▷ **Clinical Priors.** These are priors elicited from substantive experts who are taking part in the research project. This is often done because these individuals are easily captured for interviews and are motivated by a direct stake in the outcome.

▷ **Skeptical Priors.** These are priors built with the assumption that the hypothesized effect does not actually exist and are usually operationalized with a zero mean. Skeptical priors can be created because of actual skepticism or because overcoming such a prior provides stronger evidence: "... set up as representing an adversary who will need to be disillusioned by the data ..." (Spiegelhalter *et al.* 1994, Spiegelhalter, Abrams, and Myles 2004).

▷ **Enthusiastic Priors.** These are obviously the opposite of the skeptical prior. The priors are built around the positions of partisan experts or advocates and generally assume the existence of the hypothesized effect. For comparative purposes, enthusiastic priors can be specified with the same variance, but different mean, as corresponding skeptical priors.

▷ **Reference Priors.** Such priors are occasionally produced from expert sources, but they are somewhat misguided because the purpose of elicitation is to glean information that can be described formally.

The priors are restated from Spiegelhalter *et al.* (1994) in order to be less focused on the application to clinical trials. The key point is to understand the differing perspectives of experts. One approach is to contrast the posterior results obtained from divergent prior perspectives, including a formalized process of overcoming adversarial prior specifications in favor of priors more sympathetic to research questions through additional sampling (Lindley and Singpurwalla 1991), randomization strategies (Kass and Greenhouse 1989), scoring rules (Savage 1971), or other means.

4.5.2.2 Simple Elicitation Using Linear Regression

An analyst asks an expert for predictions on an expected outcome for some interval-measured event of interest. The V-method question asked is: what would be an expected low value as a 0.25 quantile (labeled $x_{0.25}$) and an expected high value as a 0.75 quantile (labeled $x_{0.75}$)? These two supplied quantile values, $x_{0.25}$ and $x_{0.75}$, correspond to normal z-scores $z_{0.25} = -0.6745$ and $z_{0.75} = 0.6745$, which specify the shape of a normal PDF since there are two equations and two unknowns:

$$z_{0.25} = \frac{x_{0.25} - \alpha}{\beta} \qquad z_{0.75} = \frac{x_{0.75} - \alpha}{\beta}. \qquad (4.19)$$

Here α and β are the mean and standard deviation parameters of the normal PDF:

$$f(x|\alpha, \beta) = (2\pi\beta^2)^{-\frac{1}{2}} \exp\left[-\frac{1}{2\beta^2}(x - \alpha)^2\right]. \qquad (4.20)$$

This notation for the parameters of a normal is different, but the reason shall soon become apparent. When we solve for α and β in (4.19), we have a fully defined prior distribution in (4.20) and the elicitation is complete.

Of course, one expert is typically not enough to produce robust prior forms, so now query experts $1, 2, \ldots, J$. This produces an over-specified series of equations since there are $J \times 2$ equations and only two unknowns (Spiegelhalter *et al.* [1994], for instance, use $J = 10$). It is necessary to assume that these experts are *exchangeable* meaning that they all provide equal quality elicitations.

Secondly, given the cost of interviewing, we are likely to ask each expert for more than just these two quantiles. Each assessor is asked to give five quantile values at $m = [0.01, 0.25, 0.5, 0.75, 0.99]$ corresponding to standard normal points z_m. At this point, (4.19) can be re-expressed for the quantile level m given by assessor j: $x_{jm} = \alpha + \beta z_{jm}$, and the total amount of expert-elicited information constitutes the following over-specification

($J \times 5$ equations and 2 unknowns) of a normal distribution:

$$x_{11} = \alpha + \beta z_{11} \qquad x_{21} = \alpha + \beta z_{21} \ldots \qquad x_{(J-1)1} = \alpha + \beta z_{(J-1)1} \qquad x_{J1} = \alpha + \beta z_{J1}$$

$$x_{12} = \alpha + \beta z_{12} \qquad x_{22} = \alpha + \beta z_{22} \ldots \qquad x_{(J-1)2} = \alpha + \beta z_{(J-1)2} \qquad x_{J2} = \alpha + \beta z_{J2}$$

$$x_{13} = \alpha + \beta z_{13} \qquad x_{23} = \alpha + \beta z_{23} \ldots \qquad x_{(J-1)3} = \alpha + \beta z_{(J-1)3} \qquad x_{J3} = \alpha + \beta z_{J3}$$

$$x_{14} = \alpha + \beta z_{14} \qquad x_{24} = \alpha + \beta z_{24} \ldots \qquad x_{(J-1)4} = \alpha + \beta z_{(J-1)4} \qquad x_{J4} = \alpha + \beta z_{J4}$$

$$x_{15} = \alpha + \beta z_{15} \qquad x_{25} = \alpha + \beta z_{25} \ldots \qquad x_{(J-1)5} = \alpha + \beta z_{(J-1)5} \qquad x_{J5} = \alpha + \beta z_{J5}$$

The approach suggested by this setup is to run a simple linear regression to estimate α as the intercept and β as the slope. There are two issues to worry about. One must check for logical inconsistencies in consistent quantile values for each assessor (see Lindley, Tversky, and Brown [1979] for a discussion of problems). Also, it is critical to apply necessary mathematical constraints such as ensuring that the estimated coefficient for β remains positive (if substantively required) since the basic linear model imposes no such restriction (Raiffa and Schlaifer 1961).

■ **Example 4.1:   Eliciting Expected Campaign Spending.** We are interested in eliciting a prior distribution for expected campaign contributions received by major-party candidates in an impending U.S. Senate election in order to specify an encompassing Bayesian model. Elicitation replaces data here since the election has not yet taken place. Eight campaign experts are queried for quantiles at levels $m = [0.1, 0.5, 0.9]$, and they provide the following values reflecting the national range of expected total intake by Senate candidates (in thousands):

$x_{11} = 400$	$x_{12} = 2500$	$x_{13} = 4000$
$x_{21} = 150$	$x_{22} = 1000$	$x_{23} = 2500$
$x_{31} = 300$	$x_{32} = 900$	$x_{33} = 1800$
$x_{41} = 250$	$x_{42} = 1200$	$x_{43} = 2000$
$x_{51} = 450$	$x_{52} = 1800$	$x_{53} = 3000$
$x_{61} = 100$	$x_{62} = 1000$	$x_{63} = 2500$
$x_{71} = 500$	$x_{72} = 2100$	$x_{73} = 4200$
$x_{81} = 300$	$x_{82} = 1200$	$x_{83} = 2000$

where x_{83} is expert 8's third quantile. None of the experts have supplied quantile values out of logical order, so these results are consistent. Using these "data" we regress x on z to obtain the intercept and slope values:

```
x <- c( 400, 2500, 4000,  150, 1000, 2500,  300,  900,
       1800,  250, 1200, 2000,  450, 1800, 3000,  100,
       1000, 2500,  500, 2100, 4200,  300, 1200 ,2000)
z <- qnorm(rep(c(0.1,0.5,0.9),8))
summary(lm(x~z))
```

which returns:

	Estimate	Std. Error	t value	Pr(>|t|)
(Intercept)	1506.3	127.0	11.858	4.99e-11
qnorm(y)	953.4	121.4	7.854	8.00e-08

Thus the normal prior mean is $\hat{\alpha} = 1506.3$ and the normal prior standard deviation is $\hat{\beta} = 953.4$.

4.5.2.3 Variance Components Elicitation

A problem with direct quantile elicitation is that assessors often misjudge the probability of unusual values because it is more difficult to visualize and estimate tail behavior than to estimate means or medians. Hora *et al.* (1992) found that when non-technical assessors are asked to estimate spread by providing high probability coverage intervals such as at 99%, then they tend to perceive this as near-certainty coverage and overstate the bounds. But this finding is not universal: in other settings people tend to think of rare events in the tails of distributions as more likely than they really are (an effect exploited by casinos and lotteries). Accordingly, O'Hagan (1998) improves elicited estimates of spread by separately requiring assessors to consider two types of uncertainty:

▷ uncertainty about an estimate relative to an assumed known summary statistic

▷ secondary uncertainty of this summary only.

The elicitee first gives a (modal) point estimate for the explanatory variable coefficient, τ, and is then asked "given your recent estimate of τ, what is the middle 50% probability interval around τ?" The elicitees must understand that this is the interval that contains the middle half of the expected values. So this (V-method) specifies a density estimate centered at the assessors modal point, and if the form of the distribution is assumed or known, then the exact value for the variance can be calculated under a distributional assumption (normal, Student's-*t*, or a log-normal if a right-skewed interval is required).

O'Hagan (1998) prefers asking for the middle 66% of the density, which he calls the "two-to-one interval" since the middle coverage is twice that of the combined tails. Now if a normal prior is used then this interval quickly yields a value for the standard deviation since it covers approximately two of them. It should actually be multiplied by $\frac{68}{66}$ to be exactly correct but analysts often do not worry about the difference since measurement error is almost certainly greater than the difference. Now the researcher calculates the implied variance from this and shows the assessor credible intervals at familiar $(1 - \alpha)$-levels. If these are deemed by the elicitee to be too large or too small, then the process is repeated.

We want to elicit prior distributions for τ_i across n cases, with unknown total $T = \sum_{i=1}^{n} \tau_i$. The assessor first provides point estimates for each case: $x_1, x_1, \ldots, x_n$, so that the estimated total is given by $x_T = \sum_{i=1}^{n} x_i$. These are useful values, but it is still necessary to get a measure of uncertainty in order to produce a variance for the full elicited prior distribution.

The individual deviance of the ith estimate from its true value can be rewritten algebraically:

$$\tau_i - x_i = \left(\tau_i - \frac{x_i}{x_T} T \right) + \frac{x_i}{x_T} \left(T - x_T \right). \tag{4.21}$$

The first quantity on the right-hand side of (4.21) is the deviance of τ_i from an estimate that would be provided if we knew T for certain:

$$E[\tau_i | T] = \frac{x_i}{x_T} T \tag{4.22}$$

(which can be considered as between-case deviance). Now the second quantity on the right-hand side of (4.21) is the weighted deviation of T, i.e., uncertainty about the true total. The expected value form (4.22) helps us obtain the variance of τ_i:

$$\text{Var}(\tau_i) = E\left[\text{Var}(\tau_i | T)\right] + \text{Var}\left(E[\tau_i | T]\right)$$

$$= E\left[\tau_i - \frac{x_i}{x_T} T\right]^2 + \left(E\left[E[\tau_i | T]^2\right] - \left(E\left[E[\tau_i | T]\right]\right)^2 \right)$$

$$= E\left[\tau_i - \frac{x_i}{x_T} T\right]^2 + \left(\frac{x_i}{x_T}\right)^2 \text{Var}(T), \tag{4.23}$$

which shows the general form of the two variance components. A better form for elicitation is achieved by dividing both sides of this equation by x_i^2:

$$\text{Var}\left(\frac{\tau_i}{x_i}\right) = \text{Var}\left(\frac{\tau_i}{x_i} - \frac{T}{x_T}\right) + \text{Var}\left(\frac{T}{x_T}\right). \tag{4.24}$$

At this point elicitees are queried about the middle spread around these two quantities individually. *First*, they give an estimate of middle spread around $\frac{T}{x_T}$, assuming accuracy of the sum x_T as an estimate of T. *Then*, they give the middle spread around each $\frac{\tau_i}{x_i}$ assuming that $\frac{T}{x_T} = 1$. This means that there is no second component in the variance to consider at this moment. Once the individual means and variances are elicited, these $\frac{x_i}{x_T}$ values are put into the assumed distribution defined over $[0:1]$ (they are proportions) to form the complete prior distribution specification. Two common distributional forms are the normal CDF and the beta distribution. In the case where $\frac{\tau_i}{T}$ is from a beta distribution, we can solve for the parameters with the beta distribution mean and variance: $E\left[\frac{\tau_i}{T}\right] = \frac{\alpha}{\alpha+\beta}$, $\text{Var}\left(\frac{\tau_i}{T}\right) = \frac{\alpha\beta}{(\alpha+\beta)^2(\alpha+\beta+1)}$.

■ **Example 4.2:   Minority Political Participation.** This example is from Gill and Walker (2005).   An expert on minority electoral participation is asked to estimate upcoming Hispanic turnout for n precincts in a given district: $\tau_1, \tau_2, \ldots, \tau_n$, with total

Hispanic turnout in the district equal to T. She first gives estimates $x_1, x_2, \ldots, x_n$ for each precinct, which give a district turnout estimate of T by summing, x_T. This result does not yet give the variance information necessary to build a prior using an assumed normal distribution.

The expert is now asked to provide the two-to-one interval for $\frac{T}{x_T}$, giving $[0.7:1.3]$: they believe that the summed estimate of Hispanic turnout is correct to plus or minus 30% with probability 0.66 (from the two-to-one interval). To confirm the expert's certainty about this, the value $\sigma_T = 0.3$ is plugged into the normal CDF at levels to give credible interval summaries:

$$50\% \, CI = [\Phi_{\mu=1, \sigma=0.3}(0.25) : \Phi_{\mu=1, \sigma=0.3}(0.75)] = [0.798 : 1.202]$$

$$99\% \, CI = [\Phi_{\mu=1, \sigma=0.3}(0.005) : \Phi_{\mu=1, \sigma=0.3}(0.995)] = [0.227 : 1.773].$$

These are then displayed to the elicitee, and if she agrees that these are reasonable summaries then the variance is $\sigma_T^2 = (0.3)^2 = 0.09$ and there is no need to iterate here. The expert is now asked to repeat this process for each of the x_i estimates under the assumption that $x_T = T$ (the estimate of the total above is correct). This temporary fixing of x_T means that the right-hand-side of (4.24) reduces to the variance of $\frac{T_i}{x_i}$ and the expert can do the same interval process as was done with $\frac{T}{x_T}$ for each of the n precincts.

Suppose that two-to-one interval for the estimate of Hispanic turnout at the first precinct ($x_1 = 0.2$) is given as $[0.5:1.5]$: she believes the estimate to be correct to plus or minus 50% with probability 0.66. This gives a variance of $\sigma_1 = (0.5)^2 = 0.25$, and we will note that the subsequent 50% and 99% credible interval summaries are approved by the expert. So the total elicited variance for the first precinct is given by (4.24) where x_1^2 is moved back to the right-hand-side: $\text{Var}(\tau_1) = x_1^2(0.25 + 0.09) = 0.0136$.

4.5.2.4 Predictive Modal Elicitation

If the outcome variable of interest is distributed Bernoulli or binomial, then it is usually straightforward to query experts *directly* for prior probabilities. For psychological reasons probabilities of binary or summed binary outcomes are more intuitively easy to visualize (Cosmides and Tooby 1996). Using the natural (conjugate) choice of a beta conjugate prior Chaloner and Duncan (1983, 1987) produce the predictive modal (PM) elicitation algorithm (see also Gavasakar [1988] for a second application). First fix a hypothetical total number of Bernoulli trials, and then ask the elicitee to specify the most likely number of successes as well as reasonable bounds on the uncertainty. These values are then worked backward into the beta-binomial parametric setup (Chapter 1) to get the beta prior distribution parameters. The elicitee is now shown the implications of their stipulated values on the shape of the beta prior on a computer terminal or with pre-prepared flip-charts. If the elicitee finds the distribution to be unlike their prior expectations, then adjustments are made in the deterministic phase and the process is repeated.

The data, $X_1, X_2, \ldots, X_n$, are distributed iid Bernoulli, so use $Y = \sum_{i=1}^{n} X_i$ distributed binomial(n, p). Interest lies in the posterior distribution of the probability of occurrence of some event of interest (bill passes/fails, treaty/no-treaty, etc.), which is p in this binomial PMF. The PM method first assumes a beta distribution prior for p with unknown parameters: $p \sim \text{beta}(A, B)$ with $A, B > 1$. The individual steps are:

▷ select a fixed number of trials for a hypothetical experiment ($n = 20$ is recommended)

▷ ask the elicitees to give the prior predictive modal value for this n: the most likely number of successes out of n trials, $m \in (1{:}n)$.

The numerator of Bayes' Law is:

$$f(y|p)f(p) = \frac{\Gamma(n+1)\Gamma(A+B)}{\Gamma(y+1)\Gamma(n-y+1)\Gamma(A)\Gamma(B)}p^{y+A-1}(1-p)^{n-y+B-1}. \qquad (4.25)$$

Now the marginal distribution for y is obtained by integrating over p:

$$\begin{aligned}
f(y|A, B) &= \int_0^1 \frac{\Gamma(n+1)\Gamma(A+B)}{\Gamma(y+1)\Gamma(n-y+1)\Gamma(A)\Gamma(B)}p^{y+A-1}(1-p)^{n-y+B-1}dp \\
&= \frac{\Gamma(n+1)\Gamma(A+B)}{\Gamma(y+1)\Gamma(n-y+1)\Gamma(A)\Gamma(B)}\frac{\Gamma(y+A)\Gamma(n-y+B)}{\Gamma(n+A+B)} \qquad (4.26)
\end{aligned}$$

(as shown before on page 49). Since m is the mode of this distribution, then $f(y = m|A, B)$ is the maximum value obtainable for the function $f(y|A, B)$. The random variable y can only take on discrete values so $f(y = m-1|A, B) < f(y = m|A, B)$ and $f(y = m+1|A, B) < f(y = m|A, B)$. Chaloner and Duncan calculate the following two ratios using properties of the binomial distribution:

$$d_\ell = \frac{f(y = m-1|A, B)}{f(y = m|A, B)} = \frac{(n-m)(m+A)}{(m+1)(n-m+B-1)}$$

$$d_r = \frac{f(y = m+1|A, B)}{f(y = m|A, B)} = \frac{m(n-m+B)}{(n-m+1)(m+A-1)}. \qquad (4.27)$$

Both terms are bounded by $(0{:}1)$. Once the elicitee has identified m for the researcher, then the prior parameters (A, B) must be constrained to lie in a cone originating at $[1, 1]$ in the A, B plane as shown by the solid lines in Figure 4.3. This cone is determined because the equations in (4.27) define linear limits in two-space from the same starting point. Any particular point within the cone represents a two-dimensional Cartesian distance from the uniform prior since the origin of the cone specifies a beta$(1, 1)$.

We do not yet have a complete answer since there are an infinite number of (A, B) pairs that could be selected and still remain inside the cone. At this point the elicitee is told to think about *spread* around the mode and is shown a histogram for binomial$(n, m/n)$, and is asked: "If we were to go one unit up (and down), how much do you think the probability of occurrence would decrease?" With these two values (up and down) we can now calculate

d_ℓ and d_r directly. Thus the equations in (4.27) define line segments for values of A and B that are necessarily bounded by the cone. The point of intersection is the (A, B) pair that satisfies both the one unit up restriction and the one unit down restriction, assuming that the following restriction is met:

$$d_\ell d_r > \frac{m(n-m)}{(m+1)(n-m+1)}. \tag{4.28}$$

If it this is not true, then the assessor is asked to provide new values of $f(y = m-1|A, B)$ and $f(y = m+1|A, B)$. The line segments and their intersection are also shown in Figure 4.3. Label this point of intersection (A_1, B_1), and calculate a new beta mode with these values:

$$m_1 = \frac{A_1 - 1}{A_1 + B_1 - 2}. \tag{4.29}$$

We then display it to the assessor with the middle 50% of the density. The assessor is then asked if this interval is too small, too large, or just right. The interval is adjusted according to the following:

"too small" $\implies h = -1$, "too large" $\implies h = +1$, "just right" $\implies h = 0$,

where h is inserted into:

$$A_2 = 1 + 2^h(A_1 - 1), \qquad B_2 = 1 + 2^h(B_1 - 1), \tag{4.30}$$

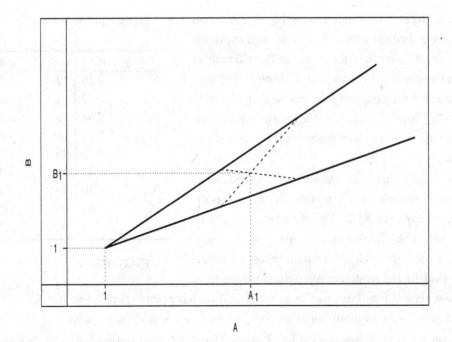

FIGURE 4.3: FINDING THE BETA PARAMETERS

for adjusted parameter values. This process is repeated until the assessor is satisfied with the interval ($h = 0$).

■ **Example 4.3:   Labor Disputes in OECD Countries.** Suppose we are interested in eliciting a prior distribution for the probability of a strike given a major labor dispute in 20 OECD countries over a certain year. Given $n = 20$ an elicitee indicates $m = 5$, and is then shown a histogram of the binomial$(20, 5/20)$ distribution. The assessor now asserts that the one unit up and down probability change is $1/50$, so $d_\ell = 0.98 = \frac{(15)(5+A)}{(6)(14+B)}$ and $d_r = 0.98 = \frac{5(15+B)}{(16)(4+A)}$ (these values are acceptable to the condition above since $d_\ell d_r = 0.9604 > \frac{m(n-m)}{(m+1)(n-m+1)} = 0.7813$). Solving the equations produces $A_1 = 1.4$, and $B_1 = 2.327$, which gives a modal value for the probability of a strike given a labor dispute of $\hat{p} = 0.232$ with the middle of 50% of the density $[0.193 : 0.537]$. The elicitee states that this interval is too large, so we set $h = +1$ and produce $A_2 = 1 + 2^h(1.4 - 1) = 1.8$, and $B_2 = 1 + 2^h(2.327 - 1) = 3.654$, which gives the middle of 50% of the density as $[0.184 : 0.456]$. The elicitee still believes that this is too large so the process is repeated with $h = +1$ now providing $[0.182 : 0.386]$. This interval is smaller than the first and is acceptable to the elicitee. Therefore the elicited prior distribution for the probability of a strike given a major labor dispute has the distribution beta$(2.6, 6.308)$.

4.5.2.5   Prior Elicitation for the Normal-Linear Model

Using the standard regression setup described on page 145, we want to elicit priors on β from experts (we can retain the uninformed approach for σ^2 or we can elicit for it as well). Kadane *et al.* (1980) as well as Kadane and Wolfson (1998) first establish m design points of the explanatory variable vector: $\tilde{\mathbf{X}}_1, \tilde{\mathbf{X}}_2, \ldots, \tilde{\mathbf{X}}_m$, where again these represent interesting cases or values spanning the range of the k variables.

These values must be chosen such that stacking the vectors into an $m \times k$ matrix, $\tilde{\mathbf{X}}$, gives a positive definite matrix $\tilde{\mathbf{X}}'\tilde{\mathbf{X}}$. The elicitees are asked to study each of the $\tilde{\mathbf{X}}_i$ row vectors and produce $\mathbf{y}_{50}$, a vector of outcome variable medians whose elements correspond to the design cases. Such values then

$\nu = df$	$a(\tilde{\mathbf{X}})$	$\nu = df$	$a(\tilde{\mathbf{X}})$
3	2.76	12	2.37
4	2.62	14	2.36
5	2.53	16	2.35
6	2.48	18	2.34
7	2.45	20	2.33
8	2.42	30	2.31
9	2.40	40	2.31
10	2.39	60	2.30
		∞	2.27

FIGURE 4.4: t RATIOS

represent typical responses to the hypothesized design points specified in the $\tilde{\mathbf{X}}_i$. Now an elicited prior point estimate for β is given by: $\hat{\mathbf{b}}_{0.50} = (\tilde{\mathbf{X}}'\tilde{\mathbf{X}})^{-1}\tilde{\mathbf{X}}'\mathbf{y}_{0.50}$.

To get a full prior description for β assume that this distribution is Student's-t around $\hat{\mathbf{b}}_{0.50}$ with greater than $\nu = 2$ degrees of freedom (we do not want to worry about the existence of moments). This is a somewhat conservative prior since large data size under weak regularity conditions leads to Bayesian posterior normality of linear model coefficients, and t-distributed forms with smaller data size (Berger 1985, p.224; Lindley and Smith 1972). There is no direct way to set the degrees of freedom for this t-distribution since the m value was established arbitrarily by the researchers.

Kadane *et al.* (1980) suggest that after eliciting $y_{0.50}$ for each $\tilde{\mathbf{X}}_i$, researchers should also elicit $y_{0.75}$ by asking for the median point above the median point just provided. Repeat this process two more times in the same direction to obtain $y_{0.875}$, and $y_{0.9375}$. For each of the m assessments calculate the ratio:

$$a(\tilde{\mathbf{X}}) = (y_{0.9375} - y_{0.50})/(y_{0.75} - y_{0.50}), \qquad (4.31)$$

where differencing makes the numerator and denominator independent of the center, and the ratio produced is now independent of the spread described. This ratio uniquely describes tail behavior for a t-distribution because it is the relative "drop-off" in quantiles. Kadane *et al.* give tabulated degrees of freedom against values of this ratio, a subset of which is given in Figure 4.4.

Values greater than 2.76 indicate that the researcher should instruct the elicitee to reconsider their responses, and values less than 2.27 imply that a standard normal prior centered at $\hat{b}_{0.50}$ is appropriate. Other distributions are feasible, but this tabulation makes the Student's-t particularly easy. Once the unique degrees of freedom are identified, the elicited prior is fully identified.

4.5.2.6 Elicitation Using a Beta Distribution

Gill and Freeman (2013) developed an elicitation process for updating social networks using beta distribution forms. The process starts with the more general form of the beta probability density function,

$$f(y) = \frac{\Gamma(\alpha + \beta)}{\Gamma(\alpha)\Gamma(\beta)} \frac{(y - a)^{\alpha-1}(b - y)^{\beta-1}}{(b - a)^{\alpha+\beta-1}}, \qquad (4.32)$$

where: $0 \le a < y < b$, and $\alpha, \beta > 0$. Now the support is over $[a:b]$ rather than just $[0:1]$, but it reduces to the standard form with the change of variable: $X = \frac{Y-a}{b-a}$, where $0 < x < 1$, but α and β are unchanged. This also means that $Y = (b - a)X + a$. This change means that qualitative experts can be queried on the more natural $[0:100]$ scale and it is converted back to a regular beta form. Now recall the following beta distribution properties:

$$\text{mean:} \quad \mu_x = \frac{\alpha}{\alpha + \beta} = \frac{\mu_y - a}{b - a}$$

$$\text{variance:} \quad \sigma_x^2 = \frac{\alpha\beta}{(\alpha + \beta)^2(\alpha + \beta + 1)} = \frac{\sigma_y^2}{(b - a)^2}.$$

Rearranging the equations above gives us the following two important relations from Law and Kelton (1982, p.205):

$$\alpha = \left[\frac{\mu_x(1 - \mu_x)}{\sigma_x^2} - 1 \right] \mu_x$$

$$\beta = \left[\frac{\mu_x(1 - \mu_x)}{\sigma_x^2} - 1 \right] (1 - \mu_x). \tag{4.33}$$

So querying for a mean and variance would produce all the necessary information to obtain the implied α and β and thus fully describe the beta distribution. Research to date demonstrates that humans are not very good subjective assessors of variance (O'Hagan 1998), necessitating other means of getting to the parameters.

Gill and Freeman propose querying the experts for their lowest but reasonable value and highest but reasonable value with the idea that these can be explained as endpoints of a 95% credible interval. Some training is therefore necessary, but not a substantial amount. Restricting $\alpha > 1$ and $\beta > 1$ guarantees that the beta distribution will be unimodal, so the credible interval is not broken up into disjoint components. Combining these, we know that $4\sigma_y \approx b - a$. We also know that

$$\sigma_y^2 = (b - a)^2 \sigma_x^2 = (b - a)^2 \frac{\alpha\beta}{(\alpha + \beta)^2(\alpha + \beta + 1)}.$$

Substituting produces the approximation: $\sigma_x \approx \frac{1}{4}$, which leads directly to preliminary estimates of the two parameters. This draws on the PERT (Program Evaluation and Review Technique) analysis for industrial and engineering project planning that uses $\sigma_x = \frac{1}{6}$ when the beta mode is between 0.13 and 0.87, and is justified by a normal approximation (Farnum and Stanton 1987, Lu 2002). More importantly, the value of 1/4 is just a starting point for a software "slide" that the elicitee is allowed to adjust on a computer screen. The elicitee is shown the resulting beta distribution graphically and given two slides: one that changes $\mu_x = \frac{\alpha}{\alpha + \beta}$ and one that changes $\sigma_x^2 = \frac{\alpha\beta}{(\alpha + \beta)^2(\alpha + \beta + 1)}$. Subsequently, there are new values of μ_x and σ_x that the elicitee prefers. This can be repeated for any reasonable number of priors.

4.5.2.7   Eliciting Some Final Thoughts on Elicited Priors

Elicited priors in Bayesian studies have been studied for quite some time, going back to the foundational papers of Kadane (1980), Kadane *et al.* (1980), Hogarth (1975), Lindley, Tversky, and Brown (1979), Savage (1971), Spetzler and Staël von Holstein (1975), Tversky (1974), Tversky and Kahneman (1974), and Winkler (1967). There have been two main impediments to widespread acceptance and use of elicited priors. *First*, there has long been a distrust by some Bayesians and others of overtly subjective priors. The failed attempt to produce a universally accepted no-information prior was fueled in part by these sentiments. *Second*, there remains worry about the cognitive process that generates elicited priors. These concerns are summed up by Hogarth (1975, p.273):

In summary, man is a selective, stepwise information processing system with limited capacity, and, as I shall argue, he is ill-equipped for assessing subjective probability distributions. Furthermore, man frequently just ignores uncertainty.

This does not mean that we should never use human experts for generating prior information, but rather that we should carefully elicit probabilistic statements, knowing that the sources have cognitive limitations. A very promising approach is the use of interactive computer queries to elicit priors (Garthwaite and Dickey 1992). The software can be written so that inconsistent, illogical, and contradictory answers are rejected at the data-entry stage rather than corrected or ignored later. Web-based implementations, though still untried, promise even greater returns given their convenience and ubiquity.

There is a close relationship between elicited priors and meta-analysis. It is possible to assemble the results from previous studies and treat these as elicitations. This is a fundamentally Bayesian idea and essentially represents a Bayesian treatment of conventional meta-analysis where the predictive distribution from previous works forms the basis of a prior on the current project (Carlin 1992). One caveat is warranted, however. Kass and Greenhouse (1989) warn that this approach implies that the previous studies pooled to form a prior are now assumed to be exchangeable and this might not be appropriate.

4.6 Hybrid Prior Forms

Some prior specifications are attempts to *mix* informative and uninformed types in such a way that gradations of information can be reflected. Technically this makes them informative priors, but we will treat them as compromises since this is how they are typically viewed. The key point is that the specification of prior distributions is very flexible, and can be tailored by researchers to reflect varying levels of qualitative or quantitative information. In this section we describe some recent designs for prior distributions that focus on addressing specific model problems rather than general prior specifications.

A number of other prior forms are common in applied settings. We will describe mixture priors in Chapter 9, how empirical Bayes uses the observed data to establish hyperpriors in a hierarchical model (prior parameters on the highest level of priors), and the idea of specifying both hierarchical structure and subjective information into the prior leads to hierarchical models in general. Much work has been done toward finding new criteria for uninformative priors. These approaches include maximally vague entropy priors (Spall and Hill 1990; Berger, Bernardo, and Mendoza 1989), indifference conjugate priors (Novick and Hall 1965, Novick 1969), and the general idea of proper but very diffuse priors (Box and Tiao 1973).

4.6.1   Spike and Slab Priors for Linear Models

Mitchell and Beauchamp (1988) introduce a species of priors designed to facilitate explanatory variable selection in linear regression. The basic idea is to specify a reasonably skeptical prior distribution by stipulating density spike at zero surrounded symmetrically by a flat slab with specified boundaries.

If desired, other non-zero values can be evaluated by simply offsetting the coefficient value. Also, this type of prior has only been applied to linear regression, but it would only be slightly more involved to apply to generalized linear models.

Thus the prior makes a reasonably strong claim about zero effect for some regression parameter, but admits the possibility of non-zero effects. Suppose we are seeking a prior for the jth regression coefficient (from $j \in 1, \ldots, k$) in a standard linear model: $\mathbf{y} = \mathbf{x}\beta + \epsilon$, with all of the usual Gauss-Markov assumptions, $k < n$, and $\epsilon \sim \mathcal{N}(0, \sigma^2)$. The prior is then:

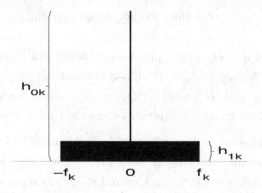

FIGURE 4.5: SPIKE AND SLAB PRIOR

$$\text{spike:} \quad p(\beta_j = 0) = h_{0j}$$
$$\text{slab:} \quad p(\beta_j \in [-f_j:f_j], \beta_j \neq 0) = 2h_{1j}f_j,$$
$$- f_j < \beta_j < f_j,$$

where $h_{0j} + 2h_{1j}f_j = 1$. This is illustrated in Figure 4.5. So obviously h_{0j} is the height of the spike and h_{1j} is the height of the slab, and varying degrees of skepticism about the veracity of the effect of $\mathbf{x}_j$ can be modeled by altering the ratio:

$$\gamma_j = \frac{h_{0j}}{h_{1j}} = 2f_j \frac{h_{0j}}{1 - h_{0j}}.$$

For terms that should be included in the model with probability one, just set $h_{0j} = 0$. The nice part of this priors setup is that we can easily reflect our prior belief in complete model specifications. Miller (2002, 203) notes that taken by itself, h_{0j} is a statement that the jth variable should not be in the model, and taken by itself, $1 - h_{0j}$ is a statement that the jth variable matters in some way. Since all prior density not accounted for by the spike falls to the slab, the prior for the mth model, A_m, is just the product given by:

$$p(A_m) = \prod_{j \in A_m} (1 - h_{0j}) \prod_{j \notin A_m} (h_{0j}),$$

which is the product of the slab density for those coefficients in model A_m, times the product of the spike density for those coefficients not in A_m (note that the jth coefficient only contributes a single term).

For the linear regression model Mitchell and Beauchamp put a prior on σ such that $\log(\sigma)$ is uniform between $-\log(\sigma_0)$ and $\log(\sigma_0)$, for some large value of σ_0. They then derive the resulting posterior for model A_m with k_m number of coefficients (including the constant) in the matrix $\mathbf{X}_m$:

$$\pi(A_m|\mathbf{X},\mathbf{y}) \propto \pi^{k_m/2}\Gamma\left(\frac{n-k_m}{2}\right)|\mathbf{X}'_m\mathbf{X}_m|^{-\frac{1}{2}}(S_m^2)^{-(n-k_m)/2}\prod_{j\notin A_m}\gamma_j,$$

where $S_m^2 = (\mathbf{y}-\mathbf{X}_m\hat{\boldsymbol{\beta}}_m)'(\mathbf{y}-\mathbf{X}_m\hat{\boldsymbol{\beta}}_m)$, $\hat{\boldsymbol{\beta}} = (\mathbf{X}'_m\mathbf{X}_m)^{-1}\mathbf{X}'_m\mathbf{y}$, and c is a normalizing constant. We can see from this how the data informs across all of the proposed model specifications. This setup also facilitates coefficient quality assessment as well since we can average coefficient probabilities across model space. So the joint posterior distribution of all k of the coefficients is given by:

$$p(\boldsymbol{\beta}|\mathbf{X},\mathbf{y}) = \sum_m \pi(A_m|\mathbf{X},\mathbf{y})p(\boldsymbol{\beta}|\boldsymbol{\alpha}_m,\mathbf{X},\mathbf{y})$$

where $p(\boldsymbol{\beta}|\boldsymbol{\alpha}_m,\mathbf{X},\mathbf{y})$ is multivariate t-distribution centered at $\hat{\boldsymbol{\beta}}$ (see Chapter 5 for more details on posterior distributions from linear models). This general approach is called Bayesian model averaging and is discussed in Chapter 6.

Such priors are easy to construct for linear models and can be applied in more general settings (see Pang and Gill 2012). Several words of caution are warranted, however. *First*, each coefficient operates on a different scale so direct comparison cannot be done without some adjustment like centering and scaling (Mitchell and Beauchamp's suggestion). *Second*, the researcher's choice of f_j and h_{0j} can have a more dramatic effect than intended.

4.6.2 Maximum Entropy Priors

Jaynes (1968, 1980, 1983) introduces the idea of an *entropy prior* to describe relative levels of uncertainty about the distribution of prior parameters. One advantage to the entropy approach is that within the same framework, uncertainty ranging from that provided by the uniform prior to that provided by absolute certainty given by a degenerate (single point) distribution can be modeled in the same way. Thus the primary advantage to entropy priors is their flexibility. Unfortunately, however, entropy priors are not invariant to reparameterization and therefore have somewhat limited applicability. Nonetheless they are worthy of study in their own right because of the link to the idea of *information* in the general study of information theory, which is closely tied to the idea of prior knowledge.

The core idea of entropy is the quantification of uncertainty of a transmission or observation (Shannon 1948, Ayres 1994), and this can be interpreted as uncertainty in a PDF or PMF (Rosenkranz 1977). Initially assume that we are interested in a discrete parameter θ.

Define the entropy of θ for a given parametric form, $p(\theta)$, as:

$$H(\theta) = -\sum_{\Theta} p(\theta_i) \log[p(\theta_i)], \tag{4.34}$$

where the sum is taken over the categories of θ. In the case of discrete distributions, we have a wide selection of parametric forms for $p(\theta)$, but consider two very different varieties for k possible values in the sample space: $\Theta = [\theta_1, \theta_2, \ldots, \theta_k]$. Following the discussion of uniform priors, if we assign each outcome a uniform prior probability of occurrence: $1/k$, then the entropy of this prior is at its maximum:

$$H(\theta) = -\sum_{\Theta} \frac{1}{k} \log\left[\frac{1}{k}\right] = \log[k]. \tag{4.35}$$

It is clear from this result that prior uncertainty increases logarithmically with increases in the number of discrete alternatives. At the other end of the uncertainty spectrum, we can stipulate that $p(\theta_i) = 1$, and $p(\theta_{\neg i}) = 0$; that is, θ equals some value with probability one and every other value with probability zero. The entropy of this prior specification is given by:

$$H(\theta) = -\sum_{\Theta[-i]} 0 \log[0] + 1 \log[1] = 0, \tag{4.36}$$

where this calculation requires the assumption that $0 \log[0] = 0$. Here we are completely certain about distribution of θ and the entropy is zero reflecting minimum prior entropy.

The utility of the entropy prior is that we can take some limited knowledge in the form of a restriction, and for a given parametric specification of the prior, produce the prior with maximum possible entropy. The goal is to include what we may know but to interject as little else as possible in the creation of the prior: minimally informative about θ for the specified $p(\theta)$ (Eaves 1985). So a distribution is produced that is as diffuse as possible given the identified "side conditions."

Suppose that we could stipulate the first two moments as constraints:

$$E[\theta] = \sum_{\Theta} p(\theta)\theta = \mu_1$$

$$E[\theta^2] = \sum_{\Theta} p(\theta)\theta^2 = \mu_2.$$

Adding the further constraint that $p(\theta)$ is proper gives the prior density:

$$\tilde{p}(\theta_i) = \frac{\exp\left[\lambda_1 \theta_i + \lambda_2 (\theta_i - \mu_1)^2\right]}{\sum_j \exp\left[\lambda_1 \theta_j + \lambda_2 (\theta_j - \mu_1)^2\right]}, \tag{4.37}$$

where the constants, λ_1 and λ_2, are determined from the constraints. This is a simplified version, and in a more general setting there are additional moments than the two estimated, say M in $E[g_m(\theta)]$. So producing the constants is more involved than the example here (Robert 2001, p.110; Zellner and Highfield 1988).

The continuous case is both more difficult and more straightforward. It is more difficult because the analog to (4.34),

$$H(\theta) = -\int_{\Theta} p(\theta) \log[p(\theta)] d\theta, \qquad (4.38)$$

leads to different answers depending on alternative definitions of the underlying reference measures (Jaynes 1968; Robert 2001, p.110). Essentially this just means that mathematical requirements for getting a usable prior are more difficult to obtain and in some circumstances are actually impossible (Berger 1985, p.93-94). For the M moment estimation case, the constraining statement is now:

$$E[g_m(\theta)] = \int_{\Theta} p(\theta) g_m(\theta) d\theta = \mu_m, \qquad m = 1, \dots, M. \qquad (4.39)$$

Consider the following simple example (O'Hagan 1994). The support of θ is $[0:\infty]$ and we specify the constraint that the expected value is equal to some constant: $E[\theta] = c$. If we further specify that the prior distribution has the least informative possible exponential PDF form, then (4.39) specifies the prior: $f(\theta|c) = c \exp(-c\theta)$.

4.6.3 Histogram Priors

Johnson and Albert (1999) suggest the "histogram prior" based on the idea that it is easy to summarize relatively vague, but informative, prior information through this graphical device. Their prior exploits the idea that it is much easier to elicit or generate a small number of binned probability statements. This is essentially a nonparametric approach in that there is no *a priori* structure on the form of the resulting histogram and is therefore attractive when the researcher is uncomfortable with specifying a standard form. The estimation process is done in segments across the range of the histogram corresponding to a bin. Within each bin a separate posterior is calculated with a bounded uniform prior producing a noncontinuous posterior weighted by the heights of the histogram bins. Although the discontinuities can be smoothed out, the resulting form may still be unreflective of a compromise between prior and likelihood information.

4.7 Nonparametric Priors

Nonparametric priors were originally introduced by Ferguson (1973) and Antoniak (1974), but not fully developed until the advent of better computing resources for estimation. Dirichlet process priors stipulate that the data are produced by a mixture distribution wherein the Bayesian prior specifications are produced by a Dirichlet process, which constitutes a distribution of distributions since each produced parameter defines a particular distribution. These distributions can be made conditional on additional parameterizations (as done in Escobar and West [1995]) and thus the models are hierarchical. This means that realizations of the Dirichlet process are discrete (with probability one), even given support over the full real line, and are thus treated like countably infinite mixtures.

This approach is substantially different from the conventional use of the Dirichlet distribution in Bayesian models as a conjugate prior distribution for the multinomial likelihood. First, let us look at the Dirichlet PDF, which is the multivariate generalization of the beta distribution. Suppose $\mathbf{X} = [x_1, x_2, \ldots, x_k]$, where each individual x is defined over the support $[0\!:\!1]$, and sum to one: $\sum_{i=1}^n x_i = 1$. The Dirichlet PDF uses a parameter vector, $\boldsymbol{\alpha} = [\alpha_1, \alpha_2, \ldots, \alpha_k]$, with the condition that all $\alpha_i > 0$. These structures are then generalizations from the beta for $\mathbf{X} = [x, 1-x]$, and $\boldsymbol{\alpha} = [\alpha, \beta]$, where $k = 2$. The PDF is:

$$f(\mathbf{X}|\boldsymbol{\alpha}) = \frac{\Gamma\left(\sum_{i=1}^n \alpha_i\right)}{\prod_{i=1}^n \Gamma(\alpha_i)} \prod_{i=1}^n x_i^{\alpha_i - 1}. \tag{4.40}$$

The resemblance to the beta is quite strong here. This distribution has expected value and variance:

$$E[X_i] = \frac{\alpha_i}{\sum_{k=1}^n \alpha_k} \tag{4.41}$$

$$\mathrm{Var}[X_i] = \frac{\alpha_i \left(\sum_{k=1}^n \alpha_k - \alpha_i\right)}{\left(\sum_{k=1}^n \alpha_k\right)^2 \left(\sum_{k=1}^n \alpha_k + 1\right)} \tag{4.42}$$

$$\mathrm{Cov}[X_i, X_j] = \frac{-\alpha_i \alpha_j}{\left(\sum_{k=1}^n \alpha_k\right)^2 \left(\sum_{k=1}^n \alpha_k + 1\right)}. \tag{4.43}$$

Consider the question of modeling dichotomous individual choices, Y_i, like turning out to vote, voting for a specific candidate, joining a social group, discontinuing education, and so on. The most common "regression-style" modeling specification is to assume that an underlying smooth utility curve dictates such preferences, providing the unobserved, but estimated threshold $\theta \in [0, 1]$. The individual's threshold along this curve then determines the zero or one outcome conditional on an additive right-hand specification, $\mathbf{X}\boldsymbol{\beta}$. Realistically, we should treat this threshold differently for each individual, but we can apply the reasonable Bayesian approach of assuming that these are different thresholds yet still generated from a single distribution G which is itself conditional on a parameter α, thus $E[nG(\theta_i|\mathbf{X}\boldsymbol{\beta}, \alpha)]$ is the expected number of affirmative outcomes. Suppose there were structures in the data such as unexplained clustering effects, unit heterogeneity, autocorrelation, or missingness that cast doubt on the notion of G as a single model. Note that this can happen in a Bayesian or non-Bayesian setting, the difference being the distributional or deterministic interpretation of θ. The choice of G is unknown by the researcher but determined by custom or intuition. We suggest, instead, a nonparametric Bayesian approach that draws θ from a mixture of appropriate prior distributions conditional on data and parameters (in this simple case a mixture of beta distributions according to $\mathcal{BE}(\alpha(\mathbf{X}\boldsymbol{\beta}), Z - \alpha(\mathbf{X}\boldsymbol{\beta}))$ for some prior parameter Z).

Gill and Casella (2007) apply this type of a prior to ordinal regression models. Such outcomes occur frequently in the social sciences, especially in survey research where instruments such as Likert and Guttman scale responses are used. Gill and Casella look at self-reported levels of stress for political executives in the United States where a five-point ordinal scale is used. Start with data $Y_1, Y_2, \ldots, Y_n$ assumed to be drawn from a mixture of distributions denoted $p(\psi)$ where the mixing over ψ is independent from distribution G, and the prior on G, $\mathcal{D}$ is a mixture of Dirichlet processes. The ordered probit model assumes first that there is a multinomial selection process:

$$Y_i \sim \text{Multinomial}(1, (p_1, p_2, \ldots, p_C)), \quad i = 1, \ldots n \qquad (4.44)$$

where $\sum_j p_j = 1$, and $Y_i = (y_{i1}, \ldots, y_{iC})$ is a $C \times 1$ vector with a 1 in one position and 0 elsewhere. The placement of the 1 denotes the class that the observation falls into. In addition, the p_j are ordered by the probit model

$$p_j = p(\theta_{j-1} \leq U_i \leq \theta_j) \qquad (4.45)$$

where these cutpoints between categories have the property that $-\infty = \theta_0 < \theta_1 < \cdots < \theta_C = \infty$. Define now the random quantity:

$$U_i \sim \mathcal{N}(X_i\beta + \psi_i, \sigma^2) \qquad (4.46)$$

where X_i are covariates associated with the ith observation, β is the coefficient vector, and ψ denotes a random effect to account for subject-specific deviation from the underlying model. Here the U_i are unobservable random variables, and we could specify the model without them, that is, from (4.45) and (4.46),

$$p_j = \Phi\left(\frac{\theta_j - X_i\beta - \psi_i}{\sigma}\right) - \Phi\left(\frac{\theta_{j-1} - X_i\beta - \psi_i}{\sigma}\right). \qquad (4.47)$$

If we do not want to require a particular structure or distribution on this random effect we can make our model semiparametric by assuming that ψ is an observation from a *Dirichlet Process*,

$$\begin{aligned} \psi_i &\sim G \\ G &\sim \mathcal{D}_{mG_{\mu,\tau^2}} \end{aligned} \qquad (4.48)$$

where G_{μ,τ^2} is a *base measure* and m is a *concentration parameter*. Thus, ψ is modeled to come from a distribution that sits in a neighborhood of G_{μ,τ^2}, with the size of the neighborhood being controlled by m. For now, we take m to be fixed, but later we will let it vary. In particular, with the mixture setup we take the prior on m to be in a discrete set and G to have root density $\mathcal{N}(\mu, \tau^2)$.

It is necessary to stipulate a full set of priors for the other terms, and the following make intuitive sense.

$$\begin{aligned} \beta &\sim \mathcal{N}(\beta_0, \sigma_\beta^2) \\ \mu &\sim \mathcal{N}(0, d\tau^2) \\ \frac{1}{\tau^2} &\sim \text{Gamma}(a, b) \end{aligned} \qquad (4.49)$$

It is common to set $\sigma_\beta^2 = \infty$, resulting in a flat prior on β. The parameters in the priors on μ and τ^2 can be chosen to make the priors sufficiently diffuse to allow the random effect to have an effect. The choice of prior mean zero for ψ does not lose generality, as the $X_i\beta$ term in (4.48) locates the distribution.

Estimation of this model requires MCMC tools, such as the Gibbs sampler. So first write:

$$c_i \sim \text{Discrete}(q_1, \ldots, q_K)$$

$$\psi_{c_i} \sim g(\psi) = \mathcal{N}(\mu, \tau^2) \qquad (4.50)$$

$$\mathbf{q} \sim \text{Dirichlet}(m/K, \ldots, m/K),$$

where the c_i serve only to group the ψ_i, resulting in a common value of $\psi_i = \psi_j$ if $c_i = c_j$. In the Gibbs sampler, the c_i are generated conditionally:

$$\mathbf{c} = (c_1, c_2, \ldots, c_n)$$

$$\mathbf{c}_{-i} = (c_1, c_2, \ldots, c_{i-1}, c_{i+1}, \ldots, c_n)$$

$$n_{-i,\ell} = \#(c_i = \ell)$$
$$f(y_i|\psi_i) = p_j \text{ see (4.45)},$$

then, for $i = 1, \ldots, n$

$$p(c_i = \ell|\mathbf{c}_{-i}) \propto \begin{cases} \frac{n_{-i,\ell}}{n-1+m} f(y_i|\psi_i) & \text{if } n_{-i,\ell} > 0 \\ \frac{m}{n-1+m} H_i & \text{if } n_{-i,\ell} = 0 \end{cases},$$

where

$$H_i = \int f(y_i|\psi)g(\psi)d\psi.$$

This then sets up the full conditional distributions for Gibbs sampling (see page 25).

4.8   Bayesian Shrinkage

The prior distribution works to move the posterior away from the likelihood and toward its own position. In cases where sharply defined priors are specified in the form of distributions with small variance, Bayesian estimates will have lower variance than corresponding classical likelihood-based estimates. Furthermore, the greater the correlation between coefficients in a given model, the greater the extent of the "shrinkage" toward the prior mean. As we shall see in Chapter 12, it is often the case that models with hierarchical specifications (multiple levels of priors) display more shrinkage due to correlations between parameters.

As a specific example of shrinkage from Hartigan (1983, p.88), consider the normal model in Chapter 3 with prior mean m and variance s^2 for μ and σ_0 known. Rewrite the posterior mean according to:

$$\hat{\mu} = \left(\frac{m}{s^2} + \frac{n\bar{x}}{\sigma_0^2} \right) \bigg/ \left(\frac{1}{s^2} + \frac{n}{\sigma_0^2} \right),$$

$$= \frac{\sigma_0^2}{\sigma_0^2 + ns^2} m + \frac{ns^2}{\sigma_0^2 + ns^2} \bar{x}$$

$$= (S_f)m + (1 - S_f)\bar{x}, \tag{4.51}$$

where $S_f = \sigma_0^2/(\sigma_0^2 + ns^2)$ is the shrinkage factor that is necessarily bounded by $[0{:}1]$. This shows that the shrinkage factor gives the *proportional* distance that the posterior mean has shrunk back to the prior mean away from the classical maximum likelihood estimate $\bar{x}$. The form of the shrinkage factor in this case highlights the fact that large data variance means that the denominator will dominate and there will be little shrinkage.

The posterior variance in the normal-normal model can also be rewritten in similar fashion:

$$\hat{\sigma}^2 = \left(\frac{1}{s^2} + \frac{n}{\sigma_0^2} \right)^{-1}$$

$$= \frac{\sigma_0^2 s^2}{\sigma_0^2 + ns^2} = \left[\frac{\sigma_0^2}{\sigma_0^2 + ns^2} \sigma_0^2 \right]^{\frac{1}{2}} \left[\frac{ns^2}{\sigma_0^2 + ns^2} \frac{s^2}{n} \right]^{\frac{1}{2}}$$

$$= (S_f \sigma_0^2)^{1/2} \left((1 - S_f) \frac{s^2}{n} \right)^{1/2}. \tag{4.52}$$

This is interesting because it shows that the posterior variance is a product of the square root of the prior variance weighted by the shrinkage factor and the square root of the data variance weighted by the complement of the shrinkage factor. So we see specifically how the shrinkage factor determines the compromise between prior uncertainty and data uncertainty.

It is noteworthy to look at both (4.51) and (4.52) when the prior distribution is very diffuse. In the extreme, if we pick an improper prior by setting $s^2 = \infty$ (or perhaps just some huge value), then it is clear that $S_f \to 0$, and the likelihood model dominates. Conversely, if s^2 is close to zero, reflecting strong prior knowledge about the distribution of μ, then S_f is close to one and the prior dominates.

Of course shrinkage is not just a feature of normal models or even models based on location-scale distributions. Returning to the Beta-Binomial conjugate setup given in Example 2.3.4:

$$Y \sim \mathcal{BN}(n, p) \qquad p \sim \mathcal{BE}(A, B) \tag{4.53}$$

where A and B are fixed prior parameters. The posterior distribution for p was shown to be:

$$p|y \sim \mathcal{BE}(y + A, n - y + B), \tag{4.54}$$

with:

$$\hat{p} = \frac{(y + A)}{(y + A) + (n - y + B)} = \left[\frac{n}{A + B + n}\right]\left(\frac{y}{n}\right) + \left[\frac{A + B}{A + B + n}\right]\left(\frac{A}{A + B}\right). \tag{4.55}$$

Here $\frac{A+B}{A+B+n}$ is the shrinkage estimate where the degree of shrinkage is determined by the magnitude of $A + B$ relative to n. So for large n the shrinkage gets small since n is in the denominator of the weight on the prior mean. While the shrinkage is overt in the example here, we can also consider the effect for the more complicated priors forms described in this chapter.

4.9   Exercises

4.1   Show that both the exponential PDF and chi-square PDF are special cases of the gamma PDF.

4.2   Suppose you have an iid sample of size n from the Rayleigh distribution:

$$f(x|\sigma) = \frac{x}{\sigma^2} \exp\left[-\frac{1}{2\sigma^2}x^2\right], \quad x \geq 0, \ \sigma > 0.$$

Produce the likelihood function for this distribution and assign an appropriate prior distribution for σ^2.

4.3   (Pearson 1920). An event has occurred p times in n trials. What is the probability that it occurs r times in the next m trials? How would your answer be interpreted differently in Bayesian terms than in standard frequentist terms?

4.4   Demonstrate that an improper prior distribution is a measure but not a probability measure.

4.5   Suppose that you had a prior parameter with the restriction: $[0 < \eta < 1]$. If you believed that η had prior mean 0.4 and variance 0.1, and wanted to specify a beta distribution, what prior parameters would you assign?

4.6   Suppose $X_1, \ldots, X_n$ are iid $\mathcal{G}(\alpha, \beta)$, $Y_1, \ldots, Y_n$ are iid $\mathcal{G}(\alpha, \gamma)$ and independent of the X_i. Produce the distribution of $\bar{X}/\bar{Y}$.

4.7   Derive the Jeffrey's prior for a normal likelihood model under three circumstances: *(1)* $\sigma^2 = 1$, *(2)* $\mu = 1$, and *(3)* both μ and σ^2 unknown (nonindependent).

4.8   For $X \sim \mathcal{N}(\mu, 1)$, show that the prior distribution $\mu \sim \mathcal{N}(m, s^2)$ gives the posterior distribution:

$$\mu | X \sim \mathcal{N} \left(\frac{1}{1+s^2} m + \frac{s^2}{1+s^2} X, \frac{s^2}{1+s^2} \right).$$

4.9   (Robert 2001) Calculate the marginal posterior distributions for the following setups:

   ▷ $x | \sigma^2 \sim \mathcal{N}(0, \sigma^2)$, $1/\sigma^2 \sim \mathcal{G}(1, 2)$.
   ▷ $x | \lambda \sim \mathcal{P}(\lambda)$, $\lambda \sim \mathcal{G}(2, 1)$.
   ▷ $x | p \sim \mathcal{NB}(10, p)$, $p \sim \mathcal{BE}(0.5, 0.5)$.

4.10 The Haldane prior for parameter $\zeta \in (0:1)$ is given by $p(\zeta) \propto \zeta^{-1}(1 - \zeta)^{-1}$. Suppose we had data $X_1, \ldots, X_n$ Bernoulli distributed according to $\mathcal{BR}(x | \zeta) = \zeta^x (1 - \zeta)^{1-x}$. Show that the resulting posterior distribution is $\pi(\zeta | \mathbf{X}) \propto \zeta^{\sum x_i - 1}(1 - \zeta)^{n - \sum x_i - 1}$. What is the resulting posterior mean? Can the Haldane prior be expressed as a special case of a beta distribution? What happens if there are no zeros or no ones in the $\mathbf{X}$ data?

4.11 Calculate the Jeffreys prior for the distributional forms in Exercise 4.9.

4.12 The triangular distribution for θ is given with limits (a, b) and mode m by the expression:

$$p(\theta | a, b, m) \begin{cases} 0 & \theta < a \\ \frac{2(\theta - a)}{(b-a)(m-a)} & a \leq \theta \leq m \\ \frac{2(b-\theta)}{(b-a)(b-m)} & m \leq \theta \leq b \\ 0 & \theta > b. \end{cases}$$

Find the mean and variance of this distribution. Produce a posterior distribution for θ assuming data that are distributed $\mathcal{N}(\theta, 1)$.

4.13 The Bayesian framework is easily adapted to problems in time-series. One of the most simple time-series specifications is the AR(1), which assumes that the previous period's outcome is important in the current period estimation.

   Given an observed outcome variable vector, $\mathbf{Y}_t$ measured at time t, AR(1) model for T periods is:

$$\mathbf{Y_t} = \mathbf{X}_t \boldsymbol{\beta} + \epsilon_t$$
$$\epsilon_t = \rho \epsilon_{t-1} + u_t, \qquad |\rho| < 1$$
$$u_t \sim \text{iid } \mathcal{N}(0, \sigma_u^2).$$

   Here $\mathbf{X}_t$ is a matrix of explanatory variables at time t, ϵ_t and ϵ_{t-1} are residuals from period t and $t-1$, respectively, u_t is an additional zero-mean error term for the autoregressive relationship, and β, ρ, σ_u^2 are unknown parameters to be estimated by the model. Backward substitution through time to arbitrary period s gives $\epsilon_t = \rho^s \epsilon_{t-s} + \sum_{j=0}^{T-1} \rho^j u_{t-j}$, and since $E[\epsilon_t] = 0$, then $\mathrm{var}[\epsilon_t] = E[\epsilon_t^2] = \frac{\sigma_u^2}{1-\rho^2}$, and the covariance between any two errors is: $\mathrm{Cov}[\epsilon_t, \epsilon_{t-j}] = \frac{\rho^j \sigma_u^2}{1-\rho^2}$. Assuming asymptotic normality, this setup leads to a general linear model with the following tridiagonal $T \times T$ weight matrix (Amemiya 1985, p.164):

$$
\Omega = \frac{1}{\sigma_u^2}
\begin{bmatrix}
1 & -\rho & & & \\
-\rho & (1+\rho^2) & -\rho & & 0 \\
& & \ddots & & \\
0 & & -\rho & (1+\rho^2) & -\rho \\
& & & -\rho & 1
\end{bmatrix}.
$$

Using the following data on worldwide fatalities from terrorism compared to some other causes per 100,000 (source: Falkenrath 2001), develop posteriors for β, ρ, and σ_u for each of the causes (Y) as a separate model stipulating the date (minus 1983) as the explanatory variable (X). Specify conjugate priors (Berger and Yang [1994] show some difficulties with nonconjugate priors here), or use a truncated normal for the prior on ρ (see Chib and Greenberg [1998]). Can you reach some conclusion about the differences in these four processes?

Year	Terrorism	Car Accidents	Suicide	Murder
1983	0.116	14.900	12.100	8.300
1984	0.005	15.390	12.420	7.900
1985	0.016	15.150	12.380	8.000
1986	0.005	15.920	12.870	8.600
1987	0.003	19.106	12.710	8.300
1988	0.078	19.218	12.440	8.400
1989	0.006	18.429	12.250	8.700
1990	0.004	18.800	11.500	9.400
1991	0.003	17.300	11.400	9.800
1992	0.001	16.100	11.100	9.300
1993	0.002	16.300	11.300	9.500
1994	0.002	16.300	11.200	9.000
1995	0.005	16.500	11.200	8.200
1996	0.009	16.500	10.800	7.400

4.14 The number of cases needed to give a $1 - \alpha$ confidence interval of width ω for a proportion p is $n = (z_\alpha)^2 p(1-p)/\omega^2$. Plot n versus ω at $\alpha = 0.05$ for $p = (0.01, 0.3, 0.5, 0.8)$.

4.15 Laplace (1774) wonders what the best choice is for a posterior point estimate. He sets three conditions for the shape of the posterior: symmetry, asymptotes, and properness (integrating to one). In addition, Laplace tacitly uses uniform priors. He proposes two possible criteria for selecting the estimate:

> ▷ La primiére est l'instant tel qu'en également probable que le véritable instant du phénomène tombe avant ou après; on pourrait appeler cet instant milieu de probabilité.

Meaning: use the median.

> ▷ Le seconde est l'instant tel qu'en le prenant pour milieu, la somme des erreurs à craindre, multipliées par leur probabilité, soit un minimum; on pourrait l'appeler milieu d'erreur ou milieu astronomique, comm étant celui auquel les astronomes doivent s'arreter de préférence.

Meaning: use the quantity at the "astronomical center of mass" that minimizes: $f(x) = \int |x - V| f(x) dx$. In modern terms, this is equivalent to minimizing the posterior expected loss: $E[L(\theta, d)|x] = \int_\Theta L(\theta, d)\pi(\theta|x)d\theta$, which is the average loss defined by the posterior distribution and d is the "decision" to use the posterior estimate of θ (see Berger [1985]).

Prove that these two criteria lead to the same point estimate.

4.16   In their popular text Gelman and Hill (2007) often specify variance components priors according to the following two steps:

$$\sigma \sim \mathcal{U}(0, 100)$$
$$\tau = \sigma^{-2},$$

where σ is a standard error and τ is a precision. Explain why this might not be a good idea for probit regression models.

4.17   Review one body of literature in your area of interest and develop a detailed plan for creating elicited priors.

4.18   Kadane and Wolfson (1996) specify a normal linear model for elicitation according to:

$$\mathbf{Y}|\mathbf{X}, \beta, \sigma^2 \sim \mathcal{N}(\mathbf{X}\beta, \sigma^2\mathbf{I})$$
$$\beta \sim \mathcal{N}(b, \sigma^2\mathbf{R}^{-1})$$
$$\frac{1}{\sigma^2} \sim \frac{1}{w\delta}\mathcal{G}(\delta/2, 2).$$

Develop a program to elicit scalars b, w, δ, and matrix $\mathbf{R}$ using a series of quantile questions.

4.19   Test a Bayesian count model for the number of times that capital punishment is implemented on a state level in the United States for the year 1997. Included in the data below (source: United States Census Bureau, United States Department of Justice) are explanatory variables for: median per capita income in dollars, the percent of the population classified as living in poverty, the percent of Black citizens in the population, the rate of violent crimes per 100,000 residents for the year before (i.e., 1996), a dummy variable to indicate whether the state is in the South, and the proportion of the population with a college degree of some kind. The data are given below and available in the BaM dataset executions.

In 1997, executions were carried out in 17 states with a national total of 74. The model should be developed from the Poisson link function, $\theta = \log(\mu)$, with the objective of finding the best β vector in:

$$\underbrace{g^{-1}(\theta)}_{17\times 1} = \exp\left[1\beta_0 + \text{INC}\beta_1 + \text{POV}\beta_2 + \text{BLK}\beta_3 + \text{CRI}\beta_4 + \text{SOU}\beta_5 + \text{DEG}\beta_6\right].$$

Specify a suitable prior with assigned prior parameters, then summarize the resulting posterior distribution.

State	Exe-cutions	Median Income	Percent Poverty	Percent Black	Violent Crime	South	Prop. Degrees
Texas	37	34453	16.7	12.2	644	1	0.16
Virginia	9	41534	12.5	20.0	351	1	0.27
Missouri	6	35802	10.6	11.2	591	0	0.21
Arkansas	4	26954	18.4	16.1	524	1	0.16
Alabama	3	31468	14.8	25.9	565	1	0.19
Arizona	2	32552	18.8	3.5	632	0	0.25
Illinois	2	40873	11.6	15.3	886	0	0.25
S. Carolina	2	34861	13.1	30.1	997	1	0.21
Colorado	1	42562	9.4	4.3	405	0	0.31
Florida	1	31900	14.3	15.4	1051	1	0.24
Indiana	1	37421	8.2	8.2	537	0	0.19
Kentucky	1	33305	16.4	7.2	321	0	0.16
Louisiana	1	32108	18.4	32.1	929	1	0.18
Maryland	1	45844	9.3	27.4	931	0	0.29
Nebraska	1	34743	10.0	4.0	435	0	0.24
Oklahoma	1	29709	15.2	7.7	597	0	0.21
Oregon	1	36777	11.7	1.8	463	0	0.25

4.20   Zellner's g-prior can be expressed most simply in a linear regression context as:

$$\beta|\tau \sim \mathcal{N}\left(\mathbb{B}, \frac{g}{\tau}(\mathbf{X}'\mathbf{X})-1\right)$$

$$p(\tau) \propto \frac{1}{\phi},$$

where $\mathbb{B}$ is the prior mean, and τ is the usual homoscedastic precision parameter, meaning that the covariance matrix of the β is given by $\mathbf{I}/\tau$. Common choices for g include $g = n$ (the data size), $g = k^2$ (the square of the number of explanatory variables in the model), and $g = \max(n, k^2)$. Rerun the model in Exercise 5.7 with this prior using your preferred value of g. How do the results differ?

Chapter 5

The Bayesian Linear Model

5.1 The Basic Regression Model

This chapter develops the Bayesian linear regression model with differing priors and assumptions. We will consider both informed and uninformed prior specifications as well as look at the common problem of heteroscedasticity. Detailed technical expositions of the Bayesian linear regression model are found in the classic article by Lindley and Smith (1972) with discussion, the follow-up article with generalizations by Smith (1973), Geweke's (1993) exposition on t-distributed errors, and the early work by Tiao and Zellner (1964b). Elsewhere Zellner and Tiao (1964) provide estimation techniques for general error models, Davis (1978) considers the Bayesian general linear model with inequality constraints, and Pollard (1986) gives a helpful chapter on Bayesian linear forms. A modern and very comprehensive volume on linear model theory is given by Ravishanker and Dey (2002).

The first treatment presented here assumes homoscedasticity. A discussion of unequal variances for the Bayesian linear model follows, and additional discussions can be found in Leonard (1975) and Boscardin and Gelman (1996). Le Cam (1986, Chapter 13) gives a detailed asymptotic analysis, and more elementary treatments of heteroscedastic linear models can be found in Goldberger (1964, p.235), Huang (1970, p.147), and Rao and Toutenburg (1995, p.101). A detailed theoretical discussion of linear model heteroscedasticity is given in Amemiya (1985, Section 6.5), and the corresponding application of the jackknife and bootstrap are outlined in Shao and Tu (1995, Chapter 7). See also Fomby, Hill, and Johnson (1980).

Start with the well-known basic multiple linear regression model, described in Appendix A, conforming to the Gauss-Markov assumptions. Define the terms conventionally:

$$\mathbf{y} = \mathbf{X}\boldsymbol{\beta} + \boldsymbol{\epsilon}, \tag{5.1}$$

where $\mathbf{X}$ is an $n \times k$, rank k matrix of explanatory variables with a leading vector of ones for the constant, $\boldsymbol{\beta}$ is a $k \times 1$ vector of coefficients to be estimated, $\mathbf{y}$ is an $n \times 1$ vector of outcome variable values, and $\boldsymbol{\epsilon}$ is a $n \times 1$ vector of errors distributed $\mathcal{N}(0, \sigma^2 I)$ for a constant σ^2. The likelihood function for a sample of size n is:

$$L(\boldsymbol{\beta}, \sigma^2 | \mathbf{X}, \mathbf{y}) = (2\pi\sigma^2)^{-\frac{n}{2}} \exp\left[-\frac{1}{2\sigma^2}(\mathbf{y} - \mathbf{X}\boldsymbol{\beta})'(\mathbf{y} - \mathbf{X}\boldsymbol{\beta})\right]. \tag{5.2}$$

So far we have the standard non-Bayesian approach to linear modeling, and once the data, $[\mathbf{X}, \mathbf{y}]$, are observed, this likelihood function (or rather its log) is maximized relative to the unknown parameter vector $\boldsymbol{\beta}$ and the unknown scalar σ. We know the values for which (5.2) is at its maximum from standard likelihood theory (bias corrected for σ^2) and ordinary least squares principles:

$$\hat{\mathbf{b}} = (\mathbf{X}'\mathbf{X})^{-1}\mathbf{X}'\mathbf{y}, \qquad \hat{\sigma}^2 = \frac{(\mathbf{y} - \mathbf{X}\hat{\mathbf{b}})'(\mathbf{y} - \mathbf{X}\hat{\mathbf{b}})}{(n-k)}, \tag{5.3}$$

and we can therefore plug these values into (5.2) and process according to:

$$
\begin{aligned}
L(\boldsymbol{\beta}, \sigma^2 | \mathbf{X}, \mathbf{y}) &\propto \sigma^{-n} \exp\left[-\frac{1}{2\sigma^2}(\mathbf{y}'\mathbf{y} - 2\boldsymbol{\beta}'\mathbf{X}'\mathbf{y} + \boldsymbol{\beta}'\mathbf{X}'\mathbf{X}\boldsymbol{\beta}) \right] \\
&= \sigma^{-n} \exp\Big[-\frac{1}{2\sigma^2}(\mathbf{y}'\mathbf{y} - 2\boldsymbol{\beta}'\mathbf{X}'\mathbf{y} + \boldsymbol{\beta}'\mathbf{X}'\mathbf{X}\boldsymbol{\beta} \\
&\qquad \underbrace{-2((\mathbf{X}'\mathbf{X})^{-1}\mathbf{X}'\mathbf{y})'\mathbf{X}'\mathbf{y} + 2((\mathbf{X}'\mathbf{X})^{-1}\mathbf{X}'\mathbf{y})'\mathbf{X}'\mathbf{X}((\mathbf{X}'\mathbf{X})^{-1}\mathbf{X}'\mathbf{y}))}_{\text{sums to zero}} \Big] \\
&= \sigma^{-n} \exp\Big[-\frac{1}{2\sigma^2}((\mathbf{y} - \mathbf{X}\hat{\mathbf{b}})'(\mathbf{y} - \mathbf{X}\hat{\mathbf{b}}) \\
&\qquad + \hat{\mathbf{b}}'\mathbf{X}'\mathbf{X}\hat{\mathbf{b}} + \boldsymbol{\beta}'\mathbf{X}'\mathbf{X}\boldsymbol{\beta} - 2\boldsymbol{\beta}'\mathbf{X}'\mathbf{X}\hat{\mathbf{b}}) \Big] \\
&= \sigma^{-n} \exp\left[-\frac{1}{2\sigma^2}(\hat{\sigma}^2(n-k) + (\boldsymbol{\beta} - \hat{\mathbf{b}})'\mathbf{X}'\mathbf{X}(\boldsymbol{\beta} - \hat{\mathbf{b}})) \right].
\end{aligned}
\tag{5.4}
$$

The "trick" used here is completing the square after inserting a quantity that is simultaneously subtracted and added (therefore adding zero), and using the property $\mathbf{X}'\mathbf{X}(\mathbf{X}'\mathbf{X})^{-1} = I$ to rearrange terms. The other mildly subtle point is the replacement of $-2\boldsymbol{\beta}'\mathbf{X}'\mathbf{y}$ with $-2\boldsymbol{\beta}'\mathbf{X}'\mathbf{X}\hat{\mathbf{b}}$ from the normal equation ($\mathbf{X}'\mathbf{X}\hat{\mathbf{b}} = \mathbf{X}'\mathbf{y}$).

Thus the result is a likelihood function expressed in terms of unknown coefficients to be estimated and observable data (Zellner 1976, p.401). This is the starting point for the Bayesian analysis of the standard linear model and we now begin to quantify existing knowledge about the mean vector and the variance parameter (a scalar due to the homoscedasticity assumption). The Bayesian complement to maximum likelihood estimation not only provides for prior information, but also easily incorporates: linear additive specifications of nonlinear functions (Bernardo and Smith 1994, pp.221-222; Leonard and Hsu 1999, Section 5.2), linear inequality constraints (Davis 1978), analysis of variance of means (Hartigan 1983, Chapter 9), polynomial regression (Halpern 1973; Lempers 1971), small area estimation (Datta and Ghosh 1991), and time-series models (Ferreira and Gamerman 2000; Kitagawa and Gersch 1996; Marín 2000; Pole, West, and Harrison 1994), among other extensions.

5.1.1 Uninformative Priors for the Linear Model

Define the improper uninformed priors $p(\boldsymbol{\beta}) \propto c$ and $p(\sigma^2) = \frac{1}{\sigma}$ over the support $[-\infty:\infty]$ and $[0:\infty]$, respectively (Tiao and Zellner 1964a, p.220). Note that we are assuming independence between $\boldsymbol{\beta}$ and σ^2. Therefore the joint posterior from the likelihood function (5.4) is provided by:

$$\pi(\boldsymbol{\beta}, \sigma^2 | \mathbf{X}, \mathbf{y}) \propto L(\boldsymbol{\beta}, \sigma^2 | \mathbf{X}, \mathbf{y}) p(\boldsymbol{\beta}) p(\sigma^2)$$

$$\propto \sigma^{-n-1} \exp\left[-\frac{1}{2\sigma^2}(\hat{\sigma}^2(n-k) + (\boldsymbol{\beta} - \hat{\mathbf{b}})'\mathbf{X}'\mathbf{X}(\boldsymbol{\beta} - \hat{\mathbf{b}}))\right]. \tag{5.5}$$

Note that the constant c drops out with proportionality.

To obtain the desired marginal distribution of $\boldsymbol{\beta}$, first make the transformation: $s = \sigma^{-2}$, with a required Jacobian: $J = |\frac{d}{ds}\sigma| = |\frac{d}{ds}s^{-\frac{1}{2}}| = \frac{1}{2}s^{-\frac{3}{2}}$. Reexpressing (5.5) in terms of s gives:

$$\pi(\boldsymbol{\beta}, s | \mathbf{X}, \mathbf{y})$$
$$\propto (s^{-\frac{1}{2}})^{-n-1} \exp\left[-\frac{1}{2}s(\hat{\sigma}^2(n-k) + (\boldsymbol{\beta} - \hat{\mathbf{b}})'\mathbf{X}'\mathbf{X}(\boldsymbol{\beta} - \hat{\mathbf{b}}))\right] \left(\frac{1}{2}s^{-\frac{3}{2}}\right)$$
$$\propto s^{\frac{n}{2}-1} \exp\left[-\frac{1}{2}s(\hat{\sigma}^2(n-k) + (\boldsymbol{\beta} - \hat{\mathbf{b}})'\mathbf{X}'\mathbf{X}(\boldsymbol{\beta} - \hat{\mathbf{b}}))\right]. \tag{5.6}$$

Now integrate with respect to s to get the marginal for $\boldsymbol{\beta}$:

$$\pi(\boldsymbol{\beta} | \mathbf{X}, \mathbf{y}) = \int_0^\infty s^{\frac{n}{2}-1} \exp\left[-\frac{1}{2}s(\hat{\sigma}^2(n-k) + (\boldsymbol{\beta} - \hat{\mathbf{b}})'\mathbf{X}'\mathbf{X}(\boldsymbol{\beta} - \hat{\mathbf{b}}))\right] ds.$$

This integral is quite easy to calculate when one notices that inside the integral is a gamma PDF kernel, with the integration performed over the appropriate support. So use the following substitution from a gamma PDF:

$$1 = \int_0^\infty \frac{q^{p+1}}{\Gamma(p+1)} s^p e^{-qs} ds, \qquad \frac{\Gamma(p+1)}{q^{p+1}} = \int_0^\infty s^p e^{-qs} ds.$$

Setting $p = \frac{n}{2} - 1$ and $q = \frac{1}{2}(\hat{\sigma}^2(n-k) + (\boldsymbol{\beta} - \hat{\mathbf{b}})'\mathbf{X}'\mathbf{X}(\boldsymbol{\beta} - \hat{\mathbf{b}}))$, and defining the degrees of freedom as $\nu = n - k$ means that:

$$\pi(\boldsymbol{\beta} | p, q) \propto q^{-(p+1)} = q^{-\frac{n}{2}}, \tag{5.7}$$

which can then be resubstituted back to get:

$$\pi(\boldsymbol{\beta} | \mathbf{X}, \mathbf{y}) \propto \left[(n-k) + (\boldsymbol{\beta} - \hat{\mathbf{b}})'\hat{\sigma}^{-2}\mathbf{X}'\mathbf{X}(\boldsymbol{\beta} - \hat{\mathbf{b}})\right]^{-\frac{n}{2}}. \tag{5.8}$$

It is easy to recognize $\pi(\boldsymbol{\beta} | \mathbf{X}, \mathbf{y})$ as the kernel of a multivariate-t distribution for $\boldsymbol{\beta} - \hat{\mathbf{b}}$, provided that the covariance matrix, $\mathbf{R} = \frac{(n-k)\hat{\sigma}^2(\mathbf{X}'\mathbf{X})^{-1}}{n-k-2}$, is positive definite (Box and Tiao 1973, p.440; Press 1989, p.135; Tong 1990, Chapter 9). Thus $E[\hat{\mathbf{b}}] = \boldsymbol{\beta}$, and the covariance between any two coefficients is given by the elements of the $\mathbf{R}$ matrix: $\text{Cov}[t_i, t_j] = R_{ij}$, where $\hat{\sigma}^2$ is defined by (5.3).

Obtaining the marginal distribution for σ^2 is considerably less involved due to an obvious shortcut. Start with the defining integral and separate terms in the exponent:

$$\pi(\sigma|\mathbf{X},\mathbf{y}) \propto \int_{-\infty}^{\infty} \sigma^{-n-1} \exp\left[-\frac{1}{2\sigma^2}(\hat{\sigma}^2(n-k) + (\boldsymbol{\beta} - \hat{\mathbf{b}})'\mathbf{X}'\mathbf{X}(\boldsymbol{\beta} - \hat{\mathbf{b}}))\right] d\boldsymbol{\beta}.$$

$$= \sigma^{-n-1} \exp\left[-\frac{1}{2\sigma^2}\hat{\sigma}^2(n-k)\right]$$

$$\times \int_{-\infty}^{\infty} \exp\left[-\frac{1}{2\sigma^2}(\boldsymbol{\beta} - \hat{\mathbf{b}})'\mathbf{X}'\mathbf{X}(\boldsymbol{\beta} - \hat{\mathbf{b}})\right] d\boldsymbol{\beta}.$$

The second exponential term is a k-dimensional kernel of a multivariate normal distribution providing the following substitution and simplification:

$$\pi(\sigma|\mathbf{X},\mathbf{y}) \propto \sigma^{-n-1} \exp\left[-\frac{1}{2\sigma^2}\hat{\sigma}^2(n-k)\right] (2\pi\sigma^2)^{\frac{k}{2}}$$

$$\propto (\sigma^2)^{-\frac{1}{2}(n-k)-\frac{1}{2}} \exp\left[-\frac{1}{2\sigma^2}\hat{\sigma}^2(n-k)\right]. \tag{5.9}$$

It is not immediately obvious, but the posterior distribution of σ is the kernel of an inverse-gamma distribution (Appendix B), and can also be parameterized as an inverse Wishart distribution (Tiao and Zellner 1964b). To see how this is really an inverse gamma form, apply the simple transformations: $\alpha = \frac{1}{2}(n-k-1)$, $\beta = \frac{1}{2}\hat{\sigma}^2(n-k)$, so that 5.9 is now:

$$\pi(\sigma^2|\alpha,\beta) \propto (\sigma^2)^{-(\alpha+1)} \exp[-\beta/\sigma^2]. \tag{5.10}$$

So far (5.10) is exactly the same inferential result that we would expect from a standard likelihood analysis of the linear model The maximum likelihood solution, which is also equivalent to minimizing the summed squared errors, is equivalent to a Bayesian solution in which improper uniform priors over the entire support of the unknown parameters are specified. Thus, the omnipresent non-Bayesian approach to linear modeling is a special case of a Bayesian model.

■ **Example 5.1:   The 2000 U.S. Election in Palm Beach County.** The 2000 U.S. election for president was marked by considerable controversy concerning the casting of ballots in the state of Florida. Because the election was so tightly contested in Florida and the state's 25 electoral college delegates were the final determining factor for electing the president in the contest between Al Gore and George W. Bush, various problems and aberrations in the voting process became magnified in importance. There was considerable evidence that the final certified outcome declaring Bush the winner by 537 votes (out of approximately 6 million) was shaped by technical problems with the voting apparatus, ballot confusion by voters, and outright discrimination against minority voters.

At the nexus of this controversy is Palm Beach County, a liberal-leaning, upper-middle class area with a considerable number of northeastern retirees where the far-right conservative candidate Pat Buchanan did suspiciously well. The data here (available in the `BaM` package associated with this text) consist of all 516 reporting precincts in Palm Beach County collected by the *Palm Beach Post* from state and federal reporting sources. The variables include party affiliation percentages, racial demographics, registration status, voting technology, and the outcome variable of interest: the number of spoiled ballots. Ballots are spoiled if the voter designates more than one presidential selection or marks the ballot in some other inappropriate way. Therefore this variable does not count so-called under-votes wherein the voter does not select any presidential candidate. For the purpose of this example, the variables described in Table 5.1 are included in the linear specification.

TABLE 5.1: PBC DATA VARIABLE DEFINITIONS

Bad Ballots	Total Number of Spoiled Ballots
Technology	0 for a datapunch machine (butterfly ballot), 1 for votomatic
New	Number of "new" voters (have not voted in the precinct for previous 6 years)
Size	Total number of precinct voters
Republican	Number of voters registered as Republican
White	Number of white (nonminority) voters

First we specify an uninformed joint prior, $p(\boldsymbol{\beta}, \sigma^2) = 1/\sigma$, and calculate the resulting posterior density according to Section 5.1.1. This joint prior comes from the simultaneous determination of the improper forms: $p(\boldsymbol{\beta}, \sigma^2) = p(\boldsymbol{\beta}|\sigma^2)p(\sigma^2)$. The resulting posterior is summarized in Table 5.2 where the posterior summaries, posterior moments, and quantiles for the coefficient estimates were produced directly from the analytical t-distribution result. Note that the mean and median are identical for the coefficient posteriors indicating symmetry.

Section 5.1.1 also demonstrated that the posterior distribution of σ^2 is an inverse-gamma specification as given by (5.9), and we can analytically determine the quantiles of interest with $\alpha = (n - k - 1)/2$, $\beta = \frac{1}{2}\hat{\sigma}^2(n - k)$ inserted into the inverse-gamma PDF (see Appendix B). However, it turns out to be much easier, and nearly equally accurate, to generate a large number of values of σ^2 and use the empirical quantiles. The R code for all estimates in Table 5.2 is included in the **Computational Addendum** for this chapter.

Some comments are necessary regarding the presentation of Table 5.2. A strictly canonical Bayesian may object to the traditional non-Bayesian reporting of the mean and standard error in the left-hand side and want more than three quantile summaries on the right-hand side. In a conventional linear regression table we would expect to see t-statistics, p-values, and "stars" instead of these quantiles. Most Bayesians publishing in the social sciences see the usefulness of giving means and standard errors to comfort reviewers and readers who may not have had substantial exposure to Bayesian results.

But, these same researchers would certainly object to t-statistics since this implies a test that is not undertaken, to p-values since there is no null distribution with which to define them, and to "stars" since this is an idiotic practice. Thus Table 5.2 represents a compromise between accessibility and the Bayesian preference for purely distributional summaries.

TABLE 5.2: POSTERIOR: LINEAR MODEL, UNINFORMATIVE PRIORS

β Covariate	Mean	Std. Error	0.025 Q.	Median	0.975 Q.
Intercept	107.351	7.759	92.108	107.351	122.593
Technology	-50.529	3.492	-57.389	-50.529	-43.670
New	-0.353	0.038	-0.427	-0.353	-0.278
Size	0.149	0.006	0.137	0.149	0.161
Republican	-0.084	0.007	-0.098	-0.084	-0.069
White	-0.048	0.006	-0.059	-0.048	-0.037
σ	22.003	0.698	20.695	21.983	23.434

What we see from the analysis in Table 5.2 is that the linear model with an uninformed prior for β and σ^2 provides an apparently very good fit to the data. Each of the marginal posterior distributions is narrow and easily statistically distinguishable from zero. Since this is still a linear model, the interpretation of these coefficients is very direct:

▷ Precincts with votomatic technology have an expected 50 less spoiled ballots.

▷ For every 2.5 new voters, there is an expected 1 less spoiled ballot.

▷ For each additional voter that turns out, there is an additional 0.15 expected increase in spoiled ballots: in other words, a turnout addition of 100 provides an expected increase in spoiled ballots of 15. This includes the new voters above; so in fact, this effect is really greater than this coefficient indicates since the new voters are actually suppressing this effect somewhat.

▷ For each additional Republican voter there is about a 9% decrease in expected spoiled ballots. That is, we expect about 1 less spoiled ballot for every 11 increased Republican voters.

▷ For each additional white voter, there is a 5% decrease in expected spoiled ballots. That means about 1 less spoiled ballot for every 20 increased white voters.

It is also important to remember that while the reported mean and standard error can be identical between those produced with Bayesian methods and ordinary least squares (equivalently maximum likelihood here), they come from fundamentally different processes and assumptions (the OECD example on page 381 gives identical coefficient estimates even though one was produced with MCMC tools and the other with minimizing squared residuals). A posterior mean is a point estimate that comes from an underlying distributional assumption about posterior quantities and the OLS point estimate comes from an assumption of a fixed underlying quantity set by nature. Furthermore, a posterior standard error also arises from the belief that unknowns should be described with distributions, whereas a maximum likelihood estimate-based standard error comes from the curvature around the mode of the likelihood function under the assumption that the mode is the optimal estimator. Obviously these values basically agree in the presence of uninfluential prior distributions, or large samples, but they do not have to.

5.1.2 Conjugate Priors for the Linear Model

We can also stipulate conjugate priors according to the principles outlined in Chapter 3. The linear regression model developed in this section is actually just a multivariate generalization of the normal model in that chapter, where the maximum likelihood estimates from the linear parametric likelihood specification are used. The well-known conjugate prior distributions are: multivariate normal for the mean vector β and inverse gamma for σ^2. We can therefore proceed in essentially the same manner as in the basic normal model, but incorporating the likelihood function from the linear regression setup.

In this case of conjugacy we also have a dependency of β on σ^2 as in the simple normal-normal model in Chapter 3:

$$p(\beta|\sigma^2) = (2\pi)^{-\frac{k}{2}}|\mathbf{\Sigma}|^{-\frac{1}{2}} \exp\left[-\frac{1}{2}(\beta - \mathbb{B})'\mathbf{\Sigma}^{-1}(\beta - \mathbb{B})\right],$$

and:

$$p(\sigma^2) \propto \sigma^{-(a-k)} \exp\left[-\frac{b}{\sigma^2}\right]. \tag{5.11}$$

This affects the joint posterior as a simple joint prior since:

$$p(\beta, \sigma^2) = p(\beta|\sigma^2)p(\sigma^2). \tag{5.12}$$

However, the dependency in (5.11) will flow through to the posterior of β. Here $\mathbf{\Sigma} = \sigma^2\mathbf{I}$ by assumption (this form is adaptable but not only makes the operations conformable, it allows us to easily transition to the general linear model later). Thus the prior for β conditional on σ^2 is a multivariate normal with mean $\mathbb{B}$ and variance σ^2. The prior for σ^2 is an inverse gamma kernel with parameters a and b (employing the form from page 73, see also Appendix B). Using such notation, assign $\alpha - 1 = a - k$, and $\beta = b$ since these are free prior parameters. The inverse gamma prior for σ^2 is not only the conjugate prior, but also the marginal posterior from the uninformed prior model. So to summarize the notation, we now have:

$$
\begin{array}{ll}
n \times k & \text{size of the } \mathbf{X} \text{ matrix} \\
\boldsymbol{\beta} & \text{the unknown linear model coefficient vector} \\
\mathbb{B} & \text{prior mean vector for } \boldsymbol{\beta} \\
\sigma^2 & \text{prior variance for } \boldsymbol{\beta}, \text{ collected in the diagonal matrix } \boldsymbol{\Sigma} \\
\hat{\mathbf{b}} & (\mathbf{X}'\mathbf{X})^{-1}\mathbf{X}'\mathbf{y} \\
\hat{\sigma}^2 & \dfrac{(\mathbf{y} - \mathbf{X}\hat{\mathbf{b}})'(\mathbf{y} - \mathbf{X}\hat{\mathbf{b}})}{(n-k)} \\
a - k & \text{assigned first parameter for } \sigma^2 \text{ prior} \\
b & \text{assigned second parameter for } \sigma^2 \text{ prior.}
\end{array}
$$

Combining the data likelihood from (5.4) with the prior specifications from (5.11) and applying Bayes' Law gives the joint posterior:

$$
\begin{aligned}
\pi(\boldsymbol{\beta}, & \sigma^2 | \mathbf{X}, \mathbf{y}) \\
&\propto \sigma^{-n} \exp\left[-\frac{1}{2\sigma^2}\left(\hat{\sigma}^2(n-k) + (\boldsymbol{\beta} - \hat{\mathbf{b}})'\mathbf{X}'\mathbf{X}(\boldsymbol{\beta} - \hat{\mathbf{b}}) \right) \right] \\
&\quad \times (2\pi)^{-\frac{k}{2}} |\boldsymbol{\Sigma}|^{-\frac{1}{2}} \exp\left[-\frac{1}{2}(\boldsymbol{\beta} - \mathbb{B})'\boldsymbol{\Sigma}^{-1}(\boldsymbol{\beta} - \mathbb{B}) \right] \sigma^{-(a-k)} \exp\left[-\frac{b}{\sigma^2} \right] \\
&\propto \sigma^{-n-a} \exp\left[-\frac{1}{2\sigma^2}\left(\hat{\sigma}^2(n-k) + (\boldsymbol{\beta} - \hat{\mathbf{b}})'\mathbf{X}'\mathbf{X}(\boldsymbol{\beta} - \hat{\mathbf{b}}) + 2b + (\boldsymbol{\beta} - \mathbb{B})'\boldsymbol{\Sigma}^{-1}(\boldsymbol{\beta} - \mathbb{B}) \right) \right],
\end{aligned}
$$

$$(5.13)$$

where σ^2 moves out of $|\boldsymbol{\Sigma}|$ from the determinant operation. This form can be simplified by coweighting the precisions and re-expressing as a Gaussian kernel. First define (Zellner 1971, Chapter 3):

$$
\tilde{\boldsymbol{\beta}} = (\boldsymbol{\Sigma}^{-1} + \mathbf{X}'\mathbf{X})^{-1}(\boldsymbol{\Sigma}^{-1}\mathbb{B} + \mathbf{X}'\mathbf{X}\hat{\mathbf{b}}) \tag{5.14}
$$

$$
\tilde{s} = 2b + \hat{\sigma}^2(n-k) + (\mathbb{B} - \tilde{\boldsymbol{\beta}})'\boldsymbol{\Sigma}^{-1}\mathbb{B} + (\hat{\mathbf{b}} - \tilde{\boldsymbol{\beta}})'\mathbf{X}'\mathbf{X}\hat{\mathbf{b}}. \tag{5.15}
$$

The initial form of the joint posterior can now be re-expressed (Exercise 5.9) as:

$$
\pi(\boldsymbol{\beta}, \sigma^2 | \mathbf{X}, \mathbf{y}) \propto (\sigma^2)^{-\frac{n+a}{2}} \exp\left[-\frac{1}{2\sigma^2}\left(\tilde{s} + (\boldsymbol{\beta} - \tilde{\boldsymbol{\beta}})'(\boldsymbol{\Sigma}^{-1} + \mathbf{X}'\mathbf{X})(\boldsymbol{\beta} - \tilde{\boldsymbol{\beta}}) \right) \right]. \tag{5.16}
$$

The advantage of this new form is that it immediately allows us to use the same marginalization trick as we did with the posterior from the uninformed priors to get the distribution of $\boldsymbol{\beta} | \mathbf{X}, \mathbf{y}$. This produces:

$$
\pi(\boldsymbol{\beta} | \mathbf{X}, \mathbf{y}) \propto \left[\tilde{s} + (\boldsymbol{\beta} - \tilde{\boldsymbol{\beta}})'(\boldsymbol{\Sigma}^{-1} + \mathbf{X}'\mathbf{X})(\boldsymbol{\beta} - \tilde{\boldsymbol{\beta}}) \right]^{-\frac{n+a}{2}+1}, \tag{5.17}
$$

which is the kernel of a multivariate-t distribution with $\nu = n + a - k - 2$ degrees of freedom (Bauwens, Lubrano, and Richard 1999, Section 2.7, Zellner 1971, Section 3.2, Box and Tiao

1973, Section 8.4). Therefore the mean and variance of the parameter estimates are given by:

$$E(\beta|\mathbf{X},\mathbf{y}) = \tilde{\beta}, \qquad \text{Cov}(\beta|\mathbf{X},\mathbf{y}) = \frac{\tilde{s}(\mathbf{\Sigma}^{-1} + \mathbf{X}'\mathbf{X})^{-1}}{n + a - k - 3}. \tag{5.18}$$

The marginal distribution of σ^2 is also produced in similar fashion to the uninformed prior derivation from before: $\pi(\sigma^2|\mathbf{X},\mathbf{y}) \propto \sigma^{-n-a+k-1} \exp\left[-\frac{1}{2\sigma^2}\hat{\sigma}^2(n+a-k)\right]$. This is the kernel of an $\mathcal{IG}(n+a-k, \frac{1}{2}\hat{\sigma}^2(n+a-k))$ distribution. For details, see Note B of Zellner (1971).

We can now compare the informed conjugate model with the uninformed model developed previously. There are some interesting similarities as well as differences in Table 5.3. Note first that the parametric form for the marginal posteriors comes from the same family but with important differences in the parameters. Specifically, conjugacy provides a in each of the possible places. Since the researcher controls $a \in [0:\infty)$, then there is temptation to say that this is manipulable to provide customary levels of statistical reliability. This is most easily seen in the posterior distribution of $\beta|\sigma^2$ where larger values of a shrink the tails of the Student's-t toward that of the normal producing smaller posterior coefficient variance. What prevents abuse of this term is the convention that authors overtly state their parameter values and the substantive reasons behind them.

TABLE 5.3: Linear Regression Model, Prior Comparison

Setup	Prior	Posterior
Conjugate	$\beta|\sigma^2 \sim \mathcal{N}(\mathbb{B}, \sigma^2)$	$\beta|\mathbf{X} \sim t(n+a-k-2)$
	$\sigma^2 \sim \mathcal{IG}(a,b)$	$\sigma^2|\mathbf{X} \sim \mathcal{IG}(n+a-k, \frac{1}{2}\hat{\sigma}^2(n+a-k))$
Uninformative	$\beta \propto c$ over $[-\infty:\infty]$	$\beta|\mathbf{X} \sim t(n-k)$
	$\sigma^2 = \frac{1}{\sigma}$ over $[0:\infty]$	$\sigma^2|\mathbf{X} \sim \mathcal{IG}(\frac{1}{2}(n-k-1), \frac{1}{2}\hat{\sigma}^2(n-k))$

The role of sample size is also highlighted in Table 5.3. For the t-distribution posteriors, increases in n will obviously push this distribution towards the normal. Since most models in the social sciences have a relatively modest number of covariates (compared to statistical genetics or some fields in engineering), k will have little effect for large sample sizes in this transition. The role of n in the inverse gamma forms is also clear from the form of $p(\sigma^2)$ in (5.11) since both are negative exponents. So asymptotically the conjugate and the uninformative linear regression models both converge to conventional large sample results with normally distributed coefficients and fixed variance.

5.1.3   Conjugate Caveats for the Cautious and Careful

Some caveats are warranted about the use of conjugate priors in the Bayesian linear regression model as described here. Conjugate priors with exponential family distributions typically provide *linear* posterior expectations (Berger 1984, Diaconis and Ylvisaker 1979, Ghosh 1969, Goel and DeGroot 1980) which are highly nonrobust to influential outliers (West 1984), although these outliers can be explicitly accounted for in a Bayesian context (Box and Tiao 1968). In the absence of credible normal assumptions about the prior, due to small samples or less well-behaved error structures, linear expectations and linear variance estimates (estimating the mean squared error) can be substituted (Hartigan 1969), but only by averaging over both the data and the parameters. Because the posterior distribution inherits its tail structure from the prior in conjugate specifications, robustness is often difficult to incorporate (Anscombe 1963; Dickey 1974; Hill 1974; Lindley 1968; Rubin 1977). Geweke (1993) shows that when the model residuals are shown to be noticeably non-normal, more complex models are required (mixtures, distributional assignment for the degrees of freedom parameter) and these sometimes require Bayesian stochastic simulation techniques (see later chapters).

The normal-normal conjugacy setup in Chapter 3 required that the prior and posterior for the mean be conditional on the variance. This was a mathematical requirement for the parameter form to flow through the likelihood function, but it seems to water down the spirit of conjugacy. For this reason it is sometimes called "pseudo-conjugacy" (or sometimes "semi-conjugacy"), which is a term that some authors dislike. This requirement on the mean also applied to the linear regression models in this chapter, and was no weaker in this case. Here, we have actually relaxed pure conjugacy even further since the normal prior for the mean produces a Student's-t posterior. However, unless the sample size, n, and the set parameter, a, are set relatively low compared to the number of covariates, k, the distribution is likely to approach normality anyway.

■ **Example 5.2:   The 2000 U.S. Election in Palm Beach County, Continued.**
Now we can impose a conjugate prior as explained in Section 5.1.2 on these data and re-perform this analysis. Suppose we now specify a "pessimistic" prior on the β coefficients with the following multivariate normal distribution and diffuse inverse-gamma distribution:

$$\beta \sim \mathcal{N}(\mathbb{B}, \Sigma)$$
$$\text{where:} \quad \mathbb{B} = [0,0,0,0,0,0], \Sigma = \text{diagonal}_{6\times 6}(2)$$
$$\sigma^2 \sim \mathcal{IG}(A, B)$$
$$\text{where:} \quad A = 3, \ B = 9.$$

and the notation "$\text{diagonal}_{6\times 6}(2)$" indicates a six-by-six square matrix with 2's on the diagonal and zeros elsewhere. The hyperparameters $A = 3, B = 9$ on σ^2 were chosen to provide a diffuse and therefore relatively uninfluential form of the prior distribution.

Later we will use better prior distributions for σ^2 based on folded distributions (i.e., normals, or Student's-t, excluding negative values), but these forms are not conjugate and require more elaborate estimation procedures. The specification above produces the model results summarized in Table 5.4 with posterior moments and quantiles.

TABLE 5.4: POSTERIOR: LINEAR MODEL, CONJUGATE PRIORS

β Covariate	Mean	Std. Error	0.025 Q.	Median	0.975 Q.
Intercept	96.337	7.347	81.904	96.337	110.770
Technology	-46.635	3.327	-53.171	-46.635	-40.099
New	-0.378	0.040	-0.456	-0.378	-0.300
Size	0.155	0.006	0.143	0.155	0.167
Republican	-0.085	0.007	-0.099	-0.085	-0.070
White	-0.049	0.006	-0.060	-0.049	-0.038
σ	15.577	0.343	14.925	15.569	16.273

From this reanalysis, we can see that it does not make a dramatic difference whether an uninformed or a conjugate prior is specified (at least with these assigned parameters). While there are slight differences, it seems to be of little substantive concern. This is partly due to the sample size of course ($n = 516$ precincts), meaning that the likelihood dominates our choice of prior here.

5.2 Posterior Predictive Distribution for the Data

We will explore the topic in greater detail in Chapter 7, but it is interesting at this point to derive the marginal distribution of future draws of the data from the Bayesian linear model. This is the *predictive* distribution of the data assumed to be generated by the model implied by the posterior distribution of the parameters. Thus we can compare the distribution of the data from this prediction with the actual data where large observed differences may indicate poor model fit.

Consider predictive data, a vector $\tilde{y}$ of length $k < n - 2$, generated from inserting actual or hypothetical covariate values into the $q \times k$ matrix $\tilde{X}$. Instead of a dataset of size n, we create our own set of explanatory values of size q. The $q \times 1$ vector of actual predictive data is assumed to be produced from the linear model,

$$\tilde{y} = \tilde{X}\beta + \epsilon, \tag{5.19}$$

but interest lies instead in the posterior distribution of the $\tilde{y}$ unconditional on parameters:

$\pi(\tilde{\mathbf{y}}|\mathbf{y}, \tilde{\mathbf{X}}, \mathbf{X})$. This means that we are interested in linear predictions that result from constructing a design matrix of covariate values and multiplying this by the estimated coefficients from the original model with this linear form, but we want a distribution for this prediction that incorporates all sources of variance, including that of the coefficient posteriors. To get this unconditional distribution, start with the relation:

$$\pi(\tilde{\mathbf{y}}, \boldsymbol{\beta}, \sigma^2|\tilde{\mathbf{X}}, \mathbf{X}, \mathbf{y}) = \pi(\tilde{\mathbf{y}}|\boldsymbol{\beta}, \sigma^2, \tilde{\mathbf{X}})\pi(\boldsymbol{\beta}, \sigma^2|\mathbf{X}, \mathbf{y}), \tag{5.20}$$

where the first term on the right-hand side is just the normal PDF of new data, given parameterization,

$$\pi(\tilde{\mathbf{y}}|\boldsymbol{\beta}, \sigma^2, \tilde{\mathbf{X}}) \propto \frac{1}{(\sigma^2)^{q/2}} \exp\left[-\frac{1}{2\sigma^2}(\tilde{\mathbf{y}} - \tilde{\mathbf{X}}\boldsymbol{\beta})'(\tilde{\mathbf{y}} - \tilde{\mathbf{X}}\boldsymbol{\beta})\right] \tag{5.21}$$

and the second term on the right-hand side is joint posterior of the parameters from the linear model,

$$\pi(\boldsymbol{\beta}, \sigma^2|\mathbf{X}, \mathbf{y}) \propto (\sigma^2)^{-\frac{n+1}{2}} \exp\left[-\frac{1}{2\sigma^2}(\hat{\sigma}^2(n-k) + (\boldsymbol{\beta} - \hat{\mathbf{b}})'\mathbf{X}'\mathbf{X}(\boldsymbol{\beta} - \hat{\mathbf{b}}))\right] \tag{5.22}$$

$(\hat{\mathbf{b}} = (\mathbf{X}'\mathbf{X})^{-1}\mathbf{X}'\mathbf{y})$, equation (5.5) from the uninformed priors $p(\boldsymbol{\beta}) \propto c$ and $p(\sigma^2) = \frac{1}{\sigma}$. We want to unwind the square in the exponent according to: $(\boldsymbol{\beta} - \hat{\mathbf{b}})\mathbf{X}'\mathbf{X}(\boldsymbol{\beta} - \hat{\mathbf{b}}) = (\boldsymbol{\beta} - (\mathbf{X}'\mathbf{X})^{-1}\mathbf{X}'\mathbf{y})'\mathbf{X}'\mathbf{X}(\boldsymbol{\beta} - (\mathbf{X}'\mathbf{X})^{-1}\mathbf{X}'\mathbf{y}) = \boldsymbol{\beta}'\mathbf{X}'\mathbf{X}\boldsymbol{\beta} - 2\boldsymbol{\beta}'\mathbf{X}'\mathbf{y} + \mathbf{y}'\mathbf{y} = (\mathbf{y} - \mathbf{X}\boldsymbol{\beta})'(\mathbf{y} - \mathbf{X}\boldsymbol{\beta})$. Putting these last two together gives:

$$\pi(\tilde{\mathbf{y}}, \boldsymbol{\beta}, \sigma^2|\tilde{\mathbf{X}}, \mathbf{X}, \mathbf{y}) \propto (\sigma^2)^{-\frac{n+q+1}{2}}$$

$$\times \exp\left[-\frac{1}{2\sigma^2}\left((\mathbf{y} - \mathbf{X}\boldsymbol{\beta})'(\mathbf{y} - \mathbf{X}\boldsymbol{\beta}) + (\tilde{\mathbf{y}} - \tilde{\mathbf{X}}\boldsymbol{\beta})'(\tilde{\mathbf{y}} - \tilde{\mathbf{X}}\boldsymbol{\beta})\right)\right]. \tag{5.23}$$

The quantity of interest is obtained by integrating out $\boldsymbol{\beta}$ and σ^2 individually, starting with σ^2:

$$\pi(\tilde{\mathbf{y}}, \boldsymbol{\beta}|\tilde{\mathbf{X}}, \mathbf{X}, \mathbf{y}) = \int_0^\infty \pi(\tilde{\mathbf{y}}, \boldsymbol{\beta}, \sigma^2|\tilde{\mathbf{X}}, \mathbf{X}, \mathbf{y})d\sigma^2$$

$$\propto \int_0^\infty (\sigma^2)^{-\frac{n+q+1}{2}} \exp\left[-\frac{1}{2\sigma^2}\left((\mathbf{y} - \mathbf{X}\boldsymbol{\beta})'(\mathbf{y} - \mathbf{X}\boldsymbol{\beta})\right.\right.$$

$$\left.\left. + (\tilde{\mathbf{y}} - \tilde{\mathbf{X}}\boldsymbol{\beta})'(\tilde{\mathbf{y}} - \tilde{\mathbf{X}}\boldsymbol{\beta})\right)\right]d\sigma^2.$$

$$\propto \left[(\mathbf{y} - \mathbf{X}\boldsymbol{\beta})'(\mathbf{y} - \mathbf{X}\boldsymbol{\beta}) + (\tilde{\mathbf{y}} - \tilde{\mathbf{X}}\boldsymbol{\beta})'(\tilde{\mathbf{y}} - \tilde{\mathbf{X}}\boldsymbol{\beta})\right]^{-\frac{n+q}{2}}, \tag{5.24}$$

where we used the same gamma PDF trick as before in (5.7). To integrate out $\boldsymbol{\beta}$ we need to collect terms in a more useful way, starting with breaking out the squares:

$$\pi(\tilde{\mathbf{y}}, \boldsymbol{\beta}|\tilde{\mathbf{X}}, \mathbf{X}, \mathbf{y}) = \left[\mathbf{y}'\mathbf{y} - 2\boldsymbol{\beta}\mathbf{X}'\mathbf{y} + \boldsymbol{\beta}'\mathbf{X}'\mathbf{X}\boldsymbol{\beta} + \tilde{\mathbf{y}}'\tilde{\mathbf{y}} - 2\boldsymbol{\beta}\tilde{\mathbf{X}}'\tilde{\mathbf{y}} + \boldsymbol{\beta}'\tilde{\mathbf{X}}'\tilde{\mathbf{X}}\boldsymbol{\beta}\right]^{-\frac{n+q}{2}}. \tag{5.25}$$

Defining $\mathbf{L} = \mathbf{X}'\mathbf{y} + \tilde{\mathbf{X}}'\tilde{\mathbf{y}}$ and $\mathbf{M} = \mathbf{X}'\mathbf{X} + \tilde{\mathbf{X}}'\tilde{\mathbf{X}}$, this last expression can be expressed as:

$$\pi(\tilde{\mathbf{y}}, \boldsymbol{\beta} | \tilde{\mathbf{X}}, \mathbf{X}, \mathbf{y}) \propto \left[\mathbf{y}'\mathbf{y} + \tilde{\mathbf{y}}'\tilde{\mathbf{y}} + \boldsymbol{\beta}'\mathbf{M}\boldsymbol{\beta} - 2\boldsymbol{\beta}'\mathbf{L} \right]^{-\frac{n+q}{2}}$$

$$= \left[\mathbf{y}'\mathbf{y} + \tilde{\mathbf{y}}'\tilde{\mathbf{y}} - \mathbf{L}'\mathbf{M}^{-1}\mathbf{L} \right.$$

$$\left. + (\boldsymbol{\beta}'\mathbf{M}\boldsymbol{\beta} - \boldsymbol{\beta}'\mathbf{M}\mathbf{M}^{-1}\mathbf{L} - \mathbf{L}'\mathbf{M}^{-1}\mathbf{M}\boldsymbol{\beta} + \mathbf{L}'\mathbf{M}^{-1}\mathbf{M}\mathbf{M}^{-1}\mathbf{L}) \right]^{-\frac{n+q}{2}}$$

$$= \left[\mathbf{y}'\mathbf{y} + \tilde{\mathbf{y}}'\tilde{\mathbf{y}} - \mathbf{L}'\mathbf{M}^{-1}\mathbf{L} + (\boldsymbol{\beta} - \mathbf{M}^{-1}\mathbf{L})'\mathbf{M}(\boldsymbol{\beta} - \mathbf{M}^{-1}\mathbf{L}) \right]^{-\frac{n+q}{2}}. \qquad (5.26)$$

Zellner (1971, p.73) integrates this form over the k length $\boldsymbol{\beta}$ vector to produce the posterior predictive distribution of interest:

$$\pi(\tilde{\mathbf{y}} | \tilde{\mathbf{X}}, \mathbf{X}, \mathbf{y}) \propto \left[\mathbf{y}'\mathbf{y} + \tilde{\mathbf{y}}'\tilde{\mathbf{y}} + \boldsymbol{\beta}'\mathbf{M}\boldsymbol{\beta} - 2\boldsymbol{\beta}'\mathbf{L} \right]^{-\frac{n+q-k}{2}}. \qquad (5.27)$$

He then defines the substitution $\mathbf{H} = (\mathbf{I} - \tilde{\mathbf{X}}\mathbf{M}^{-1}\tilde{\mathbf{X}}')/\hat{\sigma}^2$ and reintroduces $\hat{\mathbf{b}}$ to produce the form:

$$\pi(\tilde{\mathbf{y}} | \tilde{\mathbf{X}}, \mathbf{X}, \mathbf{y}) \propto \left[(n-k) + (\tilde{\mathbf{y}} - \tilde{\mathbf{X}}\hat{\mathbf{b}})'\mathbf{H}(\tilde{\mathbf{y}} - \tilde{\mathbf{X}}\hat{\mathbf{b}}) \right]^{-\frac{n+q-k}{2}}, \qquad (5.28)$$

which makes it easier to see that this is a multivariate t-distribution with $\nu = n - k$ degrees of freedom. Recalling the properties of the multivariate-t (Appendix B), this result means that:

$$E[\tilde{\mathbf{y}}] = \tilde{\mathbf{X}}\hat{\mathbf{b}}, \qquad \mathrm{Cov}[\tilde{\mathbf{y}}] = \frac{n-k}{n-k-2}\mathbf{H}^{-1}, \qquad (5.29)$$

with the obvious restriction on the size of $n - k$ from the denominator.

This result is actually quite intuitive. While the data themselves are assumed to be normally distributed, future claims about the data conditional on the model are t-distributed reflecting added uncertainty.

■ **Example 5.3: A Model of Educational Effects.** The Bayesian linear model is further illustrated through a partial replication of a Meier, Polinard, and Wrinkle (2000) study of bureaucratic effects on education outcomes in public schools (see also the reanalysis in Wagner and Gill [2005], and extensions in Meier and Gill [2000]). These authors are concerned with whether the education bureaucracy is the product or cause of poor student performance. The issue is controversial, and Chubb and Moe (1988, 1990) argue otherwise that the institutional structure of the schools, especially the overhead democratic control, resulted in the schools being ineffective. The institutional structure and the bureaucracy created a process that leads to poor performance by the public schools. This conclusion is challenged by Meier and Smith (1994), as well as in Smith and Meier (1995), who contend that bureaucracy is an adaptation to poor performance and not the cause.

Meier, Polinard, and Wrinkle develop a linear model based on panel dataset from more than 1,000 school districts for a seven-year period to test organizational theory and educational policy, producing an impressively large dataset here with $n = 7301$

cases. The question asked is whether there is a causal relationship between bureaucracy and poor performance by public schools. The central issue in this literature is one of causality through a "production function" that maps inputs to outputs in essentially an economic construct. Therefore their outcome variable is the percent of students in district/year that pass the Texas Assessment of Academic Skills (TAAS), which measures mastery of basic skills. In addition to bureaucratic causes, student and school performance can be influenced by a number of variables, some of which are causally related, including class size, state funding, teacher salary, and teacher experience. The data measure bureaucracy as the total number of full-time administrators per 100 students and lag the variable so as to create a more likely causal relationship. Other variables include three measures of financial capital, which consist of the average teacher salary, per pupil expenditures for instruction and the percentage of money each district receives from state funds. A measure of human capital was included based on teacher experience, and two policy indicators were used by measuring the average class size in the district and the percent of students in gifted classes.

The linear regression model proposed by Meier *et al.* is affected by both serial correlation and heteroscedasticity. Meier *et al.* address these concerns through a set of six dummy variables for each year as well as through the use of weighted least squares. The Meier *et al.* results are obtained by specifying diffuse normal priors on the unknown parameters. These are Gaussian normal specifications centered at zero with small precision. The model is summarized in "stacked" notation that shows the distributional assumptions (priors and likelihood):

$$Y[i] \sim \mathcal{N}(\lambda[i], \sigma^2),$$

$$\lambda[i] = \beta_0 + \beta_1 x_1[i] + \ldots + \beta_k x_k[i] + \epsilon[i]$$

$$\beta[i] \sim \mathcal{N}(0.0, 10)$$

$$\epsilon[i] \sim \mathcal{N}(0.0, \tau)$$

$$\tau \sim \mathcal{IG}(16, 6) \tag{5.30}$$

Note the hierarchical expression of these distributional and modeling assumptions here, which is the conventional way to notate models with dependent distributional features (Chapter 12). This specification above allows a close Bayesian replication of the original model since all of the coefficient prior distribution forms are highly diffuse, and the precision prior distribution is tuned to resemble $1/\sigma^2$ from Meier *et al.* This model is estimated using BUGS software, although with more agony it could be directly computed. The results are provided in Table 5.5.

For this model we can calculate the posterior predictive distribution of the data as given by (5.28) in the previous section. While there are certainly many interesting

TABLE 5.5: POSTERIOR: MEIER REPLICATION MODEL

Explanatory Variables	Mean	Std. Error	95% HPD Region
Intercept	9.172	1.358	[6.510:11.840]
Low Income Students	-0.108	0.006	[-0.119:-0.097]
Teacher Salaries	0.073	0.053	[-0.035: 0.181]
Teacher Experience	-0.009	0.046	[-0.099: 0.082]
Gifted Classes	0.097	0.023	[0.054: 0.139]
Class Size	-0.220	0.052	[-0.322:-0.118]
State Aid Percentage	-0.002	0.004	[-0.010: 0.006]
Funding Per Student ($\times 1000$)	0.065	0.174	[-0.276: 0.406]
Lag of Student Pass Rate	0.677	0.008	[0.661: 0.693]
Lag of Bureaucrats	-0.081	0.262	[-0.595: 0.431]

Posterior standard error of $\tau = 0.00072$

cases that a scholar of education policy might insert into the rows of the design matrix, $\tilde{\mathbf{X}}$, we will limit the analysis here to the predicted outcome for all of the explanatory variables set at their mean. Producing other cases to reveal important effects is much like a first difference calculation for evaluating the effect of coefficients in GLM models. The posterior predictive value for $\tilde{\mathbf{y}}$ where $\tilde{\mathbf{X}} = \bar{\mathbf{X}}$ is 48.18 (with a standard error of 0.0007202 due to the huge size of the dataset), yet the mean of the y vector is 63.84. This is interesting because it shows that the mean model outcome predicts a much poorer outcome than that observed suggesting that the model is missing some important features of the data-generating process. The other obvious suggestion that the data are right-skewed is not true.

To expand on the Meier *et al.* model, it is possible to include non-sample information for the creation of the Bayesian prior drawn from Meier's previous work on school bureaucracy and school performance with Kevin Smith (1994). Clearly, Meier *et al.* were not uninformed when specifying the model above and Bayesian inference allows for the incorporation of that knowledge. The Smith and Meier work includes data and inference on the impact of funding and other institutional variables on student achievement in Florida. These include district level data for all of the public schools in Florida. Smith and Meier note also that the Florida data provides a diverse group of students with constant measures over time. The Florida data represents both rural and urban districts as well as different ethnic and socioeconomic compositions. The prior distributions remain normal, but are now centered around values drawn from the 1995 Smith and Meier findings:

$$\beta[0] \sim \mathcal{N}(0.0, 10) \qquad \beta[1] \sim \mathcal{N}(-0.025, 10) \qquad \beta[2] \sim \mathcal{N}(0.0, 10)$$

$$\beta[3] \sim \mathcal{N}(0.23, 10) \qquad \beta[4] \sim \mathcal{N}(0.615, 10) \qquad \beta[5] \sim \mathcal{N}(-0.068, 10)$$

$$\beta[6] \sim \mathcal{N}(0.0, 10) \qquad \beta[7] \sim \mathcal{N}(-0.033, 10) \qquad \beta[8] \sim \mathcal{N}(0.299, 10)$$

$$\beta[9] \sim \mathcal{N}(0.0, 10) \qquad \beta[10] \sim \mathcal{N}(0.0, 10) \qquad \beta[11] \sim \mathcal{N}(0.0, 10)$$

$$\beta[12] \sim \mathcal{N}(0.0, 10) \qquad \beta[13] \sim \mathcal{N}(0.0, 10) \qquad \beta[14] \sim \mathcal{N}(0.0, 10)$$

$$\beta[15] \sim \mathcal{N}(0.0, 10)$$

The numbered β terms represent the prior information assigned to each explanatory variable (such BUGS statements are more elegant vectorized but the prior assignments are then less obvious to a reader). The terms are ordered as in Table 5.5, with the additional $\beta[10:15]$ values representing years 1993-97. Some of these are left relatively uninformed since the data from the Smith and Meier research were insufficient to address all of the current terms.

Interestingly, Meier *et al.* expected to find a positive relationship between teacher salaries and student performance, but did not find one. The model that includes stipulated priors generated results that were closer to the expectation of the researchers since it incorporated knowledge to which the researchers already had access. Meier *et al.* noted that economic theory expects higher salaries attract better teachers.

TABLE 5.6:   POSTERIOR: INTERACTION MODEL

Explanatory Variables	Mean	Std. Error	95% HPD Region
Intercept	4.799	2.373	[0.165: 9.516]
Low Income Students	-0.105	0.006	[-0.117:-0.094]
Teacher Salaries	0.382	0.099	[0.189: 0.575]
Teacher Experience	-0.066	0.046	[-0.156: 0.025]
Gifted Classes	0.096	0.021	[0.054: 0.138]
Class Size	0.196	0.191	[-0.180: 0.569]
State Aid Percentage	0.002	0.004	[-0.006: 0.010]
Funding Per Student (×1000)	0.049	0.175	[-0.294: 0.392]
Lag of Student Pass Rate	0.684	0.008	[0.667: 0.699]
Lag of Bureaucrats	-0.042	0.261	[-0.557: 0.469]
Class Size × Teacher Salaries	-0.015	0.007	[-0.029:-0.002]

Posterior standard error of $\tau = 0.00071$

In addition, the new model adds a multiplicative interaction between class size and teacher salary as the variables are claimed to be related in this way. The interaction coefficient was found to have a negative sign with 95% credible interval bounded away from zero, as provided in Table 5.6. So larger class sizes have an apparent dampening effect on the positive impact of increasing teacher salaries. Also, the posterior distribution for class size now shows a much less reliable effect in the interaction model (the 95% credible interval is almost centered at zero). Interaction effects can sometimes "steal" explanatory power and reliability from main effects. This finding says that the effects of class size are now only reliable in this model in the context of specified teachers' salary levels.

If this model is qualitatively different than the Meier Replication Model, then we would expect the posterior predictive distribution to show some change reflecting better fit to the data. From the mean model, applying the data mean as input $\tilde{\mathbf{X}}$, the posterior predictive value of y is 51.77 (with a standard error of 0.00071022), compared to the y data mean of 63.84. So the enhanced model predicts slightly better but not as much as we would have hoped. The point from this comparison is that one way to compare model fit in model development is to compare the predicted data to the actual. We will be much more systematic about this process in Chapter 7, but it easy to see here that posterior quantities form the basis of any such comparison.

5.3   Linear Regression with Heteroscedasticity

It is not uncommon to encounter the linear regression problem of non-constant error variance. The typical means of dealing with this problem in the non-Bayesian setting is through a more general form of least squares inserting a weighting matrix. The Bayesian approach is similar but leads to some difficult analytical issues, which we will overcome here. The definitive citation is Geweke (1993) and useful discussions can also be found in Mouchart and Simar (1984), who show that the Bayesian model can be worked out with least squares theory, and Leonard (1975), who shows the importance of exchangeability in this context.

Instead of the usual assumption about the distribution of $\mathbf{y}$ given $\mathbf{X}$, we now assert that $\mathbf{y}_i | \mathbf{X}_i \sim \mathcal{N}(\mathbf{X}_i\beta, \sigma^2\omega_i)$, where $\omega = (\omega_1, \omega_2, \ldots, \omega_n)$ is a vector of unknown regression weights (parameters), which we can also organize along the diagonal of a $n \times n$ matrix $\mathbf{\Omega}$ for convenience. The linear model is now defined as:

$$\mathbf{y} = \mathbf{X}\beta + \epsilon, \qquad \text{Var}[\epsilon] = \sigma^2\mathbf{\Omega}, \qquad (5.31)$$

which implies from the conditional distribution of the $\mathbf{y}_i$ that $\epsilon_i \sim \mathcal{N}(0, \sigma^2\omega_i)$. Thus the likelihood function from (5.4) becomes:

$$L(\beta, \sigma^2 | \mathbf{X}, \mathbf{y}) \propto \sigma^{-n} |\mathbf{\Omega}|^{-\frac{1}{2}} \exp\left[-\frac{1}{2\sigma^2}(\hat{\sigma}^2(n-k) + (\beta - \hat{\mathbf{b}})'\mathbf{X}'\mathbf{\Omega}^{-1}\mathbf{X}(\beta - \hat{\mathbf{b}})) \right]. \quad (5.32)$$

We will again use the uninformed priors $p(\boldsymbol{\beta}) \propto c$ and $p(\sigma^2) = \frac{1}{\sigma}$ over the support $[-\infty:\infty]$ and $[0:\infty]$. Geweke (1993) suggests using independent chi-square distributions, $\omega|\nu \sim \chi^2(df = \nu)$, which is also expressible as the gamma distribution $\mathcal{G}(\nu/2, 1/2)$. This ν parameter is flexible and can be fixed or estimated The resulting joint posterior distribution is:

$$\pi(\boldsymbol{\beta}, \sigma^2, \boldsymbol{\Omega}|\mathbf{X}, \mathbf{y}) \propto L(\boldsymbol{\beta}, \sigma^2|\mathbf{X}, \mathbf{y})p(\boldsymbol{\beta})p(\sigma^2)p(\boldsymbol{\Omega})$$

$$\propto \sigma^{-n-1}|\boldsymbol{\Omega}|^{-\frac{\nu+3}{2}} \exp\left[-\frac{1}{2\sigma^2}\left((\hat{\sigma}^2(n-k)\right.\right.$$

$$\left.\left. + (\boldsymbol{\beta} - \hat{\mathbf{b}})'\mathbf{X}'\boldsymbol{\Omega}^{-1}\mathbf{X}(\boldsymbol{\beta} - \hat{\mathbf{b}}) + \nu\mathrm{tr}(\boldsymbol{\Omega})^{-1}\right)\right]. \qquad (5.33)$$

This posterior turns out to be quite difficult to integrate for marginals, although $\pi(\boldsymbol{\beta}, \boldsymbol{\Omega}|\mathbf{X}, \mathbf{y})$ can be obtained with methods given previously in this chapter. Geweke instead finds three *conditional* distributions according to the following (suppressing the data in the conditional for emphasis).

▷ The conditional posterior distribution of $\boldsymbol{\beta}$:

$$\pi(\boldsymbol{\beta}|\sigma^2, \boldsymbol{\Omega}) \propto \exp\left[-\frac{1}{2\sigma^2}(\boldsymbol{\beta} - \hat{\mathbf{b}}^*)'\mathbf{X}'\boldsymbol{\Omega}^{-1}\mathbf{X}(\boldsymbol{\beta} - \hat{\mathbf{b}}^*)\right], \qquad (5.34)$$

where $\hat{\mathbf{b}}^* = (\mathbf{X}'\boldsymbol{\Omega}\mathbf{X})^{-1}\mathbf{X}'\boldsymbol{\Omega}\mathbf{y}$. Thus the conditional distribution of the $\boldsymbol{\beta}$ vector is multivariate normal according to $\mathcal{N}(\hat{\mathbf{b}}^*, \sigma^2(\mathbf{X}'\boldsymbol{\Omega}^{-1}\mathbf{X})^{-1})$.

▷ The conditional posterior distribution of σ^2:

$$\pi(\sigma^2|\boldsymbol{\Omega}) \propto (\sigma^2)^{-\frac{n+1}{2}} \exp\left[-\frac{1}{2\sigma^2}\hat{\sigma}^{2*}\right], \qquad (5.35)$$

where $\hat{\sigma}^{2*} = (\mathbf{y} - \mathbf{X}\hat{\mathbf{b}})'\boldsymbol{\Omega}^{-1}(\mathbf{y} - \mathbf{X}\hat{\mathbf{b}})$. This is clearly an inverse gamma distribution according to $\mathcal{IG}(\sigma^2|(n-1)/2, \hat{\sigma}^{2*}/2)$, and Geweke also expresses it in χ^2 terms.

▷ The conditional posterior distribution of $\boldsymbol{\Omega}$:

$$\pi(\boldsymbol{\Omega}|\boldsymbol{\beta}, \sigma^2) \propto |\boldsymbol{\Omega}|^{-\frac{\nu+3}{2}} \exp\left[-\frac{1}{2}\left(\hat{\sigma}^{2*}/\sigma^2 + \nu\mathrm{tr}(\boldsymbol{\Omega})^{-1}\right)\right] \qquad (5.36)$$

where $\hat{\sigma}^{2*}$ is defined as before. So an individual diagonal element of $\boldsymbol{\Omega}$ is conditionally distributed $\omega_i|\boldsymbol{\beta}, \sigma^2 \propto \omega_i^{-\frac{\nu+3}{2}} \exp\left[-(u_i^2/2\sigma^2 + \nu/2)/\omega_i\right]$, where $u_i = y_i - \mathbf{x}_i\boldsymbol{\beta}$. This gives an inverse gamma PDF with parameters $\frac{\nu+1}{2}$ and $(u_i^2/\sigma^2 + \nu)/2$.

Another interesting feature of this model is that the posterior distribution of the residuals are no longer normal according to $\epsilon_i \sim \mathcal{N}(0, \sigma^2\omega_i)$. In fact, they end up being Student's-t distributed with $\nu = n - k$ degrees of freedom. Geweke also points out that the model developed in this fashion produces residuals that are *independent* Student's-t (an independent χ^2 denominator of a normal), whereas previous versions produced residuals with a *joint* Student's-t (a common χ^2 denominator of a normal).

While the results above may not seem as useful as the unconditional marginal posteriors we derived earlier in this chapter, it does set us up perfectly for the Gibbs sampler as a way to get desired unconditionals. Recall from Chapter 1 that the Gibbs sampler uses iterated samples from full conditional distributions for the parameters of interest to obtain empirical estimates of marginals (page 25). In the present case we sample iteratively at the $j + 1$ step according to:

$$\boldsymbol{\beta}^{[j+1]} \sim \pi(\boldsymbol{\beta} | \sigma^{2[j]}, \boldsymbol{\Omega}^{[j]})$$

$$\sigma^{2[j+1]} \sim \pi(\sigma^2 | \boldsymbol{\Omega}^{[j]})$$

$$\boldsymbol{\Omega}^{[j+1]} \sim \pi(\boldsymbol{\Omega} | \boldsymbol{\beta}^{[j+1]}, \sigma^{2[j+1]})$$

where the $\boldsymbol{\beta}$ vector and the $\boldsymbol{\Omega}$ matrix (which is really a vector's worth of information) can be sampled individually $(\beta_1^{[j]}, \ldots, \beta_k^{[j]}, \omega_1^{[j]}, \ldots, \omega_n^{[j]})$ or as a block. Miraculously this iterative process eventually produces sample values that behave as if they were generated from the marginal distributions rather than the conditional distributions. Details on this MCMC procedure are given starting in Chapter 10 and essentially constitute the second half of this text.

■ **Example 5.4: War in Ancient China.** To demonstrate the described heteroscedastic linear model with Bayesian priors we use conflict data from West Asia for events taking place between 2700 BCE to 722 BCE. Cioffi-Revilla and Lai (1995, 2001) coded documents from multiple epigraphic and archaeological sources on war and politics in ancient China covering the Xia (Hsia), Shang, and Western Zhou (Chou) periods. These data ($n = 104$ conflicts) are available via the Murray Archive. This is the only quantitative dataset covering Chinese war during this period, but Cioffi-Revilla and Lai use the modern *Long-Range Analysis of War* definitions.

The outcome variable of interest here is an additive combination of two of the coded variables: *Political Level* (1 for internal war, 2 for interstate war) and *Political Complexity* (governmental level of the warring parties), where the first variable is multiplied by ten for scale purposes. Thus we are looking to explain the *political scope* of conflicts in terms of governmental units affected. Explanatory variables of interest are EXTENT (number of belligerents), DIVERSE (number of ethnic groups participating as belligerents), ALLIANCE (total number of alliances among belligerents), DYADS (number of alliance pairs), TEMPOR (type of war: protracted rivalry, integrative conquest, disintegrative/fracturing conflict, sporadic event), DURATION of conflict, measured in years, BALANCE (the difference in military capabilities: minor-minor, minor-major, major-major), ETHNIC (intra-group or inter-group), and POLAR (number of relatively major or great powers at the time of onset). See Cioffi-Revilla and Lai (2001) or the associated codebook for further details. Graphical investigation indicates the presence of heteroscedastic effects of concern with a homoscedastic linear model. See Figure 5.1.

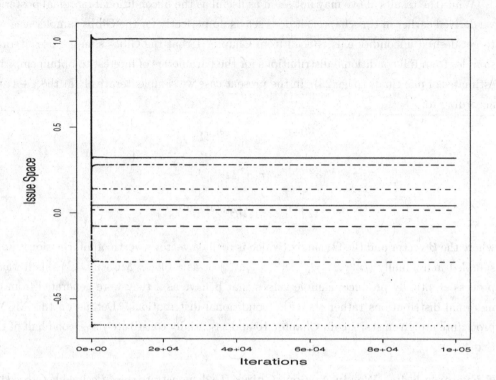

FIGURE 5.1: MCMC RUNNING MEANS, CHINESE CONFLICTS

The Bayesian heteroscedastic linear model is fit without an intercept (zero levels do not make sense here), using uninformed priors $p(\beta) \propto c$ and $p(\sigma^2) = \frac{1}{\sigma}$, as discussed above. Furthermore, fix $\nu = 50$ to elongate the distribution of weights for this example. The Gibbs sampler is run for $100,000$ iterations, throwing away the first $50,000$ (this early "burn-in" period will discussed at length starting on page 478). Table 5.7 summarizes the resulting marginal posteriors from this estimation. The R code for running this model is provided in the **Computational Addendum** to this chapter. Details about running Gibbs samplers are provided starting in Chapter 10 and diagnostics for MCMC convergence are discussed particularly in Chapter 14. For now note that the Markov chain iterations become stable in the cumulative mean plots given in Figure 5.1.

Observe from the resulting HPD regions for the coefficients that the model fits very well with no coefficient 95% HPD regions crossing zero. We see here that the number of belligerents, the number of ethnic groups, imbalance in capabilities, ethnic composition, and number of major groups all have accelerating effects on the political scope of the conflict. The type of war does as well but its interpretation is not as clear. Conversely, the number of alliances, the number of dyads, and the duration give

TABLE 5.7: Heteroscedastic Model, Ancient Chinese Conflicts

Explanatory Variables	Mean	Std. Error	95% HPD Region
EXTENT	1.0145	0.1077	[0.8034: 1.2256]
ALLIANCE	-0.2840	0.0756	[-0.4321:-0.1359]
DYADS	-0.6540	0.0739	[-0.7988:-0.5092]
TEMPOR	0.1391	0.0302	[0.0799: 0.1984]
DURATION	-0.0779	0.0353	[-0.1471:-0.0087]
BALANCE	0.2810	0.0692	[0.1454: 0.4165]
ETHNIC	0.3210	0.0574	[0.2086: 0.4335]
POLAR	0.0189	0.0078	[0.0035: 0.0343]
Mean of $\sigma^2 = 0.0454$			

the opposite effect. These are all as expected, despite using the most diffuse priors possible.

5.4 Exercises

5.1 Derive the posterior marginal for β in (5.17) from the joint distribution given by (5.16).

5.2 Write an R function that calculates R^2 and the F-statistic for the two models given in Table 5.3. Defend your choice of point estimate from the posterior distributions used for these calculations. Does the Bayesian interpretation of these values differ? Is it reasonable to use these in applied Bayesian work?

5.3 For uninformed priors, the joint posterior distribution of the regression coefficients was shown to be multivariate-t (page 147), with covariance matrix: $\mathbf{R} = \frac{(n-k)\hat{\sigma}^2(\mathbf{X}'\mathbf{X})^{-1}}{n-k-2}$. Under what conditions is this matrix positive definite (a requirement for valid inferences here).

5.4 Clogg, Petkova, and Haritou (1995) give detailed guidance for deciding between different linear regression models using the same data. In this work they define the matrices $\mathbf{X}$, which is $n \times (p+1)$ rank $p+1$, and $\mathbf{Z}$, which is $n \times (q+1)$ rank $q+1$, with $p < q$. They calculate the matrix $A = \left[\mathbf{X}'\mathbf{X} - \mathbf{X}'\mathbf{Z}(\mathbf{Z}'\mathbf{Z})^{-1}\mathbf{Z}'\mathbf{X}\right]^{-1}$. Find the dimension and rank of A.

5.5 Under standard analysis of linear models, the hat matrix is given by $\hat{y} = \mathbf{H}y$ where $\mathbf{H}$ is $\mathbf{X}(\mathbf{X}'\mathbf{X})^{-1}\mathbf{X}'$ where the diagonal values of this matrix indicate leverage, which is obviously a function of X only. Can the stipulation of strongly informed priors change data-point leverage? Influence in linear model theory depends on both hat matrix diagonal values and y_i. Calculate the influence on the `Technology` variable of each datapoint in the Palm Beach County model with uninformed priors by jackknifing out these values one at a time and re-running the analysis. Which precinct has the greatest influence?

5.6   Prove that the $\mathbf{M}$ matrix $(\mathbf{I} - \mathbf{H})$ from linear regression is symmetric and idempotent.

5.7   Meier and Keiser (1996) used a linear pooled model to examine the impact of several federal laws on state-level child-support collection policies. Calculate a Bayesian linear model and plot the posterior distribution of the parameter vector, $\boldsymbol{\beta}$, as well as σ^2, specifying an inverse gamma prior with your selection of prior parameters, and the uninformative uniform prior: $f(\sigma^2) = \frac{1}{\sigma}$. Use a diffuse normal prior for the $\boldsymbol{\beta}$, and identify outliers using the hat matrix method (try the R command `hat`).

The data are collected for the 50 states over the period 1982 to 1991, where the outcome variable, SCCOLL, is the change in child-support collections. The explanatory variables are: chapters per population (ACES), policy instability (INSTABIL), policy ambiguity (AAMBIG), the change in agency staffing (CSTAFF), state divorce rate (ARD), organizational slack (ASLACK), and state-level expenditures (AEXPEND). These data can be downloaded on the webpage for this book or from the BaM package. Additional description can be found in their original article or Meier and Gill (2000, Chapter 2).

5.8   Replicate the Meier Interaction Model in Table 5.6 on page 160, using the dataset `student.score` in BaM. Modify the prior variances for the coefficient from 10 to other numbers. Does this change the prior predictive distribution of the data (at means for $\tilde{\mathbf{X}}$)? Why or why not?

5.9   Returning to the discussion of conjugate priors for the Bayesian linear regression model starting on page 151, show that substituting (5.14) and (5.15) into (5.13) produces (5.16).

5.10   Consider a linear regression setting where the matrix $\mathbf{X}'\mathbf{X}$ is singular ($\mathbf{X}$ is $n \times k$). Clearly non-Bayesian solutions are limited, but careful stipulation of priors can lead to workable results. Using the setup of Zellner (1971, p.75) we start with the usual joint conjugate prior $p(\boldsymbol{\beta}, \sigma) = p(\boldsymbol{\beta}|\sigma)p(\sigma)$, which is essentially (5.1.2), and

$$p(\sigma) \propto \frac{1}{\sigma^{\nu_0+1}} \exp\left[-\frac{\nu_0 c_0^2}{2\sigma^2}\right]$$

$$p(\boldsymbol{\beta}|\sigma) \propto \frac{|\mathbf{A}|^{-\frac{1}{2}}}{\sigma^k} \exp\left[-\frac{1}{2\sigma^2}(\boldsymbol{\beta} - \mathbb{B})'\mathbf{A}(\boldsymbol{\beta} - \mathbb{B})\right]$$

where $\mathbb{B}$ is the prior mean for β, the prior covariance matrix, $\sigma^2 \mathbf{A}^{-1}$, is nonsingular. Using the likelihood function in (5.2), we get the joint posterior distribution:

$$p(\beta, \sigma | \mathbf{X}, \mathbf{y}) \propto \sigma^{-n'-k-1} \exp\left[-\frac{1}{2\sigma^2}(n'c^2 + (\beta - \tilde{\beta})(\mathbf{A} + \mathbf{X}'X)(\beta - \tilde{\beta})\right],$$

where $n' = n + \nu_0$, and $n'c^2 = \nu_0 c_0^2 + \mathbf{y}'\mathbf{y} + \mathbb{B}'\mathbf{A}\mathbb{B} - \tilde{\beta}'(\mathbf{A} + \mathbf{X}'X)\tilde{\beta}$ (notice that there is no need to invert $\mathbf{X}'\mathbf{X}$). Show that the marginal posterior distribution for β is

$$\pi(\beta | \mathbf{X}, \mathbf{y}) \propto \left[n'c^2 + (\beta - \tilde{\beta}'(\mathbf{A} + \mathbf{X}'\mathbf{X})(\beta - \tilde{\beta})\right]^{-\frac{n'+k}{2}},$$

and give the distributional form with parameters identified.

5.11   For the Bayesian linear regression model, prove that the posterior that results from conjugate priors is asymptotically equivalent to the posterior that results from uninformative priors. See Table 5.3 on page 153.

5.12   The following data come from the 1998 European Household Community Panel Survey (given in the R package BaM as ehcps). The two variables are: (1) the median (EU standardized) income of individuals age 65 and older as a percentage of the population age 0–64, and (2) the percentage of all age groups with income below 60% of the median (EU standardized) income of the national population. Regress the second variable on the first with a linear model, constructing conjugate and uninformative prior distributions for Over 65 Relative Income. Compare these results. Is it possible to specify highly influential priors in the conjugate case that differ markedly from the uninformative case?

Nation	Over 65 Relative Income	Total Poverty Rate
Netherlands	93.00	7.00
Luxembourg	99.00	8.00
Sweden	83.00	8.00
Germany	97.00	11.00
Italy	96.00	14.00
Spain	91.00	16.00
Finland	78.00	17.00
France	90.00	19.00
United.Kingdom	78.00	21.00
Belgium	76.00	22.00
Austria	84.00	24.00
Denmark	68.00	31.00
Portugal	76.00	33.00
Greece	74.00	33.00
Ireland	69.00	34.00

5.13   Develop a Bayesian linear model for the following data that describe the average weekly household spending on tobacco and alcohol (in pounds) for the eleven regions of the United Kingdom (Moore and McCabe 1989, originally from *Family Expenditure Survey, Department of Employment, 1981*, British Official Statistics).

Specify both an informed conjugate and uninformed prior using the level for alcoholic beverages as the outcome variable and the level for tobacco products as the explanatory variable. Do you notice a substantial difference in the resulting posteriors? Describe.

Region	Alcohol	Tobacco
Northern Ireland	4.02	4.56
East Anglia	4.52	2.92
Southwest	4.79	2.71
East Midlands	4.89	3.34
Wales	5.27	3.53
West Midlands	5.63	3.47
Southeast	5.89	3.20
Scotland	6.08	4.51
Yorkshire	6.13	3.76
Northeast	6.19	3.77
North	6.47	4.03

5.14 Using your two model results from the *Family Expenditure Survey* model above, create two graphs of predicted outcome values using the following steps for each:

(a) Create a matrix of $\mathbf{X}$ values over the range of the data, $\tilde{\mathbf{X}}$, of size greater than the data.

(b) Draw m (large) values from the posterior distributions of β: $\tilde{\beta}$.

(c) Create a vector of predicted outcomes according to: $\tilde{\mathbf{y}} = \tilde{\mathbf{X}}\tilde{\beta}$.

(d) Plot these values with a histogram and a random sample of them against the actual data with qqplot.

Observe any differences occurring from the use of different priors.

5.15 The standard econometric approach to testing parameter restrictions in the linear model is to compare $H_0 : \mathbf{R}\beta - \mathbf{q} = 0$ with $H_1 : \mathbf{R}\beta - \mathbf{q} \neq 0$ (e.g., Greene 2011), where $\mathbf{R}$ is a matrix of linear restrictions and $\mathbf{q}$ is a vector of values (typically zeros). Thus for example $\mathbf{R} = \begin{bmatrix} 1 & 0 & 0 & 0 \\ 0 & 1 & 1 & 0 \\ 0 & 0 & 1 & 1 \end{bmatrix}$ and $\mathbf{q} = [0, 1, 2]'$ indicate $\beta_1 = 0$, $\beta_2 + \beta_3 = 1$, and $\beta_4 = 2$. Derive expressions for the posteriors $\pi(\beta|\mathbf{X}, \mathbf{y})$ and $\pi(\sigma|\mathbf{X}, \mathbf{y})$ using both conjugate and uninformative priors.

5.16 The inverse of a matrix Υ^{-1} of Υ is defined as meeting five conditions: (1) $\mathbf{H}\Upsilon^{-1}\Upsilon = \mathbf{H}$, (2) $\Upsilon^{-1}\Upsilon\Upsilon^{-1} = \Upsilon^{-1}$, (3) $(\Upsilon\Upsilon^{-1})' = \Upsilon^{-1}\Upsilon$, (4) $(\Upsilon^{-1}\Upsilon) = \Upsilon\Upsilon^{-1}$, and (5) $\Upsilon^{-1}\Upsilon = \mathbf{I}$ (1–4 implied by 5, $\mathbf{H}$ a symmetric, full rank matrix). The Moore-Penrose generalized inverse matrix Υ^{-} of Υ meets only the first four conditions. A pseudo-variance matrix is calculated as $\mathbf{V}'\mathbf{V}$, where $\mathbf{V} = \text{GCHOL}(\mathbf{H}^{-})$, $\text{GCHOL}(\cdot)$ is the generalized Cholesky, and $\mathbf{H}^{-}$ is the generalized inverse of the

Hessian matrix. The result is a pseudo-variance matrix that is in most cases well conditioned (not nearly singular). Show that if the Hessian is invertible, the pseudo-variance matrix is the usual inverse of the negative Hessian.

5.17 Using the Palm Beach County electoral data (page 148, also available in the R package BaM using `data(pbc)`), calculate the posterior predictive distribution of the data using a skeptical conjugate prior (i.e., centered at zero for hypothesized effects and having large variance).

5.18 Returning to the 1998 European Household Community Panel Survey data, run a heteroscedastic specification using the Gibbs steps given on page 163 written in R. Do you see a model improvement over a regular non-Bayesian model assuming homoscedasticity? Provide evidence.

5.19 Rerun the Ancient China Conflict Model using the code in this chapter's **Computational Addendum** using informed prior specifications of your choice (you should be willing to defend these decisions though). Calculate the posterior mean for each marginal distribution of the parameters and create a set of 104 predicted outcome data values as customarily done in linear models analysis. Graph the y_i against $\hat{y}_i$ and make a statement about the quality of fit for your model. Can you improve this fit by dramatically changing the prior specification?

5.20 The "Grenander Conditions" for establishing asymptotic properties of the linear model are given by: [G1:] for each column of $\mathbf{X}$: $\mathbf{X}'_k \mathbf{X}_k \longrightarrow +\infty$ (sums of squares grow as n grows, no columns of all zeros), [G2:] no single observation dominates each explanatory variable k: $\lim_{n \to \infty} \frac{\mathbf{X}^2_{ik}}{\mathbf{X}'_k \mathbf{X}_k} = 0, i = 1, \ldots, n$, [G3:] $\mathbf{X}'\mathbf{X}$ has rank k by Gauss-Markov assumption, define $\mathbf{X}_{-0}$ as the explanatory variable matrix minus the leading column of 1s, then $\lim_{n \to \infty} \mathbf{X}_{-0'} \mathbf{X}_{-0} = C$, C a positive definite matrix. Prove that for a Bayesian linear model with conjugate priors these provide: $\hat{\beta} \underset{\sim}{\text{ asym. }} N\left[\beta, \frac{\hat{\sigma}^2}{n} \mathbf{Q}^{-1}\right]$, where $\mathbf{Q} = \lim_{n \to \infty} \frac{1}{n} \mathbf{X}'\mathbf{X}$, $\hat{\beta}$ is the posterior mean vector for the coefficients, and $\hat{\sigma}$ is the posterior mean of σ.

5.5 Computational Addendum

5.5.1 Palm Beach County Normal Model

The R code here was used to develop the linear model example with the Palm Beach County voting data.

```
# RETURNS A REGRESSION TABLE WITH CREDIBLE INTERVALS
t.ci.table <- function(coefs,cov.mat,level=0.95,degrees=Inf,
        quantiles=c(0.025,0.500,0.975))  {
```

```
    quantile.mat <- cbind( coefs, sqrt(diag(cov.mat)),
        t(qt(quantiles,degrees) %o% sqrt(diag(cov.mat)))
        + matrix(rep(coefs,length(quantiles)),
        ncol=length(quantiles)) )
    quantile.names <- c("Mean","Std. Error")
    for (i in 1:length(quantiles))
        quantile.names <- c(quantile.names,paste(quantiles[i],
        "Quantile"))
    dimnames(quantile.mat)[2] <- list(quantile.names)
    return(list(title="Posterior Quantities",round(quantile.mat,4)))
}


# READ IN THE DATA AND USE MULTIPLE IMPUTATION ON MISSING
lapply(c("BaM","mice","nnet"),library,character.only=TRUE)
data(pbc.vote)
attach(pbc.vote)
X <- cbind(tech, new, turnout, rep, whi)
Y <- badballots
detach(pbc.vote)
imp.X <- mice(X)
X <- as.matrix(cbind(rep(1,nrow(X)), complete(imp.X)))
dimnames(X)[[2]] <- c("tech", "new", "turnout", "rep", "whi")


# UNINFORMED PRIOR ANALYSIS
bhat <- solve(t(X)%*%X)%*%t(X)%*%Y
s2 <- t(Y- X%*%bhat)%*%(Y- X%*%bhat)/(nrow(X)-ncol(X))
R <- solve(t(X)%*%X)*((nrow(X)-ncol(X))*
        s2/(nrow(X)-ncol(X)-2))[1,1]
uninformed.table <- t.ci.table(bhat,R,
        degrees=nrow(X)-ncol(X))[[2]]
alpha <- (nrow(X)-ncol(X)-1)/2
beta <- 0.5*s2*(nrow(X)-ncol(X))
sort.inv.gamma.sample <- sort(1/rgamma(10000,alpha,beta))
sqrt.sort.inv.gamma.sample <- sqrt(sort.inv.gamma.sample)
uninformed.table <- rbind(uninformed.table,
        c( mean(sqrt.sort.inv.gamma.sample),
      sqrt(var(sqrt.sort.inv.gamma.sample)),
            sqrt.sort.inv.gamma.sample[250],
            sqrt.sort.inv.gamma.sample[5000],
            sqrt.sort.inv.gamma.sample[9750] ))


# CONJUGATE PRIOR ANALYSIS
A <- 3; B <- 9
BBeta <- rep(0,6); Sigma <- diag(c(2,2,2,2,2,2))
tB <- solve(solve(Sigma)
    + t(X)%*%X)%*%(solve(Sigma)%*%BBeta+t(X)%*%X%*%bhat)
```

```
ts <- 2*B + s2*(nrow(X)-ncol(X)) + (t(BBeta)-t(tB))%*%
    solve(Sigma)%*%BBeta + t(bhat-tB)%*%t(X)%*%X%*%bhat
R <- diag(ts/(nrow(X)+A-ncol(X)-3))*
    solve(solve(Sigma)+t(X)%*%X)
alpha <- nrow(x)+A-ncol(X); beta <- 0.5*S2*(nrow(X)+A-ncol(X))
conjugate.table<-t.ci.table(tB,R,
    degrees=nrow(X)+A-ncol(X)-2)[[2]]
sort.inv.gamma.sample <- sort(1/rgamma(10000,alpha,beta))
sqrt.sort.inv.gamma.sample <- sqrt(sort.inv.gamma.sample)
conjugate.table<- rbind(conjugate.table,
    c( mean(sqrt.sort.inv.gamma.sample),
        sqrt(var(sqrt.sort.inv.gamma.sample)),
            sqrt.sort.inv.gamma.sample[250],
            sqrt.sort.inv.gamma.sample[5000],
            sqrt.sort.inv.gamma.sample[9750] ))
```

5.5.2   Educational Outcomes Model

While programming in BUGS is not explained until later chapters, the code for the Meier *et al.* model and the informed extension is provided here for reference. Notice, for the time being, that BUGS code shares many features with R in terms of specifying models and distributions. The first model could be coded in a more efficient manner by making theta a vector, but leaving it as a set of scalars makes modifying the prior with substantive prior information easier to specify and evaluate. The X[] variables are out of order in the linear specification because the data matrix orders them differently than the model listing in the original article. Also, we would normally specify a theta vector below instead of a series of scalars, but the individual notation is retained to make it easier to contrast the two model specifications.

```
# MEIER, ET AL. REPLICATION MODEL
Model {
    theta0~dnorm(0.0,0.001);  theta1~dnorm(0,0.001)
    theta2~dnorm(0.0,0.001);  theta3~dnorm(0.0,0.001)
    theta4~dnorm(0.0,0.001);  theta5~dnorm(0.0,0.001)
    theta6~dnorm(0.0,0.001);  theta7~dnorm(0.0,0.001)
    theta8~dnorm(0.0,0.001);  theta9~dnorm(0.0,0.001)
    theta10~dnorm(0.0,0.001); theta11~dnorm(0.0,0.001)
    theta12~dnorm(0.0,0.001); theta13~dnorm(0.0,0.001)
    theta14~dnorm(0.0,0.001); theta15~dnorm(0.0,0.001)
    tau~dgamma(16,6); sigma.sq <- 1

    for (i in 1 : N)  {
        epsilon[i]~dnorm(0.0, sigma.sq)
        lambda[i] <- theta0 + theta1*X9[i]   + theta2*X10[i] +
            theta3*X2[i]    + theta4*X3[i]   + theta5*X4[i] +
```

```
                theta6*X5[i]    + theta7*X6[i]   + theta8*X7[i] +
                theta9*X8[i]    + theta10*X11[i] + theta11*X12[i] +
                theta12*X13[i]  + theta13*X14[i] + theta14*X15[i] +
                theta15*X16[i]  + epsilon[i]
            Y[i] ~ dnorm(lambda[i], tau)
        }
}


# MODEL WITH INFORMED PRIORS AND INTERACTION
Model {
    theta0~dnorm(0.0,0.1);    theta1~dnorm(-.025,0.1)
    theta2~dnorm(0.0,0.1);    theta3~dnorm(0.23,0.1)
    theta4~dnorm(0.615,0.1);  theta5~dnorm(-0.068,0.1)
    theta6~dnorm(0.0,0.1);    theta7~dnorm(-.033,0.1)
    theta8~dnorm(0.299,0.1);  theta9~dnorm(0.0,0.1)
    theta10~dnorm(0.0,0.1);   theta11~dnorm(0.0,0.1)
    theta12~dnorm(0.0,0.1);   theta13~dnorm(0.0,0.1)
    theta14~dnorm(0.0,0.1);   theta15~dnorm(0.0,0.1)
    tau~dgamma(16,6); sigma.sq <- 1

    for (i in 1 : N) {
        epsilon[i]~dnorm(0.0, sigma.sq)
        lambda[i] <- theta0 + theta1*X9[i]    + theta2*X10[i] +
            theta3*X2[i]    + theta4*X3[i]    + theta5*X4[i] +
            theta6*X5[i]    + theta7*X6[i]    + theta8*X7[i] +
            theta9*X8[i]    + theta10*X11[i] + theta11*X12[i] +
            theta12*X13[i]  + theta13*X14[i] + theta14*X15[i] +
            theta15*X16[i]  + theta16*X6*X3  + epsilon[i]
        Y[i] ~ dnorm(lambda[i], tau)
    }
}
```

5.5.3   Ancient China Conflict Model

This section provides the R code for running the Gibbs sampler in the Chinese wars example.

```
data(wars)
attach(wars)
X <- cbind(EXTENT,DIVERSE,ALLIANCE,DYADS,TEMPOR,DURATION)
y <- SCOPE
detach(wars)
n <- nrow(X); k <- ncol(X)
nu <- 5
num.sims <- 10000
war.samples <- matrix(NA,nrow=num.sims,(ncol=k+n+1))
```

```
beta <- rep(1,ncol(X));sigma.sq <- 3;Omega <- 3*diag(n)
b <- solve(t(X) %*% X) %*% t(X) %*% y
yXb <- y-X%*%b

for (i in 1:num.sims)  {
    Omega.inv <- solve(Omega)
    X2.Om   <- solve(t(X) %*% Omega %*% X)
    b.star <- X2.Om %*% t(X) %*% Omega %*% y
    s.sq.star <- t(yXb) %*% Omega.inv %*% (yXb)
    u <- y - X %*% beta
    beta <- as.vector( rmultinorm(1, b.star, sigma.sq *solve(t(X)
        %*% Omega.inv %*% X) ) )
    sigma.sq <- 1/rgamma(1,shape=(n-1)/2,rate=s.sq.star/2)
    for (j in 1:n) Omega[j,j] <- 1/rgamma(1, shape=(nu+1)/2,
        rate=((sigma.sq^(-1))*u^2 + nu)/2 )
    war.samples[i,] <- c(beta,sigma.sq,diag(Omega))
}
```

Chapter 6

Assessing Model Quality

6.1 Motivation

The third step in Bayesian analysis is to evaluate the model fit to the data and determine the sensitivity of the posterior distribution to the assumptions. This is typically an interactive and iterative process in which different approaches are taken until there is reasonable evidence that the conclusions are both statistically reliable *and* stable under modest changes in the assumptions. So the emphasis in this chapter is on *model adequacy*: the suitability of a single model under consideration. The subsequent chapter is really about *model testing*: determining why we should prefer one model specification over another. Both of these are critical considerations, and it is important to keep them distinct.

Model checking is a critical part of the estimation process since there is nothing in the procedures outlined in previous chapters that would *per se* prevent one from producing incorrect and overly sensitive posterior distributions for the unknown parameters of interest. In George Box's (1995) words: "Statistics has no reason for existence except as a catalyst for scientific inquiry in which only the last stage, when all the creative work has already been done, is concerned with a final fixed model and rigorous test of conclusions." Furthermore, Box (1980) also notes that "No statistical model can safely be assumed adequate." Accordingly, this chapter provides various methods for assessing the quality of "final" models in the sense that other considerations, such as data collection, parametric specification, and variable selection, have already been determined and our goal is to understand the quality of those choices.

The quality of a Bayesian posterior inference is attributable to three model assignments: the prior, the likelihood function, and the loss function (if the latter is explicitly specified). Prior specifications include the assigned parametric form as well as the prior distribution's parameter values, and in the case of hierarchical models the *hyperprior specifications*: higher level prior distributions assigned to the first level of priors. For a given prior, there is a range of possible outcomes, including: the prior is subsumed by a large data set and therefore not particularly important, the prior is moderately influential even with reasonable sample size, and the prior strongly affects the form of the posterior. The second model assignment is the specified parametric form of the likelihood function, which itself ranges from relatively unimportant to mostly determining the shape of the posterior. Finally, if a loss function is stipulated, the characteristics of this function can also affect posterior conclusions.[1]

After looking at some simple comparative methods, this chapter focuses on three related approaches to model checking in applied Bayesian work: sensitivity analysis, global robustness, and local robustness. Sensitivity analysis is the *informal* process of altering assumptions according to researcher intuition with the objective of determining the extent to which these changes modify the posterior distribution. In particular, does varying the prior parameters or modestly changing the form of the prior itself lead to vastly different conclusions from the posterior? Robustness is posterior insensitivity to broad classes of user-specified assumptions and is therefore a desirable quality of Bayesian models.[2] Robustness evaluation is the *systematic* process of determining the degree to which posterior inferences are affected by both potential misspecification of the prior and influential data points. It is actually important to remember the distinction between sensitivity analysis and robustness (Skene, Shaw, and Lee 1986). Global robustness evaluation performs a formal analysis of a large class of priors to determine the subsequent *range* of the inferences. Local robustness generally uses differential calculus to determine the volatility of specific reported results.

The distinction between global and local robustness does not need to be confined to the prior distributions under consideration for a Bayesian model. Any aspect of the specification is equally suspect and therefore deserving scrutiny. Smith (1986), for example, expresses *model criticism* distinctness in very general terms (quoting, p.97):

▷ *global* criticism, which basically asks the question "should we abandon the current framework completely, despite the fact we have nothing at all to propose in its place?"

▷ *local* criticism, which asks "should the model be modified or extended in this or that particular direction?"

The definition of global criticism may be a little bit stark here since it is almost always true that the researcher has potential alternatives. Smith actually argues that his given characterization of global robustness checking is too limited and not particularly Bayesian in spirit. Note that this is essentially the Fisherian setup where models are tested against a vague null concept that is not specified but represents a lack of systematic effect. The underlying philosophy that undergirds either approach is that a single model applied to some data produces a single conclusion, but a range of models applied to some data produces

---

[1]The implications of loss function robustness are not considered here in detail and the reader is referred to the general discussion on page 247, and to: Dey and Micheas (2000), Dey, Lou, and Bose (1998), Martín, Ríos Insua, and Ruggeri (1998), and Ramsey and Novick (1980). Kadane and Chuang (1978), extended in Kadane and Srinivasan (1996), provide an integrated way of looking at the prior and the loss function together by introducing the broader idea of posterior stability.

[2]Since robustness increases automatically with increases in sample size, then robustness evaluation is less important in models with large samples. Conversely, models with modest sample size (say less than 100, but this is data-dependent) should be thoroughly analyzed in this regard.

a range of conclusions. The degree to which this range of conclusions differs for modest changes in basic assumptions warns the researcher about the delicateness of their findings.

This chapter also covers the posterior predictive distribution as a way to investigate the quality of models through unseen data that they imply. Since all unknown quantities are treated probabilistically, Bayesian posterior results can be averaged across models to give more robust coefficient estimates as they cover multiple-model space. Bayesian model averaging can therefore be used to reduce the uncertainty inherent in presenting a single specification. Chapter 7 also shows how the Kullback-Leibler distance can be used to compare distributional distances. This allows us to measure differences between priors and posteriors.

6.1.1 Posterior Data Replication

One very simple way to evaluate model quality is to produce a replicate dataset under the estimated model and compare it to the observed data using summary statistics or graphical analysis. This can be done analytically, but is much easier with simulation. While the core ideas of Monte Carlo simulation and MCMC estimation are covered in later chapters, we have been systematically introducing some of these ideas throughout the early part of this text. Here we will see a preview of two important tools in this section: the BUGS language for estimating Bayesian models with Markov chain Monte Carlo, and the use of simulation across posterior distributions as a means of describing the implications of the model.

The process works as follows. *First*, produce a posterior distribution for all unknown quantities in the customary fashion, where the outcome variable vector of interest, $\mathbf{y}$, is modeled by observed predictors, $\mathbf{X}$, and parameters $\boldsymbol{\beta}$, according to the general specification $f(\mathbf{y}|\mathbf{X}, \boldsymbol{\beta})$. *Second*, fix $\mathbf{X}$ at these observed values and draw m replicated values of $\boldsymbol{\beta}$ from the corresponding posterior distribution: $\boldsymbol{\beta}^{(1)}, \boldsymbol{\beta}^{(2)}, \ldots, \boldsymbol{\beta}^{(m)}$. This last step can be done analytically, but is almost trivially easy if the model has been estimated with MCMC procedures described already on pages 25 and 164, since these procedures give empirical draws from the coefficient posterior distributions and we can simply sample from these. *Third*, take such a sample of posterior draws and treat them as fixed to fully define a single version of the posterior model and calculate the implied outcome variable vector, $f(\mathbf{y}^{\text{rep}(i)}|\mathbf{X}, \boldsymbol{\beta}^{(i)})$, for posterior draws $i = 1, 2, \ldots, m$. *Finally*, plot and compare the $\mathbf{y}^{\text{rep}(i)}$ with the observed $\mathbf{y}$ where large deviations imply poor model fit. There are obviously many variations to this procedure that correspond to different model characteristics (see the discussion in Gelman and Hill [2007, pp.517-524], with R and BUGS code).

■ **Example 6.1: Posterior Comparison for a Model of Abortion Attitudes in Britain.** To illustrate this model-checking procedure, we develop a model of support for abortion under different scenarios using survey data from Britain in consecutive years from 1983 to 1986. The panel data for 264 respondents is collected annually by McGrath and Waterton (1986) where seven scenarios are provided and these respondents have the option of expressing support or disagreement for abortion (see also the reanalysis by Knott *et al.* [1990]). The full collection of these seven queries do not fall into an obvious ordinal scale, so we will treat them here as nominal and judge total support for abortion as a binomial test for each respondent at each wave of the panel. The scenarios are: (1) the woman decides on her own that she does not wish to have the child, (2) the couple agree that they do not wish to have the child, (3) the woman is not married and does not wish to marry the man, (4) the couple cannot afford any more children, (5) there a strong chance that the baby has a biological defect, (6) the woman's health is seriously endangered by the pregnancy, and (7) the woman became pregnant as a result of rape. So each respondent has the ability to produce an outcome from 0 to 7. These data are available in the R package BaM.

Naturally we do not want to treat the repeated annual trials as independent since this would ignore correlation within subjects. So the model is given by the following specification for $i = 1, \ldots, 264$ respondents across $j = 1, \ldots, 4$ panel waves:

$$y_{ij} \sim \mathcal{BN}(n, p_{ij})$$

$$\text{logit}(p_{ij}) = \beta_{0,j} + \beta_{1,i} X_{1,i}$$

$$\beta_{0,j} \sim \mathcal{N}(\mu_0, \tau_0) \qquad\qquad \beta_{1,i} \sim \mathcal{N}(\mu_1, \tau_1)$$

$$\mu_0 \sim \mathcal{N}(0, 100) \qquad\qquad \mu_1 \sim \mathcal{N}(0, 100)$$

$$\tau_0 \propto \mathcal{C}_{half}(25) \qquad\qquad \tau_1 \propto \mathcal{C}_{half}(25) \qquad\qquad (6.1)$$

where the second term in the normals is a variance and half-Cauchy priors are positive-support, zero-centralized forms (location) with a scale term equal to $A = 25$, $f(\tau) = (1 + \tau/A)^{-1}$, $\tau > 0$. This is a form recommended by Gelman (2006) as an alternative to gamma distributions with small parameters for providing low-information priors for variances terms. Here $X_{1,i}$ is the ith person's self-identified religion: Catholic (1), Protestant (2), Other (3), and No Religion (4), and $n = 7$ at each wave for each person. The JAGS code is:

```
model {
    for (i in 1:PEOPLE)  {
        for (j in 1:WAVES)  {
            logit(p[i,j]) <-  b0[j] + b1[i]*x1[i];
            r[i,j] ~ dbin(p[i,j], n[i]);
        }
        b0[i] ~ dnorm(mu0, nu0);
        b1[i] ~ dnorm(mu1, nu1);
    }
    mu0    ~ dnorm(0.0,1.0E-2);
    mu1    ~ dnorm(0.0,1.0E-2);
    tau0   ~ dnorm(0,1)T(0,);
    tau1   ~ dnorm(0,1)T(0,);
    sigma  ~ dgamma(2,2);
```

```
    nu0    <- 1/(25*tau0/sqrt(sigma));
    nu1    <- 1/(25*tau1/sqrt(sigma));
}
```

While this is getting a little ahead of ourselves, it is useful to look at the JAGS code before Chapter 11. Notice that the code inside the "for" loop is essentially how we would describe the model statistically on paper. That is, p comes from a linear additive component with a link function, then the outcome variable is given a distributional assumption. The term n[i] is indexed for generality but does not need to be since all respondents are given seven statements at each wave. The Gibbs sampler is run for 20,000 iterations dispensing with the first 10,000 values. All convergence diagnostics point towards convergence. This is a hierarchical model in the sense of those described in Chapter 12, so in addition to obtaining posterior distributions for the two μ and τ parameters, there will be $n = 264$ posterior distributions for each β. Therefore we will not report the full posterior results beyond Table 6.1, and will concentrate instead on assessing fit.

TABLE 6.1: MODEL SUMMARY, ABORTION ATTITUDES IN BRITAIN

	Posterior Quantiles				
	0.025	0.25	0.50	0.75	0.975
μ_0	-0.6708	-0.2563	-0.0294	0.1960	0.6245
τ_0	0.0283	0.0500	0.0641	0.0789	0.1113
μ_1	0.3029	0.4129	0.4728	0.5291	0.6336
τ_1	0.0303	0.3136	0.6726	1.1507	2.2453

More importantly, consider Figure 6.1 where the jittered observed data values from the last wave of the panel are plotted against nine (uniformly) randomly selected jittered iterations from the recorded Markov chain. This is not a *test* for fit but indicates both positive and negative aspects of the results. The closer the results are to the upward sloping diagonal, the better the fit. Thus for high support, the model fits well and for low support the model fits less well. This makes sense since the questions are diverse and mixed support is therefore much harder to model than uniformly or nearly-uniformly strong support. Notice also the few values in the radically wrong corners of the graphs. More covariate information in the data would likely tighten the fit around the diagonal. Note that in the case where the outcome variable of interest is continuously measured, the graphical display is more straightforward with lines or points instead of these discrete categories.

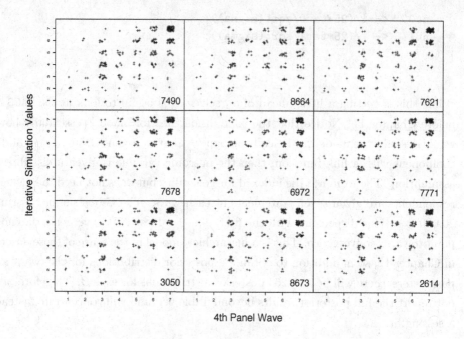

Iterative Simulation Values

4th Panel Wave

FIGURE 6.1: Outcome Comparison: Observed versus Simulated

6.1.2   Likelihood Function Robustness

The specification of the likelihood function is generally not very controversial in specifying Bayesian or non-Bayesian models because it is often narrowly defined by the form of the outcome variable and is well-studied as a contributor to model quality in the generalized linear models literature (Fahrmeir and Tutz 2001; Gill 2000; McCullagh and Nelder 1989). Conversely, frequentist criticism (and therefore Bayesian defensiveness) is often centered on the posterior implications of prior assumptions. It should be noted, however, that the likelihood function component of the model is no less suspect than the prior specification and worth attention as well since nearly all researchers "directly announce the likelihood itself without deriving it" (Poirier 1988, p.131). Also, the likelihood selection process is no less subjective than the prior selection process (de Finetti 1974, 1975). For example, the selection of a Poisson link function versus a negative binomial link function accords significantly less attention and defensive efforts than the choice of a prior for the same model.

Shyamalkumar (2000) suggests a means of checking likelihood robustness across related likelihood classes, and he shows that it is relatively simple to define a finite class of related likelihood models with greater robustness properties. For instance, it is well known that prior specifications with wider tails have better robustness properties than conjugate choices (Berger 1984, 1985). So Shyamalkumar recommends a Cauchy comparison class by matching up normal and Cauchy quantiles (for instance the interquartile range of a $\mathcal{N}(\theta, 1)$ distribution matches those of a $\mathcal{C}(\theta, 0.675)$). It is also possible to formulate this neighborhood specification nonparametrically (Lavine 1991a).

6.2 Basic Sensitivity Analysis

A very simple and helpful method for assessing posterior model quality is to vary the prior or other assumptions in some *ad hoc* but intuitive way and observe the implications with respect to posterior quantities of interest. If reasonably large changes in model assumptions are seen to have a negligible effect on the calculated posterior density, then we can comfortably establish that the data are sufficiently influential or that the varied assumption is sufficiently benign (or perhaps both) to eliminate further worry about the subjective influence of assumptions. Conversely, a far more alarming scenario occurs when one makes mild changes to the established model assumptions and dramatic changes are observed in the summary measures of the posterior.

6.2.1 Global Sensitivity Analysis

There are essentially two types of Bayesian sensitivity analysis: global and local (Leamer 1978). Global sensitivity analysis is a broad approach that evaluates a wide range of: alternative prior specifications (Berger 1984, 1990), forms of the link function (Draper 1995), missing data implications (Kong, Liu, and Wong 1994), error sensitivity (Polasek 1987), and perturbations of the likelihood and prior specifications (Kass and Raftery 1995). The purpose of global sensitivity is to vary the *widest* possible range of assumptions, although some authors have used the term to describe analysis of the sensitivity of the posterior to differences provided within a given *family* of priors (Polasek 1987).

Leamer's original objective in global sensitivity analysis is to determine the maximum amount that the model assumptions can be varied without dramatically changing the posterior inferences (1985, p.308). Thus substantive inferences are deemed to be reliable only if the range of assumptions is wide enough to be realistic and the corresponding posterior inferences are narrow enough that they provide consistent results. This is a naturally appealing idea because it would indicate which specifications are relatively "fragile" with respect to researcher-specified model assumptions, although Pagan (1987) believes that this approach does not go far enough since it is restricted to altering the parameters of the prior distribution only. Unfortunately the idea of absolute global sensitivity analysis opens a vast array of alternative specifications and tests to consider, even restricting oneself to the prior only. Often there are too many of these alternatives to consider in a reasonable amount of time, and there is also the additional challenge of clearly reporting these results to readers.

From many published works, we can see that the obvious, but not solitary, target of sensitivity analysis is the form of the prior distribution. Routinely, a flat prior of some sort is substituted for an informative prior. The observed change in the posterior can range from trivial to substantial. It should be noted, however, that if the informed prior has a substantial theoretical basis, large changes to the posterior are not *a priori* a sufficient reason to disqualify it from the analysis. Conversely, strongly informative priors that are specified purely for mathematical convenience should be evaluated carefully if they produce a radically different posterior form than a reference prior of some kind.

6.2.1.1   Specific Cases of Global Prior Sensitivity Analysis

The classical prior juxtaposition is to specify an alternative uninformative prior (typically uniform) over the support of the parameter of interest and compare it with the stipulated informative prior. If there is a substantial difference in the form of the posterior, then this is something that often needs to be explained to readers. It is not necessary to use the uniform as a comparison prior, but many researchers find it to be a convenient comparison form. An obvious question arises regarding how much of a change in the posterior results indicates strong prior influence from the informative prior. First, if the difference changes the *substantive* conclusions from the model, then the informed prior should be earnestly defended. Consider the difference in the posteriors for Teacher Salaries in the two models of educational effects in Section 5.2: a posterior mean and standard deviation of $(0.073, 0.053)$ with uninformed priors (Table 5.5), versus $(0.382, 0.099)$ with informed priors. With a sample size of 7301 cases we can safely assume that the posteriors are normally distributed for this linear model specification. Accordingly, the first posterior has 0.0842 of the density below zero while the second posterior has 0.0006 of the density below zero. Often the differences are not this stark, and sometimes the differences are merely trivial. In such settings we would be less concerned with the influence of the prior distribution since the shrinkage of the posterior to the prior is small.

If the researcher-specified prior is of the normal form, then t-distributions, varying by the degrees of freedom, can be used as wider tail alternatives (Lange, Little, and Taylor 1989). One advantage of the t-distribution approach is that it avoids problems associated with specifying improper uniform priors over the real numbers. Jeffreys (1961) recommends using the Cauchy distribution as a prior in this context since its very heavy tail structure implies a high level of conservatism, although there is often not a substantial comparative difference between an informative prior relative to a uniform prior and an informative prior relative to a Cauchy prior.

6.2.1.2   Global Sensitivity in the Normal Model Case

A form of sensitivity analysis with the normal model was performed previously with simulated data in Table 3.2. That example demonstrated that an intentionally misspecified prior affected posterior quantities of interest: HPD regions and point estimates. The obvious point was that as the sample size increased from 10 to 1,000, the misspecification of the prior parameters on the normal prior mattered considerably less.

Suppose we assign an infinitely diffuse prior for an unknown mean in the normal model discussed in Chapter 3 as a way to assess the impact of some other imposed prior. The chosen prior is $p(\mu) = 1$, $\mu \in [-\infty : \infty]$, meaning that a density of one is assigned for values of μ along the entire real line. An important question is whether it is even possible to achieve a comparative posterior distribution with such an improper prior. The posterior here is proportional to the likelihood in the following simple manner:

$$\pi(\mu|\mathbf{x}) \propto p(\mathbf{x}|\mu)p(\mu)$$

$$= \prod_{i=1}^{n} \exp\left[-\frac{1}{2\sigma_0^2}(x_i - \mu)^2\right] \times 1$$

$$= L(\mu|\mathbf{x}), \tag{6.2}$$

which can be shown to be finite over the real line:

$$\int_{-\infty}^{\infty} L(\mu|\mathbf{x})d\mu = \int_{-\infty}^{\infty} \prod_{i=1}^{n} (2\pi\sigma_0^2)^{-\frac{n}{2}} \exp\left[\frac{1}{-2\sigma_0^2}(x_i - \mu)^2\right] d\mu, \tag{6.3}$$

with σ_0^2 known. Therefore this prior leads to a posterior that is just the same conclusion that a non-Bayesian likelihood analysis would produce. This is interesting because it is one of a few cases where Bayesian analysis and non-Bayesian analysis agree perfectly, despite starting from different assumptions.

6.2.1.3 Example: Prior Sensitivity in the Analysis of the 2000 U.S. Election in Palm Beach County

Returning again to the linear regression example of the voting results from the 2000 presidential election (see page 169 forward in Chapter 5 for the setup R code), we can graphically compare the posterior implications of the uniform prior specification versus the conjugate prior developed in Section 5.1 as means of global sensitivity analysis.

Figure 6.2 shows the marginal posterior density for each posterior dimension (omitting the constant) where the dotted line results from the conjugate prior and the solid line results from the uniform prior. It is clear from these figures that there is very little difference in the marginal posteriors for the two prior specifications with the possible exception of the technology variable. Despite these similarities, it is important to note that there may be substantive reasons to prefer one over the other in terms of what is being assumed about voter behavior.

6.2.1.4 Problems with Global Sensitivity Analysis

To Leamer (1984, 1985) such analysis is "global" when the test of sensitivity to the form of the prior includes a wide variety of prior forms. The motivation is that no single prior can be assumed to fully describe prior knowledge or expert opinion about the distribution of parameters, and therefore specifying a neighborhood of priors around the original gives an important picture of the ramifications of uncertain prior information. Furthermore, if in order to get a suitably narrow HPD region, the set of prior neighborhood specifications had to be severely limited, then this would also be evidence of fragility.

Global sensitivity is often very difficult to achieve in practice because it is not always

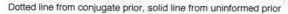

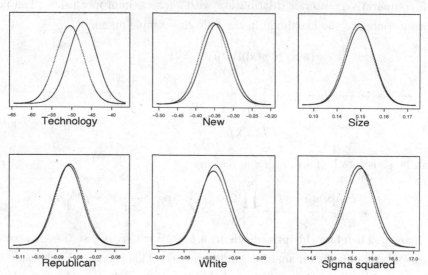

FIGURE 6.2: MARGINAL POSTERIORS, PALM BEACH COUNTY MODEL

clear what qualifies as a "neighborhood prior," and what range of these alternative priors constitutes sufficient testing criteria (O'Hagan and Berger 1988, Walley, Gurrin, and Burton 1996). This reduces to a difficult decision as to what should be considered a reasonably wide range of included prior specifications without including those that are highly unreasonable for mathematical or substantive reasons. Sometimes global sensitivity analysis is just not possible due to "an inability to sufficiently refine the usually subjective inputs of model, prior, and loss" (Berger 1986a).

Furthermore, the primary difficulty is that global sensitivity applies to simultaneous neighborhood generalization across all dimensions of the parameter vector. There can therefore be an enormous number and variety of specifications to consider and evaluate with difficult inclusion criteria. As Poirer notes, Leamer's advice is "easier to preach than to practice" (Poirer 1988, p.130). However, systematic Monte Carlo methods can considerably ease this task (Canova 1994). One very creative approach to handling this problem is the idea of "backward" sensitivity analysis or *prior partitioning* as advocated by Carlin and Louis (2009, p.188-194). The idea is to fix the posterior at some interesting or realistic configuration and see what types of priors are consistent with reaching this posterior given the data.

6.2.2   Local Sensitivity Analysis

Local sensitivity analysis is the more modest and realizable process of making minor changes in the prior parameterization while looking at the subsequent posterior effects. This has the advantage of realizing many of the benefits of global sensitivity analysis such as indicating posterior fragility, but at a much lower cost in terms of effort and reporting. The distinction between global and local sensitivity analysis is not as formal as the distinction between global and local robustness evaluation studied below, but the central point is that the basic form of the prior is not altered in local sensitivity analysis.

Generally modifying the parameters of the prior is done to produce a more diffuse form of the prior than that used in the initial analysis. The associated argument is that if the posterior does not appreciably change, then this is support for the initially specified form of the prior as one that does not interject substantial subjective prior information into the subsequent posterior. Naturally this is data-dependent and therefore application-specific.

6.2.2.1   Normal-Normal Model

Returning to the normal model example from Chapter 3 with variance known and a normal prior on the unknown mean, we specify a prior with parameters $\mathcal{N}(m, s^2)$. This produces a normal posterior according to:

$$\pi(\mu|\mathbf{X}, \sigma^2) \sim \mathcal{N}\left[\left(\frac{m}{s^2} + \frac{n\bar{x}}{\sigma_0^2}\right) \Big/ \left(\frac{1}{s^2} + \frac{n}{\sigma_0^2}\right), \left(\frac{1}{s^2} + \frac{n}{\sigma_0^2}\right)^{-1}\right]. \tag{6.4}$$

The sensitivity to the prior mean, m, can be analyzed by adding and subtracting one prior standard deviation, $m_1^* = m - s$, $m_2^* = m + s$, and recalculating the posterior distribution two more times using m_1^* and m_2^* separately as substitutes for m. It is clear from the form of (6.4) that this provides a location shift of the posterior whose scaled distance is averaged into the sample mean scaled by the known variance. In other words, if this distance is large relative to information provided by the sample, the location shift of the posterior will be large. Conversely, if this distance is small relative to sample information ($n\bar{x}$), then the mode of the posterior will be relatively stable.

From the form of the posterior distribution for μ in (6.4), we can also see that adding or subtracting some fixed quantity from the prior mean m does not affect the posterior variance in any way because the normal distribution is a location-scale family distribution, meaning that one parameter is strictly a description of the measure of centrality, and the other strictly a measure of dispersion (see Casella and Berger 2002, Chapter 3; Lehmann 1986, Chapter 1). Here conjugacy provides invariancy of this property, meaning that a conjugate location-scale prior gives a location-scale form of the posterior. There are actually situations where Bayesian invariance of location-scale family characteristics does not depend on conjugacy (Dawid 1979).

If we want to alter the variance parameter of the prior in order to assess posterior sensitivity, we need only to change the s^2 prior parameter. A typical procedure for variance parameters in this setting is to double or halve its value, $s_1^{2*} = s^2/2$, $s_2^{2*} = 2s^2$ and observe the effect on the resulting posterior. It is easy to see from the form of (6.4) that this change alters the dispersion of the posterior, subject to the relative weighting of the variance from the likelihood function; that is, the greater n is the less important the prior variance specification becomes (Lindley 1972).

6.2.2.2   Local Sensitivity Analysis Using Hyperparameter Changes

The normal-normal model described previously is actually quite a simple case in that as a location-scale family distribution, the effect of altering one parameter is independent of the other. In more general specifications the distinction is not always as clear. For example, in the beta-binomial model of Chapter 2 it is not obvious how one would alter the A and B parameters in (2.13), except to note from posterior quantities on page 50 that it can make a large difference. In such cases it is essential to employ substantive theoretical motivations for specifying ranges of the parameter space.

Figure 6.3 shows the effect of changes of the prior parameters of a model with an exponential likelihood function and a conjugate gamma prior, as described in Section 4.3.1. The posterior distribution for the unknown exponential PDF parameter is given by: $\pi(\theta|\mathbf{x}) \propto \theta^{(\alpha+n)-1} \exp\left[-\theta\left(\sum x_i + \beta\right)\right]$. In the first panel, α is set at 10 and β, the "scale parameter," is varied to show the effect on the resulting posterior distribution for θ for fixed $\sum x_i$ and n. In the second panel, β is fixed at 4 and α, the "shape parameter," is varied. Since the posterior is flattened and elongated for higher values of α and lower values of β, then it is clear that arbitrary simultaneous changes in both α and β can be difficult to interpret.

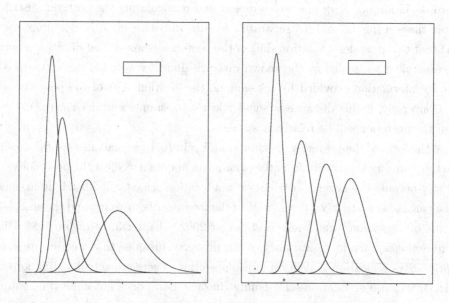

FIGURE 6.3: Exponential Model for θ with Gamma Prior

A better means of analyzing posterior sensitivity to changes in prior parameters that do not have an immediate interpretation (unlike the normal model) is to relate these parameters to the variance of the PDF or PMF of the prior. The variance of the prior distribution of θ in the exponential-gamma model is α/β^2, suggesting that we can vary this amount by some proportional quantity and view the subsequent sensitivity of the posterior. Starting with the prior parameter specification $\alpha = 10$, $\beta = 2$ we now vary the total *prior* variance by 50% in both directions by modifying either parameter singly. This exercise is demonstrated in Figure 6.4 where the posterior for $\alpha = 10$, $\beta = 2$ is located centrally in both panels with the 95% HPD region shaded. The first panel varies α and the second panel varies β, in both cases enough to alter the prior variance by 50% in either direction.

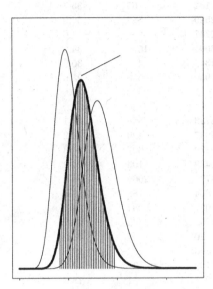 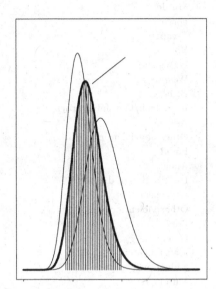

FIGURE 6.4: EXPONENTIAL MODEL SENSITIVITY

What we can see from this exercise is that the exponential-gamma posterior is relatively insensitive to modest modifications of the prior specification. If this were not true, Figure 6.4 would show dramatic changes in the location or dispersion of the posterior. This approach to local sensitivity analysis is generally very easy to perform and often provides convincing diagnostic evidence to support some researcher-generated prior specification.

6.2.3 Global and Local Sensitivity Analysis with Recidivism Data

In criminology and the public policy study of the penal system a key concern is the rate of recidivism: convicted criminals committing additional crimes after release from prison (see the running discussion in Chapter 1). This example looks at state-level recidivism data from Oklahoma over the period from January 1, 1985 through June 30, 1999. Table 6.2 shows these data (by descending percentage) as collected by the Oklahoma Department of Corrections (http://www.doc.state.ok.us), and provided in the R package BaM as recidivism.

It is clear from the table that recidivism rates differ according to the crime of original conviction, but it is not unreasonable to assert that there also exist some underlying social

TABLE 6.2: RECIDIVISM BY CRIME, OKLAHOMA, 1/1/85 TO 6/30/99

Crime Type	Released	Returned	Percentage
Unauthorized use of motor vehicle	1522	621	40.8
Burglary I	821	314	38.3
Burglary II	7397	2890	39.1
Larceny	9557	3525	36.9
DUI-2nd offense	8891	3212	36.1
Murder I	182	65	35.7
Forgery	2529	882	34.9
Escape	2747	951	34.6
Robbery	3001	1035	34.5
Bogus check	1341	414	30.9
Weapons	1950	558	28.6
Bribery	7	2	28.6
Possession/Obtaining drugs	7473	2125	28.4
Assault	3133	848	27.1
Other nonviolent crimes	1637	429	26.2
Fraud	1239	307	24.8
Arson	439	106	24.2
Distribution of drugs	8775	2066	23.5
Embezzlement	804	188	23.4
Other violent crimes	1081	234	21.7
Rape	1269	262	20.7
Gambling	20	4	20.0
Manslaughter	846	167	19.7
Kidnapping	162	31	19.1
Murder II	271	51	18.8
Sex offenses	2143	362	16.9
Drug trafficking	426	33	7.8

factors that affect all rates. Although there are several obvious ways to look at these data, we develop a simple Poisson model of the *log* of the recidivism rate (log of the Number Returned variable). Actually four models are produced, three with differently parameterized gamma priors producing the posterior: $\pi(\lambda|\mathbf{X}) \propto \lambda^{(\alpha+\sum x_i)-1} \exp[-\lambda(\beta+n)]$, and one with a bounded uniform prior producing the posterior: $\pi(\lambda|\mathbf{X}) \propto \lambda^{\sum x_i} \exp[-\lambda n]$. The posterior results are summarized in Table 6.3 with quantiles $(2.5, 50, 97.5)$ and the mean for the four prior forms.

Table 6.3 shows that there is not much of a difference across the prior specifications, including the uninformative uniform prior. This is evidence that the model is reasonably insensitive to differing conjugate prior parameterizations and to different prior forms. Also, the uniform prior can be considered a reference prior in the sense discussed in Chapter 4 and therefore it appears that the conjugate gamma prior is not overly influential in determining the posterior relative to the uniform reference. Further evidence is seen in Figure 6.5, which shows 1,000 empirical replications from each of the posterior distributions in Table 6.3 using a histogram and a smoothed density estimate.

TABLE 6.3: λ POSTERIOR SUMMARY, RECIDIVISM MODELS

Prior	2.5% Quantile	Mean	Median	97.5% Quantile
$\mathcal{G}(2,2)$	3.7765	4.4743	4.4630	5.2004
$\mathcal{G}(1,4)$	4.0534	4.7677	4.7644	5.4945
$\mathcal{G}(7,1)$	3.8328	4.4557	4.4369	5.1765
$\mathcal{U}(0,100)$	3.5177	4.1255	4.1019	4.8808

6.3 Robustness Evaluation

This section looks at two definitions of robustness: insensitivity to misspecification of the prior and insensitivity to influential data-points through the likelihood function. There is some confusion about the language of robustness, resistance, and sensitivity analysis. Classical robustness focuses on the linear model's sensitivity to influential outliers and seeks to mitigate this effect. There is now a vast literature on robust methods for identifying and handling influential outliers in the linear modeling context (Andrews 1974; Andrews *et al.* 1972; Barnett and Lewis 1978; Belsley, Kuh, and Welsch 1980; Cook and Weisberg 1982; Emerson and Hoaglin 1983; Hamilton 1992; Hampel 1974; Hampel *et al.* 1986; Huber 1972, 1973, 1981; Rousseeuw and Leroy 1987), including some from an explicitly Bayesian context (Bradlow, Weiss, and Cho 1998; Chaloner and Brant 1988; Guttman 1973; Guttman, Dutter, and Freeman 1978; Pettit and Smith 1985). The more appropriate term for this work is actually "resistance" reflecting the goal of making inferences more resistant to particular data cases, and robustness is correctly defined as low sensitivity to violations of underlying model assumptions.

Bayesian robustness is somewhat broader than the classical approach. The primary objective of Bayesian robustness is to determine the *posterior* sensitivity to substantively reasonable changes in model assumptions and thereby understand the context of uncertainty introduced by the prior, the likelihood function, and if appropriate, the loss function. The modern formal approach to Bayesian robustness, due primarily to Berger (1984), develops a mechanical framework for altering assumptions and taking these assumptions through the Bayesian inferential machinery to determine their posterior implications. Specifically, a nonrobust model specification is one in which "reasonable" changes in the prior, likelihood function, or loss function, produce large changes in the *range* of the posterior quantity of interest.

An associated goal is the identification of data values that strongly influence posterior inferences, and the complexity of this endeavor is greatly magnified as the dimension of the problem is increased. Generally computationally intensive tools are required for this analysis (Bradlow, Weiss, and Cho 1998), and it is common that time-saving summary measures are substituted for complete data expositions such as the AIC implications of subsetting (Kita-

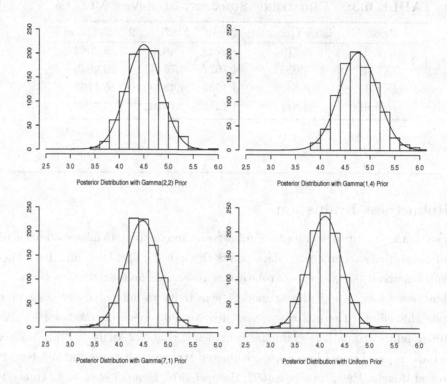

FIGURE 6.5: COMPARISON OF POSTERIORS, RECIDIVISM MODEL

gawa and Akaike 1982). A radically different, and substantially more complex, approach performs *a priori* analysis to determine robustness properties from the initial assumptions rather than as *post hoc* model adjustments. One such method by Seidenfeld, Schervish, and Kadane (1995) develops a complete Bayesian decision-theoretic axiomatic construct to derive preferences from utility and probability statements. Therefore the robustness properties are a direct result of the initial setup.

6.3.1   Global Robustness

The most common method for determining global robustness is the *range* of the posterior quantity of interest (Wasserman 1992). For instance, we can see how much the bounds of the 95% HPD region change under different prior assumptions. Other statistics of obvious interest include the posterior mean, various quantiles, and measures of dispersion. If modest changes to a specified prior distribution, say varying its parameters somewhat, produce dramatic changes in the range of these posterior quantities, then it is an indication that we should be cautious about the use of this particular prior, or at least that we should report this instability to our readers.

Berger (1984) introduced the idea of global robustness as an explicit recognition that no single prior distribution can be shown to be absolutely correct and a "class" of priors therefore provides more information about model uncertainty. The most frequent approach uses the so-called ϵ-contamination neighborhood priors. Starting with a specified prior, $p_0(\theta)$, identify a wider class of "contamination" forms that includes this prior but broadens the range of specific forms considerably: Q containing alternatives $q(\theta)$. It is also necessary to give a subjective probability indicating how *uncertain* we are about the selection of $p_0(\theta)$; label this probability ϵ. The class priors evaluation is the subset of Q given by all forms of $p(\theta)$ that satisfy:

$$\Gamma(Q, \epsilon) = \{p(\theta): p(\theta) = (1 - \epsilon)p_0(\theta) + (\epsilon)q(\theta), q(\theta) \in Q\}. \tag{6.5}$$

Thus $\Gamma(Q, \epsilon)$ is an ϵ-weighted compromise between the initial prior and the selected overarching class of alternatives. The primary question, of course, is how do we define Q? Choices include all possible contaminants (Berger and Berliner 1986), all symmetric and unimodal forms (Basu 1994, Sivaganesan and Berger 1989), unimodality-preserving forms (Moreno and González 1990), normals (Berger and Berliner 1986; see however, the warning in Lavine 1991a), more than one Q (Bose 1994a), mixtures (Bose 1994b), hierarchical classes (Moreno and Pericchi 1993; Sivaganesan 2000), and quantile restraints on Q (Moreno and Cano 1991; Moreno and Pericchi 1991, Ruggeri 1990).

Notice that ϵ is constant across the parameter space for θ, and it may be the case that particular regions such as the tails have higher or lower researcher confidence in $p_0(\theta)$. To accommodate this goal, a function, $\epsilon(\theta)$ can be defined so that prior uncertainty differs over Θ (Berger 1994; Liseo, Petrella and Salinetti 1996).

To show the effects of different contamination classes, Moreno (2000) gives the following hypothetical analysis. Suppose the data consist of a single data point distributed $\mathcal{N}(\theta, 1)$, the prior is specified as $\mathcal{N}(0, 2)$, and we set $\epsilon = 0.2$ (an atheoretic but common value in such analyses). Consider three contaminants of the form: all distributions ($q(\theta) \in Q_{\text{All}}$), all distributions with the same median as the prior ($q(\theta) \in Q_{\text{Median}}$), and all distributions with the same quartiles as the prior ($q(\theta) \in Q_{\text{Quartiles}}$). The posterior quantity of interest is the 95% HPD region for θ, and Table 6.4 shows the most extreme upper and lower bounds reached over each contaminating class for five different observed data values (calculations are found in Moreno [2000, p.50]).

This example was selected because it illustrates several aspects of robust Bayesian analysis with ϵ-contaminations. *First*, notice that the more we restrict the contamination class, the smaller the posterior range: the ranges decrease going right across the table. This means that there is a trade-off in this analysis between generalizeability of the robustness and practical utility: a very broad Q may lead to a large range and a very narrowly defined Q may lead to a small range, but neither one of these results tells as much as a broad class leading to a narrow range or a narrow class leading to a broad range. *Second*, since the data value is always greater than the prior mean here, the infimum of the posterior HPD regions is the more variable component, and this is true *even as the data value gets larger*

TABLE 6.4:   Oklahoma Recidivism 95% HPD
Regions, 1/1/85 to 6/30/99

x	Q_{All}	Q_{Median}	$Q_{Quartiles}$
0.5	[0.82:0.96]	[0.84:0.96]	[0.91:0.96]
1.0	[0.77:0.97]	[0.82:0.96]	[0.87:0.96]
2.5	[0.71:0.97]	[0.79:0.96]	[0.86:0.96]
3.0	[0.36:0.98]	[0.52:0.97]	[0.66:0.97]
4.0	[0.13:0.99]	[0.20:0.99]	[0.50:0.97]

reflecting variance scale uncertainty as well as location uncertainty. *Finally*, it is a clear (and welcome) result that the data matter in all of this: the range increases as the data point moves away from the prior mean.

Several problems often occur in performing this analysis in real data-analytic situations: the specification of the contamination class may be difficult in cases where the obvious choices are not reasonable, the calculations of posterior bounds can be foreboding in high-dimension problems, and the results may unfortunately be inconclusive.

This is an active research area and there are well-developed alternatives to ϵ-contamination forms, including those based on banding with standard CDF forms (Basu and DasGupta 1995), general quantile-restrained classes (Berger and O'Hagan 1988; O'Hagan and Berger 1988), quantile-restrained classes with robust prior properties (Moreno, Martinez, and Cano 1996), density-ratio classes (Lavine 1991b), Empirical Bayes (Maritz 1970; Morris 1983b), asymptotic Bayes risk (Berger 1984), entropy (Dey and Birmiwal 1994), and concentration functions (Regazzini 1992, Fortini and Ruggeri 2000).

6.3.2   Local Robustness

Because the scope of performing global sensitivity in a *completely* rigorous and thorough manner is beyond the resources of many research projects, there is a motivation for a more globally limited, but still informative, paradigm for assessing robustness in Bayesian models (Basu, Jammalamadaka, and Liu 1996; Cuevas and Sanz 1988; Delampady and Dey 1994; Gelfand and Dey 1991, Sivaganesan 1993). Local robustness evaluates the rate of change in posterior inferences that occur due to very small (infinitesimal) perturbations in the prior (Ruggeri and Wasserman 1993; Gustafson 1996). Because this process uses derivative quantities rather than integrals, it is often much easier to calculate than alternatives.

Define the locally perturbed prior to be a weighting of the original prior and the perturbance:

$$p^*(\theta) = (1 - \epsilon)p(\theta) + (\epsilon)q(\theta), \tag{6.6}$$

where $q(\theta)$ is a single identified disturbance prior here (rather than one of a class as in global robustness), and ϵ is not an expression of doubt about the original prior but a weighting of how much to perturb with $q(\theta)$. Given the likelihood function, $L(\theta|\mathbf{x})$, the subsequent ϵ-weighted (mixture) posterior is:

$$\pi_\epsilon(\theta|\mathbf{x}) = \frac{(1-\epsilon)p(\theta)L(\theta|\mathbf{x}) + (\epsilon)q(\theta)L(\theta|\mathbf{x})}{(1-\epsilon)m_p(\mathbf{x}) + (\epsilon)m_q(\mathbf{x})} \tag{6.7}$$

(O'Hagan 1994). This is the standard nonproportional derivation of the Bayesian posterior, but weighted by the marginal posterior densities of the data using the original prior ($m_p(\mathbf{x})$) and the disturbance prior ($m_q(\mathbf{x})$):

$$m_p(\mathbf{x}) = \int_\Theta p(\theta)L(\theta|\mathbf{x})d\theta, \qquad m_q(\mathbf{x}) = \int_\Theta q(\theta)L(\theta|\mathbf{x})d\theta. \tag{6.8}$$

Further define, according to Gustafson and Wasserman (1995), the difference between the ϵ-weighted posterior and the nondisturbed posterior as ϵ goes to zero:

$$D(q) = \lim_{\epsilon \to 0}(\pi_\epsilon(\theta|\mathbf{x}) - p(\theta)L(\theta|\mathbf{x})/m_p(\mathbf{x}))$$

$$= \left(\frac{q(\theta)L(\theta|\mathbf{x})}{m_q(\mathbf{x})} - \frac{p(\theta)L(\theta|\mathbf{x})}{m_p(\mathbf{x})}\right)\frac{m_q(\mathbf{x})}{m_p(\mathbf{x})}, \tag{6.9}$$

which gives $||D(q)||$ as the local (infinitesimal) posterior sensitivity from disturbing $p(\theta)$ by $q(\theta)$, and under very general conditions can be thought of as the derivative of the posterior with prior $p(\theta)$ in the direction of $q(\theta)$ (see Diaconis and Freedman 1986, especially p.13).

There are many possible choices for the disturbance prior, and Gustafson (2000, 75) gives a tabularized list for the best-known examples. One advantage to this approach is that it can readily be programmed in a language like R to cycle over many different alternative disturbance specifications, as well as to find worst-case behavior or average behavior across ranges of priors.

6.3.2.1 Bayesian Linear Outlier Detection

Outlier detection and robustizing is very straightforward in the Bayesian linear model. The basic idea, from Chaloner and Brant (1988), Zellner and Moulton (1985), and Zellner (1975), is to look at the modeled distribution of the error term after producing the posterior. Returning to the linear specification in Bayesian terms given in Section 5.1 with the improper prior $p(\theta, \sigma^2) = \frac{1}{\sigma}$, define the vector of residuals resulting from the insertion of maximum likelihood into (5.5):

$$\epsilon = \mathbf{y} - \mathbf{X}\hat{\theta}. \tag{6.10}$$

The "hat" matrix (Amemiya 1985, Chapter 1) is:

$$\mathbf{H} = \mathbf{X}(\mathbf{X}'\mathbf{X})^{-1}\mathbf{X}', \tag{6.11}$$

where we are principally interested in the diagonal values, h_{ii}, of the $\mathbf{H}$ matrix.[3] Outliers are defined to be those having residuals with high posterior probabilities of exceeding some arbitrary cutoff on a normalized scale, k. For a candidate point i, identify cutoff points, which identify the distance from the k-defined tail and the posterior residual value in both directions:

$$z_{1i} = \frac{k - \sqrt{\sigma}\epsilon_i}{\sqrt{h_{ii}}} \qquad z_{2i} = \frac{k + \sqrt{\sigma}\epsilon_i}{\sqrt{h_{ii}}}, \tag{6.12}$$

and the associated standard normal CDF values:

$$\Phi(z_{1i}) = \Phi\left(\frac{k - \sqrt{\sigma}\epsilon_i}{\sqrt{h_{ii}}}\right) \qquad \Phi(z_{2i}) = \Phi\left(\frac{k + \sqrt{\sigma}\epsilon_i}{\sqrt{h_{ii}}}\right). \tag{6.13}$$

The probability of interest is thus the weighted posterior area of the support of the variance term bounded by these cutoff points:

$$p(|\epsilon_i| > k\sigma|\mathbf{X}, \mathbf{y}) = \int (1 - \Phi(z_{1i}) - \Phi(z_{2i}))\pi(\sigma|\mathbf{X}, \mathbf{y})d\sigma, \tag{6.14}$$

where $\pi(\sigma|\mathbf{X}, \mathbf{y}) \propto \sigma^{-(n-k)-1} \exp\left[-\frac{1}{2\sigma^2}\hat{\sigma}^2(n-k)\right]$. This quantity can be compared for each data point to $2\Phi(-k)$ where larger values are inferred to be high probability outliers (Chaloner and Brant 1988, p.652). The probability that a given point is an outlier is therefore determined by either a large absolute model residual, $|\epsilon_i|$, or a small hat value.

Polasek (1984) provides a diagnostic for the general least squares estimator with weights given for the variance on the prior $p(\epsilon) \sim \mathcal{N}(\mathbf{0}, \mathbf{\Sigma})$, and a normal prior independent of $\mathbf{\Sigma}$: $p(\boldsymbol{\beta}) \sim \mathcal{N}(\mathbb{B}, \mathbf{S})$. Weighting in linear model analysis is very useful not only to compensate for heteroscedasticity, but also as a diagnostic tool (Birkes and Dodge 1993; Carroll and Ruppert 1988; the essays in Hoaglin, Mosteller, and Tukey 1983). The resulting setup from Polasek gives the conditional posterior for $\boldsymbol{\beta}$:

$$\pi(\boldsymbol{\beta}|\mathbf{\Sigma}, \mathbf{X}, \mathbf{y})$$
$$= \mathcal{N}\left((\mathbf{S}^{-1} + \mathbf{X}'\mathbf{\Sigma}^{-1}\mathbf{X})^{-1}(\mathbf{S}^{-1}\mathbb{B} + \mathbf{X}'\mathbf{\Sigma}^{-1}\mathbf{X}), (\mathbf{S}^{-1} + \mathbf{X}'\mathbf{\Sigma}^{-1}\mathbf{X})^{-1}\right) \tag{6.15}$$

which is not only very easy to work with, but also gives a simple way to test local sensitivity by changing the weights in the $\mathbf{\Sigma}$ matrix. Polasek gives the local sensitivity of the posterior mean and precision by taking the derivative of these with respect to the prior precision on

---

[3]The h_{ii} values are termed "leverage points" by Cook and Weisberg (1982, p.15), and are particularly useful in determining outlying values in a high-dimensional model where visualization is difficult. These values have several important characteristics: *(1)* the greater h_{ii} the more that the ith case determines the relationship between $\mathbf{y}$ and $\hat{\mathbf{y}}$, *(2)* the greater h_{ii} the smaller the variance of the ith residual, *(3)* defining the model standard error by $s^2 = \epsilon'\epsilon/(n-k)$ and therefore the ith contribution as $(s_i)^2 = (1 - h_{ii})s^2$, then the ith jackknifed residual (also called an externally studentized residual) is calculated from $s_{(i)}$, which is an estimate of σ when the regression is run omitting the ith case: $t_{(i)} = \epsilon_i/(s_{(i)}\sqrt{1 - h_{ii}})$. This statistic can also be thought of as the residual weighted inversely proportional to the jackknifed standard error.

ϵ:

$$\frac{\partial}{\partial \boldsymbol{\Sigma}^{-1}}\left[\mathbf{S}^{-1} + \mathbf{X}'\boldsymbol{\Sigma}^{-1}\mathbf{X}\right] = -\text{vec}(\mathbf{X})\text{vec}(\mathbf{X})'(\mathbf{S}^{-1} + \mathbf{X}'\boldsymbol{\Sigma}^{-1}\mathbf{X})$$

$$\frac{\partial}{\partial \boldsymbol{\Sigma}^{-1}}\left[\mathbf{S}^{-1} + \mathbf{X}'\boldsymbol{\Sigma}^{-1}\mathbf{X}\right]^{-1}(\mathbf{S}^{-1}\mathbb{B} + \mathbf{X}'\boldsymbol{\Sigma}^{-1}\mathbf{X})$$

$$= \text{vec}(\mathbf{X})(\mathbf{S}^{-1} + \mathbf{X}'\boldsymbol{\Sigma}^{-1}\mathbf{X})$$

$$\times \text{vec}(\mathbf{y} - \mathbf{X}(\mathbf{S}^{-1} + \mathbf{X}'\boldsymbol{\Sigma}^{-1}\mathbf{X})^{-1}(\mathbf{S}^{-1}\mathbb{B} + \mathbf{X}'\boldsymbol{\Sigma}^{-1}\mathbf{X}))', \qquad (6.16)$$

which is long but not complicated. The advantage of this setup is that once programmed, it is easy to run hypothetical priors through the matrix algebra and see posterior effects immediately. The unweighted local sensitivity of the ith data point is found by first replacing the $\boldsymbol{\Sigma}^{-1}$ matrix by an identity matrix substituting the ith one on the diagonal with the corresponding weight from the general linear model, w_i, to produce $\mathbf{W}_i$. The resulting unweighted estimate in the ith case is $b(w_i) = (\mathbf{X}'\mathbf{W}_i\mathbf{X})^{-1}\mathbf{X}'\mathbf{W}_i y_i$ with the local sensitivity given by $\partial b(w_i)/\partial w_i = (\mathbf{X}'\mathbf{X})^{-1}\mathbf{X}'e_i$ where e_i is the standard OLS residual for the data point i (Polasek 1984). This is a very simple setup and these priors can easily be developed in R (see the exercises).

Outlier and influence analysis in Bayesian models is certainly not restricted to the linear case. Kass, Tierney, and Kadane (1989) give outlier diagnostics for Bayesian models in general based on asymptotic principles by jackknifing out cases and looking at the subsequent change in the posterior expectation of some statistic of interest. Weiss (1996) takes a similar approach but uses the Bayes Factor (Chapter 7) to understand the resulting influence. Bradlow and Zaslavsky (1997) introduce a way of speeding up this process by importance weighting (Chapter 9) approximations of full jackknifing when the data size is large.

6.3.3 Bayesian Specification Robustness

An important question is whether it is possible to construct a prior, or class of priors, that provide desired robustness qualities before observing the data and conducting the inferential analysis. Lavine (1991a, 1991b) suggests a method for producing prior distributions that have similar support to some initially proposed prior, but possessing enhanced robustness characteristics. Suppose that the multidimensional prior vector of coefficients can be segmented according to $\theta = [\theta_1, \theta_2]$ on $\boldsymbol{\Theta}$ (the multidimensional sample space), and we specify a joint prior consisting of the product of a marginal and conditional prior: $p(\theta) = p^c(\theta_2|\theta_1)p^m(\theta_1)$. Lavine (1991b) finds four general classes of priors for the conditional prior, p^c, and the marginal or unconditional prior, p^m. These classes are designated Γ^c and Γ^m, and possess robust qualities for the forms: *(1)* quantile partitioning of prior space, *(2)* the ϵ-contaminated class (Berger and Berliner 1986), *(3)* a density ratio form of upper and lower limits, and *(4)* a density-bounded class. The primary advantage of this approach is that the problem of finding tightly bounded posterior inferences for a large class of priors reduces to a manageable sequence of linear optimizations (Lavine 1991b, p.401).

6.4   Comparing Data to the Posterior Predictive Distribution

Rubin (1984) recommends posterior predictive distributions because regardless of the complexity of the model specification, "model monitoring" is easily accomplished with simple summary statistics. This idea integrates the two previous approaches of sensitivity analysis and robust evaluation because it simultaneously tests for the two by looking at posterior reasonableness through simulated data. The basic idea is that if the data generating mechanism implied by the calculated posterior is wildly divergent from the original data or from data that we would expect given substantive knowledge, then the Bayesian analysis is suspect. Furthermore, this idea can also be extended to predictive model selection (Laud and Ibrahim 1995).

First consider a *prior predictive distribution* of a new data value, x_{new} before observing the full dataset:

$$p(x_{\text{new}}) = \int_\Theta p(x_{\text{new}}, \theta) d\theta = \int_\Theta p(x_{\text{new}}|\theta) p(\theta) d\theta. \qquad (6.17)$$

In other words the marginal distribution of an unobserved data value is the product of the prior for θ and the single variable PDF or PMF, integrating out this parameter. This makes intuitive sense as uncertainty in θ is averaged out to reveal a distribution for the data point.

More usefully, from a diagnostic perspective, is the distribution of a new data point, x_{new} *after* the full iid data set, $\mathbf{x}$, has been observed: the posterior predictive distribution. This is produced by:

$$p(x_{\text{new}}|\mathbf{x}) = \int_\Theta p(x_{\text{new}}, \theta|\mathbf{x}) d\theta = \int_\Theta \frac{p(x_{\text{new}}, \theta|\mathbf{x})}{p(\theta|\mathbf{x})} p(\theta|\mathbf{x}) d\theta$$

$$= \int_\Theta p(x_{\text{new}}|\theta, \mathbf{x}) p(\theta|\mathbf{x}) d\theta. = \int_\Theta p(x_{\text{new}}|\theta) p(\theta|\mathbf{x}) d\theta. \qquad (6.18)$$

The last simplification comes from the assumption that x_{new} and $\mathbf{x}$ are independent. The integral means that the posterior predictive distribution is the product of the single variable PDF or PMF times the full data likelihood in which we integrate over uncertainty in θ to result in a probability statement that is dependent on the observed data only.

For example, suppose we consider the now familiar example from Chapter 3 in which $X_1, X_2, \ldots, X_n$ are distributed iid $\mathcal{N}(\mu, \sigma_0^2)$, and σ_0^2 is known but μ is unknown. Placing a normal prior on μ according to $\mu \sim \mathcal{N}(m, s^2)$ gives:

$$\pi(\mu|\mathbf{x}) \propto \exp\left[-\frac{1}{2}\left(\frac{1}{s^2} + \frac{n}{\sigma_0^2}\right)\left(\mu - \frac{\left(\frac{m}{s^2} + \frac{n\bar{x}}{\sigma_0^2}\right)}{\left(\frac{1}{s^2} + \frac{n}{\sigma_0^2}\right)}\right)^2 \right]. \qquad (6.19)$$

As a simplifying procedure, re-express this posterior in terms of its mean and variance:

$$\pi(\mu|\mathbf{x}) \propto \exp\left[-\frac{1}{2\sigma_1^2}(\mu - \mu_1)^2\right]$$

$$\text{where: } \sigma_1^2 = \left(\frac{1}{s^2} + \frac{n}{\sigma_0^2}\right)^{-1} \qquad \mu_1 = \frac{\left(\frac{m}{s^2} + \frac{n\bar{x}}{\sigma_0^2}\right)}{\left(\frac{1}{s^2} + \frac{n}{\sigma_0^2}\right)}.$$

Therefore (6.18) for this model is:

$$p(x_{\text{new}}|\mathbf{x}) = \int_\mu p(x_{\text{new}}|\mu)p(\mu|\mathbf{x})d\mu$$

$$\propto \int_\mu \exp\left[-\frac{1}{2}\left(\frac{(x_{\text{new}} - \mu)^2}{\sigma_0^2} + \frac{(\mu - \mu_1)^2}{\sigma_1^2}\right)\right]d\mu. \qquad (6.20)$$

This allows us to calculate summary statistics for the posterior predictive distribution:

$$E[x_{\text{new}}|\mathbf{x}] = E[E(x_{\text{new}}|\mu, \mathbf{x})|\mathbf{x}]$$

$$= E[E(\mu|\mathbf{x})]$$

$$= E[\mu] = \mu_1.$$

$$\text{Var}[x_{\text{new}}|\mathbf{x}] = E[\text{Var}(x_{\text{new}}|\mu, \mathbf{x})|\mathbf{x}] + \text{Var}[E(x_{\text{new}}|\mu, \mathbf{x})|\mathbf{x}]$$

$$= E[\sigma_0^2|\mathbf{x}] + \text{Var}[\mu|\mathbf{x}]$$

$$= \sigma_0^2 + \sigma_1^2/n. \qquad (6.21)$$

These results demonstrate that the posterior predictive distribution in this example has the same expected value as the posterior but greater variance, reflecting additional uncertainty about an *unobserved* quantity rather than the posterior description of the *observed* quantities, $\mathbf{x}$. Furthermore, since we know that the distribution for x_{new} is normal, then we have the unconditional PDF:

$$x_{\text{new}}|\mathbf{x} \sim \mathcal{N}\left(\frac{\left(\frac{m}{s^2} + \frac{n\bar{x}}{\sigma_0^2}\right)}{\left(\frac{1}{s^2} + \frac{n}{\sigma_0^2}\right)}, \sigma_0^2 + \left(\frac{1}{s^2} + \frac{n}{\sigma_0^2}\right)^{-1}\frac{1}{n}\right), \qquad (6.22)$$

which we can easily draw from.

Since the parametric form for additional observations is fully defined by (6.18), multiple draws can be made to form a simulated data set. If the replicated data have distributional qualities noticeably different from the observed data, then this is an indication of poor model fit (Gelman *et al.* 2003, Meng, 1994a). The posterior predictive distribution of the data can also be used to make explicit model comparisons between considered alternatives (Chen, Dey, and Ibrahim 2000).

■ **Example 6.2:** **Economic Growth in Sub-Saharan Africa.** This example makes use of data provided by Bratton and Van De Walle (1997) from various authoritative and reliable sources, made available through the *Inter-University Consortium for*

Political and Social Research (ICPSR 1996). The authors collect data on regime transition for 47 sub-Saharan countries over the period from each colonial independence to 1989, with some additional variables collected for the period 1990 to 1994. These data include 99 variables describing governmental, economic, and social conditions for the 47 cases. Also included are data from 106 presidential and 185 parliamentary elections, including information about political parties, turnout, and openness. We use only the average annual rate of growth in GNP per capita, in percent. One case is missing this value, so $n = 46$ here.

TABLE 6.5: SUB-SAHARAN AFRICA AVERAGE ECONOMIC GROWTH RATE, 1990-1994

Average Economic Growth by African Country, 1965-1989

Angola	-2.1	Benin	-0.1	Botswana	8.5	Burkina Faso	1.4
Burundi	3.6	Cameroon	3.2	Cape Verde	4.0	Ctrl. Afr. Repub.	-0.5
Chad	-1.2	Comoros	0.5	Congo	3.3	Cote d'Ivoire	0.8
Djibouti	NA	Eq.Guinea	0.0	Ethiopia	-0.1	Gabon	0.9
Gambia	0.7	Ghana	-1.5	Guinea	0.0	Guinea Bissau	-3.0
Kenya	2.0	Lesotho	5.0	Liberia	-2.0	Madagascar	-1.9
Malawi	1.0	Mali	1.7	Mauritani	-0.5	Mauritius	3.0
Mozambique	-8.2	Namibia	-1.2	Niger	-2.4	Nigeria	0.2
Rwanda	1.2	Saotome	1.5	Senegal	-0.7	Seychelles	3.2
Sierra Leone	0.2	Somalia	0.3	S.Africa	3.5	Sudan	-2.0
Swaziland	2.1	Tanzania	-0.1	Togo	0.0	Uganda	-2.8
Zaire	-2.0	Zambia	-2.0	Zimbabwe	1.2		

We will start with the prior $\mathcal{N}(0, 4)$ for μ, which might be somewhat cynical, and set $\sigma_0^2 = 7$ as if it were known. In order to describe the posterior predictive distribution we plot 5000 draws from the distribution given by (6.22), as shown in Figure 6.6. In this figure the solid line is a density estimate from the draws and the dashed line shows the prior distribution.

6.5   Simple Bayesian Model Averaging

Anything not known for certain in the Bayesian framework is treated probabilistically. There is no reason that model specification, which is rarely known for certain, cannot also be

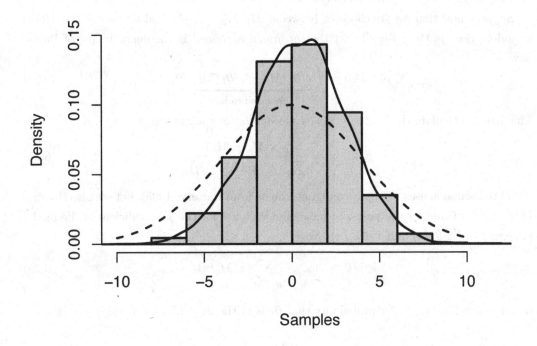

FIGURE 6.6: Posterior Predictive Distribution Draws, Africa Growth

treated as such. There are two key components to this approach: assigned priors to *models*, and producing posterior distributions that are averaged across model space. This idea of putting priors on models makes sense since we typically have information that supports some specifications over others. For recent discussions see Montgomery and Nyhan (2010) and Young (2009).

A starting point is the idea of a mixture distribution described in detail in Section 3.9 starting on page 87. We could create a likelihood based on mixtures by stipulating K distinct parametric forms that might be used to explain the data generation process, weighting these $\omega_1, \omega_2, \ldots, \omega_K$, where $\sum \omega_i = 1$. We will stipulate that each of the parametric forms is from the same parametric family, but differs in parametric assignment. Thus $f_k(x|\boldsymbol{\theta}_k)$ is the kth specification for x with values given in the parameter vector $\boldsymbol{\theta}_k$. Now the likelihood for this mixture is given by:

$$L_m(\boldsymbol{\theta}|\mathbf{x}) = \prod_{i=1}^{n} \sum_{k=1}^{K} \omega_k f_k(x_i|\boldsymbol{\theta}_k). \tag{6.23}$$

The likelihood function is therefore the product of the mixture sums for each data value. Mixture models of this type are very flexible (see Titterington, Smith, and Makov [1985]), but suffer from the problem that lots of quantities need to be specified directly by the researcher: the mixture proportions, and the full $\boldsymbol{\theta}$ vector for each mixture component. So for a mixture of ten normals, thirty constants need to be provided. For discussion of this issue and implementation concerns in more complicated scenarios, see Diebolt and Robert (1994), Kalyanam (1996), Stephens (2000), and West (1992).

Suppose now that we are choosing between: $M_1, M_2, \ldots, M_K$ and for each k, determine a model prior: $p(M_k)$. Recall that the *integrated likelihood* is the denominator of Bayes' Law:

$$p(\mathbf{x}|M_k) = \int \underbrace{p(\boldsymbol{\theta}_k|M_k, \mathbf{x})p(\boldsymbol{\theta}_k|M_k)}_{\text{likelihood} \times \text{prior}} \, d\boldsymbol{\theta}_k. \tag{6.24}$$

This lets us calculate the *posterior model probability* for each model, $\kappa = 1 \ldots K$:

$$\pi(M_\kappa|\mathbf{x}) = \frac{p(\mathbf{x}|M_\kappa)p(M_\kappa)}{\sum_{j=1}^{K} p(\mathbf{x}|M_j)p(M_j)}. \tag{6.25}$$

This discussion uses a simple averaging scheme from Raftery (1995), but see also Hoeting *et al.* (1999). Consider the posterior mean and variance for the jth coefficient of the model indexed by κ^{th}, $\boldsymbol{\theta}_j(\kappa)$ and $Var_{\boldsymbol{\theta}_j}(\kappa)$. Now:

$$p(\boldsymbol{\theta}_j \neq 0|\mathbf{x}) = \sum_{\boldsymbol{\theta}_j \in M_\kappa} \pi(M_\kappa|\mathbf{x}), \tag{6.26}$$

which is just the "posterior probability that $\boldsymbol{\theta}_j$ is in the model," as well as:

$$E[\boldsymbol{\theta}_j|\mathbf{x}] \approx \sum_{\kappa=1}^{K} \boldsymbol{\theta}_j(\kappa)\pi(M_\kappa|\mathbf{x}) \tag{6.27}$$

and:

$$\text{Var}[\boldsymbol{\theta}_j|\mathbf{x}] \approx \sum_{\kappa=1}^{K} \left[(\text{Var}_{\boldsymbol{\theta}_j}(\kappa) + \boldsymbol{\theta}_j(\kappa)^2)\pi(M_\kappa|\mathbf{x})) - E[\boldsymbol{\theta}_j|\mathbf{x}]^2 \right], \tag{6.28}$$

where we apply standard definitions to these new quantities. Note that the summation in both the expected value and the variance calculations are both averages when (as by convention) the prior model weights sum to one. Criticism of this approach, in particular generation of values such as (6.26) and the subsequent summaries, is that the researcher has made decisions about what variables to include in the various specifications and can therefore increase or decrease expected values or variances by strategical selection. However, since inclusions or exclusions are overt, readers can judge the quality of resulting averages.

6.6   Concluding Comments on Model Quality

The quality of a statistical model is a function of several criteria:

▷ Parsimony

▷ Completeness and generalizeability

▷ Clarity

▷ Robustness

▷ Resistance

▷ Predictive ability

▷ Adherence to trends in the data.

General assessment of model quality is a vast literature in both Bayesian and classical statistics; landmarks include Barnett (1973), Blalock (1961), Box (1980), Howson and Urbach (1993), Leamer (1978), Miller (2002), and Russell (1929). Regretfully, recommended criteria can be at odds with each other. For example, parsimony comes at the expense of closeness to the data in that we can never *reduce* the overall fit of the model by adding more explanatory variables, but this often creates overparameterized and theoretically confusing results (Kreuzenkamp and McAleer 1995). In a familiar example, the linear model is well-known to be fairly robust to mild deviations from the underlying Gauss-Markov assumptions (Shao 2005), but it is not very resistant to outliers in the data (Leamer 1984, Rao and Toutenburg 1995, Chapter 9).

Model checking is the general procedure for comparing final conclusions with our expectations or knowledge of reality. Obviously we will want to dismiss any model specification that produces inferences (Bayesian or otherwise) that do not comport with some sense of substantive reasonableness. Where this arises most often is when certain model features are specified for mathematical or data collection convenience. Suppose that a conjugate prior is specified for a given model without a great amount of substantive justification. A reader is less inclined to be critical if it can be shown that the consequences of the posterior distribution are in accordance with sound thinking in terms of the theoretical or situational setting. Conversely, one should be wary of assumptions like conjugate priors applied merely for analytical ease if they provide dubious substantive conclusions.

Model checking, through sensitivity analysis or robustness evaluation, is not a panacea. Bartels points out that *ad hoc* nonsystematic sensitivity analysis can produce wildly wrong statements about model quality (1997, pp.668-669). We should therefore see these tools in a more general context as components of the broader task of checking that also includes looking at the data, carefully considering the likelihood function used, and reasonableness checks with posterior predictive distributions.

The Bayesian approach to model checking also circumvents a common problem in social science data analysis. Typically when a social science model with a null hypothesis significance test is reported, it is presented as if only two models were ever considered or ever deserved to be considered: the null hypothesis and the provided research hypothesis. The quality of a research finding is then solely judged on the ability to reject the complementary null hypothesis with a sufficiently low p-value.

This process proceeds in an artificially exclusive manner regarding the selecting of a model $\mathbf{M}$: "...examining the data in x to identify the single 'best' choice $\mathbf{M}^*$ for $\mathbf{M}$, and then proceeding as if $\mathbf{M}^*$ were known to be correct in making inferences and predictions" (Draper 1995, p.46). However, during the development of the reported model, many differing alternate mixes of independent variables are tested. This is an "illusion of theory confirmation" (Greenwald 1975; Lindsay 1995) because the null hypothesis significance test is presented as evidence of the exclusivity of explanation of this single research hypothesis. Summary statistics are reported from the nth equation as if the other $n - 1$ never existed and this last model is produced from a fully controlled experiment (Leamer 1978, p.4). The null hypothesis significance test thus provides an infinitely strong bias in favor of a single research hypothesis against a huge number of other reasonable hypotheses: all other distributional alternatives are assumed to have probability zero (Rozeboom 1960, Lehmann 1986, 68, Popper 1968, 113). Worse yet are solutions that rely upon atheoretic stepwise or other mechanical search processes; for a litany and review, see Adams (1991), or Sala-I-Martin (1997).

Two entirely reasonable and statistically significant, competing models can lead to substantively different conclusions using exactly the same data (Raftery 1995). This leads to the question of determining which of the two equally plausible specifications is in some way "better." The answer is generally one in which the posterior is less sensitive to changes in the underlying assumptions (robustness) and less sensitive to influential outliers (resistance). The described process of *sensitivity analysis* makes small changes in the assumptions of the model, such as the parameterization of the prior distribution, and observes whether or not unreasonably large changes are observed in the subsequent posterior. We can therefore determine which assumptions are worth further investigation, and possibly which conclusions are more "fragile." This illustrates how the Bayesian approach is more naturally inclusive of alternatives and more overtly skeptical about assumptions.

6.7   Exercises

6.1   Derive the marginal posterior for β in

$$\pi(\beta|\mathbf{X}, \mathbf{y}) \propto \left[\tilde{s} + (\beta - \tilde{\beta})'(\mathbf{\Sigma}^{-1} + \mathbf{X}'\mathbf{X})(\beta - \tilde{\beta}) \right]^{-\frac{n+a}{2}},$$

where:

$$\tilde{\beta} = (\mathbf{\Sigma}^{-1} + \mathbf{X}'\mathbf{X})^{-1}(\mathbf{\Sigma}^{-1}\mathbb{B} + \mathbf{X}'\mathbf{X}\hat{\mathbf{b}})$$

$$\tilde{s} = 2b + \hat{\sigma}^2(n - k) + (\mathbb{B} - \tilde{\beta})'\mathbf{\Sigma}^{-1}\mathbb{B} + (\hat{\mathbf{b}} - \tilde{\beta})'\mathbf{X}'\mathbf{X}\hat{\mathbf{b}}$$

from the joint distribution given by

$$\pi(\beta, \sigma^2|\mathbf{X}, \mathbf{y})$$

$$\propto \sigma^{-n-a} \exp\left[-\frac{1}{2\sigma^2} \left(\tilde{s} + (\beta - \tilde{\beta})'(\mathbf{\Sigma}^{-1} + \mathbf{X}'\mathbf{X})(\beta - \tilde{\beta}) \right) \right].$$

See Chapter 5 (page 152) for terminology details.

6.2 Using the JAGS code in Example 6.1.1 starting on page 177, replicate Figure 6.1 on page 180 with different variance components priors. Is it possible to improve the visual fit? See Chapter 11 for details on running the JAGS software.

6.3 Derive the form of the posterior predictive distribution for the beta-binomial model. See Section 6.4 starting on page 196.

6.4 On page 180 it was noted that the IQR of $\mathcal{N}(\theta, 1)$ distribution matches that of $\mathcal{C}(\theta, 0.675)$. Find the Cauchy PDF with matching $(0.05, 0.95)$ quantiles for a $\mathcal{N}(\theta, 2.5)$.

6.5 Calculate the Kullback-Leibler distance between two gamma distributions, $f(\mathbf{X}|\alpha, \beta)$, $g(\mathbf{X}|\gamma, \delta)$, based on the same data but different parameters.

6.6 Prove that the integral in (6.3) on page 183 is finite.

6.7 Calculate the Polasek (6.15) linear model diagnostic for each of the data points in your model from Exercise 5.7. Graphically summarize your result identifying points of concern.

6.8 Given an iid random sample $\mathbf{X} \sim \mathcal{N}(\theta, \sigma^2)$, define a normal conjugate neighborhood prior for θ as $Gamma = \{\mathcal{N}(\mu, \tau^{-1}) : 0 < \tau^L < \tau < \tau^U < \infty \}$. The posterior expectation of θ is $E[\theta|\mathbf{X}, \mu, \tau] = (n\sigma^2 \bar{\mathbf{X}} + \tau\mu)/(n\sigma^2 + \tau)$. Derive the range of this expectation.

6.9 In Section 6.2.2 it was shown how to vary the prior variance 50% by manipulating the beta parameters individually. Perform this same analysis with the two parameters in the Poisson-gamma conjugate specification. Produce a graph like Figure 6.4 on page 187.

6.10 Show that the prior in (6.6) produces the posterior in (6.7) (page 192), and plot a comparison for difference values of ϵ with simulated data.

6.11 The following data (`firearm.deaths` in BaM) are annual firearm-related deaths in the United States per 100,000 from 1980 to 1997, by row (source: National Center for Health Statistics, Health and Aging Chartbook, August 24, 2001):

14.9	14.8	14.3	13.3	13.3	13.3	13.9	13.6	13.9
14.1	14.9	15.2	14.8	15.4	14.8	13.7	12.8	12.1

Specify a normal model for these data with a diffuse normal prior for μ and a diffuse inverse gamma prior for σ^2 (see Section 3.5 starting on page 76 for the more complex multivariate specification that can be simplified). In order to obtain local robustness of this specification, calculate $||D(q)||$ (see page 193) for the alternative priors: $q(\mu) = \mathcal{N}(14, 1)$, $q(\mu) = \mathcal{N}(12, 3)$, $q(\sigma^{-2}) = \mathcal{G}(1, 1)$, $q(\sigma^{-2}) = \mathcal{G}(2, 4)$.

6.12   Berger (1990) gives four important features for a class of priors from expert elicitation: (1) simple elicitation and interpretation, (2) including as many reasonable
forms as possible, (3) no unreasonable forms that are precluded by measurement or
functional form, and (4) easy calculations. Find a published example that violates
at least one of these principles.

6.13   Calculate the posterior predictive distribution of X in Exercise 6.11 assuming that
$\sigma_0^2 = 0.75$ and using the prior: $p(\mu) = \mathcal{N}(15, 1)$. Perform the same analysis using
the prior: $\mathcal{C}(15, 0.675)$. Describe the difference graphically and with quantiles.

6.14   Returning to the Africa economic growth data from Example 6.4, notice that
Botswana appears to be an outlier at 8.5% growth. Calculate the probability of
getting this value or higher from the prior distribution given in the example and
the posterior predictive distribution calculated. Make a histogram of the posterior
predictive distribution and draw a vertical bar at the value for Botswana.

6.15   Calculate the posterior predictive distribution for the Palm Beach County model
(Section 5.1.1 starting on page 148) using both the conjugate prior and the uninformative prior.

6.16   The cross-validation (also called the leave-one-out) posterior predictive distribution
is given by:

$$p(x_{\text{new}}|\mathbf{x}_{-i}) = \int_{\Theta} p(x_{\text{new}}|\theta)p(\theta|\mathbf{x}_{-i})d\theta.$$

where $\mathbf{x}_{-i}$ denotes the dataset removing the ith case (Stern and Cressie 2000). A
small value of $p(x_{\text{new}}|\mathbf{x}_{-i})$ means that this new value will occur infrequently in the
absence of $\mathbf{x}_i$. Further define:

$$A = \frac{1}{n} \sum_{i=1}^{n} \log\left(p(x_{\text{new}}|\mathbf{x}_{-i})\right),$$

which is smaller for better fitting models and asymptotically equivalent to the AIC
for iid samples (Stone 1977a, 1977b). Replicate the plot from Example 6.4 using
the cross-validation posterior predictive distribution. Observe any differences from
the original analysis. Calculate the A statistic.

6.17   Using the model specification in (6.1), produce a posterior distribution for appropriate survey data of your choosing. Manipulate the forms of the priors and indicate
the robustness of this specification for your setting. Summarize the posterior results
with tabulated quantiles.

6.18   Calculation of the posterior predictive distribution in cases where the integral cannot be calculated or standard parametric forms are not used can be challenging.
Fortunately, the common use of MCMC estimation in Bayesian modeling makes
this task easier. Suppose X and Y are bivariate normal with mean vector $(0, 0)$

and covariance matrix $\begin{bmatrix} 1 & \rho \\ \rho & 1 \end{bmatrix}$, with $\rho = 0.75$. Write a Gibbs sampler in R that draws from the full conditional distribution:

$$X|Y \sim \mathcal{N}(\rho y, 1 - \rho^2) = \rho y + \sqrt{1 - \rho^2}\mathcal{N}(0, 1)$$

$$Y|X \sim \mathcal{N}(\rho x, 1 - \rho^2) = \rho x + \sqrt{1 - \rho^2}\mathcal{N}(0, 1).$$

and show graphically that the marginal distributions are as expected. Take a bivariate sample of $n = 100$ and treat this as observed data. Now estimate ρ as if unknown and produce a sample of posterior predictive values from this result. Do they differ from the Gibbs sampling output?

6.19 Leamer (1983) proposes reporting extreme bounds for coefficient values during the specification search as a way to demonstrate the reliability of effects across model-space. Thus the researcher would record the maximum and minimum values that every coefficient took on as different model choices were attempted. Explain how this relates to Bayesian model averaging.

6.20 Observed data of size n are known to be classified into $k = 1, \ldots, K$ groups and distributed $x_k \sim \mathcal{N}(\theta_k, s_k^2)$ where the θ_k parameters are distributed $\mathcal{N}(\mu, \sigma^2)$. In this hierarchical setting (continued in Chapter 12) posterior conclusions about θ and μ are known to be sensitive to the prior treatment of the random effects standard deviation σ and many approaches have been suggested. DuMouchel and Normand (2000) suggest a uniform prior for $s_0/(s_0 + \sigma)$ where s_0^2 is the harmonic mean of the group variances: $(\frac{1}{K}\sum s_k^{-2})^{-1}$. Show that this means that $p(\sigma) = s_0/(s_0 + \sigma)^2$ and $p(\sigma^2) = s_0/(2\sigma(s_0 + \sigma)^2)$.

Chapter 7

Bayesian Hypothesis Testing and the Bayes Factor

7.1 Motivation

This chapter systematically explores Bayesian hypothesis testing, and describes the Bayes Factor as the evidence of the quality of one model specification over another. We start with a general discussion of the state of hypothesis testing in the social and behavioral sciences and outline where Bayesian posterior descriptions can replace a number of problematic practices. The emphasis will be mainly on using Bayesian tools for determining the strength of evidence, rather than the details of the many differences between Bayesian and non-Bayesian approaches. For a very clear, and succinct, discussion of the historical developments in this regard, see Marden (2000).

The material in this chapter assumes knowledge of generalized linear models, although basic likelihood theory is briefly reviewed. Appendix A contains a review of McCullagh and Nelder (1989) style GLM theory. In particular we will see their treatment of the exponential family form when discussing the structure of prior distributions. Excellent and accessible works on likelihood theory abound. Among the most commonly assigned are Berger and Wolpert (1988), Cramer (1994), Edwards (1992), Eliason (1983), and King (1989). There are also an increasing number of texts on generalized linear models. Dobson (1990) is particularly accessible. The classic and enduring work is McCullagh and Nelder (1989), and more recent additions include Lindsey (1997) and Fahrmeir and Tutz (2001).

More topically, this is a chapter about *model testing*: determining why we should prefer one model specification over another. Obviously practically inclined social science researchers have many potential specifications to choose from, and it turns out that the Bayesian paradigm has powerful and sophisticated tools for comparing alternatives. The core advantage over non-Bayesian approaches is that *everything* unknown is given a probability assessment, including model choices. This facilitates clear comparisons and stands in direct contrast to the methods in the last chapter, which focused on *model adequacy* for a single specification.

Bayesian hypothesis testing is often less formal than the non-Bayesian varieties. By far, the most common procedure for summarizing results in social science research is to simply describe the posterior distribution rather than to apply a rigid decision process. However,

Bayesian decision theory (Chapter 8) is a well-developed area, particularly in those fields where costs, risks, and consequences in general are measurable (Cyert and DeGroot 1987). More basically, the Bayes Factor described in this chapter allows a very general and directly implementable means of model comparison that is already popular in the Bayesian social sciences (Meeus, *et al.* 2010, Pang 2010, Quinn, Martin, and Whitford 1999, Raftery 1995, Sened and Schofield 2005, Zhang and Luck 2011).

The null hypothesis significance testing paradigm that currently dominates tests of statistical reliability in the social sciences is seriously defective and widely misunderstood. A fixation with arbitrary thresholds and unjustified frequentist assumptions has damaged empirical research in many fields for quite some time (Bakan 1960, Cohen 1994, Hunter 1997, Meehl 1978, Pollard 1993, Rozeboom 1960, Serlin and Lapsley 1993, Schmidt 1996). The use of p-values and "stars" (asterisks on tables) as evidence for or against the null hypothesis is flawed in many social science studies since the long-run probabilistic assumption of Neyman-Pearson, upon which these measures are built, generally does not apply to single-point cross-sectional studies with uniquely occurring data. Social scientists employ standard Neyman-Pearson frequentist standards and practices without meeting the key underlying assumption of long-run replicability, and therefore average over unobserved and unlikely events.

There is a lengthy but frustrated literature attempting to reconcile frequentist and Bayesian approaches with the idea that the resulting inferences should be the same if one could just carefully restrict the *a priori* model specifications (Bartholomew 1965; Casella and Berger 1987a; Berger, Brown, and Wolpert 1994; Berger, Boukai, and Wang 1997; De-Groot 1973; Good 1983a, 1992; Good and Crook 1974; Jeffreys 1961; Pratt 1965). Detailed and sophisticated works have still produced only limited success in a series of special cases that are not of general interest.

Also, there is much less controversy about Bayesian inference and Bayesian hypothesis testing than there was a few decades ago. This is because of an increasing recognition that the Bayesian approach, which treats all unknown quantities with distributional statements, is closer to common scientific intuition. Social and behavioral scientists typically do not have the unending stream of iid data that canonical Neyman-Pearson frequentist inference requires. Consider that the standard confidence interval means that on average 19 times out 20, for $\alpha = 0.05$, the interval covers the true parameter value. Yet we generally have only one sample of observational data, often collected by others. Also recall that maximum likelihood analysis is equivalent to a Bayesian setup with the appropriately bounded uniform distribution prior. It is also true that the two approaches lead to identical inferences asymptotically for *any* proper prior distribution specification (i.e., the data will eventually overwhelm the prior knowledge). Rather than treat the previous facts as a comforting reason to continue advocating traditional likelihoodist practices, along with the accompanying null hypothesis significance test, we observe that Bayesians have shown that uniform prior specifications can lead to a number of problematic posterior results. The fact that uniform priors are not invariant under nonlinear transformation was critical to Fisher's wholesale (and vitriolic) rejection of Bayesian inference (Fisher 1930; 1956, Chapter 2). In addition, there are many situations where uniform priors are unreasonable starting points like elections with two major party candidates and a third fringe candidate. Why would a researcher pretend that these three have equal probability of winning the election before collecting data?

7.2 Bayesian Inference and Hypothesis Testing

Bayesian modes of inference can be divided into two basic approaches: one in which basic descriptions of the posterior are provided as evidence of some effect (as we have been doing up until this point), and one in which explicit testing mechanisms are performed. The second approach is due mainly to Jeffreys (1961). Often posterior description and articulated testing are not provided as exclusive demonstrations of evidence of some effect, but are given in conjunction. When there are multiple competing model specifications arising either from theoretical propositions or from alternative specifications of the same theory, a set of posterior distributions is produced, requiring some method for comparison. The most straightforward is the ratio of the posterior probability of some specification relative to another. This posterior odds ratio gives the odds of one model relative to another and is called the Bayes Factor (discussed in detail in this chapter).

7.2.1 Problems with Conventional Hypothesis Testing

The standard process for hypothesis testing in the social sciences is an odd mix that can be called *quasi-freqentist*. Suppose we observe $X_1, X_2, \ldots, X_n$ iid $f(x|\theta)$, where θ is some unknown value on the parameter space Θ. A one-sided (non-nested) test is defined by:

$$H_0\colon \theta \leq 0 \quad vs. \quad H_1\colon \theta > 0, \tag{7.1}$$

and a two-sided (nested) test is similarly defined by:

$$H_0\colon \theta = 0 \quad vs. \quad H_1\colon \theta \neq 0. \tag{7.2}$$

In the standard setup used in empirical social and behavioral science analysis, a test statistic (T), some function of θ and the data, is calculated and compared with its known distribution under the assumption that H_0 is true. Commonly used test statistics are sample means $(\bar{X})$, chi-square statistics (χ^2), and t-statistics in linear (OLS) regression analysis. The test procedure assigns one of two decisions (D_0, D_1) to all possible values in the sample space of T, which correspond to supporting either H_0 or H_1, respectively. The p-value ("associated probability") is equal to the area in the tail (or tails; we will illustrate with a one-tailed discussion for now) of the assumed distribution under H_0 ($\theta = 0$ fixed), which starts at the point designated by the placement of T on the horizontal axis and continues to infinity:

$$p(x) = p(T(x) \geq T | \theta = 0) = \int_T^\infty f(t|\theta = 0)dt. \tag{7.3}$$

The sample space of T is segmented into two complementary regions (S_0, S_1), whereby the probability that T falls in S_1, causing decision D_1, is either a predetermined null hypothesis cumulative distribution function (CDF) level: the probability of getting this or some lower value given a specified parametric form such as normal, F, t, etc. (α = size of the test, Neyman and Pearson 1928a, 1928b, 1933a, 1933b), or the cumulative distribution function level corresponding to the value of the observed test statistic under H_0 is reported (i.e., the p-value = $\int_{S_1} P_{H_0}(T = t)dt$, Fisher [1925a]).

There are many criticisms of the use of p-values in empirical work. Hwang *et al.* (1992) point out that since the p-value is the density under the null hypothesis starting at the test statistic and continuing to infinity on the support for θ, that is an average over unlikely sample values that have not actually occurred. In addition, Berger and Wolpert (1984) note that this definition violates the likelihood principle (Birnbaum 1962) that inferences must come from observed, not hypothetical, data. Casella and Berger (1987a) observe that there is nothing "frequentist" about a p-value since it is not the probability of a Type I error. Koop (1992) shows that classic frequentist analysis with p-value evidence fails to provide evidence of unit root problems in time-series analysis. While Bayesian time-series models are not within the main scope of this text (see Pole, West, and Harrison [1994] as a good starting point), it should be mentioned that asymptotic analysis for the unit root problem is a serious problem in the non-Bayesian setting and is well-behaved for Bayesian models, a qualitative difference that is even greater when the estimated parameters occupy restricted space.

A core problem (there are many) with the null hypothesis significance test as practiced is that researchers pretend to select α levels *a priori* as in experiments based on Neyman-Pearson, but actually report p-values (or worse yet, ranges of p-values indicated by asterisks) as the strength of evidence: quasi-frequentism. This is because the social sciences are encumbered with Fisher's arbitrary thresholds (even he later recanted), despite the fact that there has *never* been a theoretical justification to support 0.01, 0.05, and 0.10 levels. Aitkin (1991) observes that atheoretic use of fixed test sizes leads to an "unreasonable test in completely specified models," and Barnard (1991, discussion of Aitkin) points out that these conventions originated from the lack of ready computing (i.e., the propagation of Fisher's tables). Because the test is typically performed once on a set of social data in time and will not reoccur in the same fashion, the reported p-value is not a long run frequentist probability. Furthermore, since only one model specification is tested (at least as far as the reader ever gets to know!), an infinite number of alternate specifications are not ruled out.

A second problem with the null hypothesis significance test that pertains directly to research in applied settings such as public policy analysis is that there is no explicitly modeled consequence of making the wrong decision (Pollard and Richardson 1987). Unlike purely academic research, decisions taking place in policy analysis and implementation have direct consequences for citizens, employees, managers, and agencies in general. Yet hypothesis testing confuses inference and decision making since it "does not allow for the costs of possible wrong actions to be taken into account in any precise way" (Barnett 1973). Decision theory (Raiffa and Schlaifer 1961) is the logical adjunct to hypothesis testing that formalizes the cost of alternatives by explicitly defining the cost of making the wrong decision by specifying a loss function and associated risk for each alternative (Berger 1985, Pollard 1986). These principles are discussed on Chapter 8. Despite the utility of this extension to settings where decision making is required, it is rare to see applications in policy studies. What makes this surprising is that loss functions have a natural role in applied settings since obvious asymmetries occur in political and social decision-making: peace versus war, election victory versus loss, social group acceptance or rejection, and so on.

7.2.1.1 One-Sided Testing

One-sided Bayesian hypothesis testing for a specified parameter is fairly basic and a Bayesian version of the standard p-value can be produced once the posterior distribution is obtained. For the simple one-sided case in (7.1), the specified prior distribution of θ provides an *a priori* probability over the two regions of the sample space of θ: $H_0: p(-\infty < \theta \leq 0) = \pi_0$, $H_1: p(0 < \theta < \infty) = \pi_1 = 1 - \pi_0$. While this can take on an obviously large number of forms, the uninformative uniform distribution is particularly useful, and many authors have suggested that lacking specific information $\pi_0 = p(H_0$ is true$) = \frac{1}{2}$ is a useful value (Berger and Sellke 1987, Jeffreys 1961).

Once prior probabilities are assigned, the Bayesian posterior probability is derived from the non-normalized region defined by the null hypothesis divided by the total non-normalized region, which can be derived as follows:

$$p(H_0|\mathbf{x}) = \int_{-\infty}^{\infty} p(H_0, \theta|\mathbf{x})d\theta$$

$$= \int_{-\infty}^{\infty} \frac{p(\mathbf{x}|H_0, \theta)p(H_0, \theta)}{p(\mathbf{x})}d\theta$$

$$= \int_{-\infty}^{\infty} p(\mathbf{x}|H_0, \theta)p(H_0, \theta)d\theta/p(\mathbf{x})$$

$$= \frac{\int_{-\infty}^{\infty} p(\mathbf{x}|H_0, \theta)p(H_0, \theta)d\theta}{\int_{-\infty}^{\infty}[p(\mathbf{x}|H_0, \theta)p(H_0, \theta) + p(\mathbf{x}|H_1, \theta)p(H_1, \theta)]d\theta}$$

$$= \frac{\int_{-\infty}^{0} p(\mathbf{x}|\theta)\pi_0 d\theta}{\int_{-\infty}^{0} p(\mathbf{x}|\theta)\pi_0 d\theta + \int_{0}^{\infty} p(\mathbf{x}|\theta)\pi_1 d\theta}$$

$$= \frac{\int_{-\infty}^{0} p(\mathbf{x}|\theta)\pi_0 d\theta}{\int_{-\infty}^{\infty} p(\mathbf{x}|\theta)\pi_0 d\theta}, \tag{7.4}$$

where the part of the integral in the numerator from 0 to ∞ contributes zero to this calculation since H_0 is on the right-hand-side of the conditionals (and the same logic holds for the H_1 part). The terms inside the integrals are modified using the definition of conditional probability: $p(\mathbf{x}|H_0, \theta)p(H_0, \theta) = p(\mathbf{x}|H_0, \theta)p(\theta|H_0)p(H_0) = p(\mathbf{x}|H_0, \theta)p(\theta|H_0)\pi_0$. In the last step, the simplification occurs only if $\pi_0 = \pi_1$ (otherwise the denominator is a weighting of the two integrals seen in the penultimate state).

More generally, this is just the slice of the density that corresponds to the one-sided restriction defining the null hypothesis calculated over the posterior distribution. This posterior probability, while slightly more difficult to construct than the standard p-value, is far more useful because it is the value that many people mistake a p-value for: the probability that the null hypothesis is true, given the data and the model. Conversely, the standard p-value is the far less revealing probability of seeing these or more extreme data, given the model and *an assumed true null hypothesis.*

Casella and Berger (1987a) showed that when $\mathbf{X}$ is generated by a symmetric location density with monotone likelihood ratio[1] (a condition greatly aided by the central limit theorem as the data set size increases), and if the prior distribution of θ is symmetric about zero, then inf $\pi(H_0|\mathbf{X}) \leq p(x)$, where the infimum (parameter or structural minimization, see Gill [2006]) is taken over the class of suitable priors. Yet there is no theoretical justification for picking the prior distribution that leads to the infimum over any other justifiable prior. Also, in the less common cases (at least with relatively large-n social science research) where the sampling density of $\mathbf{X}$ does not have a monotone likelihood ratio, then $p(H_0|\mathbf{X}) < p(x)$.

In fact, Casella and Berger's proof shows that frequentist p-values are radically biased against the null. For example (Berger and Sellke 1987, p.113, Casella and Berger 1987a, p.110), if a random variable X distributed $\mathcal{N}(\theta, 1)$ is observed to be 1.645, then the one-sided p-value is 0.05. However, for all prior distributions assigning mass of $\frac{1}{2}$ at zero and $\frac{1}{2}$ elsewhere, inf $\pi(H_0|x = 1.645) = 0.21$. This example demonstrates that concentrating non-zero mass on point null position (zero here) leads to unreasonable (and downwardly biased) posterior inferences, and are thus not "impartial" expressions of prior ignorance. And it gets worse for the p-value. Casella and Berger also show that equality of the p-value and the Bayesian posterior quantile is achieved under the same circumstances but with the extreme constant pseudo-density improper prior that gives constant density for all values in $\Re$; yet a persistent frequentist criticism of Bayesian inference is the use of these improper priors in estimation as unreasonable probability constructs.

■ **Example 7.1:   Example: One-Sided Testing with French Labor Strike Data**
   One characteristic of labor strikes in France is imitative behavior by unions: news of other strikes can stimulate additional strikes by signaling that the conditions are amenable. Conell and Cohn (1995) look at French Third Republic coal mining strikes

---

[1]Suppose we have a family of probability density functions $h(t|\theta)$ in which the random variable t is conditional on some unknown θ value to be tested. This family has a monotone likelihood ratio if for every $\theta_1 > \theta_2$, the corresponding $\frac{h(t|\theta_1)}{h(t|\theta_2)}$ is a nondecreasing function of the random variable t.

with particular attention to follow-on strike behavior by unions. Their data are given in Table 7.1

TABLE 7.1: FRENCH COAL STRIKES, BY YEAR

1902	1906	1912	1914	1919	1921	1923A	1923B	1926	1930	1933
9	8	13	23	15	23	13	6	13	15	10

There are two periods assigned to 1923 because there were two distinct "salary offensives" during this year. Since these are counts, it is natural to consider a Poisson model for the data. However, the Poisson model assumes that the mean and variance are equal and this is not the case here. Consequently we specify a negative binomial model with a Jeffreys prior.

Here we use an equivalent variant of the negative binomial PMF as that given in Appendix B:

$$\mathcal{NB}(y|r,p) = \binom{r+y-1}{y} p^r (1-p)^y, \tag{7.5}$$

where the interpretation is that y represents the number of failures before reaching the rth success. Recall that the Jeffreys prior is calculated from the negative expected value of the second derivative of the log-likelihood: $(-E_{\mathbf{X}|\theta} \frac{d^2}{d\theta^2} \log f(\mathbf{x}|\theta))^{\frac{1}{2}} = r^{\frac{1}{2}}p^{-1}(1-p)^{-\frac{1}{2}}$. This turns out here to be the kernel of a beta distribution with parameters $(0, 1/2)$, which is not strictly an allowable parameterization of the beta (both parameters are constrained to be positive). Interestingly, this does not harm the inference process in this case, although alternatively we could specify a Poisson/gamma model. In fact, from Table 4.1 we know that the beta distribution is the conjugate prior for the negative binomial. The resulting posterior is therefore also beta distributed for known r:

$$\pi(p|\mathbf{y},r) \propto r^{\frac{1}{2}}p^{-1}(1-p)^{-\frac{1}{2}}p^{nr}(1-p)^{\sum y_i} \prod_{i=1}^{n} \binom{r+y_i-1}{y_i}$$

$$\propto p^{nr-1}(1-p)^{\sum y_i - \frac{1}{2}}$$

$$\sim \mathcal{BE}\left(nr, \sum y_i + \frac{1}{2}\right). \tag{7.6}$$

An initial or follow-on strike is considered a "failure" in the model and the corresponding "success" is an end to the series of strikes for that period. Therefore for the purpose of this analysis, we set $r = 1$. The hypothesis of interest is:

$$H_0\colon p \leq 0.05$$

$$H_1\colon p > 0.05,$$

meaning that the posterior probability of a cessation to the series of strikes is one in

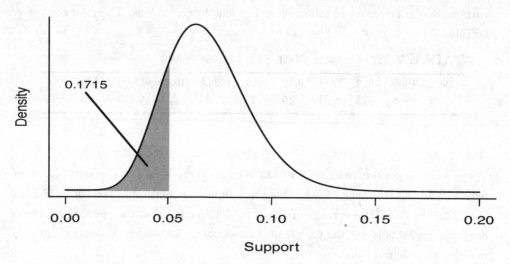

FIGURE 7.1: One-Sided Testing for the French Strikes Data

twenty or less under the null. By (7.4):

$$p(H_0|\mathbf{y}) = \frac{\int_0^{0.05} \pi(p|\mathbf{y})p(p)dp}{\int_0^1 \pi(p|\mathbf{y})p(p)dp} = 0.171. \tag{7.7}$$

This fraction is depicted in Figure 7.1 where the shaded region represents the numerator and the entire PDF is the denominator. We could integrate to get this quantity but it is actually easier, and just as accurate, to simulate the result. In fact it is nearly trivial in R: we just randomly generate 1,000,000 values distributed according to beta($nr, \sum y_i + \frac{1}{2}$) and count the proportion that are less than 0.05 (actually not this many simulated values are needed for an accurate estimate but the calculation is extremely fast).

7.2.1.2   Two-Sided Testing

While the standard, semi-frequentist approach naturally accommodates both one-sided and two-sided hypothesis tests, the Bayesian framework does not. While one-sided hypothesis testing is very straightforward in Bayesian hypothesis testing, two-sided hypothesis testing is quite difficult and remains fairly controversial in practice. This is not surprising since a Bayesian is likely to be uncomfortable placing prior mass on a point null hypothesis (H_0: $\theta = 0$, for example). While this appears to be a flaw in the Bayesian construct, it is actually an indication of how much more reasonable the approach is: *nobody* actually believes that some parameter of interest is *exactly* zero. Instead, most researchers are either truly interested in a directional conclusion (direct and regular communication reduces hostility, countries ruled by dictators are *more* likely to go to war, smaller classes lead to *better* student performance, etc.), or whether some effect size is approximately equal to zero.

Despite the evidence that point null hypothesis testing is antithetical to the Bayesian philosophy, there has been considerable effort expended trying to find a reconcilable Bayesian approach (Berger and Sellke 1987, Berger, Brown, and Wolpert 1994; Berger, Boukai, and Wang 1997, Lehmann 1993, Meng 1994a, Rubin 1984). Furthermore, unlike the non-nested case, posterior probability quantiles in the nested case are often substantially different than frequentist p-values (Lee 2004).

In testing H_0: $\theta = 0$ versus H_1: $\theta \neq 0$, we cannot assign a continuous prior distribution for θ since this would assign zero mass at the null point, thus providing an infinite bias against the nesting. One alternative is to specify a small interval around the null point, creating a focused null region: H_0: $-\epsilon \leq \theta \leq \epsilon$.

7.2.2 Attempting a Bayesian Approximation to Frequentist Hypothesis Testing

There is not a general manner in which evidence from Bayesian posterior quantiles can be calibrated with p-values since the two measures are fundamentally different in theory:

$$\int_T^\infty p(\theta)L(\theta|\mathbf{x})d\theta \neq p = p(T(\mathbf{x}) \geq T|\theta = 0) = \int_T^\infty f(\mathbf{x}|\theta = 0)dx \qquad (7.8)$$

for some α-driven critical value T and some test statistics $T(\mathbf{x})$ (Casella and Berger 1987b, p.133; Hinkley 1987, p.128). This does not mean that for a given frequentist model some Bayesian parameterization that is forced to coincide cannot be found. Particular cases include Severini (1991, 1993) for HPD regions, Stein (1965) in a repeat-sample context, DiCiccio and Stern (1994) in a multivariate setting, Thatcher (1964) for the binomial, Chang and Villegas (1986) for the multivariate normal, and Nicolaou (1993) for dealing with nuisance parameters.

Since there does not exist a default prior that is subsumed to *any* subsequently observed posterior except in the limit, then any Bayesian setup designed to agree with quasi-frequentist results is by definition a subjective assessment of the structure of the data. In addition, the notion of a Bayesian p-value analog has been described as a "paradox" (Meng, X.-L. 1994a) since the quasi-frequentist averages over data that do not exist compared to the Bayesian approach of averaging over the allowable parameter space.

The first attempt to develop a Bayesian procedure that agrees with a two-sided classical test is that of Lindley (1961). If the prior information is sufficiently vague so that one has no particular belief that $\theta = \theta_0$ versus $\theta = \theta_0 \pm \epsilon$, where ϵ is some small value, then a reference (ignorance-expressing, Chapter 4) prior can be used to obtain a posterior, and H_0 is rejected for values that fall out of the $(1 - \alpha)100$ HPD region. Highest posterior density regions are preferred over credible intervals for asymmetric distributions since credible intervals simply space out a specified distance from the mean regardless of overlying density.

7.3   The Bayes Factor as Evidence

Bayes Factors have dominated the literature on Bayesian model testing because they are often easy to calculate and have a naturally intuitive interpretation (likelihood ratio tests are a special case). The central notion is that prior and posterior information should be combined in a ratio that provides evidence of one model specification over another. Bayes Factors are also very flexible in that multiple hypotheses can be simultaneously compared. Moreover, *model nesting is not required in order to make comparisons*, addressing a major deficiency with classical approaches (Cox 1961).

The most general form of the Bayes Factor can be described as follows. Suppose we observe data $\mathbf{x}$ and wish to test two competing models, M_1 and M_2, relating these data to two different sets of parameters, $\boldsymbol{\theta}_1$ and $\boldsymbol{\theta}_2$. This is a problem of deciding between two families of density specifications:

$$M_1: f_1(\mathbf{x}|\boldsymbol{\theta}_1) \qquad M_2: f_2(\mathbf{x}|\boldsymbol{\theta}_2) \tag{7.9}$$

where $\boldsymbol{\theta}_1$ and $\boldsymbol{\theta}_2$ are either nested within a larger set of alternative parameters, $\boldsymbol{\Theta}$, or drawn from distinct parameter spaces, $\boldsymbol{\Theta_1}$ and $\boldsymbol{\Theta_2}$. The standard Bayesian setup specifies a prior unconditional distribution for the parameter vectors: $p_1(\boldsymbol{\theta}_1)$ and $p_2(\boldsymbol{\theta}_2)$, and therefore a prior probability of the two models: $p(M_1)$ and $p(M_2)$. The posterior odds ratio in favor of Model 1 versus Model 2 are therefore produced by Bayes' Law:

$$\underbrace{\frac{\pi(M_1|\mathbf{x})}{\pi(M_2|\mathbf{x})}}_{\text{posterior odds}} = \underbrace{\frac{p(M_1)/p(\mathbf{x})}{p(M_2)/p(\mathbf{x})}}_{\text{prior odds/data}} \times \underbrace{\frac{\int_{\theta_1} f_1(\mathbf{x}|\boldsymbol{\theta}_1)p_1(\boldsymbol{\theta}_1)d\boldsymbol{\theta}_1}{\int_{\theta_2} f_2(\mathbf{x}|\boldsymbol{\theta}_2)p_2(\boldsymbol{\theta}_2)d\boldsymbol{\theta}_2}}_{\text{Bayes Factor}}. \tag{7.10}$$

So the quantity of interest turns out to be the ratio of marginal likelihoods (page 41) from the two models. This expression equates the posterior odds ratio on the left-hand side to the product of the prior odds ratio and the ratio of integrated likelihoods. Note that with fairly complicated models, the integrals in (7.10) can be quite challenging to compute, even with the Markov chain Monte Carlo procedures introduced starting in Chapter 10. By rearranging we get the standard form of the Bayes Factor, which can be thought of as the magnitude of the evidence for Model 1 over Model 2, contained in the data:

$$BF_{(1,2)} = \frac{\pi(M_1|\mathbf{x})/p(M_1)}{\pi(M_2|\mathbf{x})/p(M_2)}, \tag{7.11}$$

which is also called the posterior to prior odds ratio for the obvious reason revealed in this form. In the case where we are willing to put equal prior probability on the two models ($p(M_1) = p(M_2) = \frac{1}{2}$) and the models share the same parameter space but hypothesize differing levels, then the Bayes Factor reduces to the common likelihood ratio. This is equivalent to assigning simple point mass through the priors. It is also possible to rearrange

(7.11) since the $p(\mathbf{x})$ is the same for both models:

$$\frac{\pi(M_1|\mathbf{x})/p(M_1)}{\pi(M_2|\mathbf{x})/p(M_2)} = \frac{\pi(M_1,\mathbf{x})/(p(\mathbf{x})p(M_1))}{\pi(M_2,\mathbf{x})/(p(\mathbf{x})p(M_2))}$$

$$= \frac{\pi(M_1,\mathbf{x})/p(M_1)}{\pi(M_2,\mathbf{x})/p(M_2)} = \frac{\pi(\mathbf{x}|M_1)}{\pi(\mathbf{x}|M_2)}, \qquad (7.12)$$

which gives another general form provided by some authors. Commonly, the natural log of the Bayes Factor is calculated for reasons of numerical stability.

Bayes Factors are also *transitive* in that multi-way comparisons are relative. So if we have $BF_{(1,2)}$ and $BF_{(2,3)}$, then:

$$BF_{(1,2)}BF_{(2,3)} = \frac{\pi(M_1|\mathbf{x})/p(M_1)}{\pi(M_2|\mathbf{x})/p(M_2)} \frac{\pi(M_2|\mathbf{x})/p(M_2)}{\pi(M_3|\mathbf{x})/p(M_3)} = BF_{(1,3)}, \qquad (7.13)$$

which is useful for multiple model comparisons using the same data. This property also means that if there exists a null model, then a series of alternatives can be tested against it and the resulting values are comparable on the same relative scale.

Bayes Factors do not have an inherent *scale*, exactly in the manner that likelihood ratios do not either. A fundamental criticism of Bayes Factors is that because they lack an underlying metric, *all* results are therefore arbitrary and subjective. This is not quite right since they are an overt *relative* comparison of model fit. Clearly we would rather see extremely large or extremely small values of the Bayes Factor since that indicates obvious superiority of one specification over another.

While the Bayesian approach typically eschews arbitrary decision thresholds, Jeffreys (1961, p.432) gives the following typology for comparing Model 1 versus Model 2:

$$BF_{(1,2)} > 1 \quad \text{model 1 supported}$$
$$1 > BF_{(1,2)} \geq 10^{-\frac{1}{2}} \quad \text{minimal evidence against model 1}$$
$$10^{-\frac{1}{2}} > BF_{(1,2)} \geq 10^{-1} \quad \text{substantial evidence against model 1}$$
$$10^{-1} > BF_{(1,2)} \geq 10^{-2} \quad \text{strong evidence against model 1}$$
$$10^{-2} > BF_{(1,2)} \quad \text{decisive evidence against model 1,}$$

where Model 1 is assumed to be a null model. Kass and Raftery (1995) modify these categories slightly and provide a more intuitive logarithmic scale for decision criteria (also discussed in Raftery 1996). Note that there is no explicit "acceptance" or "rejection" of hypotheses as in the Neyman-Pearson context. Instead the Bayes Factor (or the log of the Bayes Factor) is considered simply the weight of evidence for Model 1 over Model 2 provided by the data, given the prior and the model specification (Good 1985). Good points out elsewhere (1980b) that this is not a very new idea since Pierce first used "weight of evidence" in comparing hypotheses as early as 1878, and Turing (a contemporary of Jeffreys) in 1940 used the expression "factor in favor of a hypothesis" (reported in Good 1972, p.15) to mean nearly the same thing. Karl Pearson also uses this phraseology, but in a less formal comparative manner (1914 and elsewhere).

It is important to remember that Jeffreys' typology is still an arbitrary designation of levels. However, under specific circumstances the Bayes Factor relates directly to standard posterior quantities. For instance, if we are willing to take the classical stance that there are only two plausible alternative hypotheses, then it follows that $p(H_0|data) + p(H_1|data) = 1$ (we could also denote this $p(M_1|\mathbf{x}) + p(M_2|\mathbf{x}) = 1$ if we wanted to be specific that each hypothesis is represented by a model). Starting with this we can use Bayes' Law and the definition of the Bayes Factor to produce:

$$p(H_0|\text{data}) = 1 - p(H_1|\text{data})$$

$$= 1 - p(\text{data}|H_1)\frac{p(H_1)}{p(\text{data})}$$

$$= 1 - \frac{p(\text{data}|H_0)}{BF_{(1,2)}}\frac{p(H_1)}{p(\text{data})}$$

$$= 1 - \frac{1}{BF_{(1,2)}}\left[\frac{p(\text{data})}{p(H_0)}p(H_0|\text{data})\right]\frac{p(H_1)}{p(\text{data})}$$

$$= \left[1 + \frac{1}{BF_{(1,2)}}\frac{p(H_1)}{p(H_0)}\right]^{-1}. \tag{7.14}$$

Naturally other posterior quantities can be related in similar fashion to the Bayes Factor as well. This result shows that the posterior probability of the null hypothesis is a function of the Bayes Factor scaled by the ratio of priors, and highlights quite clearly the strong influence that prior specifications can have on Bayesian hypothesis testing. Lavine and Schervish (1999) provide cautionary advice when extending the Bayes Factor beyond such basic comparisons.

7.3.1   Bayes Factors for a Mean

Consider a simple setup where $X \sim \mathcal{N}(\mu, \sigma^2)$, where the population mean μ is unknown and the population variance σ^2 is known. We are interested in a two-sided test of $H_1: \mu = \mu_0$ versus $H_0: \mu \neq \mu_0$ (where μ_0 is often 0). We specify a normal prior under the research hypothesis (H_1) with mean m and variance s^2. A sample of size n is collected with mean $\bar{x}$. This leads to the Bayes Factor:

$$BF_{(H_1,H_0)} = \left(1 + \frac{ns^2}{\sigma^2}\right)\exp\left[-\frac{n}{2}\left(\frac{1}{\sigma^2} - \frac{1}{\sigma^2 + ns^2}\right)(\bar{x} - \mu_0)^2\right] \tag{7.15}$$

(Exercise 6). Despite the simplicity of this calculation, there is plenty to be uncomfortable about. First, testing a point null hypothesis is not really a Bayesian operation since μ_0 is unknown and therefore should be assigned a prior distribution as well (as noted in Section 7.3.5). As soon as we want to assign this prior the idea of a single point ceases to make obvious sense. There are ways to use the Dirac Delta function as a surrogate for such a distribution, but these are not very intuitive from an inferential sense. The next obvious alternative is to substitute a small region around the point of interest for the single point, but this leads to some additional noted challenges.

7.3.2 Bayes Factors for Difference of Means Test

This section develops the Bayesian version of the standard Student's t-test for normal data that uses a Bayes Factor. For additional details and an example, see Gönen *et al.* (2005). Surprisingly little has been done to adapt this standard tool to Bayesian use. Suppose we are interested in the two-sided test:

$$H_0: \mu_1 = \mu_2 \qquad H_1: \mu_1 \neq \mu_2,$$

with common variance σ^2 in the two groups. First we need to specify the prior distribution of the effect size (difference) to be tested. We will say that under the hypothesis of a non-zero difference, the standardized difference, $|\mu_1 - \mu_2|/\sigma$ has prior mean δ and prior variance σ_δ^2. Note that this allows great flexibility for the test to be performed. Next calculate the standard difference of means test statistic:

$$t = \frac{\bar{x}_1 - \bar{x}_2}{\left(\frac{(n_1-1)s_1^2+(n_2-1)s_2^2}{n_1+n_2-2}\right)^{\frac{1}{2}}/\sqrt{n_\delta}} \tag{7.16}$$

where $n_\delta = (n_1^{-1} + n_2^{-1})^{-1}$, and the degrees of freedom are $\nu = n_1 + n_2 - 2$. The Bayes Factor for H_0 over H_1 (large values favoring the null) is:

$$BF_{(0,1)} = \frac{T_\nu(t|0,1)}{T_\nu(t|\delta\sqrt{n_\delta}, 1 + n_\delta\sigma_\delta^2)}, \tag{7.17}$$

where $T_\nu(t|A, B)$ denotes the value that results from plugging t into a non-central t-distribution PDF with ν degrees of freedom and parameters A for location and $B^{\frac{1}{2}}$ for scale (see Johnson, Kotz, and Balakrishnan 1994). Gönen *et al.* (2005) point out that this is easily implemented in R by first determining whether the following terms can be identified:

$$pv = \sqrt{1 + n_\delta\sigma_\delta^2} \qquad ncp = \frac{\delta\sqrt{n_\delta}}{pv},$$

which is easy to implement. The Bayesian version of the difference of means test differs noticeably from the non-Bayesian variant mainly in that we get to *specify* the tested effect size prior, which is an important advantage.

7.3.3 Bayes Factor for the Linear Regression Model

An obvious and useful application for the Bayes Factor is the standard linear regression model where we want to compare two, not necessarily nested, different right-hand-side specifications in $\mathbf{y} = \mathbf{X}\boldsymbol{\beta} + \boldsymbol{\epsilon}$, where $\mathbf{X}$ is an $n \times k$, rank k matrix of explanatory variables with a leading vector of ones, $\boldsymbol{\beta}$ is a $k \times 1$ unknown vector of coefficients, $\mathbf{y}$ is an $n \times 1$

vector of outcomes, and $\boldsymbol{\epsilon}$ is a $n \times 1$ vector of residuals with $\mathcal{N}(0, \sigma^2 I)$ for a constant σ^2 (homoscedasticity). On page 145 the likelihood function for model j is:

$$L_j(\boldsymbol{\beta}_j, \sigma_j^2 | \mathbf{X}_j, \mathbf{y}) = (2\pi\sigma_j^2)^{-\frac{n}{2}} \exp\left[-\frac{1}{2\sigma_j^2}(\mathbf{y} - \mathbf{X}_j\boldsymbol{\beta}_j)'(\mathbf{y} - \mathbf{X}_j\boldsymbol{\beta}_j)\right] \quad (7.18)$$

where $j = 0, 1$ providing models M_0 and M_1. Notice that $\mathbf{y}$ is not indexed here since both models intend to explain the structure the same outcome variable. Again, make the definitions $\hat{\mathbf{b}} = (\mathbf{X}'\mathbf{X})^{-1}\mathbf{X}'\mathbf{y}$, and $\hat{\sigma}^2 = (\mathbf{y} - \mathbf{X}\hat{\mathbf{b}})'(\mathbf{y} - \mathbf{X}\hat{\mathbf{b}})/(n-k)$.

Now specify possibly different conjugate priors for each of these models with k_j columns of $\mathbf{X}$ according to:

$$p(\boldsymbol{\beta}_j | \sigma^2) = (2\pi)^{-\frac{k_j}{2}} |\boldsymbol{\Sigma}_j|^{-\frac{1}{2}} \exp\left[-\frac{1}{2}(\boldsymbol{\beta}_j - \mathbb{B}_j)'\boldsymbol{\Sigma}_j^{-1}(\boldsymbol{\beta}_j - \mathbb{B}_j)\right],$$

and:

$$p(\sigma_j^2) \propto \sigma_j^{-(a_j - k_j)} \exp\left[-\frac{b_j}{\sigma_j^2}\right] \quad (7.19)$$

as done on page 151 except for a multiplier h_j on the variance term in the normal prior for $\boldsymbol{\beta}_j$: $\boldsymbol{\Sigma}_j = h_j\sigma_j^2\mathbf{I}$. If we make the common choice of prior mean for $\boldsymbol{\beta}$ to be $\mathbb{B} = 0$ in both models, then the marginal likelihood for model j from this setup is:

$$p_j(\mathbf{y}|\mathbf{X}_j, M_j) = \frac{|\mathbf{X}_j'\mathbf{X}_j + h|^{-\frac{1}{2}}|h_j|^{\frac{1}{2}}b_j^{a_j}\Gamma(a_j + \frac{a_j}{2})}{\pi^{\frac{n}{2}}\Gamma(a_j)}\left(2b_j + (n - k_j)\hat{\sigma}_j^2\right). \quad (7.20)$$

This means that the Bayes Factor for Model 1 over Model 0 is given by:

$$BF_{(1,0)} = \frac{p_1(\mathbf{y}|\mathbf{X}_1, M_1)}{p_0(\mathbf{y}|\mathbf{X}_0, M_0)} = \frac{\frac{|\mathbf{X}_1'\mathbf{X}_1 + h|^{-\frac{1}{2}}|h_1|^{\frac{1}{2}}b_1^{a_1}\Gamma(a_1 + \frac{a_1}{2})}{\pi^{\frac{n}{2}}\Gamma(a_1)}\left(2b_1 + (n - k_1)\hat{\sigma}_1^2\right)}{\frac{|\mathbf{X}_0'\mathbf{X}_0 + h|^{-\frac{1}{2}}|h_0|^{\frac{1}{2}}b_0^{a_0}\Gamma(a_0 + \frac{a_0}{2})}{\pi^{\frac{n}{2}}\Gamma(a_0)}\left(2b_0 + (n - k_0)\hat{\sigma}_0^2\right)}. \quad (7.21)$$

This is a long expression but a relatively simple form due to the elegance of the linear model.

■ **Example 7.2: Bayes Factors for a Model of Election Surveys.** For generalized linear models with dichotomous outcome variables, it is common to specify a logit or probit link function as described in Appendix A. That is, $g^{-1}(\mathbf{X}\boldsymbol{\beta})$ is either the logit function, $\Lambda(\mathbf{X}\boldsymbol{\beta})$, or the standard normal CDF, $\Phi(\mathbf{X}\boldsymbol{\beta})$. So for outcome variable Y_i, and prior distribution $p(\boldsymbol{\beta})$ on the coefficients, we obtain the following posterior:

$$\pi(\boldsymbol{\beta}|\mathbf{X}, \mathbf{Y}) = \frac{p(\boldsymbol{\beta})\prod_{i=1}^{n}g^{-1}(\mathbf{X}\boldsymbol{\beta})^{y_i}(1 - g^{-1}(\mathbf{X}\boldsymbol{\beta}))^{1-y_i}}{\int_{\boldsymbol{\beta}}p(\boldsymbol{\beta})\prod_{i=1}^{n}g^{-1}(\mathbf{X}\boldsymbol{\beta})^{y_i}(1 - g^{-1}(\mathbf{X}\boldsymbol{\beta}))^{1-y_i}d\boldsymbol{\beta}}. \quad (7.22)$$

When the prior is some numerical constant (i.e., not a function of the $\boldsymbol{\beta}$) then it passes out of the integral in the denominator and cancels. This then becomes equivalent to the classical model described in every econometric book ever printed (only a slight exaggeration). However, in general (7.22) is not available in closed form for most prior specifications and MCMC techniques are generally required (see Chapter 12 for specific applications of this model). Assume for the moment that we specified a simple multidimensional uniform prior for $p(\beta)$ and that we wish to calculate a Bayes Factor for comparing one coefficient vector against another: H_0: $\beta = \beta_0$ vs. H_1: $\beta = \beta_1$. Determining the weight of evidence for Model 2 versus Model 1 here is performed simply by inserting β_0 and β_1 separately into (7.22) and calculating the Bayes Factor according to $BF_{(1,0)} = \pi(\beta_1|\mathbf{X}, \mathbf{Y})/\pi(\beta_0|\mathbf{X}, \mathbf{Y})$. This is actually just the likelihood ratio for Model 2 over Model 1. This process is relatively general in that we can test differing specifications as well as restricted coefficient vectors versus unrestricted estimates.

TABLE 7.2: 1964 ELECTORAL DATA

N	F	L	W	IND	DEM	WR	WD	SD
109	0.102	-2.175	9.984	0	0	0	0	0
35	0.115	-2.041	3.562	1	0	0	0	0
33	0.214	-1.301	5.551	0	1	0	0	0
75	0.258	-1.056	14.358	0	0	1	0	0
50	0.544	0.176	12.403	1	0	1	0	0
52	0.677	0.740	11.731	0	1	1	0	0
70	0.606	0.431	16.713	0	0	0	1	0
56	0.890	2.091	5.482	1	0	0	1	0
189	0.975	3.664	4.607	0	1	0	1	0
31	0.727	0.979	6.153	0	0	0	0	1
56	0.893	2.122	5.351	1	0	0	0	1
344	0.990	4.595	3.406	0	1	0	0	1

Hanushek and Jackson (1977) develop a grouped logit model for Factor data based on a clever estimation approach suggested by Theil (1970) in which the link function is applied to the outcome variable, then a grouped data form of least squares is used to obtain coefficient estimates. All generalized linear models automatically imply interaction effects because of the link function.[2] But when Theil's method is used, these automatic interaction effects are not provided, so if we believe that such effects

---

[2] To see that this is true, calculate the marginal effect of a single coefficient by taking the derivative of a GLM specification, which does not explicitly contain an interaction term, $E(Y_i) = g^{-1}(\beta_0 + \beta_1 X_{i1} + \beta_2 X_{i2})$, with regard to a variable of interest. If the form of the model implied no interactions, then we would obtain a marginal effect free of other variables, but this is clearly not so: $\frac{\partial Y_i}{\partial X_{i2}} = \frac{\partial}{\partial X_{i2}} g^{-1}(\beta_0 + \beta_1 X_{i1} + \beta_2 X_{i2}) = (g^{-1})'(\beta_0 + \beta_1 X_{i1} + \beta_2 X_{i2})\beta_2$. The calculation demonstrates that the presence of a link function requires the use of the chain rule and therefore retains other terms on the right-hand side in addition to β_2, and therefore we always get for a given variable partial effects that are dependent on the levels of the other explanatory variables.

exist, they must now be explicitly provided in the model specification. The data come from a 1964 election survey in the United States regarding the presidential election (available in BaM). The outcome variable is the log-ratio of the group proportion voting for Lyndon Johnson and the binary explanatory variables are: self-indicated indifference to the election (IND), a stated preference for Democratic party issues (DEM), and indications of party status as weak Republican (WR), weak Democrat (WD), or strong Democrat (SD). The full data set is given in Table 7.2, which is a replication from Hanushek and Jackson.

The vector $\mathbf{N}$ is the number of cases in the group, the vector $\mathbf{F}$ is the observed cell proportion voting for Johnson, $\mathbf{L}$ is log-ratio of this proportion given by $\mathbf{L} = \log[\mathbf{F}/(1-\mathbf{F})]$ (see Theil [1970, p.107]) for a justification), and $\mathbf{W}$ collects the inverse of the diagonal of the matrix for the group-weighting from $[N_i F_i (1-F_i)]$. The uniform prior coefficient estimate is produced by

$$b = (\mathbf{X'WX})^{-1}\mathbf{XWL}, \tag{7.23}$$

along with the asymptotic variance-covariance matrix $(\mathbf{X'WX})^{-1}$, and log-likelihood

$$\ell(b) = -(n/2)\log(2\pi) - (n/2)\log(s^2) - (1/(2s^2))\epsilon'\epsilon. \tag{7.24}$$

Table 7.3 gives the results for the simple model in Hanushek and Jackson and a second model that hypothesizes an interaction between indifference and weak Republican.

TABLE 7.3:   1964 ELECTION SURVEY, GROUPED DATA

	Hanushek and Jackson		Intercept Model	
	Coefficent	Std. Error	Coefficent	Std. Error
Intercept	-2.739	0.062	-2.709	0.063
IND	1.145	0.061	0.952	0.099
DEM	2.167	0.063	2.187	0.063
WR	1.598	0.080	1.477	0.095
WD	3.459	0.090	3.465	0.091
SD	4.049	0.121	4.083	0.122
(IND)(WR)			0.456	0.214

What these results show is that the interaction term is statistically reliable according to traditional levels and that the rest of the model is not substantively changed. So which specification is better? The Bayes Factor for the interaction model relative to the Hanushek and Jackson model is 0.944, indicating little support for adding the interaction according to Jeffreys' typology (given of course the unsupported assumption of uniform priors).

7.3.4 Bayes Factors and Improper Priors

From (7.10) it is easy to see that the form of the prior has a noticeable effect on the resulting Bayes Factor. This sensitivity to the prior is a main criticism of Bayes Factors in general (see Kim 1991). Interestingly, the form of the prior has a much greater effect on the Bayes Factor than other forms of Bayesian inference such as quantile descriptions of the posterior or posterior predictive distribution (Aitkin 1991; Gelman, Meng, and Stern 1996). In standard Bayesian analysis, a substantial similarity between the prior and the posterior is evidence that the data had much less of an impact than the prior. This easily detected situation would be cause for alarm. For instance, in the case where a conjugate prior is specified, if the shape of the posterior distribution is very close to the prior, then we know that the beginning assumptions are relatively unmodified by conditioning on the data.

In the case where improper priors are used, the Bayes Factor cannot be specified except under very uninteresting scenarios[3] (Berger and Mortera 1999, p.542; Kass and Wasserman 1995 [discussion], p.777). The most common improper prior setup is to specify one or both of the densities proportional to a multiplicative constant:

$$p_1(\theta_1) = c_1 g_1(\theta_1) \qquad p_2(\theta_2) = c_2 g_2(\theta_2), \tag{7.25}$$

where g_1 and/or g_2 are functions whose integrals over the respective sample spaces do not diverge. A very common improper prior specification is a constant, $p(\theta) = k$, over the Lebesgue measure on $(-\infty, \infty)$ (see Chapter 4, Section 4.4.4). One way to visualize this is as a rectangle that is $k = g(\theta)$ high and $c = \infty$ wide. Obviously c is an infinite proportionality constant, but it typically does not prevent the calculation of a proper posterior distribution since:

$$\pi(\theta|\mathbf{x}) = \frac{p(\theta)p(\mathbf{x}|\theta)}{\int_\theta p(\theta)p(\mathbf{x}|\theta)d\theta}$$

$$= \frac{cg(\theta)p(\mathbf{x}|\theta)}{c\int_\theta g(\theta)p(\mathbf{x}|\theta)d\theta}$$

$$= \frac{g(\theta)p(\mathbf{x}|\theta)}{\int_\theta g(\theta)p(\mathbf{x}|\theta)d\theta}. \tag{7.26}$$

To see that this is not true for Bayes Factors, take the form from (7.10) and substitute priors with specified proportionality constants according to (7.25) to obtain:

$$BF_{(1,2)} = \frac{c_1 \int_{\theta_1} g_1(\theta_1)p(\mathbf{x}|\theta_1)d\theta_1}{c_2 \int_{\theta_2} g_2(\theta_2)p(\mathbf{x}|\theta_2)d\theta_2}. \tag{7.27}$$

So for any two proper priors (or improper priors that are finite and proportional to proper priors) the Bayes Factor can still be calculated. However, if both of c_1, c_2 are unbounded,

---

[3]Improper priors can be assigned to unknown nuisance parameters that are common to both models. Therefore they are not applicable to the parameters that motivate the model comparison in the first place.

then the Bayes Factor is incalculable because the ratio of two different unknown unbounded quantities does not cancel out.

Returning to the problematic case of the two-sided (nested) Bayesian hypothesis test, this section treats the problem with Bayes Factors instead of posterior distribution quantiles. This case represents the widest gulf between the frequentist and the Bayesian approaches, where tests to coincide can result in differences greater than an order of magnitude (Berger and Sellke 1987, Casella and Berger 1987a, Lindley 1957). While the easiest solution to the problems of two-sided testing is to dismiss this setup, as atheoretical and impractical, avoiding such nested problems is unreasonable in practice since linear regression, easily the most commonly used statistical procedure in the social and behavioral sciences, nests a null hypothesis of a zero coefficient value within the full sample space.

7.3.4.1   Local Bayes Factor

Smith and Spiegelhalter (1980) and Spiegelhalter and Smith (1982) developed an ingenious way to solve the problem posed in (7.27) for nested linear models and known variances using an idea based loosely on Atkinson (1978), Geisser and Eddy (1979), and Lempers (1971, Chapter 6). Suppose we "imagine" a set of data as a training sample and solve for $\frac{c_1}{c_2}$ in (7.27):

$$BF_0 \frac{\int_{\theta_2} g_2(\theta_2) p(\mathbf{x_0}|\theta_2) d\theta_2}{\int_{\theta_1} g_1(\theta_1) p(\mathbf{x_0}|\theta_1) d\theta_1} = Est. \left[\frac{c_1}{c_2}\right]. \tag{7.28}$$

(the order of the fraction with integrals changes as it moves to the left-hand-side of the equality). The choice of the training data, $\mathbf{x_0}$, is done with the objective of finding the smallest data set that gives proper posterior distributions in (7.28) and so that the simplest model is most favored: $BF_0(\mathbf{x})$ is maximized for this specification over the other. So that the second criterion is not perpetuated into the next stage of the test, Smith and Spiegelhalter set $BF_0(\mathbf{x}) = 1$ with the theoretical justification that if the training model is truly minimal, then the value of one is a good estimate. Since the imaginary data set is conjured to support the model in the numerator to the greatest extent possible, this value of one is actually a conservative value as it is now the lower bound on $BF_0(\mathbf{x})$.

The final Bayes Factor is thus a product of the training "prior" for $\frac{c_1}{c_2}$ and the Bayes Factor for the rest of the data:

$$BF_{(1,2)} = Est. \left[\frac{c_1}{c_2}\right] \frac{\int_{\theta_1} g_1(\theta_1) p(\mathbf{x}|\theta_1) d\theta_1}{\int_{\theta_2} g_2(\theta_2) p(\mathbf{x}|\theta_2) d\theta_2} \tag{7.29}$$

(note that the order of the fraction is returned because it is back on the right-hand-side of the equality). Therefore by this method, we remove the non-identifiability problem associated with the undefined ratio of normalizing constants and obtain a solution for the Bayes Factor. Pettit (1992) applies this method to nested linear models with outliers to judge the sensitivity of Bayes Factors to outliers when specifying improper priors (which necessarily put more weight on the model with more outliers). Adman and Raftery (1986) apply *local Bayes Factors* (i.e., one that compares a specific model, M_1 with closely related models contained in a super-set model M_2 where closeness is defined as those alternatives that give a high weight for producing similar coefficient estimates)[4] to a nonhomogeneous Poisson process using bounded improper priors (Adman and Raftery 1986, Raftery and Adman 1986), noting that this case is more difficult with linear models since there are data conditions that can provide maximal support for the null resulting in an undefined ratio of constants.

While this idea of a minimal training sample providing a prior estimate of the incalculable quantity is very creative, two obvious problems exist. The first is the fixing of $B_0(\mathbf{x}) = 1$, a decision that lacks any theoretical justification other than the relatively vague idea that it is likely to be close in an optimal situation (for which there is no guarantee and no test). The second is that determination of the training sample is difficult both in terms of selecting the sample size (Lempers [1971] arbitrarily picked half of the full sample as the size of the training sample) and the sample components (there are $\binom{n}{m}$ ways of picking a training sample of m out of a full sample of n). For example, consider a state of nature that is maximally malevolent in that the training sample produces a premultiplier in (7.29) that is as different from the rest of the Bayes Factor calculation as possible. Since all of the integration in (7.29) is done over the sample space of θ, then the sample size of the training sample is equally weighted with the rest of the presumably much larger sample, and the ability to change the final Bayes Factor is thus substantial.

7.3.4.2 Intrinsic Bayes Factor

Berger and Pericchi (1996a, 1996b) propose a method of picking the training sample that depends on the number of possible training samples being relatively small. Their "intrinsic" Bayes Factor is an average of the Bayes Factors from every combinatorically possible training sample. In cases where the number of possible training sets is prohibitively large (i.e., almost every realistic scenario in the social sciences), then a random sample of the possible training samples can be used.

To create the intrinsic Bayes Factor, Berger and Pericchi first start with intrinsic prior densities $\nu(\theta_1)$ and $\nu(\theta_2)$. An intrinsic prior is uninformative prior in the sense that it provides the asymptotic equivalence of the Bayes Factor to the maximum likelihood ratio. For details see Berger and Pericchi (1996a) or Berger and Mortera (1999). Next define a *minimal training sample*, $\mathbf{x}_m$, the smallest possible subset of the full sample, $\mathbf{x}_n$, so that at least one of the resulting posterior densities using the intrinsic priors are proper. Calculate these minimum training set posterior distributions: $\pi_i^I(\theta_i|\mathbf{x}) \propto \nu(\theta_i) f_i(\mathbf{x}_m|\theta_i)$. The intrinsic Bayes Factor is then defined to be:

$$B^I(\mathbf{x}) = \frac{p_1^I(\theta_1|\mathbf{x})}{p_2^I(\theta_1|\mathbf{x})} AVE\left[\frac{\pi_2^I(\theta_2|\mathbf{x})}{\pi_1^I(\theta_1|\mathbf{x})}\right], \tag{7.30}$$

where $AVE[\ ]$ is some average: arithmetic mean, geometric mean, harmonic mean, median,

---

[4]Smith and Spiegelhalter (1980) demonstrate a linkage between local Bayes Factor comparisons and Akaike information criterion (AIC) comparisons, see p.232.

etc., over all possible minimal training samples. This avoids the previously mentioned problem of inadvertently selecting a maximally skewed training sample. Furthermore, since the average selected does not have to be a mean, some other robustizing measure of central location can be used so that the procedure is robust to outliers. Unfortunately, selection of the averaging procedure can also make a substantive difference in the resulting Bayes Factor, and Berger and Pericchi do not provide any theoretical justification for one form over another.

In general, intrinsic Bayes Factors are more appropriate for two-sided tests than one-sided tests, since there is no explicit incorporation of direction (see Moreno 1997). Berger and Pericchi identify intrinsic priors for large classes of nested models and specification, but found that they are not generally appropriate for developing Bayes Factors in non-nested models.

The challenge in developing the intrinsic Bayes Factor is obtaining the appropriate intrinsic prior distributions for each of the tested models. While Berger and Pericchi tabulate some common forms, most situations will require separate derivation. However, Berger *et al.* (1998) find a number of invariant specifications where the marginal density of the minimal training set is actually available in closed form analytically.

7.3.4.3   Partial Bayes Factor

O'Hagan (1995) attempts to repair the difficulties in specifying the local Bayes Factor in two ways, beginning with "partial Bayes Factors." The procedure is quite simple. First divide the sample into two components, $\mathbf{x} = (\mathbf{y}, \mathbf{z})$, where $\mathbf{y}$ is the training sample of size m, and $\mathbf{z}$ is the sample proportion of size $n - m$ used for model comparison. Use $\mathbf{y}$ to obtain training posteriors for the alternative models: $\pi_1(\theta_1|\mathbf{y})$ and $\pi_2(\theta_2|\mathbf{y})$ by assigning any desired prior for θ (including an improper prior) and using $\pi(\theta|\mathbf{x}) \propto p(\theta)p(\mathbf{x}|\theta)$. These training posteriors will necessarily be proper, and are used as priors for calculating the Bayes Factor for the rest of the sample:

$$BF(\mathbf{z}|\mathbf{y}) = \frac{\int_{\theta_1} \pi_1(\theta_1|\mathbf{y})f_1(\mathbf{z}|\theta_1,\mathbf{y})d\theta_1}{\int_{\theta_2} \pi_2(\theta_2|\mathbf{y})f_2(\mathbf{z}|\theta_2,\mathbf{y})d\theta_2}, \qquad (7.31)$$

where:

$$\int_{\theta_i} \pi_i(\theta_i|\mathbf{y})f_i(\mathbf{z}|\theta_i,\mathbf{y})d\theta_i = \int_{\theta_i} \pi_i(\theta_i)f_i(\mathbf{x}|\theta_i)d\theta_i \bigg/ \int_{\theta_i} \pi_i(\theta_i)f_i(\mathbf{y}|\theta_i)d\theta_i.$$

Since $BF(\mathbf{x}) = BF(\mathbf{y}, \mathbf{z})$, then $BF(\mathbf{z}|\mathbf{y}) = BF(\mathbf{x})/BF(\mathbf{y})$, and it is clear that the partial Bayes Factor simply divides out the undefined ratio of normalizing constants. Furthermore, O'Hagan (1995, p.105) shows that the partial Bayes Factor is asymptotically consistent in that it will choose the correct model with probability one as $n/m \to \infty$.

7.3.4.4   Fractional Bayes Factor

The partial Bayes Factor eliminates the local Bayes Factor assumptions about B_0 and the problems associated with averaging over many possible training samples as with the intrinsic Bayes Factor, but it still requires determination of the proportion of the data selected as the training sample. O'Hagan (1995) suggests a modification of the partial Bayes Factor, which makes the selection of the training sample, $\mathbf{y}$, less important. Define $\eta = m/n$. If m and n are reasonably large, then by the properties of likelihood estimators $\ell(\mathbf{y}|\theta) \approx \ell(\mathbf{x}|\theta)^\eta$, meaning that the likelihood based on the training sample approximates the full likelihood adjusted for sample size. For proper or improper priors on θ, this leads to a Bayes Factor of the following form:

$$BF_\eta(\mathbf{x}) = \frac{\int_{\theta_1} \pi_1(\theta_1) f_1(\mathbf{x}|\theta_1) d\theta_1 / \int_{\theta_1} \pi_1(\theta_1) f_1(\mathbf{x}|\theta_1)^\eta d\theta_1}{\int_{\theta_2} \pi_2(\theta_2) f_2(\mathbf{x}|\theta_2) d\theta_2 / \int_{\theta_2} \pi_2(\theta_2) f_2(\mathbf{x}|\theta_2)^\eta d\theta_2}. \tag{7.32}$$

Therefore if π_1 or π_2 are specified as improper priors, then the indeterminate constant will cancel out in either the numerator or the denominator. Also, the initial step of calculating training posteriors has been removed by the η-ratio: absolute values of the prior density are normalized out.

As with the partial Bayes Factor, determination of the proportion of the sample to commit to solving the prior problem is influential on the resulting ratio. It is obvious that η should progress toward zero as n goes to infinity: in the limit the likelihood function subsumes any prior, including improper priors. O'Hagan recommends the value m_0, which is the minimum value that provides a consistent model choice, since this provides the greatest possible proportion of the data for comparing the models. In the presence of outliers, or potential outliers, he recommends $\sqrt{n}$ or $\log(n)$ as robustizing values. In any event, the subjectivity of priors is replaced to some extent by the subjectivity of sample dichotomization.

The fractional Bayes Factor loses quite a bit of the character of Bayesian inference by mechanically removing the impact of the improper prior. The resulting quantity no longer has a Bayes Factor interpretation since it is not the ratio of alternate posteriors over priors, although it remains useful (Conigliani, Castro, and O'Hagan 2000). Rubin (1984, p.1152) sets out the following definition of *Bayesianly justifiable* calculation:

> ... it treats known values as observed values of random variables, treats unknown values as unobserved random variables, and calculates the conditional distribution of unknowns given knowns and model specifications using Bayes' theorem.

As new data are observed, the researcher will have to adjust the value of η or recognize that the test is changing criteria. Therefore new fractional Bayes Factors will not be constants of previous fractional Bayes Factors and the update from the new data. While this is only mildly inconvenient, it does violate the conditional updating of estimates of unknowns based on knowns in the Rubin statement of Bayesianly justifiable.

7.3.4.5 Redux

It is appropriate to end this section with a brief discussion of what these methods are doing relative to alternatives. Given the general problem that Bayes Factors can be sensitive to the selection of priors and the specific problem that Bayes Factors are incalculable for improper priors, these four techniques segment the sample in differing ways to cancel out the effects of indeterminate constants as seen in (7.27). While these approaches all use sample quantities to determine prior density specifications, they are not empirical Bayesian methods (Carlin and Louis 2009, Chapter 5) in the classic sense: use of the data to empirically estimate the highest level of hyperpriors in a hierarchical model. Furthermore, none of these models produces the goal of "objective Bayesianism" since subjective decisions are made in every case. Nonetheless, despite the discussed flaws, each of these methods provides a means of presenting Bayes Factor evidence for one model over another using improper priors and either a nested or non-nested test.

7.3.5   Two-Sided Hypothesis Tests and Bayes Factors

Berger and Delampady (1987) propose integrating over two sections of the prior space, $p(\theta)$, to create separate prior densities for the unknown coefficient under the alternate assumptions that H_0 and H_1 are true. This is an attempt to show that the common classical practice of testing a point null hypothesis can be approximated by a precise null *interval* hypothesis. The test begins with segmenting an unconditional prior probability across the sample space of θ, and defining two conditional priors:

$$g_0(\theta|H_0) = \frac{1}{\pi_0} p(\theta) I_\Omega(\theta)$$

$$g_1(\theta|H_1) = \frac{1}{1 - \pi_0} p(\theta) I_{\Omega\prime}(\theta)$$

$$\text{where:} \quad \Omega = \{\theta : |\theta - \theta_0| < \epsilon\}, \quad \text{and} \quad \pi_0 = \int_\Omega p(\theta)d\theta. \tag{7.33}$$

In this intuitive setup π_0 is the prior probability corresponding to H_0 and g_0 is the conditional distribution of θ assuming that this H_0 is true. Conversely, $1 - \pi_0$ is the prior density corresponding to H_1 and g_1 is the conditional distribution of θ assuming that H_1 is true. The primary difficulty is determining a reasonable interval around the point null: $\{\theta_0 - \epsilon : \theta_0 + \epsilon\}$ so that the test: $H_0 : |\theta - \theta_0| \leq \epsilon, H_1 : |\theta - \theta_0| > \epsilon$ substitutes for the true point null test: $H_0 : \theta = \theta_0, H_1 : \theta \neq \theta_0$.

The core problem is that larger values of ϵ move further away from the desired nature of the two-sided problem and that smaller values of ϵ move toward "Lindley's Paradox," alternately called "Jeffreys' Paradox" (Jeffreys 1961, Lindley 1957, Shafer 1982); this is the fact that in testing a point null hypothesis for a fixed prior, and posterior cutoff points calibrated to match some constant frequentist α value, as the sample size goes to infinity, $p(H_0) \to 1$. This happens no matter how small the value for α happens to be because parameter values of negligible likelihood are given nonzero prior weights (Aitkin 1991, p.115). Furthermore, there is a problem when the range imposed by ϵ is not theoretically driven: "Such limits would be a sheer guess and merely introduce an arbitrariness" (Jeffreys 1961, 367). Casella and Berger (1987b) object to Berger and Delampady's practice of assigning

$\pi_0 = \frac{1}{2}$ as a no-information prior since no reasonable researcher would begin some enterprise with an expectation of failure at 50%.

Berger, Brown, and Wolpert (1994) followed by Berger, Boukai, and Wang (1997) designed a conditioning statistic so that frequentist probabilities coincide with analogous Bayesian posterior density regions, where the cost of this coincidence is the introduction of a middle "no-decision" region into the hypothesis test.

Starting with a purely Bayesian perspective, if a prior probability, π_0, is assigned to H_0 with complementary prior, $1 - \pi_0$, for H_1, then the posterior probability in favor of H_0 is:

$$p(H_0|\mathbf{x}) = \left[1 + \frac{1 - \pi_0}{\pi_0} \frac{1}{BF_{(0,1)}} \right]^{-1}, \tag{7.34}$$

where $BF_{(0,1)}$ is the Bayes Factor for H_0 over H_1. If equal probability is assigned to the two hypotheses, then

$$p(H_0|\mathbf{x}) = BF_{(0,1)}/(1 + BF_{(0,1)}), \quad \text{and} \quad p(H_1|\mathbf{x}) = 1/(1 + BF_{(0,1)}). \tag{7.35}$$

This leads to the selection of H_0 if $BF_{(0,1)} \leq 1$, and H_1 if $BF_{(0,1)} > 1$.

The setup just described is deliberately rigged to resemble a frequentist decision-making process. These authors specify F_0 and F_1 as the assumed smooth and invertible cumulative distribution function of the Bayes Factor, $BF_{(0,1)}$, under the assumption of Model 0 and Model 1 respectively. Now define:

$$r = 1, \quad \alpha = F_0^{-1}(1 - F_1(1)), \quad \text{if} \quad F_0^{-1}(1 - F_1(1)) \geq 1$$
$$r = F_1^{-1}(1 - F_0(1)), \quad \alpha = 1, \quad \text{if} \quad F_0^{-1}(1 - F_1(1)) < 1. \tag{7.36}$$

This leads to the following test with three decision regions:

$$\text{Reject } H_0 \text{ if } \quad BF_{(0,1)} \leq r$$
$$\text{No decision if } \quad r < BF_{(0,1)} \leq \alpha$$
$$\text{Accept } H_0 \text{ if } \quad BF_{(0,1)} \geq \alpha \tag{7.37}$$

for values of the Bayes Factor. The frequentist conditioning statistic $S = \min[B, F_0^{-1}(1 - F_1(B))]$ leads to identical conditional error probabilities as the Bayes Factor test outlined in (7.37). Despite the authors' claims that the no-decision region is observed to be small in their empirical trials, most Bayesians are likely to be uncomfortable with the idea that some fraction of the sample space of the test statistic remains ambiguous (should a loss function be assigned to the no-decision region as well?). Non-Bayesians, however, are more accustomed to the idea of "weak evidence" for a given hypothesis. Finally, short of showing a clever intersection of Bayesian and frequentist testing, this approach may be too restrictive to both sides to be widely useful.

7.3.6 Challenging Aspects of Bayes Factors

It should also be noted that there is nothing in the setup of the Bayes Factor or the subsequent judgment about the strength of evidence for one hypothesis over another that accommodates a directional (one-sided) test. Therefore Bayes Factors are restricted to the two-sided test considered here (Berger and Mortera 1999), although some authors have found exceptions for very particular situations such as strict parameter restrictions (Dudley and Haughton 1997), or orthogonalized parameters (Kass and Vaidyanathan 1992). For discussions of one-sided Bayes Factors see Casella and Berger (1987a), Moreno (2005), Marden (2000), and Wetzels *et al.* (2009).

Occasionally authors make simplifying assumptions that are hard to support in actual data-analytic settings. For instance, some researchers use *Laplace's method*, described in Section 7.7, which makes the assumption that the posterior is approximately normally distributed to get Bayes Factors in otherwise difficult cases. See, for instance, Kass (1993), Kass, Tierney, and Kadane (1989), Kass and Raftery (1995), Tierney and Kadane (1986), and Wong and Li (1992). This assumption is far less worrisome with a large sample and in small dimensions, yet this is not always the case in social science research.

Some modifications of the Bayes Factor appear not to work in practice. The posterior Bayes Factor (Aitkin 1991) uses the ratio of the posterior means as a substitute for the standard form in order to be less sensitive to the form of the prior. Unfortunately the posterior Bayes Factor introduces some minor undesirable properties such as noninvariance to hypothesis aggregation (Lindley 1991, Comments), and that it can prefer models independently of the strength of the data (Goldstein 1991).

Sometimes the Bayes Factor is just too unstable to compute numerically due to large ratio differences and computer rounding/truncating at the register level. Machine accuracy matters, of course, when comparing very large to very small numbers. Using natural logarithms is often an effective way to deal with numerical problems, and the log of the Bayes Factor has a nice interpretation:

$$\log(BF_{(1,2)}) = \log\left(\frac{\pi(M_1|\mathbf{x})/p(M_1)}{\pi(M_2|\mathbf{x})/p(M_2)}\right) = \log\left(\frac{\pi(M_1|\mathbf{x})}{\pi(M_2|\mathbf{x})}\right) - \log\left(\frac{p(M_1)}{p(M_2)}\right). \qquad (7.38)$$

This means that the log of the Bayes Factor is the log ratio of the posteriors minus the log ratio of the priors, which contrasts posterior and prior information between the two models.

Gelman and Rubin (1995) criticize the use of Bayes Factors for the same reason that others denigrate the null hypothesis significance test: it is assumed in the comparison that one of the proposed models is the correct specification. This is really more of a problem in the way that social science research is presented in published work where often only a very limited number of specifications are tested, whereas many more were posited in earlier stages of the research. Also, Han and Carlin (2001) outline computational problems that arise from using Bayes Factors for some hierarchical models (see Chapter 12) and some of the Markov chain Monte Carlo challenges (see Chapter 10).

Occasionally the integrals in (7.10) are forbidding, but Chib (1995) provides a way of finding marginal likelihoods from Gibbs sampling output, and Chib and Jeliazkov (2001) show how this can be done with Metropolis-Hastings output. We will describe these techniques later in Chapter 14 in Section 14.5 (starting on page 515). Also Morey *et al.* (2010) use the Savage-Dickey density ratio with MCMC output to conveniently calculate the Bayes

Factor. These tools are extremely useful, but restricted to MCMC output. Additionally, the R package `MCMCpack` by Martin, Quinn and Park provides a useful function, `bayesF`, for calculating the Bayes Factor for commonly used regression models, and the R package `BayesFactor` by Morey and Rouder provides the Bayes Factor individually for a list of common models. Other package authors have included functions as well to address specific problems.

7.4   The Bayesian Information Criterion (BIC)

Kass (1993) and Kass and Raftery (1995) suggest using the Bayesian information criterion (BIC) as a substitute for the full calculation of the Bayes Factor when such calculations are difficult since the BIC can be calculated without specifying priors. It is therefore more appealing to non-Bayesians and seemingly less subjective. Alternately called the Schwarz criterion, after the author of the initial article (1978), it is given by:

$$\text{BIC} = -2\ell(\hat{\boldsymbol{\theta}}|\mathbf{x}) + p\log(n) \tag{7.39}$$

where $\ell(\hat{\boldsymbol{\theta}}|\mathbf{x})$ is the maximized log likelihood value, p is the number of explanatory variables in the model (including the constant), and n is the sample size. This is very similar to the earlier "consistent" AIC of Bozdogan (1987): $CAIC = -2\ell(\hat{\boldsymbol{\theta}}|\mathbf{x}) + p(1 + \log(n))$. To compare two models with the BIC, create a Schwarz criterion difference statistic according to:

$$S = \text{BIC}_{\text{model 1}} - \text{BIC}_{\text{model 2}} = \ell(\hat{\boldsymbol{\theta}}_1|\mathbf{x}) - \ell(\hat{\boldsymbol{\theta}}_2|\mathbf{x}) - \frac{1}{2}(p_1 - p_2)\log(n) \tag{7.40}$$

where the subscripts indicate the model source. This is extremely easy to calculate and has the following asymptotic property:

$$\frac{S - \log(BF_{(1,2)})}{\log(BF_{(1,2)})} \xrightarrow[n \to \infty]{} 0. \tag{7.41}$$

Unfortunately S is only a rough approximation to $\log(BF_{(1,2)})$ since the relative error of this approximation is $o(1)$ (although this can be improved for very specific models, e.g., Kass and Wasserman (1995) for an example using normal priors), and typically large samples are required. In fact, the log form in (7.41) is essential since $\frac{\exp[S]}{BF_{(1,2)}} \to 1$ as n goes to infinity (Kass and Wasserman 1995, p.928). Nonetheless, Kass and Raftery (1995) argue that it is often possible to use the approximation: $S \approx -2\log(BF_{(1,2)})$. Robert (2001, p.353) is critical of the use of the BIC to estimate Bayes Factors because it removes the effect of any specified priors, it is difficult or impossible to calculate in complex models with non-iid data, and maximum likelihood estimations are required for every model compared and therefore might not be as much of a shortcut as intended.

The BIC should be used with caution under certain circumstances. First, the BIC can be inconsistent as the number of parameters gets very large (Stone 1977a 1977b). Models that are this large (thousands of parameters) are rare in the social sciences, but they do occur in other areas such as statistical genetics. Berger *et al.* (2003) follow-up on Stone's brief comment with a more detailed exposition of how the BIC can be a poor approximation to the logarithm of Bayes Factor using the Stone's ANOVA example and suggest the alternative forms: Generalized Bayes Information Criterion (GBIC) and a Laplace approximation to the logarithm of the Bayes Factor. Chakrabarti and Ghosh (2006) prove some of the properties of these alternative forms.

A commonly used measure of goodness-of-fit outside of Bayesian work is the Akaike information criterion (Akaike, 1973, 1974, 1976). The idea is to select a model that minimizes the negative likelihood penalized by the number of parameters:

$$\text{AIC} = -2\ell(\hat{\boldsymbol{\theta}}|\mathbf{x}) + 2p, \tag{7.42}$$

where $\ell(\hat{\boldsymbol{\theta}}|\mathbf{x})$ is the maximized model log likelihood value and p is the number of explanatory variables in the model (including the constant). The AIC is very useful in comparing and selecting non-nested model specifications, but many authors have noted that the AIC has a strong bias toward models that overfit with extra parameters since the penalty component is obviously linear with increases in the number of explanatory variables, and the log likelihood often increases more rapidly (Carlin and Louis 2009, p.53; Neftçi 1982, p.539; Sawa 1978, p.1280).

Using the value of the log likelihood alone without some overfitting penalization as a measure of model fit is a poor strategy since the likelihood never decreases by adding more explanatory variables regardless of their inferential quality. See also the lesser known alternatives: MIC (Murata, Yoshizawa, and Amari 1994) and TIC (Takeuchi 1976), NIC (Stone 1974), and the explicitly Bayesian EAIC (Brooks 2002). Burnham and Anderson (2002) argue in their book that when the data size is small relative to the number of specified parameters, one should use a modified AIC according to:

$$AIC_c = AIC + 2p(p+1)/(n-p-1), \tag{7.43}$$

which add a further penalty.

Both the AIC and the BIC have a problem with a widened definition of parameters. Suppose we have panel data with n individuals each measured over t time-points. If individual-specific effects are drawn from a single random effects term (i.e., given a distributional assignment), then group-level variables from this specification play different roles. Such models are very common Bayesian specifications and are the subject of Chapter 12. In this simple case should this latent variable be counted as one parameter or n parameters? If it counts as n parameters, then the effect of shrinkage is ignored as well, substantially inflating the value of the AIC and BIC. On the other hand, if it counts as one parameter it appears that we are intentionally reducing the penalty component to advantage the model relative to one without such a random effect. We now turn to a tool specifically designed to handle this problem.

7.5   The Deviance Information Criterion (DIC)

A very useful tool for model assessment and model comparison is the deviance informa-tion criterion created by Spiegelhalter, Best, Carlin, and van der Linde (2002), although earlier, more narrow versions exist. The DIC has already become a popular model com-parison choice since it is integrated into the `WinBUGS` package for MCMC estimation (see Chapter 10). There are two objectives here: describing "model complexity" and model fit as in the AIC or BIC. For instance, in the BIC the term $-2\ell(\hat{\boldsymbol{\theta}}|\mathbf{x})$ is thought of as a describer of fit, whereas $p\log(n)$ shows complexity in the form of parameter vector size in the model. Suppose we have a model under consideration, defined by $p(\mathbf{y}|\boldsymbol{\theta})$ for data $\mathbf{y}$ and parameter vector $\boldsymbol{\theta}$. The first quantity to define is the "Bayesian deviance" specified by Spiegelhalter *et al.* as:

$$D(\boldsymbol{\theta}) = -2\log[p(\mathbf{y}|\boldsymbol{\theta})] + 2\log[f(\mathbf{y})], \qquad (7.44)$$

where $f(\mathbf{y})$ is some function of just the data, the "standardizing factor." Spiegelhalter *et al.* (2002) de-emphasize $f(\mathbf{y})$ for model comparison and even suggest using $f(\mathbf{y}) = 1$ (giving zero contribution above) since this term must be identical for both model calculations and therefore cancels out. Note the similarity to the AIC and BIC, except that there is no p term in the second part. We can use (7.44) in explicitly posterior terms by inserting a condition on the data and taking an expectation over $\boldsymbol{\theta}$:

$$\overline{D(\boldsymbol{\theta})} = E_{\boldsymbol{\theta}}[-2\log[p(\mathbf{y}|\boldsymbol{\theta})|\mathbf{y}]] + 2\log[f(\mathbf{y})], \qquad (7.45)$$

(the authors switch notation between $\overline{D(\boldsymbol{\theta})}$ and $\bar{D}$). This posterior mean difference is now a measure of Bayesian model fit. Now define $\tilde{\boldsymbol{\theta}}$ as a posterior estimate of $\boldsymbol{\theta}$, which can be the posterior mean or some other easily produced value (although Spiegelhalter *et al.* note that obvious alternatives such as the median may produce problems). Observe that we can also insert $\tilde{\boldsymbol{\theta}}$ into (7.44). The *effective dimension* of the model is now defined by:

$$p_D = \overline{D(\boldsymbol{\theta})} - D(\tilde{\boldsymbol{\theta}}). \qquad (7.46)$$

This is the "mean deviance minus the deviance of the means." Another way to think about effective dimensions is by counting the roles that parameters take on in Bayesian models. In this way, p_D is the sum of the parameters each of which is weighted according to: $\omega_p = 1$ for parameters unconstrained by prior information, $\omega = 0$ for parameters completely specified (fixed) by prior information, and $\omega \in [0\!:\!1]$ for parameters with specific dependencies on the data or priors.

Also, due to the subtraction, p_D is independent of our choice of $f(\mathbf{y})$. This is illustrated by expanding (7.46) and applying Bayes' Law:

$$p_D = \overline{D(\boldsymbol{\theta})} - D(\tilde{\boldsymbol{\theta}})$$

$$= \left\{ E_{\boldsymbol{\theta}|\mathbf{y}}[-2\log[p(\mathbf{y}|\boldsymbol{\theta})|\mathbf{y}] + 2\log[f(\mathbf{y})]\right\} - \left\{-2\log[p(\mathbf{y}|\tilde{\boldsymbol{\theta}})] + 2\log[f(\mathbf{y})]\right\}$$

$$= E_{\boldsymbol{\theta}|\mathbf{y}}[-2\log(p(\mathbf{y}|\boldsymbol{\theta}))] + 2\log(p(\mathbf{y}|\tilde{\boldsymbol{\theta}}))$$

$$= E_{\boldsymbol{\theta}|\mathbf{y}}\left[-2\log\left(\frac{p(\boldsymbol{\theta}|\mathbf{y})p(\mathbf{y})}{p(\boldsymbol{\theta})}\right)\right] + 2\log\left(\frac{p(\tilde{\boldsymbol{\theta}}|\mathbf{y})p(\mathbf{y})}{p(\tilde{\boldsymbol{\theta}})}\right)$$

$$= E_{\boldsymbol{\theta}|\mathbf{y}}[-2\log(p(\boldsymbol{\theta}|\mathbf{y})/p(\boldsymbol{\theta}))] + 2\log(p(\tilde{\boldsymbol{\theta}}|\mathbf{y})/p(\tilde{\boldsymbol{\theta}})). \tag{7.47}$$

We can think of the ratio $p(\boldsymbol{\theta}|\mathbf{y})/p(\boldsymbol{\theta})$ here as the gain in information provided by conditioning on the data in the model, and correspondingly $p(\tilde{\boldsymbol{\theta}}|\mathbf{y})/p(\tilde{\boldsymbol{\theta}})$ as the gain in information after plugging in the chosen estimate.

In a simulation context (subsequent chapters), p_D can be computed as the mean of simulated values of $D(\boldsymbol{\theta})$ minus $D(\boldsymbol{\theta})$ plugging the mean of the simulated values of $\boldsymbol{\theta}$, hence the interpretation as the difference between the posterior mean of the deviance and the deviance at the poster mean (or some other chosen statistic). So while $\overline{D(\boldsymbol{\theta})}$ is a Bayesian measure of model fit, p_D is designed to be a "complexity measure" for the effective number of parameters in the model. As such it is the Bayesian analogy to the second term in the AIC.

The Deviance Information Criterion is created by adding an additional model fit term to the effective dimension:

$$DIC = \overline{D(\boldsymbol{\theta})} + p_D = 2\overline{D(\boldsymbol{\theta})} - D(\tilde{\boldsymbol{\theta}}) \tag{7.48}$$

and is thus the Bayesian measure of model fit above with an extra complexity penalization. The logic here goes back to a common criticism of the AIC that using a plug-in value rather than integrating out unknown parameters leads to insufficient incentive for parsimonious models. The DIC also uses a plug-in value ($\tilde{\boldsymbol{\theta}}$), but incorporates an additional penalty to compensate. Some authors still prefer an adapted version of the AIC, such as Brooks (2002) who suggests the *expected Akaike information criterion*: $EAIC = \overline{D(\boldsymbol{\theta})} + 2p$, and $D(\bar{\boldsymbol{\theta}}) + 2p$.

■ **Example 7.3:**   **Hierarchical Models of Rural Migration in Thailand.** Garip and
   Western (2011, 2005 results shown here) use the example of village-level migration
   to contrast hierarchical model specifications for a dichotomous outcome.[5] They look
   at survey data on young adults (18-25) in 22 Northeastern Thai villages where the
   outcome variable is 1 if the respondent spent at least two months away from the village

---

[5]Note: this example is based on an earlier version of the paper which had more starkly contrasting p_D and DIC values.

in 1990 and 0 otherwise. The individual-level explanatory variables are sex, age, years of education, number of prior trips, and the village-level explanatory variables are prior trips from the villages and the Gini index of prior trips.

TABLE 7.4: AVAILABLE MODEL SPECIFICATIONS, THAI MIGRATION

Model	Specification	Prior Parameters
Pooled	$\text{logit}(p_{ij}) = \alpha + x'_{ij}\beta + z'_j\gamma$	$\alpha \sim \mathcal{N}(0, 10^6)$
		$\beta[1:4] \sim \mathcal{N}(0, 10^6)$
		$\gamma[1:2] \sim \mathcal{N}(0, 10^6)$
Fixed Effect	$\text{logit}(p_{ij}) = \alpha_j + x'_{ij}\beta$	$\alpha[1:22] \sim \mathcal{N}(0, 10^6)$
		$\beta[1:4] \sim \mathcal{N}(0, 10^6)$
Random Effect	$\text{logit}(p_{ij}) = \alpha_j + x'_{ij}\beta + z'_j\gamma$	$\mu = 0$
	$\alpha[1:22] \sim \mathcal{N}(\mu, \tau^2)$	$\tau \sim \mathcal{IG}(10^{-3}, 10^{-3})$
		$\beta[1:4] \sim \mathcal{N}(0, 10^6)$
		$\gamma[1:2] \sim \mathcal{N}(0, 10^6)$
Random Intercept and Random Slope	$\text{logit}(p_{ij}) = \alpha_j + x'_{ij}\beta_j + z'_j\gamma$	$\mu_\alpha \sim \mathcal{N}(0, 10^6)$
	$\alpha[1:22] \sim \mathcal{N}(\mu_\alpha, \tau_\alpha^2)$	$\tau_\alpha^2 \sim \mathcal{IG}(10^{-3}, 10^{-3})$
	$\beta[1:22][1:4] \sim \mathcal{N}(\mu_\beta, \tau_\beta^2)$	$\mu_\beta \sim \mathcal{N}(0, 10^6)$
		$\tau_\beta^2 \sim \mathcal{IG}(10^{-3}, 10^{-3})$
		$\gamma[1:2] \sim \mathcal{N}(0, 10^6)$

Methodological interest here centers on the utility of specifying hierarchies in the model. These are useful tools for recognizing different levels that the data affect some outcome. For classical non-Bayesian references, see Bryk and Raudenbush (2001) or Goldstein (1985). We will study Bayesian hierarchical models in considerable detail in Chapter 12, and for the moment an excellent source for background reading is Good (1980a). There are two data matrices here with rows: x_{ij} for individual i in village j, and z_j for village j. The four contrasting models for the probability of migration, p_{ij}, given by Western and Garip are given in Table 7.4 where α is a common intercept, α_j is a village-specific intercept, and $[\beta, \gamma]$ is the unknown parameter vector to be described with a posterior distribution. There are many contrasting features in these models having to do with the relative effects in the data and the role we would like these effects to play in the model. Suppose we simply wanted to test the relative quality of the models with the DIC, ignoring for now any substantive issues with different hierarchies.

Despite the notable structural differences in these models, both the individual-level and village-level coefficients are remarkably similar and do not present any substantively different stories whatsoever (except of course that the fixed effects model does not have village-level coefficients). It is a modest difference, but the posterior distributions for the random intercept and random slope/random intercept are more diffuse than for the pooled model, reflecting the random effects modeling heterogeneity.

TABLE 7.5:   Model Comparison, Thai Migration

Model:	Pooled	Fixed Effect	Random Effect	Rdm. Slope & Intercept
Intercept	-0.80 (0.12)	-0.66 (0.10)	-0.79 (0.13)	-0.76 (0.14)
Male	0.32 (0.14)	0.38 (0.14)	0.33 (0.14)	0.30 (0.15)
Age	-0.13 (0.08)	-0.15 (0.08)	-0.13 (0.08)	-0.13 (0.09)
Education	0.39 (0.07)	0.38 (0.08)	0.39 (0.07)	0.39 (0.08)
Prior trips(i)	1.08 (0.09)	1.13 (0.10)	1.08 (0.09)	1.22 (0.16)
Prior trips(j)	-0.61 (0.37)		-0.59 (0.43)	-0.69 (0.46)
Gini trips(j)	-0.62 (0.21)		-0.60 (0.24)	-0.61 (0.25)
Posterior means (posterior standard deviations in parentheses)				
p_D	7.02	26.24	10.93	30.58
DIC	1259.41	1273.29	1259.62	1247.55

We see from the $p_D = 7.02$ value (Table 7.5) that the pooled model is the most parsimonious, with the extra 0.02 above the 7 specified parameters coming from prior information. Interestingly, even though it contains $7 + 22$ parameters, the random intercept model is only slightly less parsimonious with $p_D = 10.93$. Not surprisingly, the random slope/random intercept model is the least parsimonious with 110 random effects and 2 village-level parameters specified, which is only moderately less parsimonious than the random effect model with 26 parameters, justifying the hierarchical structure imposed. Furthermore, with the lowest DIC value, we find additional support for the more complex structure.

7.5.1   Some Qualifications

Some concerns emerge with the use of the DIC (the Spiegelhalter *et al.* (2002) paper is accompanied by 23 pages of discussion). First, its lack of invariance to reparameterizations of the parameters where the subsequent differences can be large, Therefore, we should be cautious in interpreting large DIC changes that come from reparameterizations only. Secondly, there is evidence that DIC comparisons are most straightforward when the form of the likelihoods (focus) are the same and only the selected explanatory variables differ. Otherwise it is essential to make the standardizing factor the same. Likelihood functions from hierarchical models with different level structures provide difficult interpretational problems (see the example above). In fact, the DIC is strongly justified only for likelihood functions based on exponential family forms (Appendix (A)). Helpfully, Celeux *et al.* (2006) survey strengths and weakness for the DIC beyond applications to exponential families. When the posterior distributions are non-symmetric or multimodal, the use of the posterior mean may be inappropriate. It is possible for p_D to be negative, so DIC can also be negative. Clearly this provided interpretational problems. The DIC is not a function of the marginal likelihood, therefore it is not related to Bayes Factor comparisons. The DIC does not work when the likelihood depends on discrete parameters, so it does not work with mixture likelihoods (`WinBUGS` will "gray-out" the DIC button automatically). Finally, specification of the DIC implies use of the data twice: once to produce the posterior and once again with $p(\mathbf{y}|\boldsymbol{\theta})$ in the expectation. The two objections are: philosophical problems with violating canonical tenets of Bayesian inference, and practical problems with the increased tendency to overfit the data.

7.6 Comparing Posterior Distributions with the Kullback-Leibler Distance

There are many situations where it is convenient to compare distributions, posteriors in particular. The Bayes Factor (Chapter 7), in particular, provides a mechanical way to inferentially compare two model results. A different and more general method for evaluating the difference between two distributions is the *Kullback-Leibler distance* (sometimes called the *entropy distance*), which is given by:

$$I(f,g) = \int \log \left[\frac{f(x)}{g(x)} \right] f(x) dx, \tag{7.49}$$

for two candidate distributions $f(x)$ and $g(x)$ (Robert and Casella 1999, p.222, White 1996, p.9). This is the expected log-ratio of the two densities with the expectation taken relative to one selected distribution ($f(x)$ here). Despite the simple form of (7.49), it can occasionally be difficult to calculate analytically. If $f(x)$ and $g(x)$ are exponential family distributions, the resulting cancellations can considerably simplify the integral (McCulloch and Rossi 1992). For instance, suppose we are interested in comparing the Kullback-Leibler distance between two proposed univariate Poisson distributions indexed by intensity parameters λ_1

and λ_2. The form of $I(f,g)$ is given by:

$$I(f,g) = \sum_{\mathcal{X}} \log\left[\frac{e^{-\lambda_1}\lambda_1^{\mathbf{x}}/\mathbf{x}!}{e^{-\lambda_2}\lambda_2^{\mathbf{x}}/\mathbf{x}!}\right]\frac{e^{-\lambda_1}\lambda_1^{\mathbf{x}}}{\mathbf{x}!}$$

$$= \sum_{\mathcal{X}}\left[(\lambda_2 - \lambda_1) + \mathbf{x}\log\left(\frac{\lambda_1}{\lambda_2}\right)\right]\frac{e^{-\lambda_1}\lambda_1^{\mathbf{x}}}{\mathbf{x}!}$$

$$= \lambda_2 - \lambda_1 + \lambda_1 \log\left(\frac{\lambda_1}{\lambda_2}\right). \tag{7.50}$$

However, the Kullback-Leibler calculations are rarely this direct. Furthermore, the Kullback-Leibler distance is best used only comparatively among a set of alternative specifications since it is not a true Cartesian distance, the scale is still arbitrary, and it is not symmetric: $I(f,g) \neq I(g,f)$. Nonetheless, this is an excellent way to measure the closeness of a set of posteriors.

General details about using the Kullback-Leibler distance can be found in Kullback (1968) and Brown (1986), and hypothesis testing applications are given by Goel and DeGroot (1979), Janssen (1986), Ebrahimi, Habibullah, and Soofi (1992), and Robinson (1991). The relationship to the AIC model fitting criteria (Chapter 7) is explored by Hurvich and Tsai (1991) and Hurvich, Shumway, and Tsai (1990). Efron (1978) uses the "curve of constant Kullback-Leibler information" in relating the natural parameter space of coefficient with its model-induced expected parameter space, and several authors, such as Hernandez and Johnson (1980), Johnson (1984), and Sakia (1992), use Kullback-Leibler information to estimate the parameter of the popular Box-Cox transformation. There is also a wealth of studies using Kullback-Leibler information to make direct model comparisons, such as Leamer (1979), Sawa (1978), Vuong (1989), and White (1996, Chapter 9). We will return to this technique in Section 9.3.1 with a numerical example.

TABLE 7.6: MARTIKAINEN *et al.* (2005) MORTALITY
PARAMETERS

	Married	Never Married	Divorced	Widowed
Men	1.00	1.84	2.08	1.51
Women	1.00	1.59	1.62	1.28

■ **Example 7.4:** **Example: Models of Relative Mortality** Martikainen *et al.* (2005) seek to explain why the non-married population has a higher mortality rate compared to the married population, an effect seen in many countries. Their supposition is that the health-related behavior of non-married individuals contributes to their higher rates. Part of this study uses data from 1996 to 2000 in Finland for individuals aged 30-64 years. Their models produce mortality rates relative to the married populations where the values are interpreted as Poisson intensity parameters. Using their "full model" we see the relative intensity parameter estimates in Table 7.6.

These values can be evaluated as single point estimates, as the original authors do, but actually they are describing distributional differences. Given the asymmetry of the Poisson distribution, simple numerical comparison of the estimated intensity parameters may not give a complete summary. So using (7.50) we can summarize distributional differences, as given in Table 7.6.

TABLE 7.7: MARTIKAINEN *et al.* (2005) MORTALITY DIFFERENCES

Men	Never Married	Divorced	Widowed
Married	0.23023	0.34763	0.09789
Never Married		0.01441	0.03369
Divorced			0.09614
Women	Never Married	Divorced	Widowed
Married	0.12627	0.13757	0.03314
Never Married		0.00028	0.03483
Divorced			0.04162

The results in Table 7.7 reinforce the main findings that men fare worse outside of marriage than women, which supports the authors' hypothesis about behavioral factors: accidents from risky ventures, violence, and alcohol. The Kullback-Leibler distributional differences highlight more directly comparisons between the non-marriage categories. For example the distribution difference between Never Married and Divorced is much greater for Men than for Women (roughly two orders of magnitude). Since these are all relative measures, we can assert that there is virtually no distributional distance between Divorced and Never Married for Women (notice that 0.00028 is much smaller than $0.13757 - 0.12627 - 0.0113$).

7.7   Laplace Approximation of Bayesian Posterior Densities

Laplace's method (1774) is a well-known method for approximating the shape of marginal posterior densities that is very useful in the Bayesian context when direct calculations are difficult. The now standard reference to approximating Bayesian posterior densities with Laplace's method is Tierney and Kadane (1986), and theoretical details on the accuracy of the approximation can be found in Wong and Li (1992) and Kass and Vaidyanathan (1992). Kass (1993) shows how the Laplace approximation can be handy for calculating Bayes Factors. In general, the Laplace approximation is a very useful tool to have at one's disposal when a normal approximation posterior is reasonable, and can be especially useful in higher dimensions when other procedures fail.

The basic idea is to carry out a Taylor series expansion around the maximum likelihood estimate value, ignore the negligible terms, and normalize. For simplicity we will derive the approximation in only one dimension, but the generalization is obvious. Start with a posterior density of interest calculated by the likelihood times the specified prior: $\pi(\theta|\mathbf{x}) \propto L(\mathbf{x}|\theta)p(\theta)$, and assume that this distributional form is nonnegative, integrable, and single-peaked about a mode, $\hat{\theta}$, which is a reasonable set of assumptions provided that sample sizes are not small.[6]

The posterior expectation of some smooth function of θ, $g(\theta|\mathbf{x})$ (such as a mean, variance, or other desired summary quantity) is given by:

$$E[g(\theta|\mathbf{x})] = \int_\theta g(\theta|\mathbf{x})L(\mathbf{x}|\theta)p(\theta)d\theta = \int_\theta g(\theta|\mathbf{x})\pi(\theta|\mathbf{x})d\theta. \tag{7.51}$$

Using Bayes' Law we know that:

$$\pi(\theta|\mathbf{x}) = \frac{L(\mathbf{x}|\theta)p(\theta)}{\int_\theta L(\mathbf{x}|\theta)p(\theta)d\theta}. \tag{7.52}$$

So (7.51) becomes:

$$E[g(\theta|\mathbf{x})] = \int_\theta g(\theta|\mathbf{x})\frac{L(\mathbf{x}|\theta)p(\theta)}{\int_\theta L(\mathbf{x}|\theta)p(\theta)d\theta}d\theta = \frac{\int_\theta g(\theta|\mathbf{x})L(\mathbf{x}|\theta)p(\theta)d\theta}{\int_\theta L(\mathbf{x}|\theta)p(\theta)d\theta}. \tag{7.53}$$

Now define the quantities:

$$-nh_1(\theta) = \log g(\theta|\mathbf{x}) + \log L(\mathbf{x}|\theta) + \log p(\theta)$$
$$-nh_2(\theta) = \log L(\mathbf{x}|\theta) + \log p(\theta). \tag{7.54}$$

Substituting these values into (7.53) produces:

$$E[g(\theta|\mathbf{x})] = \frac{\int \exp[-nh_1(\theta)]d\theta}{\int \exp[-nh_2(\theta)]d\theta}. \tag{7.55}$$

If we have a mathematically convenient approximation for either $h_1(\theta)$ or $h_2(\theta)$, then $E[g(\theta|\mathbf{x})]$ becomes very easy to calculate. Consider a Taylor series expansion around some arbitrary point θ_0 for $h(\theta)$ where this generic identification applies to either $h_1(\theta)$ or $h_2(\theta)$:

$$h(\theta) = h(\theta_0) + (\theta - \theta_0)h'(\theta_0) + \frac{1}{2!}(\theta - \theta_0)^2 h''(\theta_0)$$
$$+ \frac{1}{3!}(\theta - \theta_0)^3 h'''(\theta_0) + R_n(\theta), \tag{7.56}$$

---

[6]Kass and Raftery (1995) recommend that for k explanatory variables one should ideally have a sample size of at least $20k$, and that less than $5k$ is "worrisome" (naturally such guidelines depend on the behavior of the likelihood function in general and substantial deviations from normality will obviously require even greater sample sizes). Carlin and Louis (2009, p.110) note that the Laplace approximation is therefore rarely helpful for higher dimensional problems such as those with more than ten explanatory variables.

where $R_n(\theta)$ is the remainder consisting of progressively smaller terms in the expansion. In fact, since $\lim_{\theta \to \theta_0} R_n(\theta)/(\theta - \theta_0)^3 = 0$ (Robert and Casella 2004), then we can safely ignore higher order terms and just use: $h_1(\theta)$ or $h_2(\theta)$:

$$h(\theta) \approx h(\theta_0) + (\theta - \theta_0)h'(\theta_0) + \frac{1}{2!}(\theta - \theta_0)^2 h''(\theta_0) + \frac{1}{3!}(\theta - \theta_0)^3 h'''(\theta_0). \qquad (7.57)$$

Recall that we are most interested in the modal point of the posterior density, so pick θ_0 to be the point where the first derivative vanishes: $\hat{\theta}$. Substituting $\hat{\theta}$ into (7.57) eliminates the linear term: $(\theta - \hat{\theta})h'(\hat{\theta})$. If we ignore the rapidly vanishing cubic term, $\frac{1}{3!}(\theta - \hat{\theta})^3 f'''(\hat{\theta})$, then this is said to be a first-order approximation. In the cases where more accuracy is required, this term can be expanded in a second Taylor series around $\hat{\theta}$ to produce second- and third-order accuracy. Often the sample size is sufficiently large so that first-order accuracy is sufficient (see Robert and Casella [2004, p.109] for details on this second-order expansion).

Take the now reduced form in (7.57) and make the substitution $\hat{\sigma}^2 = (h''(\theta_0))^{-1}$ to derive:

$$\int_\theta \exp[-nh(\theta)]d\theta$$

$$= \int_\theta \exp\left[-nh(\hat{\theta}) - \frac{n}{2\hat{\sigma}^2}(\theta - \hat{\theta})^2\right] d\theta,$$

$$= \exp[-nh(\hat{\theta})] \int_\theta \exp\left[-\frac{n}{2\hat{\sigma}^2}(\theta - \hat{\theta})^2\right] d\theta,$$

$$= \exp[-nh(\hat{\theta})](\sqrt{2\pi\hat{\sigma}^2}n^{-\frac{1}{2}}) \int_\theta \frac{n^{\frac{1}{2}}}{\sqrt{2\pi\hat{\sigma}^2}} \exp\left[-\frac{n}{2\hat{\sigma}^2}(\theta - \hat{\theta})^2\right] d\theta,$$

$$= \exp[-nh(\hat{\theta})](\sqrt{2\pi\hat{\sigma}^2}n^{-\frac{1}{2}}), \qquad (7.58)$$

using the standard trick of pushing terms into the integral so that it equals one. This is precisely the justification for a normal approximation in (7.53). The resulting approximation for $E[g(\theta|\mathbf{x})]$ is:

$$\hat{E}[g(\theta|\mathbf{x})] = \frac{\int_\theta \exp[-nh_1(\theta)]d\theta}{\int_\theta \exp[-nh_2(\theta)]d\theta}$$

$$= \frac{\exp[-nh_1(\hat{\theta}_1)](\sqrt{2\pi\sigma_1^2}n^{-\frac{1}{2}})}{\exp[-nh_2(\hat{\theta}_2)](\sqrt{2\pi\sigma_2^2}n^{-\frac{1}{2}})}$$

$$= \frac{\sigma_1}{\sigma_2} \exp[-n(h_1(\hat{\theta}_1) - h_2(\hat{\theta}_2)]. \qquad (7.59)$$

The last line of (7.59) is exactly the form found by Tierney and Kadane to be accurate to within:

$$E[g(\theta|\mathbf{x})] = \hat{E}[g(\theta|\mathbf{x})](1 + \circ(n^{-2})), \qquad (7.60)$$

where $\circ(n^{-2})$ indicates "on the order of" or "at the rate of" $1/n^2$ as n increases. Therefore Laplace's method replaces integration as a method of obtaining an estimate of $E[g(\theta|\mathbf{x})]$ with differentiation. Typically differentiation is not only easier to perform, but the algorithms are typically more numerically stable (Gill, Murray, and Wright 1981).

The variance of $g(\theta|\mathbf{x})$ is produced from standard theory: $\text{Var}[g(\theta|\mathbf{x})] = \hat{E}[g(\theta|\mathbf{x})^2] - (\hat{E}[g(\theta|\mathbf{x})])^2$, which also has relative error on the order of $\circ(n^{-2})$ (Tierney and Kadane 1986, p.83). The asymptotic properties are given in detail by several authors: de Bruijn (1981, Chapter 7), Kass, Tierney, and Kadane (1989), Tierney, Kass, and Kadane (1989b), and Schervish (1995). For an applicable, but more generalized discussion, of the asymptotic behavior of Gaussian kernel estimators, see Le Cam (1986, Section 12.4) or Barndorff-Nielsen and Cox (1989, Chapter 1).

The multivariate formulation of (7.59) follows directly by substituting $\boldsymbol{\theta}$ for θ, $|\boldsymbol{\Sigma}|^{\frac{1}{2}}$ for σ, where $\boldsymbol{\Sigma}$ is the inverse negative of the second derivative matrix (Hessian) of the likelihood function evaluated at the modal point. Occasionally the multivariate approximation imposes some additional computational difficulties, such as the conditioning of the Hessian (Albert 1988; Hsu 1995). In the case where the derivatives required to produce (7.59) are not readily calculable, MCMC simulation techniques (introduced in Chapter 10) have been developed to produce desired interim quantities (Raftery 1996). In some cases a slightly more complicated "saddle point" approximation can be substituted (Goutis and Casella 1999; Tierney, Kass, and Kadane 1989b). Tierney, Kass, and Kadane (1989a)   present a more flexible form of the basic Laplace procedure variant that reduces the need for typically more difficult higher-order derivatives.

Several implementation details are worth observing. First, note that the two modal points, $\hat{\theta}_1$ and $\hat{\theta}_2$ differ. The second point is simply the standard maximum likelihood value. Tierney and Kadane (1986) as well as Press (1989, Section 3.3.1) point out that the first value typically lives in the same neighborhood and therefore can easily be found with a simple numerical search algorithm such as Newton-Raphson, using $\hat{\theta}_2$ as a starting point. Also, under fairly general circumstances (7.59) is the *empirical Bayes estimator* for $g(\theta|\mathbf{x})$ (Lehmann and Casella 1998, p.271). It is also possible to use simulation techniques as a substitute for difficult derivatives in complex models. This approach is then called *Laplace-Metropolis* estimation (Lewis and Raftery 1997, Raftery 1996).

One convenient feature of Laplace's approximation is its flexibility with regard to new prior forms and newly observed data. Define $p_{new}(\theta)$ as an alternative prior distribution that the researcher is interested in substituting into the calculation of the statistic of interest, $E[g(\theta|\mathbf{x})]$. Rather than recalculate the Laplace approximation, we can use the following shortcut (Carlin and Louis 2009, pp.111-112). First calculate a prior odds ratio of the new prior to the one already used in the calculation:

$$b(\theta) = \frac{p_{new}(\theta)}{p(\theta)}, \qquad (7.61)$$

so the new expected value of g is:

$$\hat{E}_{new}[g(\theta|\mathbf{x})] = \frac{\int g(\theta|\mathbf{x})L(\mathbf{x}|\theta)p(\theta)b(\theta)d\theta}{\int L(\mathbf{x}|\theta)p(\theta)b(\theta)d\theta}.$$

$$= \frac{b(\theta_1)}{b(\theta_2)}\hat{E}[g(\theta|\mathbf{x})]. \qquad (7.62)$$

This means that we don't have to fully recalculate the expected value; instead we use the already maximized θ values, $\hat{\theta}_1$ and $\hat{\theta}_2$, from the previous calculations. So we can very quickly try a large number of alternate prior specifications and immediately see the implications.

7.8   Exercises

7.1   Perform a frequentist hypothesis test of H_1: $\theta = 0$ versus H_0: $\theta = 500$, where $X \sim \mathcal{N}(\theta, 1)$ (one-tail at the $\alpha = 0.01$ level). A single datapoint is observed, $x = 3$. What decision do you make? Why is this not a defensible procedure here? Does a Bayesian alternative make more sense?

7.2   A random variable X is distributed $\mathcal{N}(\mu, 1)$ with unknown μ. A standard (non-Bayesian) hypothesis test is set up as: H_1: $\mu = 0$ versus H_0: $\mu = 50$ with $\alpha = 0.05$ (one-sided such that the critical value is 1.645 for rejection). A sample of size 10 is observed, $(2.77, 0.91, 1.88, 2.28, 1.86, 1.33, 2.99, 2.07, 1.58, 2.99)$, with mean 2.07 and standard deviation 0.70. Do you decide to reject the null hypothesis? Why might this not be reasonable?

7.3   Akaike (1973) states that models with negative AIC are better than the saturated model and by extension the model with the largest negative value is the best choice. Show that if this is true, then the BIC is a better asymptotic choice for comparison with the saturated model.

7.4   For a given binomial experiment we observe x successes out of n trials. Using a conjugate beta prior distribution with parameters α and β, show that the marginal likelihood is:
$$\binom{n}{x} \frac{\Gamma(\alpha + \beta)\Gamma(x + \alpha)\Gamma(n + \beta - x)}{\Gamma(\alpha)\Gamma(\beta)\Gamma(n + \alpha + \beta)}.$$

7.5   Derive the last line of (7.14) from:
$$p(H_0|\text{data}) = 1 - p(H_1|\text{data}) = 1 - \frac{1}{BF_{(1,0)}}\left[\frac{p(\text{data})}{p(H_0)}p(H_0|\text{data})\right]\frac{p(H_1)}{p(\text{data})}.$$

7.6   Derive the Bayes Factor for a two-sided test of a mean in (7.15) on page 218.

7.7   (Berger 1985). A child takes an IQ test with the result that a score over 100 will be designated as above average, and a score of under 100 will be designated as below average. The population distribution is distributed $\mathcal{N}(100, 225)$ and the child's posterior distribution is $\mathcal{N}(110.39, 69.23)$. Test competing designations on a single test, $p(\theta \leq 100|x)$ vs. $p(\theta > 100|x)$, with a Bayes Factor using equally plausible prior notions (the population distribution), and normal assumptions about the posterior.

7.8   Recalculate the model and the Bayes Factor for including an interaction term from the Hanushek and Jackson election surveys example (page 220) using informed normal priors.

7.9   Returning to the beta-binomial model from Chapter 2, set up a Bayes Factor for: $p < 0.5$ versus $p \geq 0.5$, using a uniform prior, and the data: $[0, 1, 0, 1, 1, 1, 1, 0, 0, 0, 0, 0, 0, 0, 1]$. Perform the integration step using rejection sampling (Chapter 9).

7.10   Consider Example 15 from Appendix A (page 575). Specify normal priors for the three-dimensional vector β with mean and variance equal to the observed sample quantities in Table 15 on page 576. Bedrick, Christensen, and Johnson (1997) suggest using Bayes Factors to test competing link functions, as opposed to the more conventional idea of testing competing variable specifications. Calculate the Bayes Factor for each of the three link functions against each other. Which would you use? Why?

7.11   Demonstrate using rejection sampling the following equality from Bayes' original (1763) paper, using different values of n and p:

$$\int_0^1 \binom{n}{p} x^p (1-x)^{n-p} dx = \frac{1}{n}.$$

7.12   Using the Palm Beach County voting model in Section 5.1.1 compare the full model with a null model (intercept only) writing your own DIC function for linear specification in R.

7.13   Calculate the posterior distribution for β in a logit regression model

$$r(\mathbf{X'b}) = p(\mathbf{y} = 1 | \mathbf{X'b}) = 1/[1 + \exp(-\mathbf{X'b})]$$

with a $\mathcal{BE}(A, B)$ prior. Perform a formal test of a $\mathcal{BE}(4, 4)$ prior versus a $\mathcal{BE}(4, 1)$ prior. Calculate the Kullback-Leibler distance between the two resulting posterior distributions.

7.14   Taking the state-level obesity data from Exercise 6 on page 89, segment the cases in Southern and non-Southern states to form two groups. Do a Bayesian version of the standard Student's t-test for normal data that uses a Bayes Factor as described in Section 7.3.2, starting on page 219.

7.15   (Aitkin 1991). Model 1 specifies $y \sim \mathcal{N}(\mu_1, \sigma^2)$ with μ_1 specified and σ^2 known, Model 2 specifies $y \sim \mathcal{N}(\mu_2, \sigma^2)$ with μ_2 unknown and the same σ^2. Assign the improper uniform prior to μ_2: $p(\mu_2) = C/2$ over the support $[-C{:}C]$, where C is large enough value to make this a reasonably uninformative prior. For a predefined standard significance test level critical value z, the Bayes Factor for Model 1 versus Model 2 is given by:

$$B = \frac{2Cn^{1/2}\phi(z)}{\sigma\left[\Phi\left(n^{1/2}\frac{\bar{y}+C}{\sigma}\right) - \Phi\left(n^{1/2}\frac{\bar{y}-C}{\sigma}\right)\right]}.$$

Show that B can be made as large as one likes as $C \to \infty$ or $n \to \infty$ for any fixed value of z. This is an example of Lindley's paradox for a point null hypothesis test (Section 7.3.5 starting on page 228).

7.16　Chib (1995) introduces MCMC based tools for obtaining marginal likelihoods and therefore Bayes Factors. Using the notation of that paper, $f(\mathbf{y}|\boldsymbol{\theta}_k, M_k)$ is the density function of the observed data under model M_k with model-specific parameter vector $\boldsymbol{\theta}_k$, which has the prior distribution $\pi(\boldsymbol{\theta}_k|M_k)$ (notice that this different terminology for a prior than that used in this text). If a model is estimated with a Gibbs sampler then G iterated values, $\boldsymbol{\theta}^{(1)}, \ldots, \boldsymbol{\theta}^{(G)}$, are produced. Chib notes that the marginal likelihood under model M_k:

$$m(\mathbf{y}|M_k) = \int f(\mathbf{y}|\boldsymbol{\theta}_k, M_k)\pi(\boldsymbol{\theta}_k|M_k)d\boldsymbol{\theta}_k$$

can be estimated with the harmonic mean of the generated likelihood values:

$$\hat{m}_{NR} = \left\{\frac{1}{G}\sum_{g=1}^{G}\left(\frac{1}{(\mathbf{y}|\boldsymbol{\theta}_k^{(g)}, M_k)}\right)\right\}^{-1}.$$

For more details see Section 14.5 starting on page 515. Write an algorithm in pseudo-code to implement this process. Why is this estimate "simulation-consistent" but not numerically stable?

7.17　Given a Poisson likelihood function, instead of specifying the conjugate gamma distribution, stipulate $p(\mu) = 1/\mu^2$. Derive an expression for the posterior distribution of μ by first finding the value of μ, which maximizes the log density of the posterior, and then expanding a Taylor series around this point (i.e., Laplace approximation). Compare the resulting distribution with the conjugate result.

7.18　Returning to the example data from wars in ancient China (Section 5.3, starting on page 163), calculate the Bayes Factors from possible mixes of the covariates using the BayesFactor package in R. Start with the following code:

```
lapply(c("BaM","BayesFactor"),library,character.only=TRUE)
data(wars)
regressionBF(SCOPE ~ ., data=wars)
```

From this long list pick a baseline model where the covariate selection is substantively reasonable and determine a small set of alternative models that fit the better according to the Bayes Factor criteria.

7.19 Using the Palm Beach County voting model in Section 5.1.1 (starting on page 148) with uninformative priors, compare the full model with the nested model, leaving out the technology variable using a local Bayes Factor. Experiment with different training samples and compare the subsequent results. Do you find that the Bayes Factor is sensitive to your selection of training sample? The dataset is available as pbc.vote in BaM.

7.20 Using the capital punishments data from Exercise 19 on page 141 (in the BaM package as dataset executions), write a Gibbs sampler in R for a probit model (Appendix A) with a latent variable representation (Chib and Greenberg 1998) and uninformative priors. Use the following steps with conditional distributions at the jth iteration:

(a) Sample the latent variable according to truncated normal specification:

$$
z_i^{[j]} \sim \begin{cases} \mathcal{TN}_{(-\infty,0)} \left(\mathbf{x}_i' \beta^{[m-1]}, 1 \right) & \text{if } y_i = 0 \\ \mathcal{TN}_{[0,\infty)} \left(\mathbf{x}_i' \beta^{[m-1]}, 1 \right) & \text{if } y_i = 1 \end{cases}
$$

for $i = 1, \ldots, n$. This creates the vector $\mathbf{z}^{[j]}$.

(b) Sample the coefficient vector according to the normal specification:

$$
\beta^{[j]} \sim \mathcal{N} \left((\mathbf{X}'\mathbf{X})^{-1}\mathbf{X}'\mathbf{z}^{[j]}, (\mathbf{X}'\mathbf{X})^{-1} \right)
$$

Run this Gibbs sampler for two different specifications (mixes of covariates). Modify the code to include a calculation of the DIC within the sampler and use these values to compare the two model fits.

Chapter 8

Bayesian Decision Theory

8.1 Introducing Decision Theory

This chapter introduces the basics of *decision theory* in both a Bayesian and frequentist context. The discussion is actually a continuation of the last chapter in that we will add a "cost" dimension to model comparison and testing. The emphasis here is not on abstract theory, but rather some practical applications to problems in the social and behavioral sciences. It is important to note that decision theory is a topic that is very deep mathematically and one that touches many academic fields, including economics, management, statistics, international relations, pure mathematics, psychology, philosophy, and more. Therefore it would be hard to do justice to this vast enterprise in a single chapter. Interested readers may want to explore the cited works here, beginning with the foundational work of: Raiffa and Schlaifer (1961), Savage (1972), and Lindley (1972).

Decision theory is the logical adjunct to hypothesis testing that formalizes the cost of alternatives through explicitly defining the penalty for making the wrong decision by specifying a function that describes the loss, and therefore the associated risk, for each alternative choice. With the quasi-frequentist/quasi-Fisherian null hypothesis significance test (NHST) described in Section 7.2.1, we used a test statistic to make one of two decisions: D_0 or D_1, and only these two decisions, or actions, were allowable. So it is technically *incorrect* to make statements from the test such as "it provides modest evidence" or "it is barely significant" since that adds decisions that are not fully defined. Despite this, many authors confuse the decision process with the strength of evidence. That is, the NHST version of hypothesis testing (described in Section 7.2.1) confuses inference and decision making since it "does not allow for the costs of possible wrong actions to be taken into account in any precise way" (Barnett 1973). The cost of being wrong in the commonly applied test is completely abstract at the time of the decision.

A more reasoned approach is to assign a *loss function* to each of the two decisions in a hypothesis testing setup (Shao 1989, Leamer 1979). This is a real-valued function that explicitly provides a penalty for decision i selecting hypothesis H_i given that β is the true parameter value: $d(\beta, H_i)$. So from this we can build a *decision rule* that codifies some loss criteria that the researcher might have: minimize the maximum loss, minimize squared errors, and many others. This decision rule defines a loss function that explicitly links decisions that the research makes to costs of that decision. This extra criteria makes a lot of intuitive sense. Basic statistics courses introduce Type I and Type II errors that result from simple hypothesis testing, but do not provide the actual cost of making either of these errors. Naturally there are applications where the cost of making an incorrect statistical decision are important.

Chapter 2 introduced the key building blocks of Bayesian model specifications in order to combine prior information with data information. Here we extend that list of model components so that the researcher now must specify:

▷ A *likelihood function* that stipulates the distribution of the data and the conditionality on parameters: $L(\theta|\mathbf{X})$.

▷ A *prior distribution* that describes the distribution of the parameters of interest before seeing the data: $p(\theta)$.

▷ A *loss function* that gives the cost mis-specifying the posterior distribution away from the true form: $\mathbb{L}()$ (denoted with $\mathbb{L}$ here to distinguish it from a likelihood function).

This means that there are now three spaces over which distributions must be specified: the data space $\mathcal{X}$, the parameter space θ, and the decision space $\mathcal{T}$. This means that an additional parametric form for the loss function is required above what was required before, which can considerably add to the specification burden. Unfortunately there are many alternatives, and the complexity of these forms increases with the number of possible decisions that can be made. In the simple hypothesis testing setup described above there are only two alternatives but there can be many more (even an infinite number). Fortunately standard forms for the loss function have existed for over 200 years (Laplace 1774, Legendre 1805, Gauss 1855). Naturally this choice of loss function is a subjective decision like all model choices in statistics, regardless of paradigm. Some have argued that the choice of loss function is tied to the choice of prior and that they should be specified in conjunction (Lindley 1985). Bernardo and Smith (1994, Chapter 2) give a very detailed discussion of linking beliefs (prior distributions) to actions and utilities.

The quality of the decision rule is generally judged by a corresponding *risk function*. This is just the average loss across decisions by using this decision rule for a given value of β, where the average can be over different stochastic quantities (see below) depending on the priorities of the researcher. Risk in this context therefore combines uncertainty with cost. Decision rules will vary in quality since they stipulate differing actions for the same input. Importantly, though, good decision rules from this process have lower risk functions than known alternatives and therefore manage uncertainty more effectively. Much work has been done to assess the quality of decision rules for such criteria (DeGroot 1970).

However, some social science research applications appear to be difficult applications for assigning specific definitions of risk in inferential decision-making. It is unclear what the exact cost of being "wrong" is in analyzing second-party observational data for theoretical topics. Nonetheless this is a growth area and there are many well-developed resources (Berger 1985; Brown, Cohen, and Strawderman 1980, 1989; Carlin and Louis 2009; Ferguson

1967; Gupta and Berger 1982; Lehmann and Casella 1998; Weiss 1961). Furthermore, this is a very active research area for time-series data (Ni and Sun 2003, Clements 2004; Granger 1999; Chu and White 1982), notably in the last few years (Hong and Lee 2013, Turkov, Krasotkina, and Mottl 2012, Demir, Bovolo, and Bruzzone, 2012).

Probably the biggest concern with applied decision theory in the social sciences is that it can often be difficult (or perhaps impossible) to specify an accurate utility function. This may be either because not enough concrete information exists about costs/benefits or because the criteria are highly multidimensional with complex trade-offs. The same issues apply to the more mechanical process of defining loss functions. Another criticism is that "optimal" point estimates based on this mechanical process are not sympathetic with Bayesian reasoning that seeks to describe an entire posterior distribution. Finally, Gelman *et al.* (2003, p.567) note that it is easy to manipulate the process, through the specification of decision criteria, to obtain some desired end.

Even given these warnings, there are important circumstances when risk and loss functions should be included in a Bayesian model specification. Primarily this is true in economic, policy, and public management circumstances where gains and losses can be easily quantified in units of dollars, lives, or time. In addition, some of the concerns listed above can be mitigated by sensitivity analysis (Martin, Insua, and Ruggeri 1996) and robustness (Abraham and Cadre 2004) around the choice of loss function analogous to the investigation of alternative priors. Shao (1989) gives pre-MCMC simulations methods for dealing with difficult Bayesian decision theory calculations. Savage (1972) lays down a foundation for the Bayesian treatment of decision problems that gives principled reasons why the production of posterior quantities should be coupled with the costs of making decisions based on them, and (Herman) Rubin (1987) demonstrated that the choice of loss function and prior distribution cannot be done separately under a weak system of axioms with the assumption of rational behavior.

8.2 Basic Definitions

In this section we clarify the key terms for both Bayesian and frequentist versions of decision theory. Suppose we are interested in some unknown quantity, designated θ with distribution $f(\theta)$. Since we are going to contrast Bayesian and frequentist decision theory, we will also consider situations where θ is an unknown constant fixed by nature. In either case we are interested in making some explicit *decision* based on the data, which causes us to perform some *action*. Decisions and actions are often used synonymously in the literature, as will be done here, but sometimes it is useful to separate the deliberative process of a decision from the resulting physical action that is performed thereafter. For instance, if our model produces a *decision* that some financial security is expected to gain in value in the near-term with high probability, then a rational resulting *action* would be to purchase some amount of this security (other issues being equal).

8.2.1   Personal Preference

The first concept that needs to be defined is *preference*, which is just the notion that the decider prefers certain outcomes over others. More formally, starting with an arbitrary set of outcomes A, B, and C, the operator "$\leq$" indicates that the first outcome is "not preferred" to the second outcome. A *simple ordering* results if for every possible A, B, and C:

▷ either $A \leq B$ or $B \leq A$,

▷ if $A \leq B$ and $B \leq C$ then $A \leq C$.

Sometimes this is called a *weak ordering* due to the less-than-or-equal relation. A *strong ordering* results from the use of absolutely-less-than relations, "$<$". It is also common to use the notation $A \doteq B$ to say that we are indifferent between A and B. The full set of outcomes that can occur is either finite or infinite, although it is more common to work with finite sets. For the finite set of outcomes above:

▷ there always exists A and C in this set such for *all* B also in the set, $A \leq B \leq C$.

This just means that for a finite set of outcomes there is always a least preferred and most preferred outcome in the existence of simple ordering. So under this circumstance if outcomes are selected deterministically, then the rational actor simply picks their most preferred state. What makes decision theory interesting is that these outcomes are associated with a probability structure that does not guarantee such a deterministic choice. Now preference is more nuanced. Would a decision-maker employ a strategy that gives reasonably preferred outcomes with high probability over a most preferred outcome with low probability? Obviously this would depend on the probabilities as well as characteristics of the decision-maker.

8.2.2   Rules, Rules, Rules

To formalize how decisions are made we would like to have a codified *rule* that expresses mathematically how the decision is made in potentially repeated applications and based on some evidence in the form of data. Given a sample $\mathbf{X}$ assumed to be generated conditional on θ by $f(\mathbf{X}|\theta)$, we apply a *decision rule*, d in the set $\mathbb{D}$, that dictates a specific *decision* (action):

$$d(\mathbf{X}) = A, \qquad A \in \mathbb{A} \tag{8.1}$$

where $\mathbb{A}$ is the allowable class of actions. The decision rule is a *function* of $\mathbf{X}$ in the sense that each draw of $\mathbf{X}$ creates an estimate of θ that is mapped to a specific action A.

Consider a very simple example that is well-known. If a z-score test statistic for a difference of means test between two continuously measured data vectors is greater than 1.96 in absolute value, then we will "decide" that these two samples are not drawn from the same underlying population at the $\alpha = 0.05$ level. This is a decision rule based on characteristics of the data at hand where we will make one of two possible decisions according to:

$$d(\mathbf{X}) = \begin{cases} \text{same underlying population if} & |z| < 1.96 \\ \text{different underlying populations if} & |z| \geq 1.96 \end{cases} \tag{8.2}$$

where z is the standard z-statistic produced in a difference of means analysis. The "rule" here is determined by the magnitude of z. The necessity in this example of making a choice distinguishes decision theory from data description.

8.2.3 Lots of Loss

The decision rule is also a result of a *loss function*, such that if $\mathbf{X}$ is observed and our decision rule is $d(\mathbf{X})$, then the loss is dictated by another function:

$$\mathbb{L}(A, d(\mathbf{X})), \tag{8.3}$$

where smaller losses are preferred. Now $\mathbb{L}(A, d(\mathbf{X}))$ maps the decision/action made, A from the decision rule $d(\mathbf{X})$, to a quantifiable penalty. This idea of loss can be considered in the negative sense as a penalty for decisions that are far from the true or optimal alternative where the exactly correct decision results in a zero loss. It can also be considered in the positive sense as choosing from a set of alternative decisions, all of which provide rewards, but some rewards are higher than others, motivating a search for the highest possible return (so "losses" are unrealized gains). Bounding the loss function by $(-\infty:0)$ is convenient and avoids a host of mathematical complexities that are not necessary here. In more elaborate settings, such structures can be combined for a set of possible decisions that return absolute losses or gains. DeGroot (1970) generalizes the directional implication of loss and reward by using the term *utility*. Utility is a personalistic consideration that simply means that outcomes can be ordered in terms of preference by the relevant actor as described in Section 8.2.1. Note that these outcomes can be both positive and negative. However, bounding the loss function to $(-\infty:0)$, as commonly done, restricts utility to positive outcomes defined by $\mathbb{L}(A, d(\mathbf{X})) = -U(A, d(\mathbf{X}))$.

Since a loss is associated with a particular action, we may want to consider an overall sense of loss from the set of possible decisions. Denote $\pi(\theta|\mathbf{X}, A)$ as the posterior distribution produced from observing $\mathbf{X}$ and taking action A as inputs to the model. Then:

$$E_\pi[\mathbb{L}(A, d(\mathbf{X}))] = \int_\Theta \mathbb{L}(A, d(\mathbf{X}))dF_\pi(\theta) \tag{8.4}$$

is the *Bayesian expected loss* (sometimes called the *posterior expected loss*) over the set of possible decisions according to the associated posterior distribution $\pi(\theta|\mathbf{X}, A)$. Since this averages over the loss in θ *conditional* on an observed $\mathbf{X}$, it is the average loss resulting from having to make *some* decision for this problem. This is an overtly Bayesian statement of loss since we are averaging over uncertainty in the posterior distribution of θ, not over the *distribution* of data since it has already been observed. Thus the Bayesian version is more principled in a world where social scientists are not confronted with an endless stream of iid data. Note also that while the integration is performed over the space of the posterior distribution, the Bayesian expected loss that results is a scalar value.

Loss functions can take on many mathematical forms, mostly with stipulated penalty for moving away from some ideal point. Common examples of loss functions for the true θ and the estimated θ_A include:

▷ squared error loss: $\mathbb{L}(A, d(\mathbf{X})) = (\theta - \theta_A)^2$

▷ absolute error loss: $\mathbb{L}(A, d(\mathbf{X})) = |\theta - \theta_A|$

▷ norm loss for vectors: $\mathbb{L}(A, d(\mathbf{X}), k) = \| \boldsymbol{\theta} - \boldsymbol{\theta}_A \|^k$, for $k \in \mathcal{I}^+ k$ $(k > 1)$

▷ $0 - 1$ loss for discrete state spaces: $\mathbb{L}(A, d(\mathbf{X})) = 0$ if $\theta = \theta_A$, and 1 otherwise

▷ interval loss: $\mathbb{L}(A, d(\mathbf{X})) = 0$ if $CI_{1-\alpha}[\theta_A]$ covers θ, and 1 otherwise

▷ entropy loss: $\int_{\mathcal{X}} \log \frac{f(\mathbf{X}|\theta)}{f(\mathbf{X}|\hat{\theta})} f(\mathbf{X}|\theta) dx$

▷ LINEX loss: $\mathbb{L}(A, d(\mathbf{X}), k) = \exp[k(\theta - \hat{\theta}_A)] - k(\theta - \theta_A) - 1$, where k is some assigned constant (Varian 1975).

Note that these are frequentist-oriented definitions since there is an assumed true θ. The Bayesian analog uses these same parametric forms but only as input into an expected loss calculation as in (8.4). So θ is then treated distributionally and summarized with the Bayesian expected loss value over this distribution.

A more directly Bayesian version of the loss functions above substitutes interval decisions for point estimate decisions. Instead of $\theta = \theta_A$, consider two intervals that constitute a partition of the parameter space such that $\theta \in \Theta_A$ or $\theta \in \Theta_B$. Now set c_A as the cost of incorrectly deciding that $\theta \in \Theta_B$ ($d(\mathbf{X}) = B$), and c_B as the cost of incorrectly deciding that $\theta \in \Theta_A$ ($d(\mathbf{X}) = A$). The complete loss function is now:

$$\mathbb{L}(A, d(\mathbf{X})) = \begin{cases} c_A & \text{if } \theta \in \Theta_A \quad \text{and } d(\mathbf{X}) = B \\ c_B & \text{if } \theta \in \Theta_B \quad \text{and } d(\mathbf{X}) = A \\ 0 & \text{otherwise.} \end{cases} \tag{8.5}$$

Since c_A does not have to be set equal to c_B, this setup can have asymmetric costs for wrong decisions. An associated decision rule is picking the interval that produces the highest posterior probability from a specified model:

$$d(\mathbf{X}) = \begin{cases} A & \text{if } \pi(\theta \in \Theta_A | \mathbf{X}) > \pi(\theta \in \Theta_B) \\ B & \text{if } \pi(\theta \in \Theta_A | \mathbf{X}) < \pi(\theta \in \Theta_B) \end{cases} \tag{8.6}$$

■ **Example 8.1:   Example: Interval Decisions with the French Labor Strike Data.**   Returning to the data and analysis in Section 7.2.1.1, on page 212 ,we define

$\Theta_A = (0:0.05)$ and $\Theta_B = [0.05:\infty)$. From the calculation of the posterior in that section, we have:

$$\pi(\theta \in \Theta_A|\mathbf{X}) = 0.1715 \quad \text{and} \quad \pi(\theta \in \Theta_B|\mathbf{X}) = 0.8285. \tag{8.7}$$

Therefore according to (8.6) above $d(\mathbf{X}) = B$. Suppose it was much worse to incorrectly choose B, thus being over-optimistic about the cessation to the series of strikes. In this case we might set $c_A = 1$ and $c_B = 10$ to make the cost of B ten times higher, giving:

$$\mathbb{L}(A, d(\mathbf{X})) = \begin{cases} 1 & \text{if } \theta \in \Theta_A & \text{and } d(\mathbf{X}) = B \\ 10 & \text{if } \theta \in \Theta_B & \text{and } d(\mathbf{X}) = A \\ 0 & \text{otherwise.} \end{cases} \tag{8.8}$$

This might or might not change our decision, depending on how we determine *risk*.

8.2.4 Risky Business

The notion of frequentist *Risk* is just the average loss *across the data* for a given decision rule:

$$R_F(\theta, d(\mathbf{X})) = E_{\mathbf{X}}\left[\mathbb{L}(A, d(\mathbf{X}))\right] = \int_{\mathcal{X}} \mathbb{L}(A, d(\mathbf{X}))dF_{\mathcal{X}}(x|\theta). \tag{8.9}$$

That is, frequentist risk integrates over the distribution of the data conditional on the defined decision rule for this problem, giving the expected loss of applying the same decision rule over and over again as new datasets arrive. Thus (8.9) preserves the essential frequentist notion of a fixed underlying parameter and an endless stream of iid data. This is a different assumption about uncertainty than (8.4) since it is concerned with uncertainty from a distribution of data given a fixed parameter, rather than a uncertainty from a distribution of the parameter given a fixed set of data. Furthermore, given a single dataset at hand, this approach may not be appropriate since *the expectation is taken over the parametric form describing the data generation process not the single sample of data itself.* This distinction is only unimportant with large samples or repeated sampling from the same data-generating mechanism, both conditions that are not pervasive in the social sciences.

In specifying a risk function, the frequentist researcher may want to compare alternatives. Suppose we have two candidate decision rules, $d_1(\mathbf{X})$ and $d_2(\mathbf{X})$, then $d_1(\mathbf{X})$ is called *R-better* (for "risk better") than $d_2(\mathbf{X})$ if for every possible value of θ we have $R_F(\theta, d_1(\mathbf{X})) \leq R_F(\theta, d_2(\mathbf{X}))$, and for at least one value of θ we have $R_F(\theta, d_1(\mathbf{X})) < R_F(\theta, d_2(\mathbf{X}))$. For the frequentist this becomes an absolutist criteria such that some $R_d(\mathbf{X})$ is called *admissible* if there is no R-better alternative (not worse than every other decision rule for all possible values of the parameter of interest), and *inadmissible* otherwise. Such criteria may be overly rigid since it is unlikely that a single alternative be pairwise R-better in the same way than all alternatives with regard to the less-than or equal-to nature of the definition. Relatedly, the frequentist definition of an *unbiased decision rule* also uses expected loss compared to alternatives, but focuses on differences for true and false values of θ. A decision rule in this context is unbiased if it has no greater expected loss (frequentist risk) for the true value of θ than for all false values, θ', according to:

$$\int_{\mathcal{X}} \mathbb{L}(A, d(\mathbf{X})) dF_{\mathcal{X}}(x|\theta) \leq \int_{\mathcal{X}} \mathbb{L}(A, d(\mathbf{X})) dF_{\mathcal{X}}(x|\theta'). \qquad (8.10)$$

Notice again the hard-coded frequentist language here about true and false fixed parameter values. Unfortunately, requiring unbiased rules can lead to many pathological and illogical results (Ferguson 1967), including estimates outside of the allowed parameter space.

To further illustrate frequentist risk, suppose $X_1, \ldots, X_n$ are iid with true mean μ and true variance σ^2. Our decision rule is $d(\mathbf{X}) = \bar{X}$, for estimating μ. The chosen loss function here is squared error loss: $\mathbb{L}(A, d(\mathbf{X})) = (\mu - \bar{X})^2$, meaning that the penalty for decisions increases quadratically as this estimator moves in either direction away from the true mean. Since we have an endless stream of samples (a rarity in this book!), we will perform this estimation many times and then become concerned with the frequentist risk:

$$
\begin{aligned}
R_F(\theta, d(\mathbf{X})) &= E_{\mathbf{X}}[(\mu - \bar{X})^2] \\
&= E_{\mathbf{X}}[\mu^2 - 2\mu\bar{X} + \bar{X}^2] \\
&= \mu^2 - 2\mu E_{\mathbf{X}}[\bar{X}] + E_{\mathbf{X}}[\bar{X}^2] \\
&= \mu^2 - 2\mu E_{\mathbf{X}}[\bar{X}] + \text{Var}[\bar{X}] + (E_{\mathbf{X}}[\bar{X}])^2 \\
&= \text{Var}[\bar{X}] \\
&= \frac{\sigma^2}{n}.
\end{aligned}
$$

Frequentist risk is shown to be easy to determine given a known loss function and decision rule. The downside is, of course, the assumption of fixed parameters and infinite data, which does not comport well with most social and behavioral science settings.

Bayesian risk analysis starts with the observation that the loss as given here is a *function*, and it is therefore difficult to directly compare two losses. So we use the average of the risk function integrating (first) over the prior distribution, $p(\theta)$, instead of the distribution of the data as done before. The *Bayesian prior risk* of a decision rule (sometimes called the *preposterior risk*) is then:

$$
\begin{aligned}
R_b(\theta, d(\mathbf{X})) &= E_p[R_F(A, d(\mathbf{X}))] \\
&= \int_{\Theta} R_F(A, d(\mathbf{X})) dF_p(\theta) \\
&= \int_{\Theta} \int_{\mathcal{X}} \mathbb{L}(A, d(\mathbf{X})) dF_{\mathcal{X}}(x|\theta) dF_p(\theta), \qquad (8.11)
\end{aligned}
$$

where we use the form of the frequentist risk function but not the frequentist interpretation. In this expression $dF_{\mathcal{X}}$ denotes integration over the joint distribution of the data, and $dF_p(\theta)$ denotes integration over the prior distribution of the parameter θ. Here the model is fixed, the data is fixed but yet to be observed, and only the decision rule is to be determined. Since the data are yet to be observed, this is a hypothetical Bayesian statement. *Bayesian posterior risk* (also called *integrated risk*) is more useful, substituting the posterior for the prior by conditioning on an observed dataset:

$$R_B(\theta, d(\mathbf{X})) = E_\pi \left[R(A, d(\mathbf{X})) \right]$$
$$= \int_\Theta R_F(A, d(\mathbf{X})) dF_\pi(\theta)$$
$$= \int_\Theta \int_\mathcal{X} \mathbb{L}(A, d(\mathbf{X})) dF_\mathcal{X}(x|\theta) dF_\pi(\theta), \qquad (8.12)$$

where $dF_\pi(\theta)$ denotes integration over the posterior distribution of θ. This now reflects an update of the prior risk resulting from the obtaining data $\mathbf{X}$. Note that this is also still conditional on the form of the prior, and we have just added the effect of the data through Bayes' Law. Like the Bayesian expected loss, the Bayesian posterior risk is a single scalar so that we can directly compare risk under different decision rules where the decision rule with the smallest Bayesian risk should be preferred.

An estimator that minimizes the Bayesian posterior risk is one that for *every* $X \in \mathcal{X}$ specifies a decision rule that minimizes the posterior expected loss ($E_\pi \left[\mathbb{L}(A, d(\mathbf{X})) \right]$). The decision rule that minimizes $R_B(\theta, d(\mathbf{X}))$ for every $X \in \mathcal{X}$ is called *optimal*, and limiting the set to this choice is called a *Bayes rule*, denoted $\widehat{R_B}(\theta, d(\mathbf{X}))$ here, and defined by the value of θ_A that satisfies:

$$\widehat{R_B}(\theta, d(\mathbf{X})) = \inf_{\theta_A} \int_\Theta R_F(A, d(\mathbf{X})) dF_\pi(\theta). \qquad (8.13)$$

Analogous to the analytical calculation of maximum likelihood estimates, we can set the derivative of this expression equal to zero and solve for the resulting *minimum* value of θ_A (it is a convex rather than concave function, in contrast to many other statistical calculations). For example, if we apply squared error loss then the process for mean estimation is:

$$0 = \frac{d}{d\theta} \left[\int_\Theta (\theta - \theta_A)^2 dF_p(\theta) \right]$$
$$= \frac{d}{d\theta} \left[\int_\Theta \left(\theta^2 - 2\theta\theta_A + \theta_A^2 \right) dF_p(\theta) \right]$$
$$= \frac{d}{d\theta} \left[\int_\Theta \theta^2 dF_p(\theta) - 2\theta_A \int_\Theta \theta dF_p(\theta) + \theta_A^2 \int_\Theta dF_p(\theta) \right]$$
$$= \frac{d}{d\theta} \left[\mathrm{Var}_{f(\theta|x)}[\theta] - 2\theta_A E_{f(\theta|x)}[\theta] + \theta_A^2 \right]$$
$$= 0 - 2E_{f(\theta|x)}[\theta] + 2\theta_A. \qquad (8.14)$$

Therefore the Bayes rule estimate, θ_A, is the posterior mean, $E_{f(\theta|x)}[\theta]$. Similarly, if we specify absolute error loss, $\mathbb{L}(A, d(\mathbf{X})) = |\theta - \theta_A|$, instead, then the Bayes rule estimate is the posterior median (Exercise 6). Also, for the 0-1 error loss, $\mathbb{L}(A, d(\mathbf{X})) = 0$ if $\theta = \theta_A$, and 1 otherwise, the Bayes rule estimate is the posterior mode. Obviously with unimodal (non-uniformly so) symmetric posterior forms, these are concurrent.

■ **Example 8.2: Normal Distribution Application.** As an example of Bayesian risk analysis, consider the frequentist mean estimation process above but adding a normal prior distribution for μ: $\mathcal{N}(m, s^2)$, for hyperprior values m and s^2. We will use squared error loss again, except that in this case the posterior mean for μ, $\hat{\mu} = \frac{n\bar{x}+ms_0}{n+s_0}$ from the derivations that produced (3.18) on page 76, is an improved substitute for the data mean. This produces:

$$R_B(\theta, d(\mathbf{X})) = \int_\Theta R_F(A, d(\mathbf{X}))dF_\pi(\theta)$$

$$= \int_\mu \left(\mu - \frac{n\bar{x}+ms_0}{n+s_0}\right)^2 \left(2\pi\frac{\sigma^2}{n+s_0}\right)^{-\frac{1}{2}} \exp\left[-\frac{1}{2\sigma^2/(n+s_0)}\left(\mu - \frac{n\bar{x}+ms_0}{n+s_0}\right)^2\right]d\mu.$$

If we make the change of variable calculation $\delta_\mu = \mu - \hat{\mu} = \mu - (n\bar{x}+ms_0)/(n+s_0)$, it is easy to see that this is expression is just $R_B(\theta, d(\mathbf{X})) = E[\delta_\mu^2]$. Therefore:

$$R_B(\theta, d(\mathbf{X})) = \text{Var}[\delta_\mu] + (E[\delta_\mu])^2 = \text{Var}[\mu] - 0 = \frac{\sigma^2}{n+s_0}. \tag{8.15}$$

Where the variance of δ_μ equals the variance μ since $\hat{\mu}$ consists of all constants (observed or set), $E[\delta_\mu] = 0$ by design, and the final form of the variance comes from (3.18). We now see a practical implication for the s_0 parameter introduced on page 74. This parameter can be used by the researcher to scale the Bayesian risk relative to frequentist risk for this normal problem. Increasing values in $(0:\infty)$ decrease Bayesian risk (zero is excluded due to the functional form of $p(\mu|m, \sigma^2/s_0)$, implying useful prior information. Also, the relationship demonstrated here comes from the mean being conditional on the variance in the conjugate normal setup.

8.2.4.1 Notes on Bayes Rules

From the definition of Bayes rule above we get some very important results. First, any Bayes rule corresponding to a proper prior is admissible: there is no R-better alternative (Bernardo and Smith 1994). This is a very powerful argument for the use of proper forms since it means that we do not have to worry about a large set of alternative rules that result. It also works in reverse: all admissible rules must be Bayes rules (Exercise 14).

Unfortunately, some Bayes rules that correspond to improper priors are inadmissible (James and Stein 1961), see the discussion in Section 8.4. Restricting the set of alternatives to Bayes rules also has a computational advantage: deriving Bayes rules circumvents specifying the risk function and needs only to minimize the posterior expected loss in (8.4). Together all of these findings mean that Bayesian decision theory dominates frequentist application in both theoretical and practical ways.

■ **Example 8.3:   Nuclear Deterrence Between the United States and the Soviet Union** Patterson and Richardson (1963) consider the problem of risk in treaty observance between the two great Cold War powers (a very topical issue at the time of their writing). The key problem is whether or not to agree to a specific treaty restricting the number of thermonuclear tests, given the subjective probability that policy makers have about the probability that the other side will cheat. Specifically, should the United States pursue a negotiation strategy of insisting on a small number of tests (which is good for humanity), knowing that it increases the probability that the Soviet Union will cheat. This is critical at the time since there did not yet exist perfect detection of such events.

So consider R_T to be the risk under some negotiated treaty and R_N is the risk of no treaty and unrestricted testing. The definition of risk here is not only increasing the probability of global thermonuclear war under some treaty scenario, but also the cost of "losing face" in the presence of cheating by the other side and therefore a weakened negotiating position in the next round. Label n the actual number of tests by the other side, which is observed imperfectly with an observed value d for detections. These detections, however, are split between r real events and u unreal events (false alarms): $d = r + u$. The number of detections is assumed to be modeled as a binomial with number of possible events k and probability p, $r \sim \mathcal{BE}(n, p)$. Also, u is stochastic so we model it with a Poisson process (our protagonist side does not know the upper bound of the cheating on the other side) with intensity parameter λ: $u \sim P(\lambda)$. Since $n > d = r + u$, we also know that $u = d - r$. This leads to a convolution (summing) of $g(r)$ and $g(u)$ (Casella and Berger 2002, p.215) to get the probability of detections given events (e.g., the reliability of the verification process):

$$p(d|n) = \sum_{r=0}^{d} \left[\binom{n}{r} p^r (1-p)^{n-r} \right] \left[\frac{e^{-\lambda} \lambda^{d-r}}{(d-r)!} \right]. \qquad (8.16)$$

Therefore $E[d|n] = pn + \lambda$ and $\text{Var}[d|n] = p(1-p)n + \lambda$ (Exercise 9.6).

Of course from a gamesmanship perspective we are really more interested in estimating the probability of actual events given detections, which is obtained by first specifying a prior distribution, $p(n)$, and then applying Bayes' Law,

$$\pi(n|d) = \frac{p(n)p(d|n)}{\sum_k p(n)p(d|n)}, \qquad (8.17)$$

to get this posterior distribution.

Define $L_{r(d),n}$ as the loss from specifying response rule $r(d)$ for n actual events. This response can be from diplomatic, conciliatory, or belligerent actions. The expected loss is therefore the sum over detected events of this loss times our distribution of detections:

$$\rho(n) = \sum_d L_{r(d),n} p(d|n). \qquad (8.18)$$

That is, our actor chooses a loss function related to the probability of detections because this calculation occurs before negotiations are finished and any events can be observed. Determining $\rho(n)$ in advance means that we want to minimize this expected loss with respect to the prior distribution on n calculated by:

$$\rho = E_{p(n)}\left[\rho(n)\right]$$

$$= E_{p(n)}\left[\sum_d L_{r(d),n}p(d|n)\right]$$

$$= \sum_n \sum_d L_{r(d),n}p(d|n)p(n)$$

$$= \sum_d \left(\underbrace{\sum_n p(d|n)p(n)}_{\text{normalizing constant}} \sum_n \underbrace{L_{r(d),n}}_{\text{loss}} \underbrace{\frac{p(d|n)p(n)}{\sum_n p(d|n)p(n)}}_{\pi(n|d)} \right). \qquad (8.19)$$

So the response rule that minimizes ρ is really obtained with respect to a calculation on the posterior distribution of n. What this states in substantive terms is that the model update provides the best incorporation of risk into the calculation. This decision will determine the strategy for negotiating R_T.

8.2.5   Minimax Decision Rules

One way to narrow down the set of alternative decision rules is to apply a very conservative criteria that minimizes the worst possible risk over the sample space. If $\mathcal{D}$ is the set of alternative decision rules, we want $d^*(\mathbf{X})$ that gives the minimum maximum-possible value:

$$\sup_{\theta \in \Theta} R_F(A, d^*(\mathbf{X})) = \inf_{d(\mathbf{X}) \in \mathcal{D}} \sup_{\theta \in \Theta} R_F(A, d(\mathbf{X})). \qquad (8.20)$$

On the right-hand side, $\inf_{d \in \mathcal{D}}$ specifies the $d(\mathbf{X})$ that gives the smallest ("infimum") value of $R_F(A, d(\mathbf{X}))$ where $\sup_{\theta \in \Theta}$ requires the value of θ that gives the largest ("supremum") value of $R_F(A, d(\mathbf{X}))$, in the frequentist sense. There are several challenges with this approach:

▷ Sometimes such decision rules may be very difficult to calculate.

▷ The $d^*(\mathbf{X})$ may give very high risk over areas of the parameter space that fall just short of being the maximum, and therefore provide overall unacceptable risk. Second, minimax estimators can be barely unique, and so the choice-set can remain unhappily large.

▷ Protection against the worst possible result may be reasonable in the competitive context of game theory; it does not have an analogous motivation in standard data analysis.

▷ Minimax decision rules are generally *not* unique (Shao 2005), although unique minimax estimators are admissible!

▷ Some choices violate the transitivity of preference ordering given in Section 8.2.1 (Lindley 1972).

▷ Unique Bayes estimates from a *constant* risk function ($R_B(\theta, d(\mathbf{X})) = k$) are minimax decision rules and this may not be substantively defensible.

▷ Minimax decision rules are Bayes rules defined with respect to the prior distribution
that maximizes Bayes risk (Lehmann 1986).

The last point above is worth dwelling on. It turns out that this prior distribution is
also called the *least favorable prior* in the sense that any other prior chosen has the same
or lower Bayesian posterior risk: $R_B(\theta, d_{\text{minimax}}(\mathbf{X})) \geq R_B(\theta, d_{\text{other}}(\mathbf{X}))$. What makes this
interesting is that it ties together the form of the prior with the choice of decision rule.
Robert (2001, pp.251-252) also notes that in conventional hypothesis testing, this approach
is always biased against the null hypothesis (see also Berger and Sellke [1987] and Berger
and Delampady [1987]).

The penultimate point above is also more nuanced than it appears at first. Under
very general assumptions, a unique Bayes estimator with constant risk is minimax, *and*
admissible. However, an estimator that is not a unique Bayes form may be either admissible
or non-admissible, and a unique Bayes estimator with constant risk can be minimax but
non-admissible (Ferguson 1967). Also, Bayes estimators do not cover the set of admissible
estimators for infinite parameter space (they do for finite parameter spaces). For example
(Lehman 1986), suppose $X_1, \ldots, X_n$ is a sample from a normal distribution $\mathcal{N}(\theta, 1)$, and
we stipulate squared error loss. Then the sample mean, $\bar{X}$ is both unbiased and admissible
for θ with positive variance, but is not a Bayes estimate corresponding to any possible prior
distribution.

8.3   Regression-Style Models with Decision Theory

Applying decision theory to linear and generalized linear models usually means focusing
on making reliability decisions about estimated regression parameters, often in the process
of model choice (Brooks 1972, Brown, *et al.* 1999, Lewis and Thayer 2009). The estimation
of σ^2 in a linear context, and related quantities in hierarchical or nonlinear forms, is usually
treated as a secondary concern and not applied as a loss function. For such purposes modify
(8.3) to be:

$$\mathbb{L}(\beta_k, A, d(\mathbf{X})), \tag{8.21}$$

for the true kth regression coefficient, taking the Ath action, with data-based decision rule
$d(\mathbf{X})$. Here we temporarily suppress the k notation such that $\hat{\beta}$ is the estimate of β for this
general k variable. For most social scientists $d(\mathbf{X})$ is a dichotomous or trichotomous decision:

$$d(\mathbf{X}) = \begin{cases} \hat{\beta} \text{ is a reliable estimator if} & \left|\frac{\hat{\beta}}{se_{\hat{\beta}}}\right| \geq 1.96 \\ \hat{\beta} \text{ is a not reliable estimator if} & \left|\frac{\hat{\beta}}{se_{\hat{\beta}}}\right| < 1.96. \end{cases} \tag{8.22}$$

and,

$$d(\mathbf{X}) = \begin{cases} \hat{\beta} \text{ is a } \textit{highly} \text{ reliable estimator if} & \left|\frac{\hat{\beta}}{se_{\hat{\beta}}}\right| \geq 2.576 \\ \hat{\beta} \text{ is a reliable estimator if} & 1.96 \leq \left|\frac{\hat{\beta}}{se_{\hat{\beta}}}\right| < 2.576 \\ \hat{\beta} \text{ is a not reliable estimator if} & \left|\frac{\hat{\beta}}{se_{\hat{\beta}}}\right| < 1.96. \end{cases} \quad (8.23)$$

While these are generally not principled thresholds, as discussed in Chapter 7, they are pervasive in most social and behavioral science literatures. Furthermore, this discussion is relevant here since many Bayesian results are reported with 95% HPD regions, although Bayesians tend to see this as a convenient convention rather than prescribing any theoretical importance to the choice.

More formally for our purposes, let $\beta \in \mathcal{B}$ be the true population parameter and $\hat{\beta}$ be its estimate from some regression result. The researcher defines two non-overlapping, exhaustive regions of the support for β, $\mathcal{B}_I$, and $\mathcal{B}_{II}$ and wants to make a decision about the corresponding location of the true parameter. The first decision rule above can formalized by two actions: A_I for deciding that β is in region $\mathcal{B}_I$, and A_{II}, for deciding that β is in region $\mathcal{B}_{II}$. These are associated with negative utilities (costs) for being wrong, U_I and U_{II}, defining the well-known loss function:

$$\mathbb{L}(\beta_k, A, d(\mathbf{X})) = \begin{cases} U_I & \text{if } \beta \in \mathcal{B}_{II} \text{ and } A = A_I \\ 0 & \text{if } \beta \in \mathcal{B}_I \text{ and } A = A_I \\ U_{II} & \text{if } \beta \in \mathcal{B}_I \text{ and } A = A_{II} \\ 0 & \text{if } \beta \in \mathcal{B}_{II} \text{ and } A = A_{II}. \end{cases} \quad (8.24)$$

This means that the Bayes (optimal) action is determined probabilistically by regions of the posterior distribution:

$$\hat{A} = \begin{cases} A_I & \text{if } \pi(\mathcal{B}_I) > \frac{U_{II}}{U_I + U_{II}} \\ A_{II} & \text{if } \pi(\mathcal{B}_I) < \frac{U_{II}}{U_I + U_{II}}. \end{cases} \quad (8.25)$$

In other words, if the posterior probability of a given action is higher than the loss proportion of the alternative, then it is rational to take this action. Kempthorne (1986) contrasts this (traditional) decision-theoretic approach to coefficient hypothesis testing with the loss from simply choosing the posterior mean as the decision with squared error loss. For a selected value of the posterior mean for the regression coefficient, β^*, the posterior risk reduces to:

$$R_B(\beta, d(\mathbf{X})) = se_{\hat{\beta}} + \|\beta^* - \hat{\beta}\|, \quad (8.26)$$

meaning that Bayesian posterior risk under these circumstances is always minimized by selecting the Bayesian posterior mean. The decision-theoretic setup above based on traditional hypothesis testing steps gives the posterior risk:

$$R_B(\beta, d(\mathbf{X})) = U_I \pi(\mathcal{B}_I) I_{A_{II}} + U_{II} \pi(\mathcal{B}_{II}) I_{A_I}, \quad (8.27)$$

which is just the expected loss of a choice times the probability that it was the wrong choice, where I_{A_i} is an indicator function for making the ith choice, $i \in \{1, 2\}$. This contrast in risks rests on size of the posterior standard error for the coefficient and the costs of making wrong choices. This is also the contrast, in a Bayesian sense, from reporting posterior summaries versus making a sharp decision based on the posterior distribution.

8.3.1 Prediction from the Linear Model

In Chapter 5 (Section 5.2, page 155) the posterior predictive distribution of future y values from a Bayesian linear model was shown to have the multivariate Student's-t distribution:

$$\pi(\tilde{\mathbf{y}}|\tilde{\mathbf{X}}, \mathbf{X}, \mathbf{y}) \propto \left[(n - k) + (\tilde{\mathbf{y}} - \tilde{\mathbf{X}}\hat{\mathbf{b}})'\mathbf{H}(\tilde{\mathbf{y}} - \tilde{\mathbf{X}}\hat{\mathbf{b}}) \right]^{-\frac{n+q-k}{2}}, \qquad (8.28)$$

with mean vector and variance-covariance matrix:

$$E[\tilde{\mathbf{y}}] = \tilde{\mathbf{X}}\hat{\mathbf{b}}, \qquad\qquad \text{Cov}[\tilde{\mathbf{y}}] = \frac{n - k}{n - k - 2}\mathbf{H}^{-1} \qquad (8.29)$$

where $\mathbf{H} = (\mathbf{I} - \tilde{\mathbf{X}}\mathbf{M}^{-1}\tilde{\mathbf{X}}')/\hat{\sigma}^2$, and $\mathbf{M} = \mathbf{I} = \mathbf{X}(\mathbf{X}'\mathbf{X})^{-1}\mathbf{X}'$. The quadratic loss function for this setup is:

$$\mathbb{L}(\hat{\mathbf{b}}, A, d(\mathbf{X})) = (\tilde{\mathbf{y}} - \bar{\mathbf{y}})'\mathbf{\Upsilon}(\tilde{\mathbf{y}} - \bar{\mathbf{y}}), \qquad (8.30)$$

where $\mathbf{\Upsilon}$ is any positive definitive $k \times k$ matrix of flexible form. Then $\tilde{\mathbf{y}}$ is the single point prediction that minimizes Bayesian expected loss:

$$E_\pi\left[\mathbb{L}(\hat{\mathbf{b}}, A, d(\mathbf{X})) \right] = \int_{\boldsymbol{\beta}} \mathbb{L}(\hat{\mathbf{b}}, A, d(\mathbf{X})) dF_\pi(\boldsymbol{\beta}) \qquad (8.31)$$

(proof given in Zellner and Chetty 1965, p.610). This shows that squared error loss is directly tied to normal-linear assumptions. In fact, if y is distributed $\mathcal{N}(\mu, \sigma^2)$ and we want a prediction under the assumption of squared error loss with chosen estimator μ_A, then the frequentist risk is:

$$\begin{aligned}
E_y\left[\mathbb{L}(\mu, \mu_A) \right] &= E_y[(y - \mu_A)^2] \\
&= E_y[(y - \mu + \mu - \mu_A)^2] \\
&= E_y[(y - \mu)^2] + E_y[(\mu - \mu_A)^2] \\
&= \sigma^2 + (\mu - \mu_A)^2.
\end{aligned} \qquad (8.32)$$

Since σ^2 and μ are fixed quantities, then we actually only care about squared expected loss that results from the choice of estimator of μ_A to minimize expected loss, according to some stated criteria.

■ **Example 8.4:** **Prediction Risk for a Model of Educational Effects.** The second Bayesian linear model based on work by Meier, Polinard, and Wrinkle (2000) from Example 5.2 starting on page 157 is further analyzed here. This model used results from earlier work with educational data from Florida by Meier and Smith (1994) to inform priors for an additional analysis of the outcome: percent of students in district/year that pass the Texas Assessment of Academic Skills (TAAS).

The quantities of interest are obviously regression coefficients. So consider three scenarios: $[\beta = -0.5, \sigma_\beta^2 = 0.01]$, $[\beta = 0.0, \sigma_\beta^2 = 0.03]$, and $[\beta = 0.5, \sigma_\beta^2 = 0.01]$. We clearly would not normally apply the same criteria to each of the regression coefficients from the model, but doing so will show a range of effects for illustration here. Using the model estimates in Table 5.6 on page 160, we can calculate the frequentist risk identified in (8.32) above. This is give in Table 8.1.

TABLE 8.1: PREDICTION RISK: EDUCATION OUTCOMES MODEL

Explanatory Variables	Mean	SE	E_y [L(−0.5, 0.01)]	E_y [L(0.0, 0.03)]	E_y [L(0.5, 0.01)]
Intercept	4.799	2.373	28.080	23.031	18.482
Low Income Students	-0.105	0.006	0.156	0.012	0.366
Teacher Salaries	0.382	0.099	0.778	0.147	0.014
Teacher Experience	-0.066	0.046	0.188	0.005	0.320
Gifted Classes	0.096	0.021	0.355	0.010	0.163
Class Size	0.196	0.191	0.485	0.039	0.093
State Aid Percentage	0.002	0.004	0.252	0.001	0.248
Funding Per Student (×1000)	0.049	0.175	0.302	0.003	0.204
Lag of Student Pass Rate	0.684	0.008	1.402	0.469	0.034
Lag of Bureaucrats	-0.042	0.261	0.210	0.003	0.294
Class Size × Teacher Salaries	-0.015	0.007	0.235	0.001	0.265

Clearly the constant term fares the worst since it is the most disproportional to the criteria. Of the others, the estimate for Lag of Student Pass Rate has high frequentist risk based on its relative large coefficient estimate compared to $\{-0.5, 0.0, 0.5\}$. The range of values across rows is also interesting since even though the criteria appeared to be similar, they had a large range of effects with the middle column showing lower risk due to $\mu = 0$.

8.4   James-Stein Estimation

Stein (1955), writing in the context of a two-person zero-sum game, finds that the maximum likelihood estimator for the mean of a multivariate normal distribution is inadmissible under mean squared error risk for three or more dimensions. James and Stein (1961) then derived an alternative estimate that has lower squared error loss. This paradoxical result was very surprising at the time and still seems like an oddity. In short, it was demonstrated that if one was willing to increase bias, the variance of the estimator could be reduced dramatically. We will briefly discuss these estimators and their properties in this section, but for more detailed reviews see Lehmann and Casella (1998, Chapter 4), Carlin and Louis (2009, Chapter 5), and for nice overviews see Stein (1981), Lehmann and Casella (1998), or Heumann (2011).

First we will define a simplified multivariate normal model where an n-length data vector is distributed $\mathcal{N}(\boldsymbol{\mu}, \mathbf{I})$. This is a very basic model since all of the variances are assumed to be equal to one. Start with the mean prior defined in (3.21) on page 77:

$$\boldsymbol{\mu}|\boldsymbol{\Sigma} \sim \mathcal{N}_n\left(\mathbf{0}, \tau^2\mathbf{I}\right), \tag{8.33}$$

except we specify prior mean zero for all dimensions and τ^2 is temporarily assumed to be known. Consider this to be $\boldsymbol{\mu}_i$ mean values for $\mathbf{x}_i$ data values. Define $b = \frac{\tau^2}{\tau^2+1}$. From (3.24), this leads to the following succinct posterior distribution for vector $\boldsymbol{\mu}$:

$$\boldsymbol{\mu}|\mathbf{x} \sim \mathcal{N}_n\left(b\mathbf{x}, b\mathbf{I}\right) \tag{8.34}$$

(Exercise 17). Now suppose that we use some arbitrary point estimate $\hat{\boldsymbol{\mu}}$ from this distribution such as the posterior mean. Then the squared error loss function for the total sample is:

$$\mathbb{L}(A, d(\mathbf{X})) = \sum_{i-1}^{N}(\mu_i - \hat{\boldsymbol{\mu}}_i)^2, \tag{8.35}$$

with associated frequentist risk

$$R_F(\mu, d(\mathbf{X})) = E_{\mathbf{X}}\left[\sum_{i-1}^{N}(\mu_i - \hat{\boldsymbol{\mu}}_i)^2\right]. \tag{8.36}$$

For $\hat{\boldsymbol{\mu}} = \bar{x}$, the maximum likelihood point estimate, this gives $R_F(\mu, d(\mathbf{X})) = n$ (Exercise 15). The corresponding Bayesian risk is $R_B(\mu, d(\mathbf{X})) = bn$ from (8.15). So depending on the specified value of τ^2, the Bayesian alternative could have much lower risk. For large values of b there will not be much of a difference since b will be close to one, but for small values of b the Bayesian alternative will be superior (lower risk).

Now suppose that we want to estimate τ^2 instead of assuming that it is known a priori. One way to do this is to use the data in an additional way (Efron and Morris 1972a, 1972b). This is called *empirical Bayes* and we will cover the topic in more detail in Section 8.5. The marginal distribution of the data (integrating out $\boldsymbol{\mu}$), is $\mathbf{x} \sim \mathcal{N}(\mathbf{0}, (\tau^2+1)\mathbf{I})$. Subsequently the scaled sum of squares for the data is chi-square distributed according to: $\mathbf{x}'\mathbf{x}/(\tau^2+1) \sim \chi^2_{df=n}$ (Keating and Mason 1988). This means that $E_{\mathbf{x}}[n - 2/\mathbf{x}'\mathbf{x}] = 1/(\tau^2+1)$. Carlin and Louis (2009), and others, label $B = \tau^2/(\tau^2 + \sigma^2)$ where σ^2 is the variance of the data and τ^2 is the prior variance of μ. This now replaces the more basic $b = \tau^2/(\tau^2+1)$ used above.

We make use of the most straightforward possible estimation process where each μ_i is estimated with the corresponding x_i collected in the vector $\mathbf{x}$. By the logic in (8.13) on page 255, and in the notation above, the Bayes rule for this estimator gives the posterior distribution $\mu|\mathbf{x} \sim \mathcal{N}(B\mathbf{x}, B\mathbf{I})$ where again $B = \tau^2/(\tau^2 + \sigma^2)$. For a prior with mean vector $\mathbf{0}$, the Bayes estimator is:

$$\boldsymbol{\mu}(\mathbf{x}) = \left(1 - \frac{1}{1+\tau^2}\sigma^2\right)\mathbf{x}. \tag{8.37}$$

The maximum likelihood estimate is given simply by the vector $\boldsymbol{\mu}^{MLE}(\mathbf{x}) = \mathbf{x}$, giving squared error loss $\mathbb{L}(A, d(\mathbf{X})) = (\mathbf{x} - \boldsymbol{\mu})^2$. This means that the frequentist risk is:

$$R_F(\boldsymbol{\mu}^{MLE}, d(\mathbf{X})) = E_{\mathbf{X}}\left[\mathbb{L}(A, d(\mathbf{X}))\right] = E_{\mathbf{X}}\left[\frac{1}{n}\sum_{i=1}^{n}(\boldsymbol{\mu} - \mathbf{x})^2\right] = \sigma^2. \tag{8.38}$$

Stein's 1955 proof showed that this is inadmissible for three or more dimensions, and the 1961 paper by James and Stein gave an estimator with frequentist risk no larger than σ^2. The *James-Stein estimator* is created by modifying the usual Bayesian estimator in this context:

$$\boldsymbol{\mu}^{JS}(\mathbf{x}) = \left(1 - \frac{(n-2)}{\mathbf{x'x}}\sigma^2\right)\mathbf{x}, \tag{8.39}$$

requiring $n > 2$ to avoid additional complexities (see Lehmann and Casella [1998, p.275-277] for details). Notice that this is the Bayes estimator in (8.37) with $(n-2)/\mathbf{x'x}$ substituting for $1/(1+\tau^2)$. The James-Stein estimator is also called a "shrinkage estimator" since $(n-2)/\mathbf{x'x}$ is shrinking the posterior estimate towards the prior mean of zero. The frequentist risk of this estimator is given by:

$$R_F(\boldsymbol{\mu}^{JS}, d(\mathbf{X})) = E_{\mathbf{X}}\left[\frac{1}{n}\sum_{i=1}^{n}(\boldsymbol{\mu} - \boldsymbol{\mu}^{JS})^2\right] = \left[1 - \frac{n-2}{n}E_{\mathbf{X}}\left(\frac{(n-2)}{\mathbf{x'x}}\sigma^2\right)\right]. \tag{8.40}$$

Notice that $R_F(\boldsymbol{\mu}^{JS}, d(\mathbf{X}))$ is always less than $R_F(\boldsymbol{\mu}^{MLE}, d(\mathbf{X}))$ since the second term inside the parentheses is necessarily always positive. While discussed only in the context of a simple normal model with mean zero, the important theorem given by James and Stein (1961) shows that the James-Stein estimator dominates the maximum likelihood estimator in terms of risk in all $\boldsymbol{\mu}$ settings where $n > 2$.

Naturally, it is not required that we have prior means of zero, and specifying a different prior mean vector only slightly complicates the equations. The data are now assumed to be distributed $\mathcal{N}(\boldsymbol{\mu}, \sigma^2\mathbf{I})$, $i = 1 \ldots, n$ where $\sigma^2\mathbf{I}$ is a diagonal matrix of identical variances. We now restate the prior distribution for $\boldsymbol{\mu}$ as:

$$\boldsymbol{\mu}|\boldsymbol{\Sigma} \sim \mathcal{N}_n\left(\mathbf{m}, \tau^2\mathbf{I}\right). \tag{8.41}$$

This means that the marginal distribution of the data (integrating out $\boldsymbol{\mu}$), is

$$\mathbf{x} \sim \mathcal{N}(\mathbf{m}, (\tau^2 + \sigma^2)\mathbf{I}), \tag{8.42}$$

and the posterior distribution for $\boldsymbol{\mu}$ is:

$$\boldsymbol{\mu}|\mathbf{x} \sim \mathcal{N}_n\left(\mathbf{m}(1 - B) + B\mathbf{x}, B\sigma^2\mathbf{I}\right). \tag{8.43}$$

The Bayes rule $\mathbf{m}(1 - B) + B\mathbf{x}$ gives the James-Stein estimator for each of $i = 1, \ldots, n$:

$$\mu^{JS}(x_i) = \bar{\mathbf{x}} + \left(1 - \frac{(n-3)}{\sum_{i=1}^n (x_i - \bar{x})^2}\sigma^2\right)(x_i - \bar{\mathbf{x}}), \tag{8.44}$$

now requiring $n > 3$. This new estimator obviously generalizes the form of (8.39).

The James-Stein result means that there is unseen latent information between seemingly separate and independent decisions that is not used when modeled in isolation but appears when analyses are performed simultaneously and connected by the loss function. Essentially risk is reduced because each case borrows strength from the rest of the others, thus reducing uncertainty. A 1977 article in *Scientific American* by Efron and Morris provides a nice intuitive discussion of these ideas. Furthermore, notice that $\mu^{JS}(\mathbf{x})$ is estimated empirically and simultaneously from the data using n identical specifications. Thus the James-Stein estimator is considered an *empirical Bayes* process.

■ **Example 8.5: Government Social Spending in OECD Countries.** The following data (also provided in BaM) are the total public social expenditure as a percentage of GDP for 28 OECD countries (Organization for Economic Cooperation and Development) over the years 2006-2013 (OECD Social Expenditure Statistics Database). Chile, Japan, Korea, Mexico, and Turkey were members of the OECD during this period but are dropped from the analysis due to severe data problems. Social spending is considered an important indicator of national policy.

	2006	2007	2008	2009	2010	2011	2012	2013
Australia	0.165	0.164	0.178	0.178	0.179	0.182	0.188	0.195
Austria	0.268	0.263	0.268	0.291	0.289	0.279	0.279	0.283
Belgium	0.260	0.260	0.273	0.297	0.295	0.297	0.305	0.307
Canada	0.169	0.168	0.176	0.192	0.187	0.181	0.181	0.182
Czech Republic	0.183	0.181	0.181	0.207	0.208	0.208	0.210	0.218
Denmark	0.271	0.265	0.268	0.302	0.306	0.306	0.308	0.308
Estonia	0.127	0.127	0.158	0.200	0.201	0.182	0.176	0.177
Finland	0.258	0.247	0.253	0.294	0.296	0.292	0.300	0.305
France	0.298	0.297	0.298	0.321	0.324	0.320	0.325	0.330
Germany	0.261	0.251	0.252	0.278	0.271	0.259	0.259	0.262
Greece	0.213	0.216	0.222	0.239	0.233	0.244	0.241	0.220
Hungary	0.228	0.230	0.231	0.239	0.229	0.219	0.216	0.216
Iceland	0.159	0.153	0.158	0.185	0.180	0.181	0.176	0.172
Ireland	0.161	0.167	0.197	0.236	0.237	0.233	0.224	0.216
Israel	0.158	0.155	0.155	0.160	0.160	0.158	0.158	0.158
Italy	0.250	0.247	0.258	0.278	0.277	0.275	0.280	0.284
Luxembourg	0.218	0.203	0.208	0.236	0.230	0.226	0.232	0.234
Netherlands	0.217	0.211	0.209	0.232	0.234	0.234	0.240	0.243
New Zealand	0.189	0.186	0.198	0.212	0.213	0.214	0.220	0.224
Norway	0.203	0.205	0.198	0.233	0.230	0.224	0.223	0.229
Poland	0.208	0.197	0.203	0.215	0.218	0.205	0.206	0.209
Portugal	0.230	0.227	0.231	0.256	0.254	0.250	0.250	0.264
Slovak Republic	0.160	0.157	0.157	0.187	0.191	0.181	0.183	0.179
Slovenia	0.208	0.195	0.197	0.226	0.236	0.237	0.237	0.238
Spain	0.211	0.213	0.229	0.260	0.267	0.264	0.268	0.274
Sweden	0.284	0.273	0.275	0.298	0.283	0.276	0.281	0.286
United Kingdom	0.203	0.204	0.218	0.241	0.238	0.236	0.239	0.238
United States	0.161	0.163	0.170	0.192	0.198	0.196	0.197	0.200

TABLE 8.2: MLE AND JAMES-STEIN
ESTIMATES OF SOCIAL SPENDING

	$\mu^{MLE}(\mathbf{x})$	2007-2013 mean	$\mu^{JS}(\mathbf{x})$
Australia	0.165	0.181	0.181
Austria	0.268	0.279	0.249
Belgium	0.260	0.291	0.243
Canada	0.169	0.181	0.183
Czech Republic	0.183	0.202	0.193
Denmark	0.271	0.295	0.251
Estonia	0.127	0.174	0.156
Finland	0.258	0.284	0.242
France	0.298	0.316	0.268
Germany	0.261	0.262	0.244
Greece	0.213	0.231	0.212
Hungary	0.228	0.226	0.222
Iceland	0.159	0.172	0.177
Ireland	0.161	0.216	0.178
Israel	0.158	0.158	0.176
Italy	0.250	0.271	0.237
Luxembourg	0.218	0.224	0.216
Netherlands	0.217	0.229	0.215
New Zealand	0.189	0.210	0.197
Norway	0.203	0.220	0.206
Poland	0.208	0.208	0.209
Portugal	0.230	0.247	0.224
Slovak Republic	0.160	0.176	0.178
Slovenia	0.208	0.224	0.209
Spain	0.211	0.254	0.211
Sweden	0.284	0.282	0.259
United Kingdom	0.203	0.231	0.206
United States	0.161	0.188	0.178
Mean	0.21146	0.22971	0.21143
Variance	0.00202	0.00186	0.00087

The year 2006 proportion is used as the maximum likelihood estimate for the average over following seven years for each country. The James-Stein estimator is produced using the following R code:

```
mean.2006 <- mean(oecd[,1])
mean.vec.2006 <- oecd[,1]
var.2006 <- mean.2006*(1-mean.2006)/prod(dim(oecd))
oecd.shrink <- 1 - ((nrow(oecd)-3)*var.2006)/
    ((nrow(oecd)-1)*var(mean.vec.2006))
oecd.js <- rep(NA,length(oecd.mean))
for (i in 1:nrow(oecd))
    oecd.js[i] <- mean.2006 +
        oecd.shrink*sum(mean.vec.2006[i]-mean.2006)
mean.vec.2007.2013 <- apply(oecd[,2:ncol(oecd)],1,mean)
round(cbind(mean.vec.2006,mean.vec.2007.2013,oecd.js),3)
```

The Efron and Morris 1975 and 1977 papers do not make clear the best strategy to get the variance (they added a transformation in their running baseball batting averages example). Here a binomial-derived variance is scaled by the size of the data.

Table 8.2 gives the two sets of country estimates compared to the mean of the next seven years. There are noticeable differences since the James-Stein estimates borrow strength from other cases. Notice also the much lower variance for the James-Stein estimates than the maximum likelihood estimates. However, observe that the means are almost identical in these results.

8.5 Empirical Bayes

As noted, the James-Stein estimator is an empirical Bayes estimator since it makes use of the marginal distribution of the data. In this section we will more fully explore empirical Bayes methods. Empirical Bayes is an approach that employs the data not only at the lowest form of the hierarchy but also to estimate these highest-level hyperpriors, which can be done with or without a specific distributional form at this level (parametric versus nonparametric empirical Bayes). This is an idea that actually offends some Bayesians since the idea of "using the data twice" (hence the name "empirical") appears to controvert standard Bayesian philosophy (Lindley famously said that "...there is no one less Bayesian than an empirical Bayesian" [1969, p.421]). It also bothers some Bayesians because it implies that "regular" Bayesian inference is not *empirical* (Gelman *et al.* 2003, p.121), which clearly is not true. Furthermore, estimation of parameters at the top of the hierarchy is done with non-Bayesian techniques (maximum likelihood, method of moments, etc.), and thus strikes some as both arbitrary and counter to the core tenets of Bayesian philosophy (Robert 2001). While empirical Bayes is not an emphasis of this text, we will nonetheless explain the basic principles in this section.

For parametric empirical Bayes start with the standard setup, with a prior that is directly conditional on some parameter vector, $\mathbf{\Upsilon}$, which leads to:

$$\pi(\boldsymbol{\theta}|\mathbf{X}, \mathbf{\Upsilon}) = \frac{p(\boldsymbol{\theta}|\mathbf{\Upsilon})L(\boldsymbol{\theta}|\mathbf{X})}{\int_{\boldsymbol{\Theta}} p(\boldsymbol{\theta}|\mathbf{\Upsilon})L(\boldsymbol{\theta}|\mathbf{X})d\boldsymbol{\theta}}, \qquad (8.45)$$

and we can label the denominator as $q(\mathbf{X}|\mathbf{\Upsilon})$ for convenience. In regular cases, the $\mathbf{\Upsilon}$ parameter values are set by researchers to reflect prior information, and in the case of conjugacy $q(\mathbf{X}|\mathbf{\Upsilon})$ can be easily estimated. With empirical Bayes we instead estimate $\mathbf{\Upsilon}$ from the data, often simply using the maximum likelihood estimate of $\mathbf{\Upsilon}$ from $q(\mathbf{X}|\mathbf{\Upsilon})$. Armed with $\hat{\mathbf{\Upsilon}}$, we can plug these values into the Bayesian calculation of the posterior above. The first extension, with the odd name *Bayes empirical Bayes* (Deely and Lindley 1981), adds a hyperprior distribution, $p(\mathbf{\Upsilon}, \psi)$ so that the posterior distribution is produced by averaging over $\mathbf{\Upsilon}$: $\pi(\boldsymbol{\theta}|\mathbf{X}, \psi)$. Therefore there is a clear connection between empirical Bayes and the hierarchical models of Chapter 12. A good starting point for the parametric version is Morris (1983b), and important considerations (such as interval estimation) are covered in Carlin and Gelfand (1990) as well as Laird and Louis (1987). There are strong links between empirical Bayes and frequentist objectives (something that further offends some Bayesians), and these ideas are well-explored in works by Efron and Morris (1971, 1972a, 1972b, 1973, 1976), as well as Deely and Zimmer (1976). Commonly cited applications include: Kass and Steffey (1989), and Rubin (1980). See also the collection of papers in Ahmed and Reid (2001).

As an illustration, consider the (balanced one-way) ANOVA setup from Casella (1985) with p columns of interest having different means, $\bar{X}_i$ but the same known variance, σ_0^2. So by standard assumptions:

$$\bar{X}_i \sim \mathcal{N}(\mu_i, \sigma_0^2) \tag{8.46}$$

$$\mu_i \sim \mathcal{N}(m, s^2), \qquad i = 1, \ldots, p. \tag{8.47}$$

The resulting posterior distribution for μ_i is normal with mean:

$$\hat{\mu}_i = \frac{\sigma_0^2 m + s^2 \bar{X}_i}{\sigma_0^2 + s^2}, \tag{8.48}$$

and variance:

$$\hat{\sigma}_\mu^2 = \frac{\sigma_0^2 s^2}{\sigma_0^2 + s^2}, \tag{8.49}$$

which is the ANOVA analog to the posterior mean on page 71. Empirical Bayes proceeds by identifying the distribution of the data value of interest unconditional on μ_i,

$$q(\bar{X}_i | m, s^2) = \int_\mu f(\bar{X}_i | \mu) p(\mu | m, s^2) d\mu, \tag{8.50}$$

which easily leads to the distributional property that $\bar{X}_i \sim \mathcal{N}(m, \sigma_0^2 + s^2)$. Now, rather than assigning parameters values, we take convenient expectations over this marginal distribution:

$$E[\bar{X}_i] = m, \qquad E\left[\frac{(p-3)\sigma_0^2}{\sum(\bar{X}_i - \bar{\bar{X}})^2}\right] = \frac{\sigma_0^2}{\sigma_0^2 + s^2}, \tag{8.51}$$

where $\bar{\bar{X}}$ is the grand mean. These are convenient values in the sense that they can be plugged directly into (8.48) to produce the empirical Bayes estimate of the posterior mean:

$$\hat{\mu}_i^{EB} = \left[\frac{(p-3)\sigma_0^2}{\sum(\bar{X}_i - \bar{\bar{X}})^2}\right]\bar{\bar{X}} + \left[1 - \frac{(p-3)\sigma_0^2}{\sum(\bar{X}_i - \bar{\bar{X}})^2}\right]\bar{X}_i. \tag{8.52}$$

The interesting part about this estimator is that it balances information from the column of interest and all of the data.

Nonparametric empirical Bayes, as introduced by Robbins (1955, 1964, 1983), treats $p(\theta | \Upsilon)$ as only $p(\theta)$, which is estimated directly from the data. Important developments are contained in: van Houwelingen (1977), van Houwelingen and Stijnen (1993), as well as Maritz and Lwin (1989). Robbins' (1955) canonical example is illustrative. Start with a Poisson-distributed random variable such that $\mathcal{P}(x_i | \theta_i) = \frac{\theta_i^{x_i} e^{-\theta_i}}{x_i!}, x_i = 0, 1, \ldots, 0 \leq \theta_i < \infty$.

Under squared error loss, $\ell(\theta, T) = (\theta - T)^2$ (i.e. the cost of mis-estimation is the square of the distance from the true value), the posterior mean for θ_i is:

$$\hat{\theta}_i = \frac{\int_t \frac{t^{x_i+1}\exp[-t]}{x_i!}dp(\theta)}{\int_t \frac{t^{x_i}\exp[-t]}{x_i!}dp(\theta)} = \frac{(x_i+1)u_n(x_i+1)}{u_n(x_i)}, \qquad (8.53)$$

where $u_n()$ is Robbins' notation for the marginal distribution. The task is then reduced to finding a nonparametric replacement function for these marginals based on observed frequencies: $u_n(z) = $ (number of terms $z_1, \ldots, z_n$ equal to z). Therefore:

$$\hat{\theta}_i^{EB} = \frac{(x_i+1)(\# \text{ terms } = x_i+1)}{(\# \text{ terms } = x_i)}. \qquad (8.54)$$

Thus like the parametric-normal case above, the estimate for each θ_i borrows strength from the other components $j \in 1, \ldots, n$, $i \neq j$.

■ **Example 8.6:** **Empirical Bayes Estimates for Cabinet Duration.** Returning to the cabinet duration data in Exercise 16 on page 65, we will estimate the individual Poisson intensity parameters λ_i, $i = 1, \ldots, 11$. The data are cabinet duration (constitutional inter-election period) for eleven Western European countries from 1945 to 1980 for annualized periods given by the number of cabinets (x_i) and the average duration (δ_i). The model specification is given by:

$$x_i | \delta_i, \lambda_i \sim \mathcal{P}(\delta_i \lambda_i)$$
$$\lambda_i \sim \mathcal{G}(\alpha, \beta_{\text{rate}})$$

for $i = 1, \ldots, n$, where the δ_i are weights on the intensity parameters and the gamma distribution is specified in the rate version. This produces the posterior distribution:

$$\lambda_i \sim \mathcal{G}(\alpha + x_i, \beta + \delta_i),$$

with posterior expected value:

$$E[\lambda_i | x_i, \delta_i, \alpha, \beta] = \frac{\alpha + x_i}{\beta + \delta_i}$$

(see Exercise 18). For a multilevel model we would specify distributions for α and β, and then estimate with the principles in this chapter. Instead we will use standard maximum likelihood estimation with the data to get "plug-in" values of these unknown parameters for an empirical Bayes solution. First we need the marginal distribution of the data given these parameters:

$$p(x_i|\alpha, \beta) = \int_0^\infty \frac{(\delta_i \lambda_i)^{x_i} \exp[-\delta_i \lambda_i]}{x_i!} \times \frac{\beta^\alpha}{\Gamma(\alpha)} \exp[\beta \lambda_i] \lambda_i^{\alpha-1} d\lambda_i$$

$$= \frac{\beta^\alpha \delta_i^{x_i}}{\Gamma(\alpha) x_i!} \int_0^\infty \lambda_i^{(\alpha+x_i)-1} \exp[-(\delta_i + \beta)\lambda_i] d\lambda_i$$

$$= \frac{\beta^\alpha \delta_i^{x_i}}{\Gamma(\alpha) x_i!} \frac{\Gamma(\alpha + x_i)}{(\delta_i + \beta)^{\alpha+x_i}}$$

$$= \frac{\Gamma(\alpha + x_i)}{\Gamma(\alpha) x_i!} \beta^\alpha (\delta_i + \beta)^{-\alpha} \delta_i^{x_i} (\delta_i + \beta)^{-x_i}$$

$$= \frac{\Gamma(\alpha + x_i)}{\Gamma(\alpha) x_i!} \left(\frac{\beta}{\delta_i + \beta}\right)^\alpha \left(\frac{\delta_i}{\delta_i + \beta}\right)^{x_i}.$$

The integral is calculated from the second to the third line by noticing that the

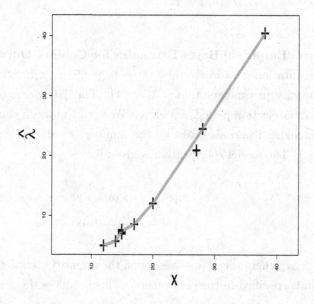

FIGURE 8.1: Empirical Bayes Estimates versus Data Values

form within is the kernel of a gamma distribution and simply multiplying this by the appropriate constants gives the integral value of one, leaving the inverse of these terms. The final expression of $p(x_i|\alpha, \beta)$ would be a negative binomial form for $y_i = \delta_i/(\delta_i+\beta)$ if $\alpha = 1 - x_i$, but it does not make sense to impose this restriction here. The joint likelihood function for α and β is simply the product of these terms:

$$L(\alpha, \beta|\mathbf{x}) = \prod_{i=1}^n \frac{\Gamma(\alpha + x_i)}{\Gamma(\alpha) x_i!} \left(\frac{\beta}{\delta_i + \beta}\right)^\alpha \left(\frac{\delta_i}{\delta_i + \beta}\right)^{x_i},$$

which we can describe with MCMC tools, but estimation with the `optim` function in

R is just as straightforward. Once this is done we can produce:

$$\hat{\lambda}_i^{EB} = \frac{\hat{\alpha} + x_i}{\hat{\beta} + \delta_i},$$

where $\hat{\alpha}$ and $\hat{\beta}$ are the MLEs from this process. The following R function performs these tasks:

```
EB.Poisson.Log.Like <- function(x.vec, d.vec) {
    poisson.ll <- function(params) {
        alpha <- exp(params[1]); beta <- exp(params[2])
        gamma.term <- sum(lgamma(alpha+x.vec) - lgamma(alpha))
        ratio.term1 <- sum(x.vec*log(d.vec/(d.vec+beta)))
        ratio.term2 <- sum(alpha*log(beta/(d.vec+beta)))
        return(gamma.term + ratio.term1 + ratio.term2)
    }
    optim.out <- optim(par=c(1,1),fn=poisson.ll,method="BFGS",
        control=list(fnscale=-1))
    alpha.hat <- exp(optim.out$par[1])
    beta.hat <- exp(optim.out$par[2])
    return((x.vec+alpha.hat)/(d.vec+beta.hat))
}
```

We can specify the data and call this function with:

```
d.vec <- c(0.833,1.070,1.234,1.671,2.065,2.080,2.114,2.168,2.274,
        2.629,2.637)
x.vec <- c(38,28,27,20,17,15,15,15,15,14,12)

( EB.out <- EB.Poisson.Log.Like(x.vec, d.vec) )
[1] 40.538450 24.541938 20.926250 12.088241  8.592669  7.640462
[7]  7.526011  7.351118  7.030418  5.775079  5.042394
```

Figure 8.1 plots the original x_i values against the produced $\hat{\lambda}_i^{EB}$ estimates with an overlayed loess smoother ($f = 0.5$). While the relationship is very near linear, the empirical Bayes estimates are mostly lower than the corresponding x_i.

8.6   Exercises

8.1   Savage (1972). For arbitrary outcomes under simple ordering prove:

▷ if $A \doteq B$ then $B \doteq A$

    ▷ $A \doteq A$ for any A

    ▷ if $A \leq B$ and $B \doteq C$ then $A \leq C$

    ▷ if $A < B$ then $B \not\leq A$

8.2   Generalize the principles in (8.1) (page 250) to an infinite number of events using efficient notation.

8.3   The decision rule is a function of $\mathbf{X}$ where each draw of $\mathbf{X}$ creates an estimate of θ which is mapped to a specific action A. Create a graphic that illustrates this function.

8.4   Equation (8.2) gave a decision rule for a normal (z-statistic) setup (difference of means). Restate this for a χ^2 test at the same α-level.

8.5   Show that the standard mean square error corresponds to a quadratic loss function. Illustrate your argument with a data vector distributed $\mathcal{N}(\mu, \sigma_n^2)$.

8.6   Prove that the Bayes rule estimator using absolute error loss ($\mathbb{L}(A, d(\mathbf{X})) = |\theta - \theta_A|$) is the posterior median.

8.7   Show that $\mu_A = \bar{y}$ is the estimator of μ that minimizes expected loss in (8.32).

8.8   Derive the mean and variance of $p(d|n)$ in (8.16).

8.9   Consider the following table of US and Soviet nuclear tests:

Year	U.S. Tests	Soviet Tests	Year	U.S. Tests	Soviet Tests
1960	0	0	1976	21	21
1961	10	1	1977	20	24
1962	98	1	1978	21	31
1963	47	0	1979	16	31
1964	47	9	1980	17	24
1965	39	14	1981	17	21
1966	48	18	1982	19	19
1967	42	17	1983	19	25
1968	56	17	1984	20	27
1969	46	19	1985	18	10
1970	39	16	1986	15	0
1971	24	23	1987	15	23
1972	27	24	1988	15	16
1973	24	17	1989	12	7
1974	23	21	1990	9	1
1975	22	19	1991	8	0

(also provided in the R package BaM). Returning to (8.19), construct a prior for Soviet tests from 1960 to 1969 ($p(n)$), and specify a reasonable loss function based on U.S. tests from 1970-1990 that allows the U.S. to test at the observed rates assuming that the two powers want to test at the same rate. Use these to construct the expected loss with respect to the prior distribution. Would it be in the U.S. interest to renegotiate at some intermediate time between 1970 and 1990?

8.10 Define $X_1, \ldots, X_n$ to be a set of iid dichotomous random variables with $p = p(X_i = 1)$ and $1 - p = p(X_i = 0)$. Show that the data mean are an admissible estimator of unknown p using squared error loss. Find another loss function that also has this property.

8.11 If $\sum t(x_i)$ is a sufficient statistic for $\boldsymbol{\theta}$ (Appendix A), then provide the Bayes (optimal) action that is a function of $\sum t(x_i)$.

8.12 For finite parameter spaces all admissible estimators are Bayesian estimators (Ferguson 1967) . Consider estimating ζ on $\Re$ from a sample distributed $\mathcal{N}(\zeta, 1)$ using $\bar{X}$. While this is admissible for ζ, show that it cannot be a Bayes action with respect to any prior distribution since it is unbiased but has non-zero variance for finite samples.

8.13 Given an iid sample, $X_1, \ldots, X_n$ from a normal distribution $\mathcal{N}(\mu, \sigma^2)$, where σ^2 is known, show that the maximum likelihood estimate for μ the frequentist risk can be expressed as:

$$\mathbb{L}(A, d(\mathbf{X}), k) = |\theta - \theta_A|^k = \Gamma\left(\frac{k+1}{2}\right) \sigma^k n^{-k/2} (2^k/\pi)^{\frac{1}{2}}.$$

8.14 Prove that all admissible rules must be Bayes rules and give an example of a Bayes rule that is inadmissible.

8.15 For the normal model described on page 78, show that for squared error loss the maximum likelihood point estimate $\hat{\mu} = \bar{x}$ has the risk function $R_F(\mu, d(\mathbf{X})) = n$.

8.16 Using the election survey data in Example 7.3.3 (page 220), calculate the maximum likelihood estimate and the James-Stein estimate for each value in IND (self-indicated indifference to the election). Compare the variances and contrast the estimated vectors graphically.

8.17 Show that the prior distribution (8.33) leads to the posterior distribution (8.34).

8.18 Derive the posterior distribution and expected value of λ_i from Example 8.5.

8.19 (Formerly Exercise 10.10) Plug (8.51) into (8.48) to obtain (8.52), the empirical Bayes estimate of the posterior mean. Using the same logic and (8.49), produce the empirical Bayes estimate of the posterior variance, assuming that it is now not known.

8.20   Given the model:

$$x_j \sim \mathcal{BN}(n_j, p_j), \qquad j = 1, \dots, J$$
$$p \sim \mathcal{BE}(\alpha, \beta)$$

with $\alpha > 0$ and $\beta > 0$ both unknown, show that the marginal distribution of the full set of $\mathbf{X}$ is:

$$p(X|\alpha, \beta) = \prod_{j=1}^{J} \binom{n_k}{x_k} \frac{\Gamma(\alpha+\beta)\Gamma(\alpha+x_j)\Gamma(n_j - x_j + \beta)}{\Gamma(\alpha)\Gamma(\beta)\Gamma(\alpha+\beta+n_j)}.$$

If $\hat{\alpha}$ and $\hat{\beta}$ are the MLEs of these parameters, show that the empirical Bayes estimator of p_j is given by:

$$E[p_j|x_j, \hat{\alpha}, \hat{\beta}] = \frac{\hat{\alpha} + x_j}{\hat{\alpha} + \hat{\beta} + n_j}.$$

Chapter 9

Monte Carlo and Related Iterative Methods

9.1 Background

This is the first chapter specifically about simulation techniques, even though some principles have already been discussed. Simulation work in applied statistics replaces analytical work on behalf of the researcher with repetitious, low-level effort by the computer. The key advantage is that when a model specification leads to a posterior form that is difficult or impossible to manipulate analytically, then it is often possible to create a set of *simulated* values that share the same distributional properties. So we describe the posterior by using empirical summaries of these simulated values rather than perform some uncomfortable integration or other operation. This is actually an old idea (see the really interesting 1951 chronicle: "Report on a Monte Carlo Calculation Performed with the Eniac," by Mayer), but one that is particularly easy to implement now that computers are ubiquitous and fast (and getting more so every day). Excellent references in addition to those specifically cited in this chapter include: Fang, Hickernell, and Niederreiter (2002), Fishman (2003), Lange (2000), Rubinstein (1981), and Sobol (1994).

To begin with, suppose we were interested in the expected value of some quantity expressed as an integral:

$$E[\boldsymbol{\theta}] = \int_a^b \boldsymbol{\theta} f(\boldsymbol{\theta}) d\boldsymbol{\theta}, \tag{9.1}$$

where $f()$ is a suitably complex form and $\boldsymbol{\theta}$ is of sufficiently high (k) dimension such that the integral is prohibitively challenging from a straight analytical standpoint. Of course, as we have seen already, we are often interested in other posterior quantities like intervals, and this chapter will address Monte Carlo simulation of the quantities of primary Bayesian interest.

If we are lucky enough to have an easy method for producing random vectors $\boldsymbol{\theta}_1, \boldsymbol{\theta}_2, \ldots, \boldsymbol{\theta}_n$ with $\boldsymbol{\theta}_i = [\theta_{i1}, \theta_{i2}, \ldots, \theta_{ik}]$, which are known to be from the correct distribution on $[a:b]$, then it is possible to use these empirical draws to summarize the unknown integral quantity by counting those that fall in this range. The idea is quite powerful: if we can generate samples from the desired sampling distribution, then we can summarize the theoretical distribution by using these simulated values rather than being required to use difficult analytical calculations.

Monte Carlo methods are digital computation methods, originally to randomly generate a large number of numerical values to perform difficult physical calculations of interest. Naturally, the definition of "large" in this context has changed considerably over time. While the varied collection of Monte Carlo techniques thus far accumulated by researchers are very important in Bayesian analysis, none of the tools provided herein are restricted to Bayesian use. In fact, most of these were developed for other purposes and eventually co-opted by Bayesians. Interestingly, the history of Monte Carlo simulation is intimately tied with the development of thermonuclear weapons, notably in the papers of Metropolis, and Ulam (1949), Kahn (1949), von Neumann (1951), Metropolis *et al.* (1953), and Pasta and Ulam (1953). The original idea came to Ulam when he was playing solitaire convalescing from an illness, and was formally developed by Ulam and von Neumann. Since the method was created in a classified environment investigating neutron diffusion in fissionable material, it was given the codename *Monte Carlo* by Metropolis who said that the iterative process of probability sampling reminded him of casino gambling. The internal report by Richtmyer (the Theoretical Division Leader at Los Alamos at the time), Ulam, and von Neumann was not declassified until 1959, partly explaining the slow pace of dissemination into general scientific use.

Common statistical computing tasks to worry about in traditional settings are optimization techniques like numerical differentiation of likelihood functions and solution finding in systems of constrained linear equations (Gill, Murray, and Wright 1981). The production of fixed-point estimates are produced by algorithmic mode-finding, where the major numerical difficulties are finding the dominating mode and the curvature around that mode. In contrast, the primary numerical challenge in Bayesian analysis is the summary of posterior distributions through *integration*. This difference is both philosophical (fixed population parameters versus population parameters with distributions) and practical (summary through point estimates and standard errors versus summary through descriptions of distributions).

Since the most frequent goal in numerical Bayesian analysis is to estimate an integral quantity of some sort based on limited information using random elements, we will concentrate in this chapter on methods for calculating integrals numerically. First we will cover basic integral quantities that are relatively well defined but require simulations to describe, then we will discuss more elaborate stochastic simulation techniques in which the simulation process contains serial correlation.

One final word of caution is necessary here. Performing statistical analysis (description, estimation, prediction) with Monte Carlo tools *is still* statistical analysis. Thus giving a Monte Carlo point estimate without a corresponding standard error is incomplete and misleading. In general, Monte Carlo techniques computationally produce samples and typically an appropriate mean, $\hat{\theta}$, is calculated from these n samples, $\theta_1, \ldots, \theta_n$. So it is trivial to calculate the *Monte Carlo standard error*: $\sqrt{s_\theta = \frac{1}{n} \sum (\theta_i - \hat{\theta})^2}$, yet this is often neglected.

Fortunately the size of this error is "controllable" since the size of n is determined by the patience of the researcher (and computers continue to increasingly cater to the impatient).

9.2 Basic Monte Carlo Integration

Suppose that we had a (normalized) probability function, $g(\theta)$, that was difficult to express or manipulate but for which we could easily generate samples on an arbitrary support of interest: $[a{:}b]$. A common quantity of interest is:

$$I[a,b] = \int_a^b g(\theta)h(\theta)d\theta, \tag{9.2}$$

that is, the expected value of some function, $h(\theta)$, of θ distributed $g(\theta)$. If $h(\theta) = \theta$, then $I[a,b]$ simply calculates the mean of θ over $[a{:}b]$. A substitute for analytically calculating (9.2) is to randomly generate n values of θ from $g(\theta)$ and calculate:

$$\hat{I}[a,b] = \frac{1}{n}\sum_{i=1}^n h(\theta_i). \tag{9.3}$$

The idea is to replace analytical integration with summation from a large number of simulated values, rejecting values outside the range of interest, $[a{:}b]$. The beauty of this approach is that by the strong law of large numbers, $\hat{I}[a,b]$ converges with probability one to the desired value, $I[a,b]$. A second positive feature is that although $\hat{I}[a,b]$ now has "simulation error," this error is easily measured by the empirical variance of the simulation estimate:

$$\mathrm{Var}(\hat{I}[a,b]) = \frac{1}{n(n-1)}\sum_{i=1}^n (h(\theta_i) - \hat{I}[a,b])^2. \tag{9.4}$$

Note that the researcher fully controls the simulation size, n, and therefore the simulation accuracy of the estimate. Furthermore, the central limit theorem applies here as long as $\mathrm{Var}(I[a,b])$ is finite, so, for instance, 95% credible intervals can be directly calculated by:

$$[95\%_{lower}, 95\%_{upper}]$$
$$= \left[\hat{I}[a,b] - 1.96\sqrt{\mathrm{Var}(\hat{I}[a,b])}, \ \hat{I}[a,b] + 1.96\sqrt{\mathrm{Var}(\hat{I}[a,b])}\right], \tag{9.5}$$

or by reporting the 0.025 and 0.975 quantiles of the set of θ_i. Obviously other credible intervals of interest can be similarly calculated.

Monte Carlo integration can be placed in more explicitly Bayesian context for our purposes by replacing $g(\theta)$ with a posterior statement $\pi(\theta|x)$ and noting that $h()$ is typically an estimate of some function of the unknown parameter: $h(\theta)$. So $I[a,b]$ is really the (posterior) expectation of $h(\theta|\mathbf{x})$:

$$E[h(\theta|x)] = \int \pi(\theta|x)h(\theta)d\theta \approx \frac{1}{n}\sum_{i=1}^n h(\theta_i). \tag{9.6}$$

Often $h(\theta) = \theta$ and this is then just the first moment of the posterior distribution.

To show how easy this process can be with modern computing consider two small problems. First, suppose we replace the following integration operation with its simulated value (ϕ indicates the standard normal PDF):

$$I[-2,1] = \int_{-2}^{1} \theta^2 \phi(\theta)d\theta \approx \hat{I}[-2,1] = \frac{1}{n}\sum_{i=1}^{n}\theta_i^2 \tag{9.7}$$

by using the R code which first samples $100,000$ random standard normal values and then applies the criteria of interest.

```
norm.sample <- rnorm(100000)
mean(norm.sample[norm.sample>-2 & norm.sample <1]^2)
```

This produces 0.5747261 (try it), saving us some analytical agony since the normal PDF is typically integrated by using polar coordinates. Now consider a much harder analytical problem:

$$I[e,\pi] = \int_{e}^{\pi} \arctan(\theta^{\frac{1}{3}})\mathcal{C}(\theta|\mu = 3, \sigma = 2)d\theta \tag{9.8}$$

where $\mathcal{C}()$ denotes the Cauchy PDF ($\mathcal{C}(\theta|\mu,\sigma) = \frac{1}{\pi\sigma}\frac{1}{1+\left(\frac{\theta-\mu}{\sigma}\right)^2}$, $-\infty < \theta, \mu < \infty, 0 < \sigma$, B). Obviously this would not be a lot of fun to work through with pencil and paper. However the Monte Carlo solution in R is trivial:

```
c.sample <- rcauchy(100000,3,2)
mean(atan(c.sample[c.sample > exp(1) & c.sample < pi]^(1/3)))
```

which produces 1.058232.

■ **Example 9.1: Monte Carlo Integration with the Pareto Distribution, European Migration Data.** The Pareto distribution (B) is often used to model phenomena that are bounded by zero and have long positive tails. Examples include city and country populations, income, and wealth. Suppose we have data at hand that are assumed to be distributed $\mathcal{PA}(x|\alpha,\beta)$, with α and β unknown. Arnold and Press (1983, 1989) develop a two-stage prior for the Pareto where there is an assumed dependence between the two parameters:

$$\alpha \sim \mathcal{G}(C,D) = \Gamma(C)D^C\alpha^{C-1}e^{-D\alpha} \tag{9.9}$$

$$\beta|\alpha \sim \mathcal{PA}(\alpha A, B) = \alpha AB(B\beta)^{-(\alpha A+1)}. \tag{9.10}$$

This idea, first suggested by Lwin (1972) and also addressed by Nigm and Handy (1987), is that independence of the two parameters is somewhat unrealistic in practice and it is better to specifically model the conditionality. We saw a normal model version of this type of prior structure in Chapter 3. The likelihood function given the data is:

$$L(\alpha,\beta|\mathbf{x}) = \alpha^n\beta^{-n\alpha}\exp\left[-(\alpha+1)\sum_{i=1}^{n}\log(x_i)\right],$$

which·leads to the following form of the joint posterior:

$$\pi(\alpha, \beta | \mathbf{x}) \propto L(\alpha, \beta | \mathbf{x}) p(\beta | \alpha) p(\alpha)$$

$$\propto \alpha^{n+C} (B\beta)^{-(n\alpha + A\alpha + 1)}$$

$$\times \exp\left[-D\alpha - \left(\sum_{i=1}^{n} \log(x_i) + A\log[B] - (n+A)\log[\min(B, \mathbf{x})]\right)\alpha\right],$$

provided that $\beta\min(B, \mathbf{x}) > 0$, and $\alpha > 0$. The conjugacy of this setup gives the marginal posteriors:

$$\alpha | \mathbf{x} \sim \mathcal{G}\left(n + C, D + \sum_{i=1}^{n} \log(x_i) + A\log[B] - (n+A)\log[\min(B, \mathbf{x})]\right)$$

$$\beta | \alpha, \mathbf{x} \sim \mathcal{PA}(n\alpha + A\alpha, \min(B, \mathbf{x})).$$

Peach (1997) looks at postwar migration into western Europe due to: the effects of a retreat from colonization, worker mobility from poorer areas of the world, and influx due to political flight. There were substantial differences in assimilation across western European countries due to local politics and culture as well as the ethnicity of the immigrants. Table 9.1 gives the 1990-1993 total ethnic minority population in western European countries, including cross-migration (excluding the "other" category).

We first set the prior parameters according to the vector: $[A, B, C, D] = [20, 100, 600, 3]$ and then evaluate the posterior according to the set forms above. See Arnold and Press (1989, p.1083) for specific guidance on empirically setting these parameters to meet moment expectations. The marginal posterior mean and variance for α are easy enough to calculate analytically since the form is a specified gamma distribution not conditioned on any unknown quantity.

Employing this method with 1,000 simulations (in actual practice this number should probably be an order of magnitude or more higher) produces a posterior mean from the α simulations of 4.754, as opposed to the theoretical value of 4.751 (not that accuracy is implied to the level of this distinction since there exists measurement error in the data as well). The presence of simulation error explains this modest difference: the standard error of the Monte Carlo statistic is 0.003.

It is often easier to use Monte Carlo integration to calculate tail probabilities and thresholds than it is to derive them analytically. It is also more flexible in that calculating different values is a trivial counting procedure using empirical draws rather than a new integral calculation. All we would have to do is to generate a set of random gamma variates and summarize numerically using simple Monte Carlo integration. Specifically, we generate N draws from the distribution of interest, sort them, and pick out the values of probabilistic interest. If we generate $N = 1,000$, then, for example, the 0.025 tails on either end begin approximately at the sorted 25th and 975th empirical values. Table 9.2 gives a set of quantiles for α from the Pareto model using exactly this procedure with the $N = 1,000$ simulated values:

TABLE 9.1:   1990-1993 W.EUROPE ETHNIC/MINORITY
POPULATIONS

Country of Origin	Estimated Total(K)	Percent of Total
Turkey	2483.9	17.79
Italy	1518.9	10.88
Former Yugoslavia	1241.1	8.89
Morocco	1178.1	8.44
Caribbean	1000.0	7.16
Portugal	925.7	6.63
India	876.6	6.28
Algeria	698.8	5.00
Ireland	641.0	4.59
Spain	572.5	4.10
Pakistan	492.8	3.53
Greece	402.9	2.89
France	331.8	2.38
Poland	328.7	2.35
Tunisia	292.2	2.09
Germany	280.7	2.01
Iran	183.8	1.32
Bangladesh	183.3	1.31
Croatia	153.1	1.10
Bosnia-Herzegovina	139.1	1.00
Afghanistan	30.0	0.21
Indonesia	9.5	0.07

9.3   Rejection Sampling

Suppose now that we could not readily produce random values from the posterior distribution of interest. This would temporarily preclude us from performing Monte Carlo integration as described above because we could not obtain the simulated sample required to calculate empirical summaries. We are not necessarily completely thwarted because if there exists an expressed form of the posterior so that for candidate values, θ, the density, $f(\theta)$, is calculable, then we can often employ *rejection sampling* to produce an empirical sample.

Rejection sampling, or the rejection method, is a means of obtaining an integral quantity using the generation of candidate random variables (the sample here) and accepting those that are determined to belong to the distribution of interest. This idea, which is only slightly more involved than Monte Carlo integration, was briefly introduced in Exercise 3.11. By comparing the quantity generated to the quantity accepted, we get a ratio value corresponding to the area we would like to measure. This principle, also called the *acceptance-rejection method*, is very general and applies to any situation where uniformly random variates can be generated over an enclosing region.

TABLE 9.2: Monte Carlo Quantiles for Pareto Parameter α

Minimum	0.01	0.25	0.50	0.75	0.99	Maximum
4.132	4.633	4.664	4.750	4.759	4.872	5.336

We could assume that $\alpha = 2$ (see Johnson, Kotz, and Balakrishnan [1997, p.575]) for a discussion of the economic history of this parameter), and get a posterior for β, but rather than "take the easy way out" we will integrate out α and simulate the subsequent marginal posterior values for β, thus incorporating uncertainty from α rather than ignoring it. This calculation starts with:

$$\pi(\beta|\mathbf{x}) = \int_\alpha \pi(\beta|\alpha, \mathbf{x})\pi(\alpha|\mathbf{x})d\alpha$$

$$\propto \int_\alpha \alpha^{n+C}(B\beta)^{-(n\alpha+A\alpha+1)}\exp\left[-\alpha(D+E)\right]d\alpha, \tag{9.11}$$

where for simplification we introduce the shorthand for constants: $E = \sum_{i=1}^n \log(x_i) + A\log[B] - (n+A)\log[\min(B, \mathbf{x})]$. Unfortunately there appears to be no convenient analytical marginalization here because our usual trick of segmenting off the kernel of a PDF is not possible. While it is possible to solve this integral with brute force mathematics, the investment of time will be nowhere near worthwhile. Instead we will employ a Monte Carlo marginalization based on *importance sampling*, which is provided in Section 9.6.

The terminology is slightly confusing in that rejection sampling is a technique that can measure difficult integral quantities as well as generate random variates from candidate distributions from which analytical sampling might otherwise be difficult (Kronmal and Peterson 1981). The original idea behind rejection sampling comes from mid-twentieth century pioneers, von Neumann (1951) and Metropolis and Ulam (1949), who developed the idea of replacing repeated computational work for analytical calculations.

9.3.1 Continuous Form with Bounded Support

Imagine that we would like to estimate some integral quantity based on a continuous random variable where we know that the support of this variable is bounded. This means only that we know the limits for simulation purposes. This does not have to be a complete probability density function, it may be a non-normalized form, a "slice" of the density, or some other composite form that does not necessarily integrate to one.

For the most basic case, posit a function $f(\theta)$ so that we can evaluate candidate values of θ through the function but we cannot easily integrate the function. If $f(\theta)$ can be evaluated for any value over the support of θ: $S_\theta = [A, B]$, then it is usually a simple process to obtain the function's maximum value (if it exists and is unique) over this support: $\tilde{\theta}$. There are two general methods for obtaining this maximum: analytical root finding and standard numerical techniques. Root finding is the basic calculus technique whereby we take the derivative of $f(\theta)$ with respect to θ, set it equal to zero and solve for θ. This produces the θ value where the derivative is zero ($\tilde{\theta}$), i.e., where the tangent line is flat at the mode, see Appendix A for details. Of course there is much more effort involved for multiple modes, minimum points, or saddle points. Tests for these conditions are very nearly trivial and involve checking the second derivative at the candidate point (the second derivative is negative at a true maximum point indicating a sloping downward in both directions away from the modal point). Numerical techniques are effective and vary in complexity from simple linear candidate testing across the range of possible θ values ("gridding") to the more sophisticated numerical root-finding algorithms discussed in this chapter.

If we have a known bounded support and maximum value for θ, then we can define a rectangle in 2-space that is guaranteed to bound all possible values for the pair: $(\theta, f(\theta))$. This is useful because it is trivial to generate a two-dimensional random variable in this rectangle by independently sampling uniformly over $[A, B]$ and $[0, \tilde{\theta}]$ and then pairing these points. Therefore once we have "boxed in" the area of interest, we can randomly generate points over the box and count the number of values that fall under the curve.

The value of the integral, the area under the curve, is just the ratio of points under the curve to the total number of points scaled by the size of the box:

$$\frac{\text{number of points under curve}}{\text{total number of points}} \times \text{size of box} \xrightarrow[n \to \infty]{} \int_A^B f(\theta) d\theta. \qquad (9.12)$$

The fact that this is a converging quantity, rather than an analytically deterministic value, should not alarm us since the degree of accuracy is entirely controlled by the researcher through the number of points generated. We can therefore be as accurate as we want just by increasing the number of simulated values to test. Discrete problems turn out to be much more straightforward as it is usually a matter of counting bin heights and taking a weighted sum.

Figure 9.1 demonstrates how rejection sampling works. In this case 100 points are sampled uniformly from the two-dimensional rectangle defined over: $[(A, B), (0, \max(f(\theta)))] = [(0, 10), (0, 0.4)]$, and observed values that are contained in the region of interest (some PDF defined on the interval (A, B)). In total 26 values fell into the area we wish to integrate, so by (9.12) we obtain the size of the interval from: $(26/100)(10 \times 0.4) = 1.04$. This makes sense since this example was constructed by using a transformed gamma PDF (lopping off a very small tail area) and the actual value is therefore very close to one. We could get much closer to one if we simulated more values than only 100.

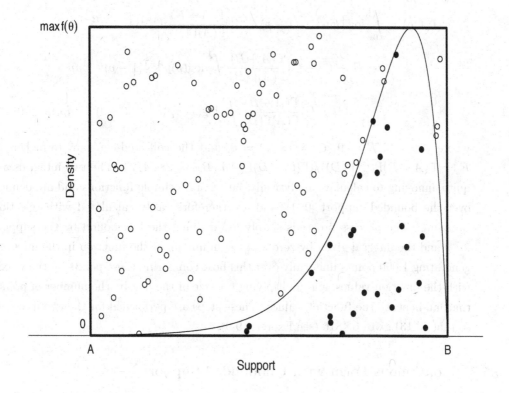

FIGURE 9.1: Rejection Sampling for Bounded Forms

This expression of rejection sampling can be particularly useful in determining the normalizing factor for non-normalized posterior distributions, provided that each dimension has bounded support that can be "boxed." Obvious examples include the beta distribution and normal forms with reasonably small tails. In the cases where this is not possible, another approach is required. Also, it is not essential that the bounding region be determined by a rectangle, and other shapes are regularly used such as the triangle, trapezoid, or various densities (see Morgan [1984, Section 5.4]).

■ **Example 9.2: Calculating the Kullback-Leibler Distance.** In Section 7.6 we introduced the Kullback-Leibler distance as a means of calculating the distance between two distributions. It was pointed out that the measure is difficult to analytically calculate for a number of distributional forms. This is true for the beta distribution, and we will use rejection sampling to produce the Kullback-Leibler distance for the two beta posteriors from the beta-binomial model of cultural consensus given in Example 2.3.4.

The two posteriors given were $f(p) = \mathcal{BE}(32, 9)$, and $g(p) = \mathcal{BE}(18, 8)$. These were plotted in Figure 2.5, and it did not appear that there was a dramatic difference despite the different approach to the priors. Using the form for the Kullback-Leibler distance given in (7.49) with the two beta posteriors, we obtain:

$$I(f,g) = \int_0^1 \log \left[Ep^{A-C}(1-p)^{B-D}\right] \frac{\Gamma(A+B)}{\Gamma(A)\Gamma(B)} p^{A-1}(1-p)^{B-1}dp$$

$$= \log(E) + (A-C)\frac{\Gamma(A+B)}{\Gamma(A)\Gamma(B)} \int_0^1 \log(p)p^{A-1}(1-p)^{B-1}dp$$

$$+ (B-D)\frac{\Gamma(A+B)}{\Gamma(A)\Gamma(B)} \int_0^1 \log(1-p)p^{A-1}(1-p)^{B-1}dp, \qquad (9.13)$$

where $A = 32$, $B = 9$, $C = 18$, $D = 8$, and the collected-constant term here is $E = (\Gamma(A+B)\Gamma(C)\Gamma(D))/(\Gamma(C+D)\Gamma(A)\Gamma(B)) = 284.4174$. The two integrals are quite annoying to calculate analytically, but contain simple functions and are defined over the bounded support $[0:1]$, and are therefore easily calculated with rejection sampling. The process proceeds simply as: defining the box bound by the support for p and the interval given by zero and the minimum of the function in the integral, generating 1,000 points uniformly over this box, comparing these points on the y-axis with the function values, and multiplying the size of the box by the number of points that fall beneath the function values. These steps are performed for the two integrals so that (9.13) gives 0.4816 (see Exercise 9.9).

9.3.2   Continuous Form with Unbounded Support

Suppose we wish to calculate the integral of some target function $f(\theta)$ in which the analytical solution is difficult or impossible, and the form of $f(\theta)$ has unbounded tails. The trick is to specify a "majorizing function," $g(\theta)$, which for every value of θ in the support of $f(\theta)$ has the property that $g(\theta) \geq f(\theta)$. This is done with the idea that we can pick some $g(\theta)$ function, such as a convenient PDF, from which it is easy to sample and therefore produce candidate values for acceptance or rejection. Thus the uniform candidate values over a two-dimensional rectangle, a nontheoretical convenience, are replaced with some distributional form with desired coverage probabilities.

If the target distribution has unbounded tails, then obviously the majorizing function must also have this property (Geweke 1989, p.1319), and simply picking a PDF for $g(\theta)$ which has heavier tails than $f(\theta)$ will not work since it will then have other regions where it is not uniformly greater than the target distribution by the fact that it must integrate to one. The solution is to use a multiplication factor so that:

$$f(\theta) \leq kg(\theta), \qquad \forall \theta, k > 1. \qquad (9.14)$$

Thus we sample θ_i from $g(\theta)$ and then make an accept/reject decision based on $f(\theta_i)$. Mechanically this means that for each θ_i point, we randomly draw a uniform(0,1) variate and accept θ_i if this uniform is less than $f(\theta_i)/kg(\theta_i)$. This process is illustrated in Figure 9.2 where the accepted points are distinguished from the rejected points (although the ratio no longer has the intuitive interpretation of that given by Figure 9.1 since the bounding

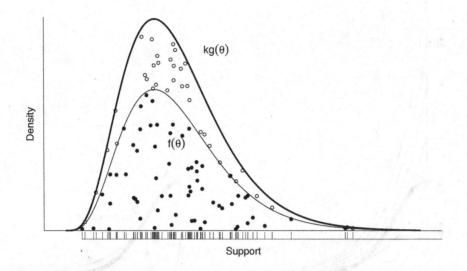

FIGURE 9.2: Rejection Sampling for Unbounded Forms

structure is no longer rectangular). The graphic underneath the distribution is called (rather cutely) a "rug," and indicates the marginal distribution of θ generated by $g(\theta)$. Note that if the majorizing distribution is very dissimilar from the target distribution, then the high-density area of the rug would be away from the mode of the target distribution, and the sampling procedure would be less efficient (more values rejected).

The key is to specify the majorizing function to have the same asymptotic properties as the target distribution in addition to its enveloping requirement. In the synthetic example in Figure 9.2, both distributions have a right-hand side asymptote. The second important characteristic is that the majorizing function resembles the target function as much as possible while still enveloping it for every value of θ in the sample space. This amounts to a performance issue rather than a mathematical requirement since an enveloping distribution that differs greatly from the target distribution will tend to generate many rejected candidate values and therefore be considerably less computationally efficient.

■ **Example 9.3: Enveloping with an Exponential Density.** Suppose that we have obtained a non-normalized posterior distribution resulting from the proportionality of Bayesian inference, of the form:

$$f(\theta) = \exp[-\theta^2/2], \qquad \theta \geq 0. \tag{9.15}$$

This is the kernel of a folded normal PDF, which is a non-normalized reflection of the standard normal defined only for positive values, leaving out constant terms. In order to normalize this posterior, we require the integral quantity in the denominator of the right-hand side of Bayes' Law (as in (2.7)): $I(\theta) = \int_0^\infty \exp[-\theta^2/2]d\theta$, which is a time-consuming integration process (although we could obtain it via normal properties).

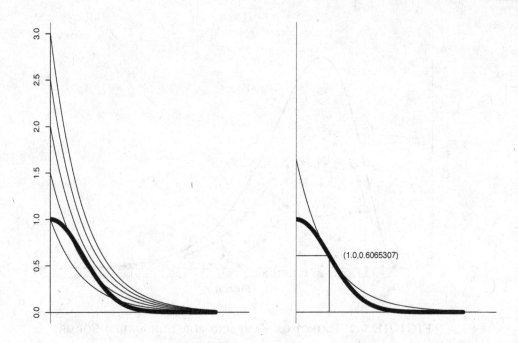

FIGURE 9.3: Rejection Sampling, Optimal Coverage

Instead of calculating an analytical solution to this integration problem, we decide to estimate the integral using the rejection method by specifying an exponential enveloping distribution: $g(\theta) = k \exp[-\theta]$. The exponential form is particularly useful here because it has the asymptotic property in the positive limit and because the constant term allows us to ensure coverage. The question remains, however, what value of k is best to ensure coverage but maximize algorithmic efficiency by minimizing rejected values.[1]

Panel 1 of Figure 9.3 shows prospective exponential functions along with the folded normal kernel (in bold). The exponential functions in the figure displayed are for $k = 1.0, 1.5, 2.0, 2.5, 3.0$ as indicated by the $\theta = 0$ point (i.e., when $k \exp[0] = k$). The first two ($k = 1.0, 1.5$) are unacceptable since they do not provide the required coverage. The others are acceptable but not optimal.

To find an optimal coverage exponential function, we would like one that has a single intersecting point with the folded normal (at its inflection point). This process turns out to be very straightforward, where we merely equate the two functions and solve

---

[1]It may seem like an inordinate amount of trouble to worry about the calculations required to reduce computing time since the computing time difference is not substantial in such a contrived example. Where it makes a significant difference is in cases we will deal with later, where such routines are called many times by higher-level Monte Carlo simulation routines and graphing functions. In these cases even small improvements in algorithmic efficiency at the lowest level of function calls can have an enormous impact.

for the k that gives the roots to the resulting quadratic equation:

$$g(\theta) = k \exp[-\theta] = \exp[-\theta^2/2] \equiv f(\theta)$$

$$k = \exp[\theta - \theta^2/2]$$

$$0 = \theta^2 - 2\theta + 2\log k.$$

Equate the two possible quadratic solutions, one of which must be 1:

$$1 = 2\log k$$

$$\therefore k = \exp(1/2).$$

Using this value, $k = 1.649$, in $g(\theta) = k \exp[-\theta]$, we get the enveloping function displayed in Panel 2 of Figure 9.3, where the single point of intersection occurs at $[1, 0.607]$.

Once we have the desired enveloping function, the next task is to sample candidate values from this function and accept or reject them based on their position relative to the folded normal. The procedure is very simple:

1. Draw n $\mathcal{EX}(1)$ random variables: $\theta_1, \theta_2, \ldots, \theta_n$.

2. For each θ_i, calculate the corresponding value of the majorizing function: $g(\theta_i) = (1.649) \exp[-\theta_i]$.

3. Draw a random uniform value for each θ_i over interval from 0 to $kg(\theta_i)$: u_{θ_i}.

4. Accept this draw as being from $f(\theta)$ if u_{θ_i} is less than $f(\theta_i)$.

5. The number of accepted values relative to the total number of draws is proportional to the ratio of the area under the target function to the area under the majorizing function. So:

$$I(\theta) = \int_0^\infty f(\theta)d\theta = k \times \frac{\text{number of accepted points}}{\text{total number of draws}}.$$

For only $n = 100$ draws, we get the surprisingly accurate result of 1.25308, which is very close to the true value of 1.25314. This process is illustrated in Figure 9.4, which shows how determining the single point of intersection (more than one point of intersection would imply a violation of the covering principle) substantially increases the efficiency of the simulation process: there are relatively few rejected points.

This section provides both the theory and the details of rejection sampling. This discussion is important in the overall context of where we are heading in that it is the first example where analytical work is replaced by computational effort. Virtually every technique discussed in subsequent chapters builds on this idea. For a further discussion and comparison of various numerical estimation techniques in Bayesian estimation, see Monahan and Genz (1996), and Stern (1997) for modern reviews.

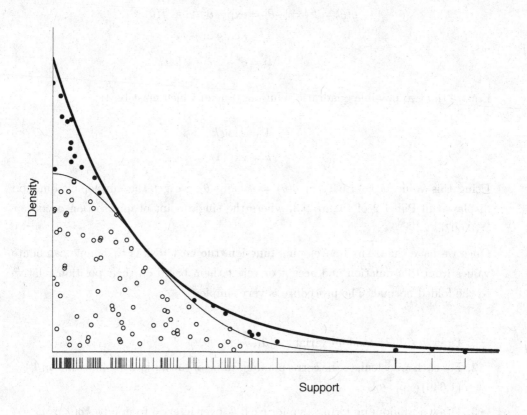

FIGURE 9.4: Folded Normal Rejection Sampling Example

9.4   Classical Numerical Integration

This section briefly reviews the traditional "quadrature" methods for numerically solving integral quantities. Long before widespread interest in Bayesian methods produced the need for a computational method of producing posterior quantities, applied mathematicians developed a wide range of tools for integrating functions that are difficult or impossible to solve analytically. It is well beyond the scope of this book to describe all of these in detail, and interested readers are directed to the excellent references of Davis and Rabinowitz (1984), Flournoy and Tsutakawa (1991), Haber (1970), Krommer and Ueberhuber (1998), and Stroud (1971), as well as the detailed practical guides on implementation of Espelid and Genz (1992), Golub and Van Loan (1996), and Press *et al.* (1986).

Difficult integrals tend to be high-dimensional where the analytic production of an associated antiderivative is unavailable, but for expositional simplicity, we will consider the general definite integral form of:

$$I(g) = \int_a^b g(\theta)d\theta = \sum_{i=1}^n \omega_i g(\theta_i). \tag{9.16}$$

Here the term ω_i is an integration weight and θ_i is an integration node in $[a:b]$. The general idea is to replace the integration process with a summation so that the weights provide the appropriate density contribution at the chosen node. Accuracy is controlled by the limit of the index i in the summation.

9.4.1 Newton-Cotes

Newton-Cotes methods involve fitting polynomial forms over equally spaced regions of the integral. The idea is to fit the shape of the integral piece-wise where the accuracy of the estimate is increased with progressively finer granularization of the support of the integral and increasing degree of the polynomial. Obviously there is a cost to increasing both of these criteria and the quality of the resulting approximation is partially a function of the skill of the researcher. While all of these tools share desirable asymptotic properties, results can differ substantially in finite simulations. There are three basic variants of Newton-Cotes: Riemann integrals, the trapezoid rule, and Simpson's rule.

9.4.1.1 Riemann Integrals

This method, sometimes called the *rectangle rule*, is the simplest, least accurate method for numerical integration, and is derived directly from basic calculus theory. The idea is to define n disjoint intervals of length $h = (b - a)/n$ so that: $\theta_0 = a$, $\theta_n = b$, and for $i = 2, \ldots, n - 1$, $\theta_i = a + ih$, to produce a histogram-like approximation. In the notation of (9.16), $\omega_i = h$ and $g(\theta_i) = g(a + ih)$. One of the reasons that this rule is so basic is that the weighting scheme is identical and equal to the spacing. The only wrinkle here is that one must select whether to employ a "left" or "right" Riemann integration:

$$\int_a^b g(\theta)d\theta = \sum_{i=1}^n \omega_i g(\theta_i) = \begin{cases} h \sum_{i=0}^{n-1} g(a + ih), & \text{left Riemann integral} \\ h \sum_{i=1}^n g(a + ih), & \text{right Riemann integral.} \end{cases} \tag{9.17}$$

Despite the obvious roughness of approximating a smooth curve with a series of rectangular bins, Riemann integrals can be extremely useful as a crude starting point since they are easily implemented. Furthermore, by calculating *both* left and right Riemann integrals, one can obtain an interval quantity *that is guaranteed to bound the true value* for monotonically increasing or decreasing functions.

9.4.1.2 Trapezoid Rule

The first and most obvious improvement to the rectangular approach of Riemann integrals is to substitute a trapezoid that uses the same binning strategy but connects the starting point of each bin to its ending point, thus creating a trapezoid. So unlike Riemann integrals, both points are on the curve and the top of the trapezoid gives a linear approximation to the curve over the bin width. The process takes a sum of trapezoids of the form: $(h/2)(g(\theta_i) + g(\theta_{i-1}))$, and simplifies through collection of terms to:

$$\int_a^b g(\theta)d\theta = \sum_{i=1}^n \omega_i g(\theta_i) = h \sum_{i=1}^{n-1} g(\theta_i) + \frac{h}{2}(g(a) + g(b)), \qquad (9.18)$$

so it remains that $\omega_i = h$, but we have an improved method for producing $g(\theta_i)$. Obviously the closer the curve is to linear within the bins, the better the trapezoid idea works. It is also clear that as the bin size gets progressively smaller, the approximation improves, albeit with a computational cost.

9.4.1.3   Simpson's Rule

The trapezoid rule does not work particularly well when the curve is very nonlinear within the bins. This occurs when the function has many local maxima and minima and when these points are sharply defined. An improved procedure, called Simpson's rule, replaces the linear approximation within each bin with a quadratic interpolating polynomial across the same space.

Suppose we wanted to fit a quadratic for the integral using only three points: a', b', and the midpoint $c' = (a' + b')/2$ over some bin within the desired range of integration: $[a:b]$. A quadratic form that is guaranteed to intersect with the integral at these three points is given by:

$$\int_{a'}^{b'} g(\theta)d\theta = \int_{a'}^{b'} \left[\frac{(\theta - c')(\theta - b')}{(a' - c')(a' - b')} g(a') + \right.$$
$$\left. \frac{(\theta - a')(\theta - b')}{(c' - a')(c' - b')} g(c') + \frac{(\theta - a')(\theta - c')}{(b' - a')(b' - c')} g(b') \right] d\theta. \qquad (9.19)$$

It is really easy to see that the quadratic form in (9.19) intersects with the integral at $\theta = a'$, $\theta = b'$, and $\theta = c'$ since in each case the other two terms will cancel out and the preceding fraction on the remaining simplifies to one (try it!). The value of this expression is that because c is chosen as the midpoint, (9.19) reduces to:

$$\int_{a'}^{b'} g(\theta)d\theta = \frac{b' - a'}{6} \left[g(a') + 4g(c') + g(b') \right]. \qquad (9.20)$$

If $g(\theta)$ is a quadratic form between a' and b', then Simpson's rule will fit exactly. In general though, it doesn't fit exactly but is much more accurate than the trapezoid rule and equally simple to code. Once again accuracy is a function of the tolerance for increased computational cost as the number of bins increases.

■ **Example 9.4:   Estimating Models with Mixture Priors.** Sometimes it is difficult to accurately describe prior information with a single PDF or PMF. In these cases it is sometimes possible to apply a mixture of distributions combined in such a way that they give a shape that is not possible otherwise. Fortunately, mixture priors possess desirable distributional and mathematical properties (O'Hagan 1994, Section 6.44-6.46). Furthermore, Dalal and Hall (1983) show that any distributional form can be approximated as accurately as we like with mixtures of conjugate distributions, although the number of components is not always easy to determine (see Carlin and Chib (1995) for a way to use Bayes Factors to test specifications with differing numbers of components). Gelman *et al.* (2003) also caution about some problems that can arise with *improper* and *non-normalized* mixture priors.

Generally though it is easy enough to normalize a mixture, say of several normals, or some other common distributional form. However, it is possible that one might want to create a more complicated mixture prior, and therefore have a need to numerically normalize it. Panel 1 of Figure 9.5 shows a prior over the support [0:12] created by:

$$p(\theta) = \begin{cases} (6-\theta)^2/200 + 0.011 & \text{for } \theta \in [0:6) \\ \mathcal{C}(11,2)/2 & \text{for } \theta \in [6:12] \end{cases} \tag{9.21}$$

where $\mathcal{C}(11,2)$ denotes a Cauchy distribution with location parameter 11 and scale parameter 2. This prior distribution is created in R by combining the functions:

```
f1 <- function(x)  ((6-x)^2)/200+0.011
f2 <- function(x)  dcauchy(x,11,2)/2
```

While it is possible to analytically calculate the normalizing constant for this prior, it is actually much easier to numerically estimate it according to the described algorithms.

Figure 9.5 displays the integration process of this unnormalized distribution by the three methods described. Only ten bins are displayed here for purely expository purposes and the integral quantities estimated, provided above the three panels, are calculated using 100 bins over this interval. While this is much more accurate, it does not provide a very revealing graph. In actual practice we might even be motivated to exceed 100 bins since the speed of calculation is very quick.

Two standard reference works on mixture distributions in general are: Titterington, Smith, and Makov (1985) and McLachlan and Basford (1988). A more Bayesian treatment of mixture models can be found in Hill and Tsai (1988), West and Turner (1994), Mengersen and Robert (1996), Ferguson (1983), and West (1993). Mixture distributions play an important role in Bayesian estimation because of their flexibility. At one extreme is the idea of nonparametric mixing functions such as the Dirichlet process prior (Dalal and Hall 1980; Ferguson 1973; Ferguson and Phadia 1979; Hjort 1996; Petrone and Raftery 1997), although these tools can be extremely complex and computationally difficult (Carlin and Louis, 2000 (51ff), Escobar 1994, Escobar and West 1995; Korwar and Hollander 1976). At the other extreme, normal mixture priors are a relatively straightforward extension (Andrews and Mallows 1974; Chib and Tiwari 1991; Geweke and Petrella 1998; Vidakovic 1999), and mixtures of *t*-distributions are only slightly more involved (Garthwaite and Dickey 1992).

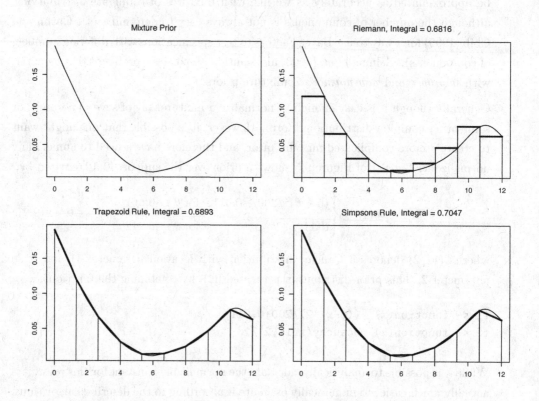

FIGURE 9.5: Illustration of Numerical Integration

9.5　Gaussian Quadrature

Gaussian quadrature replaces the analytical integration process with a very accurate form of (9.16) where the nodes of integration, the "abscissae," are no longer equally spaced. We fix an integration formula of the form:

$$\int_a^b \omega(\theta)g(\theta)d\theta \approx \sum_{i=1}^n m_i g(\theta_i) + \epsilon, \tag{9.22}$$

where: n is the order of the quadrature, m_i are nonnegative weights, θ_i are the abscissae constrained to be in the interval of integration, and ϵ is an error term. Now we are free to define a sequence of polynomials of degree k, so that they are orthogonal to the weight function on $[a:b]$:

$$\int_a^b m_i p_i(\theta) p_j(\theta) d\theta = 0, \qquad (9.23)$$

for any two polynomials $p_i(\theta)$ and $p_j(\theta)$, where $i \neq j$. Normally it would seem difficult to produce a useful set of orthogonal polynomials and the appropriate weights. Fortunately though these are already tabulated for a wide range of integral types, and the bulk of the work is reduced to fitting the quadrature formula that corresponds to the range of integration.

Before summarizing some of the standard Gaussian quadrature formulas, an artificial and commonly provided example is reviewed to illustrate the mechanics of the orthogonal polynomial manipulation. Suppose we want to integrate the standard normal PDF over the interval $[0:1]$. Obviously this quantity is easily obtained from Fisherian tables or from R: (pnorm(1) - 1/2), but using this objective allows us to check the accuracy of our procedure. The quadrature formula will be calculated with four simple orthogonal polynomials, and at only two nodes (for simplicity these will be the endpoints of the interval $[a = 0 : b = 1]$). The objective is then to find an explicit form for:

$$\int_a^b \omega(\theta) g(\theta) d\theta \approx \omega_1 g(\theta_1) + \omega_2 g(\theta_2). \qquad (9.24)$$

The four polynomials are defined over the interval of interest and evaluated with a single $m = (a + b)/2$:

▷ **Constant Function:**

$$g(\theta) = (\theta - m)^0 = \int_a^b (1) d\theta = b - a$$

$$\omega_1 (\theta_1 - m)^0 + \omega_2 (\theta_2 - m)^0 \equiv b - a.$$

▷ **Linear Function:**

$$g(\theta) = (\theta - m)^1 = \int_a^b (\theta - m) d\theta = \left. \frac{(\theta - m)^2}{2} \right|_{\theta=a}^{\theta=b} = 0$$

$$\omega_1 (\theta_1 - m)^1 + \omega_2 (\theta_2 - m)^1 \equiv 0.$$

▷ **Cubic Function:**

$$g(\theta) = (\theta - m)^3 = \int_a^b (\theta - m)^3 d\theta = \left. \frac{(\theta - m)^4}{4} \right|_{\theta=a}^{\theta=b} = 0$$

$$\omega_1 (\theta_1 - m)^3 + \omega_2 (\theta_2 - m)^3 \equiv 0.$$

▷ **Quartic Function:**

$$g(\theta) = (\theta - m)^4 = \int_a^b (\theta - m)^4 d\theta = \left. \frac{(\theta - m)^5}{5} \right|_{\theta=a}^{\theta=b} = \frac{(b - a)^3}{12}$$

$$\omega_1 (\theta_1 - m)^4 + \omega_2 (\theta_2 - m)^4 \equiv \frac{(b - a)^3}{12}.$$

We now have 4 equations and four unknowns allowing us to solve for these unknowns:

$$\theta_1 = \frac{a+b}{2} - \frac{b-a}{2\sqrt{3}}$$

$$\theta_2 = \frac{a+b}{2} + \frac{b-a}{2\sqrt{3}}$$

$$\omega_1 = \frac{b-a}{2}$$

$$\omega_2 = \frac{b-a}{2}.$$

Therefore:

$$\int_a^b \omega(\theta)g(\theta)d\theta \approx \frac{b-a}{2}g\left(\frac{a+b}{2} - \frac{b-a}{2\sqrt{3}}\right) + \frac{b-a}{2}g\left(\frac{a+b}{2} + \frac{b-a}{2\sqrt{3}}\right).$$

The integral of interest is then solved as:

$$\int_0^1 \frac{1}{\sqrt{2\pi}}e^{-\frac{1}{2}\theta^2}d\theta \approx \frac{1}{\sqrt{2\pi}}\left[\frac{1}{2}g\left(\frac{1}{2} - \frac{1}{2\sqrt{3}}\right) + \frac{1}{2}g\left(\frac{1}{2} + \frac{1}{2\sqrt{3}}\right)\right]$$

$$= 0.3412211,$$

whereas the actual answer produced by R is 0.3413447. This demonstrates that even the most simple of Gaussian quadrature schemes can produce a reasonably accurate answer.

Obviously we would not want to perform these steps for many points and more complicated integral forms. Fortunately, there are some standard forms that work very well in a wide variety of circumstances. The classic forms are expressed as follows, where the polynomials are given in the notation of Abramowitz and Stegun (1977, Chapter 22):

▷ **Legendre**

integral range: $[-1:1]$

weight function: $\omega(\theta) = 1$

$$m_j = -\frac{2}{(n+1)p_{n+1}(a_j)p'_n(a_j)}$$

polynomial: $P_n(\theta) = a_n^{-1}\sum_{m=0}^{n} c_m\theta^m.$

▷ **Jacobi**

integral range: $[-1:1]$

weight function: $\omega(\theta) = (1-\theta)^\alpha(1+\theta)^\beta$

$$m_j = \frac{-(2+2n+\alpha+\beta)\Gamma(1+n+\alpha)\Gamma(1+n+\beta)2^{(\alpha+\beta)}}{(1+n+\alpha+\beta)\Gamma(1+n+\alpha+\beta)(n+1)!p_{n+1}(a_j)p'_n(a_j)}$$

polynomial: $P_n^{(\alpha,\beta)}(\theta) = a_n^{-1}\sum_{m=0}^{n} c_m(\theta-1)^m.$

▷ **Hermite**

integral range: $[-\infty : \infty]$

weight function: $\omega(\theta) = \exp(-\theta^2)$

$$m_j = \frac{2^{n+1} n! \pi^{\frac{1}{2}}}{p_{n+1}(a_j) p_n'(a_j)}$$

polynomial: $H_n(\theta) = \sum_{m=0}^{n} c_m \theta^m.$

▷ **Laguerre**

integral range: $[0 : \infty]$

weight function: $\omega(\theta) = \exp(-\theta)$

$$m_j = \frac{n!^2}{p_{n+1}(a_j) p_n'(a_j)}$$

polynomial: $L_n(\theta) = a_n^{-1} \sum_{m=0}^{n} c_m \theta^m.$

Thisted (1988, Chapter 5) gives an extensive list of variations on these quadrature rules as well as a practical description of the theory behind orthogonal polynomials. Classic and useful references on Gaussian quadrature include Davis and Rabinowitz (1984), Galant (1969), Golub and Welsch (1969), Steen, Byrne, and Gelbard (1969), as well as Stroud and Secrest (1966). Acton (1996, Chapter 3) gives an excellent and informative analysis of what can go wrong in implementing these algorithms. There are also a number of well-known and commonly used extensions to the basic procedure. These computational tools include: Kronrod's (1965) idea of inserting an additional $n+1$ points interleaved around the original n with Legendre polynomials in order to improve the accuracy of the estimation, Patterson's (1968) similar notion of adding the maximally precision increasing p additional nodes to the original n, and the principle of concentrating more attention on problematic regions of the integral either automatically or manually (Naylor and Smith 1982, 1983; Skene 1983).

9.5.1   Redux

This subsection briefly introduced some of the very basic concepts in numerical integration. The purpose is to provide a frame of reference for the posterior estimation tools that we develop later in this chapter and in the subsequent discussion of MCMC. The presentation of the classical numerical integration procedures is far from complete as there are many variations on the ideas presented and most of these are well described in both the applied mathematics literature as well as in the cited statistics publications.

An area of importance that we have ignored here is the determination of error rates for the various techniques. In general these are extensively worked out, and detailed explanations can be found in Kennedy and Gentle (1980, Chapter 5), Pennington (1970, Chapter 7), Thisted (1988, Chapter 5), and many other references. A lot of work has been done to improve the error rates of various techniques such as the Romberg integration improvement to the trapezoid rule (Gray 1988; Pizer 1975), and composite rules (Conte and de Boor 1980, Section 7.4).

Finally, the importance of transitioning from these traditional approaches of calculating integral quantities to the stochastic simulation techniques described in later chapters cannot be overstated. While the tools just described are often sufficient, there are circumstances where they fail to produce a solution such as with high-dimensional problems, complex parameterizations, and tricky bounds (Stewart 1983). Concerning advances in the Bayesian computation of integrals, Adrian Smith observed rather presciently in 1984 that "...we are about to witness significant breakthroughs in this area, both in purpose-built Monte Carlo methodology for statistical problems and in adaptive quadrature rules exploiting statistically motivated kernels." He was certainly right, although the former has far outstripped the latter.

9.6 Importance Sampling and Sampling Importance Resampling

Importance sampling is a modification of rejection sampling that adjusts the standard procedure by placing greater emphasis on the "important" regions: those that have a higher density than others (Marshall 1956, Rubin 1988). Points that do not closely reflect the target distribution are no longer rejected (discarded), they are instead downweighted. Thus, importance sampling provides a much more efficient method by making use of every generated value, albeit with differing weights. It can also be made more flexible and adaptive (Oh and Berger 1992).

Importance sampling is motivated by the key definition underlying Monte Carlo simulation:

$$\hat{E}[h(\theta)] = \frac{1}{m} \sum_{i=1}^{m} h(\theta_i) \longrightarrow \int_{\Theta} h(\theta) f(\theta) d\theta = E[h(\theta)], \qquad (9.25)$$

where the arrow denotes almost sure convergence provided by the (strong) Law of Large Numbers (Geweke 1989). We introduce now an approximation function, $g(\theta)$, on the same support as $f(\theta)$ and hopefully resembling it over this support. Introducing the additional function allows us to express the integral in (9.25) as:

$$E[h(\theta)] = \int_{\Theta} h(\theta) f(\theta) \frac{g(\theta)}{g(\theta)} d\theta = \int_{\Theta} h(\theta) g(\theta) \left(\frac{f(\theta)}{g(\theta)} \right) d\theta. \qquad (9.26)$$

This suggests expressing the sum in (9.25) as

$$\hat{E}[h(\theta)] = \frac{1}{m} \sum_{i=1}^{m} h(\theta_i) \left(\frac{f(\theta_i)}{g(\theta_i)} \right), \tag{9.27}$$

the fundamental identity for importance sampling. As long as the ratio $f(\theta_i)/g(\theta_i)$ is easily calculated, this is a direct calculation. Now return to the familiar scenario where $f()$ is difficult to generate from, but $g()$ is not. There is no requirement in the calculation of $\hat{E}[h(\theta)]$ that the θ_i values be produced from the target distribution, so why not use the approximation distribution for this purpose? The nicest part of this approach is that we only require that the approximation function be defined on the same support and be close to the target distribution, so within such constraints, we are free to use any form that is convenient to draw from. Once the value is drawn, the calculation proceeds by simply plugging this value into three defined functions.

The first explicit application of importance sampling was due to Kloek and van Dijk (1978), and Zellner and Rossi (1984) give an early, influential application to qualitative responses. Geweke (1989) gives an extensive treatment of the theory along with proofs. For detailed summaries, see Gelman *et al.* (2003), Robert and Casella (2004, Chapter 1), and Tanner (1996, pp.54-59).

It is critical to start with an approximation distribution that is near the target distribution. Failing to do so usually requires excess computation since many points will not reflect high density areas of the target distribution and be given very low weights. One indication of a poor choice for the approximation distribution is a low average acceptance rate, and this suggests monitoring it if possible. There are a number of possible criteria for choosing the approximation distribution form, including: matching of moments or modes, Laplace approximations (normals), mixtures, and reparameterizations of convenient forms.

Since the generation of candidate values is determined by the approximation distribution and not the target distribution, these values are all characteristic of the target distribution but are on the same support. Summary statistics from the draws are adjusted by importance weights to correct for the generation process:

$$\omega_i = \frac{f(\theta_i)}{g(\theta_i)}, \tag{9.28}$$

where again $f()$ is the target distribution, $g()$ is the approximation distribution producing value θ_i. This ratio is small in cases where $f(\theta_i)$ is a low density point for the target distribution and a $g(\theta_i)$ is a high density point for the generating distribution, and vice-versa. For example, an estimate of the mean of the $f()$ function from m generated values is given by a restatement from above:

$$E[\theta] = \frac{\sum_{i=1}^{m} \theta_i \omega_i}{\sum_{i=1}^{m} \omega_i}, \tag{9.29}$$

and more generally for some function of the θs, $h(\theta)$, as defined above:

$$E[h(\theta)] = \frac{\sum_{i=1}^{m} h(\theta_i) \omega_i}{\sum_{i=1}^{m} \omega_i}. \tag{9.30}$$

Actually there is an important caveat here; this empirical quantity *converges* to the expected value above, and the rate of convergence depends on the quality of the approximation distribution (Rubin 1987a, 1988; Tanner 1996). Liu (2001, p.35) provides a handy way to measure how much the approximation distribution differs from target distribution. Define $\text{Var}_g[\omega(\theta)]$ as the variance of a sample of m importance weights over the distribution defined by $g()$, then the *effective sample size* (ESS) is defined by:

$$ESS_m = \frac{m}{1 + \text{Var}_g[\omega(\theta)]}, \tag{9.31}$$

where a large variance leads to low efficiency relative to sample size.

For calculating credible intervals and various quantiles such as tail probabilities, it is generally easier to use the sorted accepted values. So, for instance, the median statistic is produced by the sorted vector value such that the sum of the normalized sampling weights of lower values is equal to 0.5. As with rejection sampling and other similar procedures, we can easily calculate the variance of the statistics produced from importance sampling. For a statistic, $h(\theta)$, the variance from simulating is given by:

$$\text{Var}[h(\theta)] = \frac{\sum_{i=1}^m ((h(\theta_i) - E[h(\theta_i)])\omega_i)^2}{\sum_{i=1}^m \omega_i}. \tag{9.32}$$

In fact, without this adjustment, the naïve importance sampling variance taken directly from the empirical draws can sometimes be "too good": lower than with draws taken from the actual distribution (so-called super-efficiency; see Liu [1996a] for a detailed discussion).

■ **Example 9.5:   Sampling-Importance Resampling for Improving Variance Estimation.** This example, derived from a more complex model in Gill and King (2004), presents a data analysis where importance sampling is used to provide a more accurate view of the posterior variance of coefficient estimates from a simple logit model (see also the overview in Li [2004]). The application is to public policy data on poverty and its associated potential explanations, measured by state at the county-level (census "FIPS"). The data consist of 1989 county level economic and demographic variables for the 196 nonmetropolitan counties in Texas out of all 2276 nonmetropolitan U.S. counties ("ERS Typology," http://www.census.gov/). The dichotomous outcome variable indicates whether 20% or more of the county's residents live in poverty.

We place diffuse (uniform) priors on each of these explanatory variables. Since the outcome variable (Y) is dichotomous, we analyze these data using a generalized linear model with a logit link function: $p(Y_i = 1|X_i) = [1 + \exp(X_i\beta)]^{-1}$, where X_i is a set of explanatory variables for case i.

Table 9.3 gives the logit model results before and after importance sampling with a multivariate-t approximating distribution. In general there is no substantial difference, but `Transfer` becomes noticeably more reliable statistically. What is happening here is that the explanatory variable `Federal` is a poor contribution to the model and actually makes other explanatory variables less reliable than when it is excluded.

TABLE 9.3: LOGIT REGRESSION MODEL: POVERTY IN
TEXAS

	Standard Results		Importance Resampling	
Parameter	Coefficient	Std. Error	Coefficient	Std. Error
Black	15.91	3.70	15.99	3.83
Hispanic	8.66	1.48	8.46	1.64
Govt	1.16	0.78	1.18	0.74
Service	0.17	0.62	0.19	0.56
Federal	-5.78	16.20	-3.41	17.19
Transfer	1.29	0.71	1.25	0.63
Population	-0.39	0.22	-0.38	0.21
Intercept	-0.47	1.83	-0.51	1.68

The dichotomous outcome variable flags whether 20+% of the county's residents live below the poverty line. The explanatory variables selected are:

Govt	a dichotomous variable indicating whether government activities contributed a weighted annual average of 25% or more labor and proprietor income over the previous 3 years
Service	a dichotomous variable indicating whether service activities contributed a weighted annual average of 50% or more labor and proprietor income over the 3 previous years
Federal	a dichotomous variable indicating whether federally owned lands make up 30% or more of a county's land area
Transfer	a dichotomous variable indicating whether income from transfer payments (federal, state, and local) contributed a weighted annual average of 25% or more of total personal income over the past 3 years
Population	the log of the county population total for 1989
Black	the proportion of Black residents in the county
Hispanic	the proportion of Hispanic residents in the county

To demonstrate this near-singularity, Figure 9.6 provides a matrix of the bivariate profile contour plots for each pair of coefficients given at contours of $0.05, 0.15, \ldots, 0.95$, where the 0.05 contour line bounds 0.95 of the data, holding constant all other parameters at their mode. Each posterior surface is concave at the global maxima, but the curvature for Federal is very slight. This almost-flatness produces a near-ridge in the contours for each other explanatory variable paired with `Federal` where the ridge is gently sloping around maximum value in each profile plot.

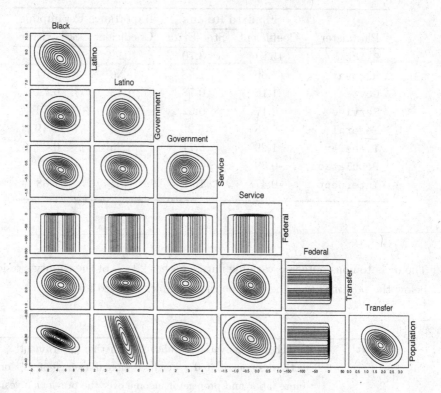

FIGURE 9.6: Contourplot Matrix, Rural Poverty in Texas

Figure 9.6 indicates that the distribution of the coefficient on `Federal` is quite asymmetric, and shows that the probability density drops as we come away from the near-ridge. This case illustrates how importance sampling can be used as a means of improving variance estimates even when less satisfactory results already exist.

■ **Example 9.6:   Monte Carlo Marginalization, European Migration Data.** We return here to the Pareto model of European migration counts from Section 9.2. Recall that it was prohibitively time-consuming to produce an analytical posterior marginal for β free from the other model parameter α: $\pi(\beta|\mathbf{x})$. The joint posterior distribution was given by:

$$\pi(\alpha, \beta|\mathbf{x}) \propto \alpha^{n+C}(B\beta)^{-(n\alpha+A\alpha+1)}\exp\left[-\alpha(D+E)\right]$$

$$\text{with } E = \sum_{i=1}^{n}\log(x_i) + A\log[B] - (n+A)\log[\min(B,\mathbf{x})]. \tag{9.33}$$

We also know that the posterior distribution of α is $\mathcal{G}(n+C, D+E)$. These two posteriors are enough to marginalize using a variant of importance sampling that will allow us to obtain the unconditional marginal posterior distribution for β. This process

(Robert and Casella [2004], Gelman *et al.* [2003]) uses $\pi(\alpha|\mathbf{x})$ as the approximation distribution since it is fully known and therefore easy to sample from. Considering the ratio of the known joint posterior to the known marginal as the importance ratio provides:

$$\pi(\beta|\mathbf{x}) = \int_\alpha \pi(\alpha,\beta|\mathbf{x})d\alpha$$

$$= \int_\alpha \frac{\pi(\alpha,\beta|\mathbf{x})}{\pi(\alpha|\mathbf{x})}\pi(\alpha|\mathbf{x})d\alpha$$

$$\approx E_{\alpha|\mathbf{x}}\left[\frac{\pi(\alpha,\beta|\mathbf{x})}{\pi(\alpha|\mathbf{x})}\right]. \tag{9.34}$$

This is implemented by the following steps:

▷ Grid the support of β into I discrete values: $[\beta_1, \beta_2, \ldots, \beta_I]$.

▷ For each value of β_i:

  1. Draw J values of α_j from $\pi(\alpha|\mathbf{x})$;
  2. Plug each of these α_j into $\frac{\pi(\alpha_j,\beta_i|\mathbf{x})}{\pi(\alpha_j|\mathbf{x})}$, and take the mean;
  3. So the result is $\pi(\beta_i|\mathbf{x})$, the posterior probability value for β at grid-point i.

The "art" involved in this process is really the determination of the grid-points for the range of β and the number of α values to sample at each iteration. Obviously there is a trade-off between computer time and accuracy, but with each iteration of new hardware, the preference for more samples should increase. The choice here is to grid the range of β into 1,000 subunits and draw 100 random values of α on each iteration. This produces an empirical marginal posterior for β. We can summarize this distribution by posterior quantiles or just calculate the weighted empirical mean and standard error: $\bar{\beta}_{\text{posterior}} = 141.96$, and $SE(\beta)_{\text{posterior}} = 62.61$.

9.6.1 Importance Sampling for Producing HPD Regions

Returning to a more appropriately Bayesian purpose, this section shows how importance sampling can be useful in producing highest posterior density intervals from simulated posterior values. Recall from Section 2.3.2.2 that HPD regions provide the region of the posterior support that meets two criteria for some chosen value α:

▷ the posterior density value for every point in the HPD region exceeds that for every point outside of the HPD region,

▷ the interval is the shortest length along the support that provides $(1 - \alpha)$ proportion of the total posterior density.

Notably this definition means that the HPD region does not need to be contiguous like the credible interval. Unfortunately, the HPD region is sometimes much more difficult to calculate since it evaluates the shape of the posterior rather than simply marching out some designated length from the mean.

Suppose that we have generated an importance sample for the posterior of interest, $\theta_1, \theta_2, \ldots, \theta_m$ from $\pi(\theta|\mathbf{X})$, using the Monte Carlo tools in this chapter or the MCMC tools in the next chapters. Denote the generating distribution as $g(\theta)$. The trick is to think in CDF terms using the empirically collected values, so that the value of interest for a specified α-level is:

$$\alpha = \int_{-\infty}^{\theta_{[\alpha]}} \pi(\theta|\mathbf{X})d\theta \approx \frac{\sum_{i=1}^{m} I_{(\theta_i \leq \theta_{[\alpha]})} \pi(\theta_i|\mathbf{X})/g(\theta_i)}{\sum_{i=1}^{m} \pi(\theta_i|\mathbf{X})/g(\theta_i)} \tag{9.35}$$

where $I_{(\theta_i \leq \theta_{[\alpha]})}$ is an indicator function equal to one if the condition is met and zero otherwise. Therefore everything in this fraction except $\sum_{i=1}^{m} I_{(\theta_i \leq \theta_{[\alpha]})}$ is really an importance weight, w_i. Intuitively, this means that we can sort the simulated values and approximate the CDF value at some arbitrarily chosen $\theta_{[\alpha]}$ with the sum $\sum_{j=1}^{i} w_{[j]}$ for $\theta_{[i]} \leq \theta_{[\alpha]} < \theta_{[i+1]}$. The precision of this approximation is simply a function of how close the consecutive sorted values are to each other (meaning really how many values were drawn). Now stated in steps, the process is:

1. generate $\theta_i \sim g(\theta)$, $i = 1, \ldots, m$,

2. sort these values, giving $\theta_{[i]}$, $\ni \theta_{[i-1]} < \theta_{[i]}$, $i = 2, \ldots, m$,

3. calculate the associated importance weights for the sorted values:

$$w_{[i]} = \frac{\pi(\theta_{[i]}|\mathbf{X})/g(\theta_{[i]})}{\sum_{i=1}^{m} \pi(\theta_{[i]}|\mathbf{X})/g(\theta_{[i]})}$$

4. for $i > 1$ and $i < m$ (since the empirical CDF is assigned to be zero and one at the endpoints),

$$\alpha_{[i]} = \sum_{j=1}^{i} w_{[j]}.$$

Once we have this last vector, the corresponding value, $\theta_{[\alpha]}$, for any chosen CDF level, $\alpha_{[i]}$ is simply selected. If we want a 0.95 HPD region for a unimodal posterior, then select the sorted $\theta_{[i]}$ values corresponding to the $(0.05/2)m$ and $(1 - 0.05/2)m$ vector locations (one reason that it is often convenient to generate samples in multiples of 100s or 1,000s). Chen and Shao (1999) explore in detail the properties of importance sampling estimates of cumulative posterior functions.

9.7   Mode Finding and the EM Algorithm

Expectation-Maximization (EM) is a very flexible and popular technique for so-called incomplete data problems. This description arises because the EM algorithm is designed to "fill in" unknown information for a specified model where the notion of what is "missing" is very general. This missing information can be unknown parameters, or missing data, or both. In fact, the missing data can even be hypothetical results, but one must be careful in interpreting these results.

The idea of the EM algorithm is marvelously and beguilingly simple. In the first step, temporary data that represent a reasonable guess are assigned to replace the missing data. Then, the parameter estimation proceeds as if we now have a complete data problem (complete in the sense that observed and missing data are now both "available" in the analysis). Once this produces a solution for the parameter estimates, then we use these to update the assignment of the temporary data values with better guesses, and again perform a full model estimation process. This two-step process is repeated as often as required until the difference in the parameter updates becomes arbitrarily close to zero and we therefore have convergence.

The EM algorithm does not just represent a practical convenience. There is strong theoretical justification for this method in that the iterated EM process gives a series of parameter estimates that are monotonically increasing on the likelihood metric and are guaranteed to converge to a maximum point under very general and nonrestrictive regularity conditions (Boyles 1983; Dempster, Laird, and Rubin 1977; Wu 1983). In the case of likelihood functions from exponential family forms (i.e., generalized linear modeling applications) there are no substantial complexities (Nelder 1977, p.23), but there do exist other cases where the first step of determining candidate values can be complicated in large dimensions or with complicated likelihood functions. Worse yet, all non-exponential family applications of the EM algorithm require the use of an asymptotic approximation to an exponential family form (Rubin 1991).

The EM algorithm was formalized by Dempster, Laird, and Rubin in their seminal 1977 article (henceforth referred to as DLR-77, see also the accompanying discussion such as Beale [1977]). However, this was not a completely new idea and actually the *ad hoc* process of iteratively filling in tentative data and estimating parameters had been developed by earlier authors. Several authors refer to a piece by McKendrick (1926) as the first known iterative EM-like technique, and Newcomb used a similar approach as early as 1886. Healy and Westmacott (1956) propose an iterative technique for least squares implemented on newly developed "automatic computers." Hartley's (1958) "simple iterative method" fills in missing tabular values due to censoring and truncation with an iterative process that is essentially a less general form of the EM algorithm restricted to discrete distributions only because his fill-in step is: "Using $_1\theta$ compute 'improved' estimates, $_1n_i$ of the 'missing frequencies' by proportional allocation based on the Poisson distribution..." (p.176). Zangwill's textbook (1969) provides some of the key proofs, including a critical one concerning conditions for monotonic convergence (Section 4.5) as well as a description of the cyclic coordinate ascent (p.111) that is very close to the general EM procedure. Haberman (1974a, Chapter 3) discusses another nearly concurrent technique, iterative proportional fitting, in the context of count data. Other notable works that develop precursors to the fully articulated EM algorithm of DLR-77 include Baum and Petrie (1966), Baum and Eagon (1967) and Baum *et al.* (1970) for Markov models, Beale and Little (1975) applied to multivariate normals, Hartley and Hocking (1971) for a special missing data case, Orchard and Woodbury (1972) for the "missing information principle," and Sundberg (1974) for the exponential family. Specialized works that predate DLR-77 are Blight (1970), Buck (1960), Carter and Myres (1973), Chen and Fienberg (1974), and Turnbull (1976).

9.7.1   Deriving the EM Algorithm

Suppose that we are interested in obtaining the posterior distribution of an unknown k-dimensional $\boldsymbol{\theta}$ coefficient vector, given an outcome variable vector $\mathbf{y}$, and an observed $\mathbf{X}$ matrix of explanatory data values assumed to be distributed iid according to $f(\mathbf{X}|\boldsymbol{\theta})$. In a complete data situation, where all of $\mathbf{X}$ is observed, we would normally proceed with standard maximum likelihood estimation by finding the maximum of the function $\ell(\boldsymbol{\theta}|\mathbf{X})$, and solving for $\hat{\boldsymbol{\theta}}$ as described in Appendix A. Analogously in a Bayesian process, we would produce a posterior distribution from $\pi(\boldsymbol{\theta}|\mathbf{X}) \propto p(\boldsymbol{\theta})p(\mathbf{X}|\boldsymbol{\theta})$. Neither of these approaches are directly possible if there happen to be missing data in $\mathbf{X}$ unless the data are considered "missing completely at random," meaning that they are safely ignorable from a parameter estimation perspective and can be removed from the analysis (Little and Rubin 2002). However, neither of these approaches are directly possible if there happen to be missing data in the $\mathbf{X}$ matrix, since deleting cases produces biased results (Schafer 1997).

The algorithm starts with first segmenting the $\mathbf{X}$-matrix into two constituent parts: $\mathbf{X} = [\mathbf{X}_{obs}, \mathbf{X}_{mis}]$, and restating the distribution function as:

$$f(\mathbf{X}|\boldsymbol{\theta}) = f(\mathbf{X}_{obs}, \mathbf{X}_{mis}|\boldsymbol{\theta}) = f(\mathbf{X}_{obs}|\boldsymbol{\theta})f(\mathbf{X}_{mis}|\mathbf{X}_{obs}, \boldsymbol{\theta}). \qquad (9.36)$$

Now similarly segment the log likelihood function into two distinct components:

$$\begin{aligned}
\ell(\boldsymbol{\theta}|\mathbf{X}) &= \ell(\boldsymbol{\theta}|\mathbf{X}_{obs}, \mathbf{X}_{mis}) \\
&= \ell(\boldsymbol{\theta}|\mathbf{X}_{obs}) + \log f(\mathbf{X}_{mis}|\mathbf{X}_{obs}, \boldsymbol{\theta}).
\end{aligned} \qquad (9.37)$$

Rearrange this form to create a statement with both unknowns collected on the right-hand side:

$$\ell(\boldsymbol{\theta}|\mathbf{X}_{obs}) = \ell(\boldsymbol{\theta}|\mathbf{X}_{obs}, \mathbf{X}_{mis}) - \log f(\mathbf{X}_{mis}|\mathbf{X}_{obs}, \boldsymbol{\theta}). \qquad (9.38)$$

Since we have an expression for the distribution of these missing values, $f(\mathbf{X}_{mis}|\mathbf{X}_{obs}, \boldsymbol{\theta})$, conditional on the quantities $\mathbf{X}_{obs}$ and $\boldsymbol{\theta}$, we can average over this uncertainty by taking expectations with respect to $\mathbf{X}_{mis}|\mathbf{X}_{obs}, \boldsymbol{\theta}$ on both sides:

$$\begin{aligned}
\int \ell(\boldsymbol{\theta}&|\mathbf{X}_{obs})f(\mathbf{X}_{mis}|\mathbf{X}_{obs}, \boldsymbol{\theta})d\mathbf{X}_{mis} \\
&= \int \ell(\boldsymbol{\theta}|\mathbf{X}_{obs}, \mathbf{X}_{mis})f(\mathbf{X}_{mis}|\mathbf{X}_{obs}, \boldsymbol{\theta})d\mathbf{X}_{mis} \\
&\quad - \int \log f(\mathbf{X}_{mis}|\mathbf{X}_{obs}, \boldsymbol{\theta})f(\mathbf{X}_{mis}|\mathbf{X}_{obs}, \boldsymbol{\theta})d\mathbf{X}_{mis}.
\end{aligned} \qquad (9.39)$$

The left-hand side here simplifies back to $\ell(\boldsymbol{\theta}|\mathbf{X}_{obs})$ because the integral ends up operating over just the isolated complete PDF for $\mathbf{X}_{mis}$:

$$\int \ell(\boldsymbol{\theta}|\mathbf{X}_{obs})f(\mathbf{X}_{mis}|\mathbf{X}_{obs},\boldsymbol{\theta})d\mathbf{X}_{mis}$$

$$= \ell(\boldsymbol{\theta}|\mathbf{X}_{obs})\int f(\mathbf{X}_{mis}|\mathbf{X}_{obs},\boldsymbol{\theta})d\mathbf{X}_{mis}$$

$$= \ell(\boldsymbol{\theta}|\mathbf{X}_{obs}). \tag{9.40}$$

Now we have an expression based only on the observed data that relates the obtainable likelihood to two quantities that can be manipulated.

$$\ell(\boldsymbol{\theta}|\mathbf{X}_{obs}) = \int \ell(\boldsymbol{\theta}|\mathbf{X}_{obs},\mathbf{X}_{mis})f(\mathbf{X}_{mis}|\mathbf{X}_{obs},\boldsymbol{\theta})d\mathbf{X}_{mis}$$

$$- \int \log f(\mathbf{X}_{mis}|\mathbf{X}_{obs},\boldsymbol{\theta})f(\mathbf{X}_{mis}|\mathbf{X}_{obs},\boldsymbol{\theta})d\mathbf{X}_{mis}. \tag{9.41}$$

But it is still not enough because we do not have values for $\boldsymbol{\theta}$ on the right-hand side of the conditionals. So start with an obviously wrong, but hopefully reasonable, value for the vector. Plug this starting value into the conditional statement for the distribution of the missing data, but not into the likelihood component since that is the object of our procedure. This produces the following expression with subfunctions labeled according to the original notation in Dempster, Laird, and Rubin (1977, p.6):

$$\ell(\boldsymbol{\theta}|\mathbf{X}_{obs}) = \underbrace{\int \ell(\boldsymbol{\theta}|\mathbf{X}_{obs},\mathbf{X}_{mis})f(\mathbf{X}_{mis}|\mathbf{X}_{obs},\boldsymbol{\theta}^{(0)})d\mathbf{X}_{mis}}_{Q(\boldsymbol{\theta}|\boldsymbol{\theta}^{(0)})}$$

$$- \underbrace{\int \log f(\mathbf{X}_{mis}|\mathbf{X}_{obs},\boldsymbol{\theta})f(\mathbf{X}_{mis}|\mathbf{X}_{obs},\boldsymbol{\theta}^{(0)})d\mathbf{X}_{mis}}_{H(\boldsymbol{\theta}|\boldsymbol{\theta}^{(0)})}. \tag{9.42}$$

The key is to evaluate the Q and H functions in (9.42) separately.

We focus exclusively on $Q(\boldsymbol{\theta}|\boldsymbol{\theta}^{(0)})$ for the moment. Treat $\boldsymbol{\theta}^{(0)}$ as known and perform the integration to obtain the conditional expectation of this missing data, given the observed data and the temporarily acceptable value for the $\boldsymbol{\theta}$ vector to produce:

$$Q(\boldsymbol{\theta}|\boldsymbol{\theta}^{(0)}) = \ell^*(\boldsymbol{\theta}|\mathbf{X}_{obs},\boldsymbol{\theta}^{(0)}). \tag{9.43}$$

This is just the expected log likelihood as if $\boldsymbol{\theta}$ were $\boldsymbol{\theta}^{(0)}$ produced by substituting expectations for the missing data. Call this part of the process the "E-Step" since we are taking an expectation.

Given that we now have a full-information log likelihood function for the unknown $\boldsymbol{\theta}$ vector, we can apply maximum likelihood estimation to produce a new estimate for $\boldsymbol{\theta}$. This is the "M-Step," and it constitutes nothing more than a straightforward application of MLE where the result, $Q(\boldsymbol{\theta}^{(1)}|\boldsymbol{\theta}^{(0)})$, is the most likely value of $\boldsymbol{\theta}$, given the observed data and the E-Step expectation over the unobserved data (which was itself conditioned on the initial values for $\boldsymbol{\theta}$). By definition, the MLE process means that:

$$Q(\boldsymbol{\theta}^{(1)}|\boldsymbol{\theta}^{(0)}) \geq Q(\boldsymbol{\theta}|\boldsymbol{\theta}^{(0)}) \tag{9.44}$$

for all other values of $\boldsymbol{\theta}$ not equal to $\boldsymbol{\theta}^{(1)}$, where equality occurs only at the maximum. Note that this is where the unimodal assumption plays a critical role. So on *each* iteration we are necessarily increasing the value of the Q function, conditional on the current value, $\boldsymbol{\theta}^{(0)}$.

Now let us worry about the $H(\boldsymbol{\theta}|\boldsymbol{\theta}^{(0)})$ component. In full form it is:

$$H(\boldsymbol{\theta}|\boldsymbol{\theta}^{(0)}) = E_{\mathbf{X}_{mis}|\boldsymbol{\theta}^{(0)}}[\log f(\mathbf{X}_{mis}|\mathbf{X}_{obs}, \boldsymbol{\theta}^{(0)})]. \tag{9.45}$$

The E-Step of going from $\boldsymbol{\theta}^{(0)}$ to estimated value $\boldsymbol{\theta}$ is characterized by expectation of the function:

$$g(\boldsymbol{\theta}^{(0)}) = \log f(\mathbf{X}_{mis}|\mathbf{X}_{obs}, \boldsymbol{\theta}^{(0)}), \tag{9.46}$$

where the expectation is concave to the x-axis and therefore has a single maximum value. By the definition of the function $g(\boldsymbol{\theta}^{(0)})$:

$$H(\boldsymbol{\theta}|\boldsymbol{\theta}^{(0)}) = E[g(\boldsymbol{\theta}^{(0)})] \quad \text{and} \quad H(\boldsymbol{\theta}^{(0)}|\boldsymbol{\theta}^{(0)}) = g(E[\boldsymbol{\theta}^{(0)}]). \tag{9.47}$$

Thus, we will subtract a smaller value of $H(\boldsymbol{\theta}|\boldsymbol{\theta}^{(0)})$ in the E-Step by moving from $\boldsymbol{\theta}^{(0)}$ to $\boldsymbol{\theta}^{(1)}$, by Jensen's inequality for *concave* functions: $E[g(\boldsymbol{\theta})] \leq g(E[\boldsymbol{\theta}])$ (Casella and Berger 2002, p.190; Shao 2005; Stuart and Ord 1994, p.67). Therefore, we have a proof that we can ignore the $H(\boldsymbol{\theta}|\boldsymbol{\theta}^{(0)})$ part since it gets smaller automatically on each iteration of the algorithm.

Actually this is more general since we now have proof that $\ell(\boldsymbol{\theta}^{(1)}|\mathbf{X}_{obs}) \geq \ell(\boldsymbol{\theta}^{(0)}|\mathbf{X}_{obs})$ for any chosen starting value of $\boldsymbol{\theta}$. This property, the "monotonicity of EM" property, holds true because:

$$Q(\boldsymbol{\theta}^{(1)}|\boldsymbol{\theta}^{(0)}) \geq Q(\boldsymbol{\theta}|\boldsymbol{\theta}^{(0)}), \tag{9.48}$$

and simultaneously

$$H(\boldsymbol{\theta}|\boldsymbol{\theta}^{(0)}) \geq H(\boldsymbol{\theta}^{(1)}|\boldsymbol{\theta}^{(0)}). \tag{9.49}$$

Now comes the best part. Since we chose $\boldsymbol{\theta}^{(0)}$ completely arbitrarily, then we can repeat the process assured of getting an equally desirable or better result.

So to summarize the steps (rather than the justification):

1. Start with...
   the data: $\mathbf{X} = [\mathbf{X}_{mis}, \mathbf{X}_{obs}]$,
   a likelihood function: $f(\mathbf{X}|\boldsymbol{\theta})$,
   and arbitrary values for the vector: $\boldsymbol{\theta}^{(k)}$.

2. **[E-Step:]**...compute:

$$Q(\boldsymbol{\theta}^{(k+1)}|\boldsymbol{\theta}^{(k)}) = \int \ell(\boldsymbol{\theta}|\mathbf{X}_{obs}, \mathbf{X}_{mis}) f(\mathbf{X}_{mis}|\mathbf{X}_{obs}, \boldsymbol{\theta}^{(k)}) d\mathbf{X}_{mis}$$

3. [**M-Step:**]... choose the value for $\boldsymbol{\theta}$ that maximizes

$$Q(\boldsymbol{\theta}^{(k+1)}|\boldsymbol{\theta}^{(k)})$$

4. repeat until the difference between $\boldsymbol{\theta}^{(k+1)}$ and $\boldsymbol{\theta}^{(k)}$ is arbitrarily small.

If in the M-Step we merely increase the Q function rather than maximize it because maximization is too difficult, then this variant is called the *generalized EM* (GEM) procedure (DLR-77, 6-7). However, this process is quite different in that now an iterative scheme such as Newton-Raphson needs to be applied so that the final result of the M-Step is a conditional maxima that can be used in the next stage of algorithm. Typically some constraints are required on this process and these can make the GEM somewhat more complex than the standard EM algorithm (McLachlan and Krishnan 1997).

9.7.2 Convergence of the EM Algorithm

Unlike most numerical maximization techniques such as Newton-Raphson, steepest descent, or Fletcher-Powell (see Thisted 1988, Chapter 4), the EM algorithm does not require the calculation of first or second derivatives. This can be a distinct advantage for complicated parametric forms. The EM algorithm is also fast relative to many competitors. The convergence rate is geometric (Laird 1978, Sundberg 1976, as well as DLR-77),[2] and there are a number of tricks for speeding up convergence further (e.g., Jamshidian and Jennrich 1993; Louis 1982, Section 5; Meilijson 1989). This rate of convergence is, however, proportional to $1 - m$, where m is the proportion of missing information (Rubin 1991, p.244). One common criticism is that convergence can be particularly slow in the absence of well-defined modes (Jamshidian and Jennrich 1993; Laird, Lange, and Stram 1987; Rubin 1991).

Wu (1983, p.98) proved that the limit points of an EM algorithm are stationary points given a continuity condition that is "very weak and should be satisfied in most practical situations," a result supported by Baum *et al.* (1970), Haberman (1977), and Boyles (1983). Unfortunately stationary points (defined as a point $\tilde{x}$ so that $\frac{d}{dx}f(\tilde{x}) = 0$ [Gill, Murray, and Wright 1981, p.62]) are not just maxima, they can be minima or inflection points as well. Luckily exponential family forms and a wide variety of the other parametric specifications used in the social sciences meet a stronger condition that assures convergence to a local or global maxima (see Wu 1983, Theorem 3). If the likelihood function is unimodal with only one stationary point, then the EM algorithm is guaranteed to find this stationary point (DLR-77, Theorem 1; Wu 1983, Property vii), and with reasonable sample size, this is rarely a difficult condition for typical parametric forms.

It should be pointed out that the fact that the EM algorithm finds local as well as

---

[2]Geometric convergence means that a series converges to its limiting value on the order of r^n as a geometric sequence defined by: $s_n = \alpha + \circ(r^n)$ as $n \to \infty$ for $|r| < 1$, and finite constant α. A process that possesses a geometric convergence rate reaches its limit approximately as fast as the series r^i, $i = 0 \to \infty$ ($|r| < 1$), reaches $1/(1-r)$.

global maxima is not necessarily a liability. In fact, there are instances when we would like to know where all maxima are located and the EM algorithm started from different points in the sample space is an ideal mode-finder. This characteristic will prove very useful later when we have procedures that describe posterior distributions by wandering around in them (i.e., MCMC), but we require some sense of the right "neighborhood" to sample. Obviously knowledge of the mode or modes of a posterior distribution will greatly facilitate such a process. However, the use of the EM algorithm as mode finding precursor to other techniques does highlight a deficiency of the procedure: it does not naturally produce a measure of uncertainty describing the curvature around the mode. However, this weakness is easily solved by adjunct procedures.

It is not necessary to furnish further proof of the convergence of the EM algorithm as the basics have just been laid out here and complete details are furnished in DLR-77, Chapter 3 of McLachlan and Krishnan (1997), and particularly carefully by Wu (1983). Boyles (1983) gives convergence results specifically for the generalized EM algorithm. The wealth of published applications of the EM algorithm in statistics and in social science fields is a testament to the robustness and durability of the algorithm across many specifications and data types.

■ **Example 9.7:   EM Applied to a Multinomial Model of U.S. Energy Consumption** This example presents a non-Bayesian model in which the consumption of energy in the United States is treated as a multinomial count by quadrillion British Thermal Units (BTUs) across different sources. The exposition here is distinct but in the spirit of the first example of DLR-77 (resummarized by other authors many, many times) in which they model genetic frequencies. A multinomial PMF with relatively few categories provides a particularly illuminating demonstration of the EM algorithm because the theoretical steps can be presented without extraneous technical material.

Consumption of energy inputs is counted annually by rounded quadrillion BTUs according to source type: fossil fuel, coal, hydroelectric, and nuclear, denoted in this order by the vector: $\mathbf{Y} = [y_1, y_2, y_3, y_4]$. These counts are assumed to arise from a multinomial$(n, p_1(\theta))$ form with cell probabilities:

$$p_1(\theta) = \left[\frac{2}{9} + \frac{2}{9}\theta, \frac{3}{9} - \frac{2}{9}\theta, \frac{2}{9}, \frac{2}{9} \right], \qquad \text{for:}\ \ 0 \leq \theta \leq 1,$$

where these cell probabilities are driven by substantive theoretical work on energy policy and politics in the United States (Bohi, Toman, and Wells 1996; Davis 1992; Sharp, Register, and Grimes 1999), and are currently a very topical issue in public policy and politics.

Therefore the specified multinomial PMF is:

$$f(y|n, \theta) = \frac{n!}{y_1! y_2! y_3! y_4!} \left(\frac{2}{9} + \frac{2}{9}\theta \right)^{y_1} \left(\frac{3}{9} - \frac{2}{9}\theta \right)^{y_2} \left(\frac{2}{9} \right)^{y_3} \left(\frac{2}{9} \right)^{y_4} \tag{9.50}$$

(Johnson, Kotz, and Balakrishnan [1997]). This is a straightforward application of the multinomial distribution and it presents no particular estimation problem as specified. Normally we would observe some data for a given year, $[y_1, y_2, y_3, y_4]$, and simply estimate θ via maximum likelihood. Now suppose that the category for fossil fuels was known to be composed of two subcategories for natural gas and petroleum with the theoretically driven cell probabilities given by:

$$z_{petroleum} = \frac{2}{9}\theta$$

$$z_{natural\ gas} = \frac{2}{9}$$

$$y_1 = z_{petroleum} + z_{natural\ gas} = \frac{2}{9} + \frac{2}{9}\theta.$$

The cell probabilities are given by the 5-category vector:

$$p_2(\theta) = \left[\frac{2}{9}\theta, \frac{2}{9}, \frac{3}{9} - \frac{2}{9}\theta, \frac{2}{9}, \frac{2}{9}\right], \qquad \text{for:} \quad 0 \le \theta \le 1.$$

Although we know that the fossil fuel category is split in terms of its theoretical contribution to the model, we do not have the data for natural gas and petroleum separately. Therefore this is a missing data problem because while we have composite data for this category, we lack the individual contributions in the data. The mapping from the 4-category incomplete data specification, $\mathbf{Y}$, to the new 5-category complete data specification, $\mathbf{Z}$, is given by:

$$y_1 = z_1 + z_2$$

$$y_2 = z_3$$

$$y_3 = z_4$$

$$y_4 = z_5.$$

So the multinomial PMF given in (9.50) is now given with the extra category:

$$g(\mathbf{z}|n, \theta) = \frac{n!}{z_1! z_2! z_3! z_4! z_5!} \left(\frac{2}{9}\theta\right)^{z_1} \left(\frac{2}{9}\right)^{z_2} \left(\frac{3}{9} - \frac{2}{9}\theta\right)^{z_3} \left(\frac{2}{9}\right)^{z_4} \left(\frac{2}{9}\right)^{z_5}. \tag{9.51}$$

This is a missing information problem because we know that the data are generated by a 5-category PMF according to (9.51) but we are only able to observe the 4 categories defined by $\mathbf{Y}$. From a public policy point of view it is important to distinguish natural gas usage from petroleum usage although both constitute critical fossil fuels.

The log likelihood corresponding to (9.51) is:

$$\ell(\theta|\mathbf{z}, n) = k + z_1 \log\left(\frac{2}{9}\theta\right) + z_3 \log\left(\frac{3}{9} - \frac{2}{9}\theta\right) \tag{9.52}$$

where the terms that do not depend on the value of θ have been swept into a constant term k. The derivative of the log likelihood function (being careful about the sign from the chain rule) is:

$$\frac{\partial}{\partial\theta}\ell(\theta|\mathbf{z}, n) = \frac{z_1}{\frac{2}{9}\theta} - \frac{z_3}{\frac{3}{9} - \frac{2}{9}\theta}. \tag{9.53}$$

Setting this quantity equal to zero easily produces a complete data maximum likelihood value of θ, given a little algebra:

$$\hat{\theta} = \frac{3z_1}{2z_1 + 2z_3}. \tag{9.54}$$

Of course we do not have a direct means of obtaining z_1 since we can only observe $y_1 = z_1 + z_2$. This is where the EM-algorithm helps out. First we will iteratively average over the missing data (the E-Step) to obtain a complete data specification for θ, then we will use maximum likelihood to obtain an interim estimate (the M-Step).

Averaging over $\mathbf{Z}$ and expressing in terms of the observed data $\mathbf{Y}$ is done by categories:

$$E[z_1|\mathbf{y}] = y_1 \frac{\frac{2}{9}\theta}{\frac{2}{9} + \frac{2}{9}\theta} = y_1 \frac{\theta}{1+\theta}$$

$$E[z_2|\mathbf{y}] = y_1 \frac{\frac{2}{9}}{\frac{2}{9} + \frac{2}{9}\theta} = y_1 \frac{1}{1+\theta}$$

$$E[z_3|\mathbf{y}] = y_2$$

$$E[z_4|\mathbf{y}] = y_3$$

$$E[z_5|\mathbf{y}] = y_4.$$

Now back these values into the full data expected log likelihood for θ, (9.52), and conditional on observed $\mathbf{y}$:

$$
\begin{aligned}
h(\theta) &= E[\log g(\mathbf{z}|n, \theta)|\mathbf{y}] \\
&= k + y_1 \frac{\theta}{1+\theta} \log\left(\frac{2}{9}\theta\right) + y_1 \frac{1}{1+\theta} \log\left(\frac{2}{9}\right) + y_2 \log\left(\frac{3}{9} - \frac{2}{9}\theta\right),
\end{aligned} \tag{9.55}
$$

where once again the terms not dependent on θ are swept into a constant (i.e., constants do not affect the calculation of the MLE). We now have a full information log likelihood, but based on only the observables since we averaged over the nonobservables. This is all we need to begin the EM algorithm. The EM steps are summarized by:

The EM algorithm is thus defined by cycling between the E-Step and the M-Step above until we are satisfied with the resulting parameter estimate (subsequent changes are minute). The data vector used is for the year 1989, which is just prior to the Gulf War (an interesting case since it is a conflict associated with petroleum availability). These values are given by:

$$\mathbf{Y}_{1989} = [y_{fossilfuels} = 44, \; y_{coal} = 19, \; y_{hydroelectric} = 3, \; y_{nuclear} = 6],$$

given in quadrillion BTUs (source: U.S. Department of Energy, http://www.energy.gov/). The magnitude of usage here along with the form of (9.50) show that on a macro-scale, energy usage is essentially a trade-off between coal and fossil fuels. Hydroelectric and nuclear power generation is fixed for a given time period due

TABLE 9.4: EM OUTPUT: U.S. ENERGY CONSUMPTION

Iteration j	$\theta^{(j)}$	log likelihood
1	0.9900000	NA
2	0.8029986	-119.9546
3	0.7615855	-119.3924
4	0.7504438	-119.2320
5	0.7472966	-119.1860
6	0.7463954	-119.1728
7	0.7461364	-119.1690
8	0.7460619	-119.1679
9	0.7460404	-119.1676
10	0.7460342	-119.1675
11	0.7460325	-119.1674
12	0.7460320	-119.1674
13	0.7460318	-119.1674
14	0.7460318	-119.1674
15	0.7460318	-119.1674
16	0.7460317	-119.1674
17	0.7460317	-119.1674
18	0.7460317	-119.1674
19	0.7460317	-119.1674
20	0.7460317	-119.1674

to the inherent cost of building generation facilities and the inflexible nature of the capacity of these facilities. Since fossil fuel sources are predominantly imported and coal fuel sources are predominantly domestically sourced, there are national security issues involved as well.

This simple R program implements the steps described above for the **Y** vector and the starting point, $\theta^{(0)} = 0.99$:

▷ **[Initialize:]** Specify the observed data, $\mathbf{Y} = [y_1, y_2, y_3, y_4]$, and some reasonable starting point for the unknown parameter, $\theta^{(0)}$.

▷ **[E-Step:]** Calculate: $z_1^{(1)} = y_1 \frac{\theta^{(0)}}{1+\theta^{(0)}}$ and $z_3^{(1)} = y_2$.

▷ **[M-Step:]** Estimate: $\hat{\theta}^{(1)} = \frac{3z_1^{(1)}}{2z_1^{(1)}+2z_3^{(1)}}$.

▷ **[Repeat:]**   Obtain $[z_1^{(2)}, z_2^{(2)}, z_3^{(2)}, \hat{\theta}^{(2)}], \ldots, [z_1^{(M)}, z_2^{(M)}, z_3^{(M)}, \hat{\theta}^{(M)}]$, until $\hat{\theta}^{(M)} - \hat{\theta}^{(M-1)}$ is sufficiently small.

```
log.like <- function(theta,z)  {
        k <- z[2]*log(2/9) + z[4]*log(2/9) + z[5]*log(2/9)
        return( k + z[1]*log((2/9)*theta)
      + z[3]*log((3/9)-(2/9)*theta))
}


y <- c(44,19,3,6); z <- rep(NA,5)
num.iterations <- 20
out.mat <- matrix(rep(NA,num.iterations*2),ncol=2)
out.mat[1,1] <- 0.99
dimnames(out.mat)[[2]] <- c("theta","loglike")

for (j in 2:num.iterations)  {
    z[1] <- y[1]*(out.mat[(j-1),1]/(1+out.mat[(j-1),1]))
    z[2] <- y[1]*(1/(1+out.mat[(j-1),1]))
    z[3] <- y[2]
    z[4] <- y[3]
    z[5] <- y[4]
    out.mat[j,1] <- 3*z[1]/(2*z[1]+2*z[3])
    out.mat[j,2] <- log.like(out.mat[j,1],z)
}
```

A couple of things are worth noting here. The log-factorial notation has not been incorporated in the k constant, but it easily could be (clearly it does not change the inference here). Secondly, several efficiencies can be obtained by setting constants, but the approach here makes the algorithm more transparent. Note also that the NA placeholder is irrelevant since there is no conditioning on this value at the starting point.

The algorithm is run for only 20 steps to produce the output given in Table 9.4. It is clear that the algorithm converges to a stable value after ten iterations, to the point $[0.7460317, -119.1674]$.

Figure 9.7 shows how the EM algorithm works in this example. We begin right after the starting point where θ values are indicated on the x-axis and the associated likelihood value designated by the first point on the lower curve. So the y-axis indicates increasing (conditional) values of the likelihood as the algorithm iterates (the scale is left off since it only matters in the relative sense). From there we condition on the observed data in the E-Step and move up the first vertical line on the right-hand side to the likelihood value that corresponds to the starting point for θ and the first conditional. Thus the E-Step moves up the vertical line to an improved estimate of the value of the likelihood function. Now we calculate the maximum likelihood value

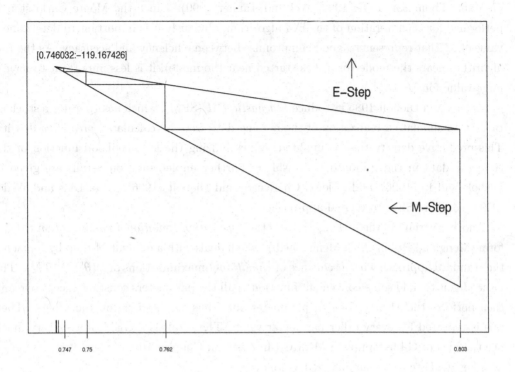

[0.746032:−119.167426]

E-Step

← M-Step

0.747 0.75 0.762 0.803

FIGURE 9.7: EM DEMONSTRATION: U.S. ENERGY CONSUMPTION

for θ to produce $\theta^{(1)}$, which moves us across the first horizontal line on the right-hand side to the first complete point. This process continues until the steps converge in the upper-left corner. This figure is quite intuitively informative because it shows how the upper line, connecting the E-Step points, converges with the lower line, connecting the M-Step points, as the number of iterations converges. Note from the "rug" along the x-axis on that the steps get progressively smaller as the algorithm converges.

9.7.3   Extensions to the EM Algorithm

In cases where the GEM E-Step is particularly difficult or time-consuming, it can be simulated without too much trouble by sampling M realizations from the distribution $\mathbf{X}^m \sim f(\mathbf{X}_{mis}|\mathbf{X}_{obs}, \theta^{(k)})$ and calculating:

$$Q(\theta^{(k+1)}|\theta^{(k)}) = \frac{1}{M} \sum_{m=1}^{M} \ell(\theta|\mathbf{X}_{obs}, \mathbf{X}^m), \qquad (9.56)$$

a substitute step that uses complete data *conditional* maximum likelihood estimation (CM), where the conditionality is over some convenient function of the parameter estimates (McCulloch 1994). This is the *Monte Carlo EM algorithm* (MCEM), which is intended to further replace analytical work with simulation (Booth and Hobert 1999; Chan and Ledolter 1995; Guo and Thompson 1991, 1994; Wei and Tanner 1990). Since the Monte Carlo step is performed for each iteration of the EM algorithm, obviously it is important to determine a value of M that represents a good compromise between efficiency and accuracy. As the EM algorithm nears the mode (or if it is started near the mode), it is less important to have a large value for M.

Celeux and Diebolt 1985 introduce *Stochastic EM* (SEM), which incorporates a stochastic step to simulate a realization of the missing data from its calculated *predictive* density. This predictive density then is updated by maximizing the log likelihood function of the complete data in the standard EM fashion. Further implementation details are given in Diebolt and Ip (1996), and Celeux, Chauveau, and Diebolt (1996). Also, Dias and Wedel (2004) compare SEM to related approaches.

Another variant of the EM algorithm is the *expectation conditional maximization* (ECM) form (Meng and Rubin 1993; Meng 1994b), which deals with a difficult M-Step by replacing the standard approach with a sequence of *conditional* maximizations of $Q(\theta^{(k+1)}|\theta^{(k)})$. The basic idea is to hold some convenient function of all the parameters constant except one and then perform the M-Step for this parameter only thus conditioning on the others. Then this is repeated for every other parameter value. The result is a GEM process very much like the idea of Gibbs sampling MCMC (discussed in Chapter 10).

The ECM steps are summarized as follows:

1. Segment the vector θ into S subvectors: $\{\theta_1, \theta_2, \ldots, \theta_S\}$.

2. (a) Strategy A: select one subvector, θ_s, and maximize, holding all others constant at previous step levels. Cycle through all subvectors.

   (b) Strategy B: select one subvector, θ_s, to hold constant at previous step levels and maximize all others. Cycle through subvectors.

There are many possible extensions, and some restrictions apply for particular problems (Meng and Rubin 1993). The primary motivation is that the resulting algorithm can be much faster than EM or ECM (Kowalski *et al.* 1997; Liu and Rubin 1998; Mkhadri 1998). One interesting variant is Liu and Rubin's (1994, 1995, 1998) ECME (Expectation/Conditional Maximization Either), which promises faster convergence than regular ECM. In this procedure, a subset of the CM steps are replaced with maximizing rather increasing processes for the constrained likelihood function implied by holding other θ values constant.

Rubin (1991) suggests that data augmentation can be thought of as a combination of EM and the multiple imputation procedure for handling missing data. Missing data values can be effectively addressed with the multiple imputation procedure (Little and Rubin 1983, 2002, Rubin 1987b), rather than being ignored in the modeling process. Essentially, this procedure creates a posterior distribution for the missing data conditional on the observed data, and draws randomly from this distribution to create multiple replications (normally

5-10) of the original dataset. The specified model is then run on each of these replicated full datasets and the results are averaged (with a slight ANOVA adjustment for the standard errors of the point estimates). This is a very simple and very common process at this point, although one with minor warnings (see Zhou and Reiter 2010). Data augmentation (details in Chapter 10) uses an added variable or variables in such a way that the estimation or optimization process is simplified rather than complicated. Consider the following modification of the EM steps:

▷ **E-Step**: take m draws of $\mathbf{X}_{mis}$ from the current form of $f(\mathbf{X}_{mis}|\mathbf{X}_{obs}, \boldsymbol{\theta}^{(k)})$ to produce $\mathbf{X}_{mis}^{(k)}$ ($k = 1, 2, \ldots, m$), or the current posterior predictive distribution in the fully Bayesian sense.

▷ **M-Step**: for each $\mathbf{X}_{mis}^{(k)}$ draw a $\boldsymbol{\theta}$ value from $\ell(\boldsymbol{\theta}|\mathbf{X}_{obs}, \mathbf{X}_{mis})$, and average as in multiple imputation.

Notice that this breaks up the two components of the function, $Q(\boldsymbol{\theta}^{(k+1)}|\boldsymbol{\theta}^{(k)})$. This is a data augmentation in the sense that $\mathbf{X}_{mis}$ need not be missing data in the strict sense, but can be any data that makes the M-Step easier. Rubin also notes that if $m = 1$, this algorithm is a Gibbs sampler (Chapter 10).

9.7.4   Additional Comments on EM

This section presented the theory and examples of the EM algorithm. Clearly two detailed cases cannot be fully representative of the applicability of the technique. Very basic explications include Greene (2011), and Jackman (2000) for latent variable models, the original DLR-77 piece, and the survey by Rubin (1991). Recently Hill and Kriesi (2001) have shown how EM can help estimate a complicated survey model specification in political science.

It is important to remember that EM does not simply provide joint maximization of the parameters and missing data together. Instead the two unknowns are treated differently since we integrate over the missing data and then calculate the maximum likelihood estimates of the unknown parameters. Treating all the unknown information, data and parameters the same is problematic since they do not share all the desirable properties of proper maximum likelihood estimation and because it is not feasible when the proportion of missing data is nontrivial (Little and Rubin 1983).

The EM algorithm is a mode finding algorithm and as developed here, does not provide the curvature around the functional mode as a means of producing variance-covariance estimates. A substantial literature has emerged to address this deficiency, including: Louis (1982), Meng and Rubin (1991), and the text by McLachlan and Krishnan (1997).

EM is easily one of the most popular statistical techniques for mode-finding and dealing with missing data. For instance, Meng and Pedlow (1992) found over 1,000 applications in the first 15 years after the publication of DLR-77, spread over a wide range of fields and subfields. While the bulk of EM citations are in applied work (and thus a testament to its status as a useful and widely accepted procedure), there exists ongoing and productive theoretical work. Heyde and Morton (1996) adapt the EM algorithm to instances where an explicit form of the likelihood function is not available using quasi-likelihood and projection methods. Laird and Louis (1982) tie EM estimation more tightly into the Bayesian world by giving a very helpful setup for calculating posteriors from conjugate priors with incomplete data sampling where this "incompleteness" comes very generally from missing values or censored, truncated, grouped, missing, or mixture data.

Often the EM algorithm is lumped together with MCMC (Chapter 10), but it is not an MCMC technique because while it does have serial nature, it does not *sample* on each iteration: each step is a deterministic decision given the present state: maximization not simulation. However, if the maximization step in EM is replaced by a simulation from the distribution identified by the E-Step, then EM is a data augmentation application of the MCMC Gibbs sampler (Robert and Casella 2004; Tanner 1996, Chapter 5). The primary linkage between EM and MCMC lies in the fact that EM is an excellent precursor to the MCMC techniques where it is important to start a chain near the high-density region (Gelman and Rubin 1992a).

9.7.5   EM for Exponential Families

Expressing the EM algorithm for problems based on exponential family forms is particularly straightforward due to the isolation of the sufficient statistic in parametric form. In practice the EM algorithm performs exceedingly well on exponential family forms due to the consistency, and asymptotic normality maximum likelihood estimates (Berk 1972).

Start with a one-parameter *complete data* conditional exponential family PMF or PDF for the random variable X in:

$$f(x|\zeta) = \exp\big[t(x)u(\zeta)\big]r(x)s(\zeta), \tag{9.57}$$

where: r and t are real-valued functions of z that do not depend on ζ, and s and u are real-valued functions of ζ that do not depend on x, and $r(x) > 0, s(\theta) > 0 \; \forall x, \zeta$. The log likelihood of independent, identically distributed (iid) $\boldsymbol{X} = \{X_1, X_2, \ldots, X_n\}$ is therefore in canonical form (Gill 2000, Section 2):

$$\ell(\theta|\mathbf{x}) = t(\mathbf{x})\theta - s'(\theta) + r'(x). \tag{9.58}$$

Here $t(\mathbf{x})$ is the complete data sufficient statistic, which we can enumerate for the jth EM algorithm iteration: $t(\mathbf{x})^{(j)}$. This notational approach greatly simplifies the EM algorithm where $\mathbf{x}_{obs}$ is the non-missing data and $\theta^{(j)}$ is the current coefficient estimate:

▷ **[E-Step:]** calculate $t(\mathbf{x})^{(j+1)} = E_{\mathbf{x}}[t(\mathbf{x})^{(j)}|\mathbf{x}_{obs}, \theta^{(j)}]$, to update the sufficient statistic.

▷ **[M-Step:]** solve for $\theta^{(j+1)}$ using the first two components in the right-hand-side of the likelihood equation in (9.58), plugging in $t(\mathbf{x})^{(j+1)}$ for $t(\mathbf{x})$.

Once the sufficient statistic is identified (a trivial exercise with exponential family forms), there is very little analytical work to be done here.

■ **Example 9.8:** EM Estimation for a Mixture Model in Demography. The example is a demography application from Li and Choe (1997) analyzing second births in China. The 1979 Chinese law allows families that accept the *one-child certificate* to get extra benefits from the government, including: cash, housing preferences, healthcare services, and school preferences. However, families that break the one-child certificate face penalties, except under special (usually tragic) circumstances. How can researchers determine the effectiveness of the law? Specifically, what other general covariates lead to a second child decision? Li and Choe attempt to answer these questions using Chinese government survey data from the Han province, 1978 to 1982, of women who had one child. It is important now to note that this law is evolving in China.

The model specification is a mixture of logistic (did a second baby occur) and hazard (how long between first and second) components. Here Y is a dichotomous variable indicating whether a woman *eventually* has a second child, which is only partly observed since the second child could occur after the period of the study (1982). Thus this is a (censored) missing data problem and appropriate for EM. The general idea is to add a weighting to the likelihood function where the weight is 1 for women observed to have a second child and a predictive weight (the probability of ever having a second child given levels of covariates) for those still with one child at the end of the survey.

Model 1 is given by:

$$p(Y|\mathbf{X}\boldsymbol{\beta}) = [1 + \exp(-\mathbf{X}\boldsymbol{\beta})]^{-1}$$

with covariates and coefficients, $\mathbf{X}$, affecting the eventual occurrence of a second birth. Define T as the random variable for the interval between births for women having the second child, with realization t. The general equation for those not having a second child by time period t:

$$S(t) = p(\text{later than } t) + p(\text{never}) = pS_1(t) + (1 - p),$$

where $S_1(t)$ is the probability of not having the second child in this period conditional on eventually having a second child, and p is the probability of ever having a second child.

Model 2 is specified by $f(T|\mathbf{ZG})$ with covariates and coefficients affecting the interval between births, giving...

the survivor function: the hazard rate:

$$S(t|\mathbf{ZG}) = 1 - F(t) = p(T \geq t) \qquad h(t|\mathbf{ZG}) = f(t|\mathbf{ZG})/S(t|\mathbf{ZG}).$$

Note that the hazard rate is zero for women who will never have a second child and thus not included in Model 2. This is therefore unlike similar hazard rate models where everyone eventually has the event (i.e., death). The "non-parametric piecewise proportional hazards" model (see Allison 1982) divides up the period after the first birth into k segments (months):

$$\xi_1 = [9{:}20] < \xi_2 = [21{:}32] < \xi_3 = [33{:}44] < \xi_4 = [45{:}56]\dots$$

$$\dots < \xi_5 = [57{:}68] < \xi_6 = [69{:}\infty]$$

and assumes constant baseline hazard rates within each segment,

$$\{\lambda_1, \lambda_2, \lambda_3, \lambda_4, \lambda_5, \lambda_6\},$$

taken from actuarial tables, and proportional changes in these hazards determined by covariates. This is done to be "simple" in that:

▷ no parametric assumptions about the shape of the survival function are required since it is now a series of steps.

▷ nonparametric flexibility can be achieved by altering the number of periods.

▷ there are fewer extra parameters to estimate relative to competitors (Cox PH, etc.).

All these assumptions provide the hazard rate for the jth period:

$$h(t|\mathbf{ZG}) = \lambda_j \exp(\mathbf{ZG}), \text{ for } t \in \xi_j,$$

with covariates and coefficients affecting the duration T. Thus negative values of $\mathbf{G}$ lengthen T by lowering the hazard rate, and positive values of $\mathbf{G}$ shorten T by raising the hazard rate. Note: some authors/literatures specify $\exp(-\mathbf{ZG})$, just switching around the above statement.

Modeling only the observed Y_i induces obvious bias, so introduce δ_i as a dichotomous variable indicating that the ith woman had the second child before the end of the study ($\delta_i = 0$), or not ($\delta_i = 1$). Note that δ_i is fully observed in the data. Specify μ_i as the expectation (probability, as in E-Step) of $Y_i = 1$ in period j, including after the survey ($j = 6$):

$$\mu_i = 1, \quad \text{if } \delta_i = 0$$

$$\mu_i = p(\text{second baby})p(\text{no second baby before time } t)/$$

$$p(\text{second baby})p(\text{no second baby before time } t)$$

$$+ p(\text{never has second baby})$$

$$= \frac{p(Y|\mathbf{X}_i\boldsymbol{\beta})S(t|\mathbf{Z}_i\mathbf{G})}{p(Y|\mathbf{X}_i\boldsymbol{\beta})S(t|\mathbf{Z}_i\mathbf{G}) + (1 - p(Y|\mathbf{X}_i\boldsymbol{\beta}))(1)}, \quad \text{if } \delta_i = 1.$$

The EM steps for Q proceed as follows:

$$Q = \sum_{\text{sample}} \sum_{\text{periods}} \Big[(1 - \delta_i) \times \log h(\text{had second baby before period } j)$$

$$+ (\text{weighting}) \times \log p(\text{has second baby at period } j)$$

$$+ (1 - \text{weighting}) \times \log p(\text{never has second baby})\Big].$$

So within each period:

$$Q = \sum_{i=1}^{n} [(1 - \delta_i) \log(h(t|\mathbf{Z}_i\mathbf{G}))$$

$$+ \mu_i \log(p(Y|\mathbf{X}_i|\boldsymbol{\beta})S(t|\mathbf{Z}_i\mathbf{G})) + (1 - \mu_i) \log(1 - p(Y|\mathbf{X}_i\boldsymbol{\beta}))]$$

$$= \underbrace{\sum_{i=1}^{n} [\mu_i \log(p(Y|\mathbf{X}_i\boldsymbol{\beta}) + (1 - \mu_i) \log(1 - p(Y|\mathbf{X}_i\boldsymbol{\beta}))]}_{Q_\beta}$$

$$+ \underbrace{\sum_{i=1}^{n} [(1 - \delta_i) \log(h(t|\mathbf{Z}_i\mathbf{G})) + \mu_i \log(S(t|\mathbf{Z}_i\mathbf{G}))]}_{Q_\mathbf{G}}$$

where the last line shows that we can separate out $\boldsymbol{\beta}$ and $\mathbf{G}$ maximum likelihood procedures. The likelihood function intuition on the observed second child event $(\delta_i = 0, \mu_i = 1)$ for a single individual (i) at time t, $\exp(Q_i)$ reduces to:

$$\exp(Q_i) = h(t|\mathbf{Z}_i\mathbf{G})[p(Y|\mathbf{X}_i\boldsymbol{\beta})S(t|\mathbf{Z}_i\mathbf{G})]$$

$$= \frac{f(t|\mathbf{Z}_i\mathbf{G})}{S(t|\mathbf{Z}_i\mathbf{G})}[p(Y|\mathbf{X}_i\boldsymbol{\beta})S(t|\mathbf{Z}_i\mathbf{G})]$$

$$= p(Y|\mathbf{X}_i\boldsymbol{\beta})f(t|\mathbf{Z}_i\mathbf{G})$$

The likelihood function intuition on unobserved second child event $(\delta_i = 1, \mu_i \neq 1)$ for a single individual at time t, $\exp(Q_i)$ reduces similarly to:

$$\exp(Q_i) = [p(Y|\mathbf{X}_i\boldsymbol{\beta})S(t|\mathbf{Z}_i\mathbf{G})]^{\mu_i}[1 - p(Y|\mathbf{X}_i\boldsymbol{\beta})]^{1-\mu_i}$$

which is an intuitive looking proportional weighting. Thus our mechanical steps are:

1. Pick MLE values using observed Y for $\boldsymbol{\beta}^{(0)}$ and $\mathbf{G}^{(0)}$.

2. kth **E-Step**:

   (a) Compute μ_i for all cases where $\delta_i = 1$ using the current values: $\boldsymbol{\beta}^{(k)}$ and $\mathbf{G}^{(k)}$.

   (b) Plug these values into $Q^{(k)}$.

3. kth **M-Step**:

   (a) Obtain MLE of $\boldsymbol{\beta}$ using $Q_\beta^{(k)}$, assign to $\boldsymbol{\beta}^{(k+1)}$.

   (b) Obtain MLE of $\mathbf{G}$ using $Q_\mathbf{G}^{(k)}$, assign to $\mathbf{G}^{(k+1)}$.

4. Repeat until $||\boldsymbol{\beta}^{(k^*+1)} - \boldsymbol{\beta}^{(k^*)}|| < \epsilon$ and $||\mathbf{G}^{(k^*+1)} - \mathbf{G}^{(k^*)}|| < \epsilon$.

The results can be stated in typical tabular form for regression models. Here the reference subject categories are: reside in what environment, education below junior high, first child's gender, and age when first child is born. Table 9.5 shows the values.

Here the reference category on residing is the countryside.

TABLE 9.5: MODEL RESULTS, MORE DEVELOPED AREAS

Covariates	Coefficient	Std. Error	Coefficient	Std. Error
Accepted Certificate	-2.322	0.313	-1.703	0.191
Reside City	-3.566	0.286	-0.021	0.220
Reside Town	-0.453	0.241	0.058	0.220
Low Education	0.825	0.373	0.002	0.272
First Child Female	1.205	0.211	0.125	0.095
Age at First Born	-0.236	0.095	-0.051	0.046
Age-Squared at First Born	-0.095	0.042	0.068	0.023

9.8   Survey of Random Number Generation

The most important simulation topic *not* detailed so far is the generation of random variables from various parametric forms. Many of the procedures in this chapter and in Chapter 10 depend on the reliable generation of values from specific PDFs and PMFs. In fact, a wide range of Bayesian and non-Bayesian simulation techniques require random number generation as a means of creating empirical sequences of interest. With the Gibbs sampler, we need to sample from full conditional distributions and with the Metropolis-Hastings algorithm, we need to sample from the jumping distribution. Unreliable production of random variables will therefore certainly affect the quality of the MCMC estimates.

To see the importance of having an adequate random number, consider the simple numerical integration described in Section 9.2. If the random numbers, as generated, favored some region of the random variable support, then it is clear that the resulting integral will be incorrect. Such inaccuracies could also occur in more complicated Monte Carlo estimation procedures including MCMC.

General descriptions of the issues involved in random number generation are found in the now-classic text of Kennedy and Gentle (1980, Chapter 6) along with Gentle's update (1998), the book by Devroye (1986), and more theoretical discussions such as Evans and Swartz (1995), Dieter (1975), Dieter and Ahrens (1971, 1973), Gentle (1990), Jagerman (1965), Jansson (1966), and Knuth's seminal work on the topic (1981, Chapter 3). Also worth reading is the foundational series of papers by Ahrens and Dieter (1972, 1973, 1974, 1980, 1982), sections of Morgan (1984, Chapters 3 to 6), the recent work of Robert and Casella (2004, Chapter 2), and the introduction by Ripley (1983).

There are many algorithms for generating different forms of random variates and these are usually cataloged by the associated PDF or PMF. Because of its basic nature and role in more complex operations, the generation of random uniforms holds special significance. There are a great many stochastic simulation techniques that require uniform variates as a means of equitably comparing multidimensional regions in a parameter space. In addition, nearly all other distributional forms can be derived by transformations or manipulations of the uniform. Consequently, there is a large and well-developed literature on the generating algorithms for uniform random numbers, including such key work as Falk (1999), MacLaren and Marsaglia (1965), Marsaglia (1961a), Mason and Lurie (1973), Müller (1959b), Walker (1974), Westlake (1967), and L'Ecuyer (1998).

A number of papers on the more modern, well-known congruential generator method for producing uniform random variates have been written by Burford (1973, 1975), Chay, Fardo, and Mazumdar (1975), Dieter (1975), Downham and Roberts (1967), Eichenauer-Herrmann (1996), Fuller (1976), Golder (1976), Hull and Dobell (1964), Lehmer (1951), Marsaglia (1968, 1972), Neave (1973), Perkins and Menzefricke (1975), Prentice and Miller (1968), and Sibuya (1961), among others. Another important algorithm is the shift register method described by Golomb (1967), Hurd (1974), Liniger (1961), Lewis and Payne (1973), and Marsaglia (1961b, 1977).

Second only to the uniform is the attention paid to generating normally random variates. Due mainly to asymptotic properties, a wide range of data-generated distributions of interest can be modeled with the normal PDF. There is therefore an extensive literature describing algorithms for generating normal random variates. Important works include Box and Müller's (1958) famous technique for generating normals from independent uniforms, along with Best (1979), Brent (1974), Cheng (1985), Gates (1978), Gebhardt (1964), George (1963), Golder and Settle (1976), Hurst and Knop (1972), Kinderman and Ramage (1976), Kinderman, Monahan, and Ramage (1975), Marsaglia (1964), Marsaglia and Bray (1964), Marsaglia, MacLaren, and Bray (1964), Müller (1958, 1959a), Niederreiter (1972, 1974, 1976), Payne (1977), and Scheuer and Stoller (1962).

Generating gamma random variables is also a fundamental task since it is a very flexible family form and the support (zero to positive infinity) makes the gamma a convenient distribution for modeling variance terms. Some of the often-cited works include Atkinson (1977), Atkinson and Pearce (1976), Best (1983), Bowman and Beauchamp (1975), Cheng (1977), Cheng and Feast (1979, 1980), Greenwood (1974), Kinderman and Monahan (1980), Phillips and Beightler (1972), Tadikamalla (1978), Tadikamalla and Ramberg (1975), Wallace (1974), Wheeler (1974, 1975), as well as Whittaker (1974).

There is an important and growing literature on testing the large number of currently available random number generators for good properties of randomness as well as computational efficiency: Altman and McDonald (2001, 2003), Atkinson (1980), Butcher (1961), Coveyou and MacPherson (1967), Downham (1970), Dudewicz (1976), Good (1957), Gorenstein (1967), Krawczyk (1992), Kronmal (1964), Learmonth and Lewis (1973), Marsaglia (1968), McArdle (1976), McCullough (1999), McCullough and Wilson (1999), Toothill, Robinson, and Adams (1971), and Whittlesey (1969). Some of these are a reaction to observed problems in serial correlation in the output stream of these produced values and the well-established problems with IBM's long-lived *RANDU* algorithm as well as other related techniques: Coveyou (1960, 1970), Fishman and Moore (1982), Hellekalek (1998).

A number of articles give informative histories and overviews of random number generation. These include Dudewicz (1975), Halton (1970), Marsaglia (1985), and Mihram and Mihram (1997). Some early-era bibliographies provide helpful background as well: Nance and Overstreet (1972), and Sowey (1972, 1978). New work worthy of attention can be found in Anderson and Louis (1996), Malov (1998), and Jones and Lunn (1996). This remains an active area of research for two main reasons. Since all "random" numbers generated are really "pseudo-random" in that they are produced deterministically by computational processes, there is always the opportunity to improve their random properties. In addition, as the use of compute-intensive techniques increases (such as Markov chain Monte Carlo), increases in efficiencies in the underlying algorithms becomes more important.

9.9   Concluding Remarks

It is important to remind readers that numerical statistical methods are not the panacea that some might expect or hope. The tools presented in this chapter, and of course the more complex simulation techniques to come, are susceptible to all of the problems associated with the misapplication of statistical tools by incautious researchers. Usually these issues occur due to misunderstanding the assumptions underlying a model or procedure as it is applied in a data-analytic setting. Furthermore, some have suggested that because the computer does the bulk of the mechanical work with these tools, there is a tendency for less human consideration of the details and an over-trusting of the results (Cleveland 1993; Higham 1996; Mooney 1997), and it is not unusual for researchers to find such faults in others' work (Siegmund 1976, p.679).

Fox (1971), also summarized in Higham (1996, p.35), indicates where researchers can go awry in numerical computing in the provocatively titled piece: "How to Get Meaningless Answers in Scientific Computation and What to Do about It." His reasons that computed answers may be useless, all applicable to statistical computing and statistical simulation, include:

▷ The problem might be ill-conditioned or unstable numerically.

▷ There might be an expectation that the computer can do more than it is capable of doing.

▷ The appearance of consistency in the simulations may be deceiving.

▷ The application may be too specialized and insufficiently generalizeable for the research objective.

As we move forward to Bayesian stochastic simulation, these are certainly important caveats to keep in the back of our minds.

9.10   Exercises

9.1   Use the trapezoidal rule with $n = 8$ to evaluate the definite integral:

$$I(g) = \int_0^\pi e^x \cos(x) dx.$$

The correct answer is $I(g) = -12.070346$; can you explain why your answer differs slightly?

9.2   The `density` function in R can be used to get a normalizing constant for non-standard distributions. Create a mixture distribution of normals with `x.vals <- c(rnorm(100,1,1),rnorm(100,8,2),rnorm(100,15,1))`, and then run the density function changing the smoothing parameter until it has a satisfactory shape: `x.dens <- density(x.vals,n=length(x.vals),adjust=...)`. Use the returned objects in `x.dens` to estimate a normalizing constant.

9.3   Use Simpson's rule with $n = 8$ to evaluate the definite integral in Exercise 9.1. Is your answer better?

9.4   Normally a histogram takes $x_1, \ldots, x_n$ produced from some distribution $f(x)$ and bins them over an equally spaced mesh along the x-axis with selected bin width h. Selection of a starting point for the bin edges is arbitrary but occasionally has an effect. The *average shifted histogram* (ASH) smooths out the bins by specifying multiple adjacent versions that are averaged (Scott 1985) to form a single estimate. Now specify m histograms instead of one where each of these has bin width h like the original specification now defining a set of bin edges given by $\{ih/m, i = 0, \ldots m\}$. This creates a finer mesh of K bins over the range of the data, $\{t_1, \ldots, t_K\}$, with widths h/m. If $I_k = [t_{k-1} : t_k)$ is the kth bin and n_k is the number of x_i values in this bin, then the value (height) of the ℓth shifted histogram is:

$$H_\ell(x) = \frac{1}{nh} \sum_{j=0}^{m-1} n_{j+i+m\lfloor(k-i)/m\rfloor} \; (x \in I_k) \text{ for } \ell = 0, \ldots, (m-1),$$

which counts cases in the larger bin of width h. The ASH is produced by the point wise average:

$$\hat{H}(x) = \frac{1}{m} \sum_{\ell=0}^{m-1} (x \in I_k).$$

Using the recidivism data on page 188 in Chapter 6, code the ASH algorithm in R and create average shifted histograms separately for the Released and Returned vectors.

9.5   It is well known that the normal is the limiting distribution of the binomial (i.e., that as the number of binomial experiments gets very large the histogram of the binomial results increasing looks like a normal). Using R, generate 25 $\mathcal{BN}(10, 0.5)$ experiments. Treat this as a posterior for $p = 0.5$ even though you know the true value and improve your estimate using importance sampling with a normal approximation density.

9.6 *Antithetic variates* are on old idea in Monte Carlo simulation (Hammersley and Morton 1956) that reduces variance by drawing negatively correlated pairs and averaging them or using both. Suppose we draw a uniformly between 0 and 1, u_{i1}, and then create $u_{i2} = 1 - u_{i1}$, for $i = 1, \ldots, m$. Show that the mean and variance of the u_{i1} and u_{i2} are the same. Derive the correlation.

9.7 Casella and Berger (2002, p.638) give the famous temperature and failure data on space shuttle launches before the *Challenger* disaster, where failure is dichotomized. Actually several o-ring failure events were multiple events. Treat the number of multiple events as missing data and write an EM algorithm to estimate the Poisson parameter given the missing data. Here are the dichotomized failure data in R form:

```
temp <- c(53,57,58,63,66,67,67,68,69,70,70,70,70,70,72,73,
          75,75,76,76,78,79,81),
fail <- c(1,1,1,1,0,0,0,0,0,0,0,0,1,1,0,0,0,1,0,0,0,0,0).
```

9.8 Generate a homogeneous Poisson process for intensity parameter λ by writing an R function that performs the following steps:

> ▷ Start with time $t_1 = 0$ and size $n = 1$
> ▷ sample $u \sim \mathcal{U}(0, 1)$
> ▷ produce $t_{n+1} = t_n - \frac{1}{\lambda} \log(u)$
> ▷ increment: $n < -n + 1$,

which continues until the desired sample size is reached. Now modify your function to create a non-homogeneous Poisson process where the time between events declines by $1/t$.

9.9 Replicate the calculation of the Kullback-Leibler distance between the two beta distributions in Section 9.3.1 using rejection sampling. In the cultural anthropology example given in Section 2.3.4 the beta priors $\mathcal{BE}(15, 2)$ and $\mathcal{BE}(1, 1)$ produced beta posteriors $\mathcal{BE}(32, 9)$ and $\mathcal{BE}(18, 8)$ with the beta-binomial specification. Are these two posterior distributions closer or farther apart in Kullback-Leibler distance than their respective priors? What does this say about the effect of the data in this example?

9.10 Monahan (2001) discusses a posterior distribution, which can be produced by the R command:

```
m.llike <- function(t1,t2)
    n1*log(2*t1*t2) + n2*log(t1*(2-t1-2*t2))
        + n3*log(t2*(2-t2-2*t1)) + 2*n4*log(t1*(1-t1-t2))
```

where $n_1 = 1, n_2 = 10, n_3 = 4, n_4 = 9$. Create a contour plot and a perspective plot over the range [0.01:0.49] in both dimensions (for both variables). Estimate the mode numerically. Use this value to calculate the Hessian analytically.

9.11   The Cauchy distribution (B) is a unimodal, symmetric density like the normal but has much heavier tails. In fact the tails are sufficiently heavy that the Cauchy distribution does not have finite moments. To understand this characteristic, perform the following experiment at least 10 times:

  ▷ Generate a sample of size 100 from $\mathcal{C}(x|\theta, \sigma)$, where you choose θ, σ.
  ▷ Perform Monte Carlo integration to attempt to calculate $E[X]$.
  ▷ Compare your estimate with the mode (θ).

What do you conclude about the moments of the Cauchy?

9.12   The *Kolmogorov-Smirnov test* uses cumulative distribution statistics test the similarity of the empirical distribution of some observed data and a specified PDF. The test statistic is created by:

$$D = \max_{i=1:n} \left\{ \frac{i}{n} - F_{(i)}, F_{(i)} - \frac{i-1}{n} \right\},$$

where $F_{(i)}$ is the ith ordered value. Large values indicate dissimilarity and the rejection of the hypothesis that the empirical distribution matches the queried theoretical distribution. The p-value is calculated from the Kolmogorov-Smirnoff CDF:

$$p(D \leq x) \frac{\sqrt{2\pi}}{x} \sum_{i=1}^{\infty} \exp\left[-(2i-1)^2 \pi^2/8x^2\right],$$

which generally requires approximation methods (see Marsaglia, Tsang, and Wang 2003). This so-called nonparametric test (this label comes from the fact that the distribution of the test statistic does not depend on the distribution of the data being tested) performs poorly in small samples, but works well in a simulation environment. Write an R function that implements this test where the reference distribution is normal. Using R generate 1,000 Cauchy random variables (`rcauchy(1000, location = 0, scale = 1)`) and perform the test.

9.13   Generate 100 standard normal random variables in R and use these values to perform rejection sampling to calculate the integral of the normal PDF from 2 to ∞. Repeat this experiment 100 times and calculate the empirical mean and variance from your replications. Now generate 10,000 Cauchy values and use this as an approximation distribution in importance sampling to obtain an estimate of the same normal interval. Which approach is more accurate?

9.14   Write a function in R to evaluate the definite integral:

$$I(g) = \int_0^\pi e^x \cos(x) dx.$$

Use Simpson's rule with $n = 8$. Give an answer with 6 digits of accuracy.

9.15   The times to failure of two groups are observed, which are assumed to be distributed exponential with unknown θ. The times to failure for the first group, $X_1, X_2, \ldots, X_m$, are fully observed, but the second group is censored at time $t > 0$ because the study grant ran out, therefore the $Y_1, Y_2, \ldots, Y_n$ values are either t or some value less than t, which can be given an indicator function: $O_i = 1$ if $Y_i = t$ and zero otherwise. The expected value of the sufficient statistic is:

$$E\left[\sum_{i=1}^n (Y_i | O_i)\right] = \left(\sum_{i=1}^n O_i\right)(t + \theta^{-1})$$
$$+ \left(n - \sum_{i=1}^n O_i\right)(\theta^{-1} - t[1 + \exp(t/\theta)]^{-1}),$$

and the complete data log likelihood is:

$$\ell(\theta) = -\sum_{i=1}^m (\log(\theta^{-1}) + x_i\theta) - \sum_{i=1}^n (\log(\theta^{-1}) + y_i\theta).$$

Substitute the value of the sufficient statistic into the log likelihood to get the $Q(\theta^{(k+1)} | \theta^{(k)})$ expression and run the EM algorithm for the following data:

X_i	1	9	4	3	11	7	2	2	-	-
Y_i	2	3	2	4	4	2	4	4	3	1

9.16   Suppose a mixture of exponential distributions is expressed as:

$$f_j(x|\theta_j) = \theta_j \exp(-\theta_j y), \ j = 1, \ldots J.$$

Each θ_j is given the a conjugate prior $\mathcal{GA}(a_j, b_j)$. Now include z_{ij} as an indicator variable for case i belonging to mixture j and $z_{ij}^{(k)}$ as the probability that case i belongs to mixture j after the kth step of the EM algorithm. If $n_j^{(k)}$ is the number of cases assigned to mixture j at the kth step, the M-Step is given by $v_j^{(k+1)} = -(\sum_{i=1}^n z_{ij}^{(k)} + b_j)/(n_j^{(k)} + a_j - 1)$, where the $\theta_j^{(k+1)} = -\left[v_j^{(k+1)}\right]^{-1}$.

9.17   One of the easier, but *very* slow, ways to generate beta random variables is *Jöhnk's method* (Jöhnk 1964; Kennedy and Gentle 1980), based on the fact that order statistics from iid random uniforms are distributed beta. The algorithm to deliver a single $\mathcal{BE}(A, B)$ random variable is given by:

(a) Generate independent:

$$u_1 \sim U(0,1), \qquad u_2 \sim U(0,1)$$

(b) Transform:

$$v_1 = (u_1)^{1/A}, \qquad v_2 = (u_2)^{1/B}$$

(c) Calculate the sum:

$$w = v_1 + v_2$$

and return v_1/w if $w \le 1$, otherwise return to step 1.

Code this algorithm in R and produce 100 random variables each according to: $\mathcal{BE}(15,2)$, $\mathcal{BE}(1,1)$, $\mathcal{BE}(2,9)$, and $\mathcal{BE}(8,8)$. For each of these beta distributions, also produce 100 random variables with the `rbeta` function in R and compare with the corresponding forms produced with Jöhnk's method using `qqplot`. Graph these four comparisons in the same window by setting up the plot with the windowing command: `par(mfrow=c(2,2))`.

9.18 Consider the following example from Wu (1983) using bivariate mean-zero data from Murray (1977) in his discussion of DLR-77:

Variable 1:	1	1	-1	-1	2	2	-2	-2	NA	NA	NA	NA
Variable 2:	1	-1	1	-1	NA	NA	NA	NA	2	2	-2	-2

where `NA` denotes a missing value. Show with an EM algorithm starting at different points that the global maxima is at $\rho = \pm 0.5, \sigma_1^2 = \sigma_2^2$. What happens when the EM algorithm is started at $\rho = 0$? What happens when the EM algorithm includes code that bounds ρ away from zero?

9.19 Hartley (1958, p.182) fits a Poisson model to the following "drastically censored" grouped data:

Number of Events	0	1	2	3-10	Total
Group Frequency	11	37	64	128	240

Develop an EM algorithm to estimate the Poisson intensity parameter.

9.20 Efron (1979) and Efron and Tibshirani (1993) introduced the bootstrap as a compute-intensive procedure to get the sampling properties of an estimator when there is not a known closed-form analytical form. It is similar to Monte Carlo simulation in that repeated computation provides information in the absence of direct calculation. For some statistic of interest, θ, and dataset $\mathbf{x}$, the steps are:

(a) Draw B "bootstrap" samples of size n (old advice: 25-200; new advice: thousands) independently, **with replacement** from the sample $\mathbf{x}$ of size n:

$$\mathbf{x}^{*1}, \mathbf{x}^{*2}, \dots, \mathbf{x}^{*B}$$

(note the notation to differentiate the bootstrap sample from the original sample).

(b) Calculate the sample statistic of interest, θ^{*b} for each bootstrap sample, and the mean of these statistics:

$$\bar{\theta}^* = \frac{1}{B} \sum_{b=1}^{B} \theta^{*b}$$

(c) Estimate the bootstrap standard error of the statistic by:

$$\mathrm{Var}(\theta) = \frac{1}{B-1} \sum_{b=1}^{B} \left(\theta^{*b} - \bar{\theta}^* \right)^2$$

where $SE(\theta) = \sqrt{\mathrm{Var}(\theta)}$.

The limit of this standard error as B goes to infinity is called the *ideal* bootstrap estimate, and this procedure is called the *nonparametric* bootstrap estimate. Consider two variables on sub-Saharan African countries: *size of military* (in thousands) and a dichotomous outcome indicating that there was a military coup during this period:

Coup	94	197	16	38	99	141	23		
No Coup	52	104	146	10	50	31	40	27	46

Note that these data are *imbalanced*. The question is whether there is a difference by size of military. Calculate the median of each group and its bootstrapped standard error.

9.11   Computational Addendum: R Code for Importance Sampling

This is a very simple importance sampling routine that provides a normal or Student's-*t* approximation function. The procedure is to draw a random sample from the multivariate normal or Student's-*t* PDF with mean equal to the MLE estimates from the input log-likelihood function, and the standard error equal to the square root of the diagonal of the inverse of the variance-covariance matrix. The function `sir` performs importance sampling with a normal approximation as a default, but one can easily change the specification by specifying a degrees of freedom parameter other than zero. It is important to note that this really implements Rubin's SIR since that way there is an empirical maximum to divide by.

```
# Call:
# sir(data.mat,     data matrix organized in columns
#   theta.vector,  the initial coefficient estimates
#   theta.matrix,  the initial vc matrix
#   M,             the number of draws
#   m,             the desired number of accepted values
#   tol,           the rounding/truncating tolerance
#   ll.func,       loglike function for empirical posterior
#   df)            the df for using the t distribution as the
#                  approx distribution, default=0 for gaussian.

sir <- function(data.mat,theta.vector,theta.matrix,M,m,
                tol=1e-06,ll.func,df=0)  {
    importance.ratio <- rep(NA,M)
    rand.draw <- mvrnorm(M,theta.vector,theta.matrix,tol=1e-04)
    if (df > 0)
        rand.draw <- rand.draw/(sqrt(rchisq(M,df)/df))
    empirical.draw.vector<-apply(rand.draw,1,ll.func,data.mat)
    if (sum(is.na(empirical.draw.vector)) == 0) {
        print("SIR: finished generating from posterior")
        empirical.draw.vector <- 1000*empirical.draw.vector
        print(summary(empirical.draw.vector))
    }
    else {
        print(paste("SIR: found",
            sum(is.na(empirical.draw.vector)),
            "NA(s) in generating from posterior density
            function, quitting"))
        return()
    }
    if (df == 0)  {
        normal.draw.vector
            <- apply(rand.draw,1,normal.posterior.ll,data.mat)
    }
    else {
        theta.matrix <- (df)/((df-2)*theta.matrix
        normal.draw.vector
            <- apply(rand.draw,1,t.posterior.ll,data.mat,df)
    }
    if (sum(is.na(normal.draw.vector)) == 0)  {
        print("SIR: finished generating from approximation
            distribution")
        print(summary(normal.draw.vector))
    }
    else {
        print(paste("SIR: found",
```

```
                    sum(is.na(normal.draw.vector)),
                    "NA(s) in generating from approximation
                    distribution, quitting"))
            return()
        }
        importance.ratio<-(empirical.draw.vector-normal.draw.vector)
        importance.ratio<-importance.ratio/max(importance.ratio)
        if (sum(is.na(importance.ratio)) == 0)  {
            print("SIR: finished calculating importance weights")
            print(summary(importance.ratio))
        }
        else  {
            print(paste("SIR: found",sum(is.na(importance.ratio)),
                "NA(s) in calculating importance weights, quitting"))
            return()
        }
        accepted.mat <- rand.draw[1:2,]
        while(nrow(accepted.mat) < m+2)  {
            rand.unif <- runif(length(importance.ratio))
            accepted.loc <-
                seq(along=importance.ratio)[(rand.unif-tol)
                <=importance.ratio]
            rejected.loc <-
                seq(along=importance.ratio)[(rand.unif-tol)
                >importance.ratio]
            accepted.mat<-rbind(accepted.mat,rand.draw[accepted.loc,])
            rand.draw <- rand.draw[rejected.loc,]
            importance.ratio <- importance.ratio[rejected.loc]
            print(paste("SIR: cycle complete,",
                (nrow(accepted.mat)-2),"now accepted"))
        }
        accepted.mat[3:nrow(accepted.mat),]
}
```

The following are log likelihood functions that can be plugged into the `sir` function above.

```
logit.posterior.ll <- function(theta.vector,X)  {
    Y <- X[,1]
    X[,1] <- rep(1,nrow(X))
    sum( -log(1+exp(-X%*%theta.vector))*Y
        -log(1+exp(X%*%theta.vector))*(1-Y) )
}


normal.posterior.ll <- function(coef.vector,X)  {
    dimnames(coef.vector) <- NULL
    Y <- X[,1]
    X[,1] <- rep(1,nrow(X))
```

```
    e <- Y - X%*%solve(t(X)%*%X)%*%t(X)%*%Y
    sigma <- var(e)
    return(-nrow(X)*(1/2)*log(2*pi)
            -nrow(X)*(1/2)*log(sigma)
            -(1/(2*sigma))*(t(Y-X%*%coef.vector)%*%
                            (Y-X%*%coef.vector)) )
}


t.posterior.ll <- function(coef.vector,X,df)  {
    Y <- X[,1]
    X[,1] <- rep(1,nrow(X))
    e <- Y - X%*%solve(t(X)%*%X)%*%t(X)%*%Y
    sigma <- var(e)*(df-2)/(df)
    d <- length(coef.vector)
    return(log(gamma((df+d)/2)) - log(gamma(df/2))
                            - (d/2)*log(df)
        -(d/2)*log(pi) - 0.5*(log(sigma))
        -((df+d)/2*sigma)*log(1+(1/df)*
                            (t(Y-X%*%coef.vector)%*%
                            (Y-X%*%coef.vector)) ))
}



probit.posterior.ll <- function (theta.vector,X,tol = 1e-05)  {
    Y <- X[,1]
    X[,1] <- rep(1,nrow(X))
    Xb <- X%*%theta.vector
    h <- pnorm(Xb)
    h[h<tol] <- tol
    g <- 1-pnorm(Xb)
    g[g<tol] <- tol
    sum( log(h)*Y + log(g)*(1-Y) )
}
```

Chapter 10

Basics of Markov Chain Monte Carlo

10.1 Who Is Markov and What Is He Doing with Chains?

The use of Markov chain Monte Carlo (MCMC) methods to evaluate integral quantities has exploded over the last two decades. Beginning with the seminal review paper by Gelfand and Smith (1990), the rate of publication of MCMC works has grown exponentially. While this is relatively a recent development, the genesis dates back to two important works: the 1953 essay by Metropolis, Rosenbluth, Rosenbluth, Teller, and Teller, as well as Geman and Geman's (1984) introduction of the Gibbs sampler as a method for obtaining difficult posterior quantities in the process of image restoration. The lack of early recognition of the importance of the 1953 contribution is a testament to the barriers that may exist between statistical physics and other fields, and the hindsight that sufficiently powerful computational resources were not widely available until sometime afterwords. This history is nicely reviewed by Robert and Casella (2011), as well as by Richey (2010). Another fundamental tool in this family is *simulated annealing* (Kirkpatrick, Gelatt, and Vecchi [1983], Černý [1985]), which is described at the begining of Chapter 15.

This chapter introduces the basic ideas behind MCMC methods with the goal of providing accessible introductions, whereas Chapter 13 covers more technical issues and Chapter 14 discusses some practical guidance on running MCMC simulations. In Chapter 15 we look at extensions and enhancements to the standard MCMC algorithms. The primary distinction made in this chapter is between standard Monte Carlo simulation methods, as covered in Chapter 9, and the *Markov chain* type of Monte Carlo methods characterized by a dependence structure between consecutive simulated values. Standard Monte Carlo methods produce a set of *independent* simulated values according to some desired probability distribution. MCMC methods produce chains in which each of the simulated values is mildly dependent on the preceding value. The basic principle is that once this chain has run long enough it will find its way to the desired posterior distribution of interest and we can summarize this distribution by letting the chain wander around, thus producing summary statistics from recorded values. The "magic" that occurs is that a process based on mechanically producing serially correlated values from joint or conditional distributions eventually gives values that can be treated as independent draws from marginals.

10.1.1   What Is a Markov Chain?

Chapter 9 paid extensive attention to the second "MC" in "MCMC" and we have yet to provide a precise definition for the first "MC." The initial definition required is that of a more primitive concept that underlies Markov chains. A *stochastic process* is a consecutive set of random quantities defined on some known state space, Θ, indexed so that the order is known: $\{\theta^{[t]} : t \in T\}$. Here the state space (which we can also refer to as the parameter space when directly referring to the support of a parameter vector of interest) is just the allowable range of values for the random vector of interest. This will be more precisely defined in Chapter 13. Frequently, but not necessarily, T is the set of positive integers implying consecutive, even-spaced time intervals: $\{\theta^{[t=0]}, \theta^{[t=1]}, \theta^{[t=2]}, \ldots\}$. With MCMC we are concerned only with this restricted type of stochastic process.

The state space, Θ, is either discrete or continuous depending on how the variable of interest is measured, but the implications for our purposes apply more to notation than to fundamental theory. Standard references on stochastic processes include Doob (1990), Hoel, Port, and Stone (1987), Karlin and Taylor (1981, 1990), and Ross (1996).

A *Markov chain* is a stochastic process with the property that at time t in the series, the probability of making a transition to any new state is dependent only on the current state of the process, $\theta^{[t]}$, and is therefore *conditionally* independent of the previous values: $\theta^{[0]}, \theta^{[1]}, \ldots, \theta^{[t-1]}$. This is stated more formally:

$$p(\theta^{[t]} \in A | \theta^{[0]}, \theta^{[1]}, \ldots, \theta^{[t-2]}, \theta^{[t-1]}) = p(\theta^{[t]} \in A | \theta^{[t-1]}), \qquad (10.1)$$

where A is any identified set (an event or range of events) on the complete state space. So a Markov chain wanders around the state space "remembering" only where it has been in the last period. This property turns out to be enormously useful in generating samples from desired limiting distributions of the chain because when the chain eventually finds the region of the state space with the highest density, it will produce a sample from this distribution that is only mildly nonindependent. These are the sample values that we will then use to describe the posterior distribution of interest.

A fundamental concern is the transition process that defines the probabilities of moving to other points in the state space, given the current location of the chain. The most convenient way to think about this structure is to define the *transition kernel*, K, as a general mechanism for describing the probability of moving to some other specified state based on the current chain status (Robert and Casella 2004, p.208). The advantage of this notation is that it subsumes both the continuous state space case as well as the discrete state space case. It is required that $K(\theta, A)$ be a defined probability measure for all θ points in the state space to the set $A \in \Theta$. Thus the function $K(\theta, A)$ maps potential transition events to their probability of occurrence.

When the state space is discrete, then K is a matrix mapping, $k \times k$ for k discrete elements in A, where each cell defines the probability of a state transition from the first term in the parentheses to all possible states:

$$P_A = \begin{bmatrix} p(\theta_1, \theta_1) & \cdots & p(\theta_1, \theta_k) \\ \vdots & & \vdots \\ p(\theta_k, \theta_1) & \cdots & p(\theta_k, \theta_k) \end{bmatrix}, \tag{10.2}$$

where the row indicates where the chain is at this period and the column indicates where the chain is going in the next period. The rows of P_A sum to one and define a conditional PMF since they are all specified for the same starting value and cover each possible destination in the state space: for row i: $\sum_{j=1}^{k} p(\theta_i, \theta_j) = 1$. Each matrix element is a well-behaved probability, $p(\theta_i, \theta_j) \geq 0$, $\forall i, j \in A$. When the state space is continuous, then K is a conditional PDF: $f(\theta|\theta_i)$, meaning a properly defined probability statement for all $\theta \in A$, given some current state θ_i.

An important feature of the transition kernel is that transition probabilities between two selected states for arbitrary numbers of steps m can be calculated multiplicatively. For instance, with a discrete state space the probability of transitioning from the state $\theta_i = x$ at time 0 to the state $\theta_j = y$ in exactly m steps is given by the multiplicative series:

$$p^m(\theta_j^{[m]} = y | \theta_i^{[0]} = x) = \underbrace{\sum_{\theta_1} \sum_{\theta_2} \cdots \sum_{\theta_{m-1}}}_{\text{all possible paths}} \underbrace{p(\theta_i, \theta_1) p(\theta_1, \theta_2) \cdots p(\theta_{m-1}, \theta_j)}_{\text{transition products}}. \tag{10.3}$$

So $p^m(\theta_j^{[m]} = y | \theta_i^{[0]} = x)$ is also a stochastic transition matrix, and this property holds for all discrete chains exactly as given, and for continuous Markov chains with only a slight modification involving integrals rather than summations. The basic idea of (10.3) is that the complete probability of transitioning from x to y is a product of all the required intermediate steps where we sum over all possible paths that reach y from x.

10.1.2   A Markov Chain Illustration

Start with a two-dimensional discrete state space, which can be thought of as discrete vote choice between two political parties, a commercial purchase decision between two brands, or some other choice. Suppose that voters/consumers who normally select θ_1 have an 80% chance of continuing to do so, and voters/consumers who normally select θ_2 have only a 40% chance of continuing to do so. Since there are only two choices, this leads to the transition matrix **P**:

$$\begin{array}{cc} & \overbrace{\begin{array}{cc} \theta_1 & \theta_2 \end{array}}^{\text{next period}} \\ \text{current period} \left\{ \begin{array}{c} \theta_1 \\ \theta_2 \end{array} \right. & \begin{bmatrix} 0.8 & 0.2 \\ 0.6 & 0.4 \end{bmatrix}. \end{array}$$

All Markov chains begin with a starting point assigned by the researcher. This two-dimensional initial state defines the proportion of individuals selecting θ_1 and θ_2 before beginning the chain. For the purposes of this example, assign the starting point:

$$S_0 = \begin{bmatrix} 0.5 & 0.5 \end{bmatrix};$$

that is, before running the Markov chain, 50% of the population select each alternative. To get to the first state, we simply multiply the initial state by the transition matrix:

$$S_1 = \begin{bmatrix} 0.5 & 0.5 \end{bmatrix} \begin{bmatrix} 0.8 & 0.2 \\ 0.6 & 0.4 \end{bmatrix} = \begin{bmatrix} 0.7 & 0.3 \end{bmatrix} = S_1.$$

So after the first iteration we have the new proportions: 70% select θ_1 and 30% select θ_2. This process continues multiplicatively as long as we like:

$$\text{Second state:} \quad S_2 = \begin{bmatrix} 0.7 & 0.3 \end{bmatrix} \begin{bmatrix} 0.8 & 0.2 \\ 0.6 & 0.4 \end{bmatrix} = \begin{bmatrix} 0.74 & 0.26 \end{bmatrix}$$

$$\text{Third state:} \quad S_3 = \begin{bmatrix} 0.74 & 0.26 \end{bmatrix} \begin{bmatrix} 0.8 & 0.2 \\ 0.6 & 0.4 \end{bmatrix} = \begin{bmatrix} 0.748 & 0.252 \end{bmatrix}$$

$$\text{Fourth state:} \quad S_4 = \begin{bmatrix} 0.748 & 0.252 \end{bmatrix} \begin{bmatrix} 0.8 & 0.2 \\ 0.6 & 0.4 \end{bmatrix} = \begin{bmatrix} 0.7496 & 0.2504 \end{bmatrix}.$$

As one might guess, the choice proportions are converging to $[0.75, 0.25]$. This is because the transition matrix is pushing toward a steady state or more appropriately "stationary" distribution of the proportions. So when we reach this distribution all future states, **S**, are constant: $\mathbf{S}^P = \mathbf{S}$.

Imagine that this stationary distribution was the articulation of some PMF or PDF that we could not analytically describe but would like to. If we could run this Markov chain sufficiently long we would eventually get the stationary distribution *for any point in the state space*. In fact, for this simple example we could solve directly for the steady state $\mathbf{S} = [s_1, s_2]$ by stipulating:

$$\begin{bmatrix} s_1 & s_2 \end{bmatrix} \begin{bmatrix} 0.8 & 0.2 \\ 0.6 & 0.4 \end{bmatrix} = \begin{bmatrix} s_1 & s_2 \end{bmatrix},$$

and solving the resulting two equations for the two unknowns (using necessarily $s_1 + s_2 = 1$). While this example is wildly oversimplified, it serves to show some basic characteristics of Markov chains. The operation of running a Markov chain until it reaches its stationary distribution is a critical part of the process employed in MCMC estimation for Bayesian models.

■ **Example 10.1:** **A Markov Chain for Card-Shuffling.** A simplistic algorithm for shuffling a deck of cards is to take the top card and insert it uniformly back into the deck, and repeat this process many times. Thus the top card has the probability $1/52$ of being placed at any position in the stack, including returning to the top position. This is clearly a stochastic process operating on a discrete state space since there is a sequential set of identifiable states from a finite number (52!) of possible states. Is this a Markov chain though? The probability of some state being the next observed state is conditional on two things: *(1)* the current arrangement of the cards in the deck, and *(2)* the placement of the top card at this step. Therefore conditional on the current state, previous states are irrelevant and the probabilistic process depends only on the transition process and this current state. So we can claim that this shuffling process is a Markov chain.

More interestingly, what is the limiting (stationary) distribution of this Markov chain? Such examples from games of chance can be quite interesting on their own, but they also show the mechanics of basic Markov chains. See Bayer and Diaconis (1992) or Diaconis (1988) for more details. Rather than analytically look at the limiting distribution, let's simulate it with R. For simplicity (without loss of generality), we will use only $n = 3$ cards. This means that the sample space (3! elements large) is the set:

$$\{[1, 2, 3], [1, 3, 2], [2, 1, 3], [2, 3, 1], [3, 1, 2], [3, 2, 1]\}$$

and the transition kernel is:

$$K = \begin{bmatrix} \frac{1}{3} & 0 & \frac{1}{3} & \frac{1}{3} & 0 & 0 \\ 0 & \frac{1}{3} & 0 & 0 & \frac{1}{3} & \frac{1}{3} \\ \frac{1}{3} & \frac{1}{3} & \frac{1}{3} & 0 & 0 & 0 \\ 0 & 0 & 0 & \frac{1}{3} & \frac{1}{3} & \frac{1}{3} \\ \frac{1}{3} & \frac{1}{3} & 0 & 0 & \frac{1}{3} & 0 \\ 0 & 0 & \frac{1}{3} & \frac{1}{3} & 0 & \frac{1}{3} \end{bmatrix}$$

where the placement of zeros on any given row indicate that element of state space is an impossible destination given the starting point indicated by the row. So on the top row, we have the starting state $[1, 2, 3]$, making $[1, 3, 2]$, $[3, 1, 2]$, and $[3, 2, 1]$ impossible from the operation of moving the top card only. We can set up a simulation with the following:

```
P <- matrix(c(1/3,0,1/3,0,1/3,0,0,1/3,1/3,0,1/3,0,1/3,
              0,1/3,0,0,1/3,1/3,0,0,1/3,0,1/3,0,1/3,0,
              1/3,1/3,0,0,1/3,0,1/3,0,1/3),nrow=6)


MC.multiply <- function(P.in,N)  {
    P1 <- c(1,0,0,0,0,0)%*%P.in
    for (i in 1:(N-1))  {
        P1 <- P1%*%P.in
        print(P1)
    }
    P1
}

MC.multiply(P,15)
```

These 15 iterations produce:

0.2222222	0.1111111	0.2222222	0.2222222	0.1111111	0.1111111
0.1851852	0.1481481	0.1851852	0.1851852	0.1481481	0.1481481
0.1728395	0.1604938	0.1728395	0.1728395	0.1604938	0.1604938
0.1687243	0.1646091	0.1687243	0.1687243	0.1646091	0.1646091
0.1673525	0.1659808	0.1673525	0.1673525	0.1659808	0.1659808
0.1668953	0.1664380	0.1668953	0.1668953	0.1664380	0.1664380
0.1667429	0.1665905	0.1667429	0.1667429	0.1665905	0.1665905
0.1666921	0.1666413	0.1666921	0.1666921	0.1666413	0.1666413
0.1666751	0.1666582	0.1666751	0.1666751	0.1666582	0.1666582
0.1666695	0.1666638	0.1666695	0.1666695	0.1666638	0.1666638
0.1666676	0.1666657	0.1666676	0.1666676	0.1666657	0.1666657
0.1666670	0.1666664	0.1666670	0.1666670	0.1666664	0.1666664
0.1666668	0.1666666	0.1666668	0.1666668	0.1666666	0.1666666
0.1666667	0.1666666	0.1666667	0.1666667	0.1666666	0.1666666
0.1666667	0.1666666	0.1666667	0.1666667	0.1666666	0.1666666

This is not a proof, but there is evidence here that the limiting distribution is uniform across the six states. In fact running `MC.multiply` for thousands or tens of thousands of iterations produces no changes in probabilities. We could therefore conclude that this algorithm produces a shuffled deck, albeit not in the most efficient fashion.

10.1.3   The Chapman-Kolmogorov Equations

The form of (10.3) also leads to a more general notion of how chain probabilities are strung together. The Chapman-Kolmogorov equations specify how successive events are bound together probabilistically. These are given here for both discrete and continuous state spaces where we abbreviate the left-hand side expression of (10.3):

$$p^{m_1+m_2}(x,y) = \sum_{\text{all } z} p^{m_1}(x,z)p^{m_2}(z,y) \qquad \text{discrete case,}$$

$$p^{m_1+m_2}(x,y) = \int_{\text{range } z} p^{m_1}(x,z)p^{m_2}(z,y)dz \qquad \text{continuous case.} \qquad (10.4)$$

The Chapman-Kolmogorov equations are particularly elegant for the discrete case because (10.4) can be represented as a series of transition matrix multiplications:

$$p^{m_1+m_2} = p^{m_1}p^{m_2} = p^{m_1}p^{m_2-1}p = p^{m_1}p^{m_2-2}p^2 = \dots . \qquad (10.5)$$

Thus iterative probabilities can be decomposed into segmented products in any way that we like, depending on the interim steps.

10.1.4   Marginal Distributions

The final basic notational characteristic of Markov chains that we will provide here is the *marginal* distribution at some step mth from the transition kernel. For the discrete case the marginal distribution of the chain at the m step is obtained by inserting the current value of the chain, $\theta_i^{[m]}$, into the row of the transition kernel for the mth step, p^m:

$$\pi^m(\theta) = [p^m(\theta_1), p^m(\theta_2), \ldots, p^m(\theta_k)]. \tag{10.6}$$

So the marginal distribution at the first step of the Markov chain is given by:

$$\pi^1(\theta) = \pi^0(\theta)p^1, \tag{10.7}$$

where π^0 is the initial starting value assigned to the chain and $p^1 = p$ is the simple transition matrix given in (10.2). A neat consequence of the defining characteristic of the transition matrix is the relationship between the marginal distribution at some (possibly distant) step and the starting value:

$$\pi^n = p\pi^{n-1} = p(p\pi^{n-2}) = p^2(p\pi^{n-3}) = \ldots = p^n\pi^0. \tag{10.8}$$

Since it is clear here that successive products of probabilities quickly result in lower probability values, the property above shows how Markov chains eventually "forget" their starting points. The marginal distribution for the continuous case is only slightly more involved since we cannot just list as a vector the quantity:

$$\pi^m(\theta_j) = \int_\theta p(\theta, \theta_j)\pi^{m-1}(\theta)d\theta, \tag{10.9}$$

which is the marginal distribution of the chain, given that the chain is currently on point θ_j at step m.

10.2   General Properties of Markov Chains

There are several properties of Markov chains that are important to us, particularly when discussing long run Markov chain stability. These properties have intimidating names that are inherited from mathematical Markov chain theory, but in reality are fairly straightforward ideas. Generally, if we can describe the mathematical status of a particular chain, then we can often determine if it is capable of producing a useful sample from the distribution of interest. The properties are only briefly summarized here and those interested in a more technical and detailed treatment should read: Gamerman and Lopes (2006), Iosifescu (1980), Norris (1997), Nummelin (1984), or Tierney (1996).

10.2.1   Homogeneity

A *homogeneous* Markov chain at step n has a transition probability that does not depend on the value of n. So the decision to move at this step is independent of this being the current point in time. Interestingly, but not importantly, at the starting point the chain cannot be homogeneous since the marginal distribution for the first step is clearly not independent of the initial values, which are hand-picked. One reason that the Gibbs sampler and the Metropolis-Hastings algorithm, both given in detail in this chapter, dominate MCMC implementations is that the chains they define possess this critical property: there is no explicit value of n that governs the transition kernel.

10.2.2   Irreducibility

There are also properties directly associated with states. A state is *absorbing* if once the chain enters this state it cannot leave: $p(A, A^c) = 0$. The obverse of absorbing is *transient*. A state is transient if, given that a chain currently occupies state A, the probability of not returning to A is non-zero. A more relevant case of absorbing is the situation where a state, A, is *closed* with regard to some other state, B: $p(A, B) = 0$.

A Markov chain is *irreducible* if every point or collection of points (a subspace, required in the continuous case), A, can be reached from every other point or collection of points. A convenient way to remember the principle behind irreducibility is the notion that you could reduce the set if you wanted to (this is obviously always possible except for the null set), but that *you do not want to* because then there will be points that cannot be reached from other points. This means that $p(\theta_i, \theta_j) \neq 0$, $\forall i, j \in A$. Notice that irreducibility is a characteristic of both the chain and the subspace. So irreducibility implies the existence of a path between any two points in the subspace. The key relationship of interest is between irreducibility and recurrence.

10.2.3   Recurrence

If a subspace is closed, finite, and irreducible, then all states within this subspace are recurrent. Recurrence is a desirable property of Markov chains. An irreducible Markov chain is called *recurrent* with regard to a given state, A, which is a single point or a defined collection of points (required for the bounded-continuous case), if the probability that the chain occupies A infinitely often over unbounded time is nonzero. This can also be restated by saying that the expected number of returns to A in the limit is infinity. The Markov chain is *positive recurrent* if the mean time to return to A is bounded, otherwise it is called *null recurrent*. This characteristic of Markov chains was introduced by Doeblin (1940) (obviously without knowing of its eventual importance in practical Bayesian applications).

If we only had to deal with discrete or finite-bounded state spaces then this form of recurrence would be enough, but with unbounded-continuous state spaces it is necessary to have a stricter definition of recurrence that guarantees that the probability of visiting

A infinitely often in the limit is now one: *Harris recurrence* (Harris 1956). First we say that a set *A* is Harris recurrent if the probability of the chain visiting *A* infinitely often in the limit is one. An irreducible Markov chain is Harris recurrent if, for a finite probability measure $\mathfrak{P}$, the chain at time *n* has for all such finite subsets *A* of the measure space a non-zero probability of reaching *A* (Athreya and Ney 1978). Formally: if there exists a σ-finite probability measure $\mathfrak{P}$ on the measure space **S** so that an irreducible Markov chain, X_n, at time *n* has the property: $p(X_n \in A) = 1$, $\forall A \in S$ where $\mathfrak{P}(A) > 0$.

The distinction between recurrence and Harris recurrence is important in demonstrating the convergence of specific Markov chain algorithms for continuous state spaces. In fact, an aperiodic, irreducible chain with an invariant distribution on an unbounded continuous state space that is *not* Harris recurrent has a positive probability of getting stuck forever in an area bounded away from convergence, given a starting point there.

Additional details on recurrence are found in the key works in this area: Meyn and Tweedie (1994), Tierney (1994), and Athreya, Doss, and Sethuraman (1996). For our purposes it is sufficient to simply consider the point: "Harris recurrence essentially says that there is no measure-theoretic pathology" (Chan 1994). The greatest concern here is that ill-chosen starting points can cause eventual problems. and Geyer

Given a set of recurrent states (nonempty, and bounded or countable) or Harris recurrent states, then the union of these states creates a new state that is closed and irreducible (Meyn and Tweedie 1993). This means that the linkage between recurrence and irreducibility is important in defining a subspace that captures a Markov chain and at the same time assures that this Markov chain will explore all of the subspace. Whenever a chain wanders into a closed, irreducible set of Harris recurrent states, it then stays there and visits every single state (eventually) with probability one (replaces almost sure convergence with convergence at every point).

10.2.4 Stationarity

Define $\pi(\theta)$ as the stationary distribution of the Markov chain for θ on the state space *A*. We denote $p(\theta_i, \theta_j)$ to indicate the probability that the chain will move from θ_i to θ_j at some arbitrary step *t* from the transition kernel, and $\pi^t(\theta)$ as the marginal distribution. This stationary distribution is then defined as satisfying:

$$\sum_{\theta_i} \pi^t(\theta_i) p(\theta_i, \theta_j) = \pi^{t+1}(\theta_j) \qquad \text{Discrete case}$$

$$\int \pi^t(\theta_i) p(\theta_i, \theta_j) d\theta_i = \pi^{t+1}(\theta_j) \qquad \text{Continuous case.} \qquad (10.10)$$

Therefore multiplication by the transition kernel and evaluating for the current point (the summation step for discrete sample spaces and the integration step for continuous sample spaces) produces the same marginal distribution: $\pi = \pi p$ in shorthand. This demonstrates that the marginal distribution remains fixed when the chain reaches the stationary distribution and we might as well drop the superscript designation for iteration number and just use $\pi(\theta)$.

Once the chain reaches its stationary distribution (also called its *invariant distribution, equilibrium distribution,* or *limiting distribution* if discussed in the asymptotic sense), it stays in this distribution and moves about, or "mixes," throughout the subspace according to marginal distribution, $\pi(\theta)$, forever. This is exactly what we want and expect from MCMC. If we can set up the Markov chain so that it reaches a stationary distribution that is the desired posterior distribution from our Bayesian model, then all we need to do is let it wander about this subspace for a while, producing empirical samples to be summarized. The good news is that the two primary forms of MCMC kernels we will use have the property that they are guaranteed to eventually reach a stationary distribution that is the desired posterior distribution.

10.2.5   Ergodicity

It is also possible to define the *period* of a Markov chain. This is simply the length of time to repeat an identical cycle of chain values. It is desirable to have an aperiodic chain, i.e., where the only length of time for which the chain repeats some cycle of values is the trivial case with cycle length equal to one. Why? It seems as though we would not necessarily care if there were some period to the chain values, particularly if the period were quite long, or perhaps in the discrete state if it included every value in the state space. The answer is that the recurrence property alone is not enough to assure that the chain reaches a state where the marginal distribution remains fixed and identical to the posterior of interest.

If a chain is irreducible, positive Harris recurrent, and aperiodic, then we call it *ergodic.* Ergodic Markov chains have the property:

$$\lim_{n \to \infty} p^n(\theta_i, \theta_j) = \pi(\theta_j), \tag{10.11}$$

for all θ_i, and θ_j in the subspace (Nummelin 1984). Therefore, in the limit, the marginal distribution at one step is identical to the marginal distribution at all other steps. Better yet, because of the recurrence requirement, this limiting distribution is now closed and irreducible, meaning that the chain will never leave it and will eventually visit every point in the subspace. Once a specified chain is determined to have reached its ergodic state, sample values behave as if they were produced by the posterior of interest from the model.

The *ergodic theorem* is analogous to the strong law of large numbers but for Markov chains. It states that any specified function of the posterior distribution can be estimated with samples from a Markov chain in its ergodic state because averages of sample values give strongly consistent parameter estimates. More formally, suppose $\theta_{i+1}, \ldots, \theta_{i+n}$ are n (not necessarily consecutive) values from a Markov chain that has reached its ergodic distribution, a statistic of interest, $h(\theta)$, can be calculated empirically:

$$\hat{h}(\theta_i) = \frac{1}{n} \sum_{j=i+1}^{i+n} h(\theta_j) \approx h(\theta), \tag{10.12}$$

and for finite quantities this converges almost surely: $p[\hat{h}(\theta_i) \to h(\theta),$ as $n \to \infty] = 1$ (Roberts and Smith 1994, p.210; Tierney 1994, p.1717). The remarkable result from ergodicity is that even though Markov chain values, by their very definition, have serial dependence, the mean of the chain values provides a strongly consistent estimate of the true parameter. For a given empirical estimator $\hat{h}(\theta_i)$ with bounded limiting variance, we get the central limit theorem results:

$$\sqrt{n} \frac{\hat{h}(\theta_i) - h(\theta)}{\sqrt{\mathrm{Var}(\hat{h}(\theta_i))}} \xrightarrow[n \to \infty]{} \mathcal{N}(0,1). \tag{10.13}$$

This is an important principle for MCMC estimation because it says that we can take the simulation values from the stationary distribution and safely ignore the serial nature of their production. For more detailed theoretical justifications, see Meyn and Tweedie, (1993, Chapter 15), or Tierney (1994).

10.3   The Gibbs Sampler

The Gibbs sampler, originating with Geman and Geman (1984), is the most widely used MCMC technique. This is a testament to its simplicity and reliability as a method for producing useful chain values. The Gibbs sampler requires specific knowledge about the conditional nature of the relationship between the variables of interest. The basic idea, which is not difficult to conceptualize, is that if it is possible to express each of the coefficients to be estimated as conditioned on the others, then by cycling through these conditional statements, we can eventually reach the true joint distribution of interest.

10.3.1   Description of the Algorithm

The Gibbs sampler is a transition kernel created by a series of full conditional distributions that is a Markovian updating scheme based on conditional probability statements. If the limiting distribution of interest is $\pi(\boldsymbol{\theta})$ where $\boldsymbol{\theta}$ is a k length vector of coefficients whose posterior distribution we want to describe, then the objective is to produce a Markov chain that cycles through these conditional statements moving toward and then around this distribution. The set of full conditional distributions for $\boldsymbol{\theta}$ are denoted $\boldsymbol{\Theta}$ and defined by $\pi(\boldsymbol{\Theta}) = \pi(\theta_i | \boldsymbol{\theta}_{-i})$ for $i = 1, \ldots, k$, where the notation $\boldsymbol{\theta}_{-i}$ indicates a specific parametric form from $\boldsymbol{\Theta}$ without the θ_i coefficient.

There must be an analytically definable full conditional statement for each coefficient in the $\boldsymbol{\theta}$ vector and these probability statements need to be completely articulated so that it is possible to draw samples from the described distribution. This requirement facilitates the iterative nature of the Gibbs sampling algorithm, which cycles through these full conditionals drawing parameter values based on the most recent version of all of the previous parameters in the list. The order does not matter, but it is essential that the most recent draws from the other samples be used. This looks like the following procedure (note the use of the most recent iteration values at each step):

1. Choose starting values: $\boldsymbol{\theta}^{[0]} = [\theta_1^{[0]}, \theta_2^{[0]}, \ldots, \theta_k^{[0]}]$

2. At the jth starting at $j = 1$ complete the single cycle by drawing values from the k distributions given by:

$$\theta_1^{[j]} \quad \sim \quad \pi(\theta_1 \quad | \quad \theta_2^{[j-1]}, \quad \theta_3^{[j-1]}, \quad \theta_4^{[j-1]}, \quad \ldots, \quad \theta_{k-1}^{[j-1]}, \quad \theta_k^{[j-1]})$$

$$\theta_2^{[j]} \quad \sim \quad \pi(\theta_2 \quad | \quad \theta_1^{[j]}, \quad \theta_3^{[j-1]}, \quad \theta_4^{[j-1]}, \quad \ldots, \quad \theta_{k-1}^{[j-1]}, \quad \theta_k^{[j-1]})$$

$$\theta_3^{[j]} \quad \sim \quad \pi(\theta_3 \quad | \quad \theta_1^{[j]}, \quad \theta_2^{[j]}, \quad \theta_4^{[j-1]}, \quad \ldots, \quad \theta_{k-1}^{[j-1]}, \quad \theta_k^{[j-1]})$$

$$\vdots$$

$$\theta_{k-1}^{[j]} \quad \sim \quad \pi(\theta_{k-1} \quad | \quad \theta_1^{[j]}, \quad \theta_2^{[j]}, \quad \theta_3^{[j]}, \quad \ldots, \quad \theta_{k-2}^{[j]}, \quad \theta_k^{[j-1]})$$

$$\theta_k^{[j]} \quad \sim \quad \pi(\theta_k \quad | \quad \theta_1^{[j]}, \quad \theta_2^{[j]}, \quad \theta_3^{[j]}, \quad \ldots, \quad \theta_{k-2}^{[j]}, \quad \theta_{k-1}^{[j]})$$

3. Increment j and repeat until convergence.

Once convergence is reached, all simulation values are from the target posterior distribution and a sufficient number should then be drawn so that all areas of the posterior are explored. Notice the important feature that during each iteration of the cycling through the $\boldsymbol{\theta}$ vector, conditioning occurs on $\boldsymbol{\theta}$ values that have already been sampled for that cycle; otherwise the $\boldsymbol{\theta}$ values are taken from the last cycle. So in the last step for a given j cycle, the sampled value for the kth parameter gets to condition on *all* j-step values.

The statements above clearly demonstrate that it is required to have the full set of conditional distributions to run the Gibbs sampling algorithm. As we will see in Chapter 12, these necessary statements often fall naturally out of the hierarchical conditional relationships. In some cases they come from classic theory. For example, a simple bivariate normal specification for θ_1 and θ_2 given by: $\mathcal{N}\left(\left[\begin{smallmatrix} 0 \\ 0 \end{smallmatrix}\right], \left[\begin{smallmatrix} 1 & \rho \\ \rho & 1 \end{smallmatrix}\right]\right)$, gives the set of full conditional distributions:

$$\theta_1^{[j]} | \theta_2^{[j-1]} \sim \mathcal{N}(\rho\theta_2^{[j-1]}, 1 - \rho^2)$$

$$\theta_2^{[j]} | \theta_1^{[j]} \sim \mathcal{N}(\rho\theta_1^{[j]}, 1 - \rho^2).$$

If the Gibbs sampler has run sufficiently long, forthcoming full cycles of the algorithm produce a complete sample of the coefficients in the $\boldsymbol{\theta}$ vector. All future iterations from this point on produce samples from the desired limiting distribution and can therefore be described empirically. The most impressive aspect of the Gibbs sampler is that these conditional distributions contain sufficient information to eventually produce a sample from the full joint distribution of interest.

10.3.2 Handling Missing Dichotomous Data with the Gibbs Sampler

Suppose that a coin is flipped 20 times and the result is recorded such that a head is fixed as 1 and a tails is fixed as 0. Of these $X_1, X_2, \ldots, X_{19}$ are observed and the last, X_{20}, is lost or unobserved. The standard question for this experiment is whether or not the coin is fair, or more generally, what is the probability that coin produces a head, θ.

A useful simplification is to treat the sum of the observed 0/1 outcomes as a single random variable as done in Example 2.3.4 (page 49), $Y = \sum_{i=1}^{19} \mathbf{X}_i$, and ignore the missing value. Thus Y is now distributed according to the binomial probability mass function:

$$p(Y|\theta) = \binom{n}{Y} \theta^Y (1-\theta)^{n-Y}$$

where $n = 19$ here. Of course this neglects the effect of the missing value (X_{20}) and may cause bias. However, we can use Gibbs sampling to estimate its value simultaneously along with the unknown θ. Clearly, the PMF of the missing value is still a Bernoulli form conditional on θ. If we were able to treat Y and X_{20} as known (observed) then the distribution of θ would be:

$$p(\theta|Y, X_{20}) = \binom{20}{Y + X_{20}} \theta^{Y + X_{20}} (1-\theta)^{20 - Y - X_{20}}$$

which gives the kernel of a beta probability density function. This means that we now have two full conditional distributions, given temporary values for each unknown, θ^* and X_{20}^*:

$$\theta|X_{20}^*, Y \sim \text{beta}(Y + X_{20}^* + 1, 20 - Y - X_{20}^* + 1)$$
$$X_{20}|Y, \theta^* \sim \text{Bernoulli}(\theta^*).$$

This is an ideal setup for Gibbs sampling since the joint distribution can be treated as a full conditional distribution for either θ or X_{20} if the other is given a temporary value. So if our fundamental interest lies in estimating θ without bias, this process explicitly incorporates the missing value rather than ignoring it.

We begin with the sample: $\{1,1,1,1,1,0,1,1,1,1,0,1,1,0,1,1,0,0,1\}$, and run the chain for 50,000 iterations. Let us consider the first 20,000 iterations as "burn-in" while the Markov chain finds its limiting distribution and dispose of them as pre-convergence values. The posterior mean of the rest of the simulated θ values is 0.714 (with variance 0.009) suggesting that coin is far from "fair." This is also different than the observed data

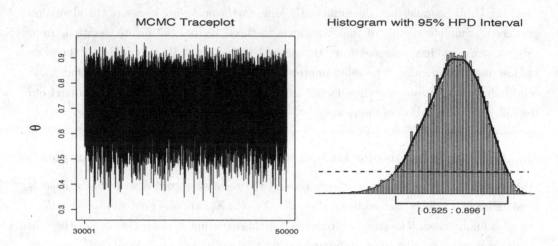

FIGURE 10.1: Gibbs Sampling Demonstration

mean of 0.737, which would have been a biased estimator. The posterior mean for the missing datapoint is 0.723 making it seem more likely to be a one than a zero, although the posterior variance of approximately 0.2 suggests some caution in the claim. However, regardless of our confidence in the posterior claim for the missing data, the estimate of θ remains unbiased.

We can, perhaps more usefully, describe the output of the Gibbs sampler graphically, as done in the two-panel Figure 10.1. Here we have a traceplot that shows the time-series of the Markov chain path and a histogram of the last 30,000 draws. One advantage to describing posterior distributions with MCMC is that summary statistics are trivial to generate since we can calculate various quantities directly on the empirical draws. The 95% highest posterior density region was produced from taking a histogram with a large number of bins (300 rather than the 100 shown in the graph) and sorting the density values observing where the thresholds of interest fall. Naturally this is just an approximation since a histogram is not a perfectly smooth density, but the parameters are sufficiently adjustable for 30,000 draws that it can be a very accurate approximation. Importance sampling can also be used to produce HPD regions as described on page 301.

■ **Example 10.2: Changepoint Analysis for Terrorism in Great Britain: 1970-2004.** This example looks at a series of terrorist events over a 35 year period in the U.K. in which there were injuries and/or fatalities. This period, 1970 to 2004, is dominated by events related to the "the troubles" in Northern Ireland and the Provisional IRA's objective of influencing British policy by bombing in England. The beginning date is selected to roughly coincide with three important events: the split of the IRA into the Official IRA and the Provisional IRA (1969), the start of internment for IRA members (1971), and the "Bloody Sunday" march in which British paratroopers killed

14, and injured 13, demonstrators. These data are subsetted from TWEED ("Terrorism in Western Europe: Events Data") compiled by Jan Oskar Engene and made available at `http://folk.uib.no/sspje/tweed.htm` as well as in BaM. The data are characterized by relatively high counts in the early era and relatively low counts in the late era. Thus the question is when was there a pronounced change in terrorism rates? The data are shown in Table 10.1.

TABLE 10.1: COUNT OF TERRORISM
INCIDENTS IN BRITAIN, 1970-2004 BY ROWS

1	21	23	9	15	17	18	5	4	8	4	1
5	2	3	0	1	2	3	3	6	5	3	2
3	0	1	1	1	1	0	0	0	0	0	

Statistically, the objective is to use this sequence to estimate the *changepoint* and also to obtain posterior estimates of the intensity parameters of the two separate Poisson processes (Poisson because these are counts). This process is addressed more generally for an unknown number of changepoints by Phillips and Smith (1996).

Specifically, $x_1, x_2, \ldots, x_n$ are a series of count data where there exists the possibility of a changepoint at some period, k, along the series. Therefore there are two Poisson data-generating processes:

$$x_i|\lambda \sim \mathcal{P}(\lambda) \qquad i = 1, \ldots, k$$
$$x_i|\phi \sim \mathcal{P}(\phi) \qquad i = k+1, \ldots, n,$$

where the determination of which to apply depends on the location of the changepoint k. So now there are three parameters to estimate: λ, ϕ, and k. This problem is distinguished by the added complexity that one of the parameters, k, operates in a different capacity on the others: determining a change in the serial data generation process, rather than as a conventional parametric input.

The three independent priors applied to this model are:

$$\lambda \sim \mathcal{G}(\alpha, \beta)$$
$$\phi \sim \mathcal{G}(\gamma, \delta)$$
$$k \sim \text{discrete uniform on}[1, 2, \ldots, n],$$

where for purposes of this example, the prior parameters are assigned according to: $\alpha = 4$, $\beta = 1$, $\gamma = 1$, $\delta = 2$. Since the mean of a gamma distribution is the ratio of its parameters (in the scale version of the gamma distribution), this assignment of parameters roughly is close to, but slightly larger than, the mean of the first 50% of the data ($\alpha\beta$), and the second 50% of the data ($\gamma\delta$). This leads to the following joint posterior and its proportional simplification:

$$\pi(\lambda, \phi, k|\mathbf{y}) \propto L(\lambda, \phi, k|\mathbf{y})\pi(\lambda|\alpha, \beta)\pi(\phi|\gamma, \delta)\pi(k)$$

$$= \left(\prod_{i=1}^{k} \frac{e^{-\lambda}\lambda^{y_i}}{y_i!}\right)\left(\prod_{i=k+1}^{n} \frac{e^{-\phi}\phi^{y_i}}{y_i!}\right)\left(\frac{\beta^{\alpha}}{\Gamma(\alpha)}\lambda^{\alpha-1}e^{-\beta\lambda}\right)$$

$$\times \left(\frac{\delta^{\gamma}}{\Gamma(\gamma)}\phi^{\gamma-1}e^{-\delta\phi}\right)\frac{1}{n}$$

$$\propto \lambda^{\alpha-1+\sum_{i=1}^{k} y_i}\phi^{\gamma-1+\sum_{i=k+1}^{n} y_i}\exp[-(k+\beta)\lambda - (n-k+\delta)\phi].$$

The last line easily provides the full conditional distribution for λ and ϕ:

$$\lambda|\phi, k \sim \mathcal{G}\left(\alpha + \sum_{i=1}^{k} y_i, \beta + k\right)$$

$$\phi|\lambda, k \sim \mathcal{G}\left(\gamma + \sum_{i=k+1}^{n} y_i, \delta + n - k\right),$$

but the full conditional distribution for k requires more work. Start with the likelihood function under the assumption that λ and ϕ are fixed and rearrange:

$$p(\mathbf{y}|k, \lambda, \phi) = \left(\prod_{i=1}^{k} \frac{e^{-\lambda}\lambda^{y_i}}{y_i!}\right)\left(\prod_{i=k+1}^{n} \frac{e^{-\phi}\phi^{y_i}}{y_i!}\right)$$

$$= \left(\prod_{i=1}^{n} \frac{1}{y_i!}\right)e^{k(\phi-\lambda)}e^{-n\phi}\lambda^{\sum_{i=1}^{k} y_i}\left(\prod_{i=k+1}^{n} \phi^{y_i}\right)\left(\prod_{i=1}^{k} \frac{\phi^{y_i}}{\phi^{y_i}}\right)$$

$$= \left(\prod_{i=1}^{n} \frac{e^{-\phi}\phi^{y_i}}{y_i!}\right)\left(e^{k(\phi-\lambda)}\left(\frac{\lambda}{\phi}\right)^{\sum_{i=1}^{k} y_i}\right)$$

$$= f(\mathbf{y}, \phi)L(\mathbf{y}|k, \lambda, \phi). \tag{10.14}$$

What this does is provide two functions, the first of which is free of k. Since our objective is simply a full conditional statement for k, we will use only the modified likelihood function that contains k, $L(\mathbf{y}|k, \lambda, \phi)$, after a cancellation below. Suppressing the conditioning on λ and ϕ for notational clarity only, apply Bayes' Law with a generic prior on k, $p(k)$:

$$p(k|\mathbf{y}) = \frac{f(\mathbf{y}, \phi)L(\mathbf{y}|k)p(k)}{\sum_{\ell=1}^{n} f(\mathbf{y}, \phi)L(\mathbf{y}|k_\ell)p(k_\ell)}$$

$$= \frac{L(\mathbf{y}|k)p(k)}{\sum_{\ell=1}^{n} L(\mathbf{y}|k_\ell)p(k_\ell)}. \tag{10.15}$$

Here we took advantage of the discrete feature of k and summed over all possible values with the index ℓ. For simple priors like our discrete uniform, this simplifies even more such that with proportionality now $p(k|\mathbf{y}) = L(\mathbf{y}|k)/\sum_{\ell=1}^{n} L(\mathbf{y}|k_\ell)$. So each iteration of the Gibbs sampler will calculate an n-length probability vector for k and draw a value accordingly. This is implemented in the following R code:

```
cp.gibbs <- function(theta.matrix,y,a,b,g,d)  {
    n <- length(y); y.bar <- mean(y)
    k.prob <- rep(0,length=n)
    for (i in 2:nrow(theta.matrix))  {
        lambda <- rgamma(1,a+sum(y[1:theta.matrix[(i-1),3]]),
            rate=b+theta.matrix[(i-1),3])
        phi    <- rgamma(1,g+sum(y[theta.matrix[(i-1),3]:n]),
            rate=d+length(y)-theta.matrix[(i-1),3])
        for (j in 1:n)  k.prob[j] <- exp( j*(phi-lambda) +
                        log(lambda/phi)*sum(y[1:j]-y.bar/j) )
        k.prob <- (lambda/phi)^y.bar*k.prob
        k      <- sample(1:n,size=1,prob=k.prob)
        theta.matrix[i,] <- c(lambda,phi,k)
    }
    theta.matrix
}
```

There are a few subtleties in the code. In the second for-loop the data y[] are mean-centered and this is undone in the next line to increase numerical stability (large count values can send entries of `k.prob` to infinity). Notice also that the function does not normalize the `k.prob` vector. This is because the `sample` function will do it automatically in cases where the vector values do not sum to one.

TABLE 10.2: Gibbs Sampler
Draws for k Parameter

Value	7	8	9	10	11	21
Count	4352	330	10	283	24	1

The function is run for 10,000 iterations retaining only the last 5,000 chain values, which are summarized in Table 10.3. This is a conservative strategy with this simple model and there is no evidence of non-convergence for the Markov chain. The quantile results for k seem confusing at first until one remembers that k is a discrete variable. The Markov chain therefore jumps from integer value to integer value. In fact, the chain had strong "preferences" as indicated by the values shown in Table 10.2. The Markov chain was started with $k = 21$, which explains that isolated value.

Figure 10.2 shows the first 100 iterations of the sampler in a pair of two-dimensional graphs where k is depicted on the y-axis in both cases plotted against λ and ϕ, respectively. Notice that the Gibbs sampler converges quickly to the region of the reported posterior.

TABLE 10.3:   Terrorism in Great Britain Model, Posterior Summary

Quantile	λ	ϕ	k
Minimum	8.61	1.39	7.00
First quartile	12.40	2.60	7.00
Median	13.30	2.85	7.00
Third quartile	14.40	3.09	7.00
Maximum	18.90	4.10	11.00
Mean	13.30	2.83	7.29

Each movement is a straight line here since the intermediate steps of the Gibbs sampler hold all other parameter values constant while sampling for a single parameter conditioned on these other values. The k dimension is jittered slightly to reveal the concentration at $k = 7$. So looking at the first step in the left-hand side panel of Figure 10.2, we begin by conditioning on $k = 21$ and move from the starting value $\lambda = 1.0$ to a sampled value of 6.447. Then the second half of the first iteration holds this value for λ constant and draws a value for k of 10. After these two steps, the first full iteration of the sampler is done. This process then continues for 4999 more iterations. Similarly, in the second panel the sampler goes from the same starting point to $\phi = 2.084$. Note that the algorithm proceeds rapidly to a more preferred region.

■ **Example 10.3:   Changepoint Analysis for Military Fatalities in Afghanistan.** A more topical example comes from looking at military fatalities among coalition forces in Afghanistan for *Operation Enduring Freedom*, the ongoing effort to depose and defeat the Taliban government forces that had sheltered Al Queda. The monthly count data covers the period from October 2001 to January 2007 and are culled from U.S. military and NATO sources, and are provided in Table 10.4 for 52 monthly periods, listed by rows.

TABLE 10.4:   NATO Fatalities in Afghanistan, 10/01 to 1/07

3	5	4	10	12	14	9	1	3	0	3	1	6	1	8	4	7	12	2	2	7
2	4	2	6	8	1	11	2	3	3	9	5	2	3	4	8	7	1	3	2	6
19	4	29	2	33	12	10	7	4	0											

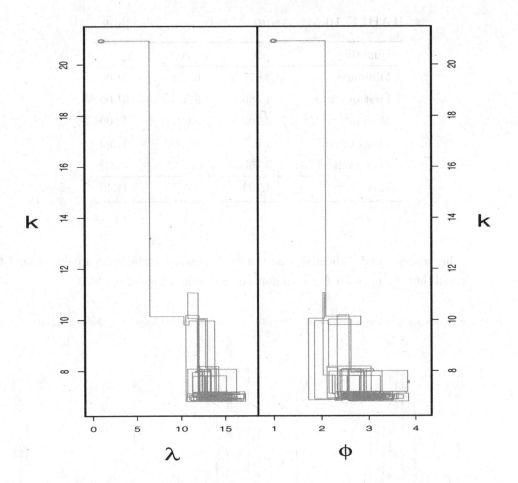

FIGURE 10.2: Gibbs Sampling Path for Terrorism Events

There is strong evidence that the resurgent Taliban made a strategic decision at some point about the possibility of winning and correspondingly escalated their activities. Naturally, we cannot know for certain the exact timing of such a decision, but it may be possible to estimate the timing by looking for a changepoint in coalition fatalities. Looking at the data for only a short period of time provides some idea that there is a measurable difference in the early periods and the later periods. In fact, the mean of the first 40 periods is about 4.95 and the last ten periods 10.66. Using these means we construct gamma priors for λ and ϕ with $\alpha = 1$, $\beta = 2$, $\gamma = 4$, and $\delta = 4$. The same algorithm from the Terrorism in Great Britain example is used, producing marginal posteriors summarized in Table 10.5.

This turns out to be quite revealing. The posterior mean for the later period intensity parameter is about twice the size of that for the early period. This supports the popular conception that the war has become more "hot" in more recent years. Also, it appears that there is strong evidence that the change occurred around the 42nd month, which is March of 2005. Figure 10.3 shows that the posterior for the changepoint is

TABLE 10.5:   AFGHAN FATALITIES POSTERIOR

Quantile	λ	ϕ	k
Minimum	3.675	6.258	31.00
First quartile	4.480	8.533	31.00
Median	4.703	9.035	42.00
Third quartile	4.913	9.590	42.00
Maximum	5.755	12.532	44.00
Mean	4.703	9.057	41.49

highly concentrated, indicating that the data are strongly asserting a point around 40 (recall that the prior for k was uniform over the full range of months).

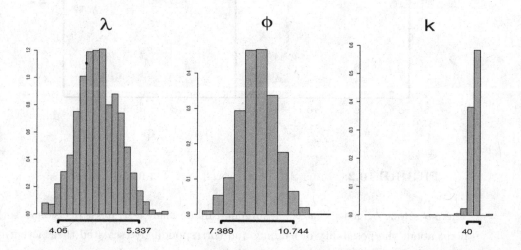

FIGURE 10.3: AFGHAN FATALITIES POSTERIOR HISTOGRAM, HPD REGION

These changepoint models have many varieties and extensions. The classic, early article by Berry and Hartigan (1993) is an excellent starting point. Other common parametric forms for the outcomes are normal (Menzefricke 1981, Skates, Pauler, and Jacobs 2001) and binomial/negative binomial (Zhao and Chu 2006, Smith 1975). While we have only covered one changepoint here in two examples, it is common to have multiple changepoints that need to be estimated and sometimes an unknown number of them (Chib 1998, Yang and Kuo 2001, Chen and Gupta 1997, Braun, Braun, and Muller 2000, Stephens 1994, Suchard *et al.* 2003, Fearnhead 2006, Green 1995). Finally, see the paper by Carlin *et al.* (1992) for a hierarchical model implementation of changepoint models.

10.3.3   Summary of Properties of the Gibbs Sampler

We finish this section with a summary of the properties of the Gibbs sampler that make it the most commonly used MCMC kernel specification. These properties are covered in greater detail in Chapter 13, for those interested in the underlying technical issues.

▷ Since the Gibbs sampler conditions on values from the last iteration of its chain values, it clearly constitutes a Markov chain.

▷ The Gibbs sampler has the true posterior distribution of the parameter vector as its limiting distribution.

▷ The Gibbs sampler is a homogeneous Markov chain: the consecutive probabilities are independent of n, the current length of the chain.

▷ The Gibbs sampler converges at a geometric rate: the total variation distance between an arbitrary time and the point of convergence decreases at a geometric rate in time (t).

▷ The Gibbs sampler is an ergodic Markov chain.

See Roberts and Polson (1994) and Tierney (1994) for the (very general) conditions. Of course the number of iterations required is still a function of the complexity of the model (hierarchies, hyperprior assignments, data characteristics, etc.), and therefore a motivation for having convergence diagnostics (Chapter 14). Casella and George (1992) provide a very clear basic introduction to the Gibbs sampler and its properties (the original title of the conference paper was "Gibbs for Kids"), a piece of which we have already seen in Chapter 1. Also, a very useful discussion of the Gibbs sampler and its relation to other MCMC techniques is provided by Smith and Roberts (1993).

Credit for introducing the Gibbs sampler on finite state spaces is usually given to Geman and Geman (1984), but Ulf Grenander (a student of Harald Cramér) actually applied it to Bayesian modeling in a well-known but unpublished paper in 1983. Early restricted versions, typically labeled as the *heatbath algorithm* in statistical physics can be found in Creutz (1979), Creutz, Jacobs and Rebbi (1983), and Ripley (1979).

10.4   The Metropolis-Hastings Algorithm

The full set of conditional distributions for the Gibbs sampler are often quite easy to specify from the hierarchy of the model since conditional relationships are directly articulated in such statements. However, the Gibbs sampler of Geman and Geman (1984) obviously does not work when the complete conditionals for the $\boldsymbol{\theta}$ parameters do not have an easily obtainable form. In these cases a chain can be produced for these parameters using the Metropolis-Hastings algorithm from statistical physics (Chib and Greenberg 1995;

Metropolis *et al.* 1953; Hastings 1970; Peskun 1973) which is often applied in fields completely unrelated to the original application (e.g., Cohen *et al.* 1998).

10.4.1 Background

The original work by Metropolis *et al.* postulated a two-dimensional enclosure with $n = 10$ molecular particles, and sought to estimate the state-dependent total energy of the system at equilibrium. The central problem that they confronted is that there is an incredibly large number of locations for the molecules in the system that must be accounted for and this number grows exponentially with time. The key contribution of Metropolis *et al.* is to model the system by generating moves that are more likely than others based on positions that are calculated using uniform probability generated candidate jump points. The moves are accepted probabilistically and likely final states are determined after a set of periods where many such decisions are made. Therefore the simulation produces an estimated force based on a statistical, rather than deterministic, arrangement of particles. The critical assumptions are already familiar to us: any molecular state can be reached from another (ergodicity), and state changes are induced probabilistically with an instrumental distribution. The result, after convergence, is a distribution of particles from which energy calculations can be made.

10.4.2 Description of the Algorithm

The simplest Metropolis-Hastings algorithm for a single selected parameter vector works as follows. Suppose we have a J-length parameter vector, $\boldsymbol{\theta} \in \boldsymbol{\Theta}^J$ to estimate, with J determining the dimension of the state space and the posterior of interest, $\pi(\theta)$. At the tth step of the Markov chain when the chain is at the position $\boldsymbol{\theta}$, draw $\boldsymbol{\theta}'$ from a distribution over the same support. This distribution is called the instrumental, jumping, or proposal distribution and is denoted $q_t(\boldsymbol{\theta}'|\boldsymbol{\theta})$. There is an obvious dependency on the data in both the posterior and the candidate-generating forms, but we will suppress this for notational clarity. Note that, unlike the Gibbs sampler, we are producing a multidimensional candidate value all at once and not serially throughout the $j = 1, \ldots, J$. It must be possible to determine the reverse function value, $q_t(\boldsymbol{\theta}|\boldsymbol{\theta}')$, and under the original constraints of Metropolis *et al.* the two conditionals need to be equal (symmetry), although we now know that this is not necessary.

Define the *acceptance ratio* (sometimes called the acceptance function) to be the following:

$$a(\boldsymbol{\theta}', \boldsymbol{\theta}) = \frac{\pi(\boldsymbol{\theta}')}{\pi(\boldsymbol{\theta})} \frac{q_t(\boldsymbol{\theta}|\boldsymbol{\theta}')}{q_t(\boldsymbol{\theta}'|\boldsymbol{\theta})}. \tag{10.16}$$

At time t, the decision that produces the $t + 1^{st}$ point in the chain is probabilistically determined according to:

$$\boldsymbol{\theta}^{[t+1]} = \begin{cases} \boldsymbol{\theta}' & \text{with probability} \quad \min(a(\boldsymbol{\theta}', \boldsymbol{\theta}^{[t]}), 1) \\ \\ \boldsymbol{\theta}^{[t]} & \text{with probability} \quad 1 - \min(a(\boldsymbol{\theta}', \boldsymbol{\theta}^{[t]}), 1). \end{cases} \qquad (10.17)$$

Several features of this algorithm are interesting. Observe that because of the ratio in the decision above, we only have to know the posterior distribution $\pi(\boldsymbol{\theta})$, up to a constant of proportionality. Note also that unlike the Gibbs sampler, the Metropolis-Hastings algorithm does not necessitate movement on every iteration: the next chosen location may be the current location. Finally, it is easy to see that in the case of symmetry in the candidate-generating density, $q_t(\boldsymbol{\theta}|\boldsymbol{\theta}') = q_t(\boldsymbol{\theta}'|\boldsymbol{\theta})$, the acceptance ratio simplifies to a ratio of just the posterior densities at the two points, which is the original 1953 construct.

We can describe a single Metropolis-Hastings iteration from the symmetric form of (10.17) in the following steps:

1. Sample $\boldsymbol{\theta}'$ from $q(\boldsymbol{\theta}'|\boldsymbol{\theta})$, where $\boldsymbol{\theta}$ is the current location.

2. Sample u from $u[0:1]$.

3. If $a(\boldsymbol{\theta}', \boldsymbol{\theta}) = \pi(\boldsymbol{\theta}')/\pi(\boldsymbol{\theta}) > u$ then accept $\boldsymbol{\theta}'$.

4. Otherwise keep $\boldsymbol{\theta}$ as the new location.

Obviously we want to choose the $q()$ distribution so that it is easy to sample from, but it is also important that $\pi(\boldsymbol{\theta}')/q(\boldsymbol{\theta}'|\boldsymbol{\theta})$ is fully known up to some arbitrary constant independent of $\boldsymbol{\theta}$. So three things can happen here: we can sample a value of higher density and move with probability one, we can sample a value of lower density but move anyway by drawing a small uniform random variable in Step 2 above, or we can draw a uniform random variable larger than the ratio of posteriors and therefore stay in the same location. One interesting feature of the Metropolis-Hastings algorithm, in contrast to EM, is that it is "okay" to move to lower density points, albeit probabilistically. So the algorithm will always move to higher density points but will move to lower density points probabilistically based on the ratio difference between the current point and the proposal. This algorithm describes the full posterior density, so it is necessary at times to move from a high-density point to a low-density point to traverse the space. Conversely, the EM algorithm never makes this kind of decision and therefore is only a mode finder rather than a method of fully sampling from the target distribution.

It can be shown that the Gibbs sampler is a special case of Metropolis-Hastings where the probability of accepting the candidate value is always one (Gelman 1992, 436; Tanner 1996, p.182). However, it is a Metropolis-Hastings algorithm where the full conditionals are required, and therefore is more restrictive in one sense (Gamerman and Lopes (2006), Besag *et al.* 1995, Tierney 1991). Furthermore, Billera and Diaconis (2001) show that the Metropolis-Hastings algorithm is the optimal variant from those in its "natural class of related algorithms," which would obviously include tools like the Gibbs sampler.

10.4.3   Metropolis-Hastings Properties

Consider a posterior of interest, $\pi(\boldsymbol{\theta})$ (suppressing conditionalities for notational convenience) with $\int \pi(\boldsymbol{\theta})d\boldsymbol{\theta} = 1$, for $\boldsymbol{\theta}$ on some state space, **S**, which is defined on a d-dimensional Lebesgue measure: $\Omega \subseteq \Re^d$. The motivation for seeking an iterative solution through MCMC techniques is that this $\pi(\boldsymbol{\theta})$ distribution is analytically complicated or unwieldy (Gelman 1992), so we want a procedure that eventually arrives at this distribution through simulation. The Metropolis-Hastings algorithm provides this function with a two-part transition kernel that has the property that it is closed with respect to the limiting distribution of $\pi(\boldsymbol{\theta})$:

$$\pi(\boldsymbol{\theta})p(\boldsymbol{\theta},\boldsymbol{\theta}') = \pi(\boldsymbol{\theta}')p(\boldsymbol{\theta}',\boldsymbol{\theta}) \qquad \forall \boldsymbol{\theta}, \boldsymbol{\theta}' \in \Omega, \tag{10.18}$$

where $p(a, b)$ defines a transition kernel from state a to state b. This is called the *reversibility* condition for $p()$ and also the *detailed balance equation*. The detailed balance equation guarantees that the Markov chain will eventually reach a single limiting distribution. The values for π are simply the posterior distribution evaluated at the two points: $\boldsymbol{\theta}$ and $\boldsymbol{\theta}'$, and $p(\boldsymbol{\theta}, \boldsymbol{\theta}')$ is the appropriately sized transition kernel from $\boldsymbol{\theta}$ to $\boldsymbol{\theta}'$. Robert and Casella (2004, Chapter 7) provide a proof that under very general conditions, virtually any conditional distribution over the appropriate support will provide a candidate jumping distribution that provides a Metropolis-Hastings chain that will eventually converge to the limiting distribution of $\pi(\boldsymbol{\theta})$ provided that (10.18) holds. This is the key theoretical importance of the algorithm because it shows that the right thing *will* happen if we let the chain run long enough.

A less general, and therefore less useful, condition is *symmetry*: $p(\boldsymbol{\theta}, \boldsymbol{\theta}') = p(\boldsymbol{\theta}', \boldsymbol{\theta})$. This was originally the specified requirement related to the distributional relationship between the two points, but the contribution of Hastings (1970) was to demonstrate that this is not strictly necessary and reversibility can be substituted.

10.4.4   Metropolis-Hastings Derivation

There are really two parts to the transition kernel, a jumping density $q(\boldsymbol{\theta}'|\boldsymbol{\theta})$, and a jumping probability $d(\boldsymbol{\theta}, \boldsymbol{\theta}')$ (or jumping decision):

$$p(\boldsymbol{\theta}, \boldsymbol{\theta}') = q(\boldsymbol{\theta}'|\boldsymbol{\theta})d(\boldsymbol{\theta}, \boldsymbol{\theta}'), \tag{10.19}$$

which determine the distribution of new $\boldsymbol{\theta}'$ values to move to and the probability of making such a move, respectively. The distribution used for $q(\boldsymbol{\theta}'|\boldsymbol{\theta})$ is arbitrary, but it must be straightforward to draw from. We sample from $q(\boldsymbol{\theta}'|\boldsymbol{\theta})$ to get potential values of the chain to jump to. The form of this jumping distribution plays only a part in determining values that we might jump to, and therefore it is not the *decision* to jump. By symmetric logic we can also define the reverse jump:

$$p(\boldsymbol{\theta}', \boldsymbol{\theta}) = q(\boldsymbol{\theta}|\boldsymbol{\theta}')d(\boldsymbol{\theta}', \boldsymbol{\theta}). \tag{10.20}$$

The decision to jump or not represents a second level of randomization as determined by the probability $d(\boldsymbol{\theta}, \boldsymbol{\theta}')$. The candidate jumping point is favored if its probability is large relative to the posterior probability associated with remaining at the current point.

A decision rule can be derived by starting with (10.19) solved for $d(\boldsymbol{\theta}, \boldsymbol{\theta}')$:

$$d(\boldsymbol{\theta}, \boldsymbol{\theta}') = \frac{p(\boldsymbol{\theta}, \boldsymbol{\theta}')}{q(\boldsymbol{\theta}'|\boldsymbol{\theta})},$$

and then inserting (10.18) solved for $p(\boldsymbol{\theta}, \boldsymbol{\theta}')$:

$$d(\boldsymbol{\theta}, \boldsymbol{\theta}') = \frac{\pi(\boldsymbol{\theta}')p(\boldsymbol{\theta}', \boldsymbol{\theta})}{\pi(\boldsymbol{\theta})q(\boldsymbol{\theta}'|\boldsymbol{\theta})}.$$

Using $p(\boldsymbol{\theta}', \boldsymbol{\theta}) = q(\boldsymbol{\theta}|\boldsymbol{\theta}')d(\boldsymbol{\theta}', \boldsymbol{\theta})$, this can be arranged as:

$$\frac{d(\boldsymbol{\theta}, \boldsymbol{\theta}')}{d(\boldsymbol{\theta}', \boldsymbol{\theta})} = \frac{\pi(\boldsymbol{\theta}')q(\boldsymbol{\theta}|\boldsymbol{\theta}')}{\pi(\boldsymbol{\theta})q(\boldsymbol{\theta}'|\boldsymbol{\theta})}. \tag{10.21}$$

An acceptance rule (Hastings 1970) that meets this criteria and accommodates the situation where we require for a jump $d(\boldsymbol{\theta}', \boldsymbol{\theta}) > d(\boldsymbol{\theta}, \boldsymbol{\theta}')$ is:

$$\alpha(\boldsymbol{\theta}', \boldsymbol{\theta}) = \min\left[\frac{\pi(\boldsymbol{\theta}')q(\boldsymbol{\theta}|\boldsymbol{\theta}')}{\pi(\boldsymbol{\theta})q(\boldsymbol{\theta}'|\boldsymbol{\theta})}, 1\right]. \tag{10.22}$$

This is exactly (10.17) restated where $q(\boldsymbol{\theta}', \boldsymbol{\theta}) = q_t(\boldsymbol{\theta}|\boldsymbol{\theta}')$ and $q(\boldsymbol{\theta}, \boldsymbol{\theta}') = q_t(\boldsymbol{\theta}'|\boldsymbol{\theta})$. If we were willing to substitute symmetry for reversibility, we would get the original simplified rule from Metropolis *et al.* which is similar, but much more intuitive:

$$\alpha(\boldsymbol{\theta}', \boldsymbol{\theta}) = \min\left[\frac{\pi(\boldsymbol{\theta}')}{\pi(\boldsymbol{\theta})}, 1\right], \tag{10.23}$$

because of cancellation. This states that the chain will move with probability one in a direction of higher posterior probability if offered by the jumping distribution, and will move with probability $a = \pi(\boldsymbol{\theta}')/\pi(\boldsymbol{\theta})$ to the new point otherwise. Therefore low values of a will often result in staying at the same chain value for that dimension. Due to the two levels of randomization, three things can now happen during each chain iteration: move to the new point with probability one, generate a uniform random variable (bounded by zero and one) that is less than $\alpha(\boldsymbol{\theta}', \boldsymbol{\theta})$ thus moving to the new point, or generate a uniform random variable greater than a and then stay at the current point.

Furthermore, Chib and Greenberg (1995, pp.328-329) give a very nice intuitive way to justify the detailed balance equation. Suppose (10.18) was "out of balance" in that:

$$\pi(\boldsymbol{\theta})p(\boldsymbol{\theta}, \boldsymbol{\theta}') > \pi(\boldsymbol{\theta}')p(\boldsymbol{\theta}', \boldsymbol{\theta}). \tag{10.24}$$

This means that the algorithm moves from $\boldsymbol{\theta}$ to $\boldsymbol{\theta}'$ more often than it should and moves from $\boldsymbol{\theta}'$ to $\boldsymbol{\theta}$ less often than it should. We know from (10.22) that the decision to jump to $\boldsymbol{\theta}'$ is really governed by $p(\boldsymbol{\theta}, \boldsymbol{\theta}') = q(\boldsymbol{\theta}'|\boldsymbol{\theta})\alpha(\boldsymbol{\theta}', \boldsymbol{\theta})$ (the probability of drawing $\boldsymbol{\theta}'$ times the probability of deciding to accept it), so we really need to make $\alpha(\boldsymbol{\theta}, \boldsymbol{\theta}')$ large enough to rectify:

$$\pi(\boldsymbol{\theta})q(\boldsymbol{\theta}'|\boldsymbol{\theta})\alpha(\boldsymbol{\theta}', \boldsymbol{\theta}) > \pi(\boldsymbol{\theta}')q(\boldsymbol{\theta}|\boldsymbol{\theta}')\alpha(\boldsymbol{\theta}, \boldsymbol{\theta}'), \tag{10.25}$$

meaning

$$\alpha(\boldsymbol{\theta}', \boldsymbol{\theta}) > \frac{\pi(\boldsymbol{\theta}')q(\boldsymbol{\theta}|\boldsymbol{\theta}')}{\pi(\boldsymbol{\theta})q(\boldsymbol{\theta}'|\boldsymbol{\theta})}\alpha(\boldsymbol{\theta}, \boldsymbol{\theta}'). \tag{10.26}$$

Since $\alpha(\boldsymbol{\theta}, \boldsymbol{\theta}')$ is a probability function, it is bounded by zero and one. Therefore the right-hand-side that makes the inequality into an equality is the minimum value of either: $\pi(\boldsymbol{\theta}')q(\boldsymbol{\theta}|\boldsymbol{\theta}')/\pi(\boldsymbol{\theta})q(\boldsymbol{\theta}'|\boldsymbol{\theta})$ (when $\alpha(\boldsymbol{\theta}, \boldsymbol{\theta}')$ has to equal one), or one (when $\alpha(\boldsymbol{\theta}, \boldsymbol{\theta}')$ has to be equal to the inverse of the fraction on the right-hand-side). Thus the defining equation of the Metropolis-Hastings decision is specified. Since $\boldsymbol{\theta}$ and $\boldsymbol{\theta}'$ are arbitrarily chosen here, this argument works for any two values in the sample space.

10.4.5   The Transition Kernel

With a little bit of trouble, we can rigorously define the properties of the 1953 Metropolis transition kernel. Start by defining an indicator function for the event that the δ' point is accepted: $I(\delta') = 1$ if δ' accepted, 0 otherwise (although this is a slightly different notation for the indicator function than was used before, it makes the explication below more clean). Thus the probability of transitioning from $\boldsymbol{\theta} = \boldsymbol{\theta}^{[k]}$ to the proposed jumping value $\boldsymbol{\theta}' = \boldsymbol{\theta}^{[k+1]}$ at the kth step is:

$$\begin{aligned}
p(\boldsymbol{\theta}, \boldsymbol{\theta}') &= p(\boldsymbol{\theta}^{[k+1]} = \boldsymbol{\theta}', I(\boldsymbol{\theta}')|\boldsymbol{\theta}^{[k]} = \boldsymbol{\theta}) \\
&= p(\boldsymbol{\theta}^{[k+1]} = \boldsymbol{\theta}'|\boldsymbol{\theta}^{[k]} = \boldsymbol{\theta})p(I(\boldsymbol{\theta}')) \\
&= q(\boldsymbol{\theta}'|\boldsymbol{\theta})\min\left[1, \frac{\pi(\boldsymbol{\theta}')}{\pi(\boldsymbol{\theta})}\right] \qquad \boldsymbol{\theta} \neq \boldsymbol{\theta}'.
\end{aligned} \tag{10.27}$$

The probability calculation for transitioning from $\boldsymbol{\theta} = \boldsymbol{\theta}^{[k]}$ to the current value (that is, not moving at all) is only slightly more complicated because it can occur two ways: a successful transition to the current state and a failed transition to a different state. The first event has probability zero in continuous state space, but is worth covering for discrete applications. This probability is:

$$p(\boldsymbol{\theta}, \boldsymbol{\theta}) = \underbrace{p(\boldsymbol{\theta}^{[k+1]} = \boldsymbol{\theta}, I(\boldsymbol{\theta})|\boldsymbol{\theta}^{[k]} = \boldsymbol{\theta})}_{\text{moving back to same point}} + \underbrace{p(\boldsymbol{\theta}^{[k+1]} \neq \boldsymbol{\theta}, \neg I(\boldsymbol{\theta})|\boldsymbol{\theta}^{[k]} = \boldsymbol{\theta})}_{\text{not moving}}$$

$$= q(\boldsymbol{\theta}, \boldsymbol{\theta}) + \sum_{\boldsymbol{\theta}' \neq \boldsymbol{\theta}} q(\boldsymbol{\theta}', \boldsymbol{\theta})\left(1 - \min\left[1, \frac{\pi(\boldsymbol{\theta}')}{\pi(\boldsymbol{\theta})}\right]\right). \tag{10.28}$$

In both of these calculations, the simpler situation of symmetry is assumed, but moving to the reversibility assumption is just a matter of substituting $(q(\boldsymbol{\theta}|\boldsymbol{\theta}')\pi(\boldsymbol{\theta}'))/(q(\boldsymbol{\theta}'|\boldsymbol{\theta})\pi(\boldsymbol{\theta}))$ for $\pi(\boldsymbol{\theta}')/\pi(\boldsymbol{\theta})$.

A critical component of the choice for the jumping distribution is the specified variance.

If this variance is too large, then the jumping distribution will be too wide relative to the target distribution and each successive step will move too far in some direction causing us to move awkwardly through the sample space in exaggerated steps. It is also possible to stipulate a jumping distribution variance that is too small causing overly cautious small steps through the sample space. In this case we will converge slowly and mix poorly through the limiting distribution once we have converged.

10.4.6   Example: Estimating a Bivariate Normal Density

This example is somewhat oversimplified in order to show the mechanics of the Metropolis-Hasting algorithm. Suppose, according to the example in Chib and Greenberg (1995, 333), we wanted to simulate the bivariate normal distribution: $\mathcal{N}\left[\begin{smallmatrix}1\\2\end{smallmatrix}\middle|\begin{smallmatrix}1.0 & -0.9\\-0.9 & 1.0\end{smallmatrix}\right]$. Obviously we could do this by more direct means, but the goal here is to demonstrate the workings of Metropolis-Hastings. The specified jumping distribution is a bivariate Student's-t and we will start from four overdispersed positions as well as at the mode of the distribution. The point here is to show that even though we deliberately disadvantage the process with regard to starting points and a heavy-tailed jumping distribution, the end-product is descriptive of our expectations.

The algorithm is given in R by:

```
metropolis <- function(theta.matrix,reps,I.mat)  {
    for (i in 2:reps)  {
        theta.star <- mvrnorm(1,theta.matrix[(i-1),],I.mat)/
                        (sqrt(rchisq(2,5)/5))
        a <-dmultinorm(theta.star[1],theta.star[2],c(0,0),I.mat)/
            dmultinorm(theta.matrix[(i-1),1],theta.matrix[(i-1),2],
                        c(0,0),I.mat)
        if (a > runif(1)) theta.matrix[i,] <- theta.star
        else theta.matrix[i,] <- theta.matrix[(i-1),]
    }
    theta.matrix
}
```

Figure 10.4 shows the last 200 iterations of the algorithm in the first panel and the path of the chain in the second. Notice that as with the Gibbs sampler example, convergence occurs rather quickly to the posterior region for this simple setup. Unlike the Gibbs sampler, each movement here is *not* a straight line parallel to one of the axes since the jumping decision takes place simultaneously in two dimensions. It is difficult to see here, but the other difference is that the chain is not required to move at each iteration.

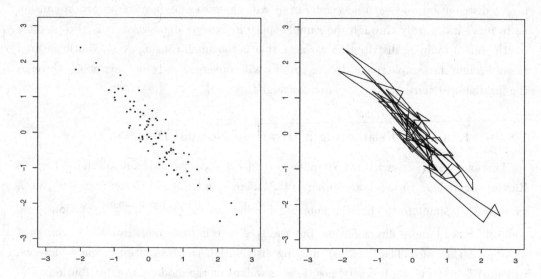

FIGURE 10.4: METROPOLIS-HASTINGS, BIVARIATE NORMAL SIMULATION

10.5   The Hit-and-Run Algorithm

The Hit-and-Run algorithm is a special case of the Metropolis-Hastings algorithm that separates the move decision into a direction decision and a distance decision. This makes it especially useful in tightly constrained parameter space because we can tune the jumping rules to be more efficient (Gelman *et al.* 1996). It is also helpful when there are several modes of nearly equal altitude. This discussion is introduced here as an example of the flexibility of the Metropolis-Hastings algorithm. Summarize the steps as following, starting from an arbitrary point $\boldsymbol{\theta}_k$, at time k:

Step 1: Generate a multidimensional direction, $\mathbf{Dr}_k$, on the surface of a
unit hypersphere from the distribution $f(\mathbf{Dr}|\boldsymbol{\theta}^{[k]})$.

Step 2: Generate a signed distance, Ds_k, from density $g(Ds|\mathbf{Dr}_k, \boldsymbol{\theta})$.

Step 3: Set the candidate jumping point to: $\boldsymbol{\theta}' = \boldsymbol{\theta}^{[k]} + Ds_k\mathbf{Dr}_k$ and
calculate:
$$a(\boldsymbol{\theta}', \boldsymbol{\theta}^{[k]}) = \frac{\pi(\boldsymbol{\theta}'|\mathbf{X})}{\pi(\boldsymbol{\theta}^{[k]}|\mathbf{X})}.$$

Step 4: Move to $\boldsymbol{\theta}^{[k+1]}$ according to:

$$\boldsymbol{\theta}_j^{[t+1]} = \begin{cases} \boldsymbol{\theta}' & \text{with probability} \quad \min(a(\boldsymbol{\theta}', \boldsymbol{\theta}^{[k]}), 1) \\ \boldsymbol{\theta}^{[k]} & \text{with probability} \quad 1 - \min(a(\boldsymbol{\theta}', \boldsymbol{\theta}^{[k]}), 1). \end{cases}$$

There are several important assumptions to consider. *First*, for all $\mathbf{Dr}_i$, $f\left(\mathbf{Dr}|\theta^{[k]}\right) > 0$. This just means that the directional distribution must be positive for all outcomes. *Second*, the distance distribution, $g(Ds|\mathbf{Dr}, \theta)$ must also be strictly greater than zero and have the property:

$$g(Ds|\mathbf{Dr}, \theta) = g(-Ds| - \mathbf{Dr}, \theta). \tag{10.29}$$

This gives a new form of the detailed balance equation:

$$g(||\theta^{[k]} - \theta'||)a(\theta', \theta^{[k]})\pi(\theta^{[k]}|\mathbf{X}) = g(||\theta' - \theta^{[k]}||)a(\theta^{[k]}, \theta')\pi(\theta'|\mathbf{X}) \tag{10.30}$$

Meeting these conditions means that the hit-and-run algorithm defines an ergodic Markov chain with stationary distribution $\pi(\theta|\mathbf{X})$.

Typically $f(\mathbf{Dr}|\theta^{[k]})$ is chosen to be uniform but others are possible, and the $a(\theta', \theta^{[k]})$ criterion can be made much more general. One advantage to this algorithm over standard Metropolis-Hastings variant is that $g(Ds|\mathbf{Dr}, \theta)$ is also flexible and disengaged from the direction decision, which makes it very tunable (Chen and Schmeiser 1993, Smith 1996). In fact, it can also be made *adaptive* as the chain matures: adaptive directional sampling.

■ **Example 10.4:** We use this algorithm to simulate again from a bivariate normal. Typically we would use hit-and-run with more problematic forms, but this example shows the integrity of the process. The problem is made more interesting specifying a strong correlation (0.95), which motivates some tuning of the algorithm. Accordingly, random directions are drawn favoring the first and third quadrants. Consider the following R code:

```
hit.run <- function(theta.matrix,reps,I.mat)  {
    for (i in 2:reps)  {
        u.vec <- c(runif(1,0,pi/2),    runif(1,pi/2,pi),
                    runif(1,pi,3*pi/2), runif(1,3*pi/2,2*pi))
        u.dr <- sample(u.vec,size=1, prob=c(1/3,1/6,1/3,1/6))
        g.ds <- rgamma(1,1,1)
        xy.theta <- c(g.ds*cos(u.dr),g.ds*sin(u.dr))
                            + theta.matrix[(i-1),]
        a <- dmultinorm(xy.theta[1],xy.theta[2],
                c(0,0),I.mat)/dmultinorm(theta.matrix[(i-1),1],
                theta.matrix[(i-1),2],c(0,0),I.mat)
        if (a > runif(1)) theta.matrix[i,] <- xy.theta
        else theta.matrix[i,] <- theta.matrix[(i-1),]
    }
    theta.matrix
}
```

Now we can run this function for our simple example and graph. Details on this step are found in the **Computational Addendum** to this chapter. This produces

Figure 10.5 where the contour bands are produced from the true target distribution, which we would ordinarily not know, and the points are the last 5,000 steps of the algorithm. There are several *tuning parameters* set here in the implementation of the algorithm: favoring positive slope direction by two-to-one in the picking of direction, a gamma distribution for the candidate-generating distribution, and the parameters of this gamma. Often when writing MCMC algorithms from scratch, these decisions deserve close inspection.

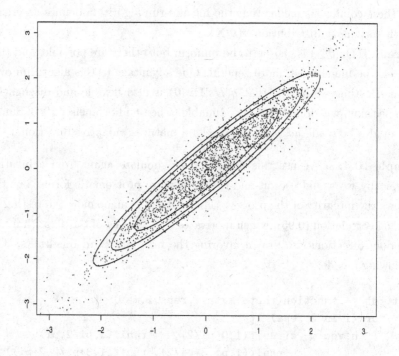

FIGURE 10.5: Hit-and-Run Algorithm Demonstration

10.6   The Data Augmentation Algorithm

Tanner and Wong (1987) introduced data augmentation (sometimes called substitution sampling) as a method for dealing with missing data or unknown parameter values by augmenting known information with candidate values much in the same way that EM does and iteratively improving the quality of these augmented quantities. In fact, data augmentation can be used instead of EM to estimate models with missing data when more than the mode of the likelihood function is required. Data augmentation is an MCMC technique that successively substitutes improved estimates conditioned on the previous state and therefore forms a Markov chain.

Much like the situation in which we applied the EM algorithm, suppose that we are interested in estimating a single-dimension (for now) parameter θ. We observe some relevant data but lack the complete set: $\mathbf{X} = [\mathbf{X}_{obs}, \mathbf{X}_{mis}]$, where all of the data (observed or not) is conditional on $\boldsymbol{\theta}$. Data augmentation requires that we know the parametric form of the posterior $p(\theta|\mathbf{X})$ corresponding to the complete data specification, and the predictive form for the missing data according to $p(\mathbf{X}_{mis}|\mathbf{X}_{obs})$. The algorithm proceeds by augmenting the observed data with simulated values of the missing data, obtained by cycling through these conditions according to the algorithm now described.

Start by defining the *posterior identity*, which is the desired quantity stated as if we could integrate out the missing data:

$$p(\theta|\mathbf{X}_{obs}) = \int_{\mathbf{X}_{mis}} p(\theta|\mathbf{X}_{obs}, \mathbf{X}_{mis}) p(\mathbf{X}_{mis}|\mathbf{X}_{obs}) d\mathbf{X}_{mis}. \qquad (10.31)$$

We can also define the *predictive identity* by asserting that there is some unknown parameter, ϕ on the sample space of θ, critical to generating the unobserved data but integrated out:

$$p(\mathbf{X}_{mis}|\mathbf{X}_{obs}) = \int_{\Phi} p(\mathbf{X}_{mis}|\phi, \mathbf{X}_{obs}) p(\phi|\mathbf{X}_{obs}) d\phi. \qquad (10.32)$$

Now insert (10.32) into (10.31) for the last term and interchange the order of integration:

$$p(\theta|\mathbf{X}_{obs}) = \int_{\mathbf{X}_{mis}} p(\theta|\mathbf{X}_{obs}, \mathbf{X}_{mis}) \left[\int_{\Phi} p(\mathbf{X}_{mis}|\phi, \mathbf{X}_{obs}) p(\phi|\mathbf{X}_{obs}) d\phi \right] d\mathbf{X}_{mis}$$

$$= \int_{\Phi} \underbrace{\int_{\mathbf{X}_{mis}} p(\theta|\mathbf{X}_{obs}, \mathbf{X}_{mis}) p(\mathbf{X}_{mis}|\phi, \mathbf{X}_{obs}) d\mathbf{X}_{mis}}_{K(\theta, \phi)} p(\phi|\mathbf{X}_{obs}) d\phi. \qquad (10.33)$$

Here Tanner and Wong use the shorthand $K(\theta, \phi)$ to make the notation cleaner, not to imply that this is a joint probability (it is really a *transition kernel*!).

The form of (10.33) implies that an iterative algorithm could be constructed that generates values of θ given an approximation for $\mathbf{X}_{mis}$, and then generates new values of $\mathbf{X}_{mis}$, given this θ. Specifically, data augmentation at the ith iteration is (beginning with candidate values for $\mathbf{X}_{mis}$):

▷ **[Imputation-Step:]**

  ▶ generate $\theta^{[i]}$ from $p^{[i-1]}(\theta|\mathbf{X}_{obs})$,
  ▶ generate m values of $\mathbf{X}_{mis}$ from $p(\mathbf{X}_{mis}|\theta^{[i]}, \mathbf{X}_{obs})$.

▷ **[Posterior-Step:]**

  ▶ update the parametric approximation using $\mathbf{X}_{mis,1}, \ldots, \mathbf{X}_{mis,m}$:

$$p^{[i]}(\theta|\mathbf{X}_{obs}) = \frac{1}{m} \sum_{j=1}^{m} p(\theta|\mathbf{X}_{obs}, \mathbf{X}_{mis,i}). \qquad (10.34)$$

In very similar fashion to importance sampling, there are two interrelated researcher-generated specifications here: the tolerance value for determining convergence ($p^{[i]}(\theta|\mathbf{X}_{obs}) - p^{[i+1]}(\theta|\mathbf{X}_{obs})$), and the number of simulation values at each step (m). The second decision is perhaps more crucial, and is a (now) familiar balance between speed and accuracy. In their original article, Tanner and Wong provide m values of 1600 and 6400 for relatively simple model specifications. It is therefore recommended that a similar value can be used as an initial parameter. With computers becoming increasingly faster, it is unlikely that this will place a heavy computation burden on the average system. Since larger values of m give better intermediate approximations, there is necessarily a trade-off between longer runs and longer calculations at each run. In fact, if $m = 1$, then data augmentation is actually Gibbs sampling, and obviously the emphasis is then purely on the length of runs.

Convergence of the data augmentation algorithm is demonstrated in Tanner and Wong. Rosenthal (1993) showed that this convergence is on the order of the log of the number of missing cases in the data for cases like this example. This is particularly encouraging because it means that even for the very largest data sets that we see in the social and behavioral sciences, data augmentation will likely be a reasonable computation process. Also, including latent data from the data augmentation procedure does not preclude model comparison or the generation of the standard model tests discussed in Chapter 7. For instance, Raftery (1996, p.182) gives the details of Bayes Factor tests when one or both of the models include such latent data.

There are enough similarities between data augmentation and EM that one might wonder when one is more applicable than the other. Both algorithms rely upon using the likelihood function under the most simple circumstances possible, and making these circumstances simple by completing the data with successively improved estimates of the missingness. EM makes more sense when the objective is simply to get the mode of the posterior because it is faster (less within-step calculations to perform). However, if the goal is to describe the complete posterior distribution, then data augmentation is more appropriate. Since data augmentation is a special case of Gibbs sampling and unlike Gibbs sampling, it is not directly implemented in BUGS (see Congdon [2001, p.114] for a nifty workaround, however), then one might have a natural preference for Gibbs sampling in practice unless one wanted to treat the missingness more explicitly.

Finally, data augmentation (Tanner and Wong 1987) can help with mixing properties even though it is most often posited as a means of dealing with missing data in Bayesian models (Liu 2001, pp.135-8). Albert and Chib (1993), Swendsen and Wang (1987), Liu, Wong, and Kong (1994), Carlin and Polson (1991), Rosenthal (1993), and more recently, Imai and van Dyk (2004), and Gelman *et al.* (2008). The basic idea behind data augmentation in the context here is that if an MCMC algorithm to marginalize $\pi(\theta|\mathbf{X})$ is mixing slowly, then it is sometimes possible to find another convenient random variable Υ such that the algorithm operates more efficiently through sampling from $\pi(\theta, \Upsilon|\mathbf{X})$ by alternating between $\pi(\theta|\Upsilon, \mathbf{X})$ and $\pi(\Upsilon|\theta, \mathbf{X})$. Chib (1992) explains how this can be quite simple and elegant when Υ is a latent feature in the model specification. Meng and van Dyk (1999), as well as van Dyk and Meng (2001), extend this idea with conditional and marginal versions of data augmentation to take advantage of unidentifiable parameters in order to improve the rate of convergence.

■ **Example 10.5: Data Augmentation for Decision-Making in the Mars Rover.**
Planetary Rovers operate in environments where human exploration is expensive, dangerous, or impossible. So NASA needs to implement *local* analysis and decision-making in Rovers due to bandwidth limitations (3.5K to 12K per second to Mars) and delay in telemetry and control signals (about 7 minutes one-way travel time to Mars, available 3 hours per day). This motivates the idea that the Rover should "learn" about its environment as a way to update and improve the quality of the decision-making process on its own. The Ames Research Center/CMU innovation is the addition of Bayesian updating within Rover circuitry to enable autonomous actions: obstacle avoidance, path planning, visual tracking, and stereo processing.

Why might this be relevant to social scientists? Proxies (agents or subordinates) need to be autonomous under some circumstances, communicating at irregular intervals and with limited information. Another tie-in is that formal models of principal/agent relationships often assume asymmetric information and divergent goals. So real-time analysis can be improved with semi-autonomous decision-making where Bayesian updating maximally leverages incomplete knowledge.

The important criteria for robotic updating are: the current status of internal systems and resources, current environmental effects, and the ramifications of actions on resources and status. Note that Rover states are discrete modes, but measured parameters are defined on continuous spaces. All diagnosis here is done as a Bayesian belief-updating system:

  ▷ begin with some prior belief over possible states,
  ▷ collect observations on status and environment,
  ▷ update the distributions to reflect new evidence.

Interestingly, the sequence of states is actually Markovian. We now formalize this in convenient notation in order to show the Bayesian learning process. For details, see the research papers from the Carnegie Mellon team: Verma, Langford, and Simmons (2001), Verma, Thrun, and Simmons (2003), and Verma, Gordon, Simmons, and Thrun (2004), as well as the technical reports from JPL/Caltech and Ames/NASA: Volpe, Nesnas, Estlin, Mutz, Petras, and Das (2001), Volpe and Peters (2003), Dearden and Clancy (2002), and Estlin, Volpe, Nesnas, Mutz, Fisher, Engelhardt, and Chien (2001).

At time t the true multivariate state of the Rover systems is denoted $\mathbf{X}_t$, but the observed state by instruments is denoted $\mathbf{O}_t$ (sensors are noisy and limited). The state of the Rover is Markovian since we know that:

$$p(\mathbf{X}_t|\mathbf{X}_0, \mathbf{X}_1, \ldots, \mathbf{X}_{t-1}) = p(\mathbf{X}_t|\mathbf{X}_{t-1}),$$

and the observations are also conditionally independent of all but the last state:

$$p(\mathbf{O}_t|\mathbf{X}_0, \mathbf{X}_1, \ldots, \mathbf{X}_{t-1}) = p(\mathbf{O}_t|\mathbf{X}_{t-1}).$$

Bayesian "filtering" (data augmentation with $\mathbf{X}_{t-1}$) for this particular state-estimating problem is done by:

$$\pi(\mathbf{X}_t|\mathbf{O}_{1:t-1}, \mathbf{O}_t) = p(\mathbf{O}_t|\mathbf{X}_t, \mathbf{O}_{1:t-1})\frac{p(\mathbf{X}_t|\mathbf{O}_{1:t-1})}{p(\mathbf{O}_t|\mathbf{O}_{1:t-1})}$$

$$\propto \underbrace{p(\mathbf{X}_t|\mathbf{O}_{1:t-1})}_{\text{prior}}\underbrace{p(\mathbf{O}_t|\mathbf{X}_t)}_{\text{likelihood}}$$

$$= \int p(\mathbf{X}_t|\mathbf{O}_{1:t-1}, \mathbf{X}_{t-1})p(\mathbf{X}_{t-1}|\mathbf{O}_{1:t-1})d\mathbf{X}_{t-1}\, p(\mathbf{O}_t|\mathbf{X}_t)$$

$$= \int \underbrace{p(\mathbf{X}_t|\mathbf{X}_{t-1})}_{\text{transition model}}\underbrace{p(\mathbf{X}_{t-1}|\mathbf{O}_{1:t-1})}_{\text{previous status}}d\mathbf{X}_{t-1}\,\underbrace{p(\mathbf{O}_t|\mathbf{X}_t)}_{\text{observation model}} \qquad (10.35)$$

This works because:

$$\int p(\mathbf{X}_t|\mathbf{O}_{1:t-1}, \mathbf{X}_{t-1})p(\mathbf{X}_{t-1}|\mathbf{O}_{1:t-1})d\mathbf{X}_{t-1}$$

$$= \int \frac{p(\mathbf{X}_t, \mathbf{O}_{1:t-1}, \mathbf{X}_{t-1})}{p(\mathbf{O}_{1:t-1}, \mathbf{X}_{t-1})}\frac{p(\mathbf{O}_{1:t-1}, \mathbf{X}_{t-1})}{p(\mathbf{O}_{1:t-1})}d\mathbf{X}_{t-1}$$

$$= \int p(\mathbf{X}_t, \mathbf{O}_{1:t-1}, \mathbf{X}_{t-1})/p(\mathbf{O}_{1:t-1})d\mathbf{X}_{t-1}$$

$$= \int p(\mathbf{X}_t, \mathbf{X}_{t-1}|\mathbf{O}_{1:t-1})d\mathbf{X}_{t-1}$$

$$= p(\mathbf{X}_t|\mathbf{O}_{1:t-1}) \qquad (10.36)$$

Often the dimensionality of this posterior, $\pi(\mathbf{X}_t|\mathbf{O}_{1:t})$, is too great for calculation in real-time, so the Rover uses a recursive *particle filter* (importance sampling):

$$\pi(\mathbf{X}_t|\mathbf{O}_{1:t}) \approx n^{-1}\sum_{i=1}^{n}\delta(\mathbf{X}_t - \mathbf{x}_t^{[i]}) \qquad (10.37)$$

where $\delta()$ is the Dirac delta function, and $\mathbf{x}_t^{[i]}$ are samples drawn from some proposal distribution. The algorithm uses $p(\mathbf{X}_t|\mathbf{X}_{t-1})$ as the proposal distribution and $p(\mathbf{O}_t|\mathbf{X}_t)$ for importance weights, and then proceeds:

1. for $i = 1 : n$, draw $\mathbf{x}_t^{[i]} \sim p(\mathbf{X}_t|\mathbf{x}_{t-1}^{[i]})$ and set $\omega_t^{[i]} = p(\mathbf{O}_t|\mathbf{X}_t^{[i]})$.

2. from this set, accept values $\mathbf{x}_t^{[i]}$ with probability proportional to $\omega_t^{[i]}$.

These transaction functions map a present mode to a PMF for future modes. So state change probabilities are estimated then updated with inherent uncertainty, and autonomous decisions are made based on the most recent posterior. So all of these calculations occur in the context of executing general instructions: "go there," "investigate that," etc. The Rover then proceeds towards its objective while Bayesianly updating the posterior for $\mathbf{X}_t$, where intermediate decisions (movement, direction, sensing) are made autonomously with criteria such as: safety, power conservation, noting interesting phenomenon. In this fashion the Bayesian characteristics are apparent.

10.7 Historical Comments

The background of the development of modern MCMC methods is interesting unto itself. The first important event was the publication of a 1953 paper by Nicholas Metropolis and his colleagues: Arianna Rosenbluth, Marshall Rosenbluth, Augusta Teller, and Edward Teller (who we know also made notable contributions in nuclear physics related to rather large explosions). Because the paper was published in the *Journal of Chemical Physics* and because it was applied exclusively to the problem of particles moving around a square, interest was restricted primarily to physics.

Metropolis *et al.* were interested in obtaining the positions and therefore the potential between all molecules in an enclosure and noted that even very modest sized setups lead to integrals of very high dimensions. Specifically, if $\hbar_{ij}$ represents the shortest distance between particles i and j, and $V(\hbar_{ij})$ is the associated potential, then the total potential energy of the whole system is given by: $E = \frac{1}{2} \sum_{i=1}^{n} \sum_{j=1, j \neq i}^{n} V(\hbar_{ij})$. Due to some simplifications this leads to an integral of "only" 200 dimensions for the force of the system. So rather than try to track and calculate future positions, they arrived at the idea of setting a molecule movement criterion in a formal model sense and then simulating a series of potential positions. In other words, it was sufficient to know where the molecules were *probabilistically* at some future point in time as opposed to *exactly*.

The Metropolis *et al.* paper was slow to permeate other disciplines including statistics partly because the authors appear not to have been aware of the widespread applicability of their technique. This is a presumed explanation for why the authors chose not to generalize it beyond the application given. The key to the dissemination of the algorithm was the refinement and generalization done by Hastings (1970) some time later. He showed that reversibility can be substituted for symmetry in the approximation distribution, applicable to continuous state spaces, and he makes the ideas accessible to statisticians. In addition, Peskun (1973) should be credited with further introducing the Metropolis algorithm to the statistics community and proving a number of important properties, including principles of reversibility.

The Geman and Geman (1984) paper introduced a new use for the Gibbs distribution in simulation and applies the tool to restoration of degraded images. This paper is very difficult to work through and most people do not persevere. Instead, the landmark Gibbs sampling paper, as far as widespread effects are concerned, is that of Gelfand and Smith (1990). They demonstrate how widely useful Gibbs sampling is in terms of setting up Markov chains to estimate posterior distributions. While nothing is entirely new in this paper, the synthesis and integration of Gibbs sampling into Markov chain Monte Carlo theory for the first time is an invaluable contribution.

One key reason for the explosion of academic attention to MCMC that occurred in the 1990s is the substantial improvement in computing power on the average desktop. This point cannot be overstated. By definition these techniques are computer-intensive, and it is hard to imagine earlier researchers being pleased with either the speed of their microcomputers or the convenience of their campus mainframes. Fortunately Moore's Law (doubling of computing power every two years) continues to apply.

The overall impact on statistics and applied statistics cannot be reasonably overstated. One long-time observer notes that "the Bayesian 'machine' together with MCMC is arguably the most powerful mechanism ever created for processing data and knowledge" (Berger 2001). Essentially the advent of MCMC freed the Bayesian analyst from the Faustian choice of accepting oversimplification of the assumptions in order to merely get a tractable answer. Because of these relatively simple tools, models of seemingly endless complexity can be estimated. There are some additional issues to worry about, such as practical considerations with Markov chain convergence (Chapter 14), but these are adequately handled with modern software.

10.7.1   Full Circle?

An inquiring mind may have realized that many of the properties used to analyze and exploit MCMC techniques are *frequentist* in nature. Principles used here such as the central limit theorem, the law of large numbers, general asymptotic analysis, and transition invariance, are all basic principles from traditional non-Bayesian statistics. Specifically, the tool that revolutionized Bayesian statistics is in fact a frequentist construction. Efron (1998) notes that Fisher's work directly implied several modern statistical computing techniques that Fisher could not have employed for purely mechanistic reasons. These include bootstrapping (the bootstrap plug-in principle is anticipated by the calculation of *Fisher Information* and Rubin [1981] shows how the bootstrap can be made explicitly Bayesian), empirical calculation of confidence/credible intervals, empirical Bayes (developing a prior using the data), and Bayes Factors.

10.8   Exercises

10.1   Find the values of α_1, α_2, and α_3 that make the following a transition matrix:

$$
\begin{bmatrix}
0.4 & \alpha_1 & 0.0 \\
0.3 & \alpha_2 & 0.6 \\
0.0 & \alpha_3 & 0.4
\end{bmatrix}.
$$

10.2 For the following simple transition matrix:

$$\begin{matrix} \theta_1 \\ \theta_2 \end{matrix} \begin{bmatrix} 0 & 1 \\ 1 & 0 \end{bmatrix}.$$

For an initial state, $[0.25, 0.75]$, describe repeated applications of the transition matrix: $\mathbf{P}^2, \mathbf{P}^3, \mathbf{P}^4, \ldots$. Is this an ergodic Markov chain?

10.3 Using the transition matrix from Section 10.1.2, run the Markov chain mechanically (step by step) in R using matrix multiplication. Start with at least two very different initial states. Run the chains for at least ten iterations.

10.4 Consider a simple Ehrenfest urn problem with two urns and only two balls. For simplicity start with one ball in each urn, then at each step one of the two balls is selected with probability 0.5 and then placed in the urn that it currently is not in. So there are only three possible states: $[2,0], [1,1], [0,2]$ Show that this is a Markov chain, provide the transition matrix, and derive the unique stationary distribution.

10.5 For the following transition matrix, which classes are closed?

$$\begin{bmatrix} 0.50 & 0.50 & 0.00 & 0.00 & 0.00 \\ 0.00 & 0.50 & 0.00 & 0.50 & 0.00 \\ 0.00 & 0.00 & 0.50 & 0.50 & 0.00 \\ 0.00 & 0.00 & 0.75 & 0.25 & 0.00 \\ 0.50 & 0.00 & 0.00 & 0.00 & 0.50 \end{bmatrix}$$

10.6 Males in three professions have the following probabilities that their primary son follows them into the same profession: Professor $p = 0.4$, Plumber $p = 0.8$, Playwright $p = 0.2$. If the son does not pick the same profession as the patriarch, he picks with even probability from the other two professions. Each protagonist possesses at least one male progeny. Produce the transition matrix for this plan and provide the steady state.

10.7 From the following transition matrix, calculate the stationary distribution for proportions:

$$\begin{bmatrix} 0.0 & 0.4 & 0.6 \\ 0.1 & 0.0 & 0.9 \\ 0.5 & 0.5 & 0.0 \end{bmatrix}.$$

What is the substantive conclusion from the zeros on the diagonal of this matrix?

10.8 An AR(1) stochastic process (autoregressive) is commonly defined as:

$$\theta^{[t+1]} = \varphi \theta^{[t]} + \Omega_t,$$

where the Ω_t values are generated iid from $\mathcal{N}(0, \tau^2)$ for finite τ^2, and the distribution of the θ values is also iid with finite variance σ^2. From the covariance of generated values $m - 1$ away from each other, $\text{Cov}(\theta^{[t+m]}, \theta^{[t]}) = \varphi^m \text{Var}(\theta)$, using the central limit theorem, show that if this process is stationary that:

$$\theta \sim \mathcal{N}(0, \sigma^2),$$

where:

$$\sigma^2 = \frac{\tau^2(1 + \varphi)}{(1 - \varphi^2)(1 - \varphi)},$$

and state the necessary restriction necessary on φ.

10.9   Develop a Bayesian specification using BUGS to model the following counts of the number of social contacts for children in a daycare center with the objective of estimating λ_i, the contact rate:

Person	1	2	3	4	5	6	7	8	9	10
Age (a_i)	11	3	2	2	7	5	6	9	7	4
Social Contacts (x_i)	22	4	2	2	9	3	4	5	2	6

Use the following specification:

$$\lambda_i \sim \mathcal{G}(\alpha, \beta)$$
$$\alpha \sim \mathcal{G}(1, 1)$$
$$\beta \sim \mathcal{G}(0.1, 1)$$

to produce a posterior summary for λ_i.

10.10   Show that for a stationary Markov chain that a set of batches (a set of consecutive draws) of the same batch length have the same joint distribution.

10.11   The table below provides the results of nine postwar elections in Italy by proportion per political party. The listed parties are:

▷ Democrazia Cristiana (**DC**),

▷ Partito Comunista Italiano (**PCI**),

▷ Partito Socialista Italiano (**PSI**),

▷ Partito Socialista Democratico Italiano (**PSDI**),

▷ Partito Repubblicano Italiano (**PRI**),

▷ Partito Liberale Italiano (**PLI**),

▷ Others.

The "Others" category is a collapsing of smaller parties: Partito Radicale (**PR**), Democrazia Proletaria (**DP**), Partito di Unità Proletaria per il Comunismo (**PdUP**), Movimento Sociale Italiano (**MSI**), South Tyrol Peoples Party (**SVP**), Sardinian

Action Party (**PSA**), Valdôtaine Union (**UV**), the Monarchists (**Mon**), and the Socialist Party of Proletarian Unity (**PSIUP**). In two cases parties presented joint election lists and the returns are split across the two parties here. The compositional data suggest a sense of stability for postwar Italian elections even though Italy has averaged more than one government per year since 1945.

Party	1948	1953	1958	1963	1968	1972	1976	1979	1983
DC	0.485	0.401	0.424	0.383	0.3910	0.388	0.387	0.383	0.329
PCI	0.155	0.226	0.227	0.253	0.2690	0.272	0.344	0.304	0.299
PSI	0.155	0.128	0.142	0.138	0.0725	0.096	0.096	0.098	0.114
PSDI	0.071	0.045	0.045	0.061	0.0725	0.051	0.034	0.038	0.041
PRI	0.025	0.016	0.014	0.014	0.0200	0.029	0.031	0.030	0.051
PLI	0.038	0.030	0.035	0.070	0.0580	0.039	0.013	0.019	0.029
Others	0.071	0.154	0.113	0.081	0.1170	0.125	0.095	0.128	0.137

Source: Instituto Centrale di Statistica, Italia

Develop a multinomial-logistic model for p_{ij} as the proportion received by party i and election j with BUGS, using time as an explanatory variable and specifying uninformative priors.

10.12 Section 10.3 described carefully how the Gibbs sampler must use the most recent draws as values on the right-hand-side of the conditional distribution for subsequent draws. Modify the R code for the Gibbs sampling in Example 10.3.2 such that the conditioned-upon values of λ and ϕ are from the previous step not the current step in the draw for k. Is this now a transient Markov chain?

10.13 Blom, Holst, and Sandell (1994) define a "homesick" Markov chain as one where the probability of returning to the starting state after $2m$ ($m > 1$) iterations is at least as large as moving to any other state: $p^{2m}(x_0, x_{\neg 0}) \leq p^{2m}(x_0, x_0)$. Does the Markov chain defined by the following transition matrix have homesickness?

$$P = \begin{bmatrix} 0.5 & 0.5 & 0.0 & 0.0 \\ 0.5 & 0.0 & 0.5 & 0.0 \\ 0.0 & 0.5 & 0.0 & 0.5 \\ 0.0 & 0.0 & 0.5 & 0.5 \end{bmatrix}$$

Do Markov chains become less homesick over time?

10.14 Using the four Metropolis-Hastings samplers from Exercise 18, modify the R plotting procedures in the **Computational Addendum** for this chapter to graph these in four partitions of the same graph.

10.15 Koppel (1999) studies political control in hybrid organizations (semi-governmental) through an analysis of government purchase of a specific type of venture capital funds: investment funds sponsored by the Overseas Private Investment Corporation (OPIC). The following data (opic.df in `BaM`provide three variables as of January 1999.

Fund	Age	Status	Size ($M)
AIG Brunswick Millennium	3	Investing	300
Aqua International Partners	2	Investing	300
Newbridge Andean Capital Partners	4	Investing	250
PBO Property	1	Investing	240
First NIS Regional	5	Investing	200
South America Private Equity Growth	4	Investing	180
Russia Partners	5	Investing	155
South Asia Capital	3	Investing	150
Modern Africa Growth and Investment	2	Investing	150
India Private Equity	4	Investing	140
New Africa Opportunity	3	Investing	120
Global Environmental Emerging II	2	Investing	120
Bancroft Eastern Europe	3	Investing	100
Agribusiness Partners International	4	Investing	95
Caucus	1	Raising	92
Asia Pacific Growth	7	Divesting	75
Global Environmental Emerging I	5	Invested	70
Poland Partners	5	Invested	64
Emerging Europe	3	Investing	60
West Bank/Gaza and Jordan	2	Raising	60
Draper International India	3	Investing	55
EnterArab Investment	3	Investing	45
Israel Growth	5	Investing	40
Africa Growth	8	Divesting	25
Allied Capital Small Business	4	Divesting	20

Develop a model using `BUGS` where the size of the fund is modeled by the age of the fund and its investment status according to the specification:

$$Y_i \sim \mathcal{N}(m_i, \tau)$$
$$m_i = \beta_0 + \beta_1 X_{1i} + \beta_2 X_{2i} + \epsilon_i$$
$$\epsilon \sim \mathcal{N}(0, k),$$

where: Y is the size of the fund, X_1 is the age of the fund, and X_2 is a dichotomous explanatory variable equal to one if the fund is investing and zero otherwise. Set a value for the constant k and an appropriate prior distribution for τ. Summarize the posterior distributions of the unknown parameters of interest.

10.16   Another special case of the Metropolis-Hastings algorithm is the multiple-try Metropolis algorithm (Liu, Liang, and Wong 2000) with local optimization steps. It is designed to be more computationally efficient with difficult posterior shapes. Assuming the current position to be θ, the steps for a single iteration are:

▷ From a proposal density, $q(\theta'|\theta)$, draw $\theta_1^*, \ldots, \theta_k^*$.

▷ Define a non-negative arbitrary function $\lambda(\theta_i^*, \theta)$.

▷ For each θ_i^* calculate $\omega(\theta_i^*, \theta) = \pi(\theta_i^*)q(\theta|\theta_i^*)\lambda(\theta_i^*, \theta)$.

▷ Draw a single θ^* proportional to these weights.

▷ Set $\theta_1 = \theta$ (current chain position), and draw $\theta_1, \ldots, \theta$ from $q(\theta|\theta^*)$

▷ Accept θ^* with the Metropolis decision probability:

$$r_q = \min\left[1, \frac{\omega(\theta_1^*, \theta) + \cdots \omega(\theta_k^*, \theta)}{\omega(\theta_1, \theta^*) + \cdots \omega(\theta_k, \theta^*)}\right]$$

or reject with probability $1 - r_q$.

Implement this algorithm in R to produce samples from a target density $\pi(\theta)$ that is a correlated central bivariate Student's-t distribution with $\rho = 0.9$ and $\nu = 5$:

$$\pi(\theta_a, \theta_b) = \frac{1}{2\pi\sqrt{1 - \rho^2}}\left[1 + \frac{\theta_a^2 + \theta_b^2 - 2\rho\theta_a\theta_b}{\nu(1 - \rho^2)}\right]^{-\frac{\nu}{2} - 1}$$

(see Appendix B for a more general multivariate non-central definition).

10.17 (Norris 1997) A Markov chain is *reversible* if the distribution of $\theta_n|\theta_{n+1} = t$ is the same as the $\theta_n|\theta_{n-1} = t$. This means that direction of time does not alter the properties of the chain. Show that the following irreducible matrix does not define a reversible Markov chain.

$$\begin{bmatrix} 0 & p & 1-p \\ 1-p & 0 & p \\ p & 1-p & 0 \end{bmatrix}.$$

See Besag *et al.* (1995) for details on reversibility.

10.18 (Chib and Greenberg 1995). To produce draws from a bivariate normal target distribution that is normal with:

$$\mu = c(1,2), \qquad \text{and} \qquad \Sigma = \begin{bmatrix} 1.0 & 0.9 \\ 0.9 & 1.0 \end{bmatrix},$$

write a Metropolis-Hastings algorithm in R with the following alternative candidate-generating strategies with the goal of a 40% to 50% acceptance rate:

▷ a random walk where the offset is a bivariate uniform with ranges $[-0.75 : 0.75]$ for the first dimension and $[-1, 1]$ for the second dimension.

▷ a random walk where the offset is a bivariate normal with mean $(0, 0)$ and variance $\begin{bmatrix} 0.6 & 0 \\ 0 & 0.4 \end{bmatrix}$.

▷ a pseudorejection scheme where independent candidates are filtered through an acceptance-rejection step using the dominating function given by $ch(x)c(2\pi)^{-1}|\mathbf{D}|^{-\frac{1}{2}}\exp\left[-\frac{1}{2}(x - \mu)'\mathbf{D}(x - \mu)\right]$, where $D = \begin{bmatrix} 2 & 0 \\ 0 & 2 \end{bmatrix}$ and $c = 0.9$ (use $\mu = c(1, 2)$ from above).

▷ an autoregressive density with $\theta^{[t+1]} = \mu - (\theta^{[t]} - \mu) + \Omega$, where Ω is an independent bivariate uniform drawn in both dimensions from $[0:1]$. Note that this reflects the draw to the other side of μ then applies an offset.

10.19   (Grimmett and Stirzaker 1992) A random walk is recurrent if the mean size of the jumps is zero. Define a random walk on the integers by the transition from integer i to either integer $i+2$ or $i-1$ with probabilities:

$$p(i, i+2) = p, \qquad p(i, i-1) = 1 - p.$$

A random walk is recurrent if the mean recurrence time, $\sum n f_{ii}(n)$, is finite, otherwise it is transient. What values of p make this random walk recurrent?

10.20   The following data give the number of casualties for 103 suicide attacks in Israel with explosives over a three-year period from November 6, 2000 to November 3, 2003 when there was a steep drop (the early period of the first "Intifada"). These data are provided by the International Policy Institute for Counter-Terrorism, and subsetted by Harrison (2006), and modeled in Kyung *et al.* (2011).

0	3	81	38	29	126	6	10	1	1
67	50	3	27	0	2	0	63	15	58
57	0	0	0	0	123	4	71	71	20
17	65	4	49	5	35	57	71	0	12
67	59	5	52	62	0	75	0	0	106
30	0	3	45	4	31	32	180	0	1
91	49	61	51	3	0	1	9	0	2
151	26	8	8	75	199	12	2	2	1
93	0	13	21	145	0	0	13	0	2
141	2	65	0	105	0	61	6	27	53
20	5	0							

There are two modes, with the larger one at zero. Fit a two-component gamma mixture model using data augmentation for p (the probability of being in the right-mode) with a beta distribution, and with a binary assignment vector as part of the Gibbs sampler. At iteration t the steps are:

▷ Draw p from $\mathcal{BE}(\alpha + n_1, \beta + n_2)$.
▷ For each case i, draw $I_i \sim \mathcal{BR}(p)$.
▷ If $I_i = 0$, draw from $\mathcal{G}(\alpha_0, \beta_0)$.
▷ If $I_i = 1$, draw from $\mathcal{G}(\alpha_1, \beta_1)$.
▷ Determine n_1 and n_2 from the I_i assignments.

Here α, β, α_0,β_0, α_0, and β_0 are hyperprior parameters and $n = n_1 + n_2$ at that point in the sampler (assign starting values).

10.9 Computational Addendum: Simple R Graphing Routines for MCMC

This addendum gives the background for the graphing R code in this chapter. In the next chapter we develop Bayesian MCMC solutions for more complex, and realistic, models The BUGS package is recommended in general for implementing MCMC estimation models.

Figure 10.2 was produced using the following function for graphing the path of a Gibbs sampler above in two chosen dimensions. You must give it values indicating which two columns of the input matrix to graph since it is a two-dimensional plot. For larger dimensional joint posteriors, modifications to this function are easy to perform.

```
plot.walk.G <- function(walk.mat,sim.rm,X=1,Y=2)  {
    plot(walk.mat[1,X],walk.mat[1,Y],type="n", xlim=range(walk.mat[,X]),
        ylim=range(walk.mat[,Y]), xlab="",ylab="")
    for(i in 1:(nrow(walk.mat)-1))  {
        segments(walk.mat[i,X],walk.mat[i,Y],
                walk.mat[(i+1),X],walk.mat[i,Y])
        segments(walk.mat[(i+1),X],walk.mat[i,Y],
                walk.mat[(i+1),X],walk.mat[(i+1),Y])
    }
}
```

```
par(mfrow=c(1,2),mar=c(2,3,1,1),oma=c(1,1,3,1))
plot.walk.G(theta.matrix[1:100,],X=1,Y=3)
mtext(outer=TRUE,side=2,cex=1.3,expression(k))
mtext(outer=FALSE,side=3,cex=1.3,expression(lambda),line=2)
plot.walk.G(theta.matrix[1:100,],X=2,Y=3)
mtext(outer=FALSE,side=3,cex=1.3,expression(phi),line=2)
```

The following R code produced Figure 10.4 for this Metropolis-Hastings output in two dimensions. It takes a matrix where the rows indicate chain iterations.

```
plot.walk.MH <- function(walk.mat)  {
    plot(walk.mat[1,1],walk.mat[1,2],type="n",
        xlim=round(range(walk.mat[,1])*1.2),
        ylim=round(range(walk.mat[,2])*1.2),
        xlab="",ylab="")
    for(i in 1:(nrow(walk.mat)-1))  {
        segments(walk.mat[i,1],walk.mat[i,2],
```

```
                    walk.mat[(i+1),1],walk.mat[(i+1),2])
   }
}


par(mfrow=c(1,2),mar=c(2,2,2,2),oma=c(1,1,3,1))
plot(theta.matrix[801:1000,],pch=".",xlab="",ylab="",
    xlim=c(-3,3), ylim=c(-3,3),cex=3)
plot.walk.MH(theta.matrix[801:1000,])
mtext(outer=TRUE,side=3,cex=1.2,
    "Metropolis-Hastings Demonstration, Bivariate Normal")
```

The previous function needs to produce bivariate normal density values, and a function for doing this follows. It is currently written only for bivariate calculations, but it can easily be vectorized to accommodate higher dimensions.

```
dmultinorm <- function(xval,yval,mu.vector,sigma.matrix) {
   normalizer <- (2*pi*sigma.matrix[1,1]*sigma.matrix[2,2]
                *sqrt(1-sigma.matrix[1,2]^2))^(-1)
   like <- exp(-(1/(2*(1-sigma.matrix[1,2]^2)))* (
            ((xval-mu.vector[1])/sigma.matrix[1,1])^2
            -2*sigma.matrix[1,2]*(((xval-mu.vector[1])
            /sigma.matrix[1,1])*
            ((yval-mu.vector[2])/sigma.matrix[2,2]))
            +((yval-mu.vector[2])/sigma.matrix[2,2])^2 ))
   normalizer*like
}
```

This last code segment implements the Hit-and-Run example along the graph on page 362.

```
num.sims <- 10000
Sig.mat <- matrix(c(1.0,0.95,0.95,1.0),2,2)
walks<-rbind(c(-3,-3),matrix(NA,nrow=(num.sims-1),ncol=2))
walks <- hit.run(walks,num.sims,Sig.mat)
z.grid <- outer(seq(-3,3,length=100),seq(-3,3,length=100),
                FUN=dmultinorm,c(0,0),Sig.mat)
contour(seq(-3,3,length=100),seq(-3,3,length=100),z.grid,
                levels=c(0.05,0.1,0.2))
points(walks[5001:num.sims,],pch=".")
```

Chapter 11

Implementing Bayesian Models with Markov Chain Monte Carlo

11.1 Introduction to Bayesian Software Solutions

While many models can be run in R and other programming environments, as demonstrated in previous chapters, it is necessary to use Markov chain Monte Carlo procedures to fit some realistic and useful social and behavioral science Bayesian models. As noted, the revolution began with the review paper of Gelfand and Smith (1990), and there were simply a class of models with hierarchies and other complex features before 1990 that couldn't be marginalized to produce a regression table. In this chapter we focus exclusively on the details of running MCMC with the BUGS language (Lunn *et al.* 2000). This will enable the attentive reader to immediately begin running MCMC procedures.

The exposition here is introductory and cannot be comprehensive due to space limitations. However, there are *many* high-quality texts dedicated to the BUGS language and its variations. The recent book by Lunn *et al.* (2012) stands out, perhaps because of the overlap of authors with the WinBUGS package. The text by Ntzoufras (2009) has detailed coverage of the language and guidance for standard models. Congdon's (2001, 2003, 2005, 2010) four texts provide an extensive library of model implementations in BUGS with accompanying descriptions. The book by Kéry (2010) is a nice basic introduction to WinBUGS, and also gives guidance on calling WinBUGS from R (although the applied focus is ecology not social science). There is also a follow-on work by Kéry and Schaub (2011) with hierarchical modeling as a focus. Kruschke (2010) produced an accessible introduction to Bayesian basics, with an orientation towards the behavioral sciences, that includes examples and exercises in BUGS.

Finally, this chapter is not meant to imply that the only way that one can estimate complex models with MCMC is to use the BUGS language. Many researchers write their own samplers directly in R (often calling C or FORTRAN for the highly repetitive parts, or increasingly in MATLAB). There is an excellent and exhaustive inventory of R resources for Bayesian analysis in the CRAN Task View on Bayesian Inference. As a reminder all R and BUGS code in this text are contained in the BaM package at CRAN. All updates to code and data are maintained in that package and therefore the guidance in this chapter will remain fairly generic and general since software changes outpace publisher changes by a considerable margin.

11.2   It's Only a Name: BUGS

In a spectacularly bad example of naming computer software, the Cambridge MRC Biostatistics Unit distributed BUGS for Unix and DOS operating systems in 1989, and corresponding research papers appeared a few years earlier. This version is now referred to as "Classic BUGS" and is no longer supported by the Cambridge team. In the early era of BUGS users were required to register annually to obtain a software key for unrestricted use (now BUGS is "immortal"). The documentation was sparse, but the software came with 41 examples fully worked out (mostly from biostatistics). These were a splendid learning tool, particularly in the late 1990s when no resources existed other than trial-and-error. There are currently 50 detailed examples in three volumes, along with valuable documentation. Although running these well-documented examples is a good way to learn the language, one caution is required. It is far better to write one's own code from scratch after learning the language than trying to modify these examples to fit the problem at hand. Inevitably the latter strategy leads to making changes in models and assumptions to fit clean running code, and this means that the software is then dictating the research rather than the researcher.

While "Classic" BUGS lived only until 1996 (R.I.P.), there are now a variety of ways to obtain the software and related packages. And while there are separate estimation "engines," they all use approximately the same language and they all interface to other environments. Many users will prefer WinBUGS,[1] which is described in this chapter, since it has a nice graphical user interface and is relatively easy to learn. As the name implies, WinBUGS is confined to Windows machines. The follow-on to WinBUGS is openbugs, which is an open-source version of the package that actually runs models from a "compound document" that encloses: model and data description text, data tables, plots, model graphs, code, and more. openbugs runs on Windows and Linux operating systems, but needs the Wine emulator to run on OS X systems. An increasingly popular alternative is the JAGS ("Just Another Gibbs Sampler") program written by Martyn Plummer. This is open source software that was developed independently of the BUGS project, but uses almost exactly the same language (differences will be described in this chapter). The main advantage of JAGS is that it runs on all commonly used operating system platforms, and Plummer has added extensible features to allow more development of user-developed functions. In this chapter both WinBUGS and JAGS models are run, where "WinBUGS" also applies to openbugs. However, by convention in this text "BUGS" refers to the general language/approach and

---

[1]Available at http://www.mrc-bsu.cam.ac.uk/software/bugs/the-bugs-project-winbugs/.

therefore all three software *engines*. Points about specific packages will be made by name as required.

Most users now run one of the packages listed above called from a standard statistical software environment, most notably R. The R package R2WinBUGS provides an interface between R and WinBUGS such that commands in R remotely direct the WinBUGS environment without the graphical interface. Similarly the R packages Rjags, runjags, bayesmix, and R2jags run JAGS from within R by commands, but with the advantage of working on all popular operating systems. Also, openbugs can be run in the same way from R with both the BRugs and the rbugs packages. John Thompson, Tom Palmer, and Santiago Moreno (2006) at the University of Leicester have developed stata ado files to run WinBUGS remotely and recover the samples. The BUGS Project homepage also offers several ways to remotely call WinBUGS from SAS. There is also an available a MATLAB helper function called mat2bugs.m that evokes a WinBUGS interface. Finally, Phil Woodward created a Microsoft Excel add-in called BugsXLA interfacing WinBUGS that accompanies his text. The discussion in this chapter focuses exclusively on running BUGS from R or in a standalone manner.

So what is BUGS? Officially it is an integrated language for specifying Bayesian models, and hierarchical models in particular (Chapter 12). It provides a library of sampling routines along with a text-based user interface for running these samplers. The syntax is intentionally R-like, but there are vastly fewer commands and some important differences in how terms are used. In BUGS the user specifies models in commands that reflect statistical thinking: variable relationships, distributions, hierarchies, and prior distributions. The software then translates these statements into a set of full conditional distributions that the user does not see. This is done in several steps that include compiling the written code and initializing the model checking for dimension mis-matches. Once this process is finished, the user then specifies a number of Gibbs sampling iterations to run and evaluates the monitored iterations. Usually the sampled values are post-processed in R, either with standard R functions or with one of two dedicated packages. This last step is quite easy since the sampled values are treated mechanically just like regular data, taking means, variances, quantiles, correlations, etc.

Because it is a Markov chain, this last process can be interrupted and recommenced at will with no implications: the only information that needs to be saved in order to restart the chain is the last chain value produced. This is important when considering the status of convergence with the diagnostics described in Chapter 14. There are other considerations during the running of the chain, including looking at traceplots that show the path of the chain in each dimension. Occasionally the BUGS variants will fail to find a full conditional distribution for one or more of the model dimensions using a Gibbs sampler. In this case the software reverts to a Metropolis-Hastings chain or slice sampling for those dimensions only, and this is done automatically. It is impossible to overstate the amazing convenience that BUGS provides. It unburdens the researcher from having to produce the full set of full conditional distributions for Gibbs sampling or the candidate-generating distribution for Metropolis-Hastings.

11.3   Model Specification with BUGS

Obviously the most important part of the model development process is understanding the data and developing a principled statistical specification that tests theories of interest and importance. This chapter focuses on what happens *after* that part and assumes that the specified model already exists. Specifying statistical models within the software production process is always a poor practice.

There are four general steps to the BUGS modeling process:

1. state the distributional features of the model, and the quantities to be estimated,

2. compile the instructions into a run-time program,

3. run this sampler that produces Markov chains,

4. and summarize the empirical distributions and assess convergence using basic diagnostics in BUGS or R.

Each of these steps will be described in detail here with developed examples. First we make a few notes about the use of the language. Unlike with other programming languages, statements are not processed serially; they constitute a full specification. This may seem strange to users used to writing code in standard programming languages where the order of statements is critical. In BUGS the order of statements within the blocks of code is not important since the software reads the blocks as a complete unit. Sometimes this also makes debugging more difficult since the point-of-failure is often not as clear as in a more serial process. Usually coding errors are detected in the compilation step and most of the frustration for a new user comes from iterating between steps one and two above. It turns out that the rest of the process is amazingly easy as we shall see.

There is a host language that comes with the use of BUGS. **Nodes** are values and variables in the model that are specified by the researcher. These come in different types and reflect the focus on hierarchical modeling. While we do not develop extensive Bayesian hierarchical modeling until Chapter 12, it is intuitive to know that some model terms in Bayesian specifications depend on other terms in the classic sense of conditional inference (e.g., $f(\theta|\alpha, \beta)$). A **parent node** is one that influences other nodes: it is higher in the hierarchy, it is on the right-hand-side of some conditional statement. A **descendant node** is the opposite of the parent node in that it is downstream from some node, meaning lower in the hierarchy. Indicative of the flexibility of hierarchical models, it can *also* be a parent node. A **founder node** is a fixed parameter rather than a variable, and it therefore has no parents. The opposite of a fixed parameter in the model is a **stochastic node**, which is a node that is assumed to have some distribution and these are both parameters and data. Lastly, a **deterministic node** is one that is a logical consequences of other nodes. A prototypical example is θ in the link from the linear structure, $\mathbf{X}\beta$, to the linear predictor, $\theta = g(\mu)$ in the specification of a GLM.

There are some additional qualifications to keep in mind here. With BUGS all priors and all likelihood functions must be either: (1) discrete, (2) conjugate, or (3) log-concave (the Gibbs sampling implications of this are discussed in Section 11.5). This is not a big deal as all GLMs with canonical link functions are well-behaved in this respect. It also means that deterministic nodes must be linear functions. Interestingly, these restrictions can be finessed with clever programming. Also BUGS likes simple, clean model structure, meaning that embedded variable transformations and heavy data processing tend to bog-down the sampler and may even lead to crashes. So if pre-processing in R can be done, it generally helps.

As mentioned, the BUGS language deliberately looks a lot like R syntax. Key identical components are: the assignment operator, <-, looping and indexing syntax, commenting beginning with #, and a host of simple functions like exp, log, mean, sd, etc. The notation for distributional statements is also consistent with R, with some exceptions. Generally ddist is the identifier for distribution "dist," and the inventory of available distributions is extensive (plus new ones can be constructed). Keep in mind that definitions inside of a ddist specification may differ from R, e.g., dnorm(mu,tau) specifies a mean and precision not a mean and variance in BUGS. Finally, all variants of BUGS (engines) can be called in batch mode from pre-constructed files.

Linear Model Example

Consider economic data from the Organization for Economic Cooperation and Development (OECD) that highlights the relationship between commitment to employment protection measured on an interval scale (0 to 4) indicating the quantity and extent of national legislation to protect jobs, and the total factor productivity difference in growth rates between 1980-1990 and 1990-1998 (see *The Economist*, September 23, 2000 for a discussion). The original data appear in Table 11.1 and these points are graphed in Figure 11.1.

TABLE 11.1: OECD PROTECTION VERSUS PRODUCTIVITY

	Prot.	Prod.		Prot.	Prod.		Prot.	Prod.
United States	0.2	0.5	Canada	0.6	0.6	Australia	1.1	1.3
New Zealand	1.0	0.4	Ireland	1.0	0.1	Denmark	2.0	0.9
Finland	2.2	0.7	Austria	2.4	-0.1	Belgium	2.5	-0.4
Japan	2.6	-0.4	Sweden	2.9	0.5	Netherlands	2.8	-0.5
France	2.9	-0.9	Germany	3.2	-0.2	Greece	3.6	-0.3
Portugal	3.9	0.3	Italy	3.8	-0.3	Spain	3.5	-1.5

We know from Gauss-Markov theory that the posterior distribution of both the intercept and the slope coefficients is Student's-t with $n - k - 1 = 16$ degrees of freedom. So why are we running BUGS on a linear model? Consider how different the estimation process really is here:

$$\hat{\mathbf{b}} = (\mathbf{X}'\mathbf{X})^{-1}\mathbf{X}'\mathbf{y}$$

versus

$$\alpha_1 \sim f(\alpha|\beta_0), \qquad\qquad \beta_1 \sim f(\beta|\alpha_1)$$
$$\alpha_2 \sim f(\alpha|\beta_1), \qquad\qquad \beta_2 \sim f(\beta|\alpha_2)$$
$$\vdots \qquad\qquad\qquad\qquad \vdots$$
$$\alpha_m \sim f(\alpha|\beta_{m-1}), \qquad\qquad \beta_m \sim f(\beta|\alpha_m).$$

not to mention the potential effect of *priors*! So this example is an implied test of the integrity of the MCMC process.

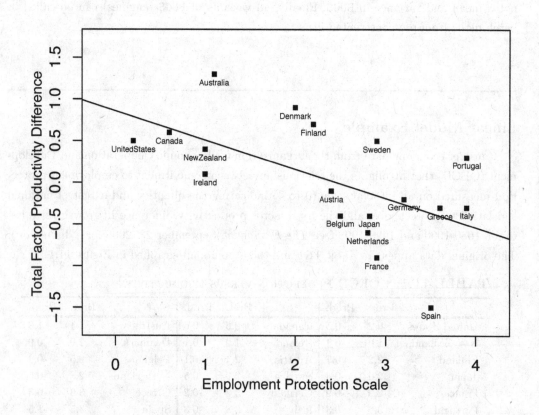

FIGURE 11.1: OECD EMPLOYMENT AND PRODUCTIVITY DATA

11.3.1 Model Specification

Two primary blocks of code need to be combined to create a full model statement, which wraps `model { }` around the whole process. The general program structure (not actual model statements) looks like this:

```
MODEL NAME
DECLARATIONS
{
    LOOPING THROUGH DATA
    DISTRIBUTIONAL STATEMENTS
}
```

where the part between the curly-braces is called the *model description*, and it has two parts where the ordering of these parts does not matter. The first line, `MODEL NAME`, consists of two parts: (1) the word `model` is necessary to indicate the start of a model specification, and (2) an optional model name. Regarding the latter, it is sometimes convenient to have a model name to distinguish individual specifications when many similar models are created. Many people ignore this feature. The second variable declaration statement identifies the set of the nodes to be used and their size. There is no variable declaration process in `WinBUGS`, whereas Classic `BUGS` required this. Both `JAGS` and `WinBUGS` use model statements (stipulated relations and inheritance from parents) and data size to determine the size of "undeclared" nodes. However, variable declaration is optional in `JAGS`. If compilation problems arise, one solution may be declaring variables, and it is also a convenient reminder when working on the rest of the model. For the OECD model, the optional `JAGS` declaration statement is:

```
var x[N], y[N], mu[N], alpha, beta, tau
```

Generally these statements are only necessary when stipulating objects of higher dimension, but giving variable declarations is also not harmful under basic circumstances. The model description is mainly contained in the `LOOPING THROUGH DATA` part, and this contains the *statistical structure* of the code. For the OECD example it is:

```
mu[i] <- alpha + beta*x[i]
y[i]  ~  dnorm(mu[i],tau)
```

where the indexing of i indicates processing through the data, $i = 1, \ldots, N$. The first line above is a GLM linear additive component collected into μ. This is a linear model so no link function is required. The second line above states that y has a normal distribution around μ with *precision* τ. It is important to remember that the second parameter in a `dnorm()` statement is a *precision* not a variance. The rest of the model description is contained in `DISTRIBUTIONAL STATEMENTS` where all of the unknown variables, except deterministic nodes, are listed along with their prior distribution assumptions. For the OECD example, these are:

```
alpha ~ dnorm(0.0,0.001)
beta  ~ dnorm(0.0,0.001)
tau   ~ dgamma(1,0.1)
```

Observe that these are parent nodes since they are higher in the hierarchy than μ, x, and y. The priors in this example are deliberately diffuse so as to mimic a regular non-Bayesian linear model as much as possible. We also need to take care of indexing with the first block above by specifying a "for" loop in exactly the same way as R. So the complete specification for the OECD model is:

```
model {
    for (i in 1:N) {
        mu[i] <- alpha + beta*x[i]
        y[i] ~ dnorm(mu[i],tau)
    }
    alpha ~ dnorm(0.0,0.001)
    beta  ~ dnorm(0.0,0.001)
    tau   ~ dgamma(1,0.1)
}
```

The looping in the BUGS model here is through the data with i. The production of chain values is not stipulated by the user in this code since chain values are generated after compilation of the model. This is an important principle to remember.

Since it is a simple linear model, the code above can be run in WinBUGS, JAGS, or openbugs exactly as written for each. In both WinBUGS and JAGS the program code is stored in a text file, but WinBUGS users often keep the model in the graphical environment model window for easy of manipulation. We still have to specify the data file. Unfortunately WinBUGS and JAGS have different types of data statements, but fortunately, they both use R language formats. WinBUGS uses R list format, so the data from Table 11.1 is specified by:

```
list(x= c(0.20, 0.60,  1.10,  1.00,  1.00,  2.00,  2.20,  2.40,  2.50,
          2.82, 2.90,  2.80,  2.90,  3.20,  3.60,  3.90,  3.90,  3.50),
     y= c(0.50, 0.60,  1.30,  0.40,  0.10,  0.90,  0.70, -0.10, -0.40,
         -0.40, 0.50, -0.60, -0.90, -0.20, -0.30,  0.30, -0.30, -1.50),
     N=18
)
```

There is an alternative matrix-structured rectangular file format for WinBUGS that some users may prefer, although constants such as N=18 must be handled separately (see Lunn *et al.* 2012, p.303 or Ntzoufras 2009, p.127). Notice that the sample size, N is in the data statement rather than inside the for loop. This is good programming practice since code is often recycled for future work with different data, and forgetting to change for (i in 1:18) is remarkably easy to do. Another good practice is writing the looping statement conditional on the data and therefore removing the possibility of future sizing mistakes:

```
for (i in 1:length(Y)) {
```

Users of WinBUGS often keep the data list in the same text window as the model statement (below it), particularly if the data size is small, since this is a convenience in the point-and-click environment of WinBUGS. However, JAGS is not an enclosed environment, so this is not possible, and the data are given in a file. A JAGS data file uses the R vector and matrix assignment convention, giving:

```
x <- c(0.20, 0.60,  1.10,  1.00,  1.00,  2.00,  2.20,  2.40,  2.50,
       2.82, 2.90,  2.80,  2.90,  3.20,  3.60,  3.90,  3.90,  3.50)
y <- c(0.50, 0.60,  1.30,  0.40,  0.10,  0.90,  0.70, -0.10, -0.40,
      -0.40, 0.50, -0.60, -0.90, -0.20, -0.30,  0.30, -0.30, -1.50)
N=18
```

which is stored in the file oecd.jags.dat. There are no commas between these data objects as required in the list format above.

We also need to stipulate starting values for the Markov chain, and there is a difference between WinBUGS and JAGS here too. The format in WinBUGS is again the R list format:

```
list(alpha = 0.0, beta = 0.0, tau = 1.0)
```

and the format in JAGS is the scalar assignment form stored in the file oecd.jags.init:

```
alpha <- 0.0
beta  <- 0.0
tau   <- 1.0
```

(although vectors can be specified too). We now have all of the pieces required to run the model. The next steps are to compile the model in BUGS and run the chain, recording values.

11.3.2 Running the Model in WinBUGS

The WinBUGS software has lots of "bells and whistles" to explore, such as running the model straight from the doodle (graphical summary, see the examples that are furnished with the software), and printing summary functions. The data from any plot can be recovered by double-clicking on it. Setting the seed may be important to you: leaving the seed as is exactly replicates chains. The WinBUGS interface uses what is called the *compound document interface*. This is an omnibus file format that holds: text, tables, code, formulae, plots, data, and initial values. The goal is to minimize cross-applications work by centralizing the model development process in one piece of software. Some useful features include: if one of the CDI elements is focused, its associated tools are made available; a built-in editor; and online documentation that has details about creating files.

The first window required is the *specification tool*, which requires that you use the following buttons:

▷ **check model**: checks the syntax of your code.

▷ **load data**: loads data from same or other file.

▷ **num of chains**: sets number of parallel chains to run.

▷ **compile**: compiles your code as specified.

▷ **load inits**: loads the starting values for the chain(s).

▷ **gen inits**: lets WinBUGS specify initial values.

It is important that these steps are performed in this order (the order of the buttons looks like a question-mark in this window). The specification tool window can be dismissed when done. The *update* window is the manner in which chain values are generated. The key buttons are:

▷ **updates**: you specify the number of chain iterations to run this cycle.

▷ **refresh**: the number of updates between screen redraws for traceplots and other displays.

▷ **update**: hit this button to begin iterations.

▷ **thin**: number of values to thin out of chain between saved values.

▷ **iteration**: current status of iterations, by UPDATE parameter.

▷ **over relax**: click in the box for option to

▷ generate multiple samples at each cycle,

▷ pick sample with greatest negative correlation to current value.

Trades cycle time for mixing qualities.

▷ **adapting**: box will be automatically clicked while the algorithm for Metropolis or slice sampling (using intentionally introduced auxiliary variables to improve convergence and mixing) is still tuning optimization parameters (4000 and 500 iterations, respectively). Other options are "greyed out" during this period.

The *sampling* window is the primary mechanism for stipulating nodes to be monitored and summarizing the resulting chains. Its buttons are:

▷ **node**: sets each node of interest for monitoring; type name and click SET for each variable of interest.

▷ Use the " * " in the window when you are done to do a full monitor.

▷ **chains**: "1 to 10" sets subsets of chains to monitor if multiple chains are being run.

▷ **beg, end**: the beginning and ending chain values current to be monitored. BEG is 1 unless you *know* the burn-in period.

▷ **thin**: yet another opportunity to thin the chain.

▷ **clear**: clear a node from being monitored.

▷ **trace**: do dynamic traceplots for monitored nodes.

▷ **history**: display a traceplot for the complete history.

▷ **density**: display a kernel density estimate.

▷ **quantiles**: displays running mean with running 95% CI by iteration number.

▷ **auto cor**: plots of autocorrelations for each node with lags 1 to 50.

▷ **coda**: display one window with the monitored chain history in "CODA" format (one long vector starting with the history of the first parameter followed by the others), and ordering information for this vector that record the location parameter boundaries in this vector.

▷ **stats**: summary statistics on each monitored node using: mean, sd, MC error, current iteration value, starting point of chain and percentiles from PERCENTILES window.

▷ Notes on **stats**:
WinBUGS regularly provides both: naive SE = sample variance$/\sqrt{n}$ and: MC Error = $\sqrt{\text{spectral density var}}/\sqrt{n}$ = asymptotic SE.

There are quite a few additional windows and pull-down features, but those described above are the essential group for running MCMC in WinBUGS. Note that WinBUGS gives two types of measures of uncertainty in the **stats** information. The "MC error" is Monte Carlo error (see the discussion starting on page 9.4 in Chapter 9), which can be reduced with additional iterations. The result labeled "sd" is the conventional posterior variance of the mean estimate, which is the quantity that should be reported in regression tables.

The following steps are given for the Classic BUGS where each command corresponds to a specific button in the WinBUGS GUI as just described. Here we run the Markov chain for 10,000 iterations without recording visited values, then monitor the nodes and run it an additional 50,000 values while recording:

▷ Compile:

```
Bugs>compile("oecd.bug")
```

▷ Run the chain for a burn-in period:

```
Bugs>update(10000)
time for    10000  updates was  00:00:01
```

▷ Turn on chain value recording:

```
Bugs>monitor(alpha)
Bugs>monitor(beta)
```

▷ Run the chain for a much longer series of values:

```
Bugs>update(50000)
time for 50000 updates was 00:00:05
```

▷ Ask for summary statistics:

```
Bugs>stats(alpha)
mean       sd        2.5% : 97.5% CI     median     sample
8.619E-1  3.402E-1  1.734E-1  1.527E+0   8.650E-1   50000
Bugs>stats(beta)
mean       sd        2.5% : 97.5% CI     median     sample
-3.507E-1  1.299E-1  -6.065E-1  -8.877E-2  -3.525E-1  50000
```

So WinBUGS goes all the way to regression-style results. As we shall see, there are good reasons to use the CODA button and analyze the results in R with the CODA package described further in Chapter 14. Note that a burn-in period of 10,000 and mixing period of 50,000 is clearly over-kill for this simple model, but since it mixes so quickly these values do not inconvenience. This issue is discussed in Chapter 14.

11.3.3 Running the Model in JAGS

Calling JAGS from R is described below, and this section shows how to run the model from a JAGS text window. After downloading and installing the software, JAGS can be called directly to produce a command window. Since we have saved the files containing code, data, and starting values, we now use these in command form in this window. The same process as done with WinBUGS above is completed by executing the commands:

```
model in "oecd.jags"
data in "oecd.jags.dat"
compile
inits in "oecd.jags.init"
initialize
update 10000
monitor set alpha
monitor set beta
update 50000
coda *
exit
```

We could also put these commands in a file (e.g., my.command.file) and simply type jags my.command.file at the prompt. Note that there may be path names required before the file names above, depending on where in the file system JAGS is started. These in-line statements are exactly analogous to the WinBUGS consecutive buttons above, except that we use the coda * statement to drop a file for R to ingest. Both WinBUGS and JAGS are capable of running multiple parallel chains at the same time. This facilitates convergence diagnostic discussions in Chapter 14. The JAGS window will display the following, indicating positive completion:

```
. Reading data file oecd.jags.dat
. Compiling model graph
    Resolving undeclared variables
    Allocating nodes
    Graph Size: 74
. Reading initial values file oecd.jags.init
. . Updating 10000
----------------------------------------| 10000
**************************************** 100%
. . . Updating 50000
```

```
------------------------------------------| 50000
**************************************** 100%
```

To run parallel chains in `JAGS` from different starting points, we repeat the initial value statements in different files and then replace the `compile` and `inits in` statements above with:

```
compile, nchains(3)
inits in "oecd-init1.R"
inits in "oecd-init2.R"
inits in "oecd-init3.R"
```

for three parallel chains. Here we would have to produce three initial value files instead of one, but it is typically trivial to copy the first and make changes to the values in the resulting copies.

TABLE 11.2: OECD MODEL RESULTS

	OLS Estimation		MCMC Posterior	
	Estimate	Std. Error	Mean	Std. Error
Intercept	0.859	0.317	0.859	0.322
Slope	-0.349	0.121	-0.349	0.123

Using the posterior mean as a point estimate from the `WinBUGS` or `JAGS` output above, we can compare with `lm` in R. This is shown in Table 11.2. Observe that the posterior means are identical (rounded to three places), but the posterior variances are slightly bigger. This makes theoretical sense: the Bayesian model has prior distributions on α and β, whereas the non-Bayesian model assumes a fixed underlying parameter.

Suppose we want to calculate some standard model summary quantities in the context of `BUGS` estimation. One set of values that we might be interested in are the linear predictions, the $\hat{y}$ values. This is done by simply monitoring the node μ in the sampler since `mu[i] <- alpha + beta*x[i]`. Another typically reported quantity is the estimate of the standard error of y, $\hat{\sigma}$. Recall that `dnorm` requires specification of the precision not the variance, so we can get the required quantity by adding `sigma <- pow(tau,-2)` to the code outside of the i loop and monitoring this new node. While the R^2 measure is a fairly blunt measure (it is quadratic in $[0:1]$ not linear, and is technically not a statistic), it is commonly reported for linear models. The easiest way to obtain it is to first create a deterministic node `y.hat[i] <- mu[i]` (we could work with μ, but this makes the code easier to read), then accumulate regression sum of squares contributions and total sum of square contributions, and outside the loop perform `R2 <- sum(SSR)/sum(SST)`. We want to remember to monitor any new nodes of interest. So with all of these enhancements, the new model is given by:

```
model {
    for (i in 1:n) {
        mu[i] <- alpha + beta*x[i];
        y[i] ~ dnorm(mu[i],tau);
        y.hat[i] <- mu[i]
        SSR[i] <- (y.hat[i] - mean(y))^2
        SST[i] <- (y[i] - mean(y))^2
    }
    sigma <- pow(tau,-2)
    R2 <- sum(SSR)/sum(SST)
    alpha ~ dnorm(0.0,0.001)
    beta  ~ dnorm(0.0,0.001)
    tau   ~ dgamma(1,0.1)
}
```

Obviously this adds some computational burden since we are making more work inside the loop, which is wholly repeated many times as we iterate the chain values. However, with a linear model this is rarely an annoying addition to run-time. With more complicated models, values that do not change across iterations, like `mean(y)`, would be taken out of the loop and calculated beforehand and used as a constant. Similarly, we could output intermediate values that allow us to perform the final R^2 calculation in R. Notice that this approach implies that R^2 is a statistic since it will have variability across chain iterations. This is not true. It is simply the calculation of a deterministic measure, *given* the current chain values for α and β, which give a $\hat{y}$ vector. Since we do not want to imply a distribution for R^2 it is better to summarize the realizations with a median ($m = 0.3763$). This is the same reason that we would not necessarily want to post-process based on the posterior mean only since that elevates it as the only considerable point summary. Finally, this model extension is intended to show that additional features can be added as desired.

■ **Example 11.1:   A Logit Model of HMO Effectiveness** Here we look at a less contrived example regarding public policy for healthcare delivery. Health Maintenance Organizations (HMOs) are private entities that provide managed care directly to patients with the goal of cost containment through contractual services. The data here include 1,180 children in Florida who visited their HMO clinic and did or did not subsequently require an emergency room visit shortly thereafter. The question is whether the HMO visit adequately addressed the child's condition. We will focus on only three variables:

> ▷ **erodd**, the dichotomous outcome variable, $[0,1]$, indicating whether or not there was an emergency room visit.

> ▷ **np**, indication of profit, $[1]$, or nonprofit, $[-1]$, status of the HMO. This is the key explanatory variable of interest.

> ▷ **metq**, a severity score, $[1,2,3]$, indicating the degree of illness diagnosed at the HMO visit. This is a required control variable.

The specification is a simple logit GLM with diffuse priors (the second parameter is a variance in these definitional statements):

$$erodd_i \sim \mathcal{BE}(p_i)$$
$$\text{logit}(p_i) = \alpha_0 + \alpha_1 np_i + \alpha_2 metq_i$$
$$\alpha_0 \sim \mathcal{N}(0, 10)$$
$$\alpha_1 \sim \mathcal{N}(0, 10)$$
$$\alpha_2 \sim \mathcal{N}(0, 10)$$

where the results of interest are the posterior distributions of α_0, α_1, and α_2 (10 is the variance above, so we will use 0.10 as the precision below). So the BUGS code is:

```
model
{
    for( i in 1 : N ) {
        logit(p[i]) <- alpha0 + alpha1 * np[i]
                        + alpha2 * metq[i]
        erodd[i] ~ dbern(p[i])
    }
    alpha0 ~ dnorm(0.0,0.1)
    alpha1 ~ dnorm(0.0,0.1)
    alpha2 ~ dnorm(0.0,0.1)
}
```

Notice that the second parameter in the normal distributional statements is now a precision. The WinBUGS data and initial value statements are:

```
list(erodd = c(1,1,1,0,1,0,1,...),
     np = c(-1,-1,-1,1,-1,-1,1,...),
     metq = c(3,3,3,3,3,2,3,...),
     N=1180)
list(alpha0 = 0, alpha1 = 0, alpha2 = 0)
```

where the dots indicate continuation of the data. The JAGS data file looks like:

```
erodd <- c(1,1,1,0,1,0,1,...)
np = c(-1,-1,-1,1,-1,-1,1,...)
metq = c(3,3,3,3,3,2,3,...)
N=1180
```

and the JAGS initial values file contains:

```
alpha0 = 0
alpha1 = 0
alpha2 = 0
```

These data can be obtained from the BaM package in R. The chain is run with a burn-in period of 1,000 cycles and then run for 9,000 more, producing the following WinBUGS output:

```
node    mean      sd  MC error     2.5%  median   97.5%  st. sample
alpha0  -1.971  0.22280  0.008254  -2.41300  -1.968  -1.540  1    9000
alpha1  0.1646  0.08042  9.194E-4  0.008893  0.1639  0.3213  1    9000
alpha2  0.2808  0.09423  0.003505  0.098070  0.2803  0.4645  1    9000
```

So there is evidence that an HMO run as a for-profit corporation (usually owned by a set of physicians and investors) has more children that need emergency room medical care shortly after their HMO visit. This is seen by the positive posterior mean for α_1, and the 95% credible interval bounded away from zero. The control variable for injury/illness severity also has a positive posterior coefficient mean, as expected.

11.4   Differences between WinBUGS and JAGS Code

We have already seen that there are differences in file formats between WinBUGS and JAGS. These differences are easy to maintain with two functions. Terry Elrod's WriteDatafileR takes an R dataframe and writes it to a file in a specified directory. This makes it easy to manipulate a dataset in R but instantly have a properly configured file for WinBUGS. Unfortunately this function is not available at CRAN, but it is available on the dedicated webpage for this text and elsewhere online. There is no corresponding function for JAGS, but the R package CODA contains the function bugs2jags that converts a WinBUGS formatted data file to a JAGS formatted data file. So consecutive use of these two functions gives JAGS users a properly formatted file after manipulation in R.

The most important difference between WinBUGS and JAGS is the handling of censoring and truncation. In WinBUGS the I(,) construct is used in WinBUGS for censoring as a posterior restriction, as well as truncation of top-level parameters as a prior restriction. The format is z ~ ddist(theta)I(lower, upper) for a restriction on an unobserved z which is distributed according to ddist (dnorm, dgamma, etc.) with a parameter θ, where this distribution exists only between the numerical values indicated by lower and upper. One-sided restrictions are achieved by omitting one of the restrictions: I(lower,) or I(,upper). If the variable z becomes observed, then these restrictions are ignored. This construct should only be used on model nodes and not to restrict the distribution of data that produces a likelihood function. Such data restrictions should be stipulated in the algebraic form of the likelihood. The WinBUGS documentation also warns that if z, θ, and the limits are unknown, then the limits cannot be modeled as functions of θ. However, I(,) can have nodes inside the function, e.g. I(,z), I(y,) to force ordering of the nodes $y < z$.

In contrast, JAGS separates censoring and truncation into two functions. For simple truncation the T(lower,upper) function is used, and it works like the I(lower,upper) function in WinBUGS. The interpretation is that the truncated variable is known a priori 100% to lie between lower and upper, which can be numeric values or other nodes for ordering. Actually, it is possible to specify I(,) as well in JAGS, but only if the distribution being truncated is fully known (no parameters to be estimated in ddist). Censoring in JAGS is represented by the novel distribution dinterval, which is not in WinBUGS. For example, suppose we had data that had left-censored values:

```
bounds.data
             x          y
[1,]  0.2470532  1.9955092
[2,]        NA  0.5612988
[3,]  1.0240813  1.1813461
⋮
⋮
```

A missing indicator (NA) is used where censored exists. If we wanted to treat the censored data as coming from a normal likelihood we would include the model statements in the data loop:

```
is.censored[i] ~ dinterval(t[i], bounds.data[i,])
t[i] ~ dnorm(mu,tau)
```

where: So is.censored[i] is a censoring indicator for the ith case, and t[i] is a censoring time. Here JAGS is more careful to distinguish unobservable due censoring, not visible to the researcher, and unobservable due to truncation, not visible because of the definition of a random quantity.

Finally, there is also an important difference in how the dedicated R functions process categorical data for WinBUGS and JAGS. None of the BUGS engines handles factors in the same natural way that R does, so factors in R must be converted to numerical values in a way that is consistent with the interpretation of the variable, particularly with regard to ordered factors. For windows systems BRugs converts to numerical values automatically, but not necessarily with the desired coding, but R2WinBUGS does not. The general advice is to take each factor variable and manually convert it to a numerical scheme in a way that makes sense. This process also includes identifying and removing one category as the reference group to create a contrast. None of the BUGS engines do the automatic contrast process that model statements in R perform. So it is up to the researcher to determine the appropriate handling of categorical data. Consider an example from Kyung *et al.* (2012) where SYS is a factor indicating classes of government structure: direct presidential elections, strong president elected by assembly, and dominant parliamentary government, summarized in R by

```
summary(SYS)
  Direct.Pres   Strong.Pres Parliamentary
        37           27           86
```

To create an ordered numeric variable `as.numeric(SYS) -1` turns these into `0,1,2` values, respectively. Note that this is a strong assumption, not just about the order, but also that the ordered distance between categories is identical. So if parliamentary government was not only the assumed highest category on this scale, but we also wanted it to be 3 times the distance as that between direct presidential elections and strong president elected by assembly, we would stipulate `SYS[SYS ==3] <- 4`. Perhaps no ordering is appropriate and this is a purely nominative measurement. Then a treatment contrast (dummy coding) must be given explicitly with the reference category identified. Suppose we want parliamentary government as the reference category, then to create two new variables to pass along to BUGS, the following R steps are necessary:

```
Direct <- abs(1-as.numeric(SYS)%%3)
Strong <- 1-abs(2-as.numeric(SYS))
```

Some people prefer sum contrast coding (Helmert and polynomial contrasts are rare in the social sciences), which can be produced for this example in R by:

```
Direct <- (as.numeric(SYS))%%3-1
Strong <- 2-as.numeric(SYS)
```

The key point is that the user is now fully responsible for coding categorical variables, ordered or unordered, in BUGS.

■ **Example 11.2:   Example: A Hierarchical Model of Lobbying Influence in the U.S. States**

The American State Administrator's Project (ASAP) survey asks administrators about the influence of a variety of external political actors including "clientele groups" in their agencies. Clientele group is arguably not perfectly synonymous with interest group, but previous studies have used these terms interchangeably (Kelleher and Yackee 2009). Gill and Witko (2013) reanalyze these data with a hierarchical Bayesian specification that accounts for the nesting of public administrators within states, and also substituting the variable `elected.board` below for their original measure of merit position, which is dropped.

Consider a 713×22 matrix $\mathbf{X}$ with a leading column of 1's for individual level explanatory variables, and a 50×3 matrix $\mathbf{Z}$ for state-level explanatory variables. These variables are:

   ▷ `contracting`: scale from $0:6$ where higher indicates more private contracting within the respondent's agency.

   ▷ `gov.influence`: respondents' assessment of the governor's influence on contracting in their agency.

▷ `leg.influence`: respondents' assessment of the legislatures' influence on contracting in their agency, ranging from $0:21$.

▷ `elect.board`: dichotomous variable coded 1 if appointed by a board, a commission or elected, and 0 otherwise.

▷ `years.tenure`: number of years that the respondent has worked at their current agency.

▷ `education`: ordinal variable for level of education possessed by the respondent.

▷ `partisan.ID`: a 5-point ordinal variable (1-5) for the respondent's partisanship (strong Democrat to strong Republican).

▷ `category_[]`: categories of agency type.

▷ `med.time`: whether the respondent spent more or less than the sample median with representatives of interest groups.

▷ `medt.contr`: interaction variable between `med.time` and `contracting`.

▷ `gov.ideology`: state government ideology from Berry et al. (1998) from 0 to 100.

▷ `lobbyists`: total state lobbying registrants in 2000-01 from Gray and Lowery (1996, 2001).

▷ `nonprofits` provides the total number of nonprofit groups in the respondents' state in the year 2008, divided by 10,000.

The outcome variable (`group.inf`) measures the respondents' perception of interest groups' influence on total budget, special budgets, and general public policies. The linear hierarchical model is given by:

$$Y_i \sim N(\alpha_{ij} + \beta \mathbf{X}_i, \sigma_y^2), \quad \text{for } i = 1, \ldots, 713$$
$$\alpha_j \sim N(\mathbf{GZ}, \sigma_\alpha^2), \quad \text{for } j = 1, \ldots, 50,$$

where α_{ij} indicates that the ith respondent is nested in the jth state to produce a state-specific random intercept. This random intercept is then parameterized at a second level by the three explanatory variables in $\mathbf{Z}$ (`gov.ideology`, `lobbyists`, `nonprofits`) and their corresponding estimated coefficients, $\mathbf{G}$. We will return to this model and the hierarchical issues in Chapter 12, but for now we will concentrate on the mechanics of estimation with MCMC.

The model uses semi-informed versions of the prior distributions for the unknown parameters since a high-quality source exists: prior distributions are diffuse normals centered at the point estimates from Kelleher and Yackee (2009), (their Model 3, 2009, p.593). This gives the specification:

$$\beta \sim N(\beta_{ky}, \Sigma_\beta)$$
$$\mathbf{G} \sim N(\mathbf{G}_{ky}, \Sigma_{\mathbf{G}}),$$

where the Σ_β and $\Sigma_{\mathbf{G}}$ matrices are diagonal forms with large variances relative to the

Kelleher and Yackee point estimates.   This leads to the following BUGS code minus the prior specifications:

```
model {
    for (i in 1:SUBJECTS) {
        mu[i] <- alpha[state.id[i]]
            + beta[1]*contracting[i]    + beta[2]*gov.influence[i]
            + beta[3]*leg.influence[i]  + beta[4]*clientappt[i]
            + beta[5]*years.tenure[i]   + beta[6]*gender[i]
            + beta[7]*education[i]       + beta[8]*partisan.ID[i]
            + beta[9]*category_2[i]      + beta[10]*category_3[i]
            + beta[11]*category_4[i]     + beta[12]*category_5[i]
            + beta[13]*category_6[i]     + beta[14]*category_7[i]
            + beta[15]*category_8[i]     + beta[16]*category_9[i]
            + beta[17]*category_10[i]    + beta[18]*category_11[i]
            + beta[19]*category_12[i]    + beta[20]*timegroupsmed[i]
            + beta[21]*timemedXcont[i]
        group.inf[i] ~ dnorm(mu[i],tau)
    }
    for (j in 1:STATES) {
        eta[j] <- gamma[1]*gov.ideology[j] + gamma[2]*lobbyists[j]
                + gamma[3]*nonprofits[j]
        alpha[j] ~ dnorm(eta[j],tau.alpha)
    }
}
```

The prior distributions are specified by:

```
beta[1]    ~ dnorm(0.070,1)      # |
beta[2]    ~ dnorm(-0.054,1)     # |
beta[3]    ~ dnorm(0.139,1)      # |
beta[4]    ~ dnorm(0.468,1)      # | PRIOR MEANS FROM KELLEHER AND
beta[5]    ~ dnorm(0.017,1)      # | YACKEE 2009, MODEL 3
beta[6]    ~ dnorm(0.207,1)      # |
beta[7]    ~ dnorm(0.056,1)      # |
beta[8]    ~ dnorm(0.039,1)      # |
beta[9]    ~ dnorm(0.0,1)
beta[10]   ~ dnorm(0.0,1)
beta[11]   ~ dnorm(0.0,1)
beta[12]   ~ dnorm(0.0,1)
beta[13]   ~ dnorm(0.0,1)
beta[14]   ~ dnorm(0.0,1)
beta[15]   ~ dnorm(0.0,1)
beta[16]   ~ dnorm(0.0,1)
beta[17]   ~ dnorm(0.0,1)
beta[18]   ~ dnorm(0.0,1)
```

```
beta[19]   ~ dnorm(0.0,1)
beta[20]   ~ dnorm(0.184,1)     # | PRIOR MEANS FROM KELLEHER AND
beta[21]   ~ dnorm(0.146,1)     # | YACKEE 2009, MODEL 3
gamma[1]   ~ dnorm(0.0,1)
gamma[2]   ~ dnorm(0.0,1)
gamma[3]   ~ dnorm(0.0,1)
tau        ~ dgamma(1.0,1)
tau.alpha  ~ dgamma(1.0,1)
```

And the JAGS formatted starting values are:

```
tau <- 10
tau.alpha <- 10
alpha <- c(1,1,1,1,1,1,1,1,1,1,1,1,1,1,1,1,1,1,1,1,1,1,1,1,1,1,1,1,1,
           1,1,1,1,1,1,1,1,1,1,1,1,1,1,1,1,1,1,1,1,1,1,1,1)
zeta <- 1
beta <- c(1,1,1,1,1,1,1,1,1,1,1,1,1,1,1,1,1,1,1,1,1)
gamma <- c(1,1,1)
```

The data are too large to present here but can be obtained from the R package BaM. We now turn our attention to processing within R. First load the necessary libraries, file, and dataset:

```
lapply(c("rjags","arm","coda","superdiag","R2WinBUGS","R2jags","lme4"),
    library, character.only=TRUE)
source("WriteDatafileR.R"); data(asap.data.list)
```

Be aware that these libraries change over time as does JAGS and R, meaning that small errors can occur because some future setup is not exactly like the one used at the time of this example. In this example we will run JAGS remotely from R using Rjags. The first step is to define the model in some file, making it an R function. Once satisfied with the model specification it can then be sourced or pasted into the R environment. The model here is given by:

```
asap.model.rjags  <- function() {
for (i in 1:SUBJECTS) {
    mu[i] <- alpha[state.id[i]] + beta[1]*contracting[i]
          + beta[2]*gov.influence[i] + beta[3]*leg.influence[i]
    :
    tau.alpha ~ dgamma(1.0,1);
}
```

where this is exactly the same model BUGS code as before (text shortened for space purposes) except that it is enclosed in `asap.model.rjags <- function() { }` in the R environment. It is important to note some features here. Since there are two levels here, there are two looping structures: `for (i in 1:SUBJECTS) { }` and `for (j in 1:STATES) { }`. In this way BUGS loops through both indices. The linkage between the individual subjects level and the states level is `mu[i] <- alpha[state.id[i]]`. Here the intercept serves to express the state-level differences since `state.id[i]` asserts that the ith individual is nested in one of the 50 states. We are then free to specify the linear model at the state level. This is called *nested indexing* as specified here and it is not the only way specify hierarchical models and data in BUGS, but it is the cleanest and closest to the theory (Lunn *et al.* [2012, pp.231-232] discuss alternatives: "padding-out" and "offsets").

It is often convenient and safe to save the model into a file, which is done by:

```
write.model(asap.model.rjags, "asap.model2.rjags")
```

This requires the R2WinBUGS package even though we are working with JAGS. Setting up the initial values and naming the parameters for the JAGS function in R is done by:

```
asap.inits <- function() list("tau.y" = 10, "tau.alpha" = 10,
    "beta" = rep(1,20), "gamma" = c(1,1,1))
asap.params <- c("beta","gamma","tau.y","tau.alpha")
```

Now that we have these values saved, we can ask JAGS to compile it with three parallel chains, and asking it to use 5,000 iterations to tune the variance of the Metropolis-Hastings candidate-generating distribution (if necessary):

```
asap.out <- jags(data=asap.jags.list, inits=asap.inits, asap.params,
    n.iter=5000, model="asap.model.rjags", DIC=TRUE)
```

Actually, DIC=TRUE is the default, but it is stated here as a reminder. Now run 200,000 iterations of the Markov chain saving all sampled values (we can remove burn-in values later:

```
asap.out3 <- update(asap.out, n.iter=200000)
```

A quick summary table is achieved with:

```
print(asap.out3)
```

A more convenient form is obtained by turning the output into an mcmc object in R:

```
asap.mcmc3 <- as.mcmc(asap.out3)
```

This lets us load the results into CODA and BOA as well as use the diagnostic suite in the superdiag packager:

```
superdiag(as.mcmc.list(asap.mcmc3), burnin=100000)
```

The results are easily summarized from the R command to produce a table. Notice from Table 11.3 that the α values are summarized by their mean. This is common in hierarchical models like this where there are 50 random intercepts and this would overwhelm readers in the context of a table.

TABLE 11.3: LOBBYING INFLUENCE RESULTS

Parameters	Mean	Std. Error	95% HPD Interval
α mean(1:50)	1.3905	0.7037	[0.0112 : 2.7698]
contracting	0.1987	0.0963	[0.0099 : 0.3874]
gov.influence	0.0481	0.0367	[-0.0239 : 0.1202]
leg.influence	0.3519	0.0397	[0.2741 : 0.4297]
elect.board	1.3436	0.3546	[0.6486 : 2.0386]
years.tenure	0.0347	0.0233	[-0.0110 : 0.0804]
education	0.1249	0.1217	[-0.1136 : 0.3634]
partisan.ID	-0.0046	0.0845	[-0.1703 : 0.1611]
category2	-0.4282	0.5423	[-1.4912 : 0.6348]
category3	-0.0596	0.5885	[-1.2131 : 1.0938]
category4	1.5501	0.4571	[0.6541 : 2.4461]
category5	-0.5473	0.5010	[-1.5292 : 0.4347]
category6	0.9227	0.5395	[-0.1348 : 1.9801]
category7	1.7014	0.4353	[0.8482 : 2.5546]
category8	1.0013	0.4986	[0.0240 : 1.9785]
category9	0.9412	0.4860	[-0.0115 : 1.8938]
category10	0.6157	0.4634	[-0.2925 : 1.5239]
category11	-0.1264	0.4265	[-0.9624 : 0.7096]
category12	-0.1592	0.5727	[-1.2816 : 0.9632]
med.time	1.1435	0.3587	[0.4405 : 1.8465]
medt.contr	-0.0869	0.1372	[-0.3559 : 0.1821]
gov.ideology	0.0182	0.0062	[0.0060 : 0.0303]
lobbyists	0.0007	0.0008	[-0.0007 : 0.0022]
nonprofits	-0.0217	0.1267	[-0.2701 : 0.2266]
τ_y	0.0763	0.0042	[0.0682 : 0.0845]
τ_α	3.1021	1.3523	[0.4517 : 5.7525]

These findings are consistent with the literature regarding interest group influence in state agencies. More contracting is positively related to perceptions of interest group influence, and this finding is statistically reliable. There is evidence that those agency heads that spent above the median amount of time with organized interests perceived groups to have more influence over their agencies. Agency heads that were elected or appointed by boards or commissions perceive interest group influence to be much greater, and this coefficient is reliable. A legislature perceived to be more influential is associated with *more* powerful clientele groups, which become legislatures' interest group allies when making decisions. Also, the more time in the current position is associated with the perception of greater interest group influence, even though 7% of the posterior density is below zero under a normal posterior assumption.

To compare this model to the null model we need to rewrite the code preserving the hierarchical component but remove all of the individual explanatory variables. Unfortunately WinBUGS requires that every variable in the data definition must be used in the model specification, which can make it awkward for the model fit process. An old trick is to assign unused variables in the data to a "dead end" child node. This is a node that does nothing but collects the unwelcome nodes and satisfies WinBUGS. Fortunately JAGS and openbugs do not worry about this match. We could easily write a "new" null model but this is irritating and can introduce new errors to debug. Instead we leave the linear additive collection but assign it to ν instead of μ, and μ gets just the mean effect:

```
asap.null.rjags <- function() {
for (i in 1:SUBJECTS) {
    nu[i] <- alpha[state.id[i]]
      + beta[1]*contracting[i]    + beta[2]*gov.influence[i]
      + beta[3]*leg.influence[i]  + beta[4]*elect.board[i]
      + beta[5]*years.tenure[i]   + beta[6]*education[i]
      + beta[7]*partisan.ID[i]    + beta[8]*category2[i]
      + beta[9]*category3[i]      + beta[10]*category4[i]
      + beta[11]*category5[i]     + beta[12]*category6[i]
      + beta[13]*category7[i]     + beta[14]*category8[i]
      + beta[15]*category9[i]     + beta[16]*category10[i]
      + beta[17]*category11[i]    + beta[18]*category12[i]
      + beta[19]*med.time[i]      + beta[20]*medt.contr[i]
    mu[i] <- alpha[state.id[i]]
    group.infl[i] ~ dnorm(mu[i],tau.y)
    for (j in 1:STATES) {
        eta[j] <- gamma[1]*gov.ideology[j] + gamma[2]*lobbyists[j]
                                           + gamma[3]*nonprofits[j]
        alpha[j] ~ dnorm(0,tau.alpha)
    }
    :
}
```

Now WinBUGS is satisfied and we can rerun the R steps from above being careful to relabel the output objects:

```
write.model(asap.null.rjags, "asap.null.rjags")
asap.null.out <- jags(data=asap.jags.list, inits=asap.inits,
    asap.params, n.iter=5000, model="asap.null.rjags", DIC=TRUE)
asap.null.out3 <- update(asap.null.out, n.iter=200000)
print(asap.null.out3)
```

```
asap.null.mcmc3 <- mcmc(asap.null.out3)
superdiag(as.mcmc.list(asap.null.mcmc3), burnin=100000)
```

The statement `print(asap.out3)` or:

```
dic.samples(asap2.model, n.iter=2500, type="pD")
```

(for more options) returns $DIC = \bar{D} + p_D = 3861 + 34.04 \approx 3895$. For the null model run above, this command returns $\bar{D} = 3964$ and $p_D = 49.26$. The null model DIC is the rounded up sum of these two: 4014. Since lower DIC values are preferred, the specified model is a better fit (3895 versus 4014). Note that these estimates are built on Monte Carlo quantities, so different runs of the same sampler will naturally produce slightly different values.

11.5   Technical Background about the Algorithm

The process of creating full conditional distributions from a model specification is the biggest contribution of the BUGS software. The mechanism for doing this and creating a run-time MCMC program is sensitive to the level of measurement of the nodes. For categorical variables CDF inversion is used, and for closed forms from interval-measured conjugate relationships, direct sampling from known analytical solutions is used. These are simple solutions that have already been discussed in other contexts. Strictly log-concave but not conjugate forms use *adaptive rejection sampling*, which is described here. Other continuous forms require Metropolis-Hastings or slice sampling (Chapter 14).

The software WinBUGS and openbugs (but not JAGS) uses an underlying engine of adaptive rejection sampling (Gilks 1992, Gilks and Wild 1992), which is an MCMC implementation that implements rejection sampling with an adaptive function. It works as follows. We want to sample from $f(\theta)$, or $g(\theta) = cf(\theta)$, defined on the support Θ. Define the *enveloping function*: $g_u(\theta) \geq g(\theta)$, $\forall \theta \in \Theta$. Now define the *squeezing function*: $g_q(\theta) \leq g(\theta)$, $\forall \theta \in \Theta$. Here is a general statement of the algorithm steps, to be repeated until the desired number of samples are obtained:

1. sample θ^* from $g_u(\theta)$

2. sample u from $\mathcal{U}(0,1)$

3. decide

If:	$u \leq g_q(\theta^*)/g_u(\theta^*)$	
Then:	accept θ^*	
Else:	**If:**	$u \leq g(\theta^*)/g_u(\theta^*)$
	Then:	accept θ^*
	Else:	reject θ^*, use it to update enveloping and squeezing functions

The adaptive rejection sampling works as follows:

▷ T_θ is a small set of points on the support of θ: $\{\theta_0, \theta_1, \ldots, \theta_{s+1}\}$, the abscissae.

▷ $h(\theta) = \log g(\theta)$ is known up to a constant (dropping conditional terms for notational convenience), and concave everywhere in Θ.

▷ Since $h(\theta)$ is concave, any non-tangent line segment that intersects the curve does so at two points corresponding to a pair of selected θ_i.

▷ Call this line $L_{i,i+1}$ determined by the points: $(\theta_i, h(\theta_i))$ and $(\theta_{i+1}, h(\theta_{i+1}))$.

▷ Now consider the triangle created by $L_{i,i+1}$ and the two tangent lines at $(\theta_i, h(\theta_i))$ and $(\theta_{i+1}, h(\theta_{i+1}))$.

▷ The triangle clearly encloses the $h(\theta)$ function, and so we also have bounds on the $f(\theta)$ metric.

▷ For each point in T_θ, calculate $h(\theta)$ and $h'(\theta)$...

▷ define the *upper hull* formed by the tangent lines to $h(\theta)$ at the abscissae in T_θ.

▷ Note that these tangent lines create the upper hull by their intersections, for the lines tangent at θ_j and θ_{j+1}, the intersection point is:

$$z_j = \frac{h(\theta_{j+1}) - h(\theta_j) - \theta_{j+1} h'(\theta_{j+1}) + \theta_j h'(\theta_j)}{h'(\theta_j) - h'(\theta_{j+1})}$$

▷ So any point on the upper hull between z_{j-1} and z_j can be found by:

$$u(\theta) = \underbrace{h(\theta_j)}_{\text{value at abscissa}} + \underbrace{(\theta - \theta_j)}_{\text{distance from abscissa}} \times \underbrace{h'(\theta_j)}_{\text{slope of tangent}}.$$

The triangles in this process are illustrated in Figure 11.2.

The squeezing function works as follows.

▷ Also for each point in T_θ, again use $h(\theta)$ and $h'(\theta)$...

▷ Define the *lower hull* formed by the chords connecting $h(\theta)$ at the abscissae in T_θ.

▷ For the chord connecting θ_j and θ_{j+1}, the line is given by:

$$\ell(\theta) = \frac{h(\theta_j)(\theta_{j+1} - \theta) + h(\theta_{j+1})(\theta - \theta_j)}{\theta_{j+1} - \theta_j}$$

▷ Log-concavity assures $\ell(\theta) \leq h(\theta) \leq u(\theta)$, where equality holds at the abscissae.

This is shown in Figure 11.3.

Now we can state the full algorithm in greater detail:

1. **Initialize** the (possibly new) abscissae, T_θ, calculate $u(\theta_i)$ and $\ell(\theta_i)$ at these points.

2. **Sample** θ^* from $\exp[u(\theta)] / \int_D \exp[u(\theta')] d\theta'$, and sample u from $\mathcal{U}(0, 1)$.

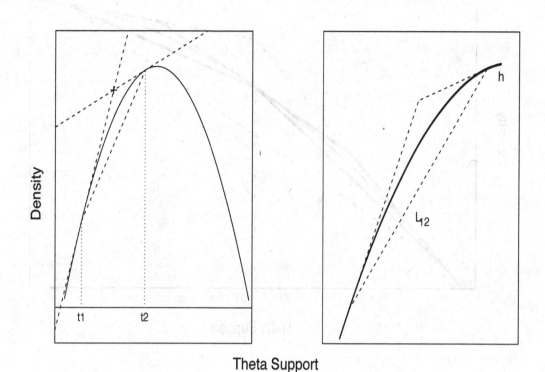

FIGURE 11.2: ADAPTIVE REJECTION SAMPLING STEP

3. **Squeezing Test:**

If: $u \leq \frac{\exp[\ell(\theta^*)]}{\exp[u(\theta^*)]}$

Then: accept θ^*

Else: calculate $h(\theta^*)$, $h'(\theta^*)$. If: $u \leq \frac{\exp[h(\theta^*)]}{\exp[u(\theta^*)]}$

Then: accept θ^*

Else: reject θ^*

---

4. **Updating Test:** If θ^* was rejected, add to the list in T_θ.

So the algorithm focuses on efficiency by trying to accept points in the most efficient method first. It also gains accuracy over time (a higher rate of accepted points) because of the increase in abscissae points. Note that this is a regular Monte Carlo implementation of an

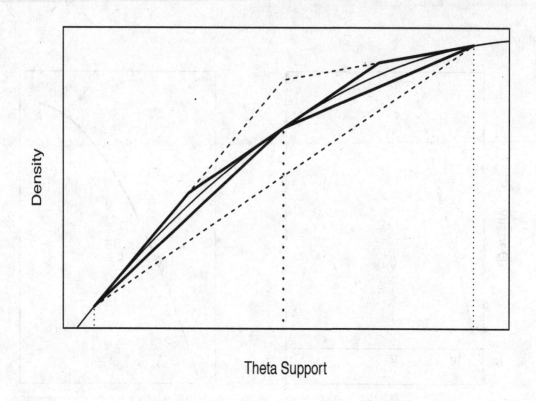

FIGURE 11.3: ADAPTIVE REJECTION SAMPLING SQUEEZING

MCMC algorithm since the $h(\theta)$ function *is* a single full conditional distribution from an inner step of the Gibbs sampler. This algorithm turns out to be quite flexible as well. For instance, Gilks, Best, and Tan (1995) show how to incorporate a Metropolis-Hastings step that allows sampling from non-log-concave target distributions.

■ **Example 11.3:   Example: An Ordered Logit Model for U.S. Election Data**
The 1960 U.S. presidential election between John Kennedy and Richard Nixon was one of the closest contests in national history, with Kennedy's margin of victory less than one percent of the popular vote. Campaigns, journalistic accounts, and social contexts can modify pre-election perceptions of the competitiveness of an uncertain outcome. This has implications for turnout and therefore may affect the election as well. This example uses the 1960 American National Election Study (subsetted in the R library BaM) to explore the link between personal characteristics and perception of closeness of the impending election. The outcome variable has four ordered categories: [one candidate will win by a lot], [one candidate will win by quite a bit], [this will be a close race–fairly even], [this will be a very close race]. The specified explanatory variables are:

▷ education, 1=8th grade or lower (233 cases), 2=highschool (428 cases), 3=some college or more (185 cases).

▷ sex, 1=male, (412 cases), 2=female (434 cases).

▷ seedebates, 1=no (141 cases), 2=yes (660 cases).

▷ importance, 1=care very much (270 cases), 2=care pretty much (308 cases), 3=pro-con/depends (6 cases), 4=don't care very much (155 cases), 5=don't care at all (85 cases).

▷ involvement, 8 categories from low to high with the distribution of cases: (158, 165, 245, 101, 78, 55, 4, 39).

▷ catholic, 0=no (623 cases), 1=yes (179 cases).

▷ partyid, 1=strong Democrat (198 cases), 2=not very strong Democrat (201 cases), 3=independent closer to Democrats (52 cases), 4=independent (67 cases), 5=independent closer to Republicans (59 cases), 6=not very strong Republican (117 cases), 7=strong Republican (139 cases).

Ordered choice models are constructed by assuming that there is a continuous latent metric dictating the categorical choices since researchers construct the scale not the respondents. So the outcome variable $\mathbf{Y}$ has C ordered categories separated by estimated thresholds (sometimes called "cutpoints" or "fences") sitting over a continuous utility metric $\mathbf{U}$ that cannot be seen:

$$\mathbf{U}_i : \ \theta_0 \underset{c=1}{\Longleftrightarrow} \theta_1 \underset{c=2}{\Longleftrightarrow} \theta_2 \underset{c=3}{\Longleftrightarrow} \theta_3 \ldots \theta_{C-1} \underset{c=C}{\Longleftrightarrow} \theta_C,$$

where the end-categories extend out to $-\infty$ and ∞, respectively. The effect of the explanatory variables is determined by a linear additive specification on the latent scale such that the ith person's utility is $U_i = \mathbf{X}_i\boldsymbol{\beta} + \epsilon_i$, where the $\boldsymbol{\beta}$ do not depend on the θ values. Note that some authors prefer a minus sign in front of $\mathbf{X}_i\boldsymbol{\beta}$, but the model defined here does not as is also the case with the R function polr. The vector of utilities across individuals is determined in the following way:

▷ the probability of the ith person choosing the kth category or less is $p(Y_i \leq k|\mathbf{X})$,

▷ this is equal to the probability that the ith person's utility is less than or equal to next threshold to the right of this category, $p(U_i \leq \theta_k)$,

▷ now substitute in the linear additive component for the utility to get $p(\mathbf{X}_i\boldsymbol{\beta} + \epsilon_i \leq \theta_k)$,

▷ rearrange to leave the stochastic component alone on the left $p(\epsilon_i \leq \theta_k - \mathbf{X}_i\boldsymbol{\beta})$,

▷ notice that this is just the CDF of ϵ_i, $F_{\epsilon_i}(\theta_k - \mathbf{X}_i\boldsymbol{\beta})$, at the point $\theta_k - \mathbf{X}_i\boldsymbol{\beta}$,

▷ specifying a logistic distribution for this distribution results in the model $p(Y_i \leq k|\mathbf{X}) = [1 + \exp(\mathbf{X}_i\boldsymbol{\beta} - \theta_k)]^{-1}$.

An ordered probit model can also be created by specifying $p(Y_i \leq k|\mathbf{X}) = \Phi(\theta_k - \mathbf{X}_i\boldsymbol{\beta})$ instead.

The ordered logit model is produced in JAGS with the following code.

```
model {
    for (i in 1:Nsub) {
        mu[i] <- beta[1]*education[i] + beta[2]*sex[i]
            + beta[3]*seedebates[i] + beta[4]*importance[i]
            + beta[5]*involvement[i]
            + beta[6]*importance[i]*involvement[i]
            + beta[7]*catholic[i] + beta[8]*partyid[i]
        for (j in 1:(Ncat-1)) { logit(Q[i,j]) <- cut[j] - mu[i] }
        p[i,1] <- Q[i,1]
        for (j in 2:(Ncat-1)) { p[i,j] <- Q[i,j] - Q[i,(j-1)] }
        p[i,Ncat] <- 1 - Q[i,(Ncat-1)]
        close[i] ~ dcat(p[i,1:Ncat])
        E.y[i] <- close[i] - mu[i]
    }
    sd.y <- sd(E.y[])
    for (k in 1:Nvar) { beta[k] ~ dt(0,1,5) }
    for (k in 1:(Ncat-1)) { cut0[k] ~ dt(0,1,5) }
    cut[1:(Ncat-1)] <- sort(cut0)
}
```

There are several features of this model that have not been introduced yet. Notice that the logistic specification is specified by `for (j in 1:(Ncat-1)) { logit(Q[i,j]) <- cut[j] - mu[i] }`, which loops through each of the first $k - 1$ categories relating the linear additive component to a cumulative probability. The last category is not necessary to calculate since it equals one. Next we want to calculate the individual category probabilities from the cumulative probabilities. The first category is easy since they are equivalent: `p[i,1] <- Q[i,1]`. Then we loop through all of the higher categories except for the last, differencing the adjacent cumulative probabilities to get the marginal probabilities: `for (j in 2:(Ncat-1)) { p[i,j] <- Q[i,j] - Q[i,(j-1)] }`. Finally the right-most category is obtained by `p[i,Ncat] <- 1 - Q[i,(Ncat-1)]`. The outcome variable is then modeled with these probabilities with: `close[i] ~ dcat(p[i,1:Ncat])`. The last line in the 1:N loop creates a residual vector, which is then summarized outside of the loop with a standard error function. The β coefficients and the cutpoints are both given Cauchy priors with 5 degrees of freedom. The last line creates the variable `cut` from the intermediate variable `cut0` to create a sorted form from the prior distribution.

As noted before in this chapter, it is good programming practice not to bury constants into the BUGS code. A common mistake is using the same code for another application (possibly modified) but not noticing that such values are hard-coded into the model statement. With good luck the program crashes leading to time debugging, but with bad luck the constant "works" but modifies the intended purpose of the new model such as looping only through part of the data. Using R the needed constants are appended to the data file:

```
system("echo 'Nsub <- 846' >> anes.jags.dat")
system("echo 'Ncat <- 4' >> anes.jags.dat")
system("echo 'Nvar <- 8' >> anes.jags.dat")
```

(although this could also be done with and editor). The system commands above require a Unix-based operating system.

TABLE 11.4: PRESIDENTIAL ELECTION MODEL RESULTS

Parameters	Mean	Std. Error	95% HPD Interval
education	0.2773	0.1067	[0.0681: 0.4864]
sex	0.3708	0.1410	[0.0944: 0.6472]
seedebates	0.4438	0.1813	[0.0884: 0.7992]
importance	-0.0432	0.1235	[-0.2852: 0.1988]
involvement	-0.3021	0.1045	[-0.5070:-0.0973]
importance×involvement	0.0534	0.0280	[-0.0015: 0.1084]
catholic	0.2484	0.1741	[-0.0928: 0.5896]
partyid	0.0403	0.0328	[-0.0240: 0.1047]
θ_1	-1.7139	0.5141	[-2.7215:-0.7064]
θ_2	-0.2879	0.4989	[-1.2658: 0.6900]
θ_2	3.0812	0.5142	[2.0734: 4.0890]

$s_y = 0.7517$, Model DIC: 1580.64, Null DIC: 1609.95

The modeling process started with using the mice package in R to create 5 complete datasets with imputed values for missing data. Then 5 chains with 5 different starting points and 5 different random seeds are run in JAGS for 100,000 iterations disposing of the first half of these. There was no indication of non-convergence using the suite of diagnostics in superdiag. The MCMC outputs are combined by first taking the mean and standard deviation of each run of the chain and combined according to standard practice with multiple imputation (mean for the coefficient mean, and weighted combination of between and within variances for the coefficient standard error). The results are summarized in Table 11.4.

We see that those higher levels of education, women, and those that watched the debates are more likely to see the race as closer. The measurement for involvement in the race combines whether the respondent cares about who wins with the degree of interest in the campaign. There is evidence from the model that higher levels of involvement lead to more likely believing one of the candidates will win convincingly. Regretfully the interaction with importance of the race does not work here, but this interaction provides better fit when included. Oddly there is no evidence that being Catholic influenced perception of closeness in an election where Kennedy's Catholicism was an important campaign issue. Finally that stalwart of U.S. electoral models, party identification, is not contributing here.

11.6   Epilogue

This is the most "vocational" chapter in this text since it is concerned with the nuts-and-bolts of estimation with popular MCMC software. The purpose is to make the mechanical part of the process as comfortable as possible. However, as we will see in subsequent chapters, there are more issues to worry about. Unfortunately, MCMC estimation will never be as automated as MLE estimation simply because there are more nuances that humans should worry about. In fact, WinBUGS documentation includes the admonition "Beware: MCMC sampling can be dangerous!"

In this chapter we also built up the level of complexity gradually going from a simple linear model to a logit model to a hierarchical model. In the process, WinBUGS and JAGS were highlighted, along with an example of the use of Rjags. This does not imply these are the only reasonable approaches and given the many R packages for calling the three BUGS engines, users can use the one that they are comfortable with. The choice between WinBUGS and openbugs is evolving. Now WinBUGS is enormously popular due to the user interface, but the developers state that new development is being done only on openbugs. The last update of WinBUGS was version 1.4.3, released August 2007 (the last of a series of patches from version 1.4 released September 2004). Finally, the software approaches in this chapter will become dated with time (hopefully not too quickly!), but the theoretical discussion in this chapter surrounding this one are based on hardened mathematical principles and will not change (although obviously new theory will be added over time).

11.7   Exercises

11.1   Rerun the model in Example 11.4. Why are your posterior summaries very slightly different?

11.2   Consider the following subset of BUGS code:

```
for (i in 1:N) {
    Y[i] ~ dnorm(eta[i], tau)
    :
}
:
```

```
tau ~ dgamma(1.0E-2, 1.0E-2)
sigma <- 1/sqrt(tau)
```

where ":" denotes additional code not essential to this question. What is σ, and how is it distributed?

11.3 Rerun the model in Example 11.3.3 using a probit link function and Cauchy priors for the three parameters.

11.4 Using the data from Example 2.3.4 (the Cultural Consensus Model in Anthropology) on page 51, write a simple logit specification with no covariates (mean effect only) in BUGS and run the model.

11.5 Write a BUGS program to calculate the posterior predictive distribution using the data in Example 6.4 concerning Economic Growth in Sub-Saharan Africa on page 197. Your model is an MCMC implementation of (6.22). Specify three different prior distributions and compare the results.

11.6 Using the Palm Beach County electoral data described on page 148, and distributed in the BaM package, specify both a linear model and a Poisson (log-normal) model in BUGS for the outcome badballots in the dataset pbc.vote. Use diffuse prior distributions in both cases. Compare the results. Is it necessary to use the Poisson GLM?

11.7 Write a Two-Way ANOVA model for the recidivism data in Example 6.2 on page 188 where the columns are determined by the variables Released and Returned, and the rows are determined by crime type. Do this with a double loop in BUGS where the values in the table are modeled according to Y[i,j] dnorm(eta[i,j], tau) where the double indexing accounts for both row and column mean effects.

11.8 In 2006 the U.S. Senate voted 56 to 44 to confirm Justice Samuel Alito to the Supreme Court. The corresponding dataset senate.vote is in the R package BaM. The variables are: PARTY indicates Democrat, Republican, or Independent, STATE is the U.S. state represented, ALITO is Yea or Nay on Alito, ADA is the Americans for Democratic Action ideology score, NUMREPS is the number of U.S. House representatives for the state, REGION is the region of the country (East, Midwest, South, West), SENIORITY is the order of service in the Senate (rank from the first senator on), and HOLMES is the vote on the district court nomination of James Leon Holmes in 2003. Construct a model in BUGS where the Alito vote is the outcome variable, picking a subset of the other variables as explanatory variables. Notice that the factors will have to be coded into numerical variables.

11.9 On page 389 the linear model in BUGS was enhanced to give an R^2 measure. Modify the linear model code to produce an F-statistic and run this code with the OECD data to get a value. Summarize the CODA output with a simple regression table (further details on CODA will be given in Chapter 14).

11.10   Replicate the French Labor Strikes model in Example 7.2.1.1, starting on page 212 by writing a BUGS program for a negative binomial model with a Jeffreys prior.

11.11   Extending the state-level obesity data from Exercise 3.6 on page 89, now add the vector of general revenue by state (divided by 1M) for 2009 to the BMI data. These are:

Alabama	21.203	Alaska	11.532	Arizona	25.548
Arkansas	15.211	California	186.315	Colorado	19.303
Connecticut	21.079	Delaware	6.700	Florida	66.734
Georgia	34.366	Hawaii	9.164	Idaho	6.452
Illinois	53.810	Indiana	29.946	Iowa	16.246
Kansas	13.576	Kentucky	21.472	Louisiana	28.080
Maine	7.891	Maryland	29.677	Massachusetts	41.573
Michigan	49.635	Minnesota	29.043	Mississippi	16.820
Missouri	24.441	Montana	5.712	Nebraska	8.403
Nevada	9.431	New Hampshire	5.721	New Jersey	49.175
New Mexico	13.1	New York	134.951	North Carolina	41.175
North Dakota	13.478	Ohio	54.383	Oklahoma	18.691
Oregon	17.752	Pennsylvania	60.726	Rhode Island	6.527
South Carolina	20.693	South Dakota	3.677	Tennessee	23.656
Texas	96.156	Utah	12.837	Vermont	4.978
Virginia	34.993	Washington	32.520	Washington DC	9.753
West Virginia	11.113	Wisconsin	29.482	Wyoming	6.041

(also available as bmi.2009 in the BaM package in R). Fit a bivariate model with BUGS where BMI is the outcome variable using appropriate prior distributions.

11.12   Returning to Exercise 4.13, construct a time-series specification in BUGS where the outcome variable is the rate of fatalities by terrorism and the other columns are explanatory variables. Use double-indexing to account for time and variables.

11.13   Consider data on depression for teenagers with the following variable definitions: anxiety ("Low","Medium","High"), behavior problems ("Present","Absent"), sex ("Male","Female"), and depression: ("Absent","Mild","Severe"). The R dataframe can be created with the statements:

```
freq <- c(9, 32, 4, 1, 8, 4, 3, 1, 40, 6, 2, 0, 1, 0, 8, 1, 9, 14,
          9, 6, 9, 41, 2, 5, 5, 7, 23, 3, 1, 7, 31, 24, 99, 2, 6, 33)
psych.df <- data.frame(freq,expand.grid(anxiety=1:3,behavioral=1:2,
                       depression=1:3,sex=1:2))
```

Format these data for BUGS and run the appropriate model for depression as the outcome.

11.14 None of the BUGS engines has a Haldane prior (Exercise 4.10 on page 139) for modeling purposes. Provide code for specifying a Haldane prior in BUGS without using dbeta.

11.15 Replicate Exercise 7.20 on page 246 in BUGS rather than R.

11.16 Exercise 3.18 analyzed Swiss suicides data using Bettina Gruen's bayesmix package. Rewrite the model directly in BUGS. Compare the results.

11.17 Merge the Senate data from Exercise 8 with the obesity data from Exercise 3.6 on page 89. Can you produce a reliable model outcome using BUGS where BMI is the outcome variable? Does this mean that there is a political component to obesity at the state level?

11.18 Using the likelihood function in Exercise 9.10 on page 324, set $t_1 = 0.25$, and write an adaptive rejection sampling algorithm (Section 11.5) in R to efficiently sample from the resulting function.

11.19 Write a module of BUGS code to calculate the conventional null and model deviance for a GLM specification (Appendix A). Apply it to the model in Exercise 8 above.

11.20 Write an R function to implement adaptive rejection sampling where the input is a function and two bounds. The function should include a check for log-concavity. Implement the function by drawing values from a log-normal distribution with $\mu = 1$ and $\sigma = \frac{1}{2}$, between the bounds [1:4].

Chapter 12

Bayesian Hierarchical Models

12.1 Introduction to Multilevel Specifications

Hierarchical data structures are regularly encountered in the social and behavioral sciences since measurement often takes place at different levels of collection. A synonymous term is "multilevel" data meaning that the data represent different levels of aggregation for the subjects of study. We will use both terms herein since both are widely used across many social science literatures. A major advantage to the Bayesian paradigm is that these multilevel/hierarchical models are described and estimated in a very clean and direct manner with distributional statements.

As an example, we can study employees in a firm by department, plant, region, or nation. Consider a typical Fortune 500 employee; they have individual attributes such as demographics, they are affected by the actions of the immediate department, they are also affected by the decisions of local senior managers and executives at corporate headquarters, and perhaps even by outside economic forces. However, the level at which these inputs matter differs widely, and individuals feel them differently depending on their position in the hierarchy. In sociological survey analysis, we might augment the collected data from individuals with historical, governmental, or economic variables measured at various geographic levels. The question then arises as to how we should treat the different levels of variables in the same statistical model. We would like predictors to enter the model at the correct level for how they affect cases and groups of cases. Ignoring the aggregate information excludes potentially important effects and treating the aggregate information as individual level effects confuses covariance in the model. The solution is to employ a hierarchical model that recognizes the different groupings or time points that information about individual observations occur, thereby specifically stipulating correlations that would not have otherwise been assumed to exist. Another common justification for specifying hierarchical models is that some distributional forms cannot adequately account for overdispersion in the outcome variable of interest (Cox 1983). Adding an additional level then *models* this attribute.

Hierarchical models are deliberately set up so that the data are assumed to be conditioned on a set of parameters, which are in turn themselves conditioned on *other* parameters, which may depend on data at other levels of aggregation. By *conditioned*, it is meant that there is a specified PDF or PMF that describes the parametric relationship between data or variables to higher-level specifications of other variables. Notice that this is a very Bayesian setup in that distributions are stipulated in the modeling process and estimated parameters are related through these distributions. So multilevel models are not only symbiotic with the Bayesian paradigm, all forms of multilevel models *are* essentially Bayesian since unknown quantities are specified with distributions.

Multilevel modeling is a very adaptable idea because there is no mathematical restriction to the number of levels of these parametric relationships or the number of relationships at each level, and as many as are practical and convenient can be used. In addition, because of these features the Bayesian hierarchical model can be used as a vehicle for making decisions about variable inclusion (Carlin and Chib 1995, George and McCulloch 1993, Green 1995, Mitchell and Beauchamp 1988, Phillips 1995).

12.2   Basic Multilevel Linear Models

A good starting point for our purposes is the most basic form of the multilevel model. Multilevel linear models take the standard linear model specification and remove the standard restriction that the estimated coefficients are constant across individual cases by specifying levels of additional effects. Start with a standard linear model specification indexed by subjects and a first level of grouping, the *context* level. Use a single explanatory variable that has the form:

$$Y_{ij} = \beta_{j0} + \beta_{j1} X_{ij} + \epsilon_{ij}. \tag{12.1}$$

Now add a second level to the model that explicitly nests effects within groups and index these groups $j = 1$ to J:

$$\beta_{j0} = \gamma_{00} + \gamma_{10} Z_{j0} + u_{j0}$$
$$\beta_{j1} = \gamma_{01} + \gamma_{11} Z_{j1} + u_{j1}, \tag{12.2}$$

where all individual level variation is assigned to groups producing department level residuals: u_{j0} and u_{j1}. These Z_{ji} are context level variables in that their effect is assumed to be measured at the group level rather than at the individual level.

The basic two-level model is now produced by inserting the context level specifications, (12.2), into the original linear expression for the outcome variable of interest, (12.1). Performing this substitution gives:

$$Y_{ij} = \gamma_{00} + \gamma_{01} X_{ij} + \gamma_{10} Z_{j0} + \gamma_{11} X_{ij} Z_{j1} + u_{j1} X_{ij} + u_{j0} + \epsilon_{ij}. \tag{12.3}$$

This equation shows that the composite error structure, $u_{j1} X_{ij} + u_{j0} + \epsilon_{ij}$, is now clearly heteroscedastic since it is conditioned on levels of the explanatory variable, causing additional estimation complexity.

The multilevel models derived from this setup have annoying synonyms, but the most commonly named forms are discussed below. Gill and Womack (2013, pp.13-14) note the pervasiveness of this problem. Here we first assume an intercept term, α, and a single explanatory variable β. For the data matrices, X_{ij} for individual i in cluster or group j, and Z_j for cluster j, there are five canonical models:

"Fully Pooled"	$Y_{ij} = \alpha + X_i'\beta + Z_j'\gamma + e_{ij}$
"Random Intercept"	$Y_{ij} = \alpha_j + X_i'\beta + e_{ij}$
"Random Slope"	$Y_{ij} = \alpha + X_{ij}'\beta + Z_j'\gamma + e_{ij}$
"Random Intercept/Random Slope"	$Y_{ij} = \alpha_j + X_{ij}'\beta_j + Z_j'\gamma + e_{ij}$
"Fully Unpooled"	$Y_{i,j=1} = \alpha_{j=1} + X_{i,j=1}'\beta_{j=1} + Z_{j=1}'\gamma_{j=1} + e_{i,j=1}$
	$Y_{i,j=2} = \alpha_{j=2} + X_{i,j=2}'\beta_{j=2} + Z_{j=2}'\gamma_{j=2} + e_{i,j=2}$
	$\vdots$
	$Y_{i,j=J} = \alpha_{j=J} + X_{i,j=J}'\beta_{j=J} + Z_{j=J}'\gamma_{j=J} + e_{i,j=J}$

The fully pooled (or just "pooled") model treats the group-level data as if it were measured at the individual level. This means that individuals will have some identical covariates as other individuals, the Z_j, since they share ignored group membership. The random intercept model, also sometimes called the "random effect" model, allows the intercept to vary in the context of a specified distribution. This means that the α_j are the same for individual i cases in the same group j, drawing from this distribution. If there is no distributional assumption, then there are $J - 1$ separate coefficients estimated using a contrast specification such as "dummy variable" coding (the treatment contrast). This is routinely called the "fixed effect" model, but the distinction is unimportant in the Bayesian hierarchical context since it simply matters whether additional distributions with hyperpriors are assigned or not. So the j subscript on α_j can mean either of these two approaches.

The problem is that "fixed" and "random" can differ in definition by literature (Kreft and De Leeuw 1988, Section 1.3.3, Gelman 2005). The random slope model puts a group distinction on the β coefficient, $j = 1, \ldots, J$ to impose group differences on the effect of the X variable. So each individual i in group j shares the same β_j effect. Like the random intercept this can come from a distribution or a contrast, and both the intercept and the slope can contribute group effects. The fully unpooled model runs J completely separate regressions for each of the J groups, under the assumption that there is commonality to be modeled between them. Given the confusion in names across fields, the best way to conceptualize these specifications is to consider them as members of a larger multilevel family where indices are turned-on turned-off systematically depending on the hierarchical purpose.

12.3   Comparing Variances

With multilevel specifications there are at least two important variance terms to evaluate. Consider the simple linear specification: $Y_{ij} = \alpha_j + X_i'\beta + e_{ij}$, with $j = 1, \ldots, J$ groups specified for the random intercept. The resulting regression output (using the `glmer` function in R or with the `BUGS` specifications in this chapter) will give σ_y^2 and σ_α^2, where their interpretation is very different. The variance of the regression, σ_y^2, describes variability from the residuals left over after the systematic effects. In basic hierarchical models this term is assumed to be constant across the data where group level differences result from differing *within*-group sizes, n_j, according to $SD_j = \sigma_y/n_j$. Obviously like any other regression setting, we prefer to minimize σ_y^2 for better model fit. Conversely, σ_α^2 gives the variance *between* groups at the higher level, and so we want this term to be as big as possible since it is an explicitly modeled term and it justifies the group distinction at the higher level. Furthermore, it contains variability that falls to σ_y^2 if the second level is eliminated from the model. In other words, the model fits best when σ_y^2 is minimized and σ_α^2 is maximized. This can also be described in ANOVA terms: σ_y^2 measures within-group variance and σ_α^2 measures between-group variance.

This comparison of variances is at the heart of determining whether the second level is justified. That is, if the grouping variable reveals no between-group differences in the form of a small σ_α^2 relative to σ_y^2, then it is clear that this grouping is not justified and the level should be eliminated from the specification. Thus a simple test is built into the process. Relatedly, it may be more difficult to justify the grouping with smaller J, but this is also revealed by the variance comparison. Therefore silly "rules of thumb" about the minimum number of groups or the minimum number of cases in each group are unnecessary as the model tests this inherently. Such rules should be completely disregarded. In fact, if σ_α^2 is comparable or large relative to σ_y^2 for small J, then the data are very clear that this group difference is important. Unfortunately there is no distributional test that makes sense when comparing these two variances. In a very general sense if σ_α is not an order of magnitude smaller than σ_y, then the group distinctions are likely to make substantive sense. However, if σ_α is vastly smaller than σ_y, then the group distinctions specified are unlikely to matter. Notice that this comparison is on the standard deviation metric. Sometimes this comparison is formally described with the *intraclass correlation coefficient* using variances:

$$ICC = \frac{\sigma_\alpha^2}{\sigma_\alpha^2 + \sigma_y^2}, \tag{12.4}$$

although this still does not supply a distributional test. Sagan (2013, p.583) suggests that the ICC can be as low as 0.05 to 0.20 and still justify the hierarchy (see Muthén and Satorra [1995] as well).

Consider the following approximation for the mean of the jth group (Gelman and Hill 2007, p.253):

$$\alpha_j \approx \left(\frac{n_j}{\sigma_y^2} \bar{y}_j + \frac{1}{\sigma_\alpha^2} \bar{y}_{\text{all}} \right) \Big/ \left(\frac{n_j}{\sigma_y^2} + \frac{1}{\sigma_\alpha^2} \right). \tag{12.5}$$

where:

$\bar{y}_j$	unpooled estimate for group j
$\bar{y}_{\text{all}}$	completely pooled estimate
σ_y^2	assumed equal within-group variance
σ_α^2	variance among the mean estimates

Although this is an approximation (we get fuller results from MCMC output but with less intuition on these issues), it reveals the important structure of multilevel models. Ignoring the denominator for the moment, the numerator reveals an important affect from the size of group j. Obviously the numerator is a weighting of the group mean and the full sample mean from all groups. If n_J is small then α_j will be more influenced by $\bar{y}_{\text{all}}$. Therefore such small groups retain their unique identity and they are also able to "borrow strength" from other groups through the overall mean in this weighting. In addition, averages from groups with smaller sample sizes contribute less to $\bar{y}_{\text{all}}$ since they are a smaller weighted contribution to the sum. Suppose a new group is discovered to exist, but there are no samples made available within this group. The α_j for this group with $n_j = 0$ (no observed values!) needs to balance its own zero weighting relative to what is known about the group $\alpha_j = \bar{y}_{\text{all}}$. This makes total sense: an unknown group that belonged in a set of groups should get the best available estimate, which is the overall mean. Conversely, groups with larger sample sizes have much greater influence over the overall model through their larger contribution to $\bar{y}_{\text{all}}$. Also, their value of α_j is less reliant on the rest of the sample due to multiplication by n_j. Furthermore, as n_j increases to some size much larger (all the way to infinity), group j will increasingly dominate the value of $\bar{y}_{\text{all}}$.

We can also consider variance effects in (12.5). First split by the numerator for additional clarity:

$$\alpha_j \approx \frac{\frac{n_j}{\sigma_y^2} \bar{y}_j}{\frac{n_j}{\sigma_y^2} + \frac{1}{\sigma_\alpha^2}} + \frac{\frac{1}{\sigma_\alpha^2} \bar{y}_{\text{all}}}{\frac{n_j}{\sigma_y^2} + \frac{1}{\sigma_\alpha^2}}. \tag{12.6}$$

As $\sigma_\alpha^2 \to 0$, the first term in the right-hand-side above goes to zero since $1/\sigma_\alpha^2$ is in the denominator and nowhere else. However, in the second term $1/\sigma_\alpha^2$ is in both the denominator and the numerator and therefore cancels out leaving $\bar{y}_{\text{all}}$ since n_j/σ_y^2 will be negligible as an addition as $1/\sigma_\alpha^2$ gets very large. Therefore as group differences disappear it makes sense simply to move to the completely pooled estimate. On the other hand, as $\sigma_\alpha^2 \to \infty$ the second term in the right-hand-side goes to zero due to the numerator, but the first term simply reduces to $\bar{y}_j$ since the n_j/σ_y^2 terms cancel out after $1/\sigma_\alpha^2$ goes to zero. In this case the best estimate of α_j is the completely unpooled estimate. So in this way multilevel model specifications are self-regulating by σ_α^2: the degree to which this value is large compared to σ_y is the degree to which the group distinctions are justified.

It is important to fully understand the role of σ_α^2. As odd as it sounds, we *want* this variance to be large since it justifies the group distinctions that form the hierarchy. Large values relative to σ_y simply mean that the groups are sufficiently different that adding a second level improves the fit of the model by moving variability from σ_y^2 to σ_α^2. This also means that if there are other non-nesting (distinct) group differences that exist, adding this feature to the model will continue to improve fit (producing another between group term, say σ_β). Conversely, if the groupings matter and are ignored then σ_y^2 will be larger, and the model will have a poorer fit. Also, as n increases (even asymptotically), the variance measured by σ_α^2 will not go to zero. This is because this inter-group set of differences still persists, if valid, even in the presence of more data in general. Obviously under this circumstance σ_y will decrease, and this difference in effects demonstrates the fundamentally different roles that these variances have in a multilevel model.

■ **Example 12.1:   Bayesian Multinomial Specifications for Employment Status.**

To illustrate how the different models above can be applied in practice, we look at an application from Pettitt *et al.* (2006) where they analyze the employment status of 5192 immigrants to Australia (the population of such persons 15 years or older, between September 1993 and August 1995), with the outcome variable being an 11-category work-status outcome measured at three time periods. In the lowest level of the hierarchical model, individual status is assumed to be distributed multinomial where the ith individual at time period t has a $J = 11$-length outcome vector with 10 zeros plus a single i1, defined as: $y_{it} \sim$ multinomial$_j(p_{it}, 1)$, where p_{itj} is the probability for person i, time t, and category j. The second level of the hierarchy connects p_{itj} to modeled causal effects through a standardized term:

$$\log(\mu_{itj}) = \mathbf{X}_{it}'\boldsymbol{\beta}_j + \mathbf{Z}_{it}'\boldsymbol{\gamma}_j + \alpha_{ij} + \epsilon_{itj}$$
$$p_{itj} = \frac{\mu_{itj}}{\sum_{j=1}^J \mu_{itj}} \tag{12.7}$$

where $\mathbf{X}_{it}$ is a matrix of explanatory variables and $\mathbf{Z}_{it}$ is a matrix of lagged outcome variables, $\boldsymbol{\beta}_j$ and $\boldsymbol{\gamma}_j$ are associated vectors of unknown parameters, and α_{ij} is a random effect term. The model is not identified without a reference category, and category 1 ("wage or salary earner") is chosen ($\beta_1, \gamma_1, \alpha_{i1}$ all set to zero) such that $\log(p_{itj}/p_{it1}) = \log(\mu_{itj}/\mu_{it1}) = \log(\mu_{itj})$ for $j \neq 1$. It is assumed that the remaining α_{ij} are multivariate normal: $\boldsymbol{\alpha}_i = (\alpha_{i2}, \alpha_{i3}, \ldots, \alpha_{ij}) \sim \mathcal{MVN}(0, \boldsymbol{\Sigma})$. Now with uninformative prior specifications (page 147), the joint posterior can be completed according to:

$$p(\boldsymbol{\beta}, \boldsymbol{\gamma}, \boldsymbol{\alpha}, \boldsymbol{\Sigma}|\mathbf{X}, \mathbf{y}) \propto p(\mathbf{X}, \mathbf{y}|\boldsymbol{\beta}, \boldsymbol{\gamma}, \boldsymbol{\alpha}, \boldsymbol{\Sigma})p(\boldsymbol{\alpha}|\boldsymbol{\Sigma})p(\boldsymbol{\beta})p(\boldsymbol{\gamma})p(\boldsymbol{\Sigma}). \tag{12.8}$$

See also the Bayesian multinomial probit specification of Imai and van Dyk (2005) for a competing approach.

The interesting part of this example from our perspective is how flexible the hierarchical model can be with only these simple components. Pettitt *et al.* specify six fundamentally different models by varying the use of the model components above across individuals, time, and response treatment:

▷ **Model 1:** $E[\log(\mu_{itj})] = \mathbf{X}'_{it}\boldsymbol{\beta}_j$.

The effect of the (unlagged only) regression parameters differs across outcome categories j, but remains constant for all individuals and over time given a state j. So employment probabilities are constant across people and constant over time.

▷ **Model 2:** $E[\log(\mu_{itj})] = \mathbf{X}'_{it}\boldsymbol{\beta}_j + \alpha_{ij}$.

The effect of the regression parameters differs across outcome categories j but not over time or individuals, and a random effect term captures within-individual clustering (i.e., varying across individuals) but is constant over time given a state j. So employment prospects are constant over time, but different for individuals where this difference does not change over time.

▷ **Model 3:** $E[\log(\mu_{itj})] = \mathbf{X}'_{it}\boldsymbol{\beta}_j + \mathbf{Z}'\boldsymbol{\gamma}_j$.

The effect of the $\boldsymbol{\beta}$ regression parameters differs across outcome categories j for the current time period, the effect of the $\boldsymbol{\gamma}$ regression parameters differs across outcome categories j for the lagged time period, and no between-individual variation is modeled, given state j. So employment prospects are constant across people again (Model 1), but now there is a lagged time effect that alters probabilities going forward.

▷ **Model 4:** $E[\log(\mu_{itj})] = \mathbf{X}'_{it}\boldsymbol{\beta}_j + \mathbf{Z}'\boldsymbol{\gamma}_j + \alpha_{ij}$.

The effect of the $\boldsymbol{\beta}$ regression parameters differs across outcome categories j for the current time period, the effect of the $\boldsymbol{\gamma}$ regression parameters differs across outcome categories j for the lagged time period, and between-individual variation is modeled with the random effect, given state j. So employment prospects are constant across people again and there is a lagged time effect that alters probabilities going forward (Model 3), but now a different effect across individuals where this difference does not change over time (Model 2).

▷ **Model 5:** $E[\log(\mu_{itj})] = \mathbf{X}'_{it}\boldsymbol{\beta}_j + \mathbf{Z}'\boldsymbol{\gamma}_{ij}$.

This is exactly the same as **Model 3**, except that the lagged effects are now assumed to be time-varying in their effect across panels. Since the panels are not equally spaced, this may be a more robust specification. So employment prospects are constant across people (Model 1), there is a lagged time effect that alters probabilities going forward (Model 3), and now this lagged time effect is individualized.

▷ **Model 6:** $E[\log(\mu_{itj})] = \mathbf{X}'_{it}\boldsymbol{\beta}_j + \mathbf{Z}'\boldsymbol{\gamma}_{ij} + \alpha_{ij}$.

This is the same as **Model 5**, but adding back the random effect to capture between-individual variation. So employment prospects are constant across people (Model 1), there is a lagged time effect that alters probabilities going forward (Model 3), this lagged time effect is individualized (Model 5), and an added different effect across individuals where this difference does not change over time (Model 2).

How many more could we have? Standard extensions include interaction effects between levels, different lagged periods, smoothing of some covariates (generalized additive models), spatial terms, and more.

Marginal posterior distributions are produced using BUGS (10,000 iterations, disposing of the first 4,000) and compare fits with the DIC (Section 7.5). The authors find poor convergence of some Σ values for models 4 and 6, leading to the conjecture that the data do not support inclusion of both γ and α in the model. Model 1 is dismissed as too basic, and the DIC points towards models 3 and 5 as fitting better than model 2.

12.4 Exchangeability

Exchangeability (sometimes also called symmetry or permutability) is an important property for hierarchical models. As mentioned in Chapter 1, exchangeability allows us to say that the data are produced conditional on the unknown model parameters in the same way for every data value. More technically, given a sample $X_1, X_2, \ldots, X_n$ and any possible permutation of these data $X_{[1]}, X_{[2]}, \ldots, X_{[n]}$, the data are exchangeable if the joint distributions are equal: $f(X_1, X_2, \ldots, X_n) = f(X_{[1]}, X_{[2]}, \ldots, X_{[n]})$, i.e., invariant to permutation. This is equivalent to saying that the subscripts (the labeling of the data) are uninformative in the sense that any subset of the data are assumed to have the same marginal distribution. Actually, as described this is *finite exchangeability*, meaning that n is fixed here. We can obtain the definition of *infinite exchangeability* by simply adding the condition that every finite subset of an infinite series of X_i is exchangeable in the sense above.

The provided definition of exchangeability has thus far conveniently omitted any dependence on parameters. The role of exchangeability becomes more obvious when we include the typical dependence of the data generation process on a parameter or parameter vector. Suppose that X_i is a series of infinitely exchangeable series of Bernoulli random variables. de Finetti's (1930) famous *representation theorem* states that there is guaranteed to exist a unique probability measure $\mathcal{Q}(\theta)$ so that:

$$f(x_1, x_2, \ldots, x_n) = \int_0^1 \prod_{i=1}^n \theta^{x_i}(1-\theta)^{1-x_i} d\mathcal{Q}(\theta), \tag{12.9}$$

where $\theta \underset{n\to\infty}{\longrightarrow} \frac{1}{n}\sum x_i$. What this implies is that conditional on θ, the exchangeable data are now iid. Diaconis and Freedman (1980) provide some technical differences between the finite and infinite applications of de Finetti's theorem. The theorem does not quite hold in the finite case and Diaconis and Freedman give the error bound for the finite case distance to the nearest iid set. Also, it is usually the case that the probability measure is sufficiently well-behaved as to define a proper PDF: $dQ(\theta) = p(\theta)d\theta$. This allows us to produce a justification for Bayesian inference based on making this substitution using the definition of conditional probability:

$$f(x_1, x_2, \ldots, x_n) = \int_0^1 p(x_1, x_2, \ldots, x_n, \theta)d\theta$$

$$= \int_0^1 p(x_1, x_2, \ldots, x_n | \theta)p(\theta)d\theta$$

$$= \int_0^1 \underbrace{\prod_{i=1}^n \theta^{x_i}(1-\theta)^{1-x_i}}_{\text{likelihood function}} \underbrace{p(\theta)}_{\text{prior}} d\theta. \qquad (12.10)$$

This is easily recognized as the denominator in Bayes' Law. More importantly it shows that we can suppose a latent random variable θ interpreted as the limiting frequency of $p(x = 1)$. Now *conditional* on this θ the x_i values are distributed iid Bernoulli with parameter θ:

$$x|\theta \sim \mathcal{BR}(\theta)$$

$$\theta \underset{iid}{\sim} p(\theta). \qquad (12.11)$$

This is now (barely) hierarchical in the sense that there is a model first for manifest zeros and ones and a higher level model for actual *rate* determining generation. So while we cannot observe θ, we are not precluded from specifically modeling it as a random quantity, where the distributional feature expresses our uncertainty about this unseen parameter. de Finetti's 1937 proof is somewhat lengthy and need not be replicated here. Bernardo and Smith (1994, Section 4.3) give a rough sketch of the proof steps, Hartigan (1983) gives a proof based on Baire functions, and Heath and Sudderth (1976) give a briefer but accessible description. Also see Haag (1924) for perhaps the first discussion of this idea.

Also, while the theorem is given so far only for Bernoulli random variables, the class of probability model to which it applies is much larger and includes the standard distributions we commonly work with (de Finetti 1937). In particular, Freedman and Diaconis (1982) proved that de Finetti's property holds true for a mixture of location symmetric interval measured random variables (symmetric about a location parameter, which itself is a random quantity, obviously an important context to Bayesians). Consider an infinitely exchangeable set of continuously measured random variables from the distribution F, $x_1, x_2, \ldots, x_n$, with probability measure $\mathfrak{P}$ such that:

$$\mathfrak{P}(x_1, x_2, \ldots, x_n) = \int_{\mathfrak{D}} \prod_{i=1}^n f(x_i)dQ(F), \qquad (12.12)$$

where $\mathcal{Q}$ is now a probability measure over $\mathfrak{D}$, the space of *all* distribution functions definable over the real line (Bernardo and Smith 1994). Furthermore $\mathcal{Q}(F)$ is the limiting distribution of the empirical distribution of the x_i, meaning that the hierarchical distribution specifies F for some distribution over all possible distributions. Clearly this is much more abstract and less practical than the definition given before, but it motivates the treatment of the prior in hierarchical terms for an extremely broad set of specifications.

The underlying implications of de Finetti's theorem are vast. As previously stated, frequentists believe that there is an unknown but fixed parameter, θ, in the data generation process. The theorem gives the asymptotic distribution of θ as a quantity from the data, not the other way around as given in introductory texts. Here θ is assigned a distribution and can therefore be thought of *subjectively* as a *belief* about the probability of success. In fact, this notion is the philosophical basis for some Bayesians' rationale for treating parameters as random quantities. For instance: "It is my tentative view that the concept of personal probability, introduced and in the preceding chapter, is except possibly for slight modifications, the only probability concept essential to science and other activities that call upon probability" (Savage 1954, p.56). Jeffreys (1961, p.401) is equally direct: "The essence of the present theory is that no probability, direct, prior, or posterior is simply a frequency." Such opinions and counter-opinions convulsed thought on Bayesian inference for at least 200 years. The interested reader is referred for details to Dale (1991), Earman (1992), Lindley, Tversky, and Brown (1979), Smith (1965), and Suppes (1974), just to start. Actually this is not just a completely theoretical or epistemological debate as the effects of researcher subjectivism can be central to assessing the validity of empirical research in the social sciences; see, for instance, the controversy over Margaret Mead's famous work in Samoa (Freeman 1983; Mead 1973; Orans 1996).

Actually, de Finetti's work is seen by some as *the* key link between subjective probability and the formulation of Bayesian models (Lindley and Novick 1981; Smith 1984), and at least one set of authors make the claim that exchangeability is actually an even more fundamental notion than that of probability because the individual psychological interpretation of similarity is more native than that of relative frequency (Draper *et al.* 1993).

Critically de Finetti (1930) points out that since exchangeability is a weaker condition than iid (iid actually implies exchangeability), then frequentist inference is a special case of Bayesian inference: the iid assumption gives a limiting frequency θ and the exchangeability assumption gives the same limiting frequency except for a region of the measure space with probability zero. This also means that the strong law of large numbers results from the theorem (Press 1989).

What does this mean to hierarchical models? We can re-express de Finetti's theorem in the following hierarchical way for $X_1, X_2, \ldots, X_n$ that are exchangeable:

$$\theta \sim p(\theta)$$
$$f(\mathbf{X}|\theta) \underset{iid}{\sim} \mathcal{BN}(n, \theta). \tag{12.13}$$

The theorem shows that exchangeability and independence are related in a hierarchical model through conditionality (as noted above). Suppose now that we couldn't be certain that every single X_i is generated from a conditional distribution with the same value of θ: $f(X_i|\theta)$. In fact, the only information that we can be absolutely sure about is that each X_i is generated from *some* θ_i. This obviously creates an impractical estimation process. However, if we impose the assumption of exchangeability on the θ_i, then by de Finetti's theorem, we can treat the joint distribution *as if* it came from a mixture of iid specifications: $f(X_1, X_2, \ldots, X_n|\theta_1, \theta_2, \ldots, \theta_n) = \prod_{i=1}^{n} f(\mathbf{X}|\boldsymbol{\theta})$.

To generalize to a broader class of hierarchical models, start with a potentially infinite exchangeable series of continuous-measured random variables: $x_{[1]}, x_{[2]}, \ldots, x_{[N]}$, with parameters $\theta_1, \theta_2, \ldots, \theta_N$. At the first stage, relate the data to parameters:

$$f(x_{[1]}, x_{[2]}, \ldots, x_{[N]}|\theta_1, \theta_2, \ldots, \theta_N) = \prod_{i=1}^{N} f(x_i|\theta_i). \qquad (12.14)$$

So the $x_{[i]}$ are independent *conditional* on the θ_i. Now define a second stage modeling the θ_i parameters as if they are a random sample from a new parametric family

$$g(\theta_1, \theta_2, \ldots, \theta_N) = \prod_{i=1}^{N} g(\theta_i|\phi). \qquad (12.15)$$

Exchangeability means that we can treat the θ_i as iid realizations from this distribution. Finally, specify a hyperprior that defines the form of ϕ:

$$\phi \sim h(\phi|A, B), \qquad (12.16)$$

where A and B are either fixed hyperparameters or random in a higher level of hyperprior hierarchy. This series of independent hierarchical specifications must end at some point and the obvious question that arises is how do we specify the top level of fixed hyperparameter values. Two approaches dominate. The *diffuse hyperprior* approach stipulates making the highest level of priors as uninformative as possible by picking hyperparameters that give a broad and flat form (Pericchi and Nazaret 1988). This is the approach taken in many of the canned BUGS examples provided by Spiegelhalter *et al.* (1996a, 1996b, 2012) and subsequently (unfortunately perhaps) retained by many users of the software.

■ **Example 12.2: Exponential-Gamma Model of Clinical Effects.** This is an example from pharmapsychology, but the clustering problem is a more general one resolved by hierarchical models. In addition, we will assess whether it is reasonable to assume exchangeability of units across different modeled groups (clinics). Stangl (1995) is concerned with modeling the time to recurrence of depression in patients served at five different clinics. The patients received antidepressant medication for eight weeks and then were randomized to either continue to take the drug (treatment, coded 1) or taken off medication (control, coded 0). This is a hazard model (see page 317 for a deeper description of the assumptions, and Dellaportas and Smith [1993] for details on a Gibbs sampler tailored for such models) because the outcome variable is the time to first recurrence of depression (censored at two years). The data values for each patient, labeled t_{ijk} for the ith patient at the jth clinic getting the kth treatment, are modeled with an exponential distribution at the first level. The hierarchy is summarized by:

$$t_{ijk}|\theta_j \sim \mathcal{EX}(\theta_j)$$
$$\theta_j|\alpha, \beta \sim \mathcal{G}(\alpha, \beta)$$
$$\alpha = 5, \quad \beta \sim \mathcal{G}(0.172, 5). \tag{12.17}$$

The rationale behind prior and hyperprior selection starts with noticing that the larger the α parameter, the greater the influence of the prior relative to the data. Several values were selected, including the rather diffuse $\alpha = 5$, during a sensitivity analysis. Holding α constant, β is interpreted as the sum of α exponential survival times, and therefore modeled gamma since the sum of independent exponential random variables is gamma distributed. The parameters of the gamma distribution at the third level are set up by setting the first equal to the pooled hazard rate before specifying a model, and the second equal to α. Therefore the hierarchy is a compromise between the assumed level of prior information and the data in a quasi-empirical Bayes fashion. See Ibrahim *et al.* (2001) for a broader discussion of Bayesian survival models.

TABLE 12.1: DEPRESSION TREATMENT BY CLINIC, CONTROL GROUP

Clinic	A	B	C	D	E
Sample Size	11	10	8	25	17
Number of Events	6	6	7	12	15
Sum of Events Times	499	391	177	1367	240
MLE Value	0.012	0.015	0.040	0.009	0.063
Mean Recurrence Time at MLE	83	67	25	111	16
Posterior Mean	0.012	0.013	0.023	0.010	0.041
Posterior Mean Recurrence Time	83	77	43	100	24

More importantly, the degree to which the hierarchical posteriors differ from separately estimated maximum likelihoods by clinic indicates the extent to which there is "sharing" in the data across clinics: evidence to support the assumption of exchangeability. Table 12.1 indicates that for the control group the Bayesian posterior modes are more compactly distributed than the MLEs, supporting the idea that there are programmatic commonalities.

12.5   Essential Structure of the Bayesian Hierarchical Model

As always, our central interest is in the generation of a posterior distribution from the likelihood times the prior:

$$\pi(\theta|\mathbf{X}) \propto L(\theta|\mathbf{X})p(\theta). \tag{12.18}$$

Following O'Hagan (1994) or Berger (1985), now suppose that the parameter θ is conditional on another unknown parameter ψ, which will be given its own prior distribution $p(\psi)$. The calculation of the posterior becomes:

$$\pi(\theta, \psi|\mathbf{X}) \propto L(\theta|\mathbf{X})p(\theta|\psi)p(\psi), \tag{12.19}$$

which can be quite straightforward if the forms of $p(\theta|\psi)$ and $p(\psi)$ are cooperative. Many times this seemingly simple extension complicates the production of (12.19) to the point where MCMC tools are required to get the marginal distribution of θ. Here θ has the prior distribution, $p(\theta|\psi)$, but it in turn is now conditional on another parameter that has its own prior, $p(\psi)$, called a "hyperprior," which can also have "hyperparameters" if desired. Inference for either parameter of interest can be obtained through looking at the respective marginal densities:

$$\pi(\theta|\mathbf{X}) = \int_{\psi} \pi(\theta, \psi|\mathbf{X})d\psi$$

$$\pi(\psi|\mathbf{X}) = \int_{\theta} \pi(\theta, \psi|\mathbf{X})d\theta, \tag{12.20}$$

where the other parameter has been integrated out of the joint posterior. What should be discernible from (12.19) is that we can continue stringing hyperpriors to the right, beginning with making ψ conditional on another parameter, $p(\psi|\zeta)$, and adding its new hyperprior distribution, $p(\zeta)$, to the calculation of the posterior:

$$\pi(\theta, \psi, \zeta|\mathbf{X}) \propto L(\theta|\mathbf{X})p(\theta|\psi)p(\psi|\zeta)p(\zeta). \tag{12.21}$$

Now ζ is the highest level parameter and therefore the only hyperprior that is unconditional.

In more realistic settings we will want to think of ψ as a parameter vector with its own hyperparameters and dependencies: $\boldsymbol{\psi} = (\psi_1, \psi_2 \ldots)$. As a simple example, consider data that are normally distributed with known variance and unknown mean, $\mathbf{X}_i \sim \mathcal{N}(\mu_i, \sigma_0^2)$, $i = 1, \ldots, n$, as described in Section 3.2 (page 70), except that instead of an unknown constant μ we assume random μ_i, $i = 1, \ldots, n$ that are drawn independently from a common, also normal, distribution: $\mathcal{N}(m_\mu, s_\mu^2)$. The model can be summarized as:

$$\begin{bmatrix} X_1 \\ X_2 \\ \vdots \\ X_n \end{bmatrix} \sim \mathcal{N}\left(\begin{bmatrix} \mu_1 \\ \mu_2 \\ \vdots \\ \mu_n \end{bmatrix}, \sigma_0^2 \mathbf{I} \right) \qquad \begin{bmatrix} \mu_1 \\ \mu_2 \\ \vdots \\ \mu_n \end{bmatrix} \sim \mathcal{N}(m_\mu, s_\mu^2). \tag{12.22}$$

The key here is that the X_i values are assumed to be generated by distinct underlying means in the normal specification, but these means are themselves drawn from a common distribution with fixed hyperparameters: m_μ, s_μ^2. This is essentially the same idea as the non-Bayesian multilevel models above except that a second stage is explicitly given in the form of a common prior distribution for the unknown μ_i. It differs, however, from the model on page 71 in that more uncertainty is imposed into the data generation process for **X**. This model can be further extended in two ways. *First*, by making m_μ and s_μ^2 additional random quantities with their own prior distributions, we add to the depth of the hierarchy. *Second*, as we add hierarchies to the model, we can specify how data at different levels enter the model specification. For instance, we might specify that the second-level model for m_μ is also normal, $\mathcal{N}(\mathbf{Z}\gamma, \Sigma)$, with data **Z** that specifically affect the mean.

While there is no *theoretical* restriction to the number of levels that can be specified in the hierarchy, the *practical* restriction is that in specifications that have greater than three or four levels, the interpretation of the estimated coefficients can be challenging. Frequently there is little reason to go beyond a two-level model in terms of nesting. Goel and DeGroot (1981) show that posterior certainty about parameters decreases at each progressive level of the hierarchical model. This is formalized by Goel (1983, Theorem 3.2) using the idea of the *conditional amount of sample information* (CASI), which is the "distance" between the prior and the posterior, indicating the relative contribution of the data. Define the f *divergence* measure of distance between two distributions $g_1(\theta)$ and $g_2(\theta)$ defined on the sample space of θ by:

$$I_F(g_1, g_2) = \int_\Theta f\left(\frac{g_1(\theta)}{g_2(\theta)}\right) g_2(\theta)d\theta, \tag{12.23}$$

for an arbitrary convex function labeled $f()$ defined on $(0{:}\infty)$, where standard measure-theoretic properties are assumed to hold. This is similar to the Kullback-Leibler distance and slightly more general in that the natural log function certainly qualifies as a candidate for $f()$ (note, however, that the order of the fraction is reversed relative to the outside term). Titterington, Smith, and Makov (1985, p.116) give a nice summary of such related distance measures. Goel's theorem means that after the second level of hierarchy, $I_F(g_1(\theta_i), g_2(\theta_{i+k}))$ is a decreasing function with increasing k. In other words, hyperparameters that are farther up the hierarchy have lower levels of information supplied by the data. This idea is actually very intuitive, given that each additional level is separated by one more parametric assumption from the root data-level of the hierarchy.

There is also a distinction between nested hierarchies and non-nested hierarchies. *Nesting* means that a hierarchy exists at a level above some hierarchy and that this hierarchy does not affect an other hierarchy at lower levels. We now assert that ψ in (12.19) had further parameterization instead of just a distribution $p(\psi)$: $\psi_k|\delta = f(\delta_{k0} + \delta_{k1}\mathbf{W}_{k1} + v)$, where δ_{k0} and δ_{k1} are given further distributions (typically conditional on further hyperparameters), $\mathbf{W}_{k1}$ is a set of covariates at this higher level of the hierarchy, and v is an additional error term. A specific example of this setup is given in the next section. *Non-nested* hierarchies do not affect each other but jointly affect a lower node. So if the jointed affected quantity is the data in a normal model, then a non-nested specification looks like:

$$y_i \sim f(\mu + \theta_{ji} + \delta_{ki}, \sigma_y^2), \quad \text{for } i = 1, \dots, n$$
$$\theta_j \sim g(\dots), \quad \text{for } j = 1, \dots, J$$
$$\delta_k \sim h(\dots), \quad \text{for } k = 1, \dots, K$$

where $f()$, $g()$, and $h()$ are user-specified distributions, $j = 1, \dots, J$ index the θ hierarchy, and $k = 1, \dots, K$ index the δ hierarchy. In this way these higher levels affect the data level without interacting in any way, although additional model features can be added to associate them. The flexibility in different arrangements of nested and non-nested hierarchies provides a dramatically greater number of possible model specifications for a given dataset.

12.5.1   A Poisson-Gamma Hierarchical Specification

Building on a specification already presented, suppose that the observed outcome variable is modeled Poisson. The obvious choice for a prior on the intensity parameter, λ, is a gamma distribution since not only is it conjugate, but also because it is a very flexible parametric form (i.e., it can be parameterized to provide many different shapes over positive real support). This gamma distribution is in turn characterized by two further parameters (often labeled α and β), which can also be assigned their own hyperpriors to reflect differences at the next level in the model specification. Of course these gamma PDF parameters are restricted to the positive real numbers just like λ, so it is logical to consider giving them gamma priors as well. In doing so, we also need to determine parameters for the gamma distributions assigned to α and β. If they are assigned fixed values, they are called hyperparameters. Otherwise we can add another level to the hierarchy and assign an additional round of hyperpriors. Figure 12.1 graphically displays the model where the gamma distributions for α and β are the final hierarchical level.

It is important to notice that the intensity parameter of the Poisson distribution is now indexed by i since it is no longer assumed to be a fixed effect that is constant in its impact on the data. Instead, we are now asserting that λ varies across cases in the data, and is therefore a random effect whose distribution is assumed here to be gamma: $\lambda_i \sim \mathcal{G}(\alpha, \beta)$. This is a fundamentally new and different idea from the types of models estimated in previous chapters. The relationship between cases in the data has now changed: instead of a sampling process that comes from a single distribution function, we have substituted a data generation process where each case is produced by a distinct but related mechanism because of the common distribution.

It is conventional to represent hierarchical models by "stacking" the parametric specifications from the data-level specification to the highest-level hyperprior specification. For

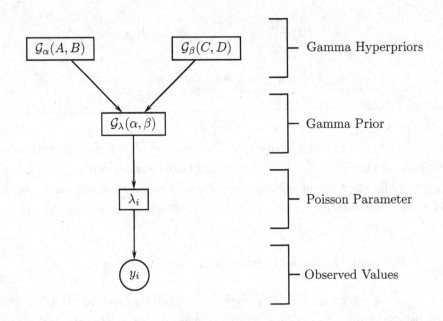

FIGURE 12.1: Poisson-Gamma Hierarchical Model

the example model, this representation is:

$$y_i \sim \mathcal{P}(\lambda_i)$$
$$\lambda_i \sim \mathcal{G}(\alpha, \beta)$$
$$\alpha \sim \mathcal{G}(A, B)$$
$$\beta \sim \mathcal{G}(C, D), \tag{12.24}$$

where the y_i are assumed conditionally independent given λ, and α and β are assumed independent. Here we can see very clearly that not only does the Poisson intensity parameter have its own specified distribution, but also the two gamma parameters that determine this Poisson distribution have their own specified distribution as well.

Other non-gamma priors are also feasible here. George, Makov, and Smith (1993) contrast four different hyperparameter specifications using this model and demonstrate the flexibility of the specification. Cohen *et al.* (1998) apply a reference prior specification $p(\alpha, \beta) \propto [\alpha(1 + \beta)^2]^{-1}$ in their study of arrest rates for drug offenders and apply a mixed Gibbs-Metropolis chain since the full conditional distributions for α and β are then no longer available.

We can write a complete specification for the posterior distribution even though there is now additional complexity from having higher levels in the model. Since each hyperprior is a prior for a lower-level prior distribution, it is possible to assemble the joint distribution multiplicatively using Bayes' Law and the definition of conditional probability includes each of the hierarchical specifications (and dropping conditionality on constants on the left-hand side):

$$p(\mathbf{y}, \boldsymbol{\lambda}, \alpha, \beta) = \prod_{i=1}^{n} p(y_i|\lambda_i)p(\lambda_i|\alpha, \beta)p(\alpha|A, B)p(\beta|C, D)$$

$$= \prod_{i=1}^{n} \left[(y_i!)^{-1}\lambda_i^{y_i} \exp(-\lambda_i)\beta^{\alpha}\Gamma(\alpha)^{-1} \exp(-\lambda_i\beta)\lambda_i^{\alpha-1} \right.$$

$$\left. \times B^A\Gamma(A)^{-1} \exp(-\alpha B)\alpha^{A-1}D^C\Gamma(C)^{-1} \exp(-\beta D)\beta^{C-1} \right],$$

so:

$$p(\lambda_i, \mathbf{y}, \alpha, \beta) \propto \lambda_i^{\sum y_i+\alpha-1}\alpha^{A-1}\beta^{\alpha+C-1}\Gamma(\alpha)^{-1} \exp[-\lambda_i(1+\beta) - \alpha B - \beta D]. \qquad (12.25)$$

This looks forbidding, especially since there are three posterior parameters of interest. Note the role of exchangeability here for the λ_i since the y_i all have the same relationship to the hyperprior specifications through the λ_i. Once the $\mathbf{y}$ data are observed, we can obtain the joint posterior distribution of interest according to:

$$\pi(\lambda_i, \alpha, \beta|\mathbf{y}) = \frac{p(\mathbf{y}, \lambda_i, \alpha, \beta)}{\iiint p(\mathbf{y}, \lambda_i, \alpha, \beta)d\lambda d\alpha d\beta}$$

$$= \frac{\lambda_i^{\sum y_i+\alpha-1}\alpha^{A-1}\beta^{\alpha+C-1}\Gamma(\alpha)^{-1} \exp[-\lambda_i(1+\beta) - \alpha B - \beta D]}{\iiint \lambda_i^{\sum y_i+\alpha-1}\alpha^{A-1}\beta^{\alpha+C-1}\Gamma(\alpha)^{-1} \exp[-\lambda_i(1+\beta) - \alpha B - \beta D]d\lambda d\alpha d\beta},$$

$$= \left(\lambda_i^{\sum y_i+\alpha-1}\alpha^{A-1}\beta^{\alpha+C-1}\Gamma(\alpha)^{-1} \exp[-\lambda_i(1+\beta) - \alpha B - \beta D] \right) \Big/$$

$$\left(\iiint \lambda_i^{\sum y_i+\alpha-1}\alpha^{A-1}\beta^{\alpha+C-1}\Gamma(\alpha)^{-1} \right.$$

$$\left. \times \exp[-\lambda_i(1+\beta) - \alpha B - \beta D]d\lambda d\alpha d\beta \right), \qquad (12.26)$$

which is unpleasant. Recall from Chapter 10 that if we can get the full conditional distributions for each of the coefficients in the posterior, we can run the Gibbs sampler and obtain the marginal posterior distributions with MCMC. The hierarchical model expressed in stacked notation immediately gives conditional expressions for the specified variables.

First consider the full conditional distribution for λ, using the properties of conditional probabilities, and the assumption that α and β are independent:

$$\pi(\lambda|\alpha, \beta) = \frac{p(\lambda, \alpha, \beta|\mathbf{y})}{p(\alpha, \beta|\mathbf{y})} = \frac{p(\lambda, \alpha, \beta)}{p(\alpha, \beta)} = \frac{p(\lambda, \alpha, \beta)}{p(\alpha)p(\beta)}$$

$$\propto \lambda^{\sum y_i+\alpha-1} \exp[-\lambda(1+\beta)]. \qquad (12.27)$$

So the random variable $\lambda|\alpha, \beta \sim \mathcal{G}(\sum y_i + \alpha, 1 + \beta)$, and given (interim) values for α and β, we can sample from the distribution for λ. This turns out to be the only real hard work required since: $p(\alpha|\beta, \lambda)$ is just $p(\alpha)$, and $p(\beta|\alpha, \lambda) = p(\beta)$, by the initial assumptions of the hierarchical model. Therefore we now have all three full conditional distributions required and the Gibbs sampler can be implemented quite easily.

There are quite a few studies that use this exact Poisson-gamma hierarchical model or some close variant of it, including: Carlin and Louis (2009, pp.32-37), Cohen *et al.* (1998), Christiansen and Morris (1997), DeSouza (1992), Hadjicostas and Berry (1999), Leonard, Hsu, and Tsui (1989), Makov, Smith, and Liu (1996), Newton and Raftery (1994), Lawson (2013, Chapter 2), Haque, Chin, and Huang (2010), De Oliveira (2013), and Neyens, Faes, and Molenberghs (2012). The subsequent posterior estimation tasks in all of these works necessitate stochastic simulation rather than analytical methods. This is common as hierarchical specifications in general lead to posteriors that are typically estimated by MCMC, and Gibbs sampling in particular since the specification usually gives easily determined full conditional distributions for the parameters. Christiansen

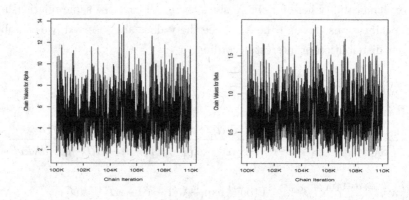

FIGURE 12.2: TRACEPLOT OF OUTPUT FOR GAMMA

■ **Example 12.3:   Marriage Rates in Italy.** This is an example of the Poisson-gamma hierarchical model just described. The data come from annual marriage counts per 1,000 of the population in Italy from 1936 to 1951 (from Columbo 1952). The substantive question addressed is whether it is practical to model marriage rates that occurred during World War II to rates just before and after. The data (y_i from above) are given by Table 12.2.

TABLE 12.2:   ITALIAN MARRIAGE RATES PER 1,000, 1936-1951

7	9	8	7	7	6	6	5	5	7	9	10	8	8	8	7

We clearly would not want to model these data with a nonhierarchical specification because that would require the Poisson intensity parameter to be constant over war and non-war years. The hierarchical model substitutes that assumption with the assumption that the λ_i are different in each year but drawn from a common distribution. Obviously this is a more flexible specification. However, if the hierarchical approach fails to fit, then there is evidence that the marriage rates are so fundamentally different that a wartime indicator parameter is needed.

The model is specified with relatively uninformed priors for α and β: both $\mathcal{G}(1,1)$. The following simple BUGS code runs this model:

```
model {
    for (i in 1:N) {
        lambda[i] ~ dgamma(alpha,beta);
        y[i] ~ dpois(lambda[i]);
    }
    alpha ~ dgamma(1,1);
    beta ~ dgamma(1,1);
}
```

These data are available in the BaM package, but the small size makes them easy to process separately as well. The WinBUGS data and initial value statements are for comparison simply:

```
list(y=c(7,9,8,7,7,6,6,5,5,7,9,10,8,8,8,7), N=16)
list(alpha=10, beta=10)
```

whereas the JAGS data file contains stored here in `italy.jags.data`:

```
N <- 16
y <- c(7,9,8,7,7,6,6,5,5,7,9,10,8,8,8,7)
```

and the JAGS initial value file, `italy.jags.inits`, contains:

```
alpha <- 10
beta <- 10
```

This is all that is needed to run this JAGS scriptfile:

```
model in "italy.bug"
data in "italy.jags.data"
compile
```

```
inits in "italy.jags.inits"
initialize
update 50000
monitor set alpha
monitor set beta
monitor set lambda
update 50000
coda *
```

The model is run for 50,000 iterations of the chain without recording the chain values. This is a very conservative burn-in period, but comes with a relatively low cost with such a simple model. Recording the chain values is turned on in BUGS and 50,000 iterations are run. Table 12.3 gives the summary statistics from these chain values. Figure 12.2 provides traceplots of the last 10,000 chain values for α and β. A traceplot is a common visual diagnostic for MCMC work in which a time-series of the chain values over the selected period are displayed. Prolonged visible upward or downward trends are generally evidence of nonconvergence.

TABLE 12.3: POSTERIOR SUMMARY STATISTICS, ITALY MODEL

Parameter	Mean	Std. Error	Median	95% HPD Region
α	5.166	1.777	4.935	[1.985: 8.644]
β	0.717	0.264	0.683	[0.240: 1.221]
1936	7.122	2.108	6.903	[3.362:11.400]
1937	8.301	2.284	8.062	[4.311:12.997]
1938	7.712	2.172	7.516	[3.785:11.982]
1939	7.100	2.123	6.861	[3.253:11.361]
1940	7.126	2.121	6.911	[3.316:11.380]
1941	6.509	1.981	6.312	[2.819:10.309]
1942	6.507	2.006	6.306	[2.887:10.506]
1943	5.930	1.904	5.734	[2.498: 9.698]
1944	5.934	1.902	5.756	[2.426: 9.654]
1945	7.080	2.073	6.840	[3.369:11.241]
1946	8.293	2.316	8.080	[4.032:12.879]
1947	8.899	2.373	8.678	[4.780:13.769]
1948	7.708	2.168	7.488	[3.889:12.164]
1949	7.721	2.147	7.511	[3.771:11.877]
1950	7.668	2.188	7.452	[3.686:11.937]
1951	7.110	2.103	6.916	[3.386:11.402]

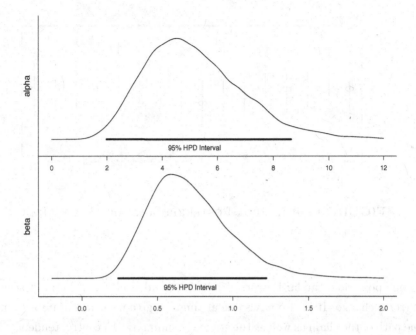

FIGURE 12.3: SMOOTHED POSTERIORS FOR α AND β

We are primarily interested in α and β since these values completely describe the distribution of the λ_i's. Figure 12.3 shows the smoothed density estimates of the last 10,000 iterates of the chain for α and β.[1] It is apparent from these density estimates that posterior distribution is unimodal and close to symmetric. We will focus extensively on MCMC convergence diagnostics in Chapter 14, but for now it is worth mentioning that two popular convergence diagnostics (from the CODA and BOA diagnostic suites in R) do not provide any evidence of nonconvergence. Geweke's diagnostic based on normal theory shows an α variable z-score of -1.1973498 and a β variable z-score of -1.337 (being in the tails of $\mathcal{N}(0, 1)$ indicates evidence of nonconvergence). The Heidelberger and Welch diagnostic gives $Est.SE_{\bar{\alpha}} = 0.225 > \text{HW.CI}_{\bar{\alpha}} = 0.128$, and $Est.SE_{\bar{\beta}} = 0.258 > \text{HW.CI}_{\bar{\beta}} = 0.019$. If the halfwidth of the credible interval for the mean is greater than a rough estimate of the variance of the sample mean given by the sample mean times some ϵ, usually defaulted to 0.1, then this test provides evidence of nonconvergence.

We also might be interested in the distribution of the λ_i parameters. Returning to the substantive question, it is clear that there is a drop in marriage rates during the war, as evidenced in Figure 12.4, which gives the posterior mean and error bars corresponding

---

[1]See Venables and Ripley (1999, pp.132-141) for specific guidance on producing smoothed density estimates in R. The frequently used `density` function is part of the base R package, and more elaborate alternatives such as `ash`, `GenKern`, and `KernSmooth` are available at CRAN.

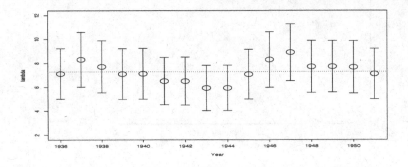

FIGURE 12.4: LAMBDA POSTERIOR SUMMARY, ITALY MODEL

to one posterior standard error. While the trend is clear, it does not seem to be an extreme change. It is also advisable at times to give measures of uncertainty through the path of the chain as well as the standard summaries of central tendency. Returning to the model of Italian marriage rates from Example 12.5.1, we now construct a slightly different kind of running mean plot of α and β across all the iterations from 1 to 10,000 with an envelope of 1.96 standard deviations from the mean on each cycle. This is shown in Figure 12.5 and strongly supports claims of convergence.

12.6   The General Role of Priors and Hyperpriors

Consider a data matrix $\mathbf{X}$ produced conditionally on an unknown parameter vector $\boldsymbol{\theta}_1$. In specifying a prior distribution for $\boldsymbol{\theta}_1$, it is possible that there is an additional parameterization to be taken into account: $\boldsymbol{\theta}_1$ is actually conditional on some second level of parameters, $\boldsymbol{\theta}_2$. For example, in the beta-binomial model given by (2.14), the data are conditional on the parameter $\boldsymbol{\theta}_1 = p$, which is itself conditional on $\boldsymbol{\theta}_2 = [A, B]$ from the beta PDF. The two coefficients can be fixed at researcher-selected constants ending the hierarchy at the second level, or an additional level of uncertainty can be specified by assigning distributions to A and B. These parameters are restricted to the support $0 < A, B < \infty$ by the form of the beta distribution. This makes the gamma or inverse gamma PDFs natural candidates for $A \sim f(A|\alpha, \beta)$, and $B \sim g(B|\gamma, \delta)$. Both of these forms are indexed by two positive parameters so we now have $\boldsymbol{\theta}_3 = [\alpha, \beta, \gamma, \delta]$. Continuing this process, it is now necessary to assign fixed values for $\boldsymbol{\theta}_3$ or assign another level of distributional assumptions.

At each level the length of the $\boldsymbol{\theta}_j$ vectors can differ, as seen in the previous example: $\boldsymbol{\theta}_1 = p$, $\boldsymbol{\theta}_2 = [A, B]$, and $\boldsymbol{\theta}_3 = [\alpha, \beta, \gamma, \delta]$. The size of the hyperprior vector for the next

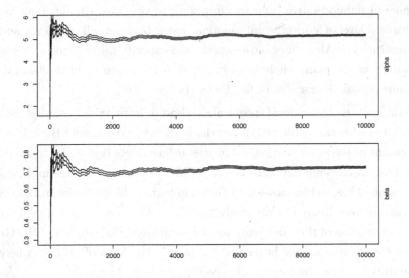

FIGURE 12.5: RUNNING MEANS, ITALY MODEL

level is determined by the number of parameters assumed to be random variables and their assigned parametric form, at the current level. Because not every parameter at a given level is required to have a specified distribution, it is common to have *mixed specification* where some parameters are assigned fixed hyperparameters and some are assumed to have additional hyperpriors. The overall specification can end up being somewhat difficult to describe, so some authors use tools such as the directed acyclic graph (DAG): a structured flowchart with parametric relationships connected by line segments in order to map out the paths of priors and hyperpriors. For use in Bayesian hierarchical models, see Gilks, Thomas, and Spiegelhalter (1994), Satten and Longini (1996), Spiegelhalter *et al.* (2000), and for general background see Lauritzen and Spiegelhalter (1988), and Whittaker (1990).

One way of thinking about hierarchical models is to consider the hierarchical structure as a complicated flat prior specification like (12.3). This works because the hierarchy of the prior can be re-expressed through a series of chained conditional probability statements: $p(\boldsymbol{\theta}_1|\boldsymbol{\theta}_2)p(\boldsymbol{\theta}_2|\boldsymbol{\theta}_3)\cdots p(\boldsymbol{\theta}_{J-1}|\boldsymbol{\theta}_J)$ for J levels of the hierarchy, so that the simplified prior is obtained by:

$$p(\boldsymbol{\theta}_1) = \int_{\Theta_2\cdots\Theta_J} p(\boldsymbol{\theta}_1|\boldsymbol{\theta}_2)p(\boldsymbol{\theta}_2|\boldsymbol{\theta}_3)\cdots p(\boldsymbol{\theta}_{J-1}|\boldsymbol{\theta}_J)d\boldsymbol{\theta}_2\cdots d\boldsymbol{\theta}_J. \qquad (12.28)$$

This shows that the hierarchical model is a correct and proper Bayesian specification since $\pi(\boldsymbol{\theta}_1|\mathbf{X}) \propto L(\boldsymbol{\theta}_1|\mathbf{X})p(\boldsymbol{\theta}_1)$. The form of (12.28) implies greater uncertainty than in those provided by the standard conjugate prior specifications studied in Chapter 4 because there is added uncertainty provided by averaging over distributions given at higher levels. Yet, except in special cases, the hierarchical form of the prior gives more parametric information than a standard uninformative prior specification. In this way, the Bayesian hierarchical model provides a level of subjective information that is between full conjugacy and uninformative specifications. Also, since most practitioners specify fairly vague forms at the top of the hierarchy at the point where fixed hyperpriors are required, hierarchical models are typically more robust than standard flat forms (Berger 1985).

In general, realistic hierarchical specifications lead to intractable posterior specifications and necessitate estimation with MCMC methods (Seltzer, Wong, and Bryk 1996). Thus the development of Bayesian hierarchical models is intimately tied to the theoretical development and increased computational ease of MCMC in practice. Interestingly, it is rare to find a discussion of hierarchical models in Bayesian texts published prior to 1990, and those that do exist are very limited. Conversely, such models are now the mainstay of applied Bayesian work because of their flexibility as well as the ready adaptation of the Gibbs sampler to such specifications. The latter point is seen in the form of (12.28), where a set of conditional distributions is automatically specified from the hierarchy.

12.7   Bayesian Multilevel Linear Regression Models

We now move on to understanding Bayesian multilevel linear models from both a theoretical and a practical standpoint. Nearly all Bayesian multilevel specifications are fit with MCMC software, so after covering some underlying theory we will proceed to BUGS solutions.

12.7.1   The Bayesian Hierarchical Linear Model of Lindley and Smith

The linear setup of Lindley and Smith (1972) is important enough to warrant a separate discussion. This model, developed before the era of widespread MCMC computing, specifies hierarchies of linear hyperpriors each of which has an associated matrix of explanatory variables. Therefore it is a Bayesian generalization of the simpler and more common hierarchical linear specification given above. Since each level has a unique conditional normal specification of differing dimension, subscripting is extensive.

Exchangeability is required here because in the estimation of the coefficients from (12.3) we need to assume that the joint distribution of each of the groups is invariant to reordering. Exchangeability is discussed in greater detail in Section 12.4 on page 420 in this chapter. In addition, we can place a prior distribution on the γ coefficients. Typically these are assigned convenient forms. The well-known and classical references on the Bayesian approach to HLMs are: Good (1980a, 1983b), Lindley and Smith (1972) plus the follow-on article by Smith (1973), the two papers by Tiao and Zellner (1964a, 1964b), and the article by Hartigan (1969). See also the extensive listing of references in Berger (1985, p.183).

Begin with a typical multivariate normal-linear model specification:

$$\mathbf{y} \sim \mathcal{N}(\mathbf{X}_1\boldsymbol{\beta}_1, \boldsymbol{\Sigma}_1) \tag{12.29}$$

for $\mathbf{X}_1$ ($n \times k_1$), $\boldsymbol{\beta}_1$ ($k_1 \times 1$), and positive definite $\boldsymbol{\Sigma}_1$ ($n \times n$). The prior on $\boldsymbol{\beta}_1$ is also normal, but according to:

$$\boldsymbol{\beta}_1 \sim \mathcal{N}(\mathbf{X}_2\boldsymbol{\beta}_2, \boldsymbol{\Sigma}_2) \tag{12.30}$$

for $\mathbf{X}_2$ ($k_1 \times k_2$), $\boldsymbol{\beta}_2$ ($k_2 \times 1$), and $\boldsymbol{\Sigma}_2$ ($k_1 \times k_1$), positive definite. The second data matrix is typically of a different dimension than the first because it is measured at a different level of aggregation or clustering. The implications of this specification are that:

$$\mathbf{y} \sim \mathcal{N}(\mathbf{X}_1\mathbf{X}_2\boldsymbol{\beta}_2, \boldsymbol{\Sigma}_1 + \mathbf{X}_1\boldsymbol{\Sigma}_2\mathbf{X}_1')$$

$$\boldsymbol{\beta}_1|\mathbf{y} \sim \mathcal{N}(\boldsymbol{\Sigma}_2^*\boldsymbol{\beta}_2^*, \boldsymbol{\Sigma}_2^*)$$

$$\text{where:} \quad \boldsymbol{\beta}_2^* = \mathbf{X}_1'\boldsymbol{\Sigma}_1^{-1}\mathbf{y} + \boldsymbol{\Sigma}_2^{-1}\mathbf{X}_2\boldsymbol{\beta}_2$$

$$\boldsymbol{\Sigma}_2^* = \left[\mathbf{X}_1'\boldsymbol{\Sigma}_1^{-1}\mathbf{X}_1 + \boldsymbol{\Sigma}_2^{-1}\right]^{-1}. \tag{12.31}$$

If the hierarchy ends here by assigning fixed hyperprior values to $\boldsymbol{\beta}_2$ and $\boldsymbol{\Sigma}_2$, then the model is equivalent to the simple one given above.

Rather than stopping at the second level of the hierarchy, $\boldsymbol{\beta}_2$ could be treated as a further random component by specifying:

$$\boldsymbol{\beta}_2 \sim \mathcal{N}(\mathbf{X}_3\boldsymbol{\beta}_3, \boldsymbol{\Sigma}_3), \tag{12.32}$$

where we specify additional data $\mathbf{X}_3$ ($k_2 \times k_3$), and fixed hyperpriors $\boldsymbol{\beta}_3$ ($k_3 \times 1$), and positive definite $\boldsymbol{\Sigma}_3$ ($k_2 \times k_2$). Winding these terms back into the prior for $\boldsymbol{\beta}_1$ gives the relatively simple form: $\boldsymbol{\beta}_1 \sim \mathcal{N}(\mathbf{X}_2\mathbf{X}_3\boldsymbol{\beta}_3, \boldsymbol{\Sigma}_2 + \mathbf{X}_2\boldsymbol{\Sigma}_3\mathbf{X}_2')$ (notice that all the second-level terms have been replaced). Of course our real interest lies in the *posterior* distribution of $\boldsymbol{\beta}_1$, which Lindley and Smith proved to be:

$$\boldsymbol{\beta}_1|\mathbf{y}, \mathbf{X}_j, \boldsymbol{\Sigma}_j, \boldsymbol{\beta}_3 \sim \mathcal{N}(\boldsymbol{\Sigma}_3^*\boldsymbol{\beta}_3^*, \boldsymbol{\Sigma}_3^*), \qquad j = 1, 2, 3$$

$$\text{where:} \quad \boldsymbol{\beta}_3^* = \mathbf{X}_1'\boldsymbol{\Sigma}_1^{-1}\mathbf{y} + (\boldsymbol{\Sigma}_2 + \mathbf{X}_2\boldsymbol{\Sigma}_3\mathbf{X}_2')^{-1}\mathbf{X}_2\mathbf{X}_3\boldsymbol{\beta}_3$$

$$\boldsymbol{\Sigma}_3^* = \left[\mathbf{X}_1'\boldsymbol{\Sigma}_1^{-1}\mathbf{X}_1 + (\boldsymbol{\Sigma}_2 + \mathbf{X}_2\boldsymbol{\Sigma}_3\mathbf{X}_2')^{-1}\right]^{-1}. \tag{12.33}$$

The nifty thing about this result is that exactly like the simple forms of Bayesian-normal inference given in Chapter 3, the posterior mean is a weighted compromise between the prior mean specification and the data mean (here the OLS estimate). To see this, note that the $\boldsymbol{\Sigma}_1$ weighted OLS estimator is $(\mathbf{X}_1'\boldsymbol{\Sigma}_1^{-1}\mathbf{X}_1)^{-1}\mathbf{X}_1'\boldsymbol{\Sigma}_1^{-1}\mathbf{y}$, and the prior mean is equal to

$\mathbf{X}_2\mathbf{X}_3\beta_3$. Rearranging the mean expression in (12.33) according to

$$\Sigma_3^*\beta_3^* = \left[\mathbf{X}_1'\Sigma_1^{-1}\mathbf{X}_1 + (\Sigma_2 + \mathbf{X}_2\Sigma_3\mathbf{X}_2')^{-1}\right]^{-1}\mathbf{X}_1'\Sigma_1^{-1}\mathbf{y}$$

$$+ \left[\mathbf{X}_1'\Sigma_1^{-1}\mathbf{X}_1 + (\Sigma_2 + \mathbf{X}_2\Sigma_3\mathbf{X}_2')^{-1}\right]^{-1}$$

$$\times (\Sigma_2 + \mathbf{X}_2\Sigma_3\mathbf{X}_2)^{-1}\mathbf{X}_2\mathbf{X}_3\beta_3 \qquad (12.34)$$

reveals that first additive term is the OLS estimate weighted by its corresponding dispersion weighted and $(\Sigma_2 + \mathbf{X}_2'\Sigma_3\mathbf{X}_2')^{-1}$, plus the prior estimate weighted by its corresponding dispersion weighted and $\mathbf{X}_1'\Sigma_1^{-1}\mathbf{X}_1$ (see Smith [1973] for discussion).

It should be clear that it is possible to continue to build the normal hierarchy in the same manner to the point where our tolerance for subscripts equals the utility of additional levels. It is rarely useful to go beyond the forms given here, however. Recall the result from Goel (1983) that we get progressively less information from the data moving up the hierarchy.

The focus in these forms has been on developing a normal hierarchy based on the mean term. It has been assumed that the variance specifications at each level of the prior could be reasonably specified. A number of alternatives have been suggested such as treating these matrices like nuisance parameters and integrating them out, perhaps specifying that each level is an arbitrary multiple of the likelihood dispersion (Lindley and Smith 1972), or putting a prior specification on these terms as well such as a Wishart (Robert 2001).

The Wishart prior (introduced in Chapter 3 on page 77, and described in Appendix B) is a flexible parametric form for modeling multivariate variance structures. Denote the PDF for a variance-covariance matrix Σ as $\mathcal{W}(\Sigma|\alpha, \beta)$, where α is a scalar and β is a symmetric nonsingular matrix. To show the flexibility of this parametric form consider a simple 2×2 case:

$$\Sigma = \begin{bmatrix} \sigma_{11} & \sigma_{12} \\ \sigma_{21} & \sigma_{22} \end{bmatrix}. \qquad (12.35)$$

Figure 12.6 shows $\sigma_{12}/(\sigma_{11}\sigma_{22})$ for three different values of α and where β is simply an identity matrix. Additional features can be imposed by specifying a more complex version of β. From the three examples we see that a variety of prior forms can be specified for various components in a hierarchical linear model.

One field that makes extensive use of such multilevel linear models is education research (Burstein, Kim, and Delandshere 1989, Bryk and Raudenbush 1989, Kreft and De Leeuw 1998). This is a natural fit because students are nested within classrooms, classrooms within schools, schools within districts, districts within states, and states within nations. Therefore data are measured at necessarily different levels of bureaucratic hierarchy and it would be naïve to treat variables as if they were produced at the same level. Standard references include Bryk and Raudenbush (2001), Goldstein (1987, 1985, 2003), Kreft (1993), Lee and Bryk (1989), and Raudenbush and Bryk (1986). A particular feature of this literature is the almost exclusive specification of distributions based on normal or t-distribution assumptions (Cocchi and Mouchart 1996; Goldstein 1986; Mason, Wong, and Entwistle 1983; Goel and

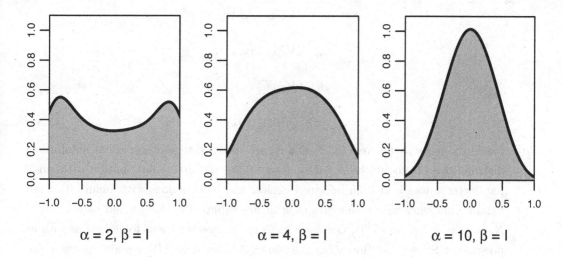

$$\alpha = 2,\ \beta = 1 \qquad\qquad \alpha = 4,\ \beta = 1 \qquad\qquad \alpha = 10,\ \beta = 1$$

FIGURE 12.6: CORRELATION DISTRIBUTIONS FROM $W(\Sigma|\alpha,\beta)$

DeGroot 1981; Smith 1973; Wakefield *et al.* 1994; Western 1998; Wong and Mason 1985, 1991), although more complex variations could include the incorporation of mixed effects (Datta and Ghosh 1991; Kackar and Harville 1984).

■ **Example 12.4:   A Normal-Normal Hierarchical Model for Economic Data.**
This example shows a minimally hierarchical model for economic indicators can be specified with BUGS (actually fitted with the JAGS package). The data come from the U.S. Department of Commerce, Survey of Current Business, and describe activity from the first quarter of 1979 to fourth quarter 1989, and therefore cover $j = 44$, also available for easy download at: http://lib.stat.cmu.edu/DASL as "Predicting Retail Sales." The variables provided are: DSB, national income wage and salary disbursements (in billions of dollars), EMP, employees on non-agricultural payrolls (in thousands), BDG building material dealer sales (in millions of dollars), CAR, retail automotive dealer sales (in millions of dollars), FRN, home furnishings dealer sales (in millions of dollars), and GMR, general merchandise dealer sales (in millions of dollars). All values are scaled for the model by dividing by one thousand.

Label as $x_j\ j = 1\ldots44$ the time variable, and list the cases (indices) by $i = 1,\ldots,5$ such that y_{ij} is the value of the ith economic value at time j. A simple normal-normal specification that accounts for the hierarchy of time within case is given by:

$$y_{ij} \sim \mathcal{N}(\beta_{0i} + \beta_{1i} x_j, \tau)$$

$$\beta_0 \sim \mathcal{N}(\mu_{\beta_0}, \tau_{\beta_0})$$

$$\beta_1 \sim \mathcal{N}(\mu_{\beta_1}, \tau_{\beta_1})$$

$$\tau \sim \mathcal{G}(0.01, 0.01), \tag{12.36}$$

and the hyperparameters for β_0 and β_1 are specified to give somewhat uninformed distributions. It should be noted that specifying the $\mathcal{G}(\epsilon, \epsilon)$ distribution with small parameter values is still an informed choice, and in situations with small estimated variance inference will be sensitive to arbitrary values of ϵ (Hodges and Sargent 2001, Natarajan and Kass 2000). Gelman (2006) recommends various folded forms such as normals or Student's-t (including the Cauchy) as more stable (see also Gelman and Hill [2007, p.433]). This specification is a (now) standard starting point in learning BUGS and the only subtlety is sweeping a mean through the x terms, and the BUGS creators note that this reduces the dependence of beta0 and beta1. In fact, this feature notably improves the fit of the model with these data. The final BUGS code is as follows:

```
model {
    for (i in 1:VALUE)  {
        for (j in 1:TIME)  {
            mu[i,j] <- beta0[i] + beta1[i]*(x1[j]-x1.mean);
            y[i,j]    ~ dnorm(mu[i,j],tau);
        }
        beta0[i]  ~ dnorm(mu.beta0,tau.beta0);
        beta1[i]  ~ dnorm(mu.beta1,tau.beta1);
    }
    tau          ~ dgamma(1.0E-2,1.0E-2);
    mu.beta0     ~ dnorm(0,1.0E-2);
    tau.beta0    ~ dgamma(1.0E-2,1.0E-2);
    mu.beta1     ~ dnorm(0,1.0E-2);
    tau.beta1    ~ dgamma(1.0E-2,1.0E-2);
    x1.mean   <- mean(x1[]);
}
```

Note the use of the double index on mu[i,j]. This model is estimated using JAGS with the supplied code and run for 100,000 iterations including a burn-in of 20,000 iterations, which are discarded. The posterior is summarized in Table 12.4 where we see that all but one of the posterior distributions have 95% HPD regions distinct from zero. So under conventional regression quality terms, this model appears to fit the data well.

TABLE 12.4: POSTERIOR SUMMARY
STATISTICS, ECONOMIC INDICATORS MODEL

Parameter	Mean	Std. Error	95% HPD
β_{01}	1.852	0.752	[0.377: 3.333]
β_{02}	96.323	0.750	[94.854:97.805]
β_{03}	17.085	0.751	[15.616:18.556]
β_{04}	66.062	0.748	[64.597:67.533]
β_{05}	16.173	0.747	[14.709:17.632]
β_{06}	37.882	0.754	[36.388:39.352]
$\beta.1_1$	0.039	0.059	[-0.077: 0.156]
β_{12}	0.496	0.059	[0.380: 0.612]
β_{13}	0.315	0.058	[0.201: 0.431]
β_{14}	1.522	0.059	[1.406: 1.639]
β_{15}	0.368	0.059	[0.253: 0.483]
β_{16}	0.548	0.059	[0.433: 0.662]

12.8 Bayesian Multilevel Generalized Linear Regression Models

The extension from Bayesian multilevel linear models to Bayesian multilevel generalized linear models is straightforward both theoretically and in practice (the latter mainly due to BUGS software solutions).

Begin by replacing the linear link function from (12.29) on page 437 with a more general form:

$$\mathbf{y} \sim g^{-1}(\mathbf{X}_1 \boldsymbol{\beta}_1, \boldsymbol{\Sigma}_1) \qquad (12.37)$$

for $\mathbf{X}_1$ ($n \times k_1$), $\boldsymbol{\beta}_1$ ($k_1 \times 1$), and positive definite $\boldsymbol{\Sigma}_1$ ($n \times n$), as described in Appendix A. In general (but not necessarily) the prior on $\boldsymbol{\beta}_1$ is again normal according to: $\boldsymbol{\beta}_1 \sim \mathcal{N}(\mathbf{X}_2 \boldsymbol{\beta}_2, \boldsymbol{\Sigma}_2)$ for $\mathbf{X}_2$ ($k_1 \times k_2$), $\boldsymbol{\beta}_2$ ($k_2 \times 1$), and $\boldsymbol{\Sigma}_2$ ($k_1 \times k_1$), positive definite. As before, the second data matrix is of a different dimension than the first because it is measured at a different level of aggregation. Because of $g^{-1}()$ this setup leads to a huge number of models and posterior forms. The resulting Bayesian hierarchical generalized linear model (BHGLM) is then a very flexible form for describing the nonlinear systematic effect on the outcome variable, but often with normal assumptions at all higher levels on the right-hand side. Such normal assumptions are not necessary, but were very convenient prior to the advent of MCMC estimation. Now there are almost no distributional restrictions give the flexibility of BUGS software and the ability to write MCMC algorithms for customized models in standard programming languages. Racine *et al.* (1986, pp.113-116) give a detailed example of model choice here, and Albert (1988) provides an overview of posterior estimation problems that can occur.

The following two examples give nonlinear (logistic) regression specifications that are typical forms. The first example uses arrest data in different cities to highlight group differences that matter to social scientists studying heterogeneous crime and drug use by geographic units. The second example specifies two non-nested hierarchies on the regression parameters.

■ **Example 12.5:   Hierarchical Model of Drug Use by Arrestees** This example uses data from *Differences in the Validity of Self-Reported Drug Use Across Five Factors in Indianapolis, Fort Lauderdale, Phoenix, and Dallas, 1994* (ICPSR Study Number 2706, Rosay and Herz (2000), from the Arrestee Drug Abuse Monitoring (ADAM) Program/Drug Use Forecasting, ICPSR Study Number 2826. The original purpose of the study was to understand the accuracy of self-reported drug use, which is a difficult problem for obvious reasons. These data provide variables measuring: gender, race, age, type of drug, and offense seriousness (misdemeanor versus felony) for 4,752 arrestees in four cities. For our purposes here, the sample is restricted to felony arrests reducing the sample size to 2,965. Confirmation of drug-use status was confirmed with past-arrest urine testing. The outcome variable used here is MJTEST, a dichotomous variable indicating a positive urine test for marijuana (2,085 negative, 880 positive). The explanatory variables selected are: AGEGRP (1 for 1,700 cases 18 through 30 years old, 2 for 1,265 cases 31 years old or over), SEX (1 for 2,213 male cases, 2 for 752 female cases), RACE (1 for 1,554 black cases, 2 for 1,411 white cases), and COCSELF indicating self-reported cocaine usage prior to arrest (0 for 2,220 negative responses, 1 for 745 positive responses). Since there are four separate cities in this study, we add a second level to account for possible differences with the variable SITE coded according to: Indianapolis = 1 (759 cases), Ft. Lauderdale = 2 (974 cases), Phoenix = 3 (646 cases), and Dallas = 4 (586 cases). The data are available in the BaM package.

Pre-processing of the R data file to BUGS and then to JAGS is accomplished with the following R code:

```
lapply(c("coda","BaM","superdiag"),library, character.only=TRUE)
source("writeDatafileR.R")
data(adam); attach(adam)
adam.df <- data.frame(SITE,MJTEST,AGEGRP,SEX,RACE,COCSELF)
writeDatafileR(adam.df,"adam.bugs.dat")
bugs2jags("adam.bugs.dat","adam.jags.dat")
system("echo 'CASES <- 2965 \n CATS <- 4 \n COVARS <- 4'
    >> adam.jags.dat")
detach(adam.df)
```

These steps are necessary only if using JAGS in a separate terminal window, as opposed to running JAGS or BUGS from R with the tools described in Chapter 11 such as with the library rjags. The system() command adds the necessary constants to the data file in a Unix context. This is better than hard-coding these into the JAGS code since those reusing the model code with other data (a common occurrence) may forget to change them resulting in misspecification. Users of WinBUGS do not need the bugs2jags statement and also need to change the system command appropriately to add these constants to the adam.bugs.dat file or add them with an editor (an easy task since they can be added right after the list() characters before the first variable). Note that path names are not specified here so the use of setwd() is required beforehand. Armed with these data the following BUGS logit model is specified and run with three parallel chains from three different starting points using JAGS:

```
model {
    for (i in 1:CASES) {
        logit(p[i]) <- alpha[SITE[i]]
            + beta[1]*AGEGRP[i] + beta[2]*SEX[i]
            + beta[3]*RACE[i] + beta[4]*COCSELF[i]
        MJTEST[i] ~ dbern(p[i])
    }
    for (j in 1:CATS) { alpha[j]  ~ dnorm(0,tau.alpha) }
    for (k in 1:COVARS) { beta[k] ~ dnorm(0,tau.beta) }
    tau.y    <- pow(sigma.y,-2)
    sigma.y ~ dt(0,1,1)T(0,)
    tau.alpha   <- pow(sigma.alpha,-2)
    sigma.alpha ~ dt(0,1,1)T(0,)
    tau.beta    <- pow(sigma.beta,-2)
    sigma.beta ~ dt(0,1,1)T(0,)
}
```

Notice in particular the alpha[SITE[i]] statement of the random effect for the four cities, providing a simple hierarchy with no covariates at that level. The variance components are specified such that the prior standard errors are each a "folded Cauchy" distribution, which gives the correct support and larger-than-normal tails to be somewhat conservative (Gelman 2006, Polson and Scott 2012). These specifications are connected to the necessary precision statement in the normal distribution by the negative square operation. The following commands (which correspond to menu selections in WinBUGS) are entered into a JAGS text window to run the model:

```
load dic
model in adam.jags
data in adam.jags.dat
compile, nchains(3)
```

```
parameters in adam1.inits
parameters in adam2.inits
parameters in adam3.inits
initialize
monitor set alpha
monitor set beta
monitor set sigma.y
monitor set sigma.alpha
monitor set sigma.beta
monitor set deviance
update 500000
coda *
```

The function `mcmc.table` provided in the **Computational Addendum** imports these samples back into R for analysis, and we have used other convenient ways to import CODA files. The output is a set of tables, depending on how many chains are specified. The use of **superdiag** shows no evidence of non-convergence from the four popular formal diagnostics introduced in Chapter 14. Being comfortable with convergence, we summarize the model in Table 12.5.

TABLE 12.5: POSTERIOR SUMMARY STATISTICS, DRUG USE BY ARRESTEES MODEL

	Mean	Std. Error	95% HPD Interval	
alpha[1]	1.0213	0.2415	0.5481	0.5481
alpha[2]	0.6248	0.2480	0.1386	0.1386
alpha[3]	0.6412	0.2751	0.1020	0.1020
alpha[4]	0.6432	0.2441	0.1648	0.1648
AGEGRP	-0.7604	0.0913	-0.9394	-0.9394
SEX	-0.4480	0.1059	-0.6556	-0.6556
RACE	0.0292	0.0907	-0.1485	-0.1485
COCSELF	-0.2129	0.0994	-0.4076	-0.4076
sigma.y	6.1765	105.3818	-200.3679	-200.3679
sigma.alpha	0.9271	0.4790	-0.0117	-0.0117
sigma.beta	0.6135	0.3024	0.0208	0.0208
deviance: 3454.8251				

First, since σ_α is not much smaller than σ_y, we know that there are important group differences accounted for in the model. In the absence of the random intercept specification, σ_y would increase non-trivially by absorbing variance accounted for in σ_{alpha}. These intercepts are also substantively interesting. It appears that Indianapolis has a noticeably larger value than the three other cities, which are relatively close in their effect. All four of the β coefficients in the model are reliable at standard levels (the 95% HPD regions do not cover zero). Younger arrestees are more likely to test positive than those over 30. Males are more likely to test positive than females. Black arrestees are more likely than white arrestees. And those that self-report recent cocaine usage are less likely to test positive for marijuana than those that did not. The last result may be due to a substitution effect.

■ **Example 12.6:   A Random Effects Logistic Model of Contraception Use.**
The hierarchical model of Wong and Mason (1985) recognizes two levels of data in cross-national surveys: the individual response level and the national aggregation level for 15 countries. The research question addressed here is what individual and national factors in lesser-developed countries affect the decision by married women, aged 40 to 44, to use modern contraception techniques (measured as 1=ever, 0=never). The explanatory variables from surveys are: the years of education for the woman (WED), a dichotomous variable for urban versus rural childhood (URC), exposure to family-planning efforts (FPE), and an interaction term specified by Wong and Mason (WED × FPE).

The hierarchy is set up to express within-country variation at the first level and across-level variation at the second level, where the outcome variable is Y_{ij} indicating the binary response for the ith observation in the jth country with n_j respondents. The hierarchy starts with a standard logit specification with $k = 4$ regressors for unit ij and then adds normal assumptions at higher levels according to:

$$Y_{ij}|p_{ij} \sim \mathcal{BR}(p_{ij})$$
$$p_{ij} = [1 + \exp(- \underset{(1\times k)(k\times 1)}{\mathbf{X}_{ij} \ \boldsymbol{\beta}_j})]^{-1}$$
$$\boldsymbol{\beta}_{jk} = \underset{(1\times l)(l\times 1)}{\mathbf{G}_{kl} \ \boldsymbol{\eta}_k} + \boldsymbol{\alpha}_{jk}$$
$$\boldsymbol{\alpha}_{jk} \sim \mathcal{N}(\mathbf{0}, \gamma_{kk})$$
$$\boldsymbol{\eta}_k \sim \mathcal{N}(\mathbf{m}_\eta, \boldsymbol{\Sigma}_\eta), \tag{12.38}$$

where $\boldsymbol{\beta}_j$ is the jth country coefficient vector, and each of these k individual coefficients $\boldsymbol{\beta}_{jk}$ is explained by a higher-level parameterization in which $\mathbf{G}_{kl}$ is a vector of second-level explanatory variables and $\boldsymbol{\eta}_k$ is the corresponding second-level coefficient, which is given a mean hyperparameter vector $\mathbf{m}_\eta$ and variance hyperparameter matrix $\boldsymbol{\Sigma}_\eta$. The second-level error term is given a normal zero mean prior, and the prior on the $\boldsymbol{\eta}_k$ is a multivariate normal with the covariance matrix parameterized so that n is contained in the denominator of the terms whereby this prior becomes deterministic in the limit.

The data values and $\boldsymbol{\eta}_k$ posterior summaries, the primary coefficient of interest, are provided in Table 12.6. Wong and Mason demonstrate that hierarchy matters, by comparing these results to a straight logit regression model (note the shrinkage). The central point in such a comparison is that a nonhierarchical model treats all the second-level units as equivalent and thus misses any cross-national effects of interest.

TABLE 12.6: CONTRACEPTION DATA, DEVELOPING
COUNTRIES (BY SIZE)

Country j	n_j	$\%(Y_{ij} = 1)$	URC	WED	FPE	WED $\times$ FPE
Indonesia	787	25	9	1.2	14	16.8
Sri Lanka	752	22	20	3.8	12	45.6
Malaysia	726	38	29	1.6	18	28.8
Korea	700	61	15	4.5	24	108.0
Peru	581	14	17	2.7	0	0.0
Fiji	517	60	15	3.7	22	81.4
Thailand	465	36	8	2.1	20	42.0
Kenya	401	9	4	0.9	6	5.4
Jordan	348	44	53	1.4	0	0.0
Costa Rica	332	59	18	4.7	21	98.7
Guyana	313	42	20	6.1	0	0.0
Panama	312	59	50	5.3	19	100.7
Columbia	294	37	47	2.7	16	43.2
Lesotho	236	6	4	3.9	0	0.0
Jamaica	210	44	8	6.9	23	158.7
η_k Posterior Mode			0.4909	0.2147	0.0850	-0.0053
η_k Posterior SD			0.1008	0.0397	0.0269	0.0024

It is almost trivial to add national-level economic and social indicator terms restricted to the second level to create even more flexible forms (Wong and Mason 1991), or to simplify this model greatly by reducing either the number of levels and complexity of the normal forms (Gilks 1996). Albert and Chib (1995) also give a direct Bayesian residuals analysis for these models.

12.9 Exercises

12.1 Show that inserting (12.2) into (12.1) produces (12.3).

12.2 Explain when you would want to run a multilevel model and when you would not want to run a multilevel model. What part of the model output from the multilevel model tells you that defining groups is justified. What output from a non-multilevel model might suggest that multilevel model should be run?

12.3   For the following simple hierarchical form,

$$X_i|\theta \sim \mathcal{N}(\theta, 1)$$
$$\theta|\sigma^2 \sim \mathcal{N}(0, \sigma^2)$$
$$\sigma^2 \sim 1/k, \quad 0 < \sigma \le k,$$

express the full joint distribution $p(\theta, \sigma, \mathbf{X})$.

12.4   Explain carefully the distinction between: *complete pooling*, *no pooling*, and *multi-level approaches*.

12.5   Using the data on firearm-related deaths per 100,000 in the United States from 1980 to 1998 (Source: National Center for Health Statistics. Health, United States, 1999, With Health and Aging Chartbook, Hyattsville, Maryland: 1999, see the online government resources at `http://www.cdc.gov/nchs/express.htm` and Chapter 6), test the hypothesis that there is a period of higher rates by specifying a normal mixture model and comparing the posterior distributions of λ_1 and λ_2 in the model:

$$x_i \sim \mathcal{N}(\lambda_{k_i}|\tau)$$
$$k_i \sim \text{Categorical}_2(K)$$
$$\tau \sim \mathcal{G}(\epsilon_{g_1}, \epsilon_{g_2})$$
$$\lambda_i \sim \mathcal{N}(0, \epsilon_{l_i})$$
$$K \sim \mathcal{D}(1, 1),$$

where K is the proportion of the observations distributed $\mathcal{N}(\lambda_{k_2}|\tau)$ and $1 - K$ is the proportion of the observations distributed $\mathcal{N}(\lambda_{k_1}|\tau)$. Specify a `BUGS` model choosing values of the $\epsilon_{..}$ parameters that specify somewhat diffuse hyperpriors. The data are repeated here:

14.9	14.8	14.3	13.3	13.3	13.3	13.9	13.6	13.9
14.1	14.9	15.2	14.8	15.4	14.8	13.7	12.8	12.1

12.6   Using equation 12.5, analytically derive the properties of α_j as $n_j \to 0$ and $n_j \to \infty$. Produce a graph of these effects from simulated data.

12.7   Describe a circumstance where data are exchangeable but not iid. How does this affect your choice of model?

12.8   Show that de Finetti's property holds true for a mixture of normals. Hint: see Freedman and Diaconis (1982).

12.9   (Carlin and Louis 2009, pp.125-128). Given the hierarchical model:

$$Y_i|\lambda_i \sim \mathcal{P}(\lambda_i t_i)$$
$$\lambda_i \sim \mathcal{G}(\alpha, \beta)$$
$$\beta \sim \mathcal{IG}(A, B), \quad \alpha \text{ known}$$

where the hyperpriors A and B are fixed, express the full conditional distributions for all of the unknown parameters.

12.10   Using the employment status data description from Example 12.3, add three additional hierarchical models to the six already listed (interactions, lags, etc.).

12.11   Hobert and Casella (1998) specify the following one-way random effects model:

$$y_{ij} \sim \mathcal{N}(\beta + u_i, \sigma_\epsilon^2)$$
$$u_i \sim \mathcal{N}(0, \sigma^2)$$
$$p(\beta) \propto 1$$
$$p(\sigma_\epsilon^2) \sim 1/\sigma_\epsilon^2), \ 0 < \sigma_\epsilon^2 < \infty$$
$$p(\sigma^2) \sim 1/\sigma^2), \ 0 < \sigma^2 < \infty,$$

where $i = 1, \ldots, K$ and $j = 1, \ldots, J$. Using the R function `rnorm()` generate contrived data for y_{ij} with $K = 50$, $J = 5$, and estimate the model in BUGS. When improper priors are specified in hierarchical models, it is important to make sure that the resulting posterior is not itself then improper. Non-detection of posterior impropriety from looking at chain values was a problem in the early MCMC literature. This model gives improper posteriors although it is not possible to observe this problem without running the chain for quite some time. Compare the standard diagnostics for a short run with a very long run. What do you see? Now specify proper priors and contrast the results.

12.12   Returning to the Palm Beach County data introduced in Example 5.1.1 on page 148 in Chapter 5 (`data(pbc.vote)`): specify a non-nested hierarchical model with levels for `technology` and your own segmentation of `size` into K groups (histogram for investigation), where there is a modeled correlation between these two hierarchies according to:

$$y_i \sim N(\beta_0 + \beta_{j1}X_1 + \beta_{k2}X_2 + \boldsymbol{\gamma}\mathbf{Z}, \sigma_y^2), \qquad i = 1, \ldots, n$$

$$\begin{pmatrix} \beta_j \\ \beta_k \end{pmatrix} \sim N\left(\begin{pmatrix} \mu_{\beta_1} \\ \mu_{\beta_2} \end{pmatrix}, \begin{pmatrix} \sigma_{\beta_1}^2 & \rho\sigma_{\beta_1}\sigma_{\beta_2} \\ \rho\sigma_{\beta_1}\sigma_{\beta_2} & \sigma_{\beta_2}^2 \end{pmatrix} \right).$$

The outcome variable is `badballots` and the matrix $\mathbf{Z}$ contains other covariates of interest. Run this model in BUGS and report the results along with convergence diagnostics from `superdiag`.

12.13   To introduce the "near-ignorance prior" Sansó and Pericchi (1992) specify the hierarchical model starting with a multivariate normal specification:

$$\mathbf{Y}|\theta \sim \mathcal{MVN}(\mathbf{X}\boldsymbol{\theta}, \sigma^2\mathbf{I})$$
$$\boldsymbol{\theta}_i|\lambda_i \sim \mathcal{N}(\mu_i, \lambda_i^2)$$
$$\lambda_i^2 \sim \mathcal{EX}(1/\tau_i^2)$$

where $i = 1, \ldots, p$ and μ_i, τ_i, σ^2 are known. Show that by substitution and integration over λ that this model is equivalent to the hierarchical specification:

$$\mathbf{Y}|\theta \sim \mathcal{MVN}(\mathbf{X}\theta, \sigma^2\mathbf{I})$$
$$\theta_i \sim \mathcal{DE}(\mu_i, \tau_i).$$

12.14 Using the general model setup in Example 12.3 starting on page 418 specify three more distinct specifications in addition to the six given.

12.15 Clyde, Müller, and Parmigiani (1995) specify the following hierarchical model for 10 Bernoulli outcomes:

$$y_i|\beta_i, \lambda_i \sim \mathcal{BR}(p(\beta, \lambda, \mathbf{x}))$$
$$p(y_i = 1|\beta, \lambda, \mathbf{x}) = (1 + \exp[-\beta_i(x - \lambda_i) - \log(0.95/0.05)])^{-1}$$
$$\begin{bmatrix} \log \beta_i \\ \log \lambda_i \end{bmatrix} \sim \mathcal{MVN}(\boldsymbol{\mu}, \boldsymbol{\Sigma})$$
$$\boldsymbol{\mu} \sim \mathcal{MVN}\left(\begin{bmatrix} 0 \\ 2 \end{bmatrix}, \begin{bmatrix} 25 & 5 \\ 5 & 25 \end{bmatrix}^{-1} \right)$$
$$\boldsymbol{\Sigma} \sim \mathcal{W}\left(10, 10 \times \begin{bmatrix} 0.44 & -0.12 \\ -0.12 & 0.14 \end{bmatrix}^{-1} \right).$$

Using BUGS, obtain the posterior distribution for $\log \beta$ and $\log \lambda$ using the data: $\mathbf{y} = [1,0,1,1,1,1,0,1,1,0]$ with $\mathbf{X}$ generated in R from $\mathcal{MVN}((1,2,3), \boldsymbol{\Sigma}), \boldsymbol{\Sigma} = 2\mathbf{I}$ along with a leading column of 1s.

12.16 Example 9.6 (page 298) discussed models of state-level poverty from Gill and King (2004). Using the dataset fips in the package BaM construct a multilevel model for the poverty outcome with a hierarchy for state: Texas versus Florida. Implement this specification in BUGS and summarize the sampled values.

12.17 Scollnik (2001) considers the following actuarial claims data for three groups of insurance policyholders, with missing values.

	Group 1		Group 2		Group 3	
Year	Payroll	Claims	Payroll	Claims	Payroll	Claims
1	280	9	260	6	NA	NA
2	320	7	275	4	145	8
3	265	6	240	2	120	3
4	340	13	265	8	105	4
5	NA	NA	285	NA	115	NA

Replicate Scollnik's hierarchical Poisson model using BUGS:

$$Y_{ij} \sim \mathcal{P}(\lambda_{ij})$$
$$\lambda_{ij} = P_{ij}\theta_j$$

$$\theta_j \sim \mathcal{G}(\alpha, \beta) \qquad\qquad P_{ij} \sim \mathcal{G}(\gamma, \delta)$$
$$\alpha \sim \mathcal{G}(5, 5) \qquad\qquad \gamma \sim \mathcal{U}(0, 100)$$
$$\beta \sim \mathcal{G}(25, 1) \qquad\qquad \delta \sim \mathcal{U}(0, 100),$$

where $i = 1, 2, 3$ and $j = 1, \ldots, 5$.

12.18   Using the data in Table 12.7 on racial composition at federal agencies, construct a two-way ANOVA model as a non-nested hierarchical model in BUGS. The data are also provided by the package BaM.

TABLE 12.7: RACIAL COMPOSITION OF U.S. FEDERAL AGENCIES (1998)

Agency	Black	Hispanic	Asian	Native	White
Agriculture	10.6	5.6	2.4	2.5	78.9
Commerce	18.3	3.4	5.2	0.6	72.5
DOD	14.2	6.2	5.4	1.0	73.3
Army	15.3	5.9	3.7	1.1	73.9
Navy	13.4	4.3	9.8	0.8	71.8
Air.Force	10.6	9.5	3.1	1.1	75.7
Education	36.3	4.7	3.3	1.0	54.7
Energy	11.5	5.2	3.8	1.3	78.2
EOP	24.2	2.4	4.2	0.3	69.0
HHS	16.7	2.9	5.1	16.9	58.5
HUD	34.0	6.7	3.2	1.1	55.0
Interior	5.5	4.3	1.6	15.9	73.2
Justice	16.2	12.2	2.8	0.8	68.1
Labor	24.3	6.6	2.9	0.7	65.6
State	14.9	4.2	3.7	0.4	76.7
Transportation	11.2	4.7	2.9	1.5	79.7
Treasury	21.7	8.4	3.3	0.8	65.8
VA	22.0	6.0	6.7	0.8	64.6
GS	28.4	5.0	3.4	0.7	62.5
NASA	10.5	4.6	4.9	0.9	79.2
EEOC	48.2	10.6	2.7	0.5	38.0

Data Source: Office of Personnel Management

12.19   One way to think of panel data models is as hierarchical models where each person/case is nested in their own group. Regular time-series model features can then be included. Rerun the BUGS model for the British abortion attitudes data in Chapter 6 (the example beginning on page 177) adding a correlation parameter over panel waves.

12.20 Obtain the `socatt` subset data from the British Social Attitudes (BSA) Survey 1983-1986 from either the `BaM` package or directly from the Centre for Multilevel Modelling, University of Bristol. The variables included are:

> ▷ `District`, identifying for geographic district
> ▷ `Respondent.Code`, respondent identifier
> ▷ `Year.Code`, 1 = 1983, 2 = 1984, 3 = 1985, 4 = 1986
> ▷ `Num.Answers`, number of positive answers to seven questions
> ▷ `Party`, 1 = Conservative, 2 = Labour, 3 = Lib/SDP/Alliance, 4 = others
> ▷ `Social.Class`, 1 = middle, 2 = upper working, 3 = lower working
> ▷ `Gender`, 1 = male, 2 = female
> ▷ `Age`, age in years 18-80
> ▷ `Religion` 1 = Roman Catholic, 2 = Protestant/Church of England, 3 = others, 4 = none.

This is a different subset from that used in the abortion attitudes example on page 177. For additional details see McGrath and Waterton (1986). Construct an ordered probit model with `BUGS` where `Num.Answers` is the outcome variable and respondents are nested in both years and social class.

12.10 Computational Addendum

12.10.1 R Function for importing BUGS output

The following R function imports MCMC output without requiring the use of the `CODA` menu structure. This is faster and more convenient to those that do not intend to use the other features of `CODA`.

```
mcmc.table <- function(n.chains=1,burnin=1/2,alpha=0.05) {
    lapply(c("coda","xtable"),library, character.only=TRUE)
    print(paste("current working directory:",getwd()))
    for (i in 1:n.chains) {
        full.out <- read.coda(paste("CODAchain",i,".txt",sep=""),
            "CODAindex.txt",quiet=TRUE)
        start <- nrow(full.out)*burnin; stop <- nrow(full.out)
        chain.mean <- apply(full.out[start:stop,],2,mean)
        chain.se <- apply(full.out[start:stop,],2,sd)
        full.tab <- cbind(chain.mean,chain.se,
            chain.mean-qnorm(1-alpha/2)*chain.se,
```

```
            chain.mean-qnorm(1-alpha/2)*chain.se)
        print(xtable(full.tab,digits=4))
    }
}
```

The output is given for each chain in a LaTeXformatted table that can then be modified. Users will want to modify the default parameters, particularly `burnin=1/2`, which stipulates one-half of the chain as a burn-in default.

Chapter 13

Some Markov Chain Monte Carlo Theory

13.1 Motivation

This chapter revisits the theoretical basis for Markov chain Monte Carlo but with greater detail and more attention to issues of convergence. Most of the technical content is intended to give a greater appreciation for the underlying mathematical process of Markov chains and therefore an understanding of the issues involved in convergence and mixing. Both convergence and mixing behavior of Markov chains affect the final inferences made with Bayesian models using MCMC. Readers who wish to move onto more practical considerations can proceed to Chapter 14 and return to this information as needed.

13.2 Measure and Probability Preliminaries

First we return to the idea of a *stochastic process*. A stochastic process is a set of observed values $\theta^{[t]}$ ($t \geq 0$) on the probability space: $(\Omega, \mathcal{F}, \mathfrak{P})$ where the superscript $[t]$ denotes the serial order of occurrence, Ω is the non-empty outcome space with the associated σ-algebra $\mathcal{F}$ and probability measure $\mathfrak{P}$ (Billingsley 1995, Doob 1990, Ross 1996). We say that $\mathcal{F}$ is the associated *field* or *algebra* of subsets of Ω if: $\Omega \in \mathcal{F}$, it is closed under complementation as well as intersections of its subsets, and a σ-field or σ-algebra if it is also closed under countable unions of subsets. These subsets can be individual elements, denoted θ, or sets of such elements, denoted $A, B, C, \ldots$. A function $\mathfrak{P}$ is a probability measure on Ω if it maps subsets of Ω to $[0\!:\!1]$ according to the Kolmogorov axioms: *(1)* $p(A) \in [0\!:\!1]$ $\forall A \in \mathcal{F}$, *(2)* $p(\Omega) = 1$, $p(\varnothing) = 0$, and *(3)* for non-overlapping multiple A_i, $P\left(\bigcup_{i=1}^n A_i\right) = \sum_{i=1}^n p(A_i)$ (even if $n = \infty$). These were stated more colloquially on page 7, but see also Billingsley (1995, Chapter 2) for an in-depth explanation or Gill (2006, Chapter 7) for an accessible introduction. Now $(\Omega, \mathcal{F}, \mathfrak{P})$ is the *probability measure space* for θ often shortened to the *state space*.

The sequence of Ω-valued $\theta^{[t]}$ random *elements*, indicated by $t = 0, 1, \ldots$, defines the Ω-valued stochastic process. Typically this is just $\Re$-valued with restrictions (Karlin and

Taylor 1981, 1990, Hoel, Port, and Stone 1987). By convention, the sequence is labeled with consecutive even-spaced intervals: $T: \{\theta^{[t=0]}, \theta^{[t=1]}, \theta^{[t=2]}, \ldots\}$. This configuration can be generalized to non-consecutive labels or non-equal time periods, although it does not provide any additional utility for current purposes.

At time t the *history* of the stochastic process is the increasing series of sub-σ-algebras defined by: $\mathcal{F}_0 \subseteq \mathcal{F}_1 \subseteq \ldots \mathcal{F}_t$ where θ is measurable on each one. A T-valued stochastic process, with transition probability p, and initial (starting) value θ_0, is a *Markov chain*, $\mathfrak{M}(\boldsymbol{\theta}_t, t \geq 0)$, if at the arbitrary $(t+1)$st time point in T, the following is true:

$$p(\theta^{[t+1]}|\mathcal{F}_t) = p(\theta^{[t+1]}|\theta^{[t]}), \ \forall t \geq 0 \tag{13.1}$$

(Zhenting and Qingfeng [1978, Chapter 6]). As stated previously in Chapter 10, the only component of the history that matters in determining movement probabilities to an arbitrary set A, which we define as a collection of individual elements, for Markov chains at the current step, is the present realization of the stochastic process:

$$p(\theta^{[t+1]} \in A|\theta^{[0]}, \theta^{[1]}, \ldots, \theta^{[t-1]}, \theta^{[t]}) = p(\theta^{[t+1]} \in A|\theta^{[t]}). \tag{13.2}$$

This is the defining *Markovian property* that distinguishes a Markov chain from the general class of stochastic processes.

Retain Ω as the space giving the support of θ, a random variable where components are denoted with subscripts: $\theta_i, \theta_j, \theta_k, \ldots$. Here $\mathcal{F}$ is again the associated σ-algebra and $\mathfrak{P}$ is a probability measure on Ω. Now use $f, g, h, \ldots$ to denote real-valued measurable functions defined on this state space, and define $\mathcal{M}$ as the full collection of signed measures on measure space $(\Omega, \mathcal{F})$ where λ and μ denote elements by notational convention. The class of *positive* measures is given by: $\mathcal{M}^+ = \{\lambda \in \mathcal{M} : \lambda(T) > 0)$, and includes $\mathfrak{P}$.

A Markov chain transition kernel, $K(\theta, A)$, is the mechanism for describing the probability structure of the Markov chain, $\mathfrak{M}(\boldsymbol{\theta}_t, t \geq 0)$: a probability measure for all θ points in the state space to the set $A \in \mathcal{F}$. Thus, it is a mapping of the potential transition events to their probability of occurrence (Robert and Casella 2004, p.208). Formally, K is a non-negative, σ-finite kernel on the state space $(\Omega, \mathcal{F})$ that provides the mapping $\Omega \times \mathcal{F} \to \Re^+$ given the following three conditions:

1. for every finite subset $A \in \mathcal{F}$, $K(\bullet, A)$ is measurable,

2. for every point $\theta_i \in \Omega$: $K(\theta_i, \bullet)$ is a measure on the state space,

3. there is a positive measurable function, $f(\theta_i, \theta_j)$, $\forall \theta_i, \theta_j \in \Omega$ where $\int K(\theta_i, d\theta_j)f(\theta_i, \theta_j) < \infty$.

This preliminary discussion sets up the discussion of Markov chain properties and characteristics. Importantly, a Markov chain is just a regular probability mechanism with two added features: time-seriality, and a memory-less decision-process.

13.3 Specific Markov Chain Properties

Having defined a stochastic process of interest, we now return to the core properties of Markov chains given in Chapter 10 but with greater attention to underlying theoretical details. These properties are important because they lead us to convergence of the Markov chain, without which no practical statistical inference can be made.

13.3.1 ψ-Irreducibility

Irreducibility is a description of accessibility. A set, A, is *irreducible* if every point or collection of points in A can be reached from every other point or collection of points in A. The associated Markov chain is irreducible if it operates on an irreducible set as its state space. Define now ψ as a positive σ-finite measure on $(\Omega, \mathcal{F})$ with arbitrary $A \in \mathcal{F}$ such that $\psi(A) > 0$. Given the transition kernel $K(\theta, A)$, if every positive ψ-subset, $A' \subseteq A$, can be reached from every part of A, then A is called ψ-*communicating*. When the full state space for the Markov chain T is ψ-communicating, then the kernel that defines the Markov chain is ψ-irreducible. Typically, we assume that ψ-irreducible here is maximally ψ-irreducible: for any *other* positive σ-finite measure on $(\Omega, \mathcal{F})$, ψ', it is necessarily true that $\psi > \psi'$ (Meyn and Tweedie 1993, pp.88-89).

13.3.2 Closed and Absorbing Sets

A non-empty set $A \in \mathcal{F}$ is *obtainable* from θ for the Markov chain defined at time t by the kernel K^t if $K^t(\theta, A) > 0$, for some $t \geq 1$, and *unobtainable* if $K^t(\theta, A) = 0$, for all $t \geq 1$. Here, A is called *closed* for K if A^c is not obtainable from A: $K^t(\theta, A^c) = 0$, for all $\theta \in A$, and all $t \geq 1$. The condition of *absorbing* (Revuz 1975) is more restrictive than closed: $K(\theta, A) = K(\theta, \Theta) = 1$, for all $\theta \in A$, since it is possible under the closed condition, but impossible under the absorbing condition, that $K(\theta, A) = K(\theta, \Theta) \neq 1$, for some $\theta \in A$. So a closed set can have subsets that are unavailable but an absorbing state fully communicates with all of its subsets.

13.3.3 Homogeneity and Periodicity

A Markov chain is *homogeneous* at the tth step if the transition probabilities at this step do not depend on the value of t. The *period* of a Markov chain is the length of time to repeat an identical cycle of values, and it is desirable that the Markov chain not have such a defined cycle. A chain that does not repeat in this fashion is called *aperiodic*. We can easily deduce that a *periodic* Markov chain is non-homogeneous because the cycle of repetitions defines transitions based on specific times.

Define now $T(A)$ as the first *hitting time* (shortest return time to an arbitrary subspace) for the Markov chain $\mathfrak{M}(\theta_t, t \geq 0)$ with invariant distribution π to the set $A \in \mathcal{F}$ not including time 0: $T(A) = \inf(t \geq 1 : \theta_t \in A)$. Athreya, Doss, and Sethuraman (1996, p.72) stipulate two conditions that lead to convergence towards the invariant distribution, which is our eventual goal.   The first condition says that for every element in A the probability that the hitting time is less than infinity is greater than zero for all of these elements, with probability 1 under the invariant distribution. The second condition forces the transition probability from one of these elements to any value in the space to be greater than or equal to the probability under the measure up to a constant.   Periodicity violates the second condition because it means that at any arbitrary step of the chain we can define a set D as the next value in the sequence specified by a period of moves.

13.3.4   Null and Positive Recurrence

Putting these previous definitions together, a homogeneous, ψ-irreducible Markov chain on a closed set (discrete or continuous and bounded) is *recurrent* or persistent with regard to the set, A, if the probability that the chain occupies each subset of A infinitely often over unbounded time is one.   When a chain moves into a recurrent state, it stays there forever and eventually visits every sub-state an infinite number of times. Furthermore, a recurrent Markov chain is *positive recurrent* if the average time to return to A is bounded, otherwise it is a *null recurrent* Markov chain (Doeblin 1940).

In the case of unbounded continuous state spaces, we have to work with a slightly more complicated version of recurrence. Define first for a set A and all elements x in A, the *expected number of visits by the Markov chain to x in the limit:* $\eta_x = \sum_{n=1}^{\infty} I_{(\theta_n \in x)}$, which is a function that counts visits to x. Now we say a ψ-irreducible Markov chain is *Harris recurrent* if there is σ-finite probability measure ψ for $(\Omega, \mathcal{F})$ such that at time n it has the property: $\psi(A) > 0$, $\forall A \in \mathcal{F}$ (Harris 1956, Athreya and Ney 1978). The new definition for unbounded continuous state spaces is required because a ψ-irreducible chain with an invariant distribution on an unbounded continuous state space that is *not* Harris recurrent has a positive probability of getting stuck indefinitely in an area bounded away from convergence, given a starting point there. The Harris definition allows us to avoid worrying about the existence of a pathological null set in $\Re^k$. The standard MCMC algorithms implemented on finite state computers are Harris recurrent (Tierney 1994).

13.3.5   Transience

Now consider the number of visits to the set itself, rather than specific elements in the set: for a set A, the *expected number of visits by chain θ_n to A in the limit:* $\eta_A = \sum_{n=1}^{\infty} I_{(\theta_n \in A)}$, where the $I()$ function counts hits on A. Transience and recurrence can both be defined in terms of expectation for this η_A: A is *uniformly transient* if there exists a scalar $M < \infty$ such that $E[\eta_A] \leq M \; \forall \theta \in A$. A single state, θ, in the discrete state space case is *transient* if: $E[\eta_\theta] < \infty$. Conversely, A is *recurrent* if: $E[\eta_A] = \infty \; \forall \theta \in A$. A single state in the discrete state space case is *recurrent* if: $E[\eta_\theta] = \infty$. For proofs see: Meyn and Tweedie (1993, pp.182-3) and Nummelin (1984, p.28).

The important theorem here is:

> **Theorem.** If $\mathfrak{M}(\boldsymbol{\theta}_t, t \geq 0)$ is a ψ-irreducible Markov chain with transition kernel $K(\theta, A)$, then it must either be transient or recurrent depending on whether it is defined on a transient or recurrent set A.

The proof is a direct consequence of Kolmogorov's zero-one law, and details are given in Billingsley (1995, 120).

This means that there is a two-state world to worry about: the chain is either recurrent and we know that it will eventually settle into a stable distribution, or it is transient and it will never achieve stability.

We can also define the *convergence parameter* of a kernel rK as the real number $0 \leq R < \infty$ on a closed set A such that: $\sum_0^\infty r^n K^n < \infty$ for every $0 \leq r < R$, and $\sum_0^\infty r^n K^n = \infty$ for every $R \geq r$. It turns out that for ψ-irreducible Markov chains there always exists a finite R that defines whether or not the kernel for this Markov chain is *R-transient* if the first condition holds, and *R-recurrent* if the second condition holds (Meyn and Tweedie 1993).

13.3.6 Markov Chain Stability

Label $\pi(\theta)$ as the stationary (limiting) distribution of the Markov chain for θ on the state space Ω, with transition probability $p(\theta_i, \theta_j)$ that gives the probability that the chain will move from arbitrary point θ_i to arbitrary point θ_j, as stipulated by the transition kernel K (i.e., from the ith row of the transition matrix for a discrete state space). Naturally, this stationary distribution, $\pi(\theta)$, is really the posterior distribution of interest from some Bayesian model where marginalization is not possible or convenient analytically. This stationary or invariant distribution satisfies the following condition from Chapter 10:

$$\sum_{\theta_i} \pi^t(\theta_i) p(\theta_i, \theta_j) = \pi^{t+1}(\theta_j) \qquad \text{Discrete State Space}$$

$$\int \pi^t(\theta_i) p(\theta_i, \theta_j) d\theta_i = \pi^{t+1}(\theta_j) \qquad \text{Continuous State Space,} \qquad (13.3)$$

(see specifically page 339). Multiplication by the transition kernel and evaluating for the current point (summation for discrete sample spaces, integration for continuous sample spaces) produces the same marginal distribution, $\pi(\theta) = \pi(\theta)P$. Therefore the marginal distribution remains fixed when the chain reaches the stationary distribution and we can ignore superscripts giving iteration number for iteration purposes. Once the chain reaches the stationary distribution (synonymously the equilibrium distribution, limiting distribution), its movement is dictated by the marginal distribution, $\pi(\theta)$ from that point on. A ψ-irreducible, aperiodic Markov chain is guaranteed to have exactly one such stationary distribution (Häggström 2002, p.37). This is the critical theoretical basis for estimation with MCMC: if the stationary distribution of the Markov chain is the posterior distribution of interest, then we are certain to eventually get samples from this posterior.

13.3.7   Ergodicity

The ergodic theorem is the key link between the mechanical process of the Monte Carlo simulation and the inferential result of MCMC. If a chain is positive (Harris if necessary) recurrent, and aperiodic on some state A, it is an *ergodic* Markov chain (Tweedie 1975). Importantly, ergodic Markov chains have the property that:

$$\lim_{n \to \infty} \left| P^n(\theta_i, \theta_j) - \pi(\theta_j) \right| = 0, \tag{13.4}$$

for all possible θ_i, and θ_j in the subspace (Norris 1997, p.53). So the transition probabilities of the chain have converged to those of the limiting distribution and therefore all future draws are treated as if from this marginal distribution of interest. This means that once a specified chain is determined to have reached this ergodic state, inference comes from running the chain for some length of time and summarizing the empirical draws. We have thus replaced analytical work (i.e., integrating over some difficult form) with empirical analysis.

Ergodic Markov chains provide the following result:

$$\lim_{t \to \infty} \frac{1}{t} \sum f(\boldsymbol{\theta}_t) = \int_{\Theta} f(\boldsymbol{\theta}) \pi(\boldsymbol{\theta}) d\boldsymbol{\theta}, \tag{13.5}$$

proven originally by Doeblin (1940). This result means that empirical averages for the function $f()$ converge to a probabilistic average of the function over the limiting distribution. In fact, it is this principle that underlies and justifies all MCMC for Bayesian stochastic simulation; it is exactly the link between "Markov chain" and "Monte Carlo."

Ergodicity gives a means of asserting eventual convergence (although not the only one), but it does not provide a firm bound on the time required to reach convergence to the limiting distribution. There are actually multiple "flavors" of ergodicity that provide differing rates of convergence for the Markov chain, and these are described in Section 13.5.

13.4   Defining and Reaching Convergence

Return to the abstract measure space $(\Omega, \mathcal{F})$ with events $A, B, C, \ldots$ in Ω, real-valued measurable functions $f, g, h, \ldots$ on $\mathcal{F}$, and signed measure $\mathcal{M}^+$ with elements λ and μ. *First*, define an appropriate norm operator. The elementary form for a bounded signed measure, λ, is:

$$\|\lambda\| \equiv \sup_{A \in \Omega} \lambda(A) - \inf_{A \in \Omega} \lambda(A) \tag{13.6}$$

which is just the total variation of λ. *Second*, assume that K is R-recurrent given by probability measure $\mathfrak{P}$, and the stationary distribution is normed such that $\pi(h) = 1$ for h on $\mathcal{F}$. In addition, assume also that

$$\mathfrak{M}(\boldsymbol{\theta}_t, t \geq 0)$$

is R-recurrent (discrete or bounded continuous space) or Harris R-recurrent (continuous unbounded space) Markov chain with transition kernel K. If K has period $p \geq 2$ then by definition the associated Markov chain cycles between the states: $\{A_0, A_1, A_2, \ldots, A_{p-1}\}$. The ψ-*null set*, $\Psi = \{A_0 \cup A_1 \cup \cdots \cup A_{p-1}\}^c$, defines the collection of such states not visited in the p-length iterations, and we want to drive this to a set of size zero.

Let the (positive) signed measures λ and μ be any two initial distributions of the Markov chain at time zero, and therefore before convergence to any other distribution. Nummelin (1984, Chapter 6) shows that if $\mathfrak{M}(\boldsymbol{\theta}_t, t \geq 0)$ is aperiodic, then:

$$\lim_{n \to \infty} ||\lambda P^n - \mu P^n|| = 0. \tag{13.7}$$

This is essentially Orey's (1961) *total variation norm theorem* applied to an aperiodic, recurrent Markov chain (Orey's result was more general but not any more useful for our endeavors; see also Athreya and Ney [1978, p.498] for a proof). However, we know that any ψ-irreducible and aperiodic Markov chain has one and only one stationary distribution and ψ-irreducibility is implied here by recurrence. Therefore we can substitute into (13.7) the stationary distribution π to get:

$$\lim_{n \to \infty} ||\lambda P^n - \pi|| = 0, \tag{13.8}$$

which gives ergodicity. This shows in greater detail than before the conditions by which convergence in distribution to the stationary distribution is justified.

Convergence to stationarity is distinct from convergence of the empirical averages, which are usually the primary substantive interest. Consider the limiting behavior of a statistic of interest, $h(\theta)$, from an aperiodic Harris recurrent Markov chain. We typically obtain empirical summaries of this statistic using the partial sums such as:

$$\bar{h} = \frac{1}{n} \sum_{i=1}^{n} h(\theta_i). \tag{13.9}$$

The expected value of the target h is $E_f h(\theta)$, so by the established properties of Harris recurrent Markov chains (Brémaud 1999, p.104), it is known that $\bar{h} \to E_f h(\theta)$ as $n \to \infty$. Equivalently, it is true that:

$$\frac{1}{n} \sum_{i=1}^{n} h(\theta_i) - E_f h(\theta) \xrightarrow[n \to \infty]{} 0. \tag{13.10}$$

We can also consider the true distribution of $h(\theta)$ at time n (even if it is not directly observed) from a chain with starting point θ_0. The interest here is in $E_{\theta_0} h(\theta)$, where the expectation is with respect to the distribution of θ_n conditional on θ_0. In the next step add and subtract this term on the left-hand side of 13.10 to obtain:

$$\left[\frac{1}{n}\sum_{i=1}^{n}h(\theta_i) - E_{\theta_0}h(\theta)\right] - \left[E_f h(\theta) - E_{\theta_0}h(\theta)\right] \xrightarrow[n\to\infty]{} 0. \tag{13.11}$$

The second bracketed term is obviously the difference between the expected value of the target $h(\theta)$ in the true distribution at time n and the expected value of $h(\theta)$ in the stationary distribution. For any ergodic Markov chain, this quantity will eventually to converge to zero. The first bracketed term is the difference between the current empirical average and its expectation at time n. Except at the uninteresting starting point, these quantities are *never* non-asymptotically equivalent, and so even in stationarity, the empirical average has not converged. This is not bad news, however, since we know for certain by the central limit theorem that:

$$\frac{\frac{1}{n}\sum_{i=1}^{n}h(\theta_i) - E_{\theta_0}h(\theta)}{\sigma/\sqrt{n}} \xrightarrow{d} \mathcal{N}(0,1). \tag{13.12}$$

(Meyn and Tweedie 1993, p.418, Jones 2004). Therefore as $\sqrt{n}\delta^n \to 0$, convergence to stationarity proceeds at a much faster rate, but does not bring along convergence of empirical averages. Note also that (13.12) explains the shape of post-convergence marginal density plots.

13.5   Rates of Convergence

So far little has been said about the actual rate of convergence, merely that chains are or are not in a state of convergence. Ergodicity, resulting from positive (Harris if necessary) recurrence and aperiodicity, is merely an asymptotic property and thus a rather indeterminate statement for Markov chains run in finite time, that is, every Markov chain ever run in actual practice (Rosenthal 1995c).

We now provide the first improvement on basic ergodicity to produce a more rapid transition to Markov chain stability. If the Markov chain has invariant distribution, $\pi(\theta)$, and $\pi(A) > 0 \,\forall A \in \mathcal{F}$ it is *ergodic of degree 2* if:

$$\int_A \pi(d\theta)E_\theta[T(A)^2] < \infty \tag{13.13}$$

(Nummelin 1984, p.118). In other words, the condition is that the second moment of the first hitting times must be finite. What does this buy us? It turns out that if the functions $f(\theta)$ (arbitrary) and $\pi(\theta)$ are regular (finite total density and finite expectations over all subregions of the support, see Billingsley [1995, p.174] for details), then the rate of convergence is proportional to n^{-2}:

$$\lim_{n\to\infty} n^2\|f(\theta_t) - \pi(\theta)\| \longrightarrow 0. \tag{13.14}$$

This is an interesting result but unfortunately it is difficult to assert degree 2 ergodicity with many practical problems.

A more useful and stronger type of ergodic convergence is *geometric ergodicity*. Make the same assumptions as those above about the Markov chain but substitute the hitting time assumption with the following requirement:

$$\|f(\theta_t) - \pi(\theta)\| \leq m(\theta)\rho^t, \ \forall \theta, \ 0 < \rho < 1, \tag{13.15}$$

where $m(\theta)$ is any finite, non-negative function. Under these conditions the tth step transition probability converges to the invariant distribution at a geometric rate, which can be very quick depending on the value of ρ. If instead of specifying the function $m(\theta)$ we find a constant m such that:

$$\|f(\theta_t) - \pi(\theta)\| \leq m\rho^t, \ \forall \theta, \ 0 < \rho < 1, \tag{13.16}$$

then the chain is *uniformly ergodic*, which means it converges even faster. The value of these properties is two-fold. *First*, knowing that a chain is geometrically or uniformly ergodic is comforting in that it is an assurance of convergence in some reasonably practical amount of time (depending of course on the complexity of the model and the structure of the data). *Second*, it allows the derivation of bounds on the number of iterations to convergence for some Markov chains. These claims are usually made by analyzing minorization and drift conditions. The *minorization* condition means that for any sub-space A, the σ-finite measure φ in Ω with $\varphi(A) > 0$ contains a *small set* C with the property that for any $\theta \in C$:

$$K_t(\theta', \theta) \geq \delta v(\theta') \tag{13.17}$$

for θ' in Ω, $\delta > 0$, and time $t > 0$, where v is a probability measure concentrated on C. See Meyn and Tweedie (1993, Chapter 5) for details on small sets for Markov chains. A Metropolis-Hastings chain always meets these conditions if $q_t(\theta|\cdot)$ and $\pi(\theta)$ are both positive and continuous. (Roberts and Rosenthal 1998).

We can now catalog some popular variants of MCMC algorithms by their ergodic properties. These are given with references for the associated proofs. It is assumed in this list that every Markov chain is at least ergodic as well as meeting the small set (minorization) condition above. As a reminder, we will indicate θ for a finding in the single-dimensional case, and $\boldsymbol{\theta}$ for a finding in the multi-dimensional case, and all statements refer to continuous state spaces unless otherwise stated. This is not a complete listing of the numerous variants, by any means, but represents most of the more important and relevant results. See also the discussion in Roberts and Smith (1994).

Gibbs Sampling

▷ A chain operating on a finite state space, with a positive invariance distribution is **geometrically ergodic** (Geman and Geman 1984, Besag 1974). The positivity condition means that the support of the invariant distribution must be the Cartesian product of the marginal supports as a way of guaranteeing irreducibility.

▷ A chain cast as two-parameter data augmentation is **geometrically ergodic** (Tanner and Wong 1987, Rosenthal 1993).

▷ A chain meeting the Geman and Geman conditions with purely systematic scan (the order of the parameter-by-parameter updating is unchanged over Gibbs iterations): $1, 2, 3, ..., d$ is **geometrically ergodic** (Schervish and Carlin 1992).

▷ A chain meeting the Geman and Geman conditions with random scan (at the t'th step only one of the $\theta_i^{[t]}$ is randomly selected, usually uniformly $p = 1/d$, and updated) is **geometrically ergodic** (Liu, Wong, and Kong 1994).

▷ Suppose there exists a non-negative function $K^*()$ on $\Re^d$ with $K^*(\theta') > 0$ on Ω, where for some $t > 0$ we have $K^{[t]}(\theta, \theta') \geq K^*(\theta'), \forall \theta$ (θ in the continuous domain of the Markov chain), and $K(\theta, \theta')$ is positive over all $(\theta, \theta') \in \Re^d \times \Re^d$. Then the chain is **uniformly ergodic** (Roberts and Polson 1994).

▷ A chain where $\pi(\theta)$ is produced from improper priors may lead to improper posteriors (Hobert and Casella 1996, 1998, Cowles 2002), so there is no longer a justification for geometric ergodicity (Chan 1993), or even degree 2 ergodicity since it may be possible to come up with sub-spaces where $\int_A \pi(d\theta) E\theta[T(A)^2] = \infty$.

Metropolis-Hastings

▷ Suppose $\pi(\theta) = h(\theta) \exp(p(\theta))$ where $p(\theta)$ is a (exponential family form) polynomial of order $m \geq 2$, and $p_m(\theta) \to -\infty$ as $|\theta| \to \infty$, where ($p_m(\theta)$ is the sub-polynomial consisting of only the terms in $p(\theta)$ of order m). There is actually a subtlety lurking here in the multivariate case. Take the term with the highest total order across terms as $m(\Theta)$, and determine if there is a case whereby setting all terms to zero but one does not result in $-\infty$. So the bivariate normal ($\propto \exp[-\frac{1}{2}(x^2 - 2xy + y^2)]$) passes but a function like $\exp[-\frac{1}{2}(x^2 + 2x^2y^2 + y^2)]$ fails. Note that this characterizes the normal distribution and those with lighter tails. Then for a symmetric candidate distribution bounded away from zero, the chain is **geometrically ergodic** (Roberts and Tweedie 1996).

▷ A chain with $\pi(\theta)$ log-concave in the tails (meaning there is an $\alpha > 0$ such that for $y \geq x$, $\log \pi(x) - \log \pi(y) \geq \alpha(y - x)$ and for $y \leq x$, $\log \pi(x) - \log \pi(y) \geq \alpha(x - y)$) and symmetric candidate distribution where $q(\theta'|\theta) = q(\theta - \theta') = q(\theta' - \theta)$ is **geometrically ergodic** (Mengersen and Tweedie 1996).

▷ A chain with σ-finite measure $\varphi < \infty$ on $(\Omega, \mathcal{F})$, where $q_t(\theta|\cdot)$ and $\pi(\theta)$ are bounded away from zero is **uniformly ergodic** (Tierney 1994). Practically, this means truncating the posterior support such that $\pi(\theta) > 0$ (strictly!), which may be challenging in high dimensions.

▷ An independence chain (defined later on page 468) with the bounded weight function, $w(\theta) = \pi(\theta)/f(\theta)$ and $\pi(\theta)$ bounded away from zero is **uniformly ergodic** with $\rho \leq 1 - \sup(w(\theta))^{-1}$ (Tierney 1994).

This listing also highlights another important point. It is not necessarily worth changing the structure of the simulation in regular practice to produce a uniformly ergodic Markov chain from a geometrically ergodic version, but it is almost always worth the trouble to obtain a geometrically ergodic setup (discarding degree 2 ergodicity as analytically difficult to assert in almost all cases). Without geometric ergodicity convergence can take dramatically longer and, for the purposes of practical MCMC work, essentially infinite time for ergodic chains. Fortunately, for many of the model types encountered in the social sciences these conditions are met with little trouble using the two standard algorithms.

Where problems may arise is in the use of hybrid chains, such as Metropolis-within-Gibbs and Variable-at-a-Time Metropolis-Hastings (Roberts and Rosenthal 1998), that combine features of more basic algorithms and are usually specified because of posterior irregularities in the first place (note, actually, that "Metropolis-within-Gibbs" is a misnomer since Gibbs sampling is a special case of Metropolis-Hastings in which a candidate is always accepted).

These can sometimes be checked with the property that a Markov chain satisfying a minorization condition and a drift condition with $V(\theta) > 2b/(1-\delta)$ is geometrically ergodic (Rosenthal 1995a). For instance, Jones and Hobert (2004, 2001: Appendix A) give specific minorization and drift conditions for a block Gibbs sampler to be geometrically ergodic (block or grouped Gibbs sampler update parameters in blocks where the joint updatings are presumed to be marginalizable post-MCMC, i.e., something like drawing $\theta_1^{[j]}, \theta_2^{[j]} \sim \pi(\theta_1, \theta_2 | \theta_3^{[j-1]}), \theta_3^{[j]} \sim \pi(\theta_3 | \theta_1^{[j]}, \theta_2^{[j]})$, see Roberts and Sahu [1997]). Also, Jarner and Roberts (2002) connect the drift condition to polynomial rate convergence measured by standard norms.

From a utilitarian standpoint knowing that the Markov chain is geometrically or uniformly ergodic is not enough. It is only part of the complete process to worry about when running the chain to obtain reliable results. While it is obviously important to demonstrate ergodic properties, it does not actually confirm any set of applied results. Rosenthal (1995b, p.741) makes this particularly clear:

> It is one thing to say that the variation distance to the true posterior distribution after k steps will be less than $A\alpha^k$ for some $\alpha < 1$ and $A > 0$. It is quite another to give some idea of how much less than 1 this α will be, and how large A is, or equivalently to give a quantitative estimate of how large k should be to make the variation distance less than some ϵ.

In other words, it is necessary to assert at least geometric ergodicity but it is not going to directly help the practitioner make decisions about the length of the runs. In the next two sections we provide findings that lead to explicit advice about how to treat convergence for the Metropolis-Hastings algorithm and the Gibbs sampler.

13.6　Implementation Concerns

In applied work there are two practical questions that a user of Markov chain Monte Carlo algorithms must ask: (1) how long should I run the chain before I can claim that it has converged to its invariant (stationary) distribution, and (2) how long do I need to run the chain in stationarity before it has sufficiently mixed throughout the target distribution? The key factor driving both of these questions is the *rate* at which the Markov chain is mixing through the parameter space: slow mixing means that the definition of "long" gets considerably worsened.

There is also a difference between *being* in the state of convergence and *measuring* the state of convergence. A Markov chain for a single dimension has converged at time t to its invariant distribution (the posterior distribution of interest for correctly set up Bayesian applications) when the transition kernel produces univariate draws arbitrarily close to this distribution and the process therefore generates only legitimate values from a distribution in proportion to the actual target density. For a given measure of "closeness" (i.e., for some specified threshold, see below), a Markov chain is either in its invariant distribution or it is not. For single-dimension chains, there is no such thing as "somewhat converged" or "approaching convergence" (this gets more complicated for chains operating in multiple dimensions). In fact, Rosenthal (1995a) gives an example Markov chain that converges in exactly one step. So diagnostics and mathematical proofs that make claims about convergence are thus analyzing only a two-state world.

To put more precision on such statements, define the vector $\boldsymbol{\theta}_t \in S \subseteq \Re^d$ as the tth empirical draw (reached point) from the chain $\mathfrak{M}(\boldsymbol{\theta}_t, t \geq 0)$, operating on d-dimensional measure space $(\Omega, \mathcal{F})$, having the transition operator f defined on the Banach space of bounded measurable functions, and having $\pi(\boldsymbol{\theta})$ as its invariant distribution. A normed vector space is called a Banach space if it is complete under this metric. Completeness means that for a given *probability* measure space, $(\Omega, \mathcal{F}, P)$, if $A \subset B$, $B \in \Omega$, $p(B) = 0$, then $A \in \Omega$ and $p(A) = 0$. This condition allows us to ignore a set of measure problems that can otherwise occur. It also provides results on a general state space as well as the easier case of a finite countable state space. Invariance in this context means that π is a probability measure on $(\Omega, \mathcal{F})$ such that $\pi(s) = \int f(\boldsymbol{\theta}, s)\pi(d\boldsymbol{\theta})$, $\forall s \in \Omega$. The transition kernel of the Markov chain, $f()$ (generalizing K above), is the mechanism that maps $\Omega \times \mathcal{F} \rightarrow [0, 1]$ such that for every $A \in \mathcal{F}$, the function $f(\cdot, A)$ is measurable and for every $\boldsymbol{\theta} \in \Omega$ the function $f(\boldsymbol{\theta}, \cdot)$ is a valid probability function, as noted previously.

A chain that is positive recurrent or positive Harris recurrent (whichever appropriately applies) and aperiodic is also α-mixing, meaning that:

$$\alpha(t) = \sup_{A,B} \left| p(\boldsymbol{\theta}_t \in B, \boldsymbol{\theta}_0 \in A) - p(\boldsymbol{\theta}_t \in B)p(\boldsymbol{\theta}_0 \in A) \right| \xrightarrow[t \to \infty]{} 0 \qquad (13.18)$$

(Rosenblatt 1971). This means that for sub-spaces A and B that produce the largest difference, the joint probability of starting at some point in A and ending at some point in B at time t converges to the product of the individual probabilities. This means that these events are asymptotically (in t) independent for any definable sub-spaces. This second result from ergodicity justifies our treatment of Markov chain iterations as iid samples (Chan 1993).

There are actually some additional measure-theoretic nuances and extensions of these properties, such as the implied assumption that $\pi(\boldsymbol{\theta})$ is not concentrated on a single point as in a Dirac delta function, but the definitions given here are sufficient for the present purposes. Also, it is important to remember that ergodicity is just one way to assert convergence. It turns out, for instance, that a *periodic* Markov chain can also converge under a different and more complicated set of assumptions (Meyn and Tweedie 1993, Chapter 13), and we can even define ergodicity without an invariant distribution (Athreya and Ney 1978).

A Markov chain that has converged has the property that repeated applications of the transition kernel produce an identical distribution: $\pi f = \pi$. By far the most commonly used method of claiming such convergence is the *total variation norm*, which is restated from (13.7):

$$\|f(\boldsymbol{\theta}_t) - \pi(\boldsymbol{\theta})\| = \frac{1}{2}\sup_{\boldsymbol{\theta} \in A} A \int_{\Theta} |f(\boldsymbol{\theta}_t) - \pi(\boldsymbol{\theta})| \, d\boldsymbol{\theta}. \qquad (13.19)$$

This is half of the well-known L_1 distance, although the L_2 "chi-square" distance is useful as well, see Diaconis and Saloff-Coste (1996). Another suggestion is the infinity norm $\|f\|_\infty = \sup_{\theta \in \Re^d} \|f(\theta)\|$ (Roberts and Polson 1994). The vector $\boldsymbol{\theta}$ is a d-dimensional random variable lying within A, the sub-space that makes the difference within the integral as great as possible. When we integrate over $\boldsymbol{\theta} \in A$ it produces a supremum over the measurable sub-space A for the set of all measurable functions on A (a set that includes $f(\boldsymbol{\theta}_t)$ and $\pi(\boldsymbol{\theta})$). So there are two important operations occurring in the statement of (13.19). *First*, there is selection of a sub-space that makes the resulting quantity as large as possible. *Second*, there is integration of the distributional difference over this sub-space. Thus one gets the most pessimistic view of the difference between $f(\boldsymbol{\theta}_t)$ and $\pi(\boldsymbol{\theta})$ as possible.

The 1/2 in (13.19) constant comes from limit theory: as $t \to \infty$, the total variation norm for A converges to twice the empirical difference for all such sub-spaces (Meyn and Tweedie 1993, p.311):

$$\lim_{t\to\infty} \|f(\boldsymbol{\theta}_t, \cdot) - \pi(\boldsymbol{\theta})\| = 2 \lim_{t\to\infty} \sup_A \|f(\boldsymbol{\theta}_t, A) - \pi(A)\|. \qquad (13.20)$$

Another way to write the total variation norm first defines $\mu(A)$ as a signed measure on the state space S for the sub-space A. In the notation above, $\mu(A)$ is the integrated difference of two distributional statements over all of A. Now the total variation norm can be expressed as:

$$\|\mu\| = \sup_{A \in S} \mu(A) - \inf_{A \in S} \mu(A), \qquad (13.21)$$

which shows the same principle as (13.19) due to the explicit statement of the integral.

Zellner and Min (1995) propose three potentially useful alternatives as well, such as: the *anchored ratio convergence criterion*, which selects two arbitrary points in the sample space to calculate a posterior ratio baseline for comparison to reached chain points, the *difference convergence criterion*, which compares different analytical equivalent conditionals from the joint distribution, and the *ratio convergence criterion*, which uses the last idea in ratio form rather than differencing (see particularly p.922). We will return to the Zellner-Min diagnostics in Chapter 14.   Thus if $\|f(\boldsymbol{\theta}_t) - \pi(\boldsymbol{\theta})\| \to 0$ as $n \to \infty$, the distribution of $\boldsymbol{\theta}$ converges to that of a random variable from $\pi(\boldsymbol{\theta})$, and this convergence is actually stronger than standard convergence in distribution (i.e., convergence of CDFs). When $\|f(\boldsymbol{\theta}_t) - \pi(\boldsymbol{\theta})\|$ reaches values close to zero (say δ for now) we are willing to assert convergence. The problem, of course, is that $\pi(\boldsymbol{\theta})$ is a difficult form to work with analytically, which is why we are using MCMC in the first place. Theoretical work that puts explicit bounds on convergence includes: Lawler and Sokal (1988), Frieze, Kannan, and Polson (1994), Ingrassia (1994), Liu (1996b), Mengersen and Tweedie (1996), Robert (1995), Roberts and Tweedie (1996), Rosenthal (1995a), as well as Sinclair and Jerrum (1988).

For discrete problems it turns out that the converge rate can be established in proportion to the absolute value of the second eigenvalue of the transition matrix (kernel) (Diaconis and Stroock 1991, Fill 1991, Fulman and Wilmer 1999, Sinclair and Jerrum 1989) but this can also be quite difficult to produce for realistic problems (Frigessi, Hwang, Di Stefano, and Sheu 1993). For examples where these approaches work in practice see: Amit (1991), Amit and Grenander (1991), Cowles and Rosenthal (1998), Goodman and Sokal (1989), Meyn and Tweedie (1994), Mira and Tierney (2001b), Polson (1996), Roberts and Rosenthal (1999), and Rosenthal (1995b, 1996). Usually these solutions are particularistic to the form of the kernel and can also produce widely varying or impractical bounds.

13.6.1   Mixing

Markov chain mixing is a related but different concern than convergence. Mixing is the rate at which a Markov chain traverses about the parameter space, before or after reaching the stationary distribution. Thus slow mixing causes two problems: it retards the advance towards the target distribution, and once there, it makes full exploration of this distribution take longer. Both of these considerations are critical to providing valid inferences since the pre-convergence distribution does not describe the desired marginals and failing to mix through regions of the final distribution biases summary statistics. Mixing problems generally come from high correlations between model parameters or weakly identified model specifications, and are often more pronounced for model precision parameters.

Detailed guidance about assessing mixing properties for particular applications is given later in Chapter 14, but several general points are worth mentioning here. When a Metropolis-Hastings chain is mixing poorly it usually has a very low acceptance rate and therefore stays in single locations for long periods. Usually this is obvious, like acceptance ratios over some period of time less that 1%. This view is unavailable for a Gibbs sampler chain since it moves on every iteration. Often, though, if one has an indication of the range of the high density area for the posterior (for instance, a rough idea of the 90% HPD region), then poor mixing is observed by reasonable chain periods that traverse a very limited subset of this interval. Often such problems with the Gibbs sampler are caused by high correlations between parameters (also discussed in Chapter 14).

13.6.2   Partial Convergence for Metropolis-Hastings

As the number of dimensions increases, the sensitivity (and complexity) of the Metropolis-Hastings algorithm increases dramatically since the measure space $(\Omega, \mathcal{F})$ is defined such that each abstract point in Ω is d-dimensional and the σ-field of subsets $\mathcal{F}$ is generated by a countable collection of sets on $\Re^d$. An ergodic Markov chain $\mathfrak{M}(\boldsymbol{\theta}_t, t \geq 0)$ has the property: $\|f(\boldsymbol{\theta}_t) - \pi(\boldsymbol{\theta})\| \to 0$ as $n \to \infty$, $\forall \boldsymbol{\theta} \in \Theta$, but the size of d is critical in determining the rate since each step is d-dimensional. The primary complexity introduced by dimensionality here has to do with the strictness by which we apply $f(\boldsymbol{\theta}_t) - \pi(\boldsymbol{\theta})$. Suppose now that there is a subset of these dimensions $e < d$ that are of primary interest and the remaining $d - e$ are essentially a result of nuisance parameters. Is it then reasonable to require only evidence of *partial convergence*? That is, at time t for some small δ:

$$\|f(\theta_t^*) - \pi(\theta^*)\| \approx \delta, \quad \forall \theta^* \in \Re^e \tag{13.22}$$

but,

$$\|f(\theta_t^\dagger) - \pi(\theta^\dagger)\| \gg \delta \quad \forall \theta^\dagger \in \Re^{d-e}, \tag{13.23}$$

where decisions are made one at a time for each of these dimensions using standard empirical diagnostics of stationarity. Even though our evidence is derived from these diagnostics it is important to note that they measure Markov chain *stability* rather than actual convergence and so the sum of each of these across dimensions is used to just *assert* convergence (convergence in total variation norm gives stationarity but the converse is not similarly guaranteed).

There is one *very* important distinction to be made here. The Markov chain $\mathfrak{M}(\boldsymbol{\theta}_t, t \geq 0)$ is assumed ergodic over all of Ω and is thus guaranteed to *eventually* converge across all of $\Re^d$. What we see by observing (13.22) and (13.23) at time point t is a lack of evidence to say that there is full dimensional convergence. Can a Markov chain operating in d dimensions be drawing from the true invariant distribution in e sub-dimensions but not in $d - e$ sub-dimensions? The standard empirical diagnostics in WinBUGS, CODA, and BOA (Brooks and Gelman, 1998a, 1998b; Geweke 1992; Heidelberger and Welch 1981a, 1981b; Raftery and Lewis, 1992, 1996, all described in Chapter 14), as well as others used in practice (Brooks, Dellaportas, and Roberts [1997] develop one based on the total variation norm discussed above), provide a series of parameter-by-parameter tests of *non*-convergence. Hence they indicate when a single dimension chain is sufficiently trending as to violate specific distributional assumptions that reflect stability, but they do assert convergence in the opposite case.

These diagnostics operate on marginal distributions individually since the output of the MCMC process is a set of *marginal* empirical draws. Unfortunately the total variation norm given above only shows us if the chain has converged simultaneously across every dimension, providing a strong disconnect between theoreticians who derive convergence properties for specific chains under specific circumstances and the masses who want to run simple empirical diagnostics in easy-to-use software environments. To see this disconnect more clearly, consider the right-hand-side of (13.19) written out with more detail:

$$\frac{1}{2}\sup_{\theta \in A} A \left| \int_{\theta_1} \cdots \int_{\theta_e} \int_{\theta_{e+1}} \cdots \int_{\theta_d} [f(\theta_1,\ldots,\theta_d)_t - \pi(\theta_1,\ldots,\theta_d)] \right.$$
$$\left. d\theta_1 \cdots d\theta_e d\theta_{e+1} \cdots d\theta_d \right|. \tag{13.24}$$

If $f()$ and $\pi()$ were able to be expressed as independent products (i.e., $f(x,y) = f(x)f(y)$) for the θ_i, then this would be a straightforward integration process. As stated this cannot be true for π since we are doing MCMC for this very reason. But what about $f()$? Consider the *actual* transition probability for the Metropolis-Hastings algorithm from $\boldsymbol{\theta}$ to $\boldsymbol{\theta}'$ from (10.28):

$$A(\boldsymbol{\theta}, \boldsymbol{\theta}') = \min \left\{ \frac{\pi(\boldsymbol{\theta}')g(\boldsymbol{\theta}|\boldsymbol{\theta}')}{f(\boldsymbol{\theta})g(\boldsymbol{\theta}'|\boldsymbol{\theta})}, 1 \right\} g(\boldsymbol{\theta}'|\boldsymbol{\theta}) + (1 - r(\boldsymbol{\theta}))\delta_{\boldsymbol{\theta}}(\boldsymbol{\theta}'), \tag{13.25}$$

where $g()$ is the proposal distribution,

$$r(\boldsymbol{\theta}) = \int \min \left\{ \frac{f(\boldsymbol{\theta}')g(\boldsymbol{\theta}|\boldsymbol{\theta}')}{f(\boldsymbol{\theta})g(\boldsymbol{\theta}'|\boldsymbol{\theta})}, 1 \right\} g(\boldsymbol{\theta}'|\boldsymbol{\theta})d\boldsymbol{\theta}',$$

and

$$\delta_{\boldsymbol{\theta}}(\boldsymbol{\theta}') = 1 \text{ if } \boldsymbol{\theta} = \boldsymbol{\theta}' \text{ and zero otherwise}$$

(the Dirac delta function). It is clear from looking at (13.25) that we cannot generally disentangle dimensions. In particular, note the conditionals that exist across $\boldsymbol{\theta}$ and $\boldsymbol{\theta}'$. What this means is that decisions to jump to a proposal point in **d**-space $(\Omega, \mathcal{F})$ are made based on the current position in every dimension for the Metropolis-Hastings algorithm. So if the chain has not converged in the ith dimension, $i \in [e + 1{:}d]$, its current placement effects the single acceptance ratio and therefore the probability of making a complete d-dimensional jump. And this is all under the assumption of ergodicity.

So now that we know that non-convergence in at least one dimension affects decisions to move in all dimensions, the natural question is how does this work? A Metropolis-Hastings chain dimension that has not converged is producing on average lower density contributions in the acceptance ratio. Therefore in cases where the conditionality on the current state is explicit (all general forms except the independence chain Metropolis-Hastings where jumping values are selected from a convenient form as in the random walk chain, but ignoring the current position completely: $g(\boldsymbol{\theta}'|\boldsymbol{\theta}) = f(\boldsymbol{\theta}'))$ it retards the mixing of the whole chain. Because the chain is ergodic, it is alpha mixing $(\sup_{A,B} |p(\boldsymbol{\theta}_t \in B, \boldsymbol{\theta}_0 \in A) - p(\boldsymbol{\theta}_t \in B)p(\boldsymbol{\theta}_0 \in A)|$ goes to zero in the limit) but inefficiently so (slowly) since non-convergence for the $d - e$

dimensions implies poorer mixing and greater distance between $p(\boldsymbol{\theta}_t \in B, \boldsymbol{\theta}_0 \in A)$ and $p(\boldsymbol{\theta}_t \in B)p(\boldsymbol{\theta}_0 \in A)$.

A chain that is completely in its stationary distribution mixes better (Robert and Casella 2004, Chapter 12). So even in the case where θ_i, the un-converged dimension here, is not a parent node in the model specified by π, there is a negative effect: a Markov chain that has not sufficiently mixed through the target distribution produces biased empirical summaries because collected chain values will be incomplete, having had insufficient time to fully explore the target.

13.6.3 Partial Convergence for the Gibbs Sampler

Robert and Richardson (1998) show that when a Markov chain, $\mathfrak{M}(\theta_t, t \geq 0)$, is derived from another Markov chain, $\mathfrak{M}(\phi_t, t \geq 0)$, by simulating from a distribution according to $\pi(\theta|\phi_t)$, the properties of the first chain inherit that of the conditional. Critically, this conditionality defines new sub-spaces of $(\Omega, \mathcal{F})$ with new measure properties.

For our purposes the important point is that if $\mathfrak{M}(\phi_t, t \geq 0)$ is geometrically ergodic, then $\mathfrak{M}(\theta_t, t \geq 0)$ is as well, which is easy to demonstrate using the data augmentation principle. The marginal distribution for the geometrically ergodic chain at time t is $\pi_t(\phi)$ with invariant distribution $\pi(\phi)$. We can now express the invariant distribution of θ in conditional terms: $\pi(\theta) = \int_\phi \pi(\theta|\phi)\pi(\phi)d\phi$, with the marginal distribution at time t: $\pi_t(\theta) = \int_\phi \pi(\theta|\phi)\pi_t(\phi)d\phi$. The conditional $\pi(\theta|\phi)$ appears in the second expression without reference to time since θ_t is simulated *at each step* from $\pi(\theta|\phi)$. These define the total variation norm for θ_t:

$$\|\pi_t(\theta) - \pi(\theta)\| = \frac{1}{2}\sup_{\theta \in A} A \left| \int_\theta \int_\phi \pi(\theta|\phi)\pi_t(\phi)d\phi d\theta - \int_\theta \int_\phi \pi(\theta|\phi)\pi(\phi)d\phi d\theta \right|$$

$$= \frac{1}{2}\sup_{\theta \in A} A \left| \int_\theta \int_\phi \pi(\theta|\phi) \left[\pi_t(\phi) - \pi(\phi) \right] d\phi d\theta \right|$$

$$\leq \|\pi_t(\phi) - \pi(\phi)\|, \tag{13.26}$$

where the inequality comes from $\pi(\theta|\phi) \leq 1$ by the integration of a probability function over the measure space for ϕ (the rate ρ carries through as well). Switching the order of integration comes from stated regularity conditions on probability functions. Note that this process is related to, but distinct from, so-called Rao-Blackwellization where intentional conditioning is imposed to reduce the variance of computed expectations or marginals (Casella and Robert 1996). The result in (13.26) is that a non-convergent dimension to the Gibbs sampler (θ here) "pushes" the others (ϕ here) away from stationarity as well, even if these pass an empirical diagnostic for convergence.

One utility of this result is that if we can intentionally augment a target chain with a simple form that is known to be geometrically ergodic, then we can *impose* this property even though we increase the dimension of $(\Omega, \mathcal{F})$ (Diaconis and Saloff-Coste 1993, Fill 1991).

Robert and Richardson (1998) point out that this is particularly useful when a target chain of unknown convergence characteristics is conditioned on a simple discrete Markov chain known to be geometrically ergodic with specific ρ and $m(\theta)$ (alternately m). Also, if the chain that is conditioned on is α-mixing, the target chain will be as well.

The big point comes from the structure of the Gibbs sampler (the default engine of BUGS). Since the kernel is an iteration of full conditionals, $\pi(\theta_j | \boldsymbol{\theta}_{-j})$ for $j = 1, \ldots, d$, then according to the logic just discussed, either the Gibbs sampler is geometrically ergodic in every dimension or it is not geometrically ergodic in any dimension. Importantly, since the sub-chains share the same geometric rate of convergence, ρ, then one should be cautious with empirical diagnostics since they provide evidence of *non-convergence* not evidence of *convergence* (see Asmussen, Glynn, and Thorisson [1992] for a detailed discussion on this point). Recall that at any given time t a Markov chain is either converged to its invariant distribution or it has not: there is a specific time when $\| f(\boldsymbol{\theta}_t) - \pi(\boldsymbol{\theta}) \| < \delta$ for some chosen δ, we just do not necessarily know the moment.

Suppose for a two-parameter Gibbs sampler θ_1 passes some empirical diagnostic and θ_2 does not. Since they share the same rate of convergence, for θ_1 to be in its invariant distribution while θ_2 is not means that you are testing the chain for convergence during the very small number of intervals where the differing results are due to probabilistic features of the chain or the test. Given the standard number iterations expected for MCMC work in the social sciences (generally tens or hundreds of thousands), the probability that you have stumbled up this exact interval is essentially zero. Conversely, the test fails for θ_2 because the Markov chain for this dimension is either not yet in stationarity, or is in stationarity but has failed to sufficiently explore the target distribution to produce a stable summary statistic for the chosen diagnostic. The latter condition only exists for a relatively short period of time, even with poor mixing. Moreover, the faster the convergence rate (i.e., geometric or uniform), the smaller the numerator in the calculation of this probability making it even less likely that the user caught the interval of intermediate results using the Gibbs samplers with listed properties above, where the size of this effect is notably a function of $m(\theta)$ (or m) and ρ. Therefore, for the Gibbs sampler, evidence of non-convergence in *any* dimension is evidence of non-convergence in *all* dimensions. So for users of the usual diagnostic packages, CODA and BOA, the standard for multidimensional convergence needs to be high. For a relatively small number of parameters, one would expect broad consensus across diagnostics. However, for large numbers of model parameters, we need to be aware that the diagnostics are built on formal hypothesis tests at selected α levels and therefore $1 - \alpha$ tests for large numbers of parameters will fail for about α proportion of dimensions, even in full convergence.

Note that this same logic applies to Metropolis-Hastings MCMC for parameters with conditions formed by hierarchies, which are a natural and common feature of Bayesian model specifications. This inheritance of convergence properties does not necessarily occur, however, for every parameter as in the perfectly symmetric case of the Gibbs sampler. It also is not reciprocal in that the conditions in a Bayesian hierarchical model flow downward from founder nodes to dependent nodes. Note that these conditionals result explicitly from the model rather than through algorithmic conditioning in the Gibbs sampler sense. In addition, parameters can be highly correlated without these structural relationships. The difficulty posed by all of these characteristics is that they generally slow the mixing of the chain, making convergence and full exploration more difficult.

13.7   Exercises

13.1   In Section 13.2, three conditions were given for $\mathcal{F}$ to be an associated field of Ω. Show that the first condition could be replaced with $\varnothing \in \mathcal{F}$ using properties of one of the other two conditions. Similarly, prove that the Kolmogorov axioms can be stated with respect to the probability of the null set or the probability of the complete set.

13.2   Given a Markov chain on Ω and two sub-states $A, B \in \Omega$, where all elements of B can be reached from A: A is called *essential for B* if all elements of A can also be reached from B, otherwise *inessential*. Show that if A is essential for B, then B is essential for A.

13.3   Suppose we have the probability space $(\Omega, \mathcal{F}, P)$, sometimes called a *triple*, and $A_1, A_2, \ldots, A_k \in \mathcal{F}$. Prove the finite sub-additive property that:

$$P\left(\bigcup_{i=1}^{k} A_i\right) \leq \sum_{i=1}^{k} p(A_i),$$

(Boole's Inequality).

13.4   Using the transition matrix from Example 10.1.2 in Chapter 10 (page 336),

$$K = \begin{bmatrix} \frac{1}{3} & 0 & \frac{1}{3} & \frac{1}{3} & 0 & 0 \\ 0 & \frac{1}{3} & 0 & 0 & \frac{1}{3} & \frac{1}{3} \\ \frac{1}{3} & \frac{1}{3} & \frac{1}{3} & 0 & 0 & 0 \\ 0 & 0 & 0 & \frac{1}{3} & \frac{1}{3} & \frac{1}{3} \\ \frac{1}{3} & \frac{1}{3} & 0 & 0 & \frac{1}{3} & 0 \\ 0 & 0 & \frac{1}{3} & \frac{1}{3} & 0 & \frac{1}{3} \end{bmatrix},$$

construct a directed graph showing the immediate paths between these six states. Does this reveal anything about moving from states that are not immediately connected.

13.5   Prove that uniform ergodicity gives a faster rate of convergence than geometric ergodicity and that geometric ergodicity gives a faster rate of convergence than ergodicity of degree 2.

13.6   Write a Metropolis-Hastings algorithm in R that is purposely transient using the data from Example 10.5 in Chapter 10 (page 361). Plot the first 500 iterations.

13.7   Construct a stochastic process without the Markovian property and construct deliberately non-homogeneous Markov chain.

13.8   A Markov chain operates on a k-cycle if it moves clockwise and counterclockwise on a loop of $1, 2, \ldots, k$ ordered values. If θ moves clockwise with probability p and counterclockwise with probability $1 - p$, then give the stationary distribution for $p = \frac{1}{2}$. Is this different for $p \neq \frac{1}{2}$?

13.9   Give an example of a maximally ψ-irreducible measure and one that it dominates.

13.10   One urn contains k black marbles and another urn contains k white marbles. At each iteration of a process one ball is uniformly randomly selected from each urn and exchanged between the urns. Define $\theta^{[t]}$ as the number of white balls in urn number 1 after the tth step. Obtain the following:

a. Prove that this is a Markov chain.

b. Derive the transition probabilities for $\theta^{[t]}$.

c. Produce the stationary distribution for $\theta^{[t]}$.

13.11   A single-dimension Markov chain, $\mathfrak{M}(\theta_t, t \geq 0)$, with transition matrix P and unique stationary distribution $\pi(\theta)$ is *time-reversible* iff:

$$\pi(\theta_i)p(\theta_i, \theta_j) = \pi(\theta_j)p(\theta_j, \theta_i).$$

for all θ_i and θ_j in Ω. Prove that the reversed chain, defined by $\mathfrak{M}(\theta_s, s \leq 0) = \mathfrak{M}(\theta_t, -t \geq 0)$, has the identical transition matrix.

13.12   Consider a Markov chain based on an AR(1) specification:

$$\theta^{[j+1]} = \varepsilon\theta^{[j]} + \epsilon^{[j]},$$

where $\epsilon^{[j]} \sim \mathcal{N}(0, 1)$, for all $j = 1, \ldots, J$ iterations, and $\varepsilon \in (0 : 1)$. See also Exercise 13 (page 139). Show that the stationary distribution of this Markov chain is $\mathcal{N}(0, 1/(1 - \varepsilon^2))$.

13.13   Meyn and Tweedie (1993, p.73) define an n-step *taboo probability* as:

$$_A P^n(x, B) = P_x(\Phi_n \in B, \tau_A \geq n), \qquad x \in \mathsf{X}, A, B \in B(\mathsf{X}),$$

meaning the probability that the Markov chain, Φ, transitions to set $\mathcal{B}$ in n steps avoiding (not hitting) set $\mathcal{A}$. Here X is a general state space with countably generated σ-field $B(\mathsf{X})$, and τ_A is the return time to $\mathcal{A}$ (their notation differs slightly from that in this chapter). Show that:

$$_A P^n(x, B) = \int_{\mathcal{A}^c} P(x, dy)_A P^{n-1}(y, B),$$

where $\mathcal{A}^c$ denotes the complementary set to $\mathcal{A}$.

13.14 For Exercise 12, implement the AR(1) Markov chain in R and graph the trajectory of the chain for three different values of ε in the same plot.

13.15 The Langevin algorithm is a Metropolis-Hastings variant where on each step a small increment is added to the proposal point in the direction of higher density. Show that making this increment an increment of the log gradient in the positive direction produces an ergodic Markov chain by preserving the detailed balance equation.

13.16 A transition matrix K is defined on the binary set $\{0, 1\}$ by:

$$K = \begin{bmatrix} 1-a & a \\ b & 1-b \end{bmatrix}$$

with $0 < a, b \leq 1$ and at least one of these less than one. Show that the stationary distribution of the Markov chain defined by this transition kernel is given by $p(0) = b/(a+b), p(1) = a/(a+b)$.

13.17 Consider a stochastic process, $\theta^{[t]}$ ($t \geq 0$), on the probability space $(\Omega, \mathcal{F}, P)$. If: $\mathcal{F}_t \subset \mathcal{F}_{t+1}$, $\theta^{[t]}$ is measurable on $\mathcal{F}$ with finite first moment, and with probability one $E[\theta^{[t+1]}|\mathcal{F}_t] = \theta^{[t]}$, then this process is called a *Martingale* (Billingsley 1995, p.458). Prove that Martingales do or do not have the Markovian property.

13.18 Replacing $E[\theta^{[t+1]}|\mathcal{F}_t] = \theta^{[t]}$ in the last exercise with $E[\theta^{[t+1]}|\mathcal{F}_t] \geq \theta^{[t]}$ produces a *supermartingale*, and replacing with $E[\theta^{[t+1]}|\mathcal{F}_t] \leq \theta^{[t]}$ produces a *submartingale*. If $t \geq 0$, show that $\theta^{[t]}$ is a supermartingale if and only if $-\theta^{[t]}$ is a submartingale, and vice versa. , and

13.19 Prove that for a Markov chain that is positive Harris recurrent, $\exists \sigma-$finite probability measure, $\mathfrak{P}$, on S such that for an irreducible Markov chain, X_t, at time t, $p(X_n \in A) = 1$, $\forall A \in S$ where $\mathfrak{P}(A) > 0$, and aperiodic is also α-mixing,

$$\alpha(t) = \sup_{A,B} \left| p(\boldsymbol{\theta}_t \in B, \boldsymbol{\theta}_0 \in A) - p(\boldsymbol{\theta}_t \in B)p(\boldsymbol{\theta}_0 \in A) \right| \xrightarrow[t \to \infty]{} 0.$$

13.20 A random walk Markov chain on $\mathcal{I}$ has the following transition kernel with fixed $0 < p < 1$:

$$K(i, j) = \begin{cases} p & \text{if } j - 1 = i \\ 1 - p & \text{if } j + 1 = i \\ 0 & \text{otherwise.} \end{cases}$$

Show that this Markov chain is irreducible with period 2. Suppose now that p is randomly drawn on each iteration of the Markov chain from the distribution $\sim \mathcal{U}(0, 1)$ (exclusive of the endpoints). Is this still an irreducible Markov chain?

Chapter 14

Utilitarian Markov Chain Monte Carlo

14.1 Objectives

This chapter has several rather practical purposes related to applied MCMC work: to introduce formal convergence diagnostic techniques, to provide tools to improve mixing and coverage, and to note a number of challenges that are routinely encountered. This is a stark contrast to the last chapter, which was concerned with theoretical properties of Markov chains and Markov chain Monte Carlo. Since applied work is generally done computationally through the convenient programs BUGS (in any of the versions) and JAGS, or by writing source code in R, C, or even Fortran, practical considerations are important to getting reliable inferences from chain values. Most of the concern centers on assessing convergence, but the speed of the sampler, and its ability to thoroughly explore the sample space are also important issues to be concerned with. This chapter also describes the two very similar R packages for analyzing MCMC output and evaluating convergence: BOA and CODA. These are merely convenient functional routines, and users will often want to go beyond their capabilities, particularly in graphics. However, the purpose here is mainly to understand the key workings of these tools rather than to function as a detailed description of the syntax of software. See Albert (2009) or Ntzoufras (2009) for recent book-length works with very detailed R and BUGS code description.

As estimation with MCMC tools becomes more common, mastery of computational mechanical challenges grows in importance. It is fundamentally import to remind ourselves that MCMC estimation is not a cookbook procedure and that it is necessary to pay attention to convergence issues. There are two critical and practical challenges in assessing MCMC reliability:

▷ For any Markov chain at some given time t, there is no absolute assurance that this chain is currently in its stationary (target) distribution.

▷ There is no way to guarantee that a Markov chain will explore all of the areas of the target distribution in a finite run time.

The bulk of the tools in this chapter are a means of adding to our confidence that a specific application has addressed these two concerns. Gelman (1996) adds three related standard problems to also consider: an inappropriately specified model, errors in programming the

Markov chain (the achieved stationary distribution of the chain may not be the desired target distribution), and *slow* convergence (distinct from mixing through the posterior). The latter can be particularly troublesome even for experienced researchers in this area as it is often difficult to anticipate problems like the chain getting stuck in low-density, high-dimension regions for long periods of time (see the litany of worries in Guan, Fleißner, and Joyce [2006]). Essentially these concerns boil down to three worries: setting up the parameters of the process correctly, ensuring good mixing through the sample space, and making a reliable decision about convergence at some point in time. These are the primary concerns of this chapter.

This preliminary discussion seems to imply that the use of MCMC algorithms is fraught with danger and despair (the WinBUGS webpage and manual contain a warning message that would impress tobacco regulators). However, we know from Chapter 13 that all ergodic Markov chains are guaranteed to eventually converge (see also Chan [1993] and Polson [1996]), and it is often quite easy to determine if a given chain has *not* converged. Consequently, convergence diagnostics have moved from fairly *ad hoc* methods (Gelfand and Smith 1990) to the reasonably sophisticated statistical tests outlined in this chapter. In addition, progressively faster processors mean that the time-dependent issues are guaranteed to diminish in the future. Furthermore, in nearly all occasions with typical generalized linear models and non-exotic priors, the Markov chain converges quickly with obvious evidence. Nonetheless, users of MCMC for Bayesian estimation should use all reasonably caution with respect to controllable features.

14.2   Practical Considerations and Admonitions

This section covers a few practical considerations in implementing MCMC algorithms. There are several critical design questions that must be answered before actually setting up the chain and running it, whether one is using a packaged resource or writing original code. These include such decisions as: the determination of where to start the chain, judging how long to "burn-in" the chain before recording values for inference, determining whether to "thin" the chain values by discarding at intervals, and setting various software implementation parameters.

14.2.1   Starting Points

Starting points as an integrated part of the simulation specification is an under-studied aspect, except perhaps in the case of one particular convergence diagnostic (Gelman and Rubin 1992a). Generally it is best to try several starting points in the state space and observe whether they lead to noticeably different descriptions of the posteriors. This is surely a sign of non-convergence of the Markov chain, although certainly not a systematic test. Unfortunately the reverse does not hold: it is not true that if one starts several Markov chains in different places in the state space and they congregate for a time in the same region that this is the region that characterizes the stationary distribution. It could be that all of the chains are seduced by the same local maxima, and will for a time mix around in its local region.

Overdispersing the starting points relative to the expected modal point is likely to provide a useful assessment (Gelman and Rubin 1992a, 1992b). We will look in detail at Gelman and Rubin's formal diagnostic in Section 14.3.3.2, but for now it is warranted to talk about strategies for determining such overdispersed points. If one can determine the mode with reasonable certainty, either by using the EM algorithm, a grid search, or some other technique (possibly analytically), then it is relatively simple to spread starting points around it at some distance. If this is not possible, or perhaps excessively complicated in high dimensions, then it is often straightforward to carefully spread starting points widely throughout the sample space.

Sometimes starting points of *theoretical interest* are available. It might be the case that a starting point can be assigned to values associated with other studies, subject-matter expertise, previous work with the same data, or even the modal point of the associated likelihood function. Often such points are close to the high density region for the posterior being explored. These strategies, of course, do not involve overdispersion and the use of the Gelman and Rubin convergence diagnostic.

Some researchers randomly distribute starting points through the state space with the idea that if little is known before the estimation process, these will at least be reasonably overdispersed points. Overdispersed starting points relative to the central region of the distribution mean that if the chains coalesce into the same region anyway, it is some evidence that they have converged to the stationary distribution.

14.2.2 Thinning the Chain

It is sometimes the case that with very long simulations, storage of the observed chain values on the computer becomes an issue, although this need has diminished substantially over time. The need for large storage files results from running the chain for extended periods, running multiple parallel chains, or monitoring a large number of parameters. Social science models, however, rarely call for hundreds or thousands of model parameters as one might see in statistical genetics, though it may be necessary in future research as the bounds between disciplines recede. More often long runs are needed in the presence of: high autocorrelation in the iterations, slow convergence of the chain to its limiting distribution combined with many parallel simultaneous runs of the chain, possibly combined with relatively high dimensionality of the model. These are addressed in this chapter. The problem introduced is that disk storage may be strained by these demands. Furthermore, many of the convergence diagnostics described in this chapter slow down considerably with very large matrix sizes. Some researchers even report software failures due to excessively large chains (Rathbun and Black 2006). However, as computer storage capabilities continue to increase rapidly in size (including cloud storage), while decreasing in cost, these problems will decline in importance.

The idea of thinning the chain is to run the chain normally but record only every kth value of the chain, k some small positive integer, thus reducing the storage demands while still preserving the integrity of the Markovian process. Importantly, note that thinning does *not* in any way improve the quality of the estimate (suggested presumably but erroneously as a way to increase the independence of the final evaluated values [Geyer 1992]), speed up the chain, or help in convergence and mixing. Instead, it is purely a device for dealing with possibly limited computer resources. In fact, the quality of the subsequent estimation always suffers because the resulting variance estimate will be higher for a given run-time (MacEachern and Berliner 1994), albeit to varying degrees depending on the length of the chain and the nature of the model. Also since the variance estimate is wrong it confounds the estimate of the serial correlation of the chain for the purpose of understanding the mixing properties. However, sometimes this variance issue is a trivial concern and thinning remains a convenience.

Given the trade-offs between storage and accuracy as well as diagnostic ability, what value of k is appropriate in a given application? The greater the amount of thinning, the more potentially important information is lost, suggesting caution. Conversely, prior to assumed convergence, thinning is irrelevant *inferentially* and may therefore be useful, provided it gives sufficient information to the diagnostics used. Many researchers pick every fourth, fifth, or tenth iteration to save, but for completely arbitrary reasons. Occasionally applications will thin to every 30th, 50th, or even 100th iteration, but this tends to decrease chain efficiency more substantially.

14.2.3   The Burn-In Period

First, one must decide the length of the burn-in period, the beginning set of runs that are discarded under the assumption that they represent pre-convergence values and are therefore not equivalent draws of the desired limiting distribution. The slower the chain is to converge, the more careful one should be about the burn-in period. Usually this involves cautiously extending the length of the chain, but this chapter also presents some customized tools for speeding up this process. Unfortunately, even starting the chain right in the area of highest density does not guarantee that the burn-in period is unimportant as it will still take the Markov chain some time to "forget" its starting region, take some time to settle into the stationary distribution, and then need some further time to fully explore the target distribution.

There is no systematic, universal, guaranteed way to calculate the length of the burn-in period, and considerable work on convergence diagnostics has been done to make specific recommendations and identify helpful tests. Raftery and Lewis (1992, 1996) suggest a running diagnostic for the length of the burn-in period, which starts with the analysis of an initial run. The idea is to solve for the number of iterations required to estimate some quantile of interest within an acceptable range of accuracy, at a specified probability level. The procedure is based on conventional normal distribution theory and implemented in the R package `mcgibbsit`, as well as in the standard diagnostic packages `BOA` and `CODA`. Unfortunately, this procedure does not always work, but it often does provide good approximate guidance (Robert and Cellier 1998, Brooks and Roberts 1997). See Section 14.3.3.4 for an extended description.

14.3 Assessing Convergence of Markov Chains

As shown in Chapter 10, the empirical results from a given MCMC analysis are not deemed reliable until the chain has reached its stationary distribution and has time to sufficiently mix throughout. Until $\mathfrak{M}(\boldsymbol{\theta}_t, t \geq 0)$ converges at time t^* (i.e. $\|f(\boldsymbol{\theta}_t^*) - \pi(\boldsymbol{\theta})\|$ is negligible), it is not possible to rely upon the effect of any variant of the central limit theorem. Therefore the single greatest risk in applied MCMC work is that the user will misjudge the required length of the burn-in period and assert convergence before the Markov chain has actually reached the stationary distribution.

Unfortunately, some convergence problems come directly from the model specification and it may not be obvious when the MCMC process fails. Natarajan and McCulloch (1995) found that it is possible to have a proper form for every full conditional distribution in a Gibbs sampler and still specify a joint posterior that is improper. Hobert and Casella (1998) in a seminal paper demonstrate that resulting improper posteriors are useless for purposes of inference. Obviously many of these issues emanate from the (sometimes desired) specification of improper priors, which are typically not a problem in simple, stylized models but can present difficult algorithmic challenges in fully-developed social science specifications. One common alternative (Chapter 4) is to use highly diffuse but proper priors. This is usually an effective alternative but can sometimes lead to slow convergence of the Markov chain, and this should be checked. Furthermore, if the model is truly non-identified in the classic sense (Manski 1995), then the MCMC estimation process is going to fail to provide useful results, even if it appears to converge (which is highly unlikely).

There are basically three approaches to determining convergence for Markov chains: assessing the theoretical and mathematical properties of particular Markov chains, diagnosing summary statistics from in-progress models, and avoiding the issue altogether with perfect sampling, which uses the idea of "coupling from the past" to produce a sample from the exact stationary distribution (Propp and Wilson 1996). The emphasis in this chapter is on the second approach, whereas perfect sampling is described in Chapter 15, Section 15.3.

The first approach is to study the mathematical properties of individual chain transition kernels, perhaps placing restrictions on data or analysis, and then determining in advance the total variation distance to the target distribution with some specified tolerance. MCMC algorithms on discrete state space converge at a rate related to how close the second largest eigenvalue of the transition matrix for discrete state spaces is to one. This is an elegant theoretical result that is often difficult to apply to actual sampling output.

The purely theoretical approach has several disadvantages: it often becomes inordinately complex mathematically, the bounds can be "weak" meaning sufficiently wide as to not be useful in practice (i.e., lacking reasonable guidance about when to stop sampling), some of the approaches are restricted to unrealistically simple models, and generally the calculations are model-specific. Nonetheless, a great deal of important work has been done here: a Gibbs sampler on variance components models for large data size and number of parameters (Rosenthal 1995b), transition kernels that satisfy certain minorization conditions on the state space (Rosenthal 1995a), log-concave functions over a discretized sample space (Frieze, Kannan, and Polson 1994), required properties of the eigenvalues (Frigessi *et al.* 1993, Ingrassia 1994), data augmentation on a finite sample space (Rosenthal 1993), the special case of Gaussian forms (Amit 1991, 1996), the conditions and rate of convergence for standard algorithms (Roberts and Polson 1994; Roberts and Smith 1994); and more generally, the models described theoretically in Meyn and Tweedie (1993, 1994). Madras and Sezer (2010) use Steinsaltz's drift functions for obtaining bounds on the rate of convergence on a general state space, Fort, Moulines, and Priouret (2011) get convergence bounds for new adaptive and interacting Markov chains, and Roberts and Rosenthal (2011) extend their work on independence samplers from geometrically ergodic Markov chains to non-geometrically ergodic Markov chains.

The second convergence assessment method involves monitoring the performance of the chain as part of the estimation process and making an often subjective determination about when to stop the chain. Also related to this process are efforts to "accelerate" convergence through various particularistic properties. There are quite a few of these techniques and several of the most popular and straightforward will be discussed here. For additional discussion of these tools, see the review essays by Brooks (1998a), Brooks and Roberts (1999), Cowles and Carlin (1996), Mengersen, Robert, and Guihenneuc-Jouyaux (1999), as well as Gelfand and Sahu (1994).

The general process is to run the chain for some conservatively large number of iterations, and then dispose of a conservatively large proportion of the early values (say half), and then to run all of the standard empirical diagnostics. If there is any indication from these diagnostics that the Markov chain is not in its stationary distribution then run the chain for an additional conservatively long period and retest. This process is iterated until the researcher is convinced that there are no long concerns from the diagnostics. There is no "magic" number of iterations that makes the process conservative, but as computers get faster something on the order of 10^4 has become something on the order of 10^5 or higher. Of course more complex models require more iterations so such numbers should not be considered as a "rule-of-thumb" (that phrase usually presages bad advice in statistics).

Finally, after one is satisfied with the length of the chain, and have moved on to writing up the results, continue the iterating chain in the background or on another machine for a very long time as an "insurance run" on the order of 10^6 and later summarize the last 10^4 as additional verification. Almost certainly the numerical values will be the same (within simulation error), but this provides additional faith in the results.

It is essential to remember that the convergence diagnostics described below, as well as others in the literature, are actually indicators of *nonconvergence*. That is, failing to find evidence of nonconvergence with these procedures is just that; it is not direct evidence of convergence. The careful practitioner should treat encouraging results from one test with continued skepticism and be willing to use multiple diagnostics on any single Markov chain, any one of which can provide sufficient evidence of failure. It is also important to remember that this is the process of evaluating the integrity of the MCMC process; it does not evaluate the overall quality of the specified statistical model.

The following subsections outline the use of popular MCMC diagnostics including those provided by `BOA` and `CODA`, and some are also integrated directly into the `WinBUGS` environment. The `R` package `superdiag` (Tsai and Gill 2012) calls all of the conventional convergence diagnostics used in typical MCMC output assessment in one convenient `R` statement. Each of these diagnostics have limitations, and therefore it is recommended that cautious users evaluate Markov chain output with each of these. Here these convergence diagnostics are illustrated with two real-data examples where one produces clean and obvious convergence assessments while the other remains problematical. In this way readers can see both positive and negative outcomes for comparison with their own results.

■ **Example 14.1: Tobit Model for Death Penalty Support.** Norrander (2000) uses Tobit models (Tobin 1958) to look at social and political influences on U.S. state decisions to impose the death penalty since the Supreme Court ruled the practice constitutional in *Furman v. Georgia* 1972. The research question is whether the ideological, racial, and religious makeup, political culture, and urbanization are causal effects for state-level death sentences from 1993 to 1995. Norrander posits a causal model whereby public opinion centrally, influenced by past policies and demographic factors, determines death penalty rates by legitimating the practice over time. The Tobit model to account for censoring is appropriate here because 15 states did not have capital punishment provisions on the books in the studied period causing the actual effect of public opinion on death penalty rates to be substantively missing. If these states had the legal ability to impose death penalty sentences, then it would be possible to observe whether there exists a relationship between the explanatory variables and the count. For instance, the death penalty in murder cases in Hawaii is recorded as zero, but is unlikely to actually be zero if observable. In addition, these data are also truncated at zero since states cannot impose a negative number of death penalty sentences.

Define now terms consistent with the discussion in Amemiya (1985, Chapter 10). If

$\mathbf{z}$ is a latent outcome variable in this context with the assumptions $\mathbf{z} = \mathbf{x}\boldsymbol{\beta} + \boldsymbol{\eta}$ and $z_i \sim \mathcal{N}(\mathbf{x}\boldsymbol{\beta}, \sigma^2)$, then the observed outcome variable is produced according to: $y_i = z_i$ if $z_i > 0$, and $y_i = 0$, if $z_i \leq 0$. The likelihood function is then:

$$L(\boldsymbol{\beta}, \sigma^2 | \mathbf{y}, \mathbf{X}) = \prod_{y_i=0} \left[1 - \Phi\left(\frac{x_i\beta}{\sigma}\right) \right] \prod_{y_i>0} (\sigma^{-1}) \exp\left[-\frac{1}{2\sigma^2}(y_i - x_i\beta)^2 \right]. \quad (14.1)$$

Chib (1992) introduces a blocked Gibbs sampling estimation process for this model using data augmentation, Albert and Chib (1993) extend this generally to discrete choice outcomes, and Chib and Greenberg (1998) focus on multivariate probit. This is a quite natural approach since augmentation (Chapter 10, Section 10.6) can be done with the computationally convenient latent variable $\mathbf{z}$. Furthermore, a flexible parameterization for the priors is given by Gawande (1998):

$$\boldsymbol{\beta}|\sigma^2 \sim \mathcal{N}(\boldsymbol{\beta}_0, \mathbf{I}\sigma^2 B_0^{-1}) \qquad \sigma^2 \sim \mathcal{IG}\left(\frac{\gamma_0}{2}, \frac{\gamma_1}{2}\right) \quad (14.2)$$

with vector hyperparameter $\boldsymbol{\beta}_0$, scalar hyperparameters B_0, $\gamma_0 > 2$, $\gamma_1 > 0$, and appropriately sized identity matrix $\mathbf{I}$. Substantial prior flexibility can be achieved with varied levels of these parameters, although values far from those implied by the data will make the algorithm run very slowly. The full conditional distributions for Gibbs sampling are given for the $\boldsymbol{\beta}$ block, σ^2, and the individual $z_i|y_i = 0$ as:

$$\boldsymbol{\beta}|\sigma^2, \mathbf{z}, \mathbf{y}, \mathbf{X} \sim \mathcal{N}\bigg((B_0 + \mathbf{X}'\mathbf{X})^{-1})(\boldsymbol{\beta}_0 B_0 + \mathbf{X}'\mathbf{z}),$$

$$(\sigma^{-2}B_0 + \sigma^{-2}\mathbf{X}'\mathbf{X})^{-1}) \bigg)$$

$$\sigma^2|\boldsymbol{\beta}, \mathbf{z}, \mathbf{y}, \mathbf{X} \sim \mathcal{IG}\left(\frac{\gamma_0 + n}{2}, \frac{\gamma_1 + (\mathbf{z} - \mathbf{X}\boldsymbol{\beta})'(\mathbf{z} - \mathbf{X}\boldsymbol{\beta})}{2} \right)$$

$$z_i|y_i = 0, \boldsymbol{\beta}, \sigma, \mathbf{X} \sim \mathcal{TN}(\mathbf{X}\boldsymbol{\beta}, \sigma^2)I_{(-\infty,0)}, \quad (14.3)$$

where $\mathcal{TN}()$ denotes the truncated normal and the indicator function $I_{(-\infty,0)}$ provides the bounds of truncation. The results from this model are sensitive to values of B_0, and this is why it is common to see very diffuse priors in this specification.

TABLE 14.1: POSTERIOR, TOBIT MODEL

	Mean	Std. Error	Median	95% HPD Interval
Intercept	-14.5451	3.6721	-14.5019	[-21.7661 : -7.3500]
Past Rates	171.1460	8.0482	171.1385	[155.2004 : 186.6079]
Political Culture	0.3461	0.1452	0.3438	[0.0596 : 0.6216]
Current Opinion	3.9738	1.0667	3.9632	[1.8575 : 6.0223]
Ideology	3.1423	1.1107	3.1436	[0.9728 : 5.3146]
Murder Rate	0.0088	0.0802	0.0095	[-0.1567 : 0.1590]

To extend Norrander's model, we add two additional explanatory variables: state average ideology and the state-level murder rate. A Gibbs sampler code in R is applied using the full conditional distributions given above for β, σ^2, and the z_i using estimates from the MLE as starting points (see the **Computational Addendum**). The prior parameters for the inverse gamma specification are stipulated to provide a relatively diffuse form: $\gamma_0 = 300$, $\gamma_1 = 100$, $B_0 = 0.02$, and $\beta = 0$. Making the gamma form less diffuse than the one specified here leads to a poor mixing chain as generated values in the truncation range become rare. The chain is run for $50,000$ iterations with the first $40,000$ values discarded as burn-in. The marginal results summarized in Table 14.1 do not generally contradict the original multiple analyses in Norrander (2000). We will see through the course of this chapter whether such results should be trusted.

TABLE 14.2: PROPORTIONAL CHANGES IN EASTERN EUROPEAN MILITARIES, 1948-1983

	Yugo.	Alb.	Bulg.	Czec.	GDR	Hung.	Poland	Rum.	USSR
1949	0.000	0.083	0.166	0.000	1.000	0.571	0.250	0.006	0.241
1950	0.000	-0.077	0.142	0.000	0.833	0.909	0.286	0.305	0.194
1951	0.498	-0.083	-0.043	0.000	0.864	0.476	0.220	0.234	0.163
1952	0.000	0.109	-0.050	0.000	0.244	0.097	0.125	0.004	0.160
1953	0.000	-0.131	-0.016	0.000	0.255	0.065	0.000	0.004	0.000
1954	0.000	-0.226	0.139	0.000	0.266	0.066	-0.333	0.004	0.000
1955	0.008	-0.049	0.000	0.000	0.296	0.057	0.000	0.000	0.000
1956	0.000	-0.051	-0.103	0.000	0.038	0.054	0.000	-0.143	-0.121
1957	0.000	-0.108	0.005	0.000	0.037	-0.977	0.000	-0.167	-0.118
1958	0.000	-0.061	-0.130	0.000	-0.124	5.000	0.000	0.057	-0.133
1959	0.000	-0.129	-0.144	0.212	-0.131	0.833	-0.333	0.054	-0.077
1960	0.000	0.000	-0.175	0.000	0.291	0.455	0.000	0.051	0.000
1961	0.032	0.037	0.017	-0.182	-0.099	0.438	0.275	-0.022	-0.167
1962	-0.102	0.071	0.125	-0.254	-0.150	0.043	0.008	0.369	0.200
1963	-0.114	0.167	0.096	0.000	0.365	0.125	0.000	-0.056	-0.083
1964	-0.089	0.086	0.115	0.270	0.034	0.030	0.058	-0.017	-0.039
1965	-0.108	0.000	0.012	0.000	0.017	0.036	0.018	-0.085	-0.076
1966	0.146	0.368	0.024	0.085	0.574	0.000	0.173	-0.027	0.159
1967	-0.085	-0.019	0.000	0.039	0.026	-0.049	-0.031	-0.112	0.022
1968	-0.077	0.000	0.012	0.000	-0.005	0.000	0.013	0.000	0.000
1969	-0.008	-0.020	-0.012	0.000	-0.036	-0.058	-0.031	-0.135	0.023
1970	0.084	0.080	-0.029	-0.234	0.069	0.062	-0.071	-0.062	-0.004
1971	-0.105	-0.222	-0.096	-0.064	-0.356	-0.270	-0.059	-0.116	-0.044
1972	0.000	-0.167	0.000	0.000	0.000	0.000	0.000	0.125	0.000
1973	0.043	0.086	0.000	0.000	0.000	0.000	0.037	-0.056	0.015
1974	-0.042	0.000	0.000	0.053	0.154	0.000	0.071	0.000	0.029
1975	0.000	0.000	0.000	0.000	-0.067	0.100	0.000	0.000	0.014
1976	0.087	0.237	0.133	-0.100	0.143	-0.091	0.000	0.059	0.020
1977	0.040	-0.043	-0.118	0.000	0.000	0.000	0.033	0.000	0.008
1978	0.038	-0.089	0.000	0.056	0.000	0.100	0.000	0.000	-0.011
1979	-0.037	0.049	0.000	0.000	0.000	-0.091	0.032	0.000	0.005
1980	0.000	-0.047	0.000	0.053	0.013	-0.070	0.000	0.028	-0.025
1981	-0.027	0.049	-0.007	-0.030	0.031	0.086	0.000	0.000	0.028
1982	-0.008	0.000	-0.007	0.015	-0.006	0.050	0.000	-0.022	0.011
1983	-0.044	-0.070	0.095	0.041	0.006	-0.009	0.063	0.050	-0.043

Proportional Change in Military Personnel

■ **Example 14.2: A Normal-Hierarchical Model of Cold War Military Personnel** These data describe changes in military personnel for seven Warsaw Pact countries plus two (Yugoslavia and Albania) during the period from 1948 to 1983, an interval that covers the height of the Cold War. These are collected by Faber (1989, ICPSR-9273) for 78 countries total, from which these 9 are taken, and include covariates for military, social, and economic conditions.

The model is a normal-gamma hierarchy with several levels and somewhat vague normal and inverse gamma hyperprior specifications. While there is a good argument to treat these data as a time-series, we will not do that here in order to produce illustrative results from a diagnostic perspective. The inverse gamma specification is articulated from the gamma statements through the BUGS convention of specifying precisions instead of variances in normal specifications. The motivation for the hierarchy is that the context level recognizes national differences from economic and cultural factors, while the primary level recognizes the political influence that comes from Warsaw Pact membership as well as the obvious pressure exerted by the USSR.

The (possibly mis-specified) model is a hierarchy of normals according to:

$$
\begin{aligned}
Y_{ij} &\sim \mathcal{N}(\alpha_i + \beta_{1i}(X_j) + \beta_{2i}(X_j^2) + \beta_{3i}\cos(X_j), \tau_c) \\
\alpha_i &\sim \mathcal{N}(\alpha_\mu, \alpha_\tau) && \alpha_\mu \sim \mathcal{N}(1, 0.1) && \alpha_\tau \sim \mathcal{G}(1, 0.1) \\
\beta_{1i} &\sim \mathcal{N}(\beta_{1,\mu}, \beta_{1,\tau}) && \beta_{1,\mu} \sim \mathcal{N}(0, 0.1) && \beta_{1,\tau} \sim \mathcal{G}(1, 0.1) \\
\beta_{2i} &\sim \mathcal{N}(\beta_{2,\mu}, \beta_{2,\tau}) && \beta_{2,\mu} \sim \mathcal{N}(0, 0.1) && \beta_{2,\tau} \sim \mathcal{G}(1, 0.1) \\
\beta_{3i} &\sim \mathcal{N}(\beta_{3,\mu}, \beta_{3,\tau}) && \beta_{3,\mu} \sim \mathcal{N}(0, 0.1) && \beta_{3,\tau} \sim \mathcal{G}(1, 0.1) \\
\tau_c &\sim \mathcal{G}(1, 0.1),
\end{aligned}
\tag{14.4}
$$

where the Y_{ij} are proportional changes in military personnel for country i at time period j and X_j is the index of the year (rows in the table). In addition to the linear treatment of the years, there are squared and cosine terms to account for systematic fluctuations over time. While the posterior distribution of the linear coefficient, β_1 is of primary interest, these secondary coefficients may reveal additional structures in the data. The τ terms and normal variances here conform to the BUGS requirement, mentioned above, of specifying precisions instead of variances. It should be clear from the hyperparameters that the prior structure is given proper distributions in this setup. Interestingly, the WinBUGS examples routinely assign very small parameter values for the gamma distribution priors ($\alpha = 0.001, \beta = 0.001$) on precisions, which puts prior density on gigantic values for the variance that exceed the typical posterior ranges for reasonable models. This practice should be replaced with other forms such as a gamma with $\alpha = 1$ and small β (Gill 2010) or a (folded) half-t (Gelman 2006).

The JAGS code for this example is given in the **Computational Addendum** for this chapter, and the data are given here (starting at 1949 since the variable has been changed from absolute numbers to annual proportional change). After compilation in JAGS, 150,000 iterations of the Markov chain are run and discarded. Monitoring is turned on for all nodes and then an additional 350,000 iterations are run. This Markov chain turns out to be relatively slow mixing and there are several lurking problems. The posterior results are summarized in Table 14.3. Clearly these results are mixed evidence of the value of the specification given in (14.4) since the standard errors indicate rather diffuse posterior forms. Interestingly, the quadratic term is the most reliable predictor, although it weakens in the absence of the linear and trigonometric contributions.

TABLE 14.3: POSTERIOR, MILITARY PERSONNEL MODEL

	Mean	Std. Error	Median	95% HPD Interval
α_μ	0.7404	3.3585	0.9814	[-5.8422 : 7.3229]
α_τ	10.0155	9.9068	7.0303	[-9.4019 : 29.4328]
$\beta_{1,\mu}$	0.0965	0.4740	0.0983	[-0.8326 : 1.0256]
$\beta_{1,\tau}$	13.3070	10.8181	10.3000	[-7.8965 : 34.5105]
$\beta_{2,\mu}$	2.0737	0.1031	2.0643	[1.8717 : 2.2758]
$\beta_{2,\tau}$	24.6509	6.9466	23.9347	[11.0357 : 38.2662]
$\beta_{3,\mu}$	-0.0011	0.0130	-0.0011	[-0.0266 : 0.0243]
$\beta_{3,\tau}$	180.0586	42.5261	176.7230	[96.7075 : 263.4097]
τ_c	0.0308	0.0038	0.0307	[0.0233 : 0.0383]

14.3.1 Autocorrelation

High correlation *between* the parameters of a chain tends to produce slow convergence, whereas high correlation within a single parameter (autocorrelation) chain leads to slow mixing and possibly individual *nonconvergence* to the limiting distribution because the chain will tend to explore less space in finite time. This is a problem since, of course, all chains are run in finite time. Furthermore, these are obviously interrelated problems for most specifications.

In addition to the empirical mean and standard deviation of the chain values (the primary inferential quantities of interest), the output from CODA gives the so-called "naïve" standard error *of the mean* (NaiveSE, which is the square root of $\sqrt{\text{sample variance}}/\sqrt{n}$, and the standard time-series adjusted standard error *of the mean* (TimeseriesSE), which is $\sqrt{\text{spectral density var}}/\sqrt{n}$ = asymptotic SE. The naïve standard error gives an overly-optimistic view (smaller) of the posterior dispersion since it ignores serial correlation that exists by definition with Markov chains. The time-series standard error is produced from an estimate of the spectral density at zero using binned chain values.

One method works around the autocorrelation problem by looking at the means of batches of the parameter. If the batch size is large enough, the batch means should be approximately uncorrelated and the normal formula for computing the standard error should work. More specifically, a quantity of interest from the simulated values, $h(\theta)$, is calculated empirically from a series $i = 1, 2, \ldots, n$:

$$\hat{h}(\theta) = \frac{1}{n} \sum_{i=1}^{n} h(\theta_i) \tag{14.5}$$

with the associated Monte Carlo variance:

$$\text{Var}[h(\theta)] = \frac{1}{n} \left(\frac{1}{n-1} \sum_{i=1}^{n} (h(\theta_i) - \hat{h}(\theta))^2 \right), \tag{14.6}$$

as described in Chapter 9. The problem with applying this to MCMC samples is that it ignores the serial nature of their generation. Instead consider specifically incorporating autocorrelation in the variance calculation. Define first for a lag t, the autocovariance:

$$\rho_t = \text{Cov} \left[h(\boldsymbol{\theta}_i), h(\boldsymbol{\theta}_{i-t}) \right]. \tag{14.7}$$

Chen *et al.* (2000, 87) show that the expected variance from MCMC samples, knowing ρ_t and σ^2 (the true variance of $h(\theta)$) is:

$$E[\sigma^2] = \sigma^2 \left[1 - \frac{2}{n-1} \sum_{t=1}^{n-1} \left(1 - \frac{t}{n} \right) \rho_t \right], \tag{14.8}$$

which shows the variance difference produced by serially dependent iterates.

In practice the lagged autocovariance is measured by:

$$\hat{\rho}_{nt} = \frac{1}{n} \sum_{t+1}^{n} \left[h(\theta_i) - \hat{h}(\theta) \right] \left[h(\theta_{i-t}) - \hat{h}(\theta) \right] \tag{14.9}$$

(Priestley 1981, p.323).

In analyzing Markov chain autocorrelation, it is helpful to identify specific lags here in order to calculate the long-run trends in correlation, and in particular whether they decrease with increasing lags. Diagnostically, though, it is rarely necessary to look beyond 30 to 50 lags. Recall that for a series of length n, the lag k autocorrelation is the sum of $n - k$ correlations according to: $\rho_k = \sum_{i=1}^{n-k}(x_i - \bar{x})(x_{i+k} - \bar{x}) / \sum_{i=1}^{n}(x_i - \bar{x})^2$.

Fortunately, both BOA and CODA give straightforward diagnostic summaries and graphical displays of autocorrelation with chains and cross-correlation matrices. The four default lag values of the correlations for the Tobit model of death penalty attitudes is given in Table 14.4. Notice that the within-chain correlations decline sharply with increasing lag, indicating no problem autocorrelation in any of these dimensions of the Markov chain. However, there are some large cross-correlations that might cause the chain to mix poorly due to constrained high-density space between two parameters. The graphical diagnostics will reveal if this is really a problem in mixing.

The model correlation structure of the path of the chain for the Eastern European military personnel example is summarized in Table 14.5 along with cross-correlations. Unfortunately, the picture is not as optimistic as in the last example. The autocorrelations for α_μ, $\beta_{1,\mu}$, $\beta_{2,\mu}$, and τ_c are unambiguously indicative of poor mixing. It is possible to

TABLE 14.4: CORRELATIONS AND AUTOCORRELATIONS, TOBIT MODEL

	Intercept	Past Rates	Political Culture	Current Opinion	Ideology	Murder Rate
		Within-Chain Correlations				
Lag 1	0.205	0.030	0.218	0.295	0.201	0.212
Lag 5	0.003	-0.006	-0.003	0.007	0.001	0.014
Lag 10	0.011	-0.005	-0.007	0.002	0.023	-0.004
Lag 50	-0.0016	-0.0156	-0.0065	-0.0183	0.0049	0.0001
		Cross-Chain Correlations				
Intercept	1.000	-0.120	0.527	-0.086	-0.975	-0.241
Past Rates		1.000	-0.167	-0.102	0.146	-0.123
Political Culture			1.000	0.022	-0.587	-0.649
Current Opinion				1.000	-0.089	0.139
Ideology					1.000	0.182
Murder Rate						1.000

look at longer lags for these variables, but the lag of 50 is sufficient evidence here from the marginals to show that the overall chain is not mixing well. The other variables show Markov chain autocorrelations that are very typical of fair to good mixing. Interestingly, although some within-chain correlations are quite high, there is little evidence that correlation between chains is slowing down the mixing. Only $\beta_{2,\mu}$ and τ_c appear to be highly (negatively) correlated. One of the main weapons for dealing with high autocorrelations is reparameterization as described in Section 14.4.1.

14.3.2 Graphical Diagnostics

Graphical diagnostics can be very useful in evaluating mixing and convergence of the chain. While they are not a formal test, as we will do in Section 14.3.3, they often show stochastic properties of importance. Figures 14.1 and 14.2 simultaneously provide two common visual diagnostics: traceplots of the path of the Gibbs sampler runs (with the burn-in period omitted to make the scale more readable), and histograms for the chains over this same period on the same vertical scale. Traceplots merely follow the path of the Markov chain over time on the x-axis, giving the consecutive mixing through the support of the posterior space on the y-axis. Because "time" is accounted for moving left to right, we can see the properties of the mixing of the chain.

In every case, the traceplots for the death penalty analysis show ideal properties: free travel up and down through the space and a flat trend across the window. This "fuzzy caterpillar" pattern is the desired outcome from looking at traceplots because it implies (not proves!) that the Markov chain is in its stationary distribution and is exploring it

TABLE 14.5: Correlations and Autocorrelations, Military Personnel Model

	α_μ	α_τ	$\beta_{1,\mu}$	$\beta_{1,\tau}$	$\beta_{2,\mu}$	$\beta_{2,\tau}$	$\beta_{3,\mu}$	$\beta_{3,\tau}$	τ_c
				Within-Chain Correlations					
Lag 1	0.998	0.896	0.981	0.871	0.883	0.304	-0.001	0.023	0.587
Lag 5	0.998	0.622	0.911	0.537	0.883	0.284	-0.001	0.003	0.534
Lag 10	0.991	0.441	0.836	0.323	0.882	0.284	-0.001	-0.004	0.533
Lag 50	0.962	0.075	0.444	0.009	0.881	0.283	0.002	-0.002	0.531
				Cross-Chain Correlations					
α_μ	1.000	0.020	-0.034	0.003	0.201	-0.023	-0.003	0.000	-0.184
α_τ		1.000	0.003	0.006	0.017	0.004	0.006	-0.002	-0.012
$\beta_{1,\mu}$			1.000	-0.008	0.024	-0.007	0.004	-0.002	-0.016
$\beta_{1,\tau}$				1.000	-0.017	0.007	0.003	0.002	0.024
$\beta_{2,\mu}$					1.000	-0.100	-0.006	0.003	-0.682
$\beta_{2,\tau}$						1.000	0.003	0.001	0.097
$\beta_{3,\mu}$							1.000	0.005	0.002
$\beta_{3,\tau}$								1.000	-0.002
τ_c									1.000

fully. The histograms show a strong central limit theorem effect from the ergodicity of the Markov chain. This complete graphical effect from Figure 14.1 is the desired result.

The graphics contained in Figure 14.2 for the Military Personnel model are not as promising. The first traceplot shows a pattern called "snaking" for α_{mu}, indicating poor mixing since the chain does not move freely through the sample space. However, the traceplots for $\beta_{1,\mu}$, $\beta_{3,\mu}$, and $\beta_{3,\tau}$ show a non-trending (flat trajectory) pattern where the chain is also making liberally wide moves through the sample space. Two variance components, α_τ and $\beta_{1,\tau}$ are somewhat long-tailed as evidenced by both the traceplots and the histograms. This suggests potential, but not certain, issues with the model specification. It is obvious that the problematic dimensions are those for $\beta_{2,\mu}$, and τ_c, where strong evidence of trending exists. This indicates that the chains are likely not to be in their stationary distribution since the time period is quite long (150,000 iterations).

One general problem with traceplots is that if the chain remains attracted to a nonoptimal mode for a long period of time, there is no visual indication that this area is not the desired high-density region. The standard solutions are to extend the run for a very long time since eventually it will leave (e.g., Geweke 1992), and to start the Markov chain from multiple widely dispersed starting points (e.g., Gelman and Rubin 1992a). Also, due to a feature in WinBUGS that keeps a dynamic running traceplot going in a separate window, it is easy to use this as a default convergence assessor. Yet WinBUGS automatically resizes the Y-axis as the chain runs for this graphic, giving an inattentive viewer the illusion of stability. So the "History" option is superior for real traceplot diagnostic purposes in the WinBUGS environment.

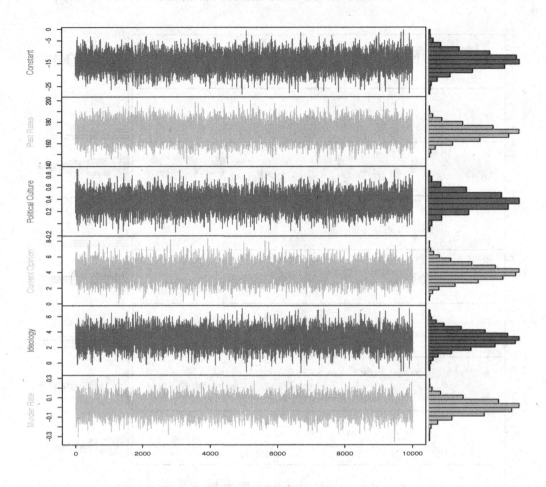

FIGURE 14.1: TRACEPLOT AND HISTOGRAM GRAPHICS, TOBIT MODEL

One important warning about traceplots is warranted here. It is common practice to run multiple parallel chains from different and dispersed starting points as a means of assessing convergence (Section 14.3.3.2). However, graphing traceplots for the full span of these chains can be very misleading because the starting, or early, values are quite far from the region where the chain will eventually settle. So the scale of the graph hides features of the chain inside a densely plotted and narrow band of line segments. As an example, Figure 14.3 shows a single dimension of the same Markov chain where the first panel covers iterations 1 through 50,000 and the second panel shows the last 500 iterations only. The longer view implies a great deal of stability and good mixing as well because the scale of the y-axis covers the large distance between the starting point at 20 and the stable region centered around 0.272. The plot of the last 500 iterations shows that there is considerable amount of snaking going on within the dense view on the left showing that the mixing is not very efficient. Unfortunately the longer view with distant starting points and small, dense

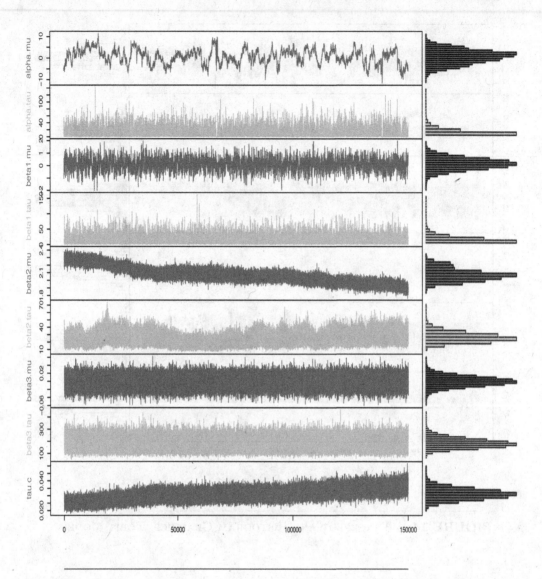

FIGURE 14.2: Traceplot and Histogram Graphics, Military Personnel
Model

stable regions can be found in published analyses and are sometimes used by researchers as
a confirmation of stability and mixing.

One very useful diagnostic is the running mean graph, which, as its name implies, gives
the mean across previous values running along the figure from left to right. Thus a stable
chain fluctuates in its early period due to a small denominator and then ends up as a flat
line. This is not iron-clad evidence of convergence because a chain may just be lingering
for a long period of time at some sub-optimal attraction, but it is strong evidence given
sufficiently lengthy evaluation periods. Figure 14.4 gives the running means for the last

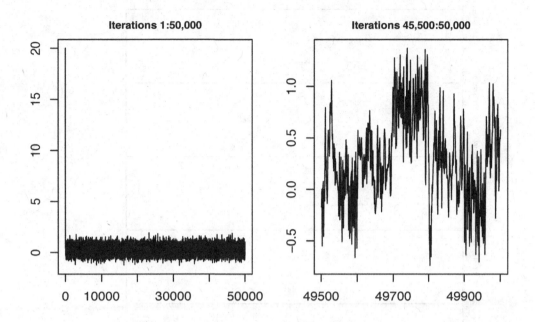

FIGURE 14.3: Traceplots of the Same Chain with Different Spans

10,000 values for the death penalty support model, and supports the claim of convergence for each parameter in the boxes on the left. Notice that each chain settles down to a stable horizontal trend. The right side gives kernel density estimates of the chain values and can be considered a smoothed version of the histograms in Figure 14.1. Underneath each density plot is a 95% HPD interval for the last half of the analyzed chain values for this plot. The values of 0.95 for the HPD interval and 0.5 for the window width can be changed in the function `mean.kernel.mcmc` to other user-selected values. This function graphical and the function for traceplots with histograms are both included in the `BaM` package in R that accompanies this book.

The idea behind an interval summary of a subset of the values used to make the density plot is that trending in the Markov chain should cause this interval to be not centered in the plot (skewed distributions will still have equal tails). These intervals in Figure 14.4 show no sign of problems.

The left-hand side of Figure 14.5 gives the same running mean diagnostic seen before with 150,000 simulated values after the burn-in period for the Military Personnel model. The graph is designed to contrast the two views: long-term stability and distributional summary. In particular $\beta_{2,\mu}$, $\beta_{,2\tau}$, and τ_c look untrustworthy in the first column of the figure since they appear not to have settled into a stable running mean. This is interesting since the Markov chain for $\beta_{2,\tau}$ appeared to be reasonably stable in the latter part of the traceplot. Robert and Mengersen (1999) use a similar diagnostic to show poor convergence properties in a normal mixture model.

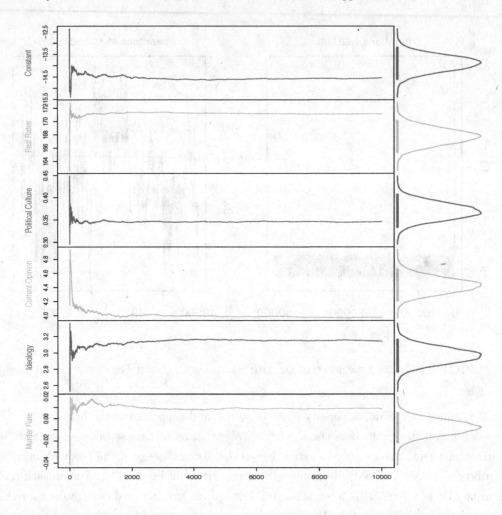

FIGURE 14.4: Running Mean Diagnostic, Tobit Model

There are a number of other graphical techniques that often give useful diagnostic information. Gelman, Gilks, and Roberts (1996) give the theoretical result that Metropolis-Hastings algorithms with a low acceptance rate will be less stable and poor at mixing through the distribution due to this inefficiency. Therefore a serial plot of the acceptance rate over time can be helpful in showing the performance of the chain.

Other graphical methods exist. Robert (1997) introduces *allocation maps* as a way of assessing convergence in mixture models. The idea is to represent each component with a different level of gray. Sudden and unambiguous shifts in color across increasing iterations is evidence of nonconvergence. Cui *et al.* (1992) and Tanner (1996, p.154) use a simple version of this idea on the θ parameter of a "witch's hat" distribution (see the discussions in Chapter 15) to show a sudden shift in chain behavior.

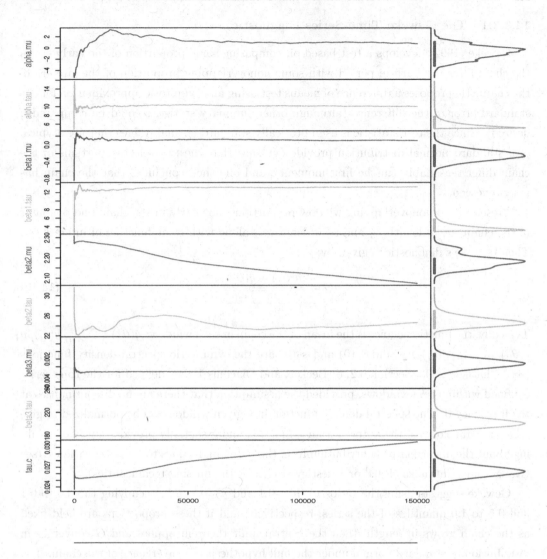

FIGURE 14.5: Running Mean Diagnostic, Military Personnel Model

14.3.3 Standard Empirical Diagnostics

This section first reviews in detail the four empirical diagnostics used most commonly in practice, giving results for our two running models of interest. These diagnostics, Geweke, Gelman-Rubin, Raftery-Lewis, and Heidelberger-Welch (all named after their creators), dominate because they are provided in easy-to-use menu form in the R packages BOA and CODA. Subsequent discussion, though, will present some interesting and useful alternatives that require user-coding.

14.3.3.1   The Geweke Time-Series Diagnostic

Geweke (1992) develops a test based on comparing some proportion of the early era of the chain after the burn-in period with some nonoverlapping proportion of the late era of the chain. He proposes a difference of means test using an asymptotic approximation for the standard error of the difference (although other summary statistics could be compared if desired). Since the test statistic is asymptotically standard normal, values that are atypical of a standard normal distribution provide evidence that the two selected portions of the chain differ reasonably (in the first moment), and one then concludes that the chain has not converged.

Preselect two nonoverlapping window proportions, one early in the chain and one later in the chain: $\boldsymbol{\theta}_1$ of length n_1 and $\boldsymbol{\theta}_2$ of length n_2 along with some function of interest $g()$. Then Geweke's diagnostic is given by:

$$G = \frac{\bar{g}(\boldsymbol{\theta}_1) - \bar{g}(\boldsymbol{\theta}_2)}{\sqrt{\frac{s_1(0)}{n_1} + \frac{s_2(0)}{n_2}}}. \tag{14.10}$$

Typically, the function choice is the mean of the chain in each window: $\bar{g}(\boldsymbol{\theta}_1) = \sum_{i=1}^{n_1} g(\theta_i)/n_1$, $\bar{g}(\boldsymbol{\theta}_2) = \sum_{i=1}^{n_2} g(\theta_i)/n_2$, and $s_1(0)$ and $s_2(0)$ are the symmetric spectral density functions (see: Chatfield 2003, Section 6.2, or Beckett and Diaconis 1994), defined for the time-series confined within these windows, provided the assumption that there are no discontinuities at the frequency 0. The spectral density function in a given window can be normalized to give a Fourier transform of the autocovariance function and essentially provides a way of thinking about the uncorrelated contribution from the individual values to the total variance (see Granger and Hatanaka [1964] or Priestley [1981] for the missing technicalities).

Geweke suggests using the ratios $n_1/n = 0.1$ and $n_2/n = 0.5$, occupying the 0.0 to 0.1 and 0.5 to 1.0 quantiles of the series, respectively, and if these proportions are held fixed as the chain grows in length, then the central limit theorem applies and G converges in distribution to standard normal under the null hypothesis of convergence of the chain. The utility is therefore that we now have a nonarbitrary method for asserting that the means of the functions of the values differ statistically. It is customary to appeal to basic normal theory and worry about values of G that are greater in absolute value than 2.

This is an overtly time-series manner of proceeding and reflects Geweke's belief that more can be learned by one very long chain since it will end up exploring areas where humans might not think to send it. Unfortunately the window proportions can greatly affect the value of G, and it is therefore important not to just rely on Geweke's recommendation (0.1/0.5), which are the default values in the diagnostic suites BOA and CODA.

■ **Example 14.3:   Geweke Diagnostic, Tobit Model.** We first perform the Geweke diagnostic for the Tobit model described above using BOA. In order to be cautious, the default settings are used and compared with windows of 20% on either side of the middle point (a more cynical approach since it includes a larger early window and smaller later window). These results are given in Table 14.6.

TABLE 14.6: GEWEKE DIAGNOSTIC, TOBIT MODEL

	0.1 vs. 0.5			0.2 vs. 0.2	
	Z-score	p-value		Z-score	p-value
Intercept	0.1901	0.4246		-0.6632	0.2536
Past Rates	0.4342	0.3321		0.5781	0.2816
Political Culture	0.0915	0.4636		-0.4210	0.3369
Current Opinion	1.0500	0.1436		0.8934	0.1858
Ideology	-0.7092	0.2391		0.4351	0.3317
Murder Rate	1.1230	0.1307		0.5769	0.2820

The output here provides little to worry about. No values are observed to be in the tail of the normal distribution at standard levels (although such points are arbitrary anyway), and so there is no evidence provided here to indicate nonconvergence. The diagnostic for `Current Opinion` shows a little bit of extra-variability, but not enough to cause alarm.

■ **Example 14.4:** **Geweke Diagnostic, Military Personnel Model.** Using the default window widths (other choices were shown to produce similar results), `CODA` returns the Geweke diagnostics here in Table 14.7. The Geweke diagnostic confirms our previous skepticism about α_μ, α_τ, $\beta_{2,\mu}$, $\beta_{2,\tau}$, $\beta_{3,\mu}$, and τ_c. The corresponding p-values are well into the tails of the standard normal for both window configurations. The similarity comes from the large sample size involved in both analyses.

TABLE 14.7: GEWEKE DIAGNOSTIC, MILITARY PERSONNEL MODEL

	0.1 vs. 0.5			0.2 vs. 0.2	
	Z-score	p-value		Z-score	p-value
α_μ	3.9268	0.0001		3.9261	0.0000
α_τ	0.4336	0.3323		0.7866	0.2158
$\beta_{1,\mu}$	0.2288	0.4095		0.2202	0.4129
$\beta_{1,\tau}$	-2.1575	0.0144		-1.1873	0.1176
$\beta_{2,\mu}$	44.0073	0.0001		53.2793	0.0001
$\beta_{2,\tau}$	-6.1943	0.0001		-3.9272	0.0001
$\beta_{3,\mu}$	-2.6983	0.0035		-2.7704	0.0028
$\beta_{3,\tau}$	0.5392	0.2949		0.9035	0.1831
τ_c	-35.9739	0.0001		-47.9269	0.0001

14.3.3.2   Gelman and Rubin's Multiple Sequence Diagnostic

Gelman and Rubin's (1992a) convergence diagnostic is based on comparing a set of chains with different starting points that are overdispersed relative to the target distribution (greater variability than the presumed limiting distribution). The method is based on normal theory approximations to the marginal posteriors using an ANOVA-based test with a normal or a Student's-t distribution-based diagnostic. Thus convergence decisions can be made *inferentially*.

The test is run according to the following steps for each scalar parameter of interest (θ):

1. Run $m \geq 2$ chains of length $2n$ from overdispersed starting points with subscripts denoting distinct chains, $(1), (2), \ldots, (m)$:

$$
\begin{array}{cccccc}
\theta_{(1)}^{[0]}, & \theta_{(1)}^{[1]}, & \cdots & \theta_{(1)}^{[n]}, & \cdots & \theta_{(1)}^{[2n-1]}, & \theta_{(1)}^{[2n]} \\
\theta_{(2)}^{[0]}, & \theta_{(2)}^{[1]}, & \cdots & \theta_{(1)}^{[n]}, & \cdots & \theta_{(2)}^{[2n-1]}, & \theta_{(2)}^{[2n]}, \\
\vdots & & & & & \\
\theta_{(m)}^{[0]}, & \theta_{(m)}^{[1]}, & \cdots & \theta_{(1)}^{[n]}, & \cdots & \theta_{(m)}^{[2n-1]}, & \theta_{(m)}^{[2n]}.
\end{array}
\tag{14.11}
$$

   The starting points can be determined by overdispersing around suspected or known modes. Discard the first n chain iterations.

2. For each parameter of interest calculate the following:

   ▷ **Within chain variance:**

   $$
   W = \frac{1}{m(n-1)} \sum_{j=1}^{m} \sum_{i=1}^{n} (\theta_{(j)}^{[i]} - \bar{\theta}_{(j)})^2
   \tag{14.12}
   $$

   where $\bar{\theta}_{(j)}$ is the mean of the n values for the jth chain.

   ▷ **Between chain variance:**

   $$
   B = \frac{n}{m-1} \sum_{j=1}^{m} (\bar{\theta}_{(j)} - \bar{\bar{\theta}})^2
   \tag{14.13}
   $$

   where $\bar{\bar{\theta}}$ is the grand mean (i.e. the mean of means since each subchain is of equal length).

   ▷ **Estimated variance:**

   $$
   \widehat{\text{Var}}(\theta) = (1 - 1/n)W + (1/n)B.
   \tag{14.14}
   $$

3. The convergence diagnostic is a single value for each dimension, called the *potential scale reduction factor* (or shrink factor):

$$
\widehat{R} = \sqrt{\frac{\widehat{\text{Var}}(\theta)}{W}}.
\tag{14.15}
$$

Along with this comes an upper confidence interval bound. The CODA output also provides an omnibus single scalar version called the multivariate potential scale reduction factor.

4. Values of $\widehat{R}$ near 1 are evidence that the m chains are all operating on the same distribution (in practice values less than roughly 1.1 or 1.2 are acceptable according to Gelman [1996]).

The logic behind this test is quite simple. Before convergence, W underestimates total posterior variation in θ because the chains have not fully explored the target distribution and W is therefore based on smaller differences. Conversely, $\widehat{\text{Var}}(\theta)$ overestimates total posterior variance because the starting points are intentionally overdispersed relative to the target. However, once the chains have converged, the difference should be incidental since the chains are exploring the same region and are therefore overlapping.

Commonly, the number of separate chains is about 5 to 10, but more complicated model specifications and more posterior complexity may require the specification of additional starting points. Note that the test here is described for only one dimension and in practice each of the m chains starts from a point in $\Re^k$ for a k-length vector of estimation parameters $\boldsymbol{\theta}$. Thus dimensionality can considerably add to the work involved in selecting overdispersed starting points.

Practically, it is very easy to monitor convergence by periodically checking the value of $\widehat{R}$ as the Markov chain runs (WinBUGS readily supplies a slightly different version of this). Therefore we can run the chain until satisfied with the statistic and then treat the most recent n values as empirical values from the desired target distribution.

The primary difficulty with the Gelman and Rubin diagnostic (noted by many observers) is that it is not always easy to obtain suitably overdispersed starting points, since determining their position requires some knowledge of the target distribution to begin with. Geyer (1992) points out that the determination of good starting points is actually *essential* in the Gelman and Rubin diagnostic because of the underdispersed/overdispersed distributional contrast. Cowles and Carlin (1996) worry about the heavy reliance on normal approximations, and it is obvious that substantial deviations from normality make the choice of starting values much more difficult (somewhat mitigated by the Brooks-Gelman modification [1998a, 1998b] which uses the more defensible Student's-t distribution assumption). Also, Brooks and Giudici (2000) give an extended split-variance refinement of this diagnostic focused on model choice, including reversible jump MCMC (see Section 15.2 in Chapter 15 starting on page 536).

In addressing the starting point issue, Gelman and Rubin (1992a) advise the use of the EM algorithm to find modes and caution that EM should be started from several different points if multimodality is suspected (1992a, 459). In practice, it is not always easy to write a customized EM algorithm for a reasonably intricate model. With the case of known multimodality Gelman and Rubin advise using a more complicated procedure that involves constructing a mixture of normals centered at the found modes, and "sharpened" with importance sampling as a means of determining the scale of overdispersion required.

■ **Example 14.5:   The Gelman and Rubin Diagnostic for the Tobit Model.** Returning to the Tobit model, we perform the Gelman and Rubin diagnostic with five separate chain paths with starting points perturbed off of the MLE from a regular GLM to overdisperse. The draw for the z_i values precluded dramatically overdispersed starting points in that dimension since the random truncated normal generator becomes extremely inefficient for highly unlikely parameter values relative to the associated data. Each chain is again run for $50,000$ iterations at these starting points with the first $40,000$ iterations discarded in all cases. The CODA diagnostic suite, with a normal distribution assumption, is used to produce the following results, which show no evidence of nonconvergence (Table 14.8). Table 14.8 also gives the multivariate potential scale reduction factor from CODA output. The MPSRF is an additional statistic from Brooks and Gelman (1998b) that seeks to give a fully multivariate view of chain convergence using the full variance-covariance matrix of the output and therefore providing a more conservative view. This is particularly useful for high-dimensional models as a convenient omnibus test, although Brooks and Gelman suggest looking at the individual PSRF statistics for important parameters.

TABLE 14.8:   GELMAN AND RUBIN DIAGNOSTIC, TOBIT MODEL

	Intercept	Past Rates	Political Culture	Current Opinion	Ideology	Murder Rate
PSRF	1.01	1.00	1.02	1.01	1.01	1.02
95% Upper CI	1.03	1.02	1.05	1.05	1.02	1.07
Multivariate Potential Scale Reduction Factor: 1.03						

■ **Example 14.6:   The Gelman and Rubin Diagnostic for the Military Personnel Model.** Again five separate Markov chains are run using overdispersed starting points. These were determined by constructing a wide grid along the sample space of each of the five parameters. Then the joint posterior is applied systematically to go through the dimensions to look for modes. This can be a very time-consuming process if the grid is reasonably granular, and it is often possible to get away with fairly large bins.

We now run the Gelman and Rubin procedure on the model of Military Personnel using BOA defaults for the five chains, including the convention of using only the second half of the recorded change values. Overdispersed starting points were determined from the posterior means and HPD intervals in Table 14.3 on a marginal basis. The results are given in Table 14.9.

TABLE 14.9: GELMAN AND RUBIN DIAGNOSTIC, MILITARY PERSONNEL MODEL

	α_μ	α_τ	$\beta_{1,\mu}$	$\beta_{1,\tau}$	$\beta_{2,\mu}$	$\beta_{2,\tau}$	$\beta_{3,\mu}$	$\beta_{3,\tau}$	τ_c
PSRF	1.00	1.00	2.28	2.01	25.32	2.49	1.04	4.70	1.36
95% Upper CI	1.01	1.00	3.58	10.11	128.56	4.13	1.04	15.49	1.82

Multivariate Potential Scale Reduction Factor: 21.7

Here all but three of the $\widehat{R}$ statistics in the first row of the table give cause for concern. Only α_μ, α_τ, and $\beta_{3,\tau}$ provide PSRF values that we would be comfortable with. Note also that the PSRF value for $\beta_{2,\mu}$ is enormous. The second line of the table provides the upper 95% credible interval for the statistics and shows large tails for the three β parameter variance components $\beta_{1,\tau}$, $\beta_{2,\tau}$, and $\beta_{3,\tau}$. All of this provides strong evidence of overall nonconvergence of the Markov chain. Markov chains with Gibbs or Metropolis kernels do not converge partially by dimension (Gill 2010). In addition, the Brooks and Gelman (1998a, 1998b) correction also gives a multivariate summary for the whole model, which is 21.7 in this case and is clearly dominated by the result for $\beta_{2,\mu}$.

14.3.3.3 The Heidelberger and Welch Diagnostic

This diagnostic works in its implemented form for individual parameters in single chains. It is easy to code on one's own and conveniently supplied in BOA and CODA. Heidelberger and Welch (1983) originally developed the algorithm in another context for simulations in operations research and based on Brownian bridge theory, using the Cramér-von Mises statistic (Heidelberger and Welch 1981a, 1981b; Schruben 1982; and Schruben, Singh, and Tierney 1983).[1]

Like the Geweke diagnostic, it has an inherently time-series orientation and uses the spectral density estimate. The null hypothesis for this test is that the chain is currently in the stationary distribution and the test starts with the full set of iterations produced from running the sampler from time zero. The general form for steps are then as follows.

1. Specify: a number of iterations (N, generally given by the current status of the chain at the proposed stopping point), an accuracy (ϵ), and an alpha level for the test.

2. Evaluate the first 10% of the chain iterations.

3. Calculate the test statistic on these samples and observe whether it indicates a tail value.

---

[1]A Brownian Bridge is a Wiener process (also called Brownian motion, Priestley [1981, pp.167-8]) over the interval [0, 1], in which the starting and stopping points are tied to the endpoints 0 and 1 and the "string" in between varies. The details are beyond the scope of this work and the interested reader is directed to Chu and White (1992), Lee (1998), as well as Nevzorov and Zhukova (1996).

4. If the test rejects the null, the first 10% of the iterations are discarded and the test is run again.

5. This process continues until either 50% of the data have been dismissed or the test fails to reject the null with the remaining iterations.

If some proportion of the data are found to be consistent with stationarity, then the *halfwidth* analysis part of the diagnostic is performed.

The process is actually a variant of the Kolmogorov-Smirnov nonparametric test for large sample sizes referred to as the Cramér-von Mises test after the form of the test statistic. A parameter, s, is defined as the proportion of the continuing sum of the chain, ranging from zero to one. We are interested in the difference between the sum of the first sT of the chain (where T is the total length of the chain), and the mean of the complete set of chain values scaled by sT. If the chain were in its stationary distribution, then this difference would be negligible. For the disparity between empirical and theoretical CDFs, the Cramér-von Mises test statistic is

$$C_n(F) = \int_X [F_n(x) - F(x)]^2 dF(x), \tag{14.16}$$

with a predefined rejection region $C_n(F) > c$, established by normal tail values under asymptotic assumptions and the hypothesis that $F_n(x) \neq F(x)$.

There are some additional aspects that must be defined to get from the Cramér-von Mises to the Heidelberger and Welch diagnostic. Define formally the following quantities:

▷ T = the total length of the "accepted" chain after discards.

▷ $s \in [0{:}1]$ = the test chain proportion.

▷ $T_{\lfloor sT \rfloor} = \sum_{i=1}^{\lfloor sT \rfloor} \theta_i$ = the sum of the chain values from one to the integer value just below sT.

▷ $\lfloor sT \rfloor \bar{\theta}$ = the chain mean times the integer value just below sT.

▷ $s(0)$ = the spectral density of the chain.

The notation $\lfloor X \rfloor$ denotes the integer value of X (the "floor" function). Using these produced quantities, for any given s, we can construct the test statistic:

$$B_T(s) = \frac{T_{\lfloor sT \rfloor} - \lfloor sT \rfloor \bar{\theta}}{\sqrt{Ts(0)}}, \tag{14.17}$$

which is the Cramér-von Mises test statistic for sums as cumulative values scaled by the spectral density. Now $B_T(s)$ can be treated as an approximate Brownian bridge and tested using normal tail values to decide on the 10% discards (where some adjustments are necessary because $s(0)$ is estimated rather than known).

The halfwidth part of the test compares two quantities:

▷ Using the proportion of the data not discarded, the halfwidth of the $(1 - \alpha)$% credible interval is calculated around the sample mean, where the estimated asymptotic standard error is the square root of spectral density divided by the remaining sample size, $s(0)/n^*$.

▷ If the mean divided by the halfwidth is lower than ϵ, then the halfwidth test is passed for this dimension.

▷ If this test fails, then longer runs are required to accurately produce an empirical estimate of the mean and assert convergence.

It should be remembered that the quality of this second part of the diagnostic is completely dependent on the ability of the first to produce a subchain in the true limiting distribution, and if it has settled for some time on a local maxima, then the halfwidth analysis can be wrong. Finally, it should be noted that most practitioners pay attention only to the Cramér-von Mises test statistic part of the test since it has an easy interpretation like the Geweke statistic.

■ **Example 14.7: Heidelberger and Welch Diagnostic for the Tobit Model.** Running the H-W diagnostic at $\epsilon = 0.1$, $\alpha = 0.05$ in BOA produces:

```
Stationarity       Test  Keep Discard C-von-M Halfwidth     Mean Halfwidth  Ratio
Intercept        passed 10000       0  0.2288    passed -14.5451    0.0971 -0.006
Past Rates       passed 10000       0  0.1926    passed 171.1460    0.1661  0.001
Political Culture passed 10000      0  0.0397    passed   0.3461    0.0038  0.011
Current Opinion  passed 10000       0  0.2211    passed   3.9738    0.0274  0.007
Ideology         passed 10000       0  0.1592    passed   3.1423    0.0299  0.009
Murder Rate      passed 10000       0  0.0511    failed   0.0088    0.0020  0.227
```

Notice that all of the chain values passed the Cramér-von Mises test component, which is not surprising given the previously observed health of this chain run and the fact that we have already discarded $40,000$ values to begin the chain. However, on the halfwidth test, the dimension for Murder Rate fails since the ratio of the halfwidth to the mean is 0.227, which is the only one greater than $\epsilon = 0.1$. Recall that this variable was the only one found to be statistically unreliable in the model given in Table 14.1 (irrespective of the unimportant constant). This is worth further investigation. The individual autocorrelations revealed no obvious stickiness problems:

```
                 Lag 1     Lag 5    Lag 10    Lag 50
Murder Rate      0.212     0.014    -0.004    0.0001
```

but there was evidence that this parameter is strongly correlated with Political Culture:

```
Past.Rates   Political Culture   Current Opinion   Ideology
  -0.123              -0.649               0.139      0.182
```

(which should explain some of the mixing). Recall, however, that slow mixing is not a serious problem if the full sample space is still fully explored. The wealth of other diagnostics certainly point us in this direction.

Figure 14.6 provides a traceplot and histogram of the last $2,000$ iterations of the marginal chain values for `Murder Rate` with the mean and 95% credible intervals indicated in each. Both figures support stability of the chain over this last period. The traceplot shows some minor snaking, possibly indicating mixing at a pace slower than the other dimensions, but actually nothing even slightly alarming is revealed about convergence. The chain is stable around its current mean, balanced between the interval ends, and the histogram shows strong evidence of the central limit theorem.

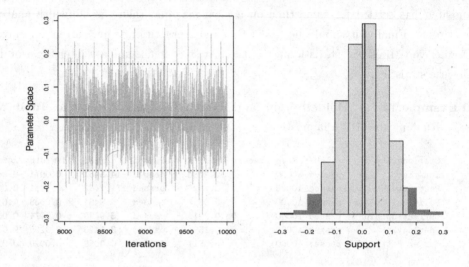

FIGURE 14.6: Traceplot and Histogram, Last 2,000 Iterations

■ **Example 14.8:   The Heidelberger and Welch Diagnostic for the Military Personnel Model.** Using the standard defaults in `BOA` (again at the default parameter values of $\epsilon = 0.1$, $\alpha = 0.05$), the H-W test gives the following output:

Stationarity	Test	Keep	Discard	C-von-M	Halfwidth	Mean	Halfwidth	Ratio
alpha.mu	passed	150000	0	0.3162	failed	0.7403	0.8404	1.1352
alpha.tau	passed	150000	0	0.2236	passed	10.0155	0.2672	0.0267
beta1.mu	passed	150000	0	0.1825	failed	0.0965	0.0238	0.2466
beta1.tau	passed	150000	0	0.2458	passed	13.3070	0.2432	0.0183
beta2.mu	failed	60000	90000	6.4482	passed	1.9996	0.0114	0.0057
beta2.tau	failed	60000	90000	2.5148	passed	25.9008	0.7142	0.0276
beta3.mu	passed	120000	30000	0.3397	passed	-0.0011	0.0001	-0.0909
beta3.tau	passed	150000	0	0.1642	passed	180.0586	0.1941	0.0011
tau.c	failed	60000	90000	6.0374	passed	0.0330	0.0004	0.0121

Obviously the results show poor convergence properties according to this diagnostic. The first test is the stationary test from the Brownian bridge approximation, and three marginal results show a failure to obtain acceptable Cramér-von Mises test statistics for discarding less than 50% of the data in 15,000 increments. The associated large values of `C-von-M` fall into the tail region past a stipulated critical value. In the second stage of the diagnostic we can ignore the results for $\beta_{2,\mu}$, $\beta_{2,\tau}$, and τ_c, since they have already failed. Now notice that α_μ and $\beta_{1,\mu}$ fail here with a small value for the halfwidth, which gives a subsequent ratio of the halfwidth to the mean that exceeds $\epsilon = 0.1$.

14.3.3.4   The Raftery and Lewis Integrated Diagnostic

The diagnostic proposed and developed in Raftery and Banfield (1991), Raftery and Lewis (1992), and summarized in Raftery and Lewis (1996) provides a rough indication of convergence for a running chain. It was originally designed to elicit desired thresholds from the user and then estimate the chain length that provides a satisfactory result. Consequently, the Raftery and Lewis diagnostic addresses the burn-in period, thinning of the chain, the posterior quantities of interest, and probabilistic reliability, as well as the number of iterations.

For a single given parameter of interest, the steps are given as follows:

1. Select the posterior quantile of interest, q, such as a tail value or some substantively interesting point:
$$q = p(f(\theta) \leq f(\theta_q)|\mathbf{X}). \tag{14.18}$$

2. Select an acceptable tolerance for this quantile, r, and the desired probability of being within that tolerance, s. For example (Raftery and Lewis 1996), suppose we want to report a 95% credible interval around the median with reliability between 92.5% and 97.5%. This is equivalent to the assignments: $q = 0.5$, $r = 0.0125$, and $s = 0.95$.

3. Select a convergence tolerance value ϵ, which is used to determine a stopping point based on a parallel Markov chain process (defaulted to 0.001).

4. Run a "pilot" sampler whose length is determined as if there is no autocorrelation in the parallel chain, given by rounding:
$$n_{\text{pilot}} = \left[\Phi^{-1}\left(\frac{s+1}{2}\right) \frac{\sqrt{q(1-q)}}{r} \right]^2, \tag{14.19}$$

where $\Phi^{-1}()$ denotes the inverse of the normal CDF. For the example above at the defaults, $n_{\text{pilot}} = 2294$.

5. The program (BOA, CODA, or mcgibbsit) will then return: the length of the burn-in period (M), the estimated number of post burn-in iterations required to meet the goals (N), and the thinning interval (k). The user then runs a Markov chain according to these parameters.

The method is based on setting up a parallel (but not Markov) chain during the pilot run, where the iterations are a binary series according to whether the generated value in the primary chain at that stage is less than the chosen quantile. Raftery and Lewis extrapolate values that give this secondary chain of zeros and ones some Markovian properties, and are then able to estimate first-chain probabilities that satisfy the user constraints (see Raftery and Lewis [1996, pp.116-118] for specifics). A nice feature of this process is that the final run of N iterations can be treated as a new pilot run, so that if the diagnostic returns a new recommended N, it would be motivation for additional sampling.

Unfortunately this process must be individually repeated for every quantile of every parameter of interest. Also, the number of iterations required (N) is different for different quantiles picked (in the example above, the number of required iterations would be only 436 here if $q = 0.95$ were of interest instead of $q = 0.5$). Since the process is inherently iterative, picking quantiles and monitoring pilot runs, then the process can be quite time-consuming (in the human sense) for large numbers of parameters.

Robert and Casella (1999, First Edition) list a number of more serious problems with the Raftery and Lewis diagnostic: the inherently single-dimension nature does not account for potential correlations between coefficients (and therefore provides evidence of collective marginal convergence rather than joint convergence, a point also mentioned by Brooks and Roberts [1997]), the dependence on the representativeness of the pilot sample, and suspect nature of the Markov approximation in the secondary chain; see Guihenneuc-Joyaux and Robert (1998) for a more rigorous but complex approach based on a *divergence evaluation* of two chains (Section 4.4). Brooks and Roberts (1998) also warn against widespread use of the default $q = 0.025$ setting since it can lead to substantial underestimation of the burn-in period. Nonetheless, the Raftery and Lewis diagnostic often works quite well with simple models and can give a general approximation of the scope of the number of iterations required in more elaborate specifications. In addition, its inclusion in BOA and CODA mean that many practitioners will make use of it regardless of any warnings that might appear in the general statistics literature.

■ **Example 14.9:   The Raftery and Lewis Diagnostic for the Tobit Model.** Running the Raftery and Lewis diagnostic at BOA defaults, $q = 0.025$, $r = \pm0.005$, $s = 0.95$, produces:

	Burn-in (M)	Total (N)	Lower Bound (Nmin)	Dependence Factor (I)
Intercept	3	4061	3746	1.08
Past Rates	3	4061	3746	1.08
Political Culture	3	4129	3746	1.10
Current Opinion	2	3802	3746	1.01
Ideology	3	4028	3746	1.08
Murder Rate	3	4061	3746	1.08

where we should not be alarmed by anything. The low burn-in numbers result from starting the diagnostic *after* burn-in $40,000$ iterations. Note that BOA gives a thinning value in the summary whereas CODA does not.

The Lower Bound is the n_{pilot} estimate of the number of iterations necessary to satisfy r and s under the assumption of independent chain values, which is less than $100,000$

so no warning was given. `Burn-in` is the estimated m, and `Thin` is the k value used in the calculations. `Total` is the estimated number of iterations required N. The last column gives the "dependence factor":

$$I_{sd} = \frac{M}{n_{\text{pilot}}}, \tag{14.20}$$

which is interpreted by Raftery and Lewis (1992) as the proportional increase in the number of iterations attributable to serial dependence (the version of I_{sd} implemented in BOA is slightly different than that given by Raftery and Lewis). Their prescriptive advice is to worry about dependence factors exceeding 5, as it might be due to an influential starting value, high correlations between coefficients, or poor mixing.

■ **Example 14.10: The Raftery and Lewis Diagnostic for the Military Personnel Model.** We now run the integrated diagnostic with the values with $q = 0.05$, $r = \pm 0.005$, and $s = 0.95$ for the Military Personnel example. The routine in BOA gives the output:

	Burn-in (M)	Total (N)	Lower Bound (Nmin)	Dependence Factor (I)
alpha.mu	864	889248	3746	237.00
alpha.tau	336	341740	3746	91.20
beta1.mu	168	178500	3746	47.70
beta1.tau	40	44616	3746	11.90
beta2.mu	608	701176	3746	187.00
beta2.tau	24	30752	3746	8.21
beta3.mu	2	3817	3746	1.02
beta3.tau	2	3819	3746	1.02
tau.c	124	146568	3746	39.10

No pilot run was necessary since 150,000 iterations were performed prior to running the diagnostic in R using BOA (if this was an insufficient period, BOA informs the user). In this second example, we should be worried about all but two of the dependence factors (`beta3.mu` and `beta3.tau`), given the length of the chain and the correlations given in Table 14.5. While this diagnostic is gives only a general indication of problems, such results as these are strong evidence that the chain has not yet reached its stationary distribution.

14.3.4 Summary of Diagnostic Similarities and Differences

The empirical examples make it clear that the four diagnostics described in detail show different characteristics of the Markov chain. Given the potential complexities of model space, the variability of posterior forms, and possibly high dimensions, it is not surprising that the stochastic behavior of the chain can be difficult to describe with a single summary statistic. It is therefore strongly recommended that researchers run all of the standard formal diagnostics with superdiag in addition to generating graphical displays. Obviously not all of these need to be included in the writeup, but they are certain to be useful in understanding chain behavior. Shockingly, there are deep misunderstandings about this process, even by producers of commercial software. For example, the Amos User's Guide in spss provides an MCMC estimation process for structural equation models. The software manual states "The default burn-in period of 500 is quite conservative, much longer than needed for most problems" (Arbuckle 2009). Obviously this is a dangerous recommendation, even for simple models.

TABLE 14.10: DIAGNOSTIC SCORECARD, TOBIT MODEL

	Geweke	Gelman-Rubin	Heidelberger-Welch	Raftery-Lewis
Intercept	✓	✓	✓	✓
Past Rates	✓	✓	✓	✓
Political Culture	✓	✓	✓	✓
Current Opinion	✓	✓	✓	✓
Ideology	✓	✓	✓	✓
Murder Rate	✓	✓	✗	✓

To summarize the convergence assessment for the two empirical examples, we can look at a complete score-card by parameter-dimension. First consider the summary for the Tobit model in Table 14.10 that summarizes each of the four diagnostics. Only the Heidelberger and Welch test for Murder Rate, and only for the halfwidth part of the test, fails here. Such a small deviation is not worth worrying about given the other results. It is important to remember that these are all *distributional* hypothesis tests and so running multiple parameters through multiple tests will lead to some tail values. To be more concrete, suppose that we ran a Geweke test on a model with 100 parameters and set $\alpha = 0.05$ for the difference of means test. Then we would *expect* about 5 parameters to fail under the null hypothesis that the complete chain had converged.

The scorecard for the Military Personnel model in Table 14.11 does not show such positive results. Overall the Raftery and Lewis diagnostic gives the most pessimistic view of convergence, even though it is the bluntest instrument. On the opposite end of the scale the Heidelberger and Welch diagnostic is the most positive with six of the nine parameter dimensions passing the test. The two most popular diagnostics disagree on α_μ, $\beta_{1,\mu}, \beta_{3,\mu}$, and $\beta_{3,\tau}$. This is actually a useful outcome because it demonstrates that the diagnostics are looking at different characteristics of stability. Such differences highlight the importance of using multiple diagnostic tools.

TABLE 14.11: DIAGNOSTIC SCORECARD, MILITARY PERSONNEL MODEL

	Geweke	Gelman-Rubin	Heidelberger-Welch	Raftery-Lewis
α_μ	✗	✓	✓	✗
α_τ	✓	✓	✓	✗
$\beta_{1,\mu}$	✓	✗	✓	✗
$\beta_{1,\tau}$	✗	✗	✓	✗
$\beta_{2,\mu}$	✗	✗	✗	✗
$\beta_{2,\tau}$	✗	✗	✗	✗
$\beta_{3,\mu}$	✗	✓	✓	✓
$\beta_{3,\tau}$	✓	✗	✓	✓
τ_c	✗	✗	✗	✗

14.3.5 Other Empirical Diagnostics

There are a wealth of additional diagnostics that are less convenient since they are not provided by BOA and CODA. These others generally require some programming or time spent understanding public domain code. Cowles and Carlin (1996, p.890) provide a very nice tabular summary of the 13 most popular methods as of 1996 (some time ago in MCMC-terms). Early debates on diagnostic strategies centered essentially on the "one long chain" versus the "many short chain" perspectives emblemized by Geweke versus Gelman and Rubin. Essentially both sides won the debate in that many of the current strategies can be characterized as "many long run" methods since computing power has become significantly more economical. What is very clear from published discussions is that it is always possible to find some example that "fools" a given convergence diagnostic (see the discussion of the Gelman and Rubin [1992b] and Geyer [1992] papers). The following are very brief descriptions of some additional proposed diagnostics that require individual coding for each model application. Because these are customized for each individual application, they cannot be "canned" in BUGS or other software.

Zellner and Min (1995). This diagnostic was first introduced here in Chapter 13, but without giving the details below. Assume that a $\boldsymbol{\theta}$ vector of unknown parameter estimates can be segmented into two parts: $(\boldsymbol{\theta}_1, \boldsymbol{\theta}_2)$, which is easy in the case of Gibbs sampling. From the definition of conditionals, we know that $\pi(\boldsymbol{\theta}_1, \boldsymbol{\theta}_2|\mathbf{D}) = \pi(\boldsymbol{\theta}_1|\boldsymbol{\theta}_2, \mathbf{D})\pi(\boldsymbol{\theta}_2|\mathbf{D}) = \pi(\boldsymbol{\theta}_2|\boldsymbol{\theta}_1, \mathbf{D})\pi(\boldsymbol{\theta}_1|\mathbf{D})$, for given $\mathbf{D}$ indicating both the data and the prior. If $\hat{\pi}_n(\boldsymbol{\theta}_1, \boldsymbol{\theta}_2, \mathbf{D})$ (note the "hat" on $\hat{\pi}$) is the vector position at the candidate stopping point of the chain at iteration n, then the "difference convergence criterion" statistic is formed by the differential Rao-Blackwellized estimate (discussed below) of the conditional times the current value, and is close to zero if the chain has converged: $\hat{\eta}_n = \pi(\boldsymbol{\theta}_1|\boldsymbol{\theta}_2, \mathbf{D})\hat{\pi}_n(\boldsymbol{\theta}_2|\mathbf{D}) - \pi(\boldsymbol{\theta}_2|\boldsymbol{\theta}_1, \mathbf{D})\hat{\pi}_n(\boldsymbol{\theta}_1|\mathbf{D})$, where the distribution $\hat{\pi}(\boldsymbol{\theta}_i|\mathbf{D})$ comes from a smoothed empirical estimate. Sometimes there is a natural segmentation of the $\boldsymbol{\theta}$ vector into the two components, but in general this is an artificial decision made by the researcher. A related quantity based on the same identity, called the "ratio convergence criterion," is given by the ratio: $\hat{\gamma}_n = (\pi(\boldsymbol{\theta}_1|\boldsymbol{\theta}_2, \mathbf{D})\hat{\pi}_n(\boldsymbol{\theta}_2|\mathbf{D}))/\pi(\boldsymbol{\theta}_2|\boldsymbol{\theta}_1, \mathbf{D})\hat{\pi}_n(\boldsymbol{\theta}_1|\mathbf{D}))$, which should be approximately one at convergence. Zellner and Min also give the procedure for obtaining the posterior odds for the hypothesis H_0: $\eta = 0$ versus the hypothesis H_1: $\eta \neq 0$, with equal probability priors for the two alternatives and a Cauchy prior on η. Using a reasonably large number of preceding values for $\hat{\eta}_i$ ($i < n$), we can calculate an empirical variance with the slightly unrealistic assumption of serial independence: $\hat{\phi}_n$. So the posterior odds of H_0 over H_1 are: $K_{OA} = \sqrt{N\pi/2}(1 + (\hat{\eta}_n^2)(n\hat{\phi}_n^2))\exp[-(\hat{\eta}_n^2)(2\hat{\phi}_n^2)]$.

Ritter and Tanner (1992). The "Gibbs stopper" of Ritter and Tanner assigns importance weights (see Chapter 9, Section 9.6) to each iteration of a Gibbs sampler so that a comparison is made between the current draw and a normalized expression for the joint posterior density. This method supports both a single chain or multiple chains (by segmenting the single chain) and assesses the bias of the full joint distribution of the Gibbs sampler (only). The Gibbs stopper, however, can be slow in complex models.

Roberts (1992, 1994). This "distance evaluation" diagnostic, initially restricted to the Gibbs sampler, gives a bias assessment of the chain simultaneously across all dimensions with a graphic summary. The idea is to cycle through multiple Gibbs samplers (generalized to other forms in 1994) in a stipulated order and then cycle back in the reverse order, creating a dual chain. Roberts suggests as many as 10 to 20 separate chains started from overdispersed starting points. The resulting statistic compares within and between behavior with a summary that has known asymptotics.

Liu, Liu, and Rubin (1992). These authors posit a "global control variable" to estimate the bias of the full joint distribution using multiple chains. The test statistic is a function of this variable whose expected value is the variance of the ratio of the posterior at the distribution implied by the current iteration to that at the limiting distribution. It is then necessary to use either graphical methods or the test setup of Gelman and Rubin to make stopping decisions.

Yu and Mykland (1998). A barely customized and particularly easy-to-implement diagnostic is the CUSUM plot (see also Brooks [1998b]). This plot is created by first calculating a running tally of the following statistic from the post-burn-in chain values $(1, \ldots, t)$: $S_t = \sum_{i=1}^{t}(\theta_t - \bar{\theta})$, and then plotting this cumulative sum against the time index. The argument is that smoothness in the resulting graph and a tendency to spend large periods away from zero are indications of slow mixing. Note that this is essentially the information provided in a plot such as the left-hand side of Figures 14.1 and 14.2.

Mykland, Tierney, and Yu (1995). This diagnostic is based on splitting a Markov chain into small sets according to renewal criteria. A Markov chain is *regenerative* (a less restrictive form of renewal) if there are times when future values are fully independent of past values and all future values are iid. Notationally, a renewal set A exists if for a given real $0 < \epsilon < 1$ and marginal probability $\eta(B)$ on another set B: $p(\theta_{n+1} \in B|\theta_n) \geq \epsilon\eta(B)$, $\forall B$ (see Robert 1995, Section 3 for additional details and examples). The diagnostic is a graphical summary of the regeneration probability from iteration to iteration (and thus implemented more directly with the Metropolis-Hastings algorithm and on discrete state spaces), where stability indicates convergence. This procedure usually involves modifying the algorithm to perform the regeneration calculations as it runs (as opposed to summarizing existing chain runs), and the resulting hybrid processes can be complex (see: Johnson [1998], and Gilks, Roberts, and Sahu [1998]). Giakoumatos (2005) produce a convergence diagnostic based on subsampling of this method and drawbacks for unobserved ARCH models, and Flegal provides a general convergence test based on subsampling bootstrap methods. Other than such specialized settings there has not been much new work in this area since the 1990s and the four main formal diagnostics described in this chapter dominate applied work.

14.3.6 Why Not to Worry Too Much about Stationarity

Considerable attention has been given to convergence diagnostics (here and elsewhere) for good reason: Markov chains that are not mixing through the target distribution do not provide useful inferential values. In this chapter we have seen a number of ways for conveniently asserting convergence or nonconvergence. However, it can be shown that these efforts are relatively cautious in their practical implications with applied work (Robert and Casella 2004, 461). While finite runs of the chain always fail to achieve asymptotic properties, each step of the chain does bring us closer, in probability, to a realization from the stationary (invariant) distribution (Robert 2001, p.303). Notably, Geyer (1991, p.158) points out that the accuracy of calculated quantities depends on the adequacy of the burn-in period, which can never be validated for certain. This does not mean that non-asymptotic results are necessarily useless. In fact, the ergodic theorem guarantees that averages from sample values give strongly consistent estimates (Meyn and Tweedie 1993), and therefore the resulting inferences are no less reliable than other sample-based procedures.

Recall the discussion in Section 13.4, where it was shown that convergence to stationarity is different from convergence of empirical averages. The arbitrary statistic of interest, $h(X)$, from a Harris recurrent Markov chain, is summarized with $\bar{h} = \frac{1}{n} \sum_{i=1}^{n} h(x_i)$, with the property: $\sum_{i=1}^{n} h(X_i)/n \longrightarrow E_f h(X)$ as $n \to \infty$. The difference between the expected value of $h(X)$ in the true distribution at time n, and the expected value of $h(X)$ in the stationary distribution, is:

$$\left[\frac{1}{n} \sum_{i=1}^{n} h(X_i) - E_{x_0} h(X) \right] - \left[E_f h(X) - E_{x_0} h(X) \right] \to 0 \text{ as } n \to \infty. \quad (14.21)$$

For a geometrically ergodic Markov chain, this quantity converges geometrically fast to zero since $\|E_f h(x) - E_{x_0} h(x)\| \le k \delta^n$ for some positive k, and $\delta \in (0 : 1)$. The first bracketed term is the difference between the empirical average and its expectation at time n. Except at the trivial point where $n = 0$, these are never non-asymptotically equivalent values *and this demonstrates that even in stationarity, the empirical average has not converged.* In fact, all we know for certain from the central limit theorem is that:

$(\frac{1}{n}\sum_{i=1}^{n} h(X_i) - E_{x_0}h(X))/(\sigma/\sqrt{n})$ converges in distribution to a standard normal, but does not bring along convergence of empirical averages.

Thus, convergence to stationarity has no bearing on the convergence of the empirical averages, which are usually of primary interest. Moreover, if stationarity is still a concern, merely start the chain in the stationary distribution (easy for Metropolis-Hastings because the stationary distribution is explicitly given) by applying an accept-reject method (a simple trial method based on testing candidate values from some arbitrary instrumental density; see Robert and Casella [2004, Chapter 1]) until a random variable from the stationary distribution is produced, then run the chain forward from there. This works because a chain started in the stationary distribution remains in the stationary distribution, and it is typically the case that the wait for the accept-reject algorithm to produce a random variable from the stationary distribution will be shorter than any reasonable burn-in period.

14.4   Mixing and Acceleration

Simple social science model specifications, even those with hierarchical features, often mix through the target distribution rather easily with standard MCMC kernels and present no estimation challenges with MCMC procedures. Conversely, mixing problems most commonly occur in high-dimension state spaces. The example models given with the BUGS software are very well-behaved in this regard, but as model specifications get larger (more parameters to estimate) and more complex, it is often the case that even when the chain has converged, it will move slowly through the stationary distribution. The problem with this is that if the chain is stopped too early, then empirical estimates will be biased.

Slow mixing chains also make the problem of convergence diagnosis more difficult because a chain that has not fully explored the limiting distribution will give biased results based on a subset of the appropriate state space.

While the problem of mixing diminishes with increasingly powerful desktop computing, it is still important to have some tools that can increase mixing rates. These typically involve "tuning" (case-specific model estimation changes) since in general they are modifications of standard algorithms that are done on a case-by-case basis. We will also provide a completely different class of approaches to solving this problem in the next chapter based on modifying the transition kernel in order to facilitate easier traversal of the chain.

14.4.1   Reparameterization

Often slow mixing through the target distribution can be attributed to high correlation between model parameters. This is particularly exacerbated with the Gibbs sampler since each sub-step is restricted to a move that is *parallel* to an axis by construction, and the full conditional distributions will specify small horizontal or vertical steps (Nandram and Chen 1996). Therefore correlations, which by definition define nonparallel structures in the posterior, will tend to restrict free movement through the parameter space, resulting in small steps and poor mixing properties. High intra-parameter correlation is also a problem with the Metropolis-Hastings algorithm in that it also slows mixing, but is made obvious by observing many rejected candidate values. So while the Gibbs sampler will chug along in a limited region without indicating such a problem to the observer, a similarly affected Metropolis-Hastings algorithm will tend to reject many candidate jumping positions and therefore be more obvious in its behavior.

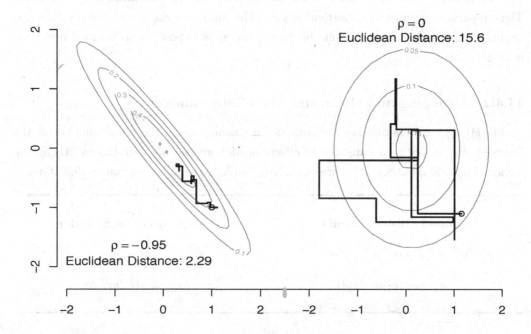

FIGURE 14.7: A COMPARISON OF GIBBS SAMPLING FOR NORMALS

As an example consider two parameters with standard normal marginal posteriors and a correlation term, ρ. In Figure 14.7 each panel shows the first 10 steps of a Gibbs sampler starting at $(1, -1)$. The total Euclidean distance traveled is an order of magnitude greater for the case where the correlation coefficient is 0.00 instead of -0.95, even though the two marginal posteriors are defined on the same support. Notice that the chain on the uncorrelated surface travels about seven times further in the same number of steps. This means that the second Markov chain will mix through the target distribution much more efficiently than the first since it is not constrained from taking big orthogonal moves in two steps.

Usually this problem is easy to detect since slow mixing leads to underestimates of the variance, an effect that appears clearly in Figure 14.7. In addition, as already shown, BUGS supplies the parameter correlation matrix. Cowles, Roberts, and Rosenthal (1999) used a simulation study of this problem, also with standard normal marginal distributions, and showed that the variance underestimation can be appreciable. Worse yet, they demonstrate that discarding iterations by thinning the chain exacerbates the problem.

A number of solutions via reparameterization have been suggested to deal with this problem. In the case of normal and related models simple algebraic forms include: $\gamma_1 = \theta_1 + \theta_2$, $\gamma_2 = 3(\theta_1 - \theta_2)$ (Gilks and Roberts 1996), centering the covariates $\mathbf{x}'_j = \mathbf{x}_j - \bar{\theta}_j$ (Hills and Smith 1992, Zuur, Garthwaite, and Fryer 2002), centering the covariates by replacing something like $Y_{ijk} = \mu + \alpha_i + \beta_{ij} + \epsilon_{ijk}$ in the normal linear model with $Y_{ijk} = \eta + \rho_{ij} + \epsilon_{ijk}$ where $\eta_i = \mu + \alpha_i$, and $\rho_{ij} = \mu + \alpha_i + \beta_{ij}$ (Gelfand, Sahu, and Carlin 1995), or just centering by "sweeping" all of the parameter means into a single term (Vines, Gilks, and Wild 1996). These reparameterizations are particularistic to the context of the model and the data and require considerable thought about the forms chosen as it possible to actually make the mixing worse.

14.4.2   Grouping and Collapsing the Gibbs Sampler

Liu (1994) suggests altering the conditional calculations of the Gibbs sampler in the following way in order to mitigate the effects of high cross-chain correlations. Using the standard notation, but for only three variables, consider the following two modifications:

Grouped Gibbs Sampler	Collapsed Gibbs Sampler
$\theta_1^{[j]}, \theta_2^{[j]} \sim \pi(\theta_1, \theta_2 \mid \theta_3^{[j-1]})$	$\theta_1^{[j]}, \theta_2^{[j]} \sim \pi(\theta_1, \theta_2)$
$\theta_3^{[j]} \sim \pi(\theta_3 \mid \theta_1^{[j]}, \theta_2^{[j]})$	$\theta_3^{[j]} \sim \pi(\theta_3 \mid \theta_1^{[j]}, \theta_2^{[j]})$

where the difference is that the collapsed Gibbs sampler first draws from the unconditional joint distribution of θ_1 and θ_2, whereas the grouped (also called blocked) Gibbs sampler first conditions on θ_3. So it must be possible to express and sample from the joint forms dictated by grouping or collapsing (the first two lines using θ_1 and θ_2 above).

The appeal of grouping and collapsing is that they shift the high correlations from the Gibbs sampling process over to a more direct random vector generation process (Seewald 1992). This principle means that if there are difficult components of the Gibbs sampler (with regard to convergence and mixing), then we can devise a "side process" to draw values in a more efficient manner, since high correlations can be "internalized" into a single joint draw.

Obviously these modifications are not a panacea; Roberts and Sahu (1997) warn that incautious specification of the blocking can even *slow down* the convergence of the Markov chain (see the example in Whittaker [1990]). It is also possible to encounter situations where it is difficult to sample from the unconditional joint form, although Chen, Shao, and

Ibrahim (2000, p.78) point out that it is possible to run sub-chains within the first step to accomplish this (that is, to draw θ_1 and θ_2 from a two-component sub-Gibbs sampler).

The choice of grouping versus collapsing is typically determined by the structure of the problem: the parametric form will frequently dictate which of the two forms are available. It is generally better in the presence of correlation problems to use grouping (Chen, Shao, and Ibrahim 2000). The real challenge is to specify the useful configuration of blocking collapsing in the sampler since this is a customized process outside the province of BUGS.

14.4.3 Adding Auxiliary Variables

Another idea for improving mixing and convergence is the strategy of introducing an additional variable into the Markov chain process as a means of improving mixing and convergence (Besag and Green 1993, Edwards and Sokal 1988, Swendsen and Wang 1987, Mira and Tierney 2001a). The principle is to augment a k-dimensional vector of coefficients, $\theta \in \Theta$, with (at least) one new variable, $u \in \mathbf{U}$, so that the $(k+1)$-dimensional vector on $\Theta \times \mathbf{U}$ space has good chain properties. The augmenting variables sometimes have direct interpretational value, but this is not necessary as the real goal is to improve the Markov chain. Actually the Metropolis-Hastings algorithm is already an auxiliary variable process since the candidate-generating distribution *is* auxiliary to the distribution of interest. Higdon (1998) points out that the Metropolis-Hastings algorithm is an auxiliary variable process because the use of the uniform random variable for acceptance decisions can be rewritten in this form.

A number of extensions have been developed to deal with circumstances that arise when one of the conditional probability statements is not directly available or when some less simple alternative to the transition kernel is required (Fredenhagen and Marcu 1987). There is a small literature on the theoretical properties of auxiliary variable Markov chains (Brooks 1998b). Hurn (1997) finds that auxiliary variable algorithms have problems with posterior distributions that are highly multimodal or have strong interaction components. More recent developments include a linkage to the slice sampler (below), (Wakefield, Gelfand, and Smith 1991; Besag and Green 1993).

14.4.4 The Slice Sampler

The slice sample is a very important tool for modern MCMC software. As noted, introducing an additional variable into the Markov chain process can improve mixing or convergence. The strategy with slice sampling is to augment a k-dimensional vector of coefficients, $\theta \in \Theta$, with an additional variable, $\psi \in \mathbf{\Psi}$, with the result that the $(k+1)$-dimensional vector on $\Theta \times \mathbf{\Psi}$ space has better mixing through this space (Neal 2003).

First consider the posterior distribution of θ is given as $\pi(\theta)$, where conditioning on the data is assumed. Now define $\pi(\psi|\theta)$ and the joint posterior distribution given by $\pi(\theta, \psi) =$

$\pi(\psi|\boldsymbol{\theta})\pi(\boldsymbol{\theta})$. Clearly the distribution $\pi(\psi|\boldsymbol{\theta})$ is picked to be easy to sample from, otherwise the implementation problems are worse. In total there are two conditional transition kernels required: one that updates $\boldsymbol{\theta}$, $P[\boldsymbol{\theta} \to \boldsymbol{\theta}'|\psi]$, and one that updates ψ, $P[\psi \to \psi'|\boldsymbol{\theta}]$. In the case of Gibbs sampling these are exactly the full conditional distributions: $\pi(\boldsymbol{\theta}'|\psi)$ and $\pi(\psi'|\boldsymbol{\theta})$, meaning that the algorithm cycles normally through the $\boldsymbol{\theta}$ statements (now conditional on one more parameter) and then adds a step for ψ. This two-step aspect of the process ensures that the stationary process remains $\pi(\boldsymbol{\theta})$ since the detailed balance equation is maintained. The nice part is that we run the chain normally, presumably with better mixing, and then just discard the ψ values.

Generally the added variable in the slice sampler is a uniform draw, so we update the marginals and the joint distribution with uniforms in the following way:

$$\psi^{[j+1]}|\boldsymbol{\theta}^{[j]} \sim \mathcal{U}(0, \pi(\boldsymbol{\theta}^{[j]}) \qquad \boldsymbol{\theta}^{[j+1]}|\psi^{[j+1]} \sim \mathcal{U}(\boldsymbol{\theta}{:}\pi(\boldsymbol{\theta}^{[j+1]}) > \psi^{[j+1]}) \qquad (14.22)$$

in going from the jth to the $(j+1)$st step. In the basic setup the target distribution is expressed as a product of the marginals, $\pi(\theta) = \prod \pi(\theta_i)$. This means that the auxiliary variable is sampled from a uniform bounded by zero and the $\boldsymbol{\theta}$, then these $\boldsymbol{\theta}$ are in turn sampled from a uniform bounded below by the auxiliary variable(s).

Roberts and Rosenthal (1998) show that the slice sampler has many good theoretical properties, such as geometric convergence, thus making it one of the more successful auxiliary variable applications (see also Mira and Tierney [2001b]). For specific applications see Damien, Wakefield, and Walker (1999), Tierney and Mira (1999), as well as Robert and Casella (2004, Chapter 8) for a list of difficulties.

A simple example given by Robert and Casella (1999), and implemented in R by Altman, Gill, and McDonald (2003), uses a uniform distribution to generate desired normals. Begin with at target $f(\theta) \propto \exp(-\theta^2/2)$, and stipulate a uniform such that:

$$\psi|\theta \sim U[0, \exp(-\theta^2/2)], \qquad \theta|\psi \sim U[-\sqrt{-2\log(\psi)}, \sqrt{-2\log(\psi)}],$$

which makes this ψ an auxiliary variable with direct sampling properties. Notice that the first statement could easily be modified to generate other forms. The R code to implement this follows directly:

```
n <- 1000; t.vals <- 0; p.vals <- 0
for (i in 2:n) {
    p.vals <- c(p.vals,runif(1,0,exp(-0.5*t.vals[(i-1)]^2)))
    t.vals <- c(t.vals,runif(1,-sqrt(-2*log(p.vals[i])),
            sqrt(-2*log(p.vals[i]))))
}
```

The resulting t.vals draws are displayed in Figure 14.8, which shows them to closely resemble normally distributed random variables. We can consider this particular example as an MCMC version of the Box-Müller algorithm (1958).

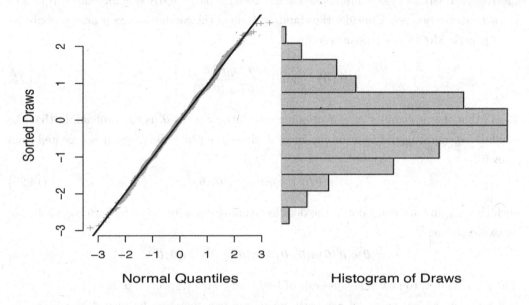

FIGURE 14.8: SLICE SAMPLER OUTPUT

14.5 Chib's Method for Calculating the Marginal Likelihood Integral

As noted in Chapter 7, the integrals in Bayes Factors can be challenging to estimate with reasonably realistic models. Usually hierarchical specifications lead to these complications, but hierarchical modeling is one of the strengths of the Bayesian paradigm. One general class of tools for getting these marginal likelihoods is provided by Chib (1995) for Gibbs sampling and Chib and Jeliazkov (2001) for Metropolis-Hastings. See also Geweke (2005, pp.257-261). The particular objective is to calculate $p(\mathbf{x}|M_i) = \int_{\boldsymbol{\theta}} f_i(\mathbf{x}|\boldsymbol{\theta}_i)p_i(\boldsymbol{\theta}_i)d\boldsymbol{\theta}_i$ in (7.10) (page 216) for the ith model. It is easier to look at this quantity if we rearrange Bayes' Law and take logs (dropping the model index since we will consider only one model here for now):

$$\log p(\mathbf{x}) = \ell(\boldsymbol{\theta}'|\mathbf{x}) + \log p(\boldsymbol{\theta}') - \log \pi(\boldsymbol{\theta}'|\mathbf{x}) \qquad (14.23)$$

where $\boldsymbol{\theta}'$ is a completely arbitrary (but acceptable) point in the sample space. Typically we will choose a point from the high density region such as the posterior mean. So if we had an estimate of $\pi(\boldsymbol{\theta}'|\mathbf{x})$ from simulation, this would be a straightforward calculation. We will now describe the method for Metropolis-Hastings. The Gibbs sampling version, a direct extension of data augmentation (Tanner and Wong 1987) is described clearly in Chib (1995), and the algorithm is described in Chapter 7, Exercise 20 (page 246). The

Metropolis-Hastings case (Chib and Jeliazkov 2001) is particularly elegant since it is tied to the acceptance process. Consider the standard form of the candidate acceptance probability for a generic Metropolis-Hastings setup:

$$\alpha(\boldsymbol{\theta}', \boldsymbol{\theta}) = \min\left[\frac{\pi(\boldsymbol{\theta}')}{\pi(\boldsymbol{\theta})}\frac{q_t(\boldsymbol{\theta}|\boldsymbol{\theta}')}{q_t(\boldsymbol{\theta}'|\boldsymbol{\theta})}, 1\right]. \tag{14.24}$$

Recall that the probability of transitioning to arbitrary point $\boldsymbol{\theta}'$ is the probability that the candidate-generating distribution produces $\boldsymbol{\theta}'$ times the probability that it is accepted from above:

$$p(\boldsymbol{\theta}, \boldsymbol{\theta}') = q(\boldsymbol{\theta}'|\boldsymbol{\theta})\alpha(\boldsymbol{\theta}', \boldsymbol{\theta}) \tag{14.25}$$

such that for any arbitrary point, the detailed balance equation, (10.18) from page 356, can be expressed as:

$$\pi(\boldsymbol{\theta})q(\boldsymbol{\theta}'|\boldsymbol{\theta})\alpha(\boldsymbol{\theta}', \boldsymbol{\theta}) = \pi(\boldsymbol{\theta}')q(\boldsymbol{\theta}|\boldsymbol{\theta}')\alpha(\boldsymbol{\theta}, \boldsymbol{\theta}'). \tag{14.26}$$

The innovation is to now take integrals of both sides with respect to $\boldsymbol{\theta}$, realizing that $\pi(\boldsymbol{\theta}')$ is a function evaluation at an arbitrary point and can therefore be moved outside of the integral. Doing this and rearranging slightly gives:

$$\pi(\boldsymbol{\theta}') = \frac{\int_{\boldsymbol{\Theta}}\pi(\boldsymbol{\theta})q(\boldsymbol{\theta}'|\boldsymbol{\theta})\alpha(\boldsymbol{\theta}', \boldsymbol{\theta})d\boldsymbol{\theta}}{\int_{\boldsymbol{\Theta}}q(\boldsymbol{\theta}|\boldsymbol{\theta}')\alpha(\boldsymbol{\theta}, \boldsymbol{\theta}')d\boldsymbol{\theta}}, \tag{14.27}$$

which is really just a ratio of two expected value calculations:

$$\pi(\boldsymbol{\theta}') = \frac{E_{\pi(\boldsymbol{\theta})}[q(\boldsymbol{\theta}'|\boldsymbol{\theta})\alpha(\boldsymbol{\theta}', \boldsymbol{\theta})]}{E_{q(\boldsymbol{\theta}|\boldsymbol{\theta}')}[\alpha(\boldsymbol{\theta}, \boldsymbol{\theta}')]}. \tag{14.28}$$

Of course what expected value calculations do is set us up for simulation solutions, and what Chib and Jeliazkov realized is that in the course of running a standard Metropolis-Hastings algorithm for marginal posterior distributions, we can get the marginal likelihood along the way with not much extra trouble. So replace (14.28) with its simulation analog:

$$\pi_{\text{sim}}(\boldsymbol{\theta}') = \frac{\frac{1}{M}\sum_{m=1}^{M}\alpha(\boldsymbol{\theta}', \boldsymbol{\theta}_m)q(\boldsymbol{\theta}'|\boldsymbol{\theta}_m)}{\frac{1}{N}\sum_{n=1}^{N}\alpha(\boldsymbol{\theta}_N, \boldsymbol{\theta}')}, \tag{14.29}$$

which of course uses known functions and values readily at hand. Typically, but not by necessity, $M = N$ here.

Recall that $\boldsymbol{\theta}'$ is chosen arbitrarily but within some high density region of the posterior. So this process substitutes a difficult integration process with simulation of the posterior density at a single point by completing (14.23) with the simulated result:

$$\log p_{\text{sim}}(\mathbf{x}) = \ell(\boldsymbol{\theta}'|\mathbf{x}) + \log p(\boldsymbol{\theta}') - \log \pi_{\text{sim}}(\boldsymbol{\theta}'|\mathbf{x}). \tag{14.30}$$

The other quantities on the right-hand side are readily available, so the marginal likelihood is calculated.

14.6 Rao-Blackwellizing for Improved Variance Estimation

Gelfand and Smith (1990) introduced the idea of "Rao-Blackwellization" as a means of exploiting the conditional nature of Gibbs sampling in order to produce improved variance estimates for statistics of interest through conditioning (the pre-MCMC simulation idea appears to originate with the text of Hammersley and Handscomb [1964]). This technique can also be applied to the Metropolis-Hastings algorithm (Casella and Robert 1996), data augmentation (Liu, Wong, and Kong 1994), EM (Meng and Schilling 1996), nonparametric Bayes (Liu 1996b), to detect outliers (Haro-López, Mallick, and Smith 2000), and also in general non-Bayesian procedures.

The Rao-Blackwell Theorem states that estimates conditioned on sufficient statistics are subsequently unaffected or improved in terms of the variance of their sampling distribution. When applied to variance calculations (the interest here), the theorem can be expressed through the relationship between conditional and unconditional variance and is not always based on sufficient statistics. Start with a statistic of interest such as a mean or quantile for one parameter of a two-parameter Markov chain: $h(\theta_1)$. The variance relationship is defined by:

$$\text{Var}[h(\theta_1)] = \text{Var}[E(h(\theta_1|\theta_2))] + E(\text{Var}[h(\theta_1|\theta_2)]). \tag{14.31}$$

So by simply rearranging we can show that the variance of the expected value of the conditional cannot be lower since variances are never negative:

$$\text{Var}[E(h(\theta_1|\theta_2)] = \text{Var}[h(\theta_1)] - E(\text{Var}[h(\theta_1|\theta_2)]), \tag{14.32}$$

provided that the method for producing the Markov chain values does not introduce exacerbating correlations between terms (see Liu, Wong, and Kong [1995] and Geyer [1995] for specifics). A standard means of comparison is the asymptotic relative efficiency (ARE), defined by the ratio of the variance of the limiting distribution of two estimates. In the case of a two-variable Gibbs sampler, these are for θ_1:

$$\underset{n\to\infty}{\text{Var}}[h(\theta_1)] = \text{Var}[h(\theta_1)] + 2\sum_{i=1}^{\infty} \text{Cov}(h(\theta_1^{[0]}), h(\theta_1^{[i]}))$$

$$\underset{n\to\infty}{\text{Var}}[E(h(\theta_1|\theta_2))] = \text{Var}[E(h(\theta_1|\theta_2))]$$

$$+ 2\sum_{i=1}^{\infty} \text{Cov}(E[h(\theta_1^{[0]}|\theta_2^{[0]})], E[h(\theta_1^{[i]}|\theta_2^{[i]})]), \tag{14.33}$$

where the brackets in the super-script denote Markov chain iterations. Levine (1996) showed that the ARE of the first unconditional variance over the Rao-Blackwellized variance is always greater than or equal to one, thus favoring use of the Rao-Blackwellized version.

Fortunately, the generalization of this last result issue is not typically a problem in that most conventional applications and a number of authors have provided suitable conditions. For instance, Liu, Wong, and Kong's (1994, 1995) identification of an *interleaving property* is satisfied by the conditions:

▷ $\theta_1^{[i]}$ and $\theta_1^{[i+1]}$ are conditionally independent given $\theta_2^{[i]}$,

▷ $\theta_2^{[i-1]}$ and $\theta_2^{[i]}$ are conditionally independent given $\theta_1^{[i]}$,

▷ and the chain has converged such that $(\theta_1^{[i]}, \theta_2^{[i-1]})$, $(\theta_1^{[i]}, \theta_2^{[i]})$ are iid.

The unconditional and Rao-Blackwellized empirical estimate of $h(\theta_1)$ from n post-convergence Markov chain values are given by the ergodic theorem as:

$$h(\theta_1)_{uc} = \frac{1}{n}\sum_{i=1}^{n} h(\theta_1^{[i]}), \qquad h(\theta_1)_{rb} = \frac{1}{n}\sum_{i=1}^{n} E(h(\theta_1^{[i]}|\theta_2^{[i]}). \tag{14.34}$$

Furthermore, both of these estimators are unbiased and are known to converge to the marginal density $h(\theta_1)$.

The general strategy of Gelfand and Smith (1990) is therefore to condition each parameter estimate on the others: if $\pi(\theta_i)$ is a marginal density estimate (smoothed or empirical), then the Rao-Blackwellized estimate can be calculated by $\hat{\pi}(\theta_i) = \frac{1}{k-1}\sum_{j=1}^{k-1}\pi(\theta_i|\theta_{j,-i})$, for k parameters as an empirical estimate of $\int_{\theta_j}\pi(\theta_i|\theta_{j,-i})\pi(\theta_{j,-i})d\theta_{j,-i}$. It turns out that this is not too hard to do and can lead to vastly improved variance estimates.

Other more elaborate schemes can also be implemented. For instance if there exists a parallel chain structure (as in the Raftery and Lewis diagnostic) conditioning on this chain at each iteration: $\hat{\pi}(\theta_i) = \frac{1}{k-1}\sum_{j=1}^{k-1}\pi(\theta_i|\mathbf{z}_{j,-i})$. Casella and Robert (1996) use the rejected values in a Metropolis-Hastings algorithm in the following way. If θ_a are the accepted values of the algorithm and θ_c are the candidate values, only some of which are accepted in the application of the Metropolis-Hastings algorithm in the process of obtaining n chain values, then the uniform random variable used to make the accept/reject decision is an ancillary statistic and can be conditioned on (integrated over). If the estimator is expressed usually using only the accepted values:

$$h(\theta_1)_{uc} = \frac{1}{n}\sum_{i=1}^{n} h(\theta_a^{[i]}), \tag{14.35}$$

then we can use an indicator function to re-express to include the rejected values:

$$h(\theta_1)_{rb} = \frac{1}{n}\sum_{i=1}^{n}\left[h(\theta_c^{[i]})\sum_{j=i}^{n} I_{(\theta_a^{[i]}=\theta_c^{[i]})}\right]. \tag{14.36}$$

Recall that the acceptance ratio was defined in Chapter 10 by $a(\theta_c, \theta_{i-1})$ and the acceptance probability of the candidate value k versus status quo value i (ρ_{ik}) is the minimum of $a(\theta_c, \theta_{i-1})$ and 1. Now define at the ith step going forth by the j:

$$\xi_{ij} = \prod_{k=i+1}^{j} (1 - \rho_{ik}) \quad i < j, \qquad \text{and} \quad \xi_{ii} = 1$$

$$\delta_i = p(\theta_a^{[i]} = \theta_c^{[i]} | \theta_c^{[1]}, \ldots, \theta_c^{[i]}) = \sum_{j=1}^{i-1} \delta_j \xi_{j(i-1)} \rho_{ji}$$

$$\phi_i = \delta_i \sum_{j=i}^{n} \xi_{ij}. \tag{14.37}$$

Then we get the following version of the Rao-Blackwellized estimator:

$$h(\theta_1)_{rb} = \frac{1}{n} \sum_{i=1}^{n} \phi_i h(\theta_c). \tag{14.38}$$

Casella and Robert subsequently show circumstances where this leads to a superior estimator over the unconditional version. This process is not limited in its application to the Metropolis-Hastings algorithm, and Chen, Shao, and Ibrahim (2000) give extensions (page 360).

Unfortunately, the application of Rao-Blackwellization is particularistic to each given model, and can require extensive customized coding of the chain since it is integrally tied into the functioning of the chain. This may deter a number of users, despite its appealing characteristics.

Utilitarian Epilogue

Some issues covered here are conceptually and managerially simple, like starting points, thinning, and the idea of a burn-in period. Most frequent users of MCMC tools fall into good habits and are aware such details need to be considered. More important is the consideration of the length of the chain to be run, conditional on the complexity of the model. Discussion ranged herein from the four dominant diagnostic tests to a range of less common alternatives and graphical approaches. At some point experienced users with relatively complex model specification leave the comfortable world of BUGS and write customized samplers. In such endeavors tools like reparameterization, grouping/collapsing, augmentation, and alternatives like the slice sampler can be enormously helpful in producing efficiently mixing Markov chains. Rao-Blackwellization from sampler output also has the potential to improve results with user-developed samplers. One purpose of the short discussion of these methods in this chapter is to direct interested researchers into the ongoing literature on sampler development.

14.7   Exercises

14.1   For the hierarchical model of firearm-related deaths in Chapter 6, perform the following diagnostic tests from Section 14.3 in CODA or BOA: Gelman and Rubin, Geweke, Raftery and Lewis, Heidelberger and Welsh.

14.2   (Robert and Casella 2004, Chapter 8). In Chapter 7, Exercise 20 (page 246), samples from the truncated normal distribution $f(x) = k \exp[-\frac{1}{2}(x+3)^2]I_{0,1}$ (k a normalizing constant) were generated. Rejection sampling here can be inefficient, so write a slice sampler in R according to the following steps at iteration t:

   (a) draw $u^{[t+1]} \sim \mathcal{U}_u(0:f(x^{[t]}))$ (the vertical step)

   (b) draw $x^{[t+1]} \sim \mathcal{U}_{f(x)}(u^{[t+1]}:1)$ (the horizontal step)

   where the second step means that a value of x is drawn uniformly such that $k \exp[-\frac{1}{2}(x+3)^2]I_{0,1} > uf(x^{[t]})$.

14.3   Plot in the same graph a time sequence of 0 to 100 versus the annealing cooling schedules: logarithmic, geometric, semi-quadratic, and linear. What are the trade-offs associated with each with regard to convergence and processing time? (*This question is left here to be compatible with the answer key to the Second Edition. Read Section annealing.section before attempting.*)

14.4   Explain how dramatic increases in computational power has made the 1990s debate about "one long run" versus "many short runs" for assessing convergence unimportant.

14.5   Show how the full conditional distributions for Gibbs sampling in the Bayesian Tobit model on page 482 can be changed into grouped or collapsed Gibbs sampler forms (Section 14.4.2).

14.6   Write a sampler in R to implement the grouped or collapsed Gibbs sampler from the problem above and use it to estimate the marginal posteriors for the death penalty support data (Example 14.3, page 481) by modifying the code in this chapter's **Computational Addendum**.

14.7   BUGS includes a quick diagnostic command, diag, which implements a version of Geweke's (1992) time-series diagnostic for the first 25% of the data and the last 50% of the data. Rather than calculate spectral densities for the denominator of the G statistic, BUGS divides the two periods into 25 equal-sized bins and then takes the variance of the means from within these bins. Run the example in Section 14.3 using the BUGS code in the **Computational Addendum** with 10,000 iterations, and load the data into R with the command . military.mcmc <-

`boa.importBUGS("my.bugs.dir/bugs")`. Write a function to calculate the short-cut for Geweke's diagnostic for each of the five chains (columns here). Compare your answer with BUGS.

14.8   David C. Baldus, Charles Pulaski, and George Woodworth (1983, 1990), i.e., the "Baldus Study," looked at the potential disparity in the imposition of the death sentence in Georgia based on the race of the murder victim and the race of the defendant. Using the dataset `baldus` in BaM, write a model in BUGS or JAGS with the variable `sentence` (death sentence imposed: 325 0 cases, 127 1 cases) and specify a linear model including a multivariate normal prior (`dmnorm`). Modify this prior specification to make it a multivariate version of Zellner's g-prior (Chapter 4, Exercise 20, on page 142).

14.9   For the model of marriage rates in Italy given in Section 12.5.1, calculate the Zellner and Min diagnostic with the segmentation: $\theta_1 = \alpha$, $\theta_2 = \beta$, for the posteriors defined at the iteration points: $\hat{\pi}_{1,000}(\alpha, \beta, \boldsymbol{\lambda}, \mathbf{X})$ and $\hat{\pi}_{10,000}(\alpha, \beta, \boldsymbol{\lambda}, \mathbf{X})$. Test H_0: $\eta = 0$ versus H_1: $\eta \neq 0$ with the statistic K_{OA} assuming serial independence.

14.10 Write an R function that implements the Ritter and Tanner diagnostic (page 508) and apply it to the Tobit death penalty example (starting on 481).

14.11 (Raftery and Lewis 1996). Write a Gibbs sampler in R to produce a Markov chain to sample for θ_1, θ_2, and θ_3, which are distributed according to the trivariate normal distribution

$$
\begin{bmatrix} \theta_1 \\ \theta_2 \\ \theta_3 \end{bmatrix} \sim \mathcal{N} \left(\begin{bmatrix} 0 \\ 0 \\ 0 \end{bmatrix}, \begin{bmatrix} 99 & -7 & -7 \\ -7 & 1 & 0 \\ -7 & 0 & 0 \end{bmatrix} \right)
$$

by cycling through the normal conditional distributions with variances: $V(\theta_1|\theta_2,\theta_3) = 1$, $V(\theta_2|\theta_1,\theta_3) = 1/50$ and $V(\theta_3|\theta_1,\theta_2) = 1/50$. Contrast the Geweke diagnostic with the Gelman and Rubin diagnostic for 1,000 iterations.

14.12 Write a dynamic animation that shows the witch's hat distribution melting with simulated annealing.

14.13 Prove that data augmentation is a special case of Gibbs sampling.

14.14 The military personnel model in Section 14.3 fits poorly because of lack of independence in the countries (columns) and the years (rows). Write the appropriate model in BUGS/JAGS that accounts for one very influential country and the time series effect. Compare your results to those in Table 14.3.

14.15 For the following simple bivariate normal model

$$
\begin{bmatrix} \theta_1 \\ \theta_2 \end{bmatrix} \sim \mathcal{N} \left(\begin{bmatrix} 0 \\ 0 \end{bmatrix}, \begin{bmatrix} 1 & \rho \\ \rho & 1 \end{bmatrix} \right),
$$

the Gibbs sampler draws iteratively according to:

$$\theta_1|\theta_2 \sim \mathcal{N}(\rho\theta_2, 1 - \rho^2)$$
$$\theta_2|\theta_1 \sim \mathcal{N}(\rho\theta_1, 1 - \rho^2).$$

Write a simple implementation of the Gibbs sampler in R, then calculate the unconditional and conditional (Rao-Blackwellized) posterior standard error, for both $\rho = 0.05$ and $\rho = 0.95$. Compare these four values and comment.

14.16   The Lehmann-Scheffé theorem states that an unbiased estimate that is conditioned on a complete *and* sufficient statistic is the unique most efficient estimator: if T is a complete and sufficient statistic and $E(f(T)) = \theta$, then $f(T)$ is the minimum variance (unbiased) estimator of θ. This is therefore a stronger statement than the Rao-Blackwell theorem. Why then is the Rao-Blackwellized posterior standard error more useful for MCMC purposes than a "Lehmann-Scheffé" posterior standard error?

14.17   Produce the ARE result for the Rao-Blackwellization from (14.33) with a two-variable Metropolis-Hastings algorithm instead of the Gibbs sampler that was used.

14.18   The Behrens-Fisher problem is an old problem that is awkward in non-Bayesian settings and straightforward with a Bayesian treatment. Suppose there are two independent samples that are both normally distributed according to:

$$x_{11}, \ldots, x_{1n_1} \sim \mathcal{N}(\mu_1, \sigma_1^2), \qquad x_{21}, \ldots, x_{2n_1} \sim \mathcal{N}(\mu_2, \sigma_2^2)$$

where all four parameters are unknown. Following the advice of Berger (1985) specify the following prior distributions:

$$p(\mu_1) \propto 1 \, p(\mu_2) \propto 1$$
$$p(\sigma_1^2) \propto 1/\sigma_1^2 \, p(\sigma_2^2) \propto 1/\sigma_2^2.$$

Using two selected countries from military personnel data in Section 14.3, write a sampler to produce posterior values from $\mu_1 - \mu_2$.

14.19   Modify the Metropolis-Hastings R code on page 359 to produce the marginal likelihood with Chib's method.

14.20   Write an implementation of Chib and Jeliazkov's (2001) method for obtaining the marginal likelihood from Gibbs sampling output in R.

14.8 Computational Addendum: Code for Chapter Examples

14.8.1 R Code for the Death Penalty Support Model

```
# PACKAGES AND DATA LOAD
lapply(c("BaM","msm","survival","LearnBayes"),library,character.only=TRUE)
# msm FOR THE rtnorm FUNCTION, survival FOR THE survreg FUNCTION
# LearnBayes FOR THE rigamma FUNCTION
data(norr)
# SETUP REDUCED DATA STRUCTURE FOR GIBBS
attach(norr.df)
Y <- c(ds9395p)
X <- cbind(rep(1,nrow(norr.df)),ep4089n,polcul,d8892r2,id8892m2,murder90)
detach(norr.df)
dimnames(X)[[2]] <- c("Intercept","Past Rates","Political Culture",
                      "Current Opinion","Ideology","Murder Rate")

# RUN BIVARIATE REGRESSIONS TO GET PRIOR MEANS FOR BETAS
beta0 <- rep(NA,ncol(X)); beta0[1] <- 1
for (i in 2:ncol(X))   beta0[i] <- summary(lm(Y~X[,i]))$coef[2,1]

# SET PARAMETERS FOR GIBBS
gamma0 <- 300; gamma1 <- 100               # HYPERPARAMETERS
B0 <- 0.02                                 # SCALING ON SIGMA^0
mc.start <- 1; mc.stop <- 50000            # MCMC CONTROL
norr.tob <- survreg(Surv(Y,Y>0,type='left') ~ X[,-1],
                    dist='gaussian',data=norr.df)
B.mat   <- matrix(summary(norr.tob)$coef,nrow=1) # STARTING POINTS
s.sq    <- matrix(rep(10,6), nrow=1)       # PRIOR on SIGMA^2 OF BETA
y.star <- Y                                # LATENT DATA START

# GIBBS SAMPLER
for (i in mc.start:mc.stop)  {
    for (j in 1:length(Y)) {
        if (Y[j] == 0)  y.star[j] <- rtnorm(1,X[j,]%*%B.mat[i,],
                                         s.sq[i,],lower=-Inf,upper=0)
    }
    delta <- t(y.star - X %*% B.mat[i,]) %*% (y.star - X %*% B.mat[i,])
    s.temp <- rep(Inf,ncol(s.sq)); while(!is.finite(sum(s.temp)))
                    s.temp <- rigamma(ncol(s.sq), (gamma0+length(Y))/2,
                                    (gamma1+delta)/2)
    s.sq <- rbind(s.sq,s.temp)
    B.hat <- solve( B0 + t(X)%*%X ) %*% (B0*beta0 + t(X)%*%y.star)
    B.var <-  make.symmetric( solve( s.sq[(i+1)]^(-1)*B0
                         + s.sq[(i+1)]^(-1)*t(X)%*%X ) )
```

```
        B.mat <- rbind( B.mat, rmultinorm(1,B.hat,B.var,tol=1e-06) )
        if (i %% 100 == 0)  print(paste("iteration:",i))
}
```

14.8.2   JAGS Code for the Military Personnel Model

```
model {
    for (i in 1:YEAR) {
        for (j in 1:COUNTRY) {
            mu[i,j] <- alpha[i] + beta1[i]*(x[j]) + beta2[i]*(x[j]^2)
                       + beta3[i]*cos(x[j])
            y[i,j]    ~ dnorm(mu[i,j],tau.c)
        }
        alpha[i] ~ dnorm(alpha.mu,alpha.tau)
        beta1[i] ~ dnorm(beta1.mu,beta1.tau)
        beta2[i] ~ dnorm(beta2.mu,beta2.tau)
        beta3[i] ~ dnorm(beta3.mu,beta3.tau)
    }
    alpha.mu   ~ dnorm(1,0.1)
    alpha.tau ~ dgamma(1,0.1)
    beta1.mu   ~ dnorm(0,0.1)
    beta1.tau ~ dgamma(1,0.1)
    beta2.mu   ~ dnorm(0,0.1)
    beta2.tau ~ dgamma(1,0.1)
    beta3.mu   ~ dnorm(0,0.1)
    beta3.tau ~ dgamma(1,0.1)
    tau.c      ~ dgamma(1,0.1)
}
```

Chapter 15

Markov Chain Monte Carlo Extensions

This chapter reviews some recent extensions and modifications of the classic MCMC algorithms described in previous chapters. Generally, researchers develop these methods to deal with problematic estimation conditions such as multimodality, extremely high dimension, and difficult convergence issues. This chapter also discusses some more recent developments that speed-up the process of MCMC estimation in ways that extend standard Bayesian stochastic estimation. These deviate from the normal orthodoxy but perhaps point towards future development. One very promising direction described here is Hamiltonian Monte Carlo, which is a variant of the Metropolis-Hastings algorithm. This literature is incredibly dynamic will remain so for the next decade or more, so some of these sections merely point at emerging literatures.

15.1 Simulated Annealing

Kirkpatrick, Gelatt, and Vecchi (1983) and Černý (1985) proposed the MCMC traversal technique *simulated annealing*, which, like Metropolis-Hastings, has roots in statistical physics but is also useful in general stochastic simulation for Bayesian modeling. Simulated annealing is a flexible stochastic procedure for iteratively traversing highly textured sample spaces where conditions have been changed to make movement easier. Correctly set up and run, the annealing algorithm is guaranteed to explore the global maxima (Lundy 1985), unlike the EM algorithm, which seeks the nearest mode and remains there. This makes it an enormously useful approach in MCMC work with multimodal posterior forms, although it can also be very slow both in movement and with necessary human involvement (van Laarhoven and Aarts 1987). There is also a related literature on simulated annealing for pure optimization (Eglese 1990, Fleischer 1995) rather than posterior description which is our interest here.

We have seen that it is possible for a Markov chain to get "stuck" for an inordinately long period of time in a nonoptimal region of the sample space exploring attractive modal features that are not the primary density areas of interest. This is particularly troublesome when such a region separates two high-density areas and the chain finds it difficult to fully explore both. Usually this problem is more common in higher dimensions, but imagine a two-dimensional, bimodal structure with a large mode, a small mode, and a wide gulf in the middle. It may be the case in this example that the trough in the middle seriously impedes the chain from easily passing from one mode to the other.

The analogous idea behind simulated annealing is the principle that less brittle solid metals are produced when allowed to cool slowly from a liquid state. In fact the metallurgic process *is* called annealing. Imagine that we can increase the "temperature" of the state space, melting down prominent features: flattening out modes and valleys such that the chain finds it easier to move through previously low-density regions. Then when the temperature is reduced back to normal levels the chain has hopefully passed through impediments to free travel. This process is done by temporarily altering all of the probabilities in the transition kernel for a Metropolis-Hastings chain, running the Markov chain for a time, and then returning the probabilities to their previous values.

The primary decision is the determination of the temperature schedule that dictates the process by which the chain returns to its original "cold" state. Start by defining a temperature parameter at time t: $T_t > 1$, and modifying the transition kernel according to $\pi^*(\theta) = \pi(\theta)^{\frac{1}{T}}$, being careful to renormalize. So heating the kernel by making T large flattens out its probability structure toward a uniform distribution. If there are multiple modes, they will melt into the surface and therefore no longer be strong attractions. It is always necessary to define an initial temperature, T_0, that provides sufficient melting, then determine a rate that slowly decreases T_t until reaching one and therefore returning to the original cold transition kernel probabilities, which are those of inferential value. The cooling schedule recognizes the competing phenomenon: *(1)* slow cooling enables greater coverage of the sample space, and *(2)* faster cooling gives more reasonable simulation times.

Consider what happens to a Metropolis-Hastings algorithm in this scenario. As the jumping distribution generates candidate positions, very few of these will be rejected and the Markov chain will rarely stay in place. This is "good;" it means that the chain can freely explore the sample space without impediments. It is also "bad" in that there is obviously much less of a tendency to remain in the (previous) high density areas. This is where the researcher-specified cooling process comes in.

The following modified Metropolis-Hastings setup, which was the original idea behind simulated annealing as first proposed by Metropolis *et al.* (1953), gives a simple Markov chain for example purposes. Suppose that at time (step) k we are at temperature T_k, having started at step 0, and the heated up temperature T_0. Then the next value is chosen according to the algorithm:

▷ At the tth step draw θ' from a uniform distribution around the current position, $\theta^{[t]}$.

▷ Define: $a(\theta', \theta) = \exp[-(\pi(\theta'_j) - \pi(\theta^{[t]}_j))/T_t]$, and make the decision:

$$\theta_j^{[t+1]} = \begin{cases} \theta'_j & \text{with probability} \quad \min\left(a(\theta', \theta), 1\right) \\[2mm] \theta_j^{[t]} & \text{with probability} \quad 1 - \min\left(a(\theta', \theta), 1\right). \end{cases} \tag{15.1}$$

▷ After sampling sufficiently that convergence is concluded at this temperature, move up the temperature schedule from T_t to T_{t+1}.

▷ Repeat steps 1-3 until the temperature schedule has been completed.

Therefore the transition matrix is first heated up such that the chain converges weakly over a near-uniform distribution. Once the cooling begins, the chain is observed to converge at progressively cooler temperatures until the transition matrix returns to its original state. This process can be repeated as many times as necessary to collect a sufficient number of cold chain values. This iterative process, which requires human interaction in this basic case, can be generalized in several ways described below to reduce the required management by the researcher.

The hottest jumping distribution for generating candidate values does not need to be uniform, but often is specified to be. While the Markov chain described here is not homogeneous as is the standard Metropolis-Hastings algorithm, Hàjek (1988) showed that the discrete state space still has required convergence properties by considering the recorded "cold" chain values to be the only expression of the Markovian process similar (but different) than the idea of thinning with an appeal to a classic theorem about grouping Markovian iterations in Meyn and Tweedie (1993). Similar properties hold for the continuous case and are given by Duflo (1996), Geman and Hwang (1986), Holley, Kusuoka, and Stroock (1989), and Jeng and Woods (1990) in more complex settings.

In general the temperature (cooling) schedule should gradually cool down by decreasing T_t very slightly on each iteration after T_0. A number of schemes have been proposed such as a logarithmic scale (Geman and Geman 1984): $T_t = kT_1/\log(t)$, a geometric-type process: $T_t = \epsilon T_{t-1}, 0 < \epsilon < 1$ (Mitra, Romeo, and Sangiovanni-Vincentelli 1986), as well as more complex algorithms from statistical physics (Aarts and Kors 1989). Hàjek (1988) showed that a logarithmic cooling schedule proportional to the height of the global maximum (often obtainable with EM or some other search tool) converges asymptotically to the full set of maxima.

The choice of a cooling schedule is necessarily a compromise between slow cooling to ensure greater coverage of the sample space and faster cooling back to zero in order to give a reasonable simulation time. This is not necessarily a stark trade-off. For instance, in the case of the Metropolis-Hastings algorithm the bounds of the jumping distribution can be adjusted so that as the temperature cools down the number of rejected jumping points does not go up dramatically (Gelfand and Mitter 1993).

■ **Example 15.1: An Illustrative Simple Simulated Annealing Setup.** This is a contrived and pathological example designed purely to show the characteristics of the simulated annealing algorithm. Consider a 21×21 matrix representing a two-dimensional discrete posterior state space. This matrix of unnormalized posterior density values is design to deter a standard Metropolis-Hastings chain from reaching and describing the global maxima values in the corners by putting an attractive mode in the middle, ringing it with compelling but nonoptimal points:

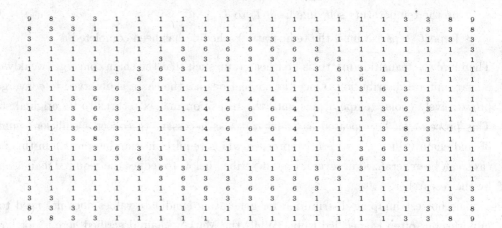

As a demonstration, we first run the simulated annealing chain without any temperature schedule (in other words a Metropolis-Hastings chain), so that the decision criteria is based on $\exp[-(\pi(\theta'_j) - \pi(\theta_j^{[t]}))]$. At each iteration the eight adjoining cells for any presently occupied location (fewer for locations on boundaries) is equally likely as a candidate point: a discrete uniform jumping distribution. The starting point is randomly selected within the state space and the algorithm is run for 10,000 iterations. The resulting distribution of visits is summarized by the matrix:

```
0 0 0   0 0 0   0 0   0   0   0   0   0   0 0 0   0   0   0 0 0
0 0 0   0 0 0   0 0   0   0   0   0   1   0 0 0   0   0   0 0 0
0 0 0   0 0 0   0 0  11   4   3   4   6   0 0 0   0   0   0 0 0
0 0 0   0 0 0   0 13 132 169 166  92  93   1 0 0   0   0   0 0 0
0 0 0   0 0 0   1 87   5   7   6   7   9  66 0 0   0   0   0 0 0
0 0 0   0 0 2  54  3   0   0   2   4   0   4 106 2  0   0   0 0 0
0 0 0   0 69 4  0   2   3   5   5   2   0   4 128 3  0   0   0 0 0
0 0 1   4 77 1  1   1   0   3   5   4   3   0   0 1 123  0   0 0 0
1 2 5 100  4 1  1   0  27  22  39  25  18   1   1 1   6  82   2 0 0
2 0 4  68  9 2  1   1  34 1681 1539 1590  25  1   1 0   6 103   3 0 0
2 0 3  62  5 2  1   1  23 1624 1573 1637  30  1   1 0   8 133   7 0 0
1 1 0 135  4 0  0   0  27 1525 1561 1588  20  0   2 2   4 179   2 1 0
1 1 9 182  3 1  0   1  36  52  28  21  16   2   2 2   9 127   4 0 0
1 1 3   6 226 6 0   1   2   2   2   0   1   1   1 5 130   5   0 0 0
1 1 0   0  4 189 6  1   1   2   3   1   2   0   1 66  6   1   0 0 0
3 1 0   0  0 10 151 6   3   1   1   1   0   2  71 6   1   0   0 0 0
0 2 0   0  1 1  1 162  12  16   5   0   0 118   1 0   0   0   0 0 0
0 0 1   1  2 1  1  14 195 175 106  81 102   0   0 0   0   0   0 0 0
0 0 0   0  1 0  1   0   3   6   6   7   1   0   0 0   0   0   0 0 0
0 0 0   0  0 0  0   0   0   0   2   0   1   0   0 0   0   0   0 0 0
0 0 0   0  0 0  0   0   0   0   3   1   0   0   0 0   0   0   0 0 0
```

It is clear that as intended, the Metropolis-Hastings algorithm missed the high modal values in the corners but spent quite some time following the ring around the center and exploring the nonoptimal mode area in the center, despite the fact that all of these points are lower than the corner values. The rejection rate was 0.6014, meaning that the algorithm was fairly inefficient as well. This is an indication that the chain chose to avoid low-density areas rather than having had the opportunity to mix well.

Now we add a temperature schedule according to the popular logarithmic choice:

$T_i = \log(2)/\log(i)$ for $i = 1, \ldots, 100$, and sample 100 times at each of these cooling levels. In real applications we would not be so cavalier about the number of chain values at each temperature, and we would instead take care to observe evidence of convergence each time (hence the reason that simulated annealing can take a long time). This setup produces the following matrix of visits for the first temperature schedule point:

9	3	4	6	5	7	8	17	18	19	15	14	17	18	22	25	36	45	33	34	24
2	5	5	4	10	9	15	30	32	26	22	23	27	38	25	55	53	32	52	46	46
2	1	4	7	4	14	16	20	40	28	41	38	45	31	39	41	47	36	44	58	24
5	2	4	6	16	14	13	15	62	77	46	47	49	45	33	34	51	44	38	31	22
6	4	3	4	13	9	14	48	22	22	39	42	28	44	50	33	38	37	35	18	13
5	15	10	11	11	25	64	35	20	18	26	23	30	29	44	54	24	27	23	19	11
9	14	11	6	18	49	23	21	19	21	24	18	18	20	32	35	36	14	19	19	9
9	17	20	20	56	16	16	14	23	19	17	14	20	19	23	33	32	22	22	23	10
11	18	14	30	20	23	16	25	29	25	30	20	26	23	26	30	32	47	24	15	9
8	16	29	49	23	17	22	19	34	44	39	31	39	22	20	23	38	31	26	14	8
12	15	25	27	18	16	18	24	37	50	35	47	39	20	27	23	35	28	48	24	10
5	9	15	29	19	18	24	25	58	54	44	45	40	23	14	35	26	46	21	26	9
8	9	13	25	22	18	24	39	53	55	33	34	27	21	25	26	28	25	21	18	9
8	10	14	24	39	24	12	35	30	26	28	20	18	17	26	35	39	26	20	15	8
11	14	9	20	25	27	22	17	22	30	29	23	17	22	40	44	32	20	16	16	14
7	14	9	16	14	14	32	21	21	23	23	22	29	35	45	32	12	12	19	21	24
8	12	16	13	13	15	25	33	36	39	28	22	33	38	20	23	16	14	23	28	17
8	21	20	12	13	25	18	35	49	40	31	33	39	26	27	20	15	14	23	23	18
15	18	27	19	20	23	20	28	42	39	23	26	25	31	24	28	21	15	12	11	22
18	19	14	30	32	29	37	31	34	30	21	17	21	12	22	20	28	19	23	31	25
37	20	18	23	24	24	19	18	12	15	17	12	14	15	12	22	12	20	20	19	35

Not only does the annealing algorithm do a much better job of exploring the sample space, it also had a rejection rate of only 0.0676 (okay here due to the small sample space), an order of magnitude better than the version without a temperature schedule. In addition, rejections remained rare even as the cooling levels reached back to normal levels.

15.1.1 General Points on Simulated Annealing

A more general simulated annealing algorithm is given by Kirkpatrick, Gelatt, and Vecchi (1983), where the procedure is the same as that given above, but cast in more Bayesian statistics than physics terms. Specifically, we generate candidates in general Metropolis-Hastings fashion from any desired distribution, and we do not need the exponential metric. The cold distribution is again given by $\pi(\boldsymbol{\theta})$, and we will assume a general temperature schedule $f(T_t|t)$. So at iteration t and temperature T_t we have the current position $\boldsymbol{\theta}^{[t]}$, and perform the steps:

▷ Generate multivariate $\boldsymbol{\theta}'$ from a candidate-generating distribution $q(\boldsymbol{\theta}'|\boldsymbol{\theta})$ that does or does not account for the current position.

▷ Define:

$$a(\boldsymbol{\theta}', \boldsymbol{\theta}^{[t]}) = \left[\frac{q(\boldsymbol{\theta}^{[t]}|\boldsymbol{\theta}')}{q(\boldsymbol{\theta}'|\boldsymbol{\theta}^{[t]})} \frac{\pi(\boldsymbol{\theta}')^{\frac{1}{T_t}}}{\pi(\boldsymbol{\theta}^{[t]})^{\frac{1}{T_t}}} \right].$$

$$\triangleright \ \boldsymbol{\theta}^{[t+1]} = \begin{cases} \boldsymbol{\theta}' & \text{with probability} \quad \min(a(\boldsymbol{\theta}', \boldsymbol{\theta}^{[t]}), 1) \\ \boldsymbol{\theta}^{[t]} & \text{with probability} \quad 1 - \min(a(\boldsymbol{\theta}', \boldsymbol{\theta}^{[t]}), 1) \end{cases}$$

$\triangleright$ Update T_t to T_{t+1} from the temperature schedule.

Notice that this is a straightforward extension of the Metropolis-Hastings algorithm.

Good overviews of simulated annealing are given by Bertsimas and Tsitsiklis (1993), Brooks and Morgan (1995), and Hàjek (1985). Neal (2001) connects importance sampling (Chapter 9) to simulated annealing as a way to create importance sampling distributions in high dimensions where it would otherwise be difficult to get reasonable acceptance rates. Other extensions exist and simulated annealing continues to be an active research area in terms of both theoretical development and applied work.

A number of authors have worried about the various conditions necessary to assert convergence (Ferrari, Frigessi, and Schonmann 1993; Gidas 1985; Locatelli 2000; Lundy and Mees 1986, Szu and Hartley 1987) generally by restricting the form of the domain or specifying a uniform minorization condition for a specific chain. As mentioned, the time to run useful simulated annealing programs can be long. Several works look at estimating this time and how it might be improved (Chiang and Chow 1988; Ingber 1992; Tsitsiklis 1988; Mitra, Romeo, and Sangiovanni-Vincentelli 1986).

Applications are usually motivated by a need to optimize ill-behaved functions, often as a reasonable alternative to exhaustive searches of the sample space (e.g., Bohachevsky, Johnson, and Stein 1986; Kollman, Miller, and Page 1997). Explicitly Bayesian works include Bernardo (1992), van Laarhoven *et al.* (1989), Geyer and Thompson (1995), and Press *et al.* (1986).

15.1.2 Metropolis-Coupling

With high-dimensional and multimodal objective functions (posteriors) of interest, the candidate distribution and the temperature schedule in simulated annealing must be chosen with great care to allow adequate exploration of the space. Unfortunately it is possible to stipulate wildly inappropriate choices of both, thus preventing convergence or mixing with simulated annealing. One early solution to this problem is *Metropolis-Coupled Markov chain Monte Carlo* (MCMCMC) (Geyer 1991). This algorithm is characterized by the steps:

$\triangleright$ Run N parallel chains at different heat levels from m^1 to m^{1/β_N}, where the temperature values have the characteristic: $\beta_1 = 1 < \beta_2 < \ldots < \beta_N$.

$\triangleright$ Thus N transition kernels are defined, $K_1, K_2, \ldots, K_N$.

$\triangleright$ At time t select two chains, i and j, and attempt to swap states:

$$\boldsymbol{\theta}_i^{[t]} \Leftarrow \boldsymbol{\theta}_j^{[t]}, \qquad \boldsymbol{\theta}_j^{[t]} \Leftarrow \boldsymbol{\theta}_i^{[t]},$$

$\triangleright$ with a Metropolis decision, probability:

$$\min\left\{ 1, \frac{m_i(\boldsymbol{\theta}_j^{[t]}) m_j(\boldsymbol{\theta}_i^{[t]})}{m_i(\boldsymbol{\theta}_i^{[t]}) m_j(\boldsymbol{\theta}_j^{[t]})} \right\}.$$

▷ Record only the cold chain, m^1, for inferential purposes.

The key advantage to MCMCMC is that chains that get stuck in non-optimal maxima will eventually get swapped-out to some other, presumably more free, state. A notable disadvantage, though, is the need to possibly run *many* parallel chains for problems with highly complex targets. Also, it is possible to run parallel chains concurrently using additional software such as snow (Olivera and Gill 2011).

15.1.3 Simulated Tempering

Marinari and Parisi (1992) and (independently) Geyer and Thompson (1995) propose an alternative algorithm called *simulated tempering*, which reduces the MCMCMC algorithm to a single chain. Essentially the temperature itself becomes a random variable so the system can heat *and* cool as time proceeds. Why would one want do this in a simulated annealing process? Now elderly chains can still avoid being trapped at local maxima by getting more general Metropolis-Hastings candidate positions. That is, step 1 of the simulated annealing algorithm above is replaced with:

1*a*. Generate β from some distribution of temperature, $f(\beta)$.

1*b*. Generate $\boldsymbol{\theta}'_j \sim \left[\boldsymbol{\theta}_j^{[t']} + \mathcal{N}(0, \sigma^2) \right]$.

The number of $f(\beta)$ choices is obviously vast, but this decision can be simplified by using a discrete distribution that resembles some desired, but not implemented, cooling schedule. Often this can be rigged to provide a large number of cold draws. The algorithm above can also be thought of as an augmented sampler in the context of Tanner and Wong (1987) where temperature is the augmentation variable.

Temperature T=1   Temperature T=25   Temperature T=300

FIGURE 15.1: SIMULATED ANNEALING APPLIED TO THE WITCH'S HAT DISTRIBUTION

Figure 15.1 shows the infamous witch's hat distribution centered at $[0.6, 0.4]$ (see Exercise 15.5 for details) which has been suggested as an MCMC diagnostic since it causes serious mixing problems with various algorithms, particularly in high dimensions. Geyer and Thompson (1995) apply simulated tempering to improve the mixing problem caused by the difficulty of moving from the brim area to the peak, and obtain an acceptance rate improvement from 7% to 72%. The panels of Figure 15.1 show the "cold" distribution at T=1 and heated up versions at T=25 and T=300 plotted on the unit square. Notice how the troublesome crease at the base of the peak is melted down.

15.1.4   Tempored Transitions

Neal (1996) builds on simulated tempering with *tempered transitions* to heat up the posterior distribution in-place so that a random walk can move more freely, but also to preserve the detailed balance equation (reversibility condition) at each step.   See also Celeux, Hurn, and Robert (2000) for an application to mixture distributions, and Liu and Sabatti (1999) for the "simulated simpering" variant. The basic idea is to "ladder" up and down in heat at each time t with random walk steps. Each ladder step specifies a (non-normalized) stationary distribution defined on the same state space but at progressively hotter temperatures going up.   Finally the last (bottom) ladder value is accepted or discarded with a Metropolis decision. This process is summarized by:

▷ τ_1 is the target joint density,

▷ β_i is the temperature value at the ith ladder step,

▷ tempered transitions: define a sequence of candidate densities $m_i, i = 1, \ldots, N$, where as i increases the m_i get "flatter" going up the ladder, then again more peaked going down the ladder.

▷ parameterize: $m_i = m^{1/\beta_i}$,

▷ where: $1 < \beta_1 < \beta_2 < \cdots < \beta_{N-1} < \beta_N$

▷ then: $\beta_N > \beta_{N+1} > \cdots > \beta_{2N-2} > \beta_{2N-1} > 1$.

▷ Starting from the original candidate m, at each step we cycle through the m_i as follows:

1. If we let $K(\boldsymbol{\theta}, m)$ denote an MCMC kernel with position $\boldsymbol{\theta}$ and stationary distribution m,

2. then we use the following transitions starting at iteration t:

$$
\begin{array}{ll}
\textbf{step 0:} & \boldsymbol{\theta}'_{1,0} \sim \mathrm{K}(\boldsymbol{\theta}_1^{[t]}, \tau_1) \\[4pt]
\textbf{step 1:} & \boldsymbol{\theta}'_{1,1} \sim \mathrm{K}(\boldsymbol{\theta}'_{1,0}, m_1) \\[4pt]
& \vdots \\[4pt]
\textbf{step N:} & \boldsymbol{\theta}'_{1,N} \sim \mathrm{K}(\boldsymbol{\theta}'_{1,N-1}, m_N) \\[4pt]
\textbf{step N+1:} & \boldsymbol{\theta}'_{1,N+1} \sim \mathrm{K}(\boldsymbol{\theta}'_{1,N}, m_{N-1}) \\[4pt]
& \vdots \\[4pt]
\textbf{step 2N-1:} & \boldsymbol{\theta}'_{1,2N-1} \sim \mathrm{K}(\boldsymbol{\theta}'_{1,2N-2}, m_1).
\end{array}
$$

3. The sequence of θ_1 values is then input into a final Metropolis-Hastings acceptance step, accepting $\theta'_{1,2N-1}$ as $\theta_1^{[t+1]}$ with probability:

$$\min\left\{1, \frac{m_1(\theta_1^{[t]})}{\tau_1(\theta_1^{[t]})} \cdots \frac{m_N(\theta'_{1,N-1})}{m_{N-1}(\theta'_{1,N-1})} \cdots \frac{\tau_1(\theta'_{1,2N-1})}{m_1(\theta'_{1,2N-1})}\right\},$$

which preserves the detailed balance condition.

To look at this in a slightly different way, we can also substitute the β_i parameterization back in. Now accept $\theta_{1,2N-1}^{[t]}$ as $\theta_1^{[t+1]}$ with probability:

$$\min\left\{1, \left(\frac{m^{1/\beta_1}(\theta_{1,0}^{[t]})}{m^1(\theta_{1,0}^{[t]})}\right)\left(\frac{m^{1/\beta_2}(\theta_{1,1}^{[t]})}{m^{1/\beta_1}(\theta_{1,1}^{[t]})}\right)\left(\frac{m^{1/\beta_3}(\theta_{1,2}^{[t]})}{m^{1/\beta_2}(\theta_{1,2}^{[t]})}\right)\right.$$

$$\cdots \left(\frac{m^{1/\beta_{N-1}}(\theta_{1,N-2}^{[t]})}{m^{1/\beta_{N-2}}(\theta_{1,N-2}^{[t]})}\right)\left(\frac{m^{1/\beta_N}(\theta_{1,N-1}^{[t]})}{m^{1/\beta_{N-1}}(\theta_{1,N-1}^{[t]})}\right)\left(\frac{m^{1/\beta_{N+1}}(\theta_{1,N}^{[t]})}{m^{1/\beta_N}(\theta_{1,N}^{[t]})}\right)$$

$$\cdots \left.\left(\frac{m^{1/\beta_2}(\theta_{1,2N-3}^{[t]})}{m^{1/\beta_3}(\theta_{1,2N-3}^{[t]})}\right)\left(\frac{m^{1/\beta_1}(\theta_{1,2N-2}^{[t]})}{m^{1/\beta_2}(\theta_{1,2N-2}^{[t]})}\right)\left(\frac{m^1(\theta_{1,2N-1}^{[t]})}{m^{1/\beta_1}(\theta_{1,2N-1}^{[t]})}\right)\right\}.$$

Note that $\theta_{1,0}^{[t]} = \theta_1^{[t]}$. We can also add a weighting function within each term above: $\frac{w(\theta_{1,0}^{[t]})}{w(\theta_{1,2N-1}^{[t]})}$ (sometimes called a *pseudo-prior* in this context). The original stationary distribution of the Markov chain is maintained as long as the m_i satisfy a detailed balance condition, which is given in Neal (1996) with proof.

This sequence of transitions allows excellent exploration of the parameter space, as the density m_N is typically chosen as very "hot," for example, uniform on the entire space. Setting $\beta_i - \beta_{i+1}$ as small gives higher acceptance rates but poorer mixing. Conversely a large difference between m and $m^{1/\beta}$ is good for mixing around the space but may lead to inordinately high rejection rates. Both criteria can be satisfied with taller ladders: more steps and a higher maximum temperature.

15.1.5 Comparison of Algorithms

It is interesting to compare these approaches with a deliberately "ugly" example objective function on $[-1,1]^2$:

$$f(x,y) = \text{abs}((x\sin(20y-90) - y\cos(20x+45))^3 a\cos(\sin(90y+42)x) \quad (15.2)$$

$$+ (x\cos(10y+10) - y\sin(10x+15))^2 a\cos(\cos(10x+24)y)). \quad (15.3)$$

This function is displayed in Figure 15.2. While the vast majority of posterior forms will not exhibit such challenging characteristics, it still serves as a benchmark for algorithm performance. Furthermore, increasingly complex model specifications *are* much more prevalent in

FIGURE 15.2: A HIGHLY MULTIMODAL BOUNDED SURFACE

recent Bayesian and non-Bayesian work in the social sciences, leading to potentially similar forms. This function, while bounded, presents problems for conventional MCMC algorithms since it has multiple modes concentrated at the corners and a wide, flat plain in the middle that must be traversed to fully describe the functional form.

We now apply a regular random walk Metropolis-Hastings algorithm, simulated annealing, and tempered transitions to this function. Each Markov chain is run for 5,000 iterations (an insufficient but illustrative period), and the chain visits are given in Figure 15.3. Such a number of iterations is revealing here because, while all of these algorithms are ergodic and will therefore *eventually* explore the full target form, our concern is with the rate at which they do so. Specifically, do the algorithms differ in the efficiency by which they mix through the space?

It is clear that the standard algorithm fails during this period to break out of the diagonal, and spends the bulk of its time visiting two of the four corners where large modes exist. The simulated annealing algorithm appears to be traversing the space much better but still fails to break out of the diagonal area. Conversely, the algorithm based on tempered transitions manages to explore the full state space, even in this small number of iterations. So the leverage that we gain from using the tempered transitions is greater assurance that we have mixed through the target form in a fixed amount of time.

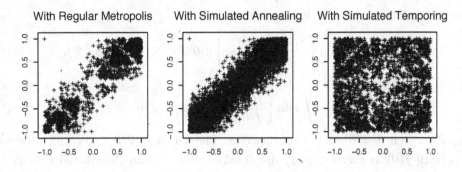

FIGURE 15.3: A COMPARISON OF TRAVELS

15.1.6 Dynamic Tempered Transitions

Gill and Casella (2004) extend the idea of tempered transitions to account for current placement of the chain in the state space. The objective is to escape the necessary trade-offs in ladder height (maximum heating level) and spacing between rungs (number of steps). When the area around the chain is highly irregular, it is better to have a lower (cooler maximum temperature) ladder in order to get to the top of the mode and fully explore this area. When the area around the chain is smooth, it is better to have a longer (hotter maximum temperature) ladder in order to avoid being trapped in the low density region. Also, setting the number of rungs for a given ladder height as too small can reduce the acceptance rate because high-quality candidates may not be offered. Conversely, setting the number of rungs as too large can also reduce the acceptance rate because of the product in the Metropolis-Hastings decision step.

The general strategy in these regards is to specify a distribution of ladders all having the same number of rungs, but differing heights (differing maximum temperatures). We observe the multidimensional curvature at the current Markov chain location and specify a greater probability of selecting a cooler ladder when this curvature is high, and a greater probability of selecting a hotter ladder when the curvature is low. The number of rungs is essentially a nuisance parameter, which can be fixed at the beginning of the chain or tuned during the early runs by comparing acceptance probabilities. The actual goal is not to determine the exact optimal number of rungs, but to pick a reasonable number that is not overly affecting the acceptance probabilities.

The key challenge is that by making the behavior of the Markov chain adjust to its surroundings (i.e., conditional on changes in the posterior for θ), we run the risk of creating a non-homogeneous Markov chain, and also losing the detailed balance equation. This would then deny the ability to assert ergodicity of the chain. We can solve this problem by taking advantage of the structure of the Metropolis algorithm.

Let $f(\theta)$ be the stationary distribution (objective function), let $g(\theta'|\theta)$ be a candidate distribution, and let $K(\theta, \theta')$ be the associated transition kernel. By the construction of the Metropolis algorithm, $K(\boldsymbol{\theta}, \boldsymbol{\theta}')$ is given by:

$$K(\boldsymbol{\theta}, \boldsymbol{\theta}') = \min\left\{\frac{f(\boldsymbol{\theta}')g(\boldsymbol{\theta}|\boldsymbol{\theta}')}{f(\boldsymbol{\theta})g(\boldsymbol{\theta}'|\boldsymbol{\theta})}, 1\right\} g(\boldsymbol{\theta}'|\boldsymbol{\theta}) + (1 - r(\boldsymbol{\theta}))\delta_{\boldsymbol{\theta}}(\boldsymbol{\theta}'), \qquad (15.4)$$

where:

$$r(\boldsymbol{\theta}) = \int \min\left\{\frac{f(\boldsymbol{\theta}')g(\boldsymbol{\theta}|\boldsymbol{\theta}')}{f(\boldsymbol{\theta})g(\boldsymbol{\theta}'|\boldsymbol{\theta})}, 1\right\} g(\boldsymbol{\theta}'|\boldsymbol{\theta})d\boldsymbol{\theta}' \qquad (15.5)$$

and $\delta_{\boldsymbol{\theta}}(\boldsymbol{\theta}') = 1$ if $\boldsymbol{\theta} = \boldsymbol{\theta}'$ and zero otherwise. The kernel $K(\boldsymbol{\theta}, \boldsymbol{\theta}')$ now satisfies detailed balance with $f(\boldsymbol{\theta})$ as the stationary distribution (exactly from Robert and Casella 1999, Theorem 6.2.3).

Now, for each $\boldsymbol{\theta}$, let $\rho(\lambda|\boldsymbol{\theta})$ be a probability distribution, that is, $\rho(\lambda|\boldsymbol{\theta}) \geq 0$ and $\int \rho(\lambda|\boldsymbol{\theta})d\lambda = 1$. Here we are considering λ to be continuous, to be general, but usually λ will be discrete. Our candidate distribution is:

$$g_\lambda^*(\boldsymbol{\theta}'|\boldsymbol{\theta}) = \rho(\lambda|\boldsymbol{\theta}), \qquad (15.6)$$

and forms the Metropolis kernel based on $g_\lambda^*(\boldsymbol{\theta}'|\boldsymbol{\theta})$ and $f(\boldsymbol{\theta})$. By construction, detailed balance is satisfied and we have an ergodic Markov chain.

As an example, suppose that there are $i = 1, \ldots, k$ ladders, and as i increases the ladders get hotter. If $|f''(\boldsymbol{\theta})|$ is big (so we are near a mode) we might want to favor the cooler ladders. To do this we can take $\rho(\lambda|\boldsymbol{\theta})$ to be a binomial mass function with k trials and success probability $p(\boldsymbol{\theta})$, where

$$\text{logit}p(\boldsymbol{\theta}) = a - b|f''(\boldsymbol{\theta})|, \quad b > 0. \qquad (15.7)$$

So big values of $|f''(\boldsymbol{\theta})|$ would result in small $p(\boldsymbol{\theta})$, which would favor the smaller values of i and the cooler ladders. This way, on average, we spend sufficient time exploring the modal area. Eventually, since there is always a positive probability of getting a hot ladder, the chain eventually escapes from every mode. In flat areas, since hot ladders are more probable, this happens on average sooner.

15.2   Reversible Jump Algorithms

Green (1995) introduces *reversible jump Markov chain Monte Carlo* (RJMCMC) where the model specification (i.e., variable selection) is part of the estimation process. The key idea is to modify a Metropolis-Hastings kernel to jump not only between points in the parameter space but also between different model specifications (although the Gibbs sampler has been used in at least one RJMCMC algorithm, Keith *et al.* [2004]). Recently Geyer (2011) argues that RJMCMC is a fundamental feature of MCMC and regular implementations are really special cases.

Reversible jump MCMC is a form of Bayesian model averaging (page 198) where the posterior probability of a model is provided by the proportion of the total MCMC (post-convergence) run-time that the chain spends in the space defined by that model. Interestingly, the number of models considered does not need to be specified in advance, but model priors are still required. For example, Richardson and Green (1997) express this idea with "births" and "deaths" of model alternatives (see also Zhang *et al.* [2004]). We will very generally describe the algorithm here, and for details, see the original article, or the reviews by Waagepetersen and Sorensen (2001), Clyde (1999), Chen *et al.* (2000, pp.300-303), Andrieu, Djurić, and Doucet (2001), or Richardson and Green (1997).

Suppose we have a countable set of K alternative model specifications to evaluate, denoted $\{\mathcal{M}_k, k \in K\}$. Naturally these models have different numbers of parameters, which may or may not provide nesting, such model $\mathcal{M}_k$ has a n_k-length ($n_k \geq 1$) parameter vector $\theta^{(k)}$ measured on $\Re^{(n_k)}$. For observed data $\mathbf{y}$, the joint distribution of interest is formed from the product of the prior on model choice, $p(k)$, the prior for the coefficient vector given model choice, $p(\theta^{(k)}|k)$, and the likelihood for $\mathbf{y}$ given model choice, $L(\mathbf{y}|k, \theta^{(k)})$, producing:

$$p(k, \theta^{(k)}, \mathbf{y}) = p(k)p(\theta^{(k)}|k)L(\mathbf{y}|k, \theta^{(k)}). \tag{15.8}$$

Since the algorithmic focus is on moving between model specifications, start with the pair $(k, \theta^{(k)})$, and denote it with simply x, which for a specific case of k must lie in the parameter space $\mathcal{C}_k = \{k\} \times \Re^{n_k}$. Therefore in the course of the algorithm x varies over the combined parameter space defined by $\mathcal{C} = \cup_{k \in K}\mathcal{C}_k$. Since the pairing is inseparable, the posterior of interest is the joint expression:

$$\pi(x|\mathbf{y}) = \pi(k, \theta^{(k)}|\mathbf{y}) = p(k|\mathbf{y})p(\theta^{(k)}|k, \mathbf{y}). \tag{15.9}$$

If $(k, \theta^{(k)})$ is the current position of the Markov chain at time t, then denote $x' = (m, \theta^{(m)}) \in \{m\} \times \Re^{n_m}$ as a proposed destination for time $t+1$ produced by the candidate-generating distribution with probability $q_m(x, x')$. One useful way to generate such candidates is by the inclusion of a random component, $\mathbf{U}$ in $\Re^{n_{km}}$, $km \geq 1$, independent of x with density $q_{km}(\theta^{(km)}, \cdot)$. So define the deterministic mapping $g_{mk} : \Re^{n_k + n_{km}} \rightarrow \Re^{n_m}$ such that we can rewrite: $x' = g_{mk}(x, u)$. Now g_{mk} is called a *bijection* since it gives a one-to-one mapping between $(\theta^{(k)}, u)$ and $(\theta^{(m)}, u')$, where u' is the reverse analog of u from x' instead of x.

This is a setup to meet the *dimension matching condition*, which is a necessary part of ensuring that the detailed balance equation holds. The potential move under consideration is:

$$(k, \theta^{(k)}) \longrightarrow (m, \theta^{(m)}) = (m, g_{mk}(x, u)), \tag{15.10}$$

with the corresponding reverse move:

$$(m, \theta^{(m)}) \longrightarrow (k, \theta^{(k)}) = (k, g_{km}(x', u')). \tag{15.11}$$

This now makes it straightforward to match the dimensions:

$$n_k + n_{km} = n_k + n_{mk} \tag{15.12}$$

so that

$$\pi(k, \theta^{(k)}, \mathbf{y}) g_{mk}(x, u) = \pi(m, \theta^{(m)}, \mathbf{y}) g_{km}(x', u') \tag{15.13}$$

holds. This leads directly to the Metropolis-Hastings acceptance criteria:

$$\alpha(x, x') = \min\left\{1, \frac{\pi(m, \theta^{(m)}, \mathbf{y})}{\pi(k, \theta^{(k)}, \mathbf{y})} \frac{q_m(x, x')}{q_k(x', x)} \frac{q_{mk}(\theta^{(mk)}, u')}{q_{km}(\theta^{(km)}, u)} \left| \frac{\partial g_{mk}(x, u)}{\partial x \partial u} \right| \right\}, \tag{15.14}$$

where the Jacobian is necessary because $x' = g_{mk}(x, u)$ is a deterministic function in the proposal process for the change in variable from (x, u) to x'. Under a wide range of applied circumstances $\alpha(x, x')$ is more simple than the general form given here.

Finally, suppose we have a (post-convergence) MCMC sample indexed $i = 1, 2, \ldots, I$, and define the following indicator function:

$$\infty_i(i_k) = \begin{cases} 1, & \text{if } \mathcal{M}_i = \mathcal{M}_k \\ 0, & \text{if } \mathcal{M}_i \neq \mathcal{M}_k. \end{cases} \tag{15.15}$$

This function produces a 1 if the ith model in the series is the kth specification in the countable set: $k \in K$. Now the posterior probability of model k is simply:

$$\pi(k|\mathbf{y}) = \frac{1}{I} \sum_{i=1}^{I} \infty_i(i_k), \tag{15.16}$$

where the posterior variance can be calculated according to the tools in Section 14.3.1.

The mechanics of RJMCMC are sometimes difficult to implement and must be customized for each application, although Hastie (2005) has developed the `AutoMix` package based on normal mixtures as a reasonably general approach. Like all advanced MCMC techniques there are a number of tuning parameters that researchers need to pay close attention to. In particular the candidate-generating distribution can be challenging in that proposals now have two criteria for acceptance: parameter values and model space. Nonetheless, this is an active research area because the idea of assessing model quality within the context of the sampler is attractive.

15.3   Perfect Sampling

As discussed, one obvious worry is the length of the pre-convergence era of a particular Markov chain. While we can be comforted by knowing that MCMC algorithms used in general practice are ergodic, an eventual decision about cessation is required. An ingenious solution to this problem is provided by Propp and Wilson (1996) who give a way to automatically obtain *perfect* samples immediately from an MCMC algorithm that are guaranteed to be from the stationary distribution of interest. Their method, called *coupling from the past*, (CFTP) is a particular kind of perfect sampling, although the terms are often (but incorrectly) used synonymously.

The idea behind CFTP is that if a chain had been started at step $t = -\infty$ and run forward in time, then at time $t = 0$ it must be in its stationary distribution. Their innovation is that they found a way to obtain this same sample without the inconvenience (or impossibility!) of dealing with the infinite past. Define terms as in previous chapters where the Markov chain operates on a finite state space with the transition matrix $K(\theta, \theta')$ and the stationary (target) distribution $\pi(\theta)$. Since this is a finite state space we can consider all the ways to transition forward from $\theta^{[-1]}$ to $\theta^{[0]}$ using the expression in (10.3):

$$p(\theta^{[0]} = j | \theta^{[-1]} = i) = K(i, j) \tag{15.17}$$

and summing over these possible ways to produce the cumulative transition probability:

$$p(\theta^{[0]} \leq j | \theta^{[-1]} = i) = \sum_{k=1}^{j} K(i, k). \tag{15.18}$$

Call this last equation $C(i, j)$ for clarity to distinguish it from $K(i, j)$. Now draw a uniform random number (u_0) between zero and one and use the cumulative function to specify a move:

$$\theta^{[0]} = j \quad \text{if} \quad C(i, j - 1) < u_0 \leq C(i, j). \tag{15.19}$$

This is fairly routine MCMC so far but now perform this operation using the same u_0 for every state at time $t = -1$, not just $\theta^{[-1]} = i$. This means we have a parallel set of Markov chains equal in number to the set of unique states at time $t - 1$. We can apply the transition rule defined by:

$$\theta^{(0)} = \phi(u_0, \theta^{(-1)}). \tag{15.20}$$

Now comes the innovative part. If

$$\phi(u_0, \theta^{(-1)} = i) = j, \quad \forall i \tag{15.21}$$

then we say that the chain has *coupled* and the value given by j is an exact (perfect) sample from $\pi(\theta)$. Why is this true? If, hypothetically, the Markov chain had been run from $t = -\infty$ to $t = -1$, then there is no question that it is in stationarity at time $t = -1$. If the Markov chain is in stationarity at time $t = -1$ then it is clearly in stationarity at time $t = 0$ from the condition imposed by $\phi(u_0, \theta^{(-1)})$, and of course for any time thereafter. All this certainty makes this result sound frequent and easy, but the truth is that the probability that all of the parallel Markov chains couple at this one step is inconveniently low. This is not a difficult impediment since the selection of the zero time point was arbitrary, thus:

$$\theta^{(-t)} = \phi(u_{-t}, \theta^{(-t-1)}). \tag{15.22}$$

Applying this general idea recursively through the history of the chain gives:

$$\theta^{(0)} = \phi(u_0, \phi(u_{-1}, (\phi_{-2}, \ldots, \phi(u_{-T+1}(\theta^{(-T)}) \cdots)))), \tag{15.23}$$

where T is some large number of our choosing, possibly very large, and we have a series of uniform draws identified: $u_{-T}, u_{-T+1}, \ldots, u_0$. If we start our M parallel Markov chains at time $-T$, for each of the M possible discrete locations in the state space, then at step $-T+1$ we will have a number of unique chain locations equal to or less than M: a *coalescence* of chains. Since such chains remain coalesced permanently, over time the collection of M chains will eventually coalesce to a single chain, and when this happens we will call it time $t = 0$. Note that any collection of M Markov chains started at $t = -\infty$ will have come through the time $t = -T$ with M or less unique states, then the sample at $t = 0$ must be that defined by $\theta^{(0)}$ above.

Various implementations of CFTP use the coupling strategy in incremental steps. The objective is to find the time $-T$ that accomplishes the full coalescence above, and such that $\theta^{(0)}$ is independent of $\theta^{(0)}$. Suppose again that there are M states in the sample space and $\phi(u_j, \theta^{(j)})$ is a transition rule at time j. Then the steps proceed as follows:

1. Start M parallel chains in each of the M states at time $t = -1$, and generate $u_0 \sim \mathcal{U}(0, 1)$.

2. Apply the transition rule, $\phi(u_0, \theta^{(-1)})$, to each of the chains. If all of the chains have coalesced at time $t = 0$, then set $-T = -1$. The value $\theta^{(0)}$ is a perfect draw from $\pi(\theta)$.

3. If all the chains have not coalesced, then go to time $t = -2$, generate $u_{-1} \sim \mathcal{U}(0, 1)$, and apply the transition rule, $\phi(u_{-1}, \theta^{(-2)})$, to each of the chains. If all of the chains have coalesced at time $t = -1$, then set $-T = -2$. The value $\theta^{(0)}$ is a perfect draw from $\pi(\theta)$.

4. If all the chains have not coalesced, then go to time $t = -2$, and repeat.

5. Continue as necessary back in time until full coalescence.

It is important to remember that no matter how far we go back, the first value that is a perfect draw from the stationary distribution is still the common value at time $t = 0$. From that point on we can then collect perfect simulations without worrying about convergence. One other reminder is warranted. Moving backward it is critical to record the uniform draws so that we can use *these exact values* going forward again: $u_{-T}, u_{-T+1}, \ldots, u_{-1}, u_0$.

It should be clear that the time required, T, and $\theta^{(0)}$ are random variables dependent on each other such that the full coalescence process must complete before sampling can move forward with draws from the stationary distribution. This may be a problem in practical applications if the initial process is interrupted. Fill (1998) addresses this with a version of perfect sampling (perfect rejection sampling) that is independent of T. It proceeds in similar fashion for the same setup:

1. Choose a time T and a state at this time $\theta^{(T)} = \theta$ at convenience.

2. Generate a conditional series backward to the zero point: $\theta^{(T-1)}|\theta^{(T)}, \theta^{(T-2)}|\theta^{(T-1)}, \ldots, \theta^{(1)}|\theta^{(2)}, \theta^{(0)}|\theta^{(1)}$.

3. Generate a series of associated uniform draws:

   $(u_1|\theta^{(0)}, \theta^{(1)})$, $(u_2|\theta^{(1)}, \theta^{(2)})$, ..., $(u_{T-1}|\theta^{(T-2)}, \theta^{(T-1)})$, $(u_T|\theta^{(T-1)}, \theta^{(T)})$.

4. Run M parallel chains starting at time $T = 0$ and use the same uniform draws to update all chains.

5. If all the chains have coalesced by the time the series runs out at T, then accept $\theta^{(0)}$ as a draw from $\pi(\theta)$.

6. If all the chains have not coalesced by this time then run the algorithm again, perhaps with a larger value of T.

Relatively large values of T in general are recommended by Fill (1998) and others.

Fill *et al.* (1999) and Casella *et al.* (2001) provide the necessary proofs that this time reversal strategy removing the dependence between the backward path length and the sample value generated at time $T = 0$ provides samples from the stationary distribution. Clearly the second step of conditioning given the initial path is the most challenging part here. However, the independence between the length of the path generation and the production of $\theta^{(0)}$ values combined with the reuse of the uniform draws makes this process relatively efficient for users. Notable extensions have been developed, including the general version of this algorithm provided by Fill *et al.* (1999). Murdoch and Rosenthal (1998) remove the required assumption of stochastic monotonicity of the chain. Møller and Schladitz (1999) are also able remove this assumption in the context of stochastic repulsive sequences, leading new applications.

Several mechanical challenges remain with perfect sampling. Unless the number of states is relatively small, then the size of the process at each step can be awkward and tracking coalescence can be burdensome. For non-trivial state spaces (i.e., excluding the toy examples that dominate published introductions), it can be very difficult to show that a large number of paths have all coalesced. However, if the states are ordered with a monotone transition rule, then one convenient shortcut is to track coalescence of just the maximum and minimum values at each step since these will squeeze the other values towards complete coupling. By monotone here we mean that $\theta_i^{(t)} \geq \theta_j^{(t)}$ implies $\theta_i^{(t+1)} \geq \theta_j^{(t+1)}$ (Fill 1998, Definition 4.2), i.e., paths on lower starting points remain below paths on higher starting points until full coalescence. Furthermore, setting up and running unbiased algorithms is labor-intensive and specific to each application. Therefore we are not likely to see commercial software support in the foreseeable future. Finally, it is critical to reuse the uniform random draws correctly for unbiased properties to hold.

Perfect sampling is one of the more exciting new research areas in MCMC. Good reviews of perfect sampling include Casella *et al.* (2001) and Craiu and Meng (2011). The biggest challenge, of course, is to generalize the state space definition (see Green and Murdoch [1998] as well as Murdoch and Green [1998]). Meng (2000) produced an algorithm based on a multistage version of the CFTP backward coupling scheme using cluster sampling to reduce time to coalescence. Hobert *et al.* (1999) are able to impose the monotonicity requirement on cases with two and three component mixtures, but apparently with a high computational cost. Corcoran and Tweedie (1998) show that perfect sampling based on the independent Metropolis-Hastings kernel has useful monotonicity properties and is therefore a good algorithmic choice. Brooks *et al.* (2006) link perfect sampling with simulated tempering discussed earlier in this chapter. Casella *et al.* (2002) develop perfect slice samplers and apply them to mixtures of distributions. Murdoch and Takahara (2006) develop applications to queueing theory and networks. However, there are more applications in statistical physics and point processes (stochastic processes with binary outcomes occurring over continuous time) than there are in the social sciences.

15.4   Hamiltonian Monte Carlo

Hamiltonian Monte Carlo (also called Hybrid Monte Carlo) is a variant of the Metropolis-Hastings algorithm that uses physical system dynamics as a means of generating candidates for a Metropolis decision. When properly implemented it is faster than regular Metropolis-Hastings because it incorporates more information about posterior topography. This idea was originally presented by Duane *et al.* (1987), and further developed by Neal (1993). This section will very generally outline the method and the most detailed description for MCMC purposes is currently Neal (2011).

As noted by every description to date, a comprehension of Hamiltonian dynamics is a necessary qualification for understanding hybrid Monte Carlo. This is a core topic in physics, so there exist many relevant textbooks such as that by Meyer and Hall (1992). Hamiltonian dynamics is the model whereby physicists describe an object's trajectory within a defined system. First define $\boldsymbol{\vartheta}_t$ as a k-dimensional location vector and $\mathbf{p}_t$ as a k-dimensional momentum (mass times velocity) vector, both recorded at time t. The Hamiltonian system at time t with $2k$ dimensions is described by the joint Hamiltonian function:

$$H(\boldsymbol{\vartheta}_t, \mathbf{p}_t) = U(\boldsymbol{\vartheta}_t) + K(\mathbf{p}_t) \qquad (15.24)$$

(sometimes just abbreviated H), where $U(\boldsymbol{\vartheta}_t)$ is the function describing the *potential energy* at the point $\boldsymbol{\vartheta}_t$, and $K(\mathbf{p}_t)$ is the function describing the *kinetic energy* for momentum $\mathbf{p}_t$. Neal (2011) gives the simple 1-dimensional example:

$$U(\vartheta_t) = \frac{\vartheta_t^2}{2} \qquad\qquad K(p_t) = \frac{p_t^2}{2}, \qquad (15.25)$$

which is equivalent to a standard normal distribution for ϑ. Commonly the kinetic energy function is defined as:

$$K(\mathbf{p}_t) = \mathbf{p}_t' \boldsymbol{\Sigma}^{-1} \mathbf{p}_t, \qquad (15.26)$$

where $\boldsymbol{\Sigma}$ is a symmetric and positive-definite matrix that can be as simple as an identity matrix times some scalar that can serve the role of a variance: $\boldsymbol{\Sigma} = \sigma^2\mathbf{I}$. This simple form is equivalent to the log PDF of the multivariate normal with mean vector zero and variance-covariance matrix $\boldsymbol{\Sigma}$.

Hamiltonian dynamics describe the gradient-based way that potential energy changes to kinetic energy and kinetic energy changes to potential energy as the object moves over time throughout the system (multiple objects require equations for gravity, but that is fortunately not our concern here). The mechanics of this process are given by Hamilton's equations, which are the set of simple differential equations:

$$\frac{\partial \boldsymbol{\vartheta}_{it}}{\partial t} = \frac{\partial H}{\partial \mathbf{p}_{it}} = \frac{K(\partial \mathbf{p}_{it})}{\partial \mathbf{p}_{it})} \tag{15.27}$$

$$\frac{\partial \mathbf{p}_{it}}{\partial t} = -\frac{\partial H}{\partial \boldsymbol{\vartheta}_{it}} = -\frac{U(\partial \boldsymbol{\vartheta}_{it})}{\partial \boldsymbol{\vartheta}_{it})} \tag{15.28}$$

for dimension i at time t. For continuously measured time these equations give a mapping from time t to time $t + \tau$, meaning that from some position $\boldsymbol{\vartheta}_t$ and momentum $\mathbf{p}_t$ at time t we can predict $\boldsymbol{\vartheta}_\tau$ and $\mathbf{p}_\tau$. Returning to the one-dimensional standard normal case, these equations are simply $d\vartheta_t/dt = p$ and $dp/dt = -\vartheta$.

There are three important properties of Hamiltonian dynamics that are actually *required* if we are going to use them to construct an MCMC algorithm (Neal 2011). First, Hamiltonian dynamics is *reversible*, meaning that the mapping from $(\boldsymbol{\vartheta}_t, \mathbf{p}_t)$ to $(\boldsymbol{\vartheta}_{t+\tau}, \mathbf{p}_{t+\tau})$ is one-to-one and therefore also defines the reverse mapping from $(\boldsymbol{\vartheta}_{t+\tau}, \mathbf{p}_{t+\tau})$ to $(\boldsymbol{\vartheta}_t, \mathbf{p}_t)$. Second, *total* energy is conserved over time t and dimension k, and the Hamiltonian is invariant, as shown by:

$$\frac{\partial H}{\partial t} = \sum_{i=1}^{k}\left[\frac{\partial \boldsymbol{\vartheta}_i}{\partial t}\frac{\partial H}{\partial \boldsymbol{\vartheta}_i} + \frac{\partial \mathbf{p}_i}{\partial t}\frac{\partial H}{\partial \mathbf{p}_i}\right] = \sum_{i=1}^{k}\left[\frac{\partial H}{\partial \mathbf{p}_i}\frac{\partial H}{\partial \boldsymbol{\vartheta}_i} - \frac{\partial H}{\partial \boldsymbol{\vartheta}_i}\frac{\partial H}{\partial \mathbf{p}_i}\right] = 0. \tag{15.29}$$

This provides detailed balance (reversibility) for the MCMC algorithm. Second, Hamiltonian dynamics preserve volume in the $2k$ dimensional space. In other words, elongating some region in a direction requires withdrawing another region as the process continues over time. This ensures that there is no change in the scale of Metropolis-Hastings acceptance probability. Finally, Hamiltonian dynamics provides a *symplectic mapping* in $\mathcal{R}^{2k}$ space. Define first the smooth mapping $\psi : \mathcal{R}^{2k} \rightarrow \mathcal{R}^{2k}$ with respect to some constant and invertible matrix $\mathbf{J}$ with $\mathbf{J}' = -\mathbf{J}$ and $\det(\mathbf{J}) \neq 0$, along with having Jacobian $\psi(z)$ for some $z \in \mathcal{R}^{2k}$. The mapping ψ is symplectic if:

$$\psi(z)'\mathbf{J}^{-1}\psi(z) = \mathbf{J}^{-1}. \tag{15.30}$$

Leimkuhler and Reich (2005, p.53) give the following mapping in 2-dimensional space $z = (\vartheta, p)$:

$$\psi(\vartheta, p) = \begin{bmatrix} p \\ 1 + b\vartheta + ap^2 \end{bmatrix}, \tag{15.31}$$

with constants $a, b \neq 0$. The Jacobian of $\psi(\vartheta, p)$ is calculated by:

$$\frac{\partial}{\partial \vartheta} \frac{\partial}{\partial p} \psi(\vartheta, p) = \begin{bmatrix} 0 & 1 \\ b & 2ap \end{bmatrix}. \tag{15.32}$$

We check symplecticness by:

$$\left[\frac{\partial}{\partial \vartheta} \frac{\partial}{\partial p} \psi(\vartheta, p)\right]' \mathbf{J}^{-1} \left[\frac{\partial}{\partial \vartheta} \frac{\partial}{\partial p} \psi(\vartheta, p)\right] = \begin{bmatrix} 0 & 1 \\ b & 2ap \end{bmatrix}' \begin{bmatrix} 0 & -1 \\ 1 & 0 \end{bmatrix} \begin{bmatrix} 0 & 1 \\ b & 2ap \end{bmatrix} = -b\mathbf{J}^{-1}. \tag{15.33}$$

Thus we say that $\psi(\vartheta, p)$ is symplectic for $b = -1$ and any $a \neq 0$.

Everything discussed so far assumed continuous time, but obviously for a computer implementation in a Markov chain Monte Carlo context we need to discretize time. So we will grid $t + \tau$ time into intervals of size v: $v, 2v, 3v, \ldots, mv$. We need a way to obtain this discretization while preserving volume, and so we use a tool called the *leapfrog methods*. The notation is more clear if we now move t from the subscript to functional notation: $\vartheta(t)$ and $\mathbf{p}(t)$, which is also a reminder that time is now discrete rather than continuous. To complete a single step starting at time t, first update each of the momentum dimensions by $v/2$ with the following:

$$\mathbf{p}_i\left(t + \frac{v}{2}\right) = p_i(t) - \frac{v}{2} \frac{\partial U(\boldsymbol{\vartheta}_t)}{\partial \vartheta_i(t)}. \tag{15.34}$$

Now take a full v-length step to update each of the position dimensions to leapfrog over the momentum:

$$\boldsymbol{\vartheta}_i(t + v) = \vartheta_i(t) + v \frac{\partial K(\mathbf{p}_t)}{\partial \mathbf{p}_i(t + \frac{v}{2})}, \tag{15.35}$$

and finish with the momentum catching up in time the step:

$$\mathbf{p}(t + v) = \mathbf{p}_i\left(t + \frac{v}{2}\right) - \frac{v}{2} \frac{U(\boldsymbol{\vartheta}_t)}{\partial \vartheta_i(t + v)}. \tag{15.36}$$

Notice that the leapfrog method is reversible since it is a one-to-one mapping from t to $t+v$. Obviously, running these steps M times completes the Hamiltonian dynamics for $M \times v$ period of total time. The determination of v is a key tuning parameter in the algorithm since smaller values give a closer estimation to continuous time but also add more steps to the algorithm.

A Metropolis-Hastings algorithm is configured such that the Hamiltonian function serves as the candidate-generating distribution. This requires connecting the regular posterior density function, $\pi(\theta)$, to a potential energy function, $U(\boldsymbol{\vartheta}_t)$, where a kinetic energy function, $K(\mathbf{p}_t)$, serves as a (multidimensional and necessary) auxiliary variable in the manner discussed in Section 14.4.3 starting on 513. This connection is done via the *distribution!canonical* commonly used in physics:

$$p(x) = \frac{1}{Z} \exp\left[-\frac{E(x)}{T}\right], \tag{15.37}$$

where $E(x)$ is the energy function of some system at state x, T is the temperature of the system (which can simply be set at 1), and Z is just a normalizing constant so that $p(x)$ is a regular density function. In the Hamiltonian context (15.37) is:

$$p(\boldsymbol{\vartheta}, \mathbf{p}) = \frac{1}{Z} \exp\left[-\frac{H(\boldsymbol{\vartheta}, \mathbf{p})}{T}\right]$$

$$= \frac{1}{Z} \exp\left[-\frac{U(\boldsymbol{\vartheta}_t) + K(\mathbf{p}_t)}{T}\right]$$

$$= \frac{1}{Z} \exp\left[-\frac{U(\boldsymbol{\vartheta}_t)}{T}\right] \exp\left[-\frac{K(\mathbf{p}_t)}{T}\right], \qquad (15.38)$$

demonstrating that $\boldsymbol{\vartheta}$ and $\mathbf{p}$ are independent. Finally we connect the energy function metric with the regular posterior density metric with the function:

$$E(\boldsymbol{\vartheta}) = -\log(\pi(\boldsymbol{\theta})), \qquad (15.39)$$

thus completing the connection. Notice that the $\boldsymbol{\theta}$ variables must all be continuous in the model, although Hamiltonian Monte Carlo can be combined with other MCMC strategies in a hybrid algorithm.

The Hamiltonian Monte Carlo algorithm uses two general steps at time t:

▷ generate, independent of the current $\boldsymbol{\vartheta}_t$, the momentum $\mathbf{p}_t$ from the multivariate normal distribution implied by $K(\mathbf{p}_t) = \mathbf{p}_t' \boldsymbol{\Sigma}^{-1} \mathbf{p}_t$ with mean vector zero and variance-covariance matrix $\sigma^2 \mathbf{I}$ (or some other desired symmetric and positive-definite form).

▷ run the leapfrog method M times with v steps to produce the candidate $(\tilde{\boldsymbol{\vartheta}}, \tilde{\mathbf{p}})$.

▷ accept this new location or accept the current location as the $t+1$ step with a standard Metropolis decision using the H function:

$$\min\left[1, \exp(-H((\tilde{\boldsymbol{\vartheta}}, \tilde{\mathbf{p}})) + H(\boldsymbol{\vartheta}, \mathbf{p})\right]. \qquad (15.40)$$

While this process looks simple, there are several complications to consider. We must be able to take the partial derivatives of the log-posterior distribution, which might be hard. Also the chosen values of the leapfrog parameters, M and v are critical. If v is too small then exploration of the posterior density will be very gradual with small steps, and if v is too big then many candidates will be rejected. Choosing M is important because this parameter allows the Hamiltonian process to explore strategically with respect to gradients. Excessively large values of M increase compute time, but excessively small values of M also lead to many rejected candidates. In both cases where the parameters are too small we lose the advantages of the gradient calculations and produce an inefficient random walk. Finally, σ^2 affects efficiency of the algorithm in the conventional sense of appropriating tuning the variance of the multivariate normal for the momentum. These can be difficult parameter decisions and Neal (2011) gives specific guidance on trial runs and analysis of the results.

■ **Example 15.2: An Illustrative Simple Simulated Hamiltonian Setup.** This is a basic illustration of Hamiltonian Monte Carlo applied to the same target distribution as the Hit-and-Run sampler in Example 10.5 (page 361). The target distribution is a bivariate normal PDF with a correlation of 0.95. Similar to the data structures in the Hit-and-Run code and the basic Metropolis-Hastings code in Example 10.4.6 (starting on page 359), a matrix is created with all NA values to fill in. Here there are four stored values on each row: x, y, ϑ_1, and ϑ_2. Consider the following R code:

```
ham.norm <- function(theta.matrix,reps,I.mat,upsilon,M,Sigma)  {
    for (i in 2:reps)  {
        vartheta <- theta.matrix[(i-1),3:4]
        p <- rnorm(length(vartheta),0,1)
        p.half <- p - (upsilon/2)*vartheta
        for (j in 1:M)  {
            vartheta <- vartheta + upsilon*p.half
            p.half <- p.half - upsilon*vartheta
        }
        p.full <- -(p.half + (upsilon/2)*vartheta)
        new.U <- t(vartheta) %*% solve(I.mat) %*% vartheta/2
        new.K <- t(p.full) %*% solve(Sigma) %*% p.full
        old.U <- t(theta.matrix[(i-1),1:2]) %*% solve(I.mat)
                    %*% theta.matrix[(i-1),1:2]/2
        old.K <- t(-p) %*% solve(Sigma) %*% -p
        a <- exp(old.U - new.U - new.K + old.K)
        if (a > runif(1)) theta.matrix[i,1:2] <- vartheta
        else theta.matrix[i,1:2] <- theta.matrix[(i-1),1:2]
        theta.matrix[i,3:4] <- vartheta
    }
    theta.matrix
}
```

This code is also provided in the BaM package in R. The following lines code give the setup and the function call for 10,000 iterations. The last 5,000 iterations are displayed in Figure 15.4. Notice the values $\upsilon = 0.001$ and $M = 1000$. These were set up by trial-and-error, and even a model as simple as this requires some tuning of these parameters.

```
num.sims <- 10000
Sig.mat <- matrix(c(1.0,0.95,0.95,1.0),2,2)
upsilon.in <- 0.001
M.in <- 1000
Sigma.in <- matrix(c(3,0,0,3),2,2)
walks<-rbind(c(-3,-3,1,1),matrix(NA,nrow=(num.sims-1),ncol=4))
walks <- ham.norm(walks,num.sims,Sig.mat,upsilon.in,M.in,Sigma.in)
```

This is intended to be an illustrative example only since a bivariate normal does not require estimation with MCMC. For a fully developed package that implements Hamiltonian Monte Carlo see the `Stan` package developed by Andrew Gelman and his colleagues at `http://mc-stan.org`.

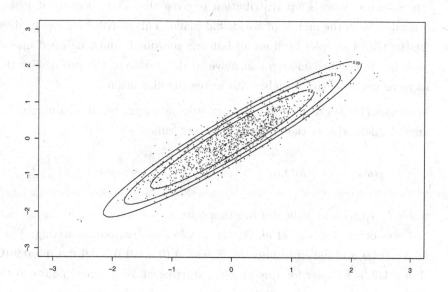

FIGURE 15.4: SAMPLES FROM HAMILTONIAN MONTE CARLO

15.5 Exercises

15.1 Plot in the same graph a time sequence of 0 to 100 versus the annealing cooling schedules: logarithmic, geometric, semi-quadratic, and linear. What are the trade-offs associated with each with regard to convergence and processing time?

15.2 Plot a bivariate normal distribution at 4 different temperature levels with enough variation that the subplots are distinct.

15.3 Replicate the contrived simulated annealing example from Section 15.1 using each of the cooling schedules from Exercise 1, checking for convergence before moving on to the next temperature. What differences do you observe?

15.4 Using the HMO data in Example 11.3.3 on page 390 (available in the `BaM` package) implement a Metropolis-coupling algorithm (MCMCMC) at 10 different temperature levels. The lowest temperature defines the regular joint posterior and the highest temperature defines a joint uniform density. Run this in R to produce a posterior estimate from the cold chain values to compare with the BUGS code in Chapter 11.

15.5   The so-called *witch's hat* distribution is given this name because it looks like a conical spike in the middle of a wide flat plane. This distribution can be disastrous for the Gibbs sampler because all but one coordinate must be lined up with the peak before a subchain step can move to the peak and the probability that this happens reduces exponentially with increasing dimensions.

Matthews (1993) gives the following multivariate form based on a normal/uniform mixture, defined over the d-dimensional unit cube:

$$p(\theta|\mathbf{x}) = (1-\delta)[2\pi\sigma^2]^{-d/2} \exp\left[-\sum_{i=1}^{d} \frac{1}{2\sigma^2}(x_i - \theta_i)^2\right] + \delta I_{(0,1)}(x_i),$$

where $I_{(0,1)}(x_i)$ is an indicator function equal to one when x_i is in the interval $(0,1)$ and zero otherwise. Cui *et al.* (1992) sample this distribution setting: $\delta = 10^{-11}$, $\sigma^2 = 0.0009$, and the nine-dimensional peak at $(0.9, 0.9, 0.9, 0.9, 0.9, 0.9, 0.9, 0.9, 0.9)$. Run a Gibbs sampler for this problem starting at 10 different points in the unit cube, and test convergence with the Gelman and Rubin diagnostic. What do you conclude about convergence?

15.6   Using again the witch's hat distribution from the last exercise, write a Metropolis-Hastings variant that fully explores the "brim" of the hat efficiently (in reasonable time).

15.7   (Bohachevsky, Johnson, and Stein 1986). Use simulated annealing to find the global minima of the function

$$f(x, y) = x^2 + 2y^2 - 0.3\cos(3\pi x) - 0.4\cos(4\pi y) + 0.7,$$

on the support $-1 \leq x, y \leq 1$. Using the same function from Bohachevsky, Johnson, and Stein (1986), write a simple Metropolis-Hastings algorithm in R to explore the sample space. Show the problems associated with this approach by graphing the acceptance rate over time.

15.8   **Simulated Tunneling.** Wenzel and Hamacher (1999) design a stochastic search related to simulated annealing that randomly "hops" from point to point to accommodate efficient search of complex multimodal functions without being trapped for long periods of time at local minima. Replace the target function $f(x)$ with $f_{\text{STUN}} = 1 - \exp[-\zeta(f(x) - f_{\min})]$, where $f_{\min}$ is the lowest minimum yet encountered and $\zeta > 0$ is a tunable "tunneling parameter" that regulates the steepness of descents. Write and run a Metropolis-Hastings implementation using the same function as the last exercise.

15.9 One of the difficulties with slice sampling is that in high dimensions, the Markov chain may be restricted to a set of unusual subspaces of the parameter space: nonlinear and varying dramatically in size. This can obviously lead to poor convergence and mixing properties. Develop a simulated annealing algorithm that alleviates this problem in at least one application.

15.10 Consider the surface (3D function) created by the following R code:

```
f.xyz <- function(x, y)  {
    0.75*exp( -((10*x-1)^2 + (10*y-1)^2)/5 ) +
    0.50*exp( -((10*x-7)^2 + (10*y-5)^2)/5 ) -
    0.25*exp( -((10*x-4)^2 + (10*y-7)^2)/5 )
}
set.seed(pi);   n <- 15;
x2 <- x1 <- seq(0,1,length=n)
y <- outer(x1, x2, f.xyz)
y <- y + rnorm(n^2,0,0.05*max(abs(y)))
```

Write a simulated tunneling application for this surface that accommodates maxima instead of minima. Graph the function and 200 post-convergence visits on the graph.

15.11 It is possible that not all full conditional probability statements can be identified to set up a Gibbs sampler. One way to solve this problem is to embed a Metropolis-Hastings algorithm within the Gibbs sampler, called "Metropolis-within-Gibbs," for the parameter or parameters that are causing the difficulty. Write out the full statement of this algorithm and derive the detailed balance equation.

15.12 Using the terrorism in Great Britain example (Example 10.3.2, starting on page 346) implement a Metropolis-with-Gibbs example where λ and ϕ get a Gibbs draw as done before but k is sampled using Metropolis-Hastings.

15.13 In his original article Green (1995) analyzes the well-known coal mining data used in many papers and texts, except that he compares models with one, two, and three possible changepoints using reversible jump MCMC and uses days instead of years. Green's analysis supports the notion of two changepoints (his Figure 2). Replicate that analysis using R and make a decision about which model should eventually be preferred with a Bayes Factor calculation. This is the data vector:

```
coal.mining.disasters <- c(4,5,4,0,1,4,3,4,0,6,3,3,4,0,2,6,
                          3,3,5,4,5,3,1,4,4,1,5,5,3,4,2,5,
                          2,2,3,4,2,1,3,2,2,1,1,1,1,3,0,0,
```

$$1,0,1,1,0,0,3,1,0,3,2,2,0,1,1,1,$$
$$0,1,0,1,0,0,0,2,1,0,0,0,1,1,0,2,$$
$$3,3,1,1,2,1,1,1,1,2,4,2,0,0,1,4,$$
$$0,0,0,1,0,0,0,0,0,1,0,0,1,0,1)$$

(also included in the BaM package).

15.14   Using the death penalty data in Example 14.3 on page 481, write a RJMCMC procedure in R for the Tobit model.

15.15   Show that the acceptance criteria given in (15.14) satisfies the detailed balance equation.

15.16   **Umbrella Sampling.** Torrie and Valleau (1977) suggested a generalization of importance sampling, solidified by Geyer (2011). Here the approximation distribution is replaced with a mixture distribution: $g(\theta) \equiv \sum_{j=1}^{J} g(\theta_j)c_j$ for a J-component mixture where c_j is a set of normalizing weights (using the language of Section 9.6 starting on page 296, rather than that of Geyer). This makes the importance weight $\omega_i = f(\theta_i)/\sum_{j=1}^{J} g(\theta_j)c_j$. The choice of mixture distribution is problem-dependent and must be made carefully. Apply this sampler to the function given in (15.2).

15.17   Using R run a coupling from the past algorithm with the following transition matrix:

$$\mathbf{K} = \begin{bmatrix} 0.9 & 0.1 \\ 0.5 & 0.5 \end{bmatrix}.$$

Print out the reached values as part of the code, indicate when coalescence occurs, and determine stationary distribution.

15.18   A state space has four ordered states, $\{A, B, C, D\}$, where D wraps around back to A. For any given state the probability of moving "forward" is $2/3$ and the probability of moving "backward" is $1/3$. Implement a CFTP algorithm in R to obtain the stationary distribution.

15.19   Rerun the CFTP algorithm from Exercise 15.17 but instead of re-using the uniform draws, generate new ones at each step. What difference do you see? Why did this go wrong?

15.20   The Hoff (2005, 2009) social networking model starts with the $n \times n$ symmetric matrix $\mathbf{Y}$ of links between n individuals, where $y_{ij} = 1$ indicates a known link between node i and node j, and $y_{ij} = 0$ indicates the absence of evidence of a link. The $n \times n \times K$ array $\mathbf{X}$ defines for each $n \times n$ relationship between individual i and individual j with a K-length vector of covariate information. Relate $\mathbf{X}$ and $\mathbf{Y}$ with the random effects logistic regression specification:

$$p(\mathbf{Y}|\boldsymbol{\theta}_{ij}) = \prod_{i \neq j} \frac{\exp(\boldsymbol{\theta}_{ij})}{1 + \exp(\boldsymbol{\theta}_{ij})} \qquad \boldsymbol{\theta}_{ij} = \boldsymbol{\beta}'\mathbf{x}_{ij} + z_{ij} \qquad z_{ij} = \mathbf{u}_i'\mathbf{G}\mathbf{v}_j + \epsilon_{ij}$$

where β is a K-length vector of coefficients to estimate, and z_{ij} is a random effects term to account for dependencies between attribute relationships. The z_{ij} term has two components: a $\mathbf{u}_i'$ vector of sender-specific latent factors, a $\mathbf{v}_j$ vector of receiver-specific latent factors, a $\mathbf{G}$ diagonal matrix of unknown coefficients, plus a ϵ_{ij} scalar error specific to the ij edge. Assign priors and write a Metropolis-Hastings variant using the Roethlisberger and Dickson (1939) Hawthorne Plant wiring room dataset in the R package BaM.

Appendix A

Generalized Linear Model Review

A.1   Terms

Consider a one-parameter conditional probability density function or probability mass function for the random variable Z of the form: $f(z|\zeta)$. This family form is classified as an exponential family form if it can be reparameterized as: $f(z|\zeta) = \exp[t(z)u(\zeta)]r(z)s(\zeta)$, where: r and t are real-valued functions of z that do not depend on ζ, and s and u are real-valued functions of ζ that do not depend on z, and $r(z) > 0, s(\zeta) > 0 \; \forall z, \zeta$. The canonical form is the result of one-to-one transformation that reduces the complexity of the symbolism and reveals structure. If $t(z) = z$, then we say that this PDF or PMF is in its canonical form for the random variable Z. Otherwise, we can make the simple transformation: $y = t(z)$ to force a canonical form. Similarly, if $u(\zeta) = \zeta$ in this expression, then this PDF or PMF is in its canonical form for the parameter ζ. If not, we can again force a canonical form by transforming: $\theta = u(\zeta)$, and call θ the canonical parameter. This produces the final form: $f(y|\theta) = \exp[y\theta - b(\theta) + c(y)]$. Here $b(\theta)$ plays a key role in calculating the moments of the distribution and other important quantities. If the form of θ, the *canonical link* between the original form and the θ parameterized form, is the source of the link function in generalized linear models (Fahrmeir and Kaufmann 1985; Jørgensen 1983; Wedderburn 1976), then this process is equivalent to finding a k-dimensional global modal point.

Our real interest lies in obtaining the posterior distribution of the unknown k-dimensional $\boldsymbol{\theta}$ coefficient vector, given an observed $\mathbf{X}$ matrix of data values: $p(\boldsymbol{\theta}|\mathbf{X})$. This allows us to determine the "most likely" values of the $\boldsymbol{\theta}$ vector using the k-dimensional mode (maximum likelihood inference, Fisher 1925b), or alternatively to describe the resulting distribution (as in Bayesian inference). This posterior is produced by Bayes' Law:

$$p(\boldsymbol{\theta}|\mathbf{X}) = p(\mathbf{X}|\boldsymbol{\theta})\frac{p(\boldsymbol{\theta})}{p(\mathbf{X})} \tag{A.1}$$

where $p(\mathbf{X}|\boldsymbol{\theta})$ is the n-dimensional joint probability function for data (the *probability of the sample* for a fixed $\boldsymbol{\theta}$) under the assumption that the data are independent and identically distributed according to $p(X_i|\boldsymbol{\theta}) \; \forall \; i = 1, \ldots, n$, and $p(\boldsymbol{\theta})$, $p(\mathbf{X})$ are the corresponding unconditional probabilities.

The Bayesian approach integrates to obtain $p(\mathbf{X})$ (or ignores it using proportionality) and specifies an assumed (prior) distribution on $\boldsymbol{\theta}$, thus allowing fairly direct computation of $p(\boldsymbol{\theta}|\mathbf{X})$ from (A.1). If we regard $p(\mathbf{X}|\boldsymbol{\theta})$ as a function of $\boldsymbol{\theta}$ for some given observed data $\mathbf{X}$, then $L(\boldsymbol{\theta}|\mathbf{X}) = \prod_{i=1}^{n} p(\mathbf{X}|\boldsymbol{\theta})$ is called a likelihood function (DeGroot 1986, p.339). The maximum likelihood principle states that an admissible[1] $\boldsymbol{\theta}$ that maximizes the likelihood function probability (discrete case) or density (continuous case), relative to alternative values of $\boldsymbol{\theta}$, provides the $\boldsymbol{\theta}$ that is most "likely" to have generated the observed data, $\mathbf{X}$, given the assumed parametric form. Restated, if $\hat{\boldsymbol{\theta}}$ is the maximum likelihood estimator for the unknown parameter vector, then it is necessarily true that $L(\hat{\boldsymbol{\theta}}|\mathbf{X}) \geq L(\boldsymbol{\theta}|\mathbf{X}) \; \forall \; \boldsymbol{\theta} \in \Theta$, where Θ is the admissible set of $\boldsymbol{\theta}$.

The likelihood function differs from the inverse probability, $p(\boldsymbol{\theta}|\mathbf{X})$, in that it is necessarily a *relative* function since it is not a normalized probability measure ($\mathfrak{P}()$) bounded by zero and one.[2] Thus maximum likelihood estimation substitutes the unbounded notion of likelihood for the bounded definition of probability (Casella and Berger 2002, p. 316; Fisher 1922, p.327; King 1989, p.23). This is an important theoretical distinction, but of little significance in applied practice.

Typically it is mathematically more convenient to work with the natural log of the likelihood function. This does not change any of the resulting parameter estimates because the likelihood function and the log likelihood function have identical modal points. Using a probability density function for a single parameter of interest the basic *log* likelihood function is very simple:

$$\ell(\boldsymbol{\theta}|\mathbf{X}) = \log(L(\boldsymbol{\theta}|\mathbf{X})), \tag{A.2}$$

where we use $\ell(\boldsymbol{\theta}|\mathbf{X})$ as shorthand to distinguish the log likelihood function from the likelihood function: $L(\boldsymbol{\theta}|\mathbf{X})$.

The score function is the first derivative of the log likelihood function with respect to the parameters of interest:

$$\dot{\ell}(\boldsymbol{\theta}|\mathbf{X}) = \frac{\partial}{\partial\boldsymbol{\theta}}\ell(\boldsymbol{\theta}|\mathbf{X}). \tag{A.3}$$

Setting $\dot{\ell}(\boldsymbol{\theta}|\mathbf{X})$ equal to zero and solving gives the maximum likelihood estimate, $\hat{\boldsymbol{\theta}}$. This is now the "most likely" value of $\boldsymbol{\theta}$ from the parameter space Θ treating the observed data as given: $\hat{\boldsymbol{\theta}}$ maximizes the likelihood function at the observed values. The *Likelihood Principle* (Birnbaum 1962) states that once the data are observed, and therefore treated as given, all of the available evidence for estimating $\hat{\boldsymbol{\theta}}$ is contained in the (log) likelihood function, $\ell(\boldsymbol{\theta}|\mathbf{X})$. This is a very handy data reduction tool because it tells us exactly what treatment

---

[1] *Admissible* here means values of θ are taken from the valid parameter space (Θ): values of θ that are unreasonable according to the form of the sampling distribution of θ are not considered (integrated over). A Bayesian decision rule is called *admissible* if it is no worse (equal or lower risk) than every other decision rule and has at least one better alternative for all possible values of the parameter as discussed in Chapter 8 See also: Berger (1985, Section 1.3; Ghosh and Meeden (1997, Section 2.2).

[2] From a frequentist standpoint, the probabilistic uncertainty is a characteristic of the random variable $\mathbf{X}$, not the unknown but fixed $\boldsymbol{\theta}$. Barnett (1973, p.131) clarifies this distinction: "Probability remains attached to X, not θ; it simply reflects inferentially on θ."

of the data is important to us and allows us to ignore an infinite number of alternates (Poirer 1988, p.127).

Setting the score function from the joint PDF or PMF equal to zero and rearranging gives the likelihood equation:

$$\sum t(x_i) = n \frac{\partial}{\partial \boldsymbol{\theta}} E[\mathbf{x}] \tag{A.4}$$

where $\sum t(x_i)$ is the remaining function of the data, depending on the form of the probability density function (PDF) or probability mass function (PMF), and $E[\mathbf{x}]$ is the expectation over the kernel of the density function for $\mathbf{x}$.[3] The underlying theory is remarkably strong. Solving (A.4) for the unknown coefficient produces an estimator that is unique (a unimodal posterior distribution), consistent (converges in probability to the population value),[4] and asymptotically efficient (the variance of the estimator achieves the lowest possible value as the sample size becomes adequately large: the Cramér-Rao lower bound, see Shao 2005). This result combined with the central limit theorem gives the asymptotic normal form for the estimator: $\sqrt{n}(\hat{\boldsymbol{\theta}} - \boldsymbol{\theta}) \overset{\mathcal{D}}{\to} \mathcal{N}(\mathbf{0}, \Sigma_{\boldsymbol{\theta}})$. This means that as the sample size gets large, the difference between the estimated value of $\boldsymbol{\theta}$ and the true value of $\boldsymbol{\theta}$ gets progressively close to zero, with a variance governed by $\frac{1}{\sqrt{n}}\Sigma_{\boldsymbol{\theta}}$, where $\Sigma_{\boldsymbol{\theta}}$ is the $k \times k$ variance-covariance matrix for $\boldsymbol{\theta}$. Furthermore, $\sum t(x_i)$ is a sufficient statistic for $\boldsymbol{\theta}$, meaning that all of the relevant information about $\boldsymbol{\theta}$ in the data is contained in $\sum t(x_i)$. For example, the normal log likelihood expressed as a joint exponential family form is $\ell(\boldsymbol{\theta}|\mathbf{X}) = \left(\mu \sum x_i - \frac{n\mu^2}{2} \right)/\sigma^2 - \frac{1}{2\sigma^2} \sum x_i^2 - \frac{n}{2} \log(2\pi\sigma^2)$. So $t(\mathbf{x}) = \sum x_i$, $\frac{d}{d\mu} \frac{n\mu^2}{2} = n\mu$, and equating gives the maximum likelihood estimate of μ to be the sample average that we know from basic texts: $\frac{1}{n} \sum x_i$.

A.1.1 The Linear Regression Model

The classic linear model dates back to the early nineteenth century (see Stigler 1999, Chapter 17) and is the undeniable workhorse of statistical work in the social and behavioral sciences. The linear model, as elegant as it is, requires a relatively strict set of assumptions. The Gauss-Markov Theorem states that if one can assume that:

1. the relationship between each explanatory variable and the outcome variable is approximately linear in structure,

---

[3]The kernel of a PDF or PMF is the component of the parametric expression that directly depends on the form of the random variable, i.e., what is left when normalizing constants are omitted. We can often work with kernels of distributions for convenience and recover all probabilistic information at the last stage of analysis by renormalizing (ensuring summation or integration to one). The kernel is the component of the distribution that assigns *relative* probabilities to levels of the random variable (see Gill 2000, Chapter 2). For example the kernel of a gamma distribution is just the part $x^{\alpha-1} \exp[-x\beta]$, without $\beta^\alpha/\Gamma(\alpha)$ (see A for gamma distribution details).

[4]In one's enthusiasm for the maximum likelihood estimator it is easy to forget that it is asymptotically unbiased, but not necessarily unbiased in finite sample situations. For instance the maximum likelihood estimate for the variance of a normal model, $\hat{\sigma}^2 = \frac{1}{n} \sum (x_i - \bar{x})^2$ is biased by $n/(n-1)$. This difference is rarely of significance and clearly the bias disappears in the limit, but it does illustrate that unbiasedness of the maximum likelihood estimate is guaranteed only in asymptotic circumstances.

2. the residuals are independent with mean zero and constant variance,

3. there is no correlation between any regressor and disturbance,

then the solution produced by selecting coefficient values that minimize the sum of the squared residuals is unbiased and has the lowest total variance amongst unbiased linear alternatives. This is sometimes called BLUE for *Best Linear Unbiased Estimate.*

The linear model can be expressed as follows:

$$\underset{(n\times 1)}{\mathbf{Y}} = \underset{(n\times p)(p\times 1)}{\mathbf{X}\boldsymbol{\beta}} + \underset{(n\times 1)}{\boldsymbol{\epsilon}}, \qquad \underset{(n\times 1)}{E[\mathbf{Y}]} = \underset{(n\times p)(p\times 1)}{\mathbf{X}\boldsymbol{\beta}}. \qquad (A.5)$$

The right-hand sides of the two equations are very familiar: $\mathbf{X}$ is the model matrix of observed data or the design matrix of stipulated values (organized by column), $\boldsymbol{\beta}$ is the column vector of unknown coefficients to be estimated, $\mathbf{X}\boldsymbol{\beta}$ is called the "linear structure vector," and $\boldsymbol{\epsilon}$ is the column vector of (assumed) independent, normally distributed error terms with constant variance: the stochastic component. In the expectation component of (A.5), $E[\mathbf{Y}] = \boldsymbol{\theta}$ is the column vector of means: the systematic component. The vector, $\mathbf{Y}$, is distributed iid normal with mean $\boldsymbol{\theta}$, and constant variance σ^2. This is exactly the multivariate linear model described in basic statistics texts, but provided here in matrix form. Gauss (1809) showed that a good way to estimate the unknown linear slope term(s) is to minimize the sum of the squared errors. The squaring is a convenience since otherwise positive and negative residuals will cancel each other out. Actually there are many ways to accomplish this such as taking the absolute value of the residuals: L_1 regression (Legendre 1805). The principle behind L-regression (also called quantile regression) is that a selected quantile, $\kappa \in [0,1]$, is selected and the following quantity is minimized:

$$\underset{\hat{\beta}\in\mathbb{R}}{\min}\left[\sum_{i=1}^{n}\kappa|y_i - x_i\hat{\beta}| + \sum_{i=1}^{n}(1-\kappa)|y_i - x_i\hat{\beta}|\right]. \qquad (A.6)$$

Obviously if $\kappa = \frac{1}{2}$, this is greatly simplified and in fact becomes a well-known alternative to least squares estimates called the least absolute errors (LAE) estimator, the L_1 estimator, or just the median regression estimator. In the notation of (A.5), minimizing squared error is equivalent to:

$$\min_{\mathbf{b}} \sum_{i=1}^{n}\varepsilon_i^2 = \min_{\mathbf{b}} \sum_{i=1}^{n}(y_i - \mathbf{X}_i\mathbf{b})^2. \qquad (A.7)$$

Or in matrix notation, this is the process of minimizing the quantity $S(\mathbf{b})$ over the range of possible values of $\mathbf{b}$:

$$S(\mathbf{b}) = \boldsymbol{\epsilon}'\boldsymbol{\epsilon} = (\mathbf{y} - \mathbf{Xb})'(\mathbf{y} - \mathbf{Xb}) = \mathbf{y}'\mathbf{y} - 2\mathbf{b}\mathbf{X}'\mathbf{y} + \mathbf{b}\mathbf{X}'\mathbf{Xb}. \qquad (A.8)$$

This is a quadratic form with a positive sign on the squared term, so there is a unique value that minimizes $S(\mathbf{b})$. We can therefore differentiate with respect to $\mathbf{b}$ and solve for zero to get:

$$\frac{\partial}{\partial b} S(\mathbf{b}) = -2\mathbf{X}'\mathbf{y} + 2\mathbf{X}'\mathbf{X}\mathbf{b} \equiv 0$$

$$\mathbf{X}'\mathbf{X}\mathbf{b} = \mathbf{X}'\mathbf{y} \quad \text{(the "normal equation")}$$

$$\hat{\mathbf{b}} = (\mathbf{X}'\mathbf{X})^{-1}\mathbf{X}'\mathbf{y}.$$

In addition, it is also possible to derive an estimator of β using maximum likelihood estimation. By the Gauss-Markov *along with* the central limit theorem, the residuals are normally distributed with mean zero and constant variance. The likelihood equation for the residuals is therefore:

$$L(\epsilon) = (2\pi\sigma^2)^{-\frac{n}{2}} \exp\left[-\frac{1}{2\sigma^2}\epsilon'\epsilon\right]$$

$$= (2\pi\sigma^2)^{-\frac{n}{2}} \exp\left[-\frac{1}{2\sigma^2}(\mathbf{y} - \mathbf{X}\mathbf{b})'(\mathbf{y} - \mathbf{X}\mathbf{b})\right].$$

The function $L(\epsilon)$ is concave for this equation, although concavity is not a guaranteed feature of the likelihood equation.[5]

The log of $L(\epsilon)$ is maximized at the same point as the function itself, allowing us to use the easier form, taking the derivative with respect to $\mathbf{b}$, and solving for zero:

$$\log L(\epsilon) = -\frac{n}{2}\log(2\pi) - \frac{n}{2}\log(2\sigma^2) - \frac{1}{2\sigma^2}(\mathbf{y} - \mathbf{X}\mathbf{b})'(\mathbf{y} - \mathbf{X}\mathbf{b})$$

$$\frac{\partial}{\partial \mathbf{b}}\log L(\epsilon) = \frac{1}{\sigma^2}\mathbf{X}'(\mathbf{y} - \mathbf{X}\mathbf{b}) \equiv 0$$

$$\hat{\mathbf{b}} = (\mathbf{X}'\mathbf{X})^{-1}\mathbf{X}'\mathbf{y}.$$

This demonstrates that the least squares estimator is identical to the maximum likelihood estimator for the parameter vector $\mathbf{b}$ in linear regression. In general, most statistical procedures do not provide such a nice linkage between disparate theories, and it is a reflection of both the elegance and the fundamental nature of the linear model that this is true here.

A.2 The Generalized Linear Model

To look at a broader application of maximum likelihood estimation, we now turn to the generalized linear model construct for specifying nonlinear regression specifications. This will demonstrate the utility of maximizing the likelihood function relative to some observed data in a more complex setting.

---

[5]In the absence of concavity over the allowable range of parameter values multiple modes may exist or the maximum may not be unique (Shao 2005). A number of authors have shown the existence and uniqueness of a maximum value not on the boundaries of the parameter space (and therefore concavity) for commonly specified likelihood functions: Haberman (1974b), Kaufmann (1988), Lesaffre and Kaufmann (1992), Makelainen, Schmidt, and Styan (1981), Silvapulle (1981), Wedderburn (1976), and Fahrmeir and Tutz (2001, p.43) discuss the properties for generalized linear models.

A.2.1   Defining the Link Function

Consider the standard linear model meeting the Gauss-Markov conditions:

$$\underset{(n\times1)}{\mathbf{V}} = \underset{(n\times p)(p\times1)}{\mathbf{X}\boldsymbol{\beta}} + \underset{(n\times1)}{\boldsymbol{\epsilon}}, \qquad E(\mathbf{V}) = \underset{(n\times1)}{\boldsymbol{\theta}} = \underset{(n\times p)(p\times1)}{\mathbf{X}\boldsymbol{\beta}}, \qquad (\text{A.9})$$

where $E(\mathbf{V}) = \boldsymbol{\theta}$ is the vector of means (the systematic component), and $\mathbf{V}$ is distributed iid normal with mean $\boldsymbol{\theta}$, and constant variance σ^2. We use $\mathbf{V}$ instead of $\mathbf{Y}$ (as opposed to (A.5)), because $\mathbf{V}$ is not the real outcome variable of interest; it is merely a description of the linear component on the right-hand-side. Now generalize this slightly with a function based on the mean of the outcome variable, which is no longer required to be normally distributed or even continuous:

$$\underset{(n\times1)}{g(\boldsymbol{\mu})} = \underset{(n\times1)}{\boldsymbol{\theta}} = \underset{(n\times p)(p\times1)}{\mathbf{X}\boldsymbol{\beta}}. \qquad (\text{A.10})$$

Here $g()$ is required to be an invertible, *smooth* function of the mean vector $\boldsymbol{\mu}$ of the y_i.[6]

Information from the explanatory variables is now expressed in the model only through the link from the linear structure, $\mathbf{X}\boldsymbol{\beta}$, to the linear predictor, $\boldsymbol{\theta} = g(\boldsymbol{\mu})$, controlled by the form of the link function, $g()$. This link function connects the linear predictor to the *mean* of the outcome variable, not directly to the expression of the outcome variable itself, so the outcome variable can now take on a variety of non-normal forms. The link function connects the stochastic component that describes some response variable from a wide variety of forms to all the standard normal theory supporting the systematic component through the mean function:

$$g(\boldsymbol{\mu}) = \boldsymbol{\theta} = \mathbf{X}\boldsymbol{\beta}$$
$$g^{-1}(g(\boldsymbol{\mu})) = g^{-1}(\boldsymbol{\theta}) = g^{-1}(\mathbf{X}\boldsymbol{\beta}) = \boldsymbol{\mu} = E(\mathbf{Y}).$$

So the inverse of the link function ensures that $\mathbf{X}\boldsymbol{\beta}$ maintains the Gauss-Markov assumptions for linear models and all the standard theory applies even though the outcome variable no longer meets the required assumptions.

The generalization of the linear model now has four components derived from the expressions above.

I. **Stochastic Component:** $\mathbf{Y}$ is the random or stochastic component that remains distributed iid according to a specific parametric family distribution with mean $\boldsymbol{\mu}$.

II. **Systematic Component:** $\boldsymbol{\theta} = \mathbf{X}\boldsymbol{\beta}$ is the systematic component with an associated Gauss-Markov normal basis.

III. **Link Function:** the stochastic component and the systematic component are linked by a function of $\boldsymbol{\theta}$, which is *exactly the canonical link function*, summarized in Table

---

[6] More specifically, $g()$ must be a one-to-one function that is everywhere differentiable over the support of $\boldsymbol{\mu}$.

A.1 below. We can consider $g(\mu)$ as "tricking" the linear model into thinking that it is still acting upon normally distributed outcome variables.

IV. **Residuals:** Although the residuals can be expressed in the same manner as in the standard linear model, observed outcome variable value minus predicted outcome variable value, a more useful quantity is the deviance residual described in detail below.

This setup is much more powerful than it initially appears. The outcome variable described by the outcome family form is affected by the explanatory variables strictly through the link function applied to the systematic component, $g^{-1}(\mathbf{X}\beta)$, and nothing else.

Table A.1 summarizes the link functions for common distributions.[7] Note that $g()$ and $g^{-1}()$ are both included.

TABLE A.1: NATURAL LINK FUNCTIONS FOR COMMON SPECIFICATIONS

Distribution	Canonical Link $\theta = g(\mu)$	Inverse Link $\mu = g^{-1}(\theta)$
Poisson	$\log(\mu)$	$\exp(\theta)$
Normal	μ	θ
Gamma	$-\frac{1}{\mu}$	$-\frac{1}{\theta}$
Negative binomial	$\log(1-\mu)$	$1-\exp(\theta)$
Binomial logit	$\log\left(\frac{\mu}{1-\mu}\right)$	$\frac{\exp(\theta)}{1+\exp(\theta)}$
Binomial probit	$\Phi^{-1}(\mu)$	$\Phi(\theta)$
Binomial cloglog	$\log(-\log(1-\mu))$	$1-\exp(-\exp(\theta))$

A substantial advantage of the generalized linear model is its freedom from the standard Gauss-Markov assumption that the residuals have mean zero and constant variance. Yet this freedom comes with the price of interpreting more complex stochastic structures. Currently, the dominant philosophy is to assess this stochastic element by looking at (summed) discrepancies: a function that describes the difference between observed and expected outcome data for some specified model: $D = \sum_{i=1}^{n} d(\theta, y_i)$. This definition is intentionally left vague for the moment to stress that the format of D is widely applicable. For instance, if the discrepancy in D is measured as the squared arithmetic difference from a single mean, then this becomes the standard form for the variance. In terms of generalized linear models, the squared difference from the mean will prove to be an overly restrictive definition of discrepancy, and a likelihood-based measure will be shown to be far more useful.

---

[7]In the case of the binomial, only the logit link function qualifies as the canonical link function. The other two forms are provided because of their substitutabillity and widespread use. See also the robit model for robust estimation of dichotomous outcomes (Gelman and Hill 2007, pp.124-5).

A.2.2   Deviance Residuals

Starting with the log likelihood for a proposed model from the (A.2) notation, add the '$\hat{\ }$' notation as a reminder that it is evaluated at the maximum likelihood values:

$$\ell(\hat{\boldsymbol{\theta}}, \psi|\mathbf{y}) = \sum_{i=1}^{n} \frac{y_i\hat{\theta} - b(\hat{\theta})}{a(\psi)} + c(\mathbf{y}, \psi),$$

where ψ is a scale parameter and the exponential family form is given by its "canonical" form: $f(y|\theta) = \exp\left[\frac{y\theta - b(\theta)}{a(\psi)} + c(y, \psi)\right]$. When a given PDF or PMF does not have a scale parameter, then $a(\psi) = 1$ and this reduces to the previous form. Often we need to explicitly treat these nuisance parameters instead of ignoring them or assuming they are known, and the most important case of a two-parameter exponential family is this form here where the second parameter is a scale parameter.

Also consider the same log likelihood function with the same data and the same link function, except that it now has n coefficients for the n data points, i.e., the saturated model log likelihood function with the '$\tilde{\ }$' function to denote the n-length $\boldsymbol{\theta}$ vector:

$$\ell(\tilde{\boldsymbol{\theta}}, \psi|\mathbf{y}) = \sum_{i=1}^{n} \frac{y_i\tilde{\theta} - b(\tilde{\theta})}{a(\psi)} + c(\mathbf{y}, \psi).$$

This is the highest possible value for the log likelihood function achievable with the given data, $\mathbf{y}$. Yet it is also often analytically unhelpful except as a benchmark. The deviance function is defined as minus twice the log likelihood ratio (that is, the arithmetic difference since both terms are already written on the log metric):

$$D(\boldsymbol{\theta}, \mathbf{y}) = -2\left[\ell(\hat{\boldsymbol{\theta}}, \psi|\mathbf{y}) - \ell(\tilde{\boldsymbol{\theta}}, \psi|\mathbf{y})\right]$$

$$= -2\sum_{i=1}^{n}\left[y_i(\hat{\theta} - \tilde{\theta}) - \left(b(\hat{\theta}) - b(\tilde{\theta})\right)\right]a(\psi)^{-1}. \qquad (A.11)$$

A utility of the deviance function is that it also allows a look at the individual deviance contributions in an analogous way to linear model residuals. The single point deviance function is just the deviance function for the y_i^{th} point (i.e., without the summation):

$$d(\boldsymbol{\theta}, y_i) = -2\left[y_i(\hat{\theta} - \tilde{\theta}) - \left(b(\hat{\theta}) - b(\tilde{\theta})\right)\right]a(\psi)^{-1}.$$

To define the deviance residual at the y_i point, we take the square root:

$$R_{Deviance} = \frac{(y_i - \mu_i)}{|y_i - \mu_i|}\sqrt{|d(\boldsymbol{\theta}, y_i)|},$$

where $\frac{(y_i - \mu_i)}{|y_i - \mu_i|}$ is a sign-preserving function. For example, the natural link function for the binomial is the logit function: $\theta = \log(\mu/(1 - \mu))$, and the log likelihood contribution from a single datum, in exponential family form, is given by $\ell(p|y_i, n_i) = y_i\log(p/(1 - p)) - (-n_i\log(1 - p)) + \log\binom{n_i}{y_i}$. Substituting in the maximum likelihood result for p, $\mu_i = n_i\hat{p}$, and the saturated model result for p, $\mathbf{y}_i = n_i\tilde{p}$ into (A.11) gives:

$$D(p, \mathbf{y}) = -2\sum_{i=1}^{n}\left[y_i\log\left(\frac{y_i}{\mu_i}\right) + (n_i - y_i)\log\left(\frac{n_i - y_i}{n_i - \mu_i}\right)\right]. \qquad (A.12)$$

Comparisons are made between: the *null model*, a common mean μ for all y meaning $y = g^{-1}(\mu + \epsilon)$ (data is modeled as all random variation). the *saturated* or *full* model, where the data are explained exactly but no data reduction or underlying trend information is obtained. This is typically n parameters for n datapoints (data is modeled as all systematic). the proposed model where we have partitioned the data into systematic structures *and* a random component according to some theoretical consideration. The log-likelihood for the full model versus the research model can be compared in ratio terms:

$$2(\ell(y, \phi|y) - \ell(\hat{\mu}, \phi|y)). \tag{A.13}$$

Assuming iid data and $a(\phi_i) = \phi/w_i$, this becomes:

$$\sum_i 2w_i(y_i(\tilde{\theta}_i - \hat{\theta}_i) - b(\tilde{\theta}_i) + b(\hat{\theta}_i))/\phi \tag{A.14}$$

Call $D(y, \hat{\mu})$ the *deviance*, and the above forms the scaled deviance $(D(y, \hat{\mu})/\phi)$. For common models, these deviances are summarized in Table A.2.

TABLE A.2: SUMMED DEVIANCES FOR COMMON SPECIFICATIONS

GLM	Unscaled Deviance, $D(y, \hat{\mu})$
Gaussian	$\sum_i(y_i - \hat{\mu}_i)^2$
Poisson	$2\sum_i [y_i \log(y_i/\hat{\mu}_i) + (y_i - \hat{\mu}_i)]$
Binomial	$2\sum_i [y_i \log(y_i/\hat{\mu}_i) + (m - \hat{\mu}_i) \log(((m - y_i)/(m - \hat{\mu}_i)))]$
	where m is the sample size so μ is the count not the proportion.
Gamma	$2\sum_i [-\log(y_i/\hat{\mu}_i) + (y_i - \hat{\mu}_i)/\hat{\mu}_i]$
Negative Binomial	$\sum_i(y_i - \hat{\mu}_i)^2/(\hat{\mu}_i^2 y_i)$

In standard practice, the two main tools: the deviances, as described, and Pearson's statistic:

$$X^2 = \sum_i \mathbf{R}^2_{Pearson} = \sum_i \frac{(\mathbf{y}_i - \hat{\mu}_i)^2}{var(\hat{\mu}_i)} \tag{A.15}$$

which lead to asymptotic χ^2 tests with the degrees of freedom equal to the difference in the number of parameters. Paradigm: We compare two *nested* models where the more parsimonious model is one that puts linear restrictions (usually $\beta_j = 0$) on the parameters. Intuition: an unrestricted model versus a restricted model. The goodness of fit test to the data is just a nested test where the nesting model is the saturated model. Two caveats: if

$\phi \neq 1$, more elaborate testing required (estimates of ϕ), and sample sizes need to be large for less-granular responses (huge for dichotomous).

For the deviance comparison we are comparing a large (possibly saturated) model Ω to a restricted research model of interest ω. The difference in the scaled deviances, $D_\omega - D_\Omega$ is asymptotically χ^2 with df the number of restrictions. The restricted model will have larger deviances because we are making theoretical statements away from just trending through the data. General test: Model 1 has p parameters and model 2 has $q > p$ parameters. Then

$$D_p - D_q \underset{\sim}{asym} \chi^2_{df=q-p}. \tag{A.16}$$

If this difference is "small" then the restrictions make sense. If this difference is "large" then they take us far from what the data want to say. Our claim is that the $p - q$ parameters all have coefficients equal to zero (this is the restriction). The number of parameters in each model determine the corresponding degrees of freedom: the saturated model has n parameters, the research model with p parameters (counting the intercept), and the null model has 1 parameter to account for the mean. To test the research model versus null model: the research model gives D_Ω, the null model gives D_ω, and the degrees of freedom are $\#params(\Omega) - \#params(\omega) = p - 1$. So large values of $D_\omega - D_\Omega$ support the saturated model. In this case we want to be in the tail of the associated chi-square test to distinguish the research model from the simple null model. To test the saturated model versus research model: the saturated model gives D_Ω, the research model gives D_ω, and the degrees of freedom are $\#params(\Omega) - \#params(\omega) = n - p$ So small values of $D_\omega - D_\Omega$ support the research model by indicating that it is statistically close to the saturated model. Note that failing this test does not discredit any specification as it merely means more information may be available to use.

■ **Example A.1:   Poisson Model of Military Coups.** Sub-Saharan Africa has experienced a disproportionately high proportion of regime changes due to the military takeover of government for a variety of reasons, including ethnic fragmentation, arbitrary borders, economic problems, outside intervention, and poorly developed governmental institutions. These data, selected from a larger set collected by Bratton and Van De Walle (1994), look at potential causal factors for counts of military coups (ranging from 0 to 6 events) in 33 sub-Saharan countries over the period from each country's colonial independence to 1989. Seven explanatory variables are chosen here to model the count of military coups: Military Oligarchy (the number of years of this type of rule); Political Liberalization (0 for no observable civil rights for political expression, 1 for limited, and 2 for extensive); Parties (number of legally registered political parties); Percent Legislative Voting; Percent Registered Voting; Size (in one thousand square kilometer units); and Population (given in millions).

A generalized linear model for these data with the Poisson link function is specified as:

$$g^{-1}(\boldsymbol{\theta}) = g^{-1}(\mathbf{X}\boldsymbol{\beta}) = \exp\left[\mathbf{X}\boldsymbol{\beta}\right] = E[\mathbf{Y}] = E[\textbf{Military Coups}].$$

In this specification, the systematic component is $\mathbf{X}\beta$, the stochastic component is $\mathbf{Y} = \mathbf{Military\ Coups}$, and the link function is $\theta = \log(\mu)$. We can re-express this model by moving the link function to the left-hand side exposing the linear predictor: $g(\mu) = \log(E[\mathbf{Y}]) = \mathbf{X}\beta$ (although this is now a less intuitive form for understanding the outcome variable). The R language GLM call for this specification is:

```
data(africa)
africa.out <- glm(MILTCOUP ~ MILITARY + POLLIB + PARTY93 + PCTVOTE
                  + PCTTURN + SIZE*POP + NUMREGIM*NUMELEC,
                  family=poisson, data=africa)
summary(africa.out)
```

This gives the model results in Table A.3.

TABLE A.3: POISSON MODEL OF MILITARY COUPS

	Coefficient	Std. Error	95% CI
Intercept	2.9209	1.3368	[0.3008: 5.5410]
Military Oligarchy	0.1709	0.0509	[0.0711: 0.2706]
Political Liberalization	-0.4654	0.3319	[-1.1160: 0.1851]
Parties	0.0248	0.0109	[0.0035: 0.0460]
Percent Legislative Voting	0.0613	0.0218	[0.0187: 0.1040]
Percent Registered Voting	-0.0361	0.0137	[-0.0629:-0.0093]
Size	-0.0018	0.0007	[-0.0033:-0.0004]
Population	-0.1188	0.0397	[-0.1965:-0.0411]
Regimes	-0.8662	0.4571	[-1.7621: 0.0298]
Elections	-0.4859	0.2118	[-0.9010:-0.0709]
(Size)(Population)	0.0001	0.0001	[0.0001: 0.0002]
(Regimes)(Elections)	0.1810	0.0689	[0.0459: 0.3161]

Note that the two interaction terms are specified by using the multiplication character. The iteratively weighted least squares algorithm converged in only four iterations using Fisher scoring, and the results are provided in the table. The model appears to fit the data quite well: an improvement from the null deviance of 62 on 46 degrees of freedom to a residual deviance of 7.5 on 35 degrees of freedom (evidence that the model does not fit would be supplied by a deviance value in the tail of a χ^2_{n-k} distribution). Nearly all the coefficients have 95% confidence intervals bounded away from zero and therefore appear reliable in the model. Since this is a Poisson link function, we might also want to check for overdispersion:

```
sum(residuals(africa.out,type="pearson")^2)
[1] 6.823595
```

```
1-pchisq(sum(residuals(africa.out,type="pearson")^2),
    df=africa.out$df.residual, lower.tail=FALSE)
[1] 0.9984635
```

This gives no evidence that we have extra-Poisson variance.

A.3   Numerical Maximum Likelihood

Generally the maximum likelihood estimate for the unknown parameter vector in a generalized linear model does not have a closed-form analytical solution. So this section explains the method for applying maximum likelihood estimation to generalized linear models using numerical techniques. The primary iterative root-finding procedure is *iteratively weighted least squares* (IWLS), created by Nelder and Wedderburn (1972). In order to fully understand the numerical aspects of this technique, this section first discusses finding coefficient estimates in nonlinear models (i.e., simple root finding), followed by weighted regression, and finally the iterating algorithm. This section can be skipped on first reading as it addresses the lower level computational aspects of statistical software only.

A.3.1   Newton-Raphson and Root Finding

The problem of finding the best possible estimate for some coefficient value typically reduces to the problem of finding the mode of the likelihood function, given some observed data and a parametric specification. In textbook examples, this is the challenge of finding the point where the first derivative of the likelihood function is equal to zero (including the secondary consideration that the second derivative be negative to ensure that the point is not a minima). However, in realistic work, the problem is that of finding a high-dimensional mode with likelihood functions that are considerably less simple and even have suboptimal modes.

Rather than using analytical solutions we are driven to using numerical techniques, which are algorithms that manipulate the data and the specified model to produce a mathematical solution for the modal point. Because the solutions are produced by iterating/searching procedures using the data, they are considerably more messy with regard to machine-generated round-off and truncation in intermediate steps of the applied algorithm.

Consider the problem of numerical maximum likelihood estimation as that of finding the top of an "ant hill" in the parameter space. It is easy then to see that this is equivalent to finding the parameter vector value where the derivative of the likelihood function is equal to zero: where the tangent line is horizontal. The most widely used procedure is called Newton-Raphson, based on Newton's method for finding the roots of polynomial equations.

Newton's method exploits the properties of a Taylor series expansion around some given point. The Taylor series expansion gives the relationship between the value of a mathematical function at point, x_0, and the function value at another point, x_1, given as:

$$f(x_1) = f(x_0) + (x_1 - x_0)f'(x_0) + \frac{1}{2!}(x_1 - x_0)^2 f''(x_0)$$
$$+ \frac{1}{3!}(x_1 - x_0)^3 f'''(x_0) + \ldots,$$

where f' is the first derivative with respect to x, f'' is the second derivative with respect to x, and so on. Note that it is required that $f()$ have continuous derivatives over the relevant support of the function here: the allowable parameter space of x. Otherwise the algorithm fails.

Infinite precision is achieved with the infinite extending of the series into higher order derivatives and higher order polynomials (of course the factorial component in the denominator means that these are rapidly decreasing increments). This process is both unobtainable and unnecessary, and only the first two terms are required as a step in an iterative process.

The point of interest is x_1 such that $f(x_1) = 0$. This value is a root of the function, $f()$ in that it provides a solution to the polynomial expressed by the function. It is also the point where the function crosses the x-axis in a graph of x versus $f(x)$. This point could be found in one step with an infinite Taylor series: $0 = f(x_0) + (x_1 - x_0)f'(x_0) + \frac{1}{2!}(x_1 - x_0)^2 f''(x_0) + \ldots + \frac{1}{\infty!}(x_1 - x_0)^\infty f^{(\infty)}(x_0) + \ldots$. While this is impossible, it is true that we could use just the first two terms to get closer to the desired point: $0 \cong f(x_0) + (x_1 - x_0)f'(x_0)$. Now rearrange to produce at the $(j+1)^{\text{th}}$ step: $x^{(j+1)} = x^{(j)} - \frac{f(x^{(j)})}{f'(x^{(j)})}$, so that progressively improved estimates are produced until $f(x^{(j+1)})$ is sufficiently close to zero. It has been shown that this method converges quadratically to a solution provided that the selected starting point is reasonably close to the solution, although the results can be very bad if this condition is not met.

■ **Example A.2: A Simple Application of Newton-Raphson.** Suppose that we wanted a numerical routine for finding the square root of a number, μ. This is equivalent to finding the root of the simple equation $f(x) = x^2 - \mu = 0$. The first derivative is just $\frac{d}{dx}f(x) = 2x$. If we insert these functions into $(j+1)^{\text{th}}$ step: $x^{(j+1)} = x^{(j)} - \frac{f(x^{(j)})}{f'(x^{(j)})}$, we get:

$$x^{(j+1)} = x^{(j)} - \frac{(x^{(j)})^2 - \mu}{2x^{(j)}} = \frac{1}{2}(x^{(j)} + \mu(x^{(j)})^{-1}). \tag{A.17}$$

A very basic R function for implementing (A.17) is:

```
newton.raphson.ex <- function(mu,x,iterations)  {
    for (i in 1:iterations)
        x <- 0.5*(x + mu/x)
    return(x)
}
```

This function is run with the command `newton.raphson.ex(99,2,3)` to find the square root of 99 starting at 2, but only taking 3 steps. The result is the (incorrect) value 10.74386. However, running the function according to `newton.raphson.ex(99,2,6)`, that is six iterations instead of three, gives the correct value of 9.949874.

A.3.1.1   Newton-Raphson for Statistical Problems

The Newton-Raphson algorithm, when applied to mode finding in a statistical setting, substitutes $\theta^{(j+1)}$ for $x^{(j+1)}$ and $\theta^{(j)}$ for $x^{(j)}$ (where the θ values are iterative estimates of the parameter vector) and $f()$ is the score function (A.3). The real goal is to estimate a k-dimensional $\hat{\theta}$ estimate, given data and a model. The applicable multivariate likelihood updating equation is now provided by:

$$\boldsymbol{\theta}^{(j+1)} = \boldsymbol{\theta}^{(j)} - \frac{\partial}{\partial\boldsymbol{\theta}}\ell(\boldsymbol{\theta}^{(j)}|\mathbf{x})\left(\frac{\partial^2}{\partial\boldsymbol{\theta}\partial\boldsymbol{\theta}'}\ell(\boldsymbol{\theta}^{(j)}|\mathbf{x})\right)^{-1}. \tag{A.18}$$

When $\boldsymbol{\theta}^{(j)}$ is the maximum likelihood coefficient vector, the quantity in the denominator is a matrix called the *Hessian*. For exponential family distributions and natural link functions, the observed and expected Hessian are identical (Fahrmeir and Tutz 2001, p.42; Lehmann and Casella, 1998, pp.124-8), although Efron and Hinkley (1978) give situations where the observed information is preferable. So it is common to replace this calculation with forms that are equivalent for the exponential family, such as the Fisher Information:

$$\mathbf{A}_F = -E\left(\frac{\partial^2}{\partial\boldsymbol{\theta}\partial\boldsymbol{\theta}'}\ell(\boldsymbol{\theta}^{(j)}|\mathbf{x})\right)$$

(Fisher 1925), or the square of the score function:

$$\mathbf{A}_B = E\left[\frac{\partial}{\partial\boldsymbol{\theta}}\ell(\boldsymbol{\theta}^{(j)}|\mathbf{x})'\frac{\partial}{\partial\boldsymbol{\theta}}\ell(\boldsymbol{\theta}^{(j)}|\mathbf{x})\right].$$

which is sometimes called the *BHHH* method (Berndt, Hall, Hall, and Hausman 1974).

At each step of these Newton-Raphson steps, a system of equations determined by the multivariate normal equations must be solved:

$$\underbrace{\mathbf{A}}_{\text{angle}}\underbrace{(\boldsymbol{\theta}^{(j+1)} - \boldsymbol{\theta}^{(j)})}_{\text{direction: }\delta\mathbf{b}} = \underbrace{\frac{\partial}{\partial\boldsymbol{\theta}}\ell(\boldsymbol{\theta}^{(j)}|\mathbf{x})}_{\text{size of direction: }\mathbf{u}}, \tag{A.19}$$

which builds a linear structure in the parameter vector, and leads to estimates from the system of linear equations: $\delta\mathbf{b} = \mathbf{A}^{-1}\mathbf{u}$. It is computationally convenient to solve on each iteration by least squares, so that the problem of mode finding reduces to a repeated weighted least squares application in which the inverse of the diagonal values of the second derivative matrix in the denominator are the appropriate weights (this is a diagonal matrix by the iid assumption).

A.3.1.2   Weighted Least Squares

A standard compensating technique for nonconstant error variance (so-called heteroscedasticity) is to insert a diagonal matrix of weights, $\boldsymbol{\Omega}$, into the calculation of $\hat{\beta}$ such that this heteroscedasticity is removed by design. The Ω matrix is created by taking the error variance of the ith case (estimated or known), ν_i, and assigning it to the ith diagonal $\Omega_{ii} = \frac{1}{\nu_i}$, leaving the off-diagonal elements as zero. So large error variances are reduced by premultiplying the model terms by this reciprocal.

We can premultiply each term in the standard linear model setup, by the square root of the Ω matrix (that is, by the standard deviation). This "square root" is actually produced from a Cholesky factorization: if $\mathbf{A}$ is a positive definite symmetric ($\mathbf{A}' = \mathbf{A}$) matrix, then there must exist a matrix $\mathbf{G}$ such that: $\mathbf{A} = \mathbf{GG}'$. A matrix, $\mathbf{A}$, is positive definite if for any nonzero $p \times 1$ vector $\mathbf{x}$, $\mathbf{x}'\mathbf{Ax} > 0$. In our case, this decomposition is greatly simplified because the Ω matrix has only diagonal values (all off-diagonal values equal to zero). Therefore the Cholesky factorization is produced simply from the square root of these diagonal values. Premultiplying gives:

$$\Omega^{\frac{1}{2}}\mathbf{Y} = \Omega^{\frac{1}{2}}\mathbf{X}\beta + \Omega^{\frac{1}{2}}\epsilon. \tag{A.20}$$

Instead of minimizing squared errors in the usual manner, we now minimize $(\mathbf{Y} - \mathbf{X}\beta)'\Omega(\mathbf{Y} - \mathbf{X}\beta)$, and the subsequent weighted least squares estimator is found by $\hat{\beta} = (\mathbf{X}'\Omega\mathbf{X})^{-1}\mathbf{X}'\Omega\mathbf{Y}$. The weighted least squares estimator gives the best linear unbiased estimate (BLUE) of the coefficient estimator in the presence of heteroscedasticity.

A.3.1.3 Iterative Weighted Least Squares

It is more common that the individual variances used to make the reciprocal diagonal values for Ω are unknown, and cannot be easily estimated, but are known to be a function of the mean of the outcome variable: $\nu_i = f(E[Y_i])$. So if the expected value of the outcome variable, $E[Y_i] = \mu$ and the form of the relation function, $f()$ are known, then this is a very straightforward estimation process. Unfortunately, although it is common for the variance structure to be dependent on the mean function, it is relatively rare to know the exact form of the dependence.

Nelder and Wedderburn (1972) provide a solution to this problem that iteratively re-estimates the weights, improving the estimate on each cycle using the mean function. Because $\mu = g^{-1}(\mathbf{X}\beta)$, then the coefficient estimate, $\hat{\beta}$, provides a mean estimate and vice versa. So the algorithm iteratively estimates these values using progressively better weights. This proceeds as follows. First assign starting values to the weights, generally equal to one: $\frac{1}{\nu_1^{(1)}} = 1$, and specify the diagonal matrix Ω, guarding against division by zero. Then iterate the following steps:

▷ Define the current (or starting) point of the linear predictor by:

$$\underset{(n \times 1)}{\hat{\eta}_0} = \underset{(n \times p)(p \times 1)}{\mathbf{X}\beta_0}$$

with fitted value $\hat{\mu}_0$ from $g^{-1}(\hat{\eta}_0)$.

▷ Form the "adjusted dependent variable" according to:

$$\underset{(n\times 1)}{z_0} = \underset{(n\times 1)}{\hat{\eta}_0} + \underset{diag(n\times n)}{\left(\left.\frac{\partial\eta}{\partial\mu}\right|_{\hat{\mu}_0}\right)}\underset{(n\times 1)}{(\mathbf{y} - \hat{\mu}_0)},$$

which is a linearized form of the link function applied to the data. As an example of this derivative function, the Poisson form looks like $\eta = \log(\mu) \implies \frac{\partial\eta}{\partial\mu} = \frac{1}{\mu}$.

▷ Form the *quadratic weight matrix*, which is the variance of z:

$$\underset{(n\times n)}{w_0^{-1}} = \left(\left.\frac{\partial\eta}{\partial\mu}\right|_{\hat{\mu}_0}\right)^2 v(\mu)|_{\hat{\mu}_0}$$

where $v(\mu)$ is the variance function: $\frac{\partial}{\partial\theta}b'(\theta) = b''(\theta)$.

▷ So the general iteration scheme is:

1. Construct z, w. Regress z on the covariates with weights to get a new interim estimate:

$$\underset{(p\times 1)}{\hat{\beta}_1} = (\underset{(p\times n)}{\mathbf{X}'}\underset{(n\times n)}{w_0}\underset{(n\times p)}{\mathbf{X}})^{-1}\underset{(p\times n)}{\mathbf{X}'}\underset{(n\times n)}{w_0}\underset{(n\times 1)}{z_0}$$

2. Use the coefficient vector estimate to update the linear predictor:

$$\hat{\eta}_1 = \mathbf{X}'\hat{\beta}_1$$

3. Iterate: $z_1, w_1 \Rightarrow \hat{\beta}_2, \hat{\eta}_2, z_2, w_2 \Rightarrow \hat{\beta}_3, \hat{\eta}_3, z_3, w_3 \Rightarrow \hat{\beta}_4, \hat{\eta}_4, \ldots$.

These steps are repeated until convergence (i.e., $\mathbf{X}\hat{\beta}_j - \mathbf{X}\hat{\beta}_{j+1}$ is close to zero). Under very general conditions, satisfied by the exponential family of distributions, the IWLS procedure finds the mode of the likelihood function, thus producing the maximum likelihood estimate of the unknown coefficient vector, $\hat{\beta}$, and the variance matrix, $\hat{\sigma}^2(\mathbf{X}'\mathbf{\Omega X})^{-1}$ (Gill 2000, Green 1984, Del Pino 1989).

The purpose of this section is to illustrate the means by which generalized linear models are estimated in practice. This process is the core methodological practice in non-Bayesian statistical analysis in the social and behavioral sciences, even though many practitioners believe they are using some particularistic model (logit, Poisson, duration, etc.). It should not be inferred from this discussion that this process always works without computational difficulties. For extensive discussions of maximum likelihood computational issues and what can go wrong, see Altman and McDonald (2001), McCullough (1998, 1999), and Gill, Murray, and Wright (1981, Chapter 8).

A.4   Quasi-Likelihood

Wedderburn (1974) introduced the concept of "quasi-likelihood" estimation to address circumstances when either the parametric form of the likelihood is known to be misspecified, or only the first two moments[8] are definable. Albert (1988) and Albert and Pepple (1989) detail the utility of applying quasi-likelihood models to hierarchical Bayesian estimation problems where the posterior calculation is difficult. This is before the revolution brought on by the widespread use of MCMC techniques, but quasi-likelihood remains useful as a way of relaxing distributional assumptions in model specifications. General applications of quasi-likelihood specifications in the social sciences include: White's (1982) look at econometric model misspecification, Goldstein and Rasbash's (1996) application to public policy, the use of quasi-likelihood in analyzing multinational corporate decision-making by Hannan *et al.* (1995), Western's (1995) application to studying union decline, Sampson and Raudenbush's (1999) study of behavior in public spaces, and Mebane's (1994) evaluation of the linkage between taxation policy and subsequent elections.

Suppose that we know something about the parametric form of the distribution generating the data, but not in complete detail. Obviously this precludes the standard maximum likelihood estimation of unknown parameters since we cannot specify a full likelihood equation. Wedderburn's idea was to develop an estimation procedure that only requires specification for the mean function of the data and a stipulated relationship between this mean function and the variance function. This is useful in a Bayesian context when we have prior information readily at hand but only a vague idea of the form of the likelihood.

Re-express the exponential family form as a joint distribution function of observed data:

$$f(\mathbf{y}|\theta) = \exp\left[\sum_{i=1}^{n} y_i\theta - nb(\theta) + \sum_{i=1}^{n} c(y_i)\right], \tag{A.21}$$

and with the more realistic assumption of a multiparameter model with k parameters: $f(y|\boldsymbol{\theta}) = \exp\left[\sum_{i=1}^{n}\sum_{j=1}^{k} y\theta_j - nb(\theta_j) + \sum_{i=1}^{n} c(y)\right]$. Adopting the canonical form with a scale parameter gives: $f(y|\theta) = \exp\left[\frac{y\theta - b(\theta)}{a(\psi)} + c(y, \psi)\right]$.

We can also define a variance *function* for a given exponential family expression, which is used in generalized linear models to indicate the dependence of the variance of Y on location and scale parameters: $v(\mu) = \frac{\partial^2}{\partial\theta^2} b(\theta)$, meaning that $\text{Var}[Y] = a(\psi)v(\mu)$ indexed by θ. Note that the dependence on $b(\theta)$ explicitly states that the variance function is conditional on the mean function, whereas there was no such stipulation with the $a(\psi)$ form. It is conventional to leave the variance function in terms of the canonical parameter, θ, rather than return it to the parameterization in the original probability function as was done for the variance of Y. Table A.4 summarizes some common variance functions.

The log likelihood function in this context is now written as:

$$\ell(\boldsymbol{\theta}) = \sum_{i=1}^{n}\sum_{j=1}^{k} \frac{y\theta_j - nb(\theta_j)}{a(\psi)} + \sum_{i=1}^{n} c(y, \psi), \tag{A.22}$$

---

[8]Moments are characterizations of distributions based on expectations. The nth moment of a random variable X is given by: $\mu_n = E[X^n]$. So the mean of X is the first moment: $E[X] = \mu_1$, and the variance of X is the second moment minus the square of the first moment: $\text{Var}[X] = \mu_2 - (\mu_1)^2$. See Stuart and Ord (1994, Chapter 3) for a lengthy discussion of moments and related quantities.

TABLE A.4: Normalizing Constants and Variance Functions

Distribution	$b(\theta)$	$v(\mu) = \frac{\partial^2}{\partial \theta^2} b(\theta)$
Poisson	$\exp(\theta)$	$\exp(\theta)$
Binomial	$n \log(1 + \exp(\theta))$	$n \exp(\theta)(1 + \exp(\theta))^{-2}$
Normal	$\frac{\theta^2}{2}$	1
Gamma	$-\log(-\theta)$	$\frac{1}{\theta^2}$
Negative binomial	$r \log(1 - \exp(\theta))$	$r \exp(\theta)(1 - \exp(\theta))^{-2}$

where $E[\mathbf{y}] = \frac{\partial b(\theta_j)}{\partial \theta}$. The expected value of the first derivative of the joint distribution is equal to zero since this is the slope of the tangent line at the mode:

$$E\left[\frac{\partial \ell(\boldsymbol{\theta})}{\partial \boldsymbol{\theta}}\right] = E\left[\sum_{i=1}^{n} \frac{y}{a(\psi)} - \frac{\partial nb(\theta_j)}{\partial \boldsymbol{\theta}} \frac{1}{a(\psi)}\right] = 0,$$

and

$$\text{Var}\left[\frac{\partial \ell(\boldsymbol{\theta})}{\partial \boldsymbol{\theta}}\right] = -E\left[\frac{\partial^2 \ell(\boldsymbol{\theta})}{\partial \boldsymbol{\theta}^2}\right] = -E\left[-\frac{\partial^2 nb(\theta_j)}{\partial \boldsymbol{\theta}^2} \frac{1}{a(\psi)}\right] = \frac{\partial^2 nb(\theta_j)}{\partial \boldsymbol{\theta}^2} \frac{1}{a(\psi)}.$$

So again $\text{Var}[Y] = a(\psi)v(\mu)$, where $v(\mu) = \frac{\partial^2 nb(\theta_j)}{\partial \theta^2}$ is the variance function.

Instead of taking the first derivative of log likelihood with respect to the parameter vector, $\boldsymbol{\theta}$, suppose we take this derivative with respect to the mean function in a generalized linear model, $\boldsymbol{\mu}$, with the analogous property that: $E\left[\frac{\partial \ell(\boldsymbol{\theta})}{\partial \boldsymbol{\mu}}\right] = 0$. Thus

$$\text{Var}\left[\frac{\partial \ell(\boldsymbol{\theta})}{\partial \boldsymbol{\mu}}\right] = -E\left[\frac{\partial^2 \ell(\boldsymbol{\theta})}{\partial \boldsymbol{\mu}^2}\right] = \frac{1}{a(\psi)v(\mu)}.$$

Therefore what we have here is a linkage between the mean function and the variance function that does not depend on the form of the likelihood function. If we combine this with a form of $\frac{\partial \ell(\boldsymbol{\theta})}{\partial \boldsymbol{\mu}}$ that satisfies the condition above that the expected value of its first derivative is zero, and the variance property above holds, then we have a replacement for the unknown specific form of the likelihood function that still provides the desired properties of maximum likelihood estimation as described. Thus we imitate these three criteria of the score function with a function that contains significantly less parametric information: only the mean and variance.

A substitution function that satisfies all of these conditions is: $\mathbf{q} = \frac{\mathbf{y} - \boldsymbol{\mu}}{a(\psi)v(\mu)}$ (McCullagh and Nelder 1989, 325; McCulloch and Searle 2001, 152; Shao 2005). The contribution to the log likelihood function from the ith point is defined by:

$$Q_i = \int_{y_i}^{\mu_i} \frac{y_i - t}{a(\psi)v(\mu)} dt,$$

so finding the maximum likelihood estimator for this setup, $\hat{\boldsymbol{\theta}}$ is equivalent to solving:

$$\frac{\partial}{\partial \boldsymbol{\theta}} \sum_{i=1}^{n} Q_i = \sum_{i=1}^{n} \frac{y_i - \mu_i}{a(\psi)v(\mu)} \frac{\partial \mu_i}{\partial \boldsymbol{\theta}} = \sum_{i=1}^{n} \frac{y_i - \mu_i}{a(\psi)v(\mu)} \frac{\mathbf{x}_i}{g(\mu)} = \mathbf{0},$$

where $g(\mu)$ is the canonical link function for a generalized linear model specification. In other words we can use the usual maximum likelihood engine for inference with complete asymptotic properties such as consistency and normality (McCullagh 1983), by only specifying the relationship between the mean and variance functions as well as the link function (which actually comes directly from the form of the outcome variable data).

As an example, suppose we assume that the mean and variance function are related by stipulating that $a(\psi) = \sigma^2 = 1$, and $b(\theta) = \frac{\theta^2}{2}$, so $v(\mu) = \frac{d^2 b(\theta)}{d\theta^2} = 1$. Then it follows that:

$$Q_i = \int_{y_i}^{\mu_i} \frac{y_i - t}{a(\psi)v(\mu)}dt = -\frac{(y_i - \mu_i)^2}{2}.$$

The quasi-likelihood solution for $\hat{\theta}$ comes from solving the quasi-likelihood equation:

$$\frac{d}{d\theta} \sum_{i=1}^{n} Q_i = \frac{d}{d\theta} \sum_{i=1}^{n} \frac{y_i - \theta}{2} = -\sum_{i=1}^{n} y_i + n\theta = \mathbf{0}.$$

In other words, $\hat{\theta} = \bar{y}$, because this example was set up with the same assumptions as a normal maximum likelihood problem but without specifying a normal likelihood function.

Quasi-likelihood models drop the requirement that the true underlying density of the outcome variable belongs to a particular exponential family form. Instead, all that is required is the identification of the first and second moments and an expression for their relationship up to a proportionality constant. It is assumed that the observations are independent and that mean function describes the mean effect of interest. Even given this generalization of the likelihood assumptions, it can be shown that quasi-likelihood estimators are consistent asymptotically equal to the true estimand (Fahrmeir and Tutz 2001, pp.55-60, Firth 1987; McCullagh 1983). However, a quasi-likelihood estimator is often less efficient than a corresponding maximum likelihood estimator (McCullagh and Nelder 1989, pp.347-348; Shao 2005).

Despite this drawback with regard to variance, there are often times when it is convenient or necessary to specify a quasi-likelihood model. A number of authors have extended the quasi-likelihood framework to: *extended quasi-likelihood* models to compare different variance functions for the same data (Nelder and Pregibon 1987), *pseudo-likelihood* models, which build upon extended quasi-likelihood models by substituting a χ^2 component instead of a deviant component in dispersion analysis (Breslow 1990; Carroll and Ruppert 1982; Davidian and Carroll 1987), and models where the dispersion parameter is dependent on specified covariates (Smyth 1989). Nelder and Lee (1992) provide an informative overview of these variations. It is also the case that quasi-likelihood models are not more difficult to compute (Nelder 1985), and the R package has preprogrammed functions that make the process routine (see below).

■ **Example A.3:   A Quasi-Likelihood Model of Military Coups.** We can develop a quasi-likelihood model for counts in the Africa Coups data as an alternative to stipulating a regular Poisson GLM as done in the previous example. First stipulate that:

$$\mu_i = v(\mu_i),\qquad(A.23)$$

so that

$$Q_i = \int_{y_i}^{\mu_i} \frac{y_i - t}{a(\psi)v(\mu)}dt = (y_i \log\mu_i - \mu_i - y_i \log +y_i)/a(\psi).\qquad(A.24)$$

Therefore we get the form:

$$\frac{d}{d\mu}\sum_{i=1}^{n}Q_i = \frac{d}{d\mu}\sum_{i=1}^{n}(y_i \log\mu_i - \mu_i),\qquad(A.25)$$

which is the same as the regular GLM model for a Poisson link function except that the scale parameter is no longer fixed to be equal to one and is instead estimated by:

$$a(\hat\psi) = (n-k)^{-1}\sum(y_i - \mu_i)^2/v(\mu_i).\qquad(A.26)$$

This means that the coefficient estimates will be exactly as they were before but the corresponding errors will differ here because the scale parameter is now free to vary.

Constructing this model is very easy in R:

```
africa.quasi.out <- glm(formula = MILTCOUP ~ MILITARY + POLLIB
        + PARTY93 + PCTVOTE + PCTTURN + SIZE*POP + NUMREGIM*NUMELEC,
        data=africa, family=quasipoisson(link="log"))
summary(africa.quasi.out)
```

Running this we notice that the coefficient estimates are identical to the regular Poisson GLM specification, as are the deviance summaries. The estimated scale parameter is estimated to be 0.325, justifying use of the more flexible form. The only differences found, as expected, are the resulting standard errors:

Intercept	MILITARY	POLLIB	PARTY93
0.76200	0.02901	0.18920	0.00619
PCTVOTE	PCTTURN	SIZE	POP
0.01240	0.00780	0.00041	0.02260
NUMREGIM	NUMELEC	SIZE:POP	NUMREGIM:NUMELEC
0.26057	0.12070	0.00002	0.03929

A.5   Exercises

A.1   Suppose $X_1,\ldots,X_n$ are iid exponential: $f(x|\theta) = \theta e^{-\theta x}$, $\theta > 0$. Find the maximum likelihood estimate of θ by constructing the joint distribution, express the log likelihood function, take the first derivative with respect to θ, set this function equal to zero, and solve for $\hat\theta$ the maximum likelihood value.

A.2   If the variance of the residuals in the linear model is not constant, then the regression model is heteroscedastic. The *general linear model* can be used when the form of the heteroscedasticity is known. Assuming the residuals are uncorrelated, the new $n \times n$ variance matrix is given by:

$$
\sigma\Omega = \begin{bmatrix}
\sigma_1^2 & 0 & \cdots & \cdots & 0 \\
0 & \sigma_2^2 & 0 & \cdots & 0 \\
\vdots & \cdots & \ddots & \cdots & \vdots \\
\vdots & \cdots & \cdots & \ddots & \vdots \\
0 & 0 & 0 & \cdots & \sigma_n^2
\end{bmatrix}.
$$

Using this matrix, calculate the new maximum likelihood estimate for the unknown parameter vector **b**.

A.3   Below are two sets of data each with least square regression lines calculated ($\hat{y} = 6 + 0x$). Answer the following questions by looking at the plots.

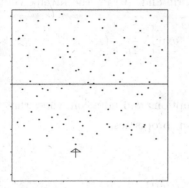 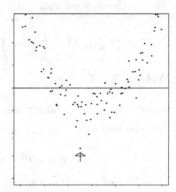

(a) Does the construction of the least squares line in panel 1 violate any of the Gauss-Markov assumptions?

(b) Does the construction of the least squares line in panel 2 violate any of the Gauss-Markov assumptions?

(c) Does the identified point (identically located in both panels) have a substantively different interpretation?

A.4   Calculate the maximum likelihood estimate of the intensity parameter of the Poisson distribution, $f(y|\mu) = \frac{e^{-\mu}\mu^y}{y!}$, $\mu > 0$, for the data: $[7, 4, 3, 4, 7, 6, 9, 11, 21, 3]$.

A.5   Consider the bivariate normal PDF:

$$
f(x_1, x_2) = \left(2\pi\sigma_1\sigma_2\sqrt{1-\rho^2}\right)^{-1} \times
$$
$$
\exp\left[-\frac{1}{2(1-\rho^2)}\left(\frac{(x_1-\mu_1)^2}{\sigma_1^2} - \frac{2\rho(x_1-\mu_1)(x_2-\mu_2)}{\sigma_1\sigma_2} + \frac{(x_2-\mu_2)^2}{\sigma_2^2}\right)\right].
$$

for $-\infty < \mu_1, \mu_2 < \infty$, $\sigma_1, \sigma_2 > 0$, and $\rho \in [-1:1]$.

For $\mu_1 = 3, \mu_2 = 2, \sigma_1 = 0.5, \sigma_2 = 1.5, \rho = 0.75$, calculate a grid search using R for the mode of this bivariate distribution on $\mathbb{R}^2$. A grid search bins the parameter space into equal space intervals on each axis and then systematically evaluates each resulting subspace. First set up a two dimensional coordinate system stored in a matrix covering 99% of the support of this bivariate density, then do a systematic analysis of the density to show the mode without using "for" loops. Hint: see the R help menu for the `outer` function. Use the `contour` function to make a figure depicting bivariate contour lines at 0.05, 0.1, 0.2, and 0.3 levels.

A.6   For the simplified Pareto PDF:

$$f(x|\theta) = \theta x^{-\theta-1}, \qquad 0 \le x, \theta < \infty,$$

find the maximum likelihood estimate for an iid sample: $X_1, \ldots, X_n$.

A.7   Derive the exponential family form and $b(\theta)$ for the Inverse Gaussian distribution:

$$f(x|\mu,\lambda) = \left(\frac{\lambda}{2\pi x^3}\right)^{1/2} \exp\left[-\frac{\lambda}{2\mu^2 x}(x-\mu)^2\right], \qquad x > 0, \mu > 0.$$

Assume $\lambda = 1$.

A.8   Using exponential family assumptions and notation, show that the score function has the following three important properties:

▷ $E_Y[\frac{\partial}{\partial \mu_i}\ell(\boldsymbol{\theta}(\mu_i)|y_i,\phi)] = 0$,

▷ $\text{Var}_Y\left[\frac{\partial}{\partial \mu_i}\ell(\boldsymbol{\theta}(\mu_i)|y_i,\phi)\right] = \frac{1}{\phi v(\mu_i)}$,

▷ $-E_Y\left[\frac{\partial^2}{\partial \mu_i^2}\ell(\boldsymbol{\theta}(\mu_i)|y_i,\phi)\right] = \frac{1}{\phi v(\mu_i)}$.

Now prove that the quasi-likelihood substitution function, $q_i = \frac{y_i-\mu_i}{\phi v(\mu_i)}$, also satisfies these three properties for likelihood estimation.

A.9   Show that the Weibull distribution, $w(x|\gamma,\beta) = \frac{\gamma}{\beta}x^{\gamma-1}\exp\left(-\left(\frac{x}{\beta}\right)^\gamma\right)$ is an exponential family form when $\gamma = 2$, labeling each part of the final answer.

A.10   A well-known alternative to Newton-Raphson that does not require the calculation of derivatives is the secant method, where at the $(j+1)^{\text{th}}$ step we calculate: $x^{(j+1)} = x^{(j)} - f(x^{(j)})\frac{x^{(j)}-x^{(j-1)}}{f(x^{(j)})-f(x^{(j-1)})}$. Write an algorithm (R or pseudo-code) to apply the secant method to the problem in the root of the simple equation $f(x) = x^2 - \mu = 0$ (i.e. a function that finds the square root of μ).

A.11   For normally distributed data a robust estimate of the standard deviation can be calculated by (Devore 1999):

$$\tilde{\sigma} = \frac{1}{0.6745n}\sum_{i=1}^n |X_i - \bar{X}|.$$

Write an R function to calculate $\tilde{\sigma}$ for the FARM variable in Exercise A.15 below. Compare this value to the standard deviation. Is there evidence of influential outliers in this variable?

A.12 Cigarette smoking among elderly males has decreased substantially whereas cigarette smoking among elderly females has remained fairly constant. Given the following data, calculate the maximum likelihood estimate of the mean and variance for each group under a normal assumption. Test for a significant difference.

	Proportion of Smokers, 65 and Over											
Year	1965	1974	1979	1983	1985	1990	1992	1993	1994	1995	1997	1998
Male	28.5	9.6	24.9	12.0	20.9	13.2	22.0	13.1	19.6	13.5	14.6	11.5
Female	16.6	12.4	13.5	10.5	13.2	11.1	14.9	11.5	12.8	11.5	10.4	11.2

Source: Centers for Disease Control and Prevention, National Center for Health Statistics. National Health Interview Survey, respective years.

A.13 Derive the generalized hat matrix from $\hat{\beta}$ produced from the final (n^{th}) IWLS step. Specifically, give the quantities of interest from the nth and $(n-1)^{th}$ step, and show how the hat matrix is produced. Secondly, give the form of Cook's D appropriate to a GLM and show that it contains hat matrix values.

A.14 An exponential family form has *quadratic variance* if the variance term can be expressed as $v(\mu) = \phi''(\theta) = \eta_0 + \eta_1\mu + \eta_2\mu^2$, where μ is the mean function. Show that there exists an exponential family form conjugate distribution and that it also has quadratic variance.

A.15 The likelihood function for dichotomous choice regression is given by:

$$L(\mathbf{b}, \mathbf{y}) = \prod_{i=1}^{n} [F(\mathbf{x}_i\mathbf{b})]^{y_i} [1 - F(\mathbf{x}_i\mathbf{b})]^{1-y_i}.$$

There are several common choices for the $F()$ function:

Logit:	$\Lambda(\mathbf{x}_i\mathbf{b}) = \frac{1}{1+\exp[\mathbf{x}_i\mathbf{b}]}$
Probit:	$\Phi(\mathbf{x}_i\mathbf{b}) = \int_{-\infty}^{\mathbf{x}_i\mathbf{b}} \frac{1}{\sqrt{2\pi}}\exp[-t^2/2]dt$
Cloglog:	$CLL(\mathbf{x}_i\mathbf{b}) = 1 - \exp(-\exp(\mathbf{x}_i\mathbf{b}))$.

Using the depression era economic and electoral data, calculate a dichotomous regression model for using each of these link functions according to the following model specification:

$$FDR \sim f[(POST.DEP - PRE.DEP) + FARM]$$

where: FDR indicates whether or not Roosevelt carried that state in the 1932 presidential elections, $PRE.DEP$ is the mean per-state income before the onset of the Great Depression (1929) in dollars, $POST.DEP$ is the mean per-state income after the onset of the great depression (1932) in dollars, and $FARM$ is the total farm wage and salary disbursements in thousands of dollars per state in 1932.

State	FDR	PRE.DEP	POST.DEP	FARM
Alabama	1	323	162	4067
Arizona	1	600	321	6100
Arkansas	1	310	157	8134
California	1	991	580	83371
Colorado	1	634	354	10167
Connecticut	0	1024	620	10167
Delaware	0	1032	590	3050
District of Columbia	1	1269	1054	0
Florida	1	518	319	14234
Georgia	1	347	200	10167
Idaho	1	507	274	7117
Illinois	1	948	486	27451
Indiana	1	607	310	11184
Iowa	1	581	297	28468
Kansas	1	532	266	22368
Kentucky	1	393	211	8134
Louisiana	1	414	241	10167
Maine	0	601	377	6100
Maryland	1	768	512	14234
Massachusetts	1	906	613	14234
Michigan	1	790	394	14234
Minnesota	1	599	363	25418
Mississippi	1	286	127	4067
Missouri	1	621	365	15251
Montana	1	592	339	10167
Nebraska	1	596	307	17284
Nevada	1	868	550	4067
New Hampshire	0	686	427	3050
New Jersey	1	918	587	18301
New Mexico	1	410	208	5084
New York	1	1152	1676	38635
North Carolina	1	332	187	8134
North Dakota	1	382	176	14234
Ohio	1	771	400	18301
Oklahoma	1	455	216	9150
Oregon	1	668	379	11184
Pennsylvania	0	772	449	25418
Rhode Island	1	874	575	2033
South Carolina	1	271	159	7117
South Dakota	1	426	189	8134
Tennessee	1	378	198	6100
Texas	1	479	266	33552
Utah	1	551	305	4067
Vermont	0	634	365	5084
Virginia	1	434	284	15251
Washington	1	741	402	14234
West Virginia	1	460	257	6100
Wisconsin	1	673	362	21351
Wyoming	1	675	374	5084

(also provided in the R package BaM). Do you find substantially different results with the three link functions? Explain. Which one would you use to report results? Why?

A.16   Suppose you have an iid sample of size n from the exponential distribution:

$$f(y|\theta) = \frac{1}{\theta} \exp\left[-\frac{y}{\theta}\right], \quad y \geq 0, \; \theta > 0.$$

Derive the MLE for θ^2. For the data $\mathbf{y} = [5, 3, 2, 1, 4]$, calculate the MLE.

A.17   Tobit regression (censored regression) deals with an interval-measured outcome variable that is censored such that all values that would have naturally been observed as negative are reported as zero, generalizeable to other values (Tobin 1958, Amemiya (1985, Chapter 10), Chib (1992). There can be left censoring and right censoring at any arbitrary value, single and double censoring, mixed truncating and censoring. If $\mathbf{z}$ is a latent outcome variable in this context with the assumed relation:

$$\mathbf{z} = \mathbf{x}\boldsymbol{\beta} + \boldsymbol{\eta} \qquad \text{and} \qquad z_i \sim \mathcal{N}(\mathbf{x}\boldsymbol{\beta}, \sigma^2),$$

then for *left censoring at zero*, the observed outcome variable is produced according to:

$$y_i = \begin{cases} z_i \text{ if } z_i > 0 \\ 0 \text{ if } z_i \leq 0. \end{cases}$$

The resulting likelihood function is:

$$L(\boldsymbol{\beta}, \sigma^2 | \mathbf{y}, \mathbf{x}) = \prod_{y_i=0} \left[1 - \Phi\left(\frac{x_i\boldsymbol{\beta}}{\sigma} \right) \right] \prod_{y_i>0} (\sigma^{-1}) \exp\left[-\frac{1}{2\sigma^2}(y_i - x_i\boldsymbol{\beta})^2 \right],$$

where σ^2 is called the *scale*. Explain the structure of the likelihood function: what parts accommodate observed values and what parts accommodate unobserved values? Replicate Tobin's original analysis in R using the `survreg` function in the `survival` package. The data are obtainable by:

```
tobin <- read.table(
        "http://artsci.wustl.edu/~jgill/data/tobin.dat",header=TRUE)
```

or in the BaM package.

A.18   An effective way of compensating for heteroscedasticity in probit models is to identify explanatory variables that introduce widely varying response patterns and specifically associate them in the model with the dispersion term. With this method, based on Harvey (1976), the functional form is changed from: $P(Y = 1) = \Phi(\mathbf{X}\boldsymbol{\beta})$ to $P(Y = 1) = \Phi\left(\frac{\mathbf{X}\boldsymbol{\beta}}{e^{\mathbf{Z}\boldsymbol{\alpha}}} \right)$, where the exponent function in the denominator is just a convenience that prevents division by zero. This approach distinguishes between the standard treatment of the estimated coefficients $\boldsymbol{\beta}$ with corresponding matrix of explanatory observations $\mathbf{X}$, and the set of dispersion determining estimated coefficients $\boldsymbol{\alpha}$ with a corresponding matrix of explanatory observations $\mathbf{Z}$. By this means the dispersion term is reparameterized to be a function of a set of coefficients and observed values that are suspected of causing the differences in error standard deviations: $\sigma_i^2 = e^{\mathbf{Z}_i\boldsymbol{\alpha}}$. Derive the log likelihood of the heteroscedastic probit and give a formal likelihood ratio test for the existence of heteroscedasticity.

A.19   Sometimes when estimation is problematic, *Restricted Maximum Likelihood* (REML) estimation is helpful (Bartlett 1937). REML uses a likelihood function calculated from a transformed set of data so that some parameters have no effect on the estimation on the others. For a One-Way ANOVA, $y|\mathbf{G} \sim \mathcal{N}(\mathbf{X}\boldsymbol{\beta} + \mathbf{Z}\mathbf{G}, \sigma^2\mathbf{I})$, $\mathbf{G} \sim \mathcal{N}(0, \sigma^2\mathbf{D})$, such that $\mathrm{Var}[y] = \mathrm{Var}[\mathbf{Z}\mathbf{G}] + \mathrm{Var}[\epsilon] = \sigma^2\mathbf{Z}\mathbf{D}\mathbf{Z}' + \sigma^2\mathbf{I}$. The unconditional distribution is now: $y \sim \mathcal{N}(\mathbf{X}\boldsymbol{\beta}, \sigma^2(\mathbf{I} + \mathbf{Z}\mathbf{D}\mathbf{Z}'))$ where $\mathbf{D}$ needs to be estimated. If $\mathbf{V} = \mathbf{I} + \mathbf{Z}\mathbf{D}\mathbf{Z}'$, then the likelihood for the data is: $\ell(\boldsymbol{\beta}, \sigma, \mathbf{D}|\mathbf{y}) = -\frac{n}{2}\log(2\pi) - \frac{1}{2}\log|\sigma^2 V| - \frac{1}{2\sigma^2}(\mathbf{y} - \mathbf{X}\boldsymbol{\beta})'\mathbf{V}^{-1}(\mathbf{y} - \mathbf{X}\boldsymbol{\beta})$. The REML steps for this model are: (1) find a linear transformation k such that $k'\mathbf{X} = 0$, so $k'\mathbf{y} \sim \mathcal{N}(0, k'\sigma k)$, (2) run MLE on this model to get $\hat{\mathbf{D}}$, which no longer has any fixed effects, (3) then estimate the fixed effects with ML in the normal way. Code this procedure in R and run the function on the depression era economic and electoral data in Exercise 15.

A.20   Maltzman and Wahlbeck (1996) look at the reasons that Supreme Court justices switch their votes. Consider the following distances between a justice and the author of the opposing opinion for four cases:

case	1	1	1	1	1	1	1
distance	-22	22	-22	-22	22	12	-22
case	1	2	2	2	2	2	2
distance	-22	-42.1	-42.1	14.3	-32.1	-44	-13.9
case	2	2	3	3	3	3	3
distance	-42.1	-42.1	-2.2	2.2	-2.2	2.2	2.2
case	3	3	3	4	4	4	4
distance	-2.2	2.2	2.2	-57.6	16.6	-39.8	-57.6
case	4	4	4	4	4		
distance	20.4	-39.4	-57.6	-57.6	-19.6		

Run a One-Way ANOVA model in R for these data according to:

```
library(lme4)
scd.out <- lmer(dist2 ~ 1+(1|case), scd)
summary(scd.out)
```

and interpret the REML results.

Appendix B

Common Probability Distributions

This appendix serves as a reference for the parametric forms used in the text. Considerably more detail can be found in the standard references: Johnson *et al.* (2005) for univariate forms on the counting measure; Johnson, Kotz, and Balakrishnan (1997) for multivariate forms on the counting measure; Johnson, Kotz, and Balakrishnan (1994, 1995) for univariate forms on the Lebesgue measure; Johnson, Kotz, and Balakrishnan (2000) for multivariate forms on the Lesbegue measure; Fang, Kotz, and Ng (1990) concentrating on symmetric forms; Kotz and Nadarajah (2000) for extreme value distributions; and more generally, Evans, Hastings, and Peacock (2000), Balakrishnan and Nevzorov (2003), and Krishnamoorthy (2006).

▷ **Bernoulli**

▶ PMF: $\mathcal{BR}(x|p) = p^x(1-p)^{1-x}, x = 0, 1, \quad 0 < p < 1.$

▶ $E[X] = p.$

▶ $\mathrm{Var}[X] = p(1-p).$

▷ **Beta**

▶ PDF: $\mathcal{BE}(x|\alpha, \beta) = \frac{\Gamma(\alpha+\beta)}{\Gamma(\alpha)\Gamma(\beta)} x^{\alpha-1}(1-x)^{\beta-1}, \quad 0 < x < 1, 0 < \alpha, \beta.$

▶ $E[X] = \frac{\alpha}{\alpha+\beta}.$

▶ $\mathrm{Var}[X] = \frac{\alpha\beta}{(\alpha+\beta)^2(\alpha+\beta+1)}.$

▷ **Binomial**

▶ PMF: $\mathcal{BN}(x|n, p) = \binom{n}{x} p^x (1-p)^{n-x}, x = 0, 1, \ldots, n, \quad 0 < p < 1.$

▶ $E[X] = np.$

▶ $\mathrm{Var}[X] = np(1-p).$

▷ **Cauchy**

▶ PDF: $\mathcal{C}(x|\theta, \sigma) = \frac{1}{\pi\sigma} \frac{1}{1+\left(\frac{x-\theta}{\sigma}\right)^2}, \quad -\infty < x, \theta < \infty, 0 < \sigma.$

▶ $E[X] = $ Does not exist.

▶ $\mathrm{Var}[X] = $ Does not exist.

▶ Note: sometimes θ and σ are labeled location and scale, respectively.

▷ **Dirichlet**

▶ PDF: $\mathcal{D}(\mathbf{x}|\alpha_1, \ldots, \alpha_k) = \frac{\Gamma(\alpha_1 + \ldots + \alpha_k)}{\Gamma(\alpha_1) \cdots \Gamma(\alpha_k)} x_1^{\alpha_1-1} \cdots x_k^{\alpha_k-1} \quad 0 \le x_i \le 1, \sum_{i=1}^{k} x_i = 1, 0 < \alpha_i, \forall i \in [1, 2, \ldots, k].$

▶ $E[X_i] = \frac{\alpha_i}{\alpha_0}$, where $\alpha_0 = \sum_{j=1}^{k} \alpha_j.$

- ▶ $\text{Var}[X_i] = \frac{\alpha_i(\alpha_0 - \alpha_i)}{\alpha_0^2(\alpha_0 + 1)}$.
- ▶ $\text{Cov}[X_h, X_i] = -\frac{\alpha_h \alpha_i}{\alpha_0^2(\alpha_0 + 1)}$.

▷ **Double Exponential**

- ▶ PDF: $\mathcal{DE}(x|\mu, \ldots, \tau) = \frac{1}{2\tau}\exp[-|x - \mu|/\tau]$ $-\infty < \mu, x < \infty, 0 < \tau$.
- ▶ $E[X] = \mu$.
- ▶ $\text{Var}[X] = 2\tau^2$.

▷ **F**

- ▶ PDF: $\mathcal{F}(x|\nu_1, \nu_2) = \frac{\Gamma\left(\frac{\nu_1 + \nu_2}{2}\right)}{\Gamma\left(\frac{\nu_1}{2}\right)\Gamma\left(\frac{\nu_2}{2}\right)}\left(\frac{\nu_1}{\nu_2}\right)^{\nu_1/2}\frac{x^{(\nu_1 - 2)/2}}{\left(1 + \frac{\nu_1}{\nu_2}x\right)^{\frac{\nu_1 + \nu_2}{2}}}$,

  $0 \le x < \infty, \nu_1, \nu_2 \in \mathbb{I}^+$.
- ▶ $E[X] = \frac{\nu_2}{\nu_2 - 2}, \nu_2 > 2$.
- ▶ $\text{Var}[X] = 2\left(\frac{\nu_2}{\nu_2 - 2}\right)^2\frac{\nu_1 + \nu_2 - 2}{\nu_1(\nu_2 - 4)}, \quad \nu_2 > 4$.

▷ **Gamma**

- ▶ PDF rate version: $\mathcal{G}(x|\alpha, \beta) = \frac{\beta^\alpha}{\Gamma(\alpha)}x^{\alpha - 1}\exp[-x\beta]$, $0 \le x < \infty$, $0 \le \alpha, \beta$.
- ▶ PDF scale version: $\mathcal{G}(x|\alpha, \beta) = \frac{\beta^{-\alpha}}{\Gamma(\alpha)}x^{\alpha - 1}\exp[-x/\beta]$, $0 \le x < \infty$, $0 < \alpha, \beta$.
- ▶ $E[X] = \frac{\alpha}{\beta}$, rate version.
- ▶ $\text{Var}[X] = \frac{\alpha}{\beta^2}$, rate version.
- ▶ $E[X] = \alpha\beta$, scale version.
- ▶ $\text{Var}[X] = \alpha\beta^2$, scale version.
- ▶ Note: the χ^2 distribution is $\mathcal{G}\left(\frac{\nu}{2}, \frac{1}{2}\right)$ (ν is the degrees of freedom parameter), and the exponential distribution comes from setting the shape parameter to one: $\mathcal{EX}(\beta)$ is $\mathcal{G}(1, \beta)$ (rate version).

▷ **Geometric**

- ▶ PMF: $\mathcal{GEO}(x|p) = p(1 - p)^{x - 1}$, $x = 1, 2, \ldots$, $0 \le p \le 1$.
- ▶ $E[X] = \frac{1}{p}$.
- ▶ $\text{Var}[X] = \frac{1 - p}{p^2}$.

▷ **Hypergeometric**

- ▶ PMF: $\mathcal{HG}(x|n, m, k) = \frac{\binom{m}{x}\binom{n - m}{k - x}}{\binom{n}{k}}$, $m - n + k \le x \le m$, $n, m, k \ge 0$.
- ▶ $E[X] = \frac{km}{n}$.
- ▶ $\text{Var}[X] = \frac{km(n - m)(n - k)}{n^2(n - 1)}$.

▷ **Inverse Gamma**

- ▶ PDF: $\mathcal{IG}(x|\alpha, \beta) = \frac{\beta^\alpha}{\Gamma(\alpha)}(x)^{-(\alpha + 1)}\exp[-\beta/x]$, $0 < x, \alpha, \beta$.
- ▶ $E[X] = \frac{\beta}{\alpha - 1}, \alpha > 1$.
- ▶ $\text{Var}[X] = \frac{\beta^2}{(\alpha - 1)^2(\alpha - 2)}, \alpha > 2$.

▷ **Lognormal**

- ▶ PDF: $\mathcal{LN}(x|\mu, \sigma) = (2\pi\sigma^2)^{-\frac{1}{2}}x^{-1}\exp[-(\log(x) - \mu)^2/2\sigma^2]$,

  $-\infty < \mu, x < \infty, 0 < \sigma^2$
- ▶ $E[X] = \exp[\mu + \sigma^2/2]$

▶ $\text{Var}[X] = \exp[2(\mu + \sigma^2)] - \exp[2\mu + \sigma^2]$.

▷ **Multinomial**

- ▶ PMF: $\mathcal{MN}(x|n, p_1, \ldots, p_k) = \frac{n!}{x_1! \cdots x_k!} p_1^{x_1} \cdots p_k^{x_k}, \quad x_i = 0, 1, \ldots, n,$
  $0 < p_i < 1, \quad \sum_{i=1}^{k} p_i = 1$.
- ▶ $E[X_i] = np_i$.
- ▶ $\text{Var}[X_i] = np_i(1 - p_i)$.
- ▶ $\text{Cov}[X_i, X_j] = -np_i p_j$.

▷ **Negative Binomial**

- ▶ PMF: $\mathcal{NB}(x|r, p) = \binom{x-1}{r-1} p^r (1-p)^{x-r}, \quad x = 0, 1, \ldots, 0 < p < 1, r \in \mathcal{I}^+$.
- ▶ $E[X] = \frac{r(1-p)}{p}$.
- ▶ $\text{Var}[X] = \frac{r(1-p)}{p^2}$.

▷ **Normal**

- ▶ PDF: $\mathcal{N}(x|\mu, \sigma^2) = (2\pi\sigma^2)^{-\frac{1}{2}} \exp\left[-\frac{1}{2\sigma^2}(x-\mu)^2\right]$,
  $-\infty < \mu, x < \infty, 0 < \sigma$.
- ▶ $E[X] = \mu$.
- ▶ $\text{Var}[X] = \sigma^2$.

Multivariate case: $\mathcal{N}_k(\mathbf{x}|\boldsymbol{\mu}, \boldsymbol{\Sigma}^2) = (2\pi)^{-k/2} |\boldsymbol{\Sigma}|^{-1/2} \exp\left[-\frac{1}{2}(\mathbf{x}-\boldsymbol{\mu})' \boldsymbol{\Sigma}^{-1}(\mathbf{x}-\boldsymbol{\mu})\right]$

▷ **Pareto**

- ▶ PDF: $\mathcal{PA}(x|\alpha, \beta) = \alpha\beta^\alpha x^{-(\alpha+1)}, \quad \beta < x, 0 < \alpha, \beta$.
- ▶ $E[X] = \frac{\beta\alpha}{\alpha-1}$, exists provided $\alpha > 1$.
- ▶ $\text{Var}[X] = \frac{\beta^2\alpha}{(\alpha-1)^2(\alpha-2)}$, exists provided $\alpha > 2$.

▷ **Poisson**

- ▶ PMF: $\mathcal{P}(x|\lambda) = \frac{\lambda^x e^{-\lambda}}{x!}, \quad x = 0, 1, \ldots, \quad 0 \le \lambda < \infty$.
- ▶ $E[X] = \lambda$.
- ▶ $\text{Var}[X] = \lambda$.

▷ **Student's-t**

- ▶ PDF: $\mathcal{T}(x|\nu) = \frac{\Gamma\left(\frac{\nu+1}{2}\right)}{\Gamma\left(\frac{\nu}{2}\right)} \frac{1}{(\pi\nu)^{\frac{1}{2}}(1+x^2/\nu)^{(\nu+1)/2}}, \quad -\infty < \mu, x < \infty, \nu \in \mathbb{I}^+$.
- ▶ $E[X] = 0, 1 < \nu$.
- ▶ $\text{Var}[X] = \frac{\nu}{\nu-2}, 2 < \nu$.

▷ **Student's-t, Multivariate**

- ▶ PDF: $\mathcal{MVT}(\mathbf{x}|\mathbf{M}, \nu) = |\mathbf{M}|^{-\frac{1}{2}} (\pi\nu)^{-k/2} \frac{\Gamma\left(\frac{\nu+k}{2}\right)}{\Gamma\left(\frac{\nu}{2}\right)} \left(1 + \frac{(\mathbf{x}-\boldsymbol{\mu})'\mathbf{M}(\mathbf{x}-\boldsymbol{\mu})}{\nu}\right)^{-\frac{\nu+k}{2}}$, where $\mathbf{x}$ is a k-length vector, $\mathbf{M}$ is a $k \times k$ positive definite matrix, and ν is a positive scalar.
- ▶ $E[\mathbf{X}] = \boldsymbol{\mu}$.
- ▶ $\text{Var}[\mathbf{X}] = \frac{\nu}{\nu-2}\mathbf{M}^{-1}$.

▷ **Uniform**

k-Category Discrete Case PMF:

► $\mathcal{U}(x) = p(X = x) = \begin{cases} \frac{1}{k}, & \text{for } x = 1, 2, \ldots, k \\ 0, & \text{otherwise} \end{cases}$

► $E[X] = \frac{k+1}{2}$.

► $\text{Var}[X] = \frac{(k+1)(k-1)}{12}$.

Continuous Case PDF:

► $\mathcal{U}(x) = f(x) = \begin{cases} \frac{1}{b-a}, & \text{for } a = 0 \leq x \leq b = 1 \\ 0, & \text{otherwise} \end{cases}$

► $E[X] = \frac{b-a}{2}$.

► $\text{Var}[X] = \frac{(b-a)^2}{12}$.

▷ **Weibull**

► PDF: $w(x|\gamma, \beta) = \frac{\gamma}{\beta} x^{\gamma-1} \exp\left(-\left(\frac{x}{\beta}\right)^\gamma\right)$ if $x \geq 0$ and 0 otherwise, where: $\gamma, \beta > 0$.

► $E[\mathbf{X}_{ij}] = \beta \Gamma\left[1 + \frac{1}{\gamma}\right]$.

► $\text{Var}[X_{ij}] = \beta^2 \left(\Gamma\left[1 + \frac{2}{\gamma}\right] - \gamma\left[1 + \frac{1}{\gamma}\right]^2\right)$

▷ **Wishart**

► PDF: $\mathcal{W}(\mathbf{X}|\alpha, \beta) = \frac{|\mathbf{X}|^{(\alpha-(k+1))/2}}{\Gamma_k(\alpha)|\beta|^{\alpha/2}} \exp[-\text{tr}(\beta^{-1}\mathbf{X})/2]$
where: $\Gamma_k(\alpha) = 2^{\alpha k/2} \pi^{k(k-1)/4} \prod_{i=1}^k \Gamma\left(\frac{\alpha+1-i}{2}\right)$, $2\alpha > k - 1$, β symmetric nonsingular, and $\mathbf{X}$ symmetric positive definite.

► $E[\mathbf{X}_{ij}] = \alpha\beta_{ij}$

► $\text{Var}[X_{ij}] = \alpha(\beta_{ij}^2 + \beta_{ii}\beta_{jj})$

► $\text{Cov}[X_{ij}, X_{kl}] = \alpha(\beta_{ik}\beta_{jl} + \beta_{il}\beta_{jk})$

References

Aarts, E. H. L. and Kors, T. J. (1989). *Simulated Annealing and Boltzmann Machines: A Stochastic Approach to Combinatorial Optimization and Neural Computing.* New York: John Wiley & Sons.

Abraham, C. and Cadre, B. (2004). Asymptotic Global Robustness in Bayesian Decision Theory. *The Annals of Statistics* **32**, 1341-1366.

Abramowitz, M. and Stegun, I. A. (eds.). (1977). *Handbook of Mathematical Functions: With Formulas, Graphs, and Mathematical Tables.* Mineola, NY: Dover Publications.

Acton, F. S. (1996). *Real Computing Made Real: Preventing Errors in Scientific and Engineering Calculations.* Princeton: Princeton University Press.

Adams, J. L. (1991). A Computer Experiment to Evaluate Regression Strategies. *Proceedings of the Computational Statistics Section of the American Statistical Association* **14**, 55-62.

Adman, V. E. and Raftery, A. E. (1986). Bayes Factors for Non-Homogeneous Poisson Processes with Vague Prior Information. *Journal of the Royal Statistical Society, Series B* **48**, 322-329.

Agresti, A. (2002). *Categorical Data Analysis.* Second Edition. New York: John Wiley & Sons.

Ahmed, S. E. and Reid, N. (2001). *Empirical Bayes and Likelihood Inference.* New York: Springer-Verlag.

Ahrens, J. H. and Dieter, U. (1972). Computer Methods for Sampling from the Exponential and Normal Distributions. *Communications of the ACM* **15**, 873-882.

Ahrens, J. H. and Dieter, U. (1973). Extensions of Forsythe's Method for Random Sampling from the Normal Distribution. *Mathematics for Computation* **27**, 927-937.

Ahrens, J. H. and Dieter, U. (1974). Computer Methods for Sampling from Gamma, Beta, Poisson, and Binomial Distributions. *Computing* **12**, 223-246.

Ahrens, J. H. and Dieter, U. (1980). Sampling from Binomial and Poisson Distributions: A Method with Bounded Computation Times. *Computing* **25**, 193-208.

Ahrens, J. H. and Dieter, U. (1982). Generation of Poisson Deviates from Modified Normal Distributions. *ACM Transactions on Mathematical Software* **8**, 163-179.

Aitkin, M. (1991). Posterior Bayes Factor. With Discussion. *Journal of the Royal Statistical Society, Series B* **53**, 111-142.

Akaike, H. (1973). Information Theory and an Extension of the Maximum Likelihood Principle. In *Proceedings of the 2nd International Symposium on Information Theory*, N. Petrov and F. Caski (eds.). Budapest: Akadémiai Kiadó, pp.176-723.

Akaike, H. (1974). A New Look at Statistical Model Identification. *IEEE Transactions Automatic Control* **AU-19**, 716-722.

Akaike, H. (1976). Canonical Correlation Analysis of Time Series and the Use of an Information Criterion. In *System Identification: Advances and Case Studies*, R. K. Mehra and D. G. Lainiotis (eds.). San Diego: Academic Press, pp.52-107.

Akaike, H. (1980). The Interpretation of Improper Prior Distributions as Limits of Data-Dependent Proper Prior Distributions. *Journal of the Royal Statistical Society, Series B* **42**, 46-52.

Albert, J. H. (1988). Computational Methods Using a Bayesian Hierarchical Generalized Linear Model. *Journal of the American Statistical Association* **83**, 1037-1044.

Albert, J. H. (1990). *Bayesian Computation with R (Use R!).* New York: Springer-Verlag.

Albert, J. H. and Chib, S. (1993).   Bayesian Analysis of Binary and Polychotomous Response Data. *Journal of the American Statistical Association* **88**, 669-679.

Albert, J. H. and Chib, S. (1995).   Bayesian Residual Analysis for Binary Response Regression Models *Biometrika* **82**, 747-759.

Albert, J. H. and Pepple, P. A. (1989).   A Bayesian Approach to Some Overdispersion Models. *Canadian Journal of Statistics* **17**, 333-344.

Allison, P. D. (1982). Discrete-Time Methods for the Analysis of Event Histories. *Sociological Methodology* **13**, 61-98.

Altman, M., Gill, J. and McDonald, M. P. (2003).   *Numerical Issues in Statistical Computing for the Social Scientist.* New York: John Wiley & Sons.

Altman, M. and McDonald, M. P. (2001).   Choosing Reliable Statistical Software. *PS: Political Science and Politics* **34**, 681-687.

Altman, M. and McDonald, M. P. (2003).   Replication with Attention to Numerical Accuracy. *Political Analysis.* **11**, 302-307.

Amemiya, T. (1985). *Advanced Econometrics.* Cambridge, MA: Harvard University Press.

Amit, Y. (1991). On Rates of Convergence of Stochastic Relaxation for Gaussian and non-Gaussian Distributions. *Journal of Multivariate Analysis* **38**, 82-99.

Amit, Y. (1996). Convergence Properties of the Gibbs Sampler for Perturbations of Gaussians. *Annals of Statistics* **24**, 122-140.

Amit, Y. and Grenander, U. (1991).   Comparing sweep strategies for stochastic relaxation. *Journal of Multivariate Analysis* **37**, 197-222.

Anderson, E. B. (1970). Sufficiency and Exponential Families for Discrete Sample Spaces. *Journal of the American Statistical Association* **65**, 1248-1255.

Anderson, J. E. and Louis, T. A. (1996).   Generating Pseudo-Random Variables From Mixture Models by Exemplary Sampling. *Journal of Statistical Computation and Simulation* **54**, 45-53.

Andrews, D. F. (1974). A Robust Method for Multiple Linear Regression. *Technometrics* **16**, 523-531.

Andrews, D. F., Bickel, P. J., Hampel, F. R., Huber, P. J., Rogers, W. H. and Tukey, J. W. (1972). *Robust Estimates of Location.* Princeton: Princeton University Press.

Andrews, D. F. and Mallows, C. L. (1974).   Scale Mixtures of Normality. *Journal of the Royal Statistical Society, Series B* **36**, 99-102.

Andrieu, C., Djurić, P. M. and Doucet, A. (2001).   Model Selection by MCMC Computation. *Signal Processing* **81**, 19-37.

Anscombe, F. J. (1963). Bayesian Inference Concerning Many Parameters with Reference to Supersaturated Designs. *Bulletin of the International Statistical Association* **40**, 733-741.

Antoniak, C. E. (1974). Mixtures of Dirichlet Processes with Applications to Bayesian Nonparametric Problems. *Annals of Statistics* **2**, 1152-1174.

Arbuckle, J. L. (2009). Amos 18 User's Guide. Chicago: SPSS Inc.

Arnold, B. C. and Press, S. J. (1983).   Bayesian Inference for Pareto Populations. *Journal of Econometrics* **21**, 287-306.

Arnold, B. C. and Press, S. J. (1989).   Bayesian Estimation and Prediction for Pareto Data. *Journal of the American Statistical Association* **84**, 1079-1084.

Asmussen, S. P., Glynn, P. and Thorisson, H. (1992).   Stationarity Detection in the Initial Transient Problem. *ACM Transactions on Modeling and Computer Simulation* **2**, 130-57.

Athreya, K. B. and Ney, P. (1978).   A New Approach to the Limit Theory of Recurrent Markov Chains. *Transactions of the American Mathematical Society* **245**, 493-501.

Athreya, K. B., Doss, H. and Sethuraman, J. (1996). On the Convergence of the Markov Chain Simulation Method. *Annals of Statistics* **24**, 69-100.

Atkinson, A. C. (1977). An Easily Programmed Algorithm for Generating Gamma Random Variates. *Journal of the Royal Statistical Society, Series A* **140**, 232-234.

Atkinson, A. C. (1978). Posterior Probabilities for Choosing a Regression Model. *Biometrika* **65**, 39-48.

Atkinson, A. C. (1980). Tests of Pseudo-Random Numbers. *Applied Statistics* **29**, 164-171.

Atkinson, A. C. and Pearce, M. C. (1976). The Computer Generation of Beta, Gamma, and Normal Random Variables. *Journal of the Royal Statistical Society, Series A* **139**, 431-461.

Ayres, D. (1994). *Information, Entropy, and Progress.* New York: American Institute of Physics Press.

Bakan, D. (1960). The Test of Significance in Psychological Research. *Psychological Bulletin* **66**, 423-437.

Balakrishnan, N. and Nevzorov, V. B. (2003). *A Primer on Statistical Distributions.* New York: John Wiley & Sons.

Baldus, D. C., Pulaski, C. and Woodworth, G. (1983). Comparative Review of Death Sentences: An Empirical Study of the Georgia Experience. *Journal of Criminal Law and Criminology* **74**, 661-753.

Baldus, D. C., Pulaski, C. and Woodworth, G. (1990). *Equal Justice and the Death Penalty: A Legal and Empirical Analysis.* Boston: Northeastern University Press.

Barankin, E. W. and Maitra, A. P. (1963). Generalization of the Fisher-Darmois-Koopman-Pittman Theorem on Sufficient Statistics. *Sankhyā, Series A* **25**, 217-244.

Barnard, G. A. (1991). Discussion of Aitkin. *Journal of the Royal Statistical Society, Series B* **53**, 128-130.

Barndorff-Nielsen, O. E. (1978). *Information and Exponential Families in Statistical Theory.* New York: John Wiley & Sons.

Barndorff-Nielsen, O. E. and Cox, D. R. (1989). *Asymptotic Techniques for Use in Statistics.* London: Chapman & Hall.

Barnett, V. (1973). *Comparative Statistical Inference.* New York: John Wiley & Sons.

Barnett, V. and Lewis, T. (1978). *Outliers in Statistical Data.* New York: Wiley & Sons.

Bartels, L. M. (1997). Specification Uncertainty and Model Averaging. *American Journal of Political Science* **41**, 641-674.

Bartholomew, D. J. (1965). A Comparison of Some Bayesian and Frequentist Inferences. *Biometrika* **52**, 19-35.

Bartlett, M. S. (1937). Maximum Likelihood Approaches to Variance Component Estimation and to Related Problems. *Proceedings of the Royal Society A: Mathematical, Physical, and Engineering Sciences* **160**, 268-278.

Basu, S. (1994). Variations of Posterior Expectations for Symmetric Unimodal Priors in a Distribution Band. *Sankhyā, Series A* **31**, 320-334.

Basu, S. and DasGupta, A. (1995). Robust Bayesian Analysis with Distribution Bands. *Statistics and Decisions* **13**, 333-349.

Basu, S., Jammalamadaka, S. R. and Liu, W. (1996). Local Posterior Robustness with Parametric Priors: Maximum and Average Sensitivity. In *Maximum Entropy and Bayesian Methods*, G. Heidlbreder (ed.). Dordrecht: Kluwer, pp.97-106.

Baum, L. E. and Eagon, J. A. (1967). An Inequality with Applications to Statistical Estimation for Probabilistic Functions of Markov Processes and to a Model for Ecology. *Bulletin of the American Mathematical Society* **73**, 360-363.

Baum, L. E. and Petrie, T. (1966). Statistical Inference for Probabilistic Functions of Finite Markov Chains. *Annals of Mathematical Statistics* **37**, 1554-1563.

Baum, L. E., Petrie, T., Soules, G. and Weiss, N. (1970). A Maximization Technique Occurring in the Statistical Analysis of Probabilistic Functions of Markov Chains. *Annals of Mathematical Statistics* **41**, 164-171.

Bauwens, L., Lubrano, M. and Richard, J-F. (1999). *Bayesian Inference in Dynamic Econometric Models.* Oxford, UK: Oxford University Press.

Bayer, D. and Diaconis, P. (1992). Trailing the Dovetail Shuffle to its Lair. *Annals of Applied Probability* **2**, 294-313.

Bayes, T. (1763). An Essay Towards Solving a Problem in the Doctrine of Chances. *Philosophical Transactions of the Royal Society of London* **53**, 370-418.

Beale, E. M. L. (1977). Discussion of the Paper by Professor Dempster, Professor Laird and Dr. Rubin. *Journal of the Royal Statistical Society, Series B* **139**, 22.

Beale, E. M. L. and Little, R. J. A. (1975). Missing Values in Multivariate Analysis. *Journal of the Royal Statistical Society, Series B* **37**, 129-145.

Beckett, L. and Diaconis, P. (1994). Spectral Analysis for Discrete Longitudinal Data. *Advances in Mathematics* **103**, 107-128.

Bedrick, E. J., Christensen, R. and Johnson, W. (1997). Bayesian Binomial Regression: Predicting Survival at a Trauma Center. *The American Statistician* **51**, 211-218.

Belsley, D. A., Kuh, E. and Welsch, R. E. (1980). *Regression Diagnostics.* New York: John Wiley & Sons.

Berger, J. O. (1984). The Robust Bayesian Viewpoint. In *Robustness of Bayesian Analysis*, Joseph B. Kadane (ed.). Amsterdam: North Holland, pp.63-144.

Berger, J. O. (1985). *Statistical Decision Theory and Bayesian Analysis.* Second Edition. New York: Springer-Verlag.

Berger, J. O. (1986a). Discussion: On the Consistency of Bayes Estimates. *Annals of Statistics* **14**, 30-37.

Berger, J. O. (1986b). Bayesian Salesmanship. In *Bayesian Inference and Decision Techniques with Applications: Essays in Honor of Bruno de Finetti*, Arnold Zellner (ed.). Amsterdam: North Holland, pp.473-488.

Berger, J. O. (1990). Robust Bayesian Analysis: Sensitivity to the Prior. *Journal of Statistical Planning and Inference* **25**, 303-28.

Berger, J. O. (1994). An Overview of Robust Bayesian Analysis. *Test* **3**, 5-124.

Berger, J. O. (2001). Bayesian Analysis: A Look at Today and Thoughts of Tomorrow. In *Statistics in the 21st Century*, Adrian E. Raftery, Martin A. Tanner and Martin T. Wells (eds.). New York: Chapman & Hall, pp.275-290.

Berger, J. O. (2003). Could Fisher, Jeffreys and Neyman Have Agreed on Testing? *Statistical Science* **18**, 1-32.

Berger, J. and Berliner, L. M. (1986). Robust Bayes and Empirical Bayes Analysis with ϵ-Contaminated Priors. *The Annals of Statistics* **14**, 461-486.

Berger, J. O. and Bernardo, J. M. (1989). Estimating the Product of Means: Bayesian Analysis with Reference Priors. *Journal of the American Statistical Association* **84**, 200-207.

Berger, J. O. and Bernardo, J. M. (1992). On the Development of the Reference Prior Method. In *Bayesian Statistics 4*, James O. Berger, José M. Bernardo, A. P. Dawid and Adrian F. M. Smith (eds.). Oxford: Oxford University Press, pp.35-49.

Berger, J. O., Brown, L. D. and Wolpert, R. L. (1994). A Unified Conditional Frequentist and Bayesian Test for Fixed and Sequential Simple Hypothesis Testing. *Annals of Statistics* **22**, 1787-1807.

Berger, J. O. and Delampady, M. (1987). Testing Precise Hypotheses. With Discussion. *Statistical Science* **2**, 317-352.

Berger, J. O., Ghosh, J. K. and Mukhopadhyay, N. (2003). Approximations to the Bayes Factor In Model Selection Problems and Consistency Issues. *Journal of Statistical Planning and Inference* **112**, 241-258.

Berger, J. O. and Mortera, J. (1999). Default Bayes Factors for Nonnested Hypothesis Testing. *Journal of the American Statistical Association* **94**, 542-554.

Berger, J. O. and O'Hagan, A. (1988). Ranges of Posterior Probabilities for Unimodal Priors with Specified

Quantiles. In *Bayesian Statistics 3*, J. M. Bernardo, M. H. DeGroot, D. V. Lindley and A. F. M. Smith (eds.). Oxford: Oxford University Press, pp.45-65.

Berger, J. O. and Pericchi, L. R. (1996a). The Intrinsic Bayes Factor for Model Selection and Prediction. *Journal of the American Statistical Association* **91**, 109-122.

Berger, J. O. and Pericchi, L. R. (1996b). The Intrinsic Bayes Factor for Linear Models. In *Bayesian Statistics 5*, J. O. Berger, J. M. Bernardo, A. P. Dawid, D. V. Lindley and A. F. M. Smith (eds.). Oxford: Oxford University Press, pp.23-42.

Berger, J. O., Pericchi, L. R. and Varshavsky, J. A. (1998). Bayes Factors and Marginal Distributions in Invariant Situations. *Sankhyā, Series A (Special Issue on Bayesian Analysis)* **60**, 307-321.

Berger, J. O. and Sellke, T. (1987). Test of a Point Null Hypothesis: the Irreconcilability of Significance Levels and Evidence. With Discussion. *Journal of the American Statistical Association* **82**, 112-139.

Berger, J. O. and Wolpert, R. L. (1984). *The Likelihood Principle.* Hayward, CA: Institute of Mathematical Statistics Monograph Series.

Berger, J. O. and Wolpert, R. L. (1988). *The Likelihood Principle.* Second Edition. Hayward, CA: Institute of Mathematical Statistics Monograph Series.

Berger, J. O. and Yang, R. (1994). Noninformative Priors and Bayesian Testing for the AR(1) Model. *Econometric Theory* **10**, 461-482.

Berger, J. O., Bernardo, J. M. and Mendoza, M. (1989). On Priors That Maximize Expected Information. In *Recent Developments in Statistics and Their Applications*, J. Klein and J. Lee (eds.). Seoul, ROK: Freedom Academy, pp.1-20.

Berger, J. O., Boukai, B. and Wang, Y. (1997) Unified Frequentist and Bayesian Testing of a Precise Hypothesis. *Statistical Science* **12**, 133-160.

Berk, R. H. (1972). Consistency and Asymptotic Normality of MLE's for Exponential Models. *Annals of Mathematical Statistics* **43**, 193-204.

Bernardo, J. M. (1979). Reference Prior Distributions for Bayesian Inference. *Journal of the Royal Statistical Society, Series B* **41**, 113-147.

Bernardo, J. M. (1984). Monitoring the 1982 Spanish Socialist Victory: A Bayesian Analysis. *Journal of the American Statistical Association* **79**, 510-515.

Bernardo, J. M. (1992). Simulated Annealing in Bayesian Decision Theory. *Computational Statistics. Proceedings of 10th Symposium on Computational Statistics, Volume 1.* Heidelberg: Physica-Verlag. 547-552.

Bernardo, J. M. and Smith, A. F. M. (1994). *Bayesian Theory.* New York: John Wiley & Sons.

Berndt, E., Hall, B., Hall, R. and Hausman, J. (1974). Estimation and Inference in Nonlinear Structural Models. *Annals of Economic and Social Measurement* **3/4**, 653-665.

Berry, D. and Hartigan, J. A. (1993). A Bayesian Analysis for Changepoint Problems. *Journal of the American Statistical Association* **88**, 309-319.

Berry, W. D., Rinquist, E. J., Fording, R. C. and Hanson, R. L. (1998). Measuring Citizen and Government Ideology in the American States, 1960-93. *American Journal of Political Science* **42**, 327-348.

Bertsimas, D. and Tsitsiklis, J. N. (1993). Simulated Annealing. *Statistical Science* **8**, 10-15.

Besag, J. (1974). Spatial Interaction and the Statistical Analysis of Lattice Systems. *Journal of the Royal Statistical Society, Series B* **36**, 192-236.

Besag, J. and Green, P. J. (1993). Spatial Statistics and Bayesian Computation. *Journal of the Royal Statistical Society, Series B* **55**, 1-52.

Besag, J., Green, P. J., Higdon, D. M. and Mengersen, K. L. (1995). Bayesian Computation and Stochastic Systems (with discussion). *Statistical Science* **10**, 3-66.

Best, D. J. (1979). Some Easily Programmed Pseudo-Random Normal Generators. *Australian Computing Journal* **11**, 60-62.

Best, D. J. (1983). A Note on Gamma Variate Generators With Shape Parameter Less Than Unity. *Computing, Archives for Informatics and Numerical Computation* **30**, 185-188.

Best, N. G., Spiegelhalter, D. J., Thomas, A. and Brayne, C. E. G. (1996). Bayesian Analysis of Realistically Complex Models. *Journal of the Royal Statistical Society, Series A* **159**, 323-342.

Bickel, P. and Doksum, K. (1977). *Mathematical Statistics*. San Francisco: Holden-Day.

Billera, L. J. and Diaconis, P. (2001). A Geometric Interpretation of the Metropolis-Hastings Algorithm. *Statistical Science* **16**, 335-339.

Billingsley, P. (1995). *Probability and Measure*. Third Edition. New York: John Wiley & Sons.

Birkes, D. and Dodge, Y. (1993). *Alternative Methods of Regression*. New York: Wiley & Sons.

Birnbaum, A. (1962). On the Foundations of Statistical Inference. *Journal of the American Statistical Association* **57**, 269-306.

Blalock, H. M. (1961). *Causal Inferences in Nonexperimental Research*. Chapel Hill, NC: University of North Carolina Press.

Blight, B. J. N. (1970). Estimation from a Censored Sample for the Exponential Family. *Biometrika* **57**, 389-395.

Blom, G., Holst, L. and Sandell, D. (1994). *Problems and Snapshots from the World of Probability*. New York: Springer-Verlag.

Blyth, S. (1995). The Dead of the Gulag: an Experiment in Statistical Investigation. *Applied Statistic*, 307-21.

Boehmke, F. J. (2009). Approaches to Modeling the Adoption and Modification of Policies with Multiple Components. *State Politics and Policy Quarterly* **9**, 229-252.

Bohachevsky, I. O., Johnson, M. E. and Stein, M. L. (1986). Generalized Simulated Annealing for Function Optimization. *Technometrics* **28**, 209-217.

Bohi, D. R., Toman, M. A. and Wells, M. A. (1996). *The Economics of Energy Security*. Boston: Kluwer Academic Publishers.

Boole, G. (1854). *An Investigation of the Laws of Thought on Which Are Founded the Mathematical Theories of Logic and Probabilities*. London: Macmillan.

Booth, J. G. and Hobert, J. P. (1999). Maximizing Generalized Linear Mixed Model Likelihoods with an Automated Monte Carlo EM Algorithm. *Journal of the Royal Statistical Society, Series B* **61**, 265-285.

Boscardin, J. W. and Gelman, A. (1996). Bayesian Computation for Parametric Models of Heteroscedasticity in the Linear Model. In *Advances in Econometrics, Volume 11 (Part A)*, R. C. Hill (ed.). Connecticut: JAI Press Inc., pp.87-109.

Bose, S. (1994a). Bayesian Robustness with Mixture Classes of Priors. *Annals of Statistics* **22**, 652-667.

Bose, S. (1994b). Bayesian Robustness with More Than One Class of Contaminations. *Journal of Statistical and Inference* **40**, 177-187.

Bowman, K. O. and Beauchamp, J. J. (1975). Pitfalls with Some Gamma Variate Simulation Routines. *Journal of Statistical Computation and Simulation* **4**, 141-154.

Box, G. E. P. (1980). Sampling and Bayes' Inference in Scientific Modeling and Robustness. *Journal of the Royal Statistical Society, Series A* **143**, 383-430.

Box, G. E. P. (1995). Discussion. *Journal of the Royal Statistical Society, Series B* **57**, 77.

Box, G. E. P. and Müller, M. E. (1958). A Note on Generation of Normal Deviates. *Annals of Mathematical Statistics* **28**, 610-611.

Box, G. E. P. and Tiao, G. C. (1968). A Bayesian Approach to Some Outlier Problems. *Biometrika* **55**, 119-129.

Box, G. E. P. and Tiao, G. C. (1973). *Bayesian Inference in Statistical Analysis*. New York: John Wiley & Sons.

Boyles, R. A. (1983). On the Convergence of the EM Algorithm. *Journal of the Royal Statistical Society, Series B* **45**, 47-50.

Bozdogan, H. (1987). Model Selection and Akaike's Information Criterion (AIC): The General Theory and Its Analytical Extensions. *Psychometrika* **52**, 345-370.

Bradlow, E. T. and Zaslavsky, A. M. (1997). Case Influence Analysis in Bayesian Inference. *Journal of Computational and Graphical Statistics* **6**, 314-331.

Bradlow, E. T., Weiss, R. E. and Cho, M. (1998). Bayesian Identification of Outliers in Computerized Adaptive Tests. *Journal of the American Statistical Association* **93**, 910-919.

Braithwaite, R. B. (1953). *Scientific Explanation.* New York: Harper & Brothers.

Brandstätter, Eduard. (1999). Confidence Intervals As An Alternative To Significance Testing. *Methods of Psychological Research* **Online 4.2**, 33-46.

Bratton, M. and Van De Walle, N. (1994). Neopatrimonial Regimes and Political Transitions in Africa. *World Politics* **46**, 453-489.

Bratton, M. and Van De Walle, N. (1997). Political Regimes and Regime Transitions in Africa, 1910-1994. *ICPSR Study Number I06996.* Ann Arbor: Inter-University Consortium for Political and Social Research.

Braun J. V., Braun R. K. and Muller H. G. (2000). Multiple Changepoint Fitting Via Quasilikelihood, With Application to DNA Sequence Segmentation. *Biometrika* **87**, 301-314.

Brémaud, P. (1999). *Markov Chains: Gibbs Fields, Monte Carlo Simulation, and Queues.* New York: Springer-Verlag.

Brent, R. P. (1974). A Gaussian Pseudo-Random Number Generator. *Communications of the Association for Computing Machinery* **17**, 704-706.

Breslow, N. (1990). Tests of Hypotheses in Overdispersed Poisson Regression and Other Quasi-Likelihood Models. *Journal of the American Statistical Association* **85**, 565-571.

Broemeling, L. D. (1985). *Bayesian Analysis of Linear Models.* New York: Marcel Dekker.

Brooks, R. J. (1972). Decision Theory Approach to Optimal Regression Designs. *Biometrika* **59**, 563-571.

Brooks, S. P. (1998a). Markov Chain Monte Carlo and its Applications. *The Statistician* **47**, 69-100.

Brooks, S. P. (1998b). Quantitative Convergence Assessment for Markov Chain Monte Carlo Via Cusums. *Statistics and Computing* **8**, 267-274.

Brooks, S. P. (2002). Discussion on the Paper by Spiegelhalter, Best, Carlin, and van der Linde. *Journal of the Royal Statistical Society, Series B* **64**, 616-618.

Brooks, S. P., Fan, Y. and Rosenthal, J. S. (2006). Perfect Forward Simulation via Simulated Tempering. *Communications in Statistics, Simulation and Computation* **35**, 683-713.

Brooks, S. P. and Roberts, G. O. (1999). On Quantile Estimation and Markov Chain Monte Carlo Convergence. *Biometrika* **86**,710-717.

Brooks, S. P. and Gelman, A. (1998a). Convergence Assessment Techniques for Markov Chain Monte Carlo. *Statistics and Computing* **8**, 319-335.

Brooks, S. P. and Gelman, A. (1998b). General Methods for Monitoring Convergence of Iterative Simulations. *Journal of Computational and Graphical Statistics* **7**, 434-455.

Brooks, S. P. and Giudici, P. (2000). MCMC Convergence Assessment via Two-Way ANOVA. *Journal of Computational and Graphical Statistics* **9**, 266-285.

Brooks, S. P. and Morgan, B. J. T. (1995). Optimization Using Simulated Annealing. *The Statistician* **44**, 241-257.

Brooks, S. P. and Roberts, G. O. (1998). Convergence Assessment Techniques for Markov Chain Monte Carlo. *Statistics and Computing* **8**, 319-335.

Brooks, S. P., Dellaportas, P. and Roberts, G. O. (1997). An Approach to Diagnosing Total Variation Convergence of MCMC Algorithms. '*Journal of Computational and Graphical Statistics* **6**, 251-265.

Brooks, S. and Roberts, G. (1997). On Quantile Estimation and MCMC Convergence. *Biometrika* **86**, 710-717.

Brown, L. D. (1964). Sufficiency Statistics in the Case of Independent Random Variables. *Annals of Mathematical Statistics* **35**, 1456-1474.

Brown, L. D. (1986). *Foundations of Exponential Families.* Hayward, CA: Institute of Mathematical Statistics Monograph Series 6.

Brown, L. D., Cohen, A. and Strawderman, W. E. (1980). Complete Classes for Sequential Tests of Hypotheses. *Annals of Statistics* **8**, 377-398.

Brown, L. D., Cohen, A. and Strawderman, W. E. (1989). *Correction of:* Complete Classes for Sequential Tests of Hypotheses. (1980, *Annals of Statistics* **8**, 377-398). *Annals of Statistics* **17**, 1414-1416.

Brown, P. J., Fearn, T. and Vannucci, M. (1999). The Choice of Variables in Multivariate Regression: A Non-Conjugate Bayesian Decision Theory Approach. *Biometrika* **86**, 635-648.

Browne, E. C., Frendreis, J. P. and Gleiber, D. W. (1986). The Process of Cabinet Dissolution: An Exponential Model of Duration and Stability in Western Democracies. *American Journal of Political Science* **30**, 628-650.

Bryk, A. S. and Raudenbush, S. W. (1989). Toward a More Appropriate Conceptualization of Research on School Level Effects: A Three Level Hierarchical Linear Model. In *Multilevel Analysis of Educational Data*, R. Darrell Bock (ed.). San Diego: Academic Press, pp.159-204.

Bryk, A. S. and Raudenbush, S. W. (2001). *Hierarchical Linear Models.* Second Edition. Beverly Hills: Sage.

Buck, S. F. (1960). A Method of Estimation of Missing Values in Multivariate Data Suitable for Use with an Electronic Computer. *Journal of the Royal Statistical Society, Series B* **22**, 302-306.

Buck, C. E., Cavanaugh, W. G. and Litton, C. D. (1996). *Bayesian Approach to Interpreting Archaeological Data.* New York: John Wiley & Sons.

Burford, R. L. (1973). A Better Additive Congruential Random Number Generator. *Decision Sciences* **4**, 190-193.

Burford, R. L. (1975). A Better Additive Congruential Random Number Generator? A Reply. *Decision Sciences* **6**, 199-201.

Burkhardt, H. and Schoenfeld, A. H. (2003). Improving Educational Research: Toward a More Useful, More Influential, and Better-Funded Enterprise. *Educational Researcher*,**32**, 3-14.

Burnham, K. P. and Anderson, D. R. (2002). *Model Selection and Multimodal Inference: A Practical Information-Theoretical Approach.* Second Edition. New York: Springer-Verlag.

Burstein, L., Kim, K-S. and Delandshere, G. (1989). Multilevel Investigations of Systematically Varying Slopes: Issues, Alternatives, and Consequences. In *Multilevel Analysis of Educational Data*, R. Darrell Bock (ed.). San Diego: Academic Press, pp.233-279.

Butcher, J. C. (1961). A Partition Test for Pseudo-Random Numbers. *Mathematics of Computation* **15**, 198-199.

Campbell, A. and Converse, P. (1999). American National Election Study, 1960. ICPSR07216-v3. Ann Arbor, MI: *Inter-university Consortium for Political and Social Research* [distributor].

Carver, Ronald P. (1978). The Case Against Statistical Significance Testing. *Harvard Education Review* **48**, 378-99.

Carver, Ronald P. (1993). The Case Against Statistical Significance Testing, Revisited. *Journal of Experimental Education* **61**, 287-92.

Canes-Wrone, B., Brady, D. W. and Cogan, J. F. (2002). Out of Step, Out of Office: Electoral Accountability and House Members' Voting. *American Political Science Review* **96**, 127-140.

Canova, F. (1994). Statistical Inference in Calibrated Models. *Journal of Applied Econometrics* **9**, S123-S144.

Carlin, B. P. and Chib, S. (1995). Bayesian Model Choice Via Markov Chain Monte Carlo Methods. *Journal of the Royal Statistical Society, Series B* **57**, 473-484.

Carlin, B. P. and Gelfand, A. E. (1990). Approaches for Empirical Bayes Confidence Intervals. *Journal of the American Statistical Association* **409**, 105-114.

Carlin, B. P. and Louis, T. A. (1996). Identifying Prior Distributions That Produce Specific Decisions, With Application to Monitoring Clinical Trials. In *Bayesian Analysis in Statistics and Econometrics: Essays in Honor of Arnold Zellner*, D. Berry, K. Chaloner and J. Geweke (eds.). New York: John Wiley & Sons, pp.493-503.

Carlin, B. P. and Louis, T. A. (2001). *Bayes and Empirical Bayes Methods for Data Analysis.* Second Edition. New York: Chapman & Hall.

Carlin, B. P. and Louis, T. A. (2009). *Bayes and Empirical Bayes Methods for Data Analysis.* Third Edition. New York: Chapman & Hall.

Carlin, B. P. and Polson, N. G. (1991). Inference for Nonconjugate Bayesian Models Using the Gibbs Sampler. *Canadian Journal of Statistics* **19**, 399-405.

Carlin, B. P., Chaloner, K., Church, T., Louis, T. A. and Matts, J. P. (1993). Bayesian Approaches for Monitoring Clinical Trials with an Application to Toxoplasmic Encephalitis Prophylaxis. *The Statistician* **42**, 355-367.

Carlin, B. P., Chaloner, K., Louis, T. A. and Rhame, F. S. (1995). Elicitation, Monitoring, and Analysis for an AIDS Clinical Trial. In *Case Studies in Bayesian Statistics, Volume II*, Constantine Gatsonis, James S. Hodges, Robert E. Kass and Nozer D. Singpurwalla (eds.). New York: Springer-Verlag, pp.48-78.

Carlin, B. P., Gelfand, A. E. and Smith, A. F. M. (1992). Hierarchical Bayesian Analysis of Changepoint Problems. *Applied Statistics* **41**, 389-405.

Carlin, J. B. (1992). Meta-analysis for 2×2 Tables: A Bayesian Approach. *Statistics in Medicine* **11**, 141-158.

Carroll, R. J. and Ruppert, D. (1982). Robust Estimation in Heteroscedastic Linear Models. *Annals of Statistics* **10**, 429-441.

Carroll, R. J. and Ruppert, D. (1988). *Transforming and Weighting in Regression.* New York: Chapman & Hall.

Carter, W. J., Jr. and Myres, R. H. (1973). Maximum Likelihood Estimation from Linear Combinations of Discrete Probability Functions. *Journal of the American Statistical Association* **68**, 203-206.

Casella, G. (1985). An Introduction to Empirical Bayes Data Analysis. *The American Statistician* **39**, 83-87.

Casella, G. (2005). James-Stein Estimator. In Encyclopedia of Biostatistics 4. Peter Armitage and Theodore Colton (eds.). New York: Wiley & Sons.

Casella, G. and Berger, R. L. (1987a). Reconciling Bayesian and Frequents Evidence in the One-Sided Testing Problem. *Journal of the American Statistical Association* **82**, 106-111.

Casella, G. and Berger, R. L. (1987b). Rejoinder. *Journal of the American Statistical Association* **82**, 133-135.

Casella, G. and Berger, R. L. (2002). *Statistical Inference.* Second Edition. Belmont, CA: Duxbury Advanced Series.

Casella, G. and George, E. I. (1992). Explaining the Gibbs Sampler. *The American Statistician* **46**, 167-174.

Casella, G., Lavine, M. and Robert, C. (2001). Explaining the Perfect Sampler. *The American Statistician* **55**, 299-305.

Casella, G., Mengersen, K. L., Robert, C. P. and Titterington, D. M. (2002) Perfect Slice Samplers for Mixtures of Distributions. *Journal of the Royal Statistical Society, Series B*, **64**, 777-790.

Casella, G. and Robert, C. P. (1996). Rao-Blackwellization of Sampling Schemes. *Biometrika* **93**, 81-94.

Celeux, G. and Diebolt, J. (1985). The SEM Algorithm: A Probabilistic Teacher Algorithm Derived from the EM Algorithm for the Mixture Problem. *Computational Statistics* **2**, 73-82.

Celeux, G., Chauveau, D. and Diebolt, J. (1996). Stochastic Versions of the EM Algorithm: an Experimental Study in the Mixture Case. *Journal of Statistical Computation and Simulation* **55**, 287-314.

Celeux, G., Forbes, F., Robert, C. P. and Titterington, D. M. (2006). Deviance Information Criteria for Missing Data Models. *Bayesian Analysis* **1**, 1-24.

Celeux, G., Hurn, M. and Robert, C. P. (2000). Computational and Inferential Difficulties with Mixture Posterior Distributions. *Journal of the American Statistical Association* **95**, 957-70.

Černý, V. (1985). A Thermodynamic Approach to the Traveling Salesman Problem: An Efficient Simulation. *Journal of Optimization Theory and Applications* **45**, 41-51.

Chakrabarti, A. and Ghosh, J. K. (2006). A Generalization of BIC for the General Exponential Family. *Journal of Statistical Planning and Inference* **136**, 2847-2872.

Chaloner, K. and Brant, R. (1988). A Bayesian Approach to Outlier Detection and Residual Analysis. *Biometrika* **75**, 651-659.

Chaloner, K. and Duncan, G. T. (1983). Assessment of a Beta Prior Distribution: PM Elicitation. *The Statistician* **27**, 174-180.

Chaloner, K. and Duncan, G. T. (1987). Some Properties of the Dirichlet-Multinomial Distribution and Its Use in Prior Elicitation. *Communications in Statistics, Part A, Theory and Methods* **16**, 511-523.

Chan, K. S. (1993). Asymptotic Behavior of the Gibbs Sampler. *Journal of the American Statistical Association* **88**, 320-326.

Chan, K. S. and Geyer, C. J. (1994). Discussion of "Markov Chains for Exploring Posterior Distributions." *Annals of Statistics* **22**, 1747-1758.

Chan, K. S. and Ledolter, J. (1995). Monte Carlo EM Estimation for Time Series Models Involving Counts. *Journal of the American Statistical Association* **90**, 242-252.

Chang, T. and Villegas, C. (1986). On a Theorem of Stein Relating Bayesian and Classical Inferences in Group Models. *Canadian Journal of Statistics* **14**, 289-296.

Chao, M. T. (1970). The Asymptotic Behavior of Bayes' Estimators. *Annals of Mathematical Statistics* **41**, 601-608.

Chatfield, C. (2003). *The Analysis of Time Series*. Sixth Edition. New York: Chapman & Hall.

Chay, S. C., Fardo, R. D. and Mazumdar, M. (1975). On Using the Box-Müller Transformation with Multiplicative Congruential Pseudo-Random Number Generators. *Applied Statistics* **24**, 132-135.

Chen, J. and Gupta A. K. (1997). Testing and Locating Changepoints With Application To Stock Prices. *Journal of the American Statistical Association* **92**, 739-747.

Chen, J. and Ibrahim, J. G. (2006). The Relationship Between the Power Prior and Hierarchical Models. *Bayesian Analysis* **1**, 551-574.

Chen, M-H. and Shao, Q-M. (1999). Monte Carlo Estimation of Bayesian Credible and HPD Intervals. *Journal of Computational and Graphical Statistics* **8**, 69-92.

Chen, M-H. and Schmeiser, B. (1993). Performance of the Gibbs, Hit-and-Run, and Metropolis Samplers. *Journal of Computational and Graphical Statistics* **2**, 251-72.

Chen, M-H., Dey, D. and Ibrahim, J. G. (2000). Bayesian Criterion Based Model Assessment for Categorical Data. *Biometrika* **91**, 45-63.

Chen, M-H., Shao, Q-M. and Ibrahim, J. G. (2000). *Monte Carlo Methods in Bayesian Computation*. New York: Springer-Verlag.

Chen, T. T. and Fienberg, S. E. (1974). Two-dimensional Contingency Tables with Both Completely and Partially Cross-Classified Data. *Biometrics* **30**, 629-642.

Cheng, R. C. H. (1977). The Generation of Gamma Variables with Non-Integral Shape Parameter. *Applied Statistics* **26**, 71-75.

Cheng, R. C. H. (1985). Generation of Multivariate Normal Samples with Given Mean and Covariance Matrix. *Journal of Statistical Computation and Simulation* **21**, 39-49.

Cheng, R. C. H. and Feast, G. M. (1979). Some Simple Gamma Variate Generators. *Applied Statistics* **28**, 290-295.

Cheng, R. C. H. and Feast, G. M. (1980). Gamma Variate Generators with Increased Shape Parameter Range. *Communications of the ACM* **23**, 389-393.

Chiang, T-S. and Chow, Y. (1988). On Eigenvalues and Optimal Annealing Rate. *Mathematical Operations Research* **13**, 508-511.

Chib, S. (1992). Bayes Inference in the Tobit Censored Regression Model. *Journal of Econometrics* **51**, 79-99.

Chib, S. (1995). Marginal Likelihood from the Gibbs Output. *Journal of the American Statistical Association* **90**, 1313-1321.

Chib S. (1998). Estimation and Comparison of Multiple Change-point Models. *Journal of Econometrics* **86**, 221-241.

Chib, S. and Greenberg, E. (1995). Understanding the Metropolis-Hastings Algorithm. *The American Statistician.* **49**, 327-335.

Chib, S. and Greenberg, E. (1998). Analysis of Multivariate Probit Models. *Biometrika* **85**, 347-361.

Chib, S. and Jeliazkov, I. (2001). Marginal Likelihood from the Metropolis-Hastings Output. *Journal of the American Statistical Association* **96**, 270-281.

Chib, S. and Tiwari, R. C. (1991). Robust Bayes Analysis in Normal Linear Regression With An Improper Mixture Prior. *Communications in Statistics, Part A, Theory and Methods* **20**, 807-829.

Christiansen, C. L. and Morris, C. N. (1997). Hierarchical Poisson Regression Modeling. *Journal of the American Statistical Association* **92**, 618-632.

Chu, C-S. J. and White, H. (1992). A Direct Test for Changing Trend. *Journal of Business and Economic Statistics* **10**, 298-299.

Chubb, J. E. and Moe, T. (1988). Politics, Markets and the Organization of American Schools. *American Political Science Review* **82**, 1065-1089.

Chubb, J. E. and Moe, T. (1990). *Politics, Markets and America's Schools.* Washington, DC: Brookings Institution.

Chung, K. L. (1974). *A Course in Probability Theory.* San Diego: Academic Press.

Cioffi-Revilla, C. and Lai, D. (1995). War and Politics in Ancient China, 2700 B.C. to 722 B.C.: Measurement and Comparative Analysis. *Journal of Conflict Resolution* **39**, 467-494.

Cioffi-Revilla, C. and Lai, D. (2001). The Second International System: China and East Asia, 5000 B.C. to 200 B.C. Presented at the Hong Kong Meeting of the International Studies Association, July 26-28, 2001.

Clements, Michael. (2004). Evaluating the Bank of England Density Forecasts of Inflation. *Economic Journal* **114**, 844-866.

Cleveland, W. S. (1979). Robust Locally Weighted Regression and Smoothing Scatterplots. *Journal of the American Statistical Association* **74**, 829-36.

Cleveland, W. S. (1981). LOWESS: A Program for Smoothing Scatterplots by Robust Locally Weighted Regression. *The American Statistician* **35**, 54.

Cleveland, W. S. (1993). *Visualizing Data.* Murray Hill, NJ: Hobart Press.

Clogg, C. C., Petkova, E. and Haritou, A. (1995). Statistical Methods for Comparing Regression Coefficients Between Models. *American Journal of Sociology* **100**, 1261-1293.

Clyde, M. (1999). Bayesian Model Averaging and Model Search Strategies. In *Bayesian Statistics 6*, J. O. Berger, J. M. Bernardo, A. P. Dawid, D. V. Lindley and A. F. M. Smith (eds.). Oxford: Oxford University Press, pp.157-185.

Clyde, M., Müller, P. and Parmigiani, G. (1995). Optimal Designs for Heart Defibrillators. *Case Studies in Bayesian Statistics II. Lecture Notes in Statistics* **105**, 278-292. New York: Springer-Verlag.

Cocchi, D. and Mouchart, M. (1996). Quasi-Linear Bayes Estimation in Stratified Finite Populations. *Journal of the Royal Statistical Society, Series B* **58**, 293-300.

Cohen, J. (1962). The Statistical Power of Abnormal-Social Psychological Research: A Review. *Journal of Abnormal and Social Psychology* **65**, 145-153.

Cohen, J. (1977). *Statistical Power Analysis for the Behavioral Sciences* (revised edition). New York: Academic Press.

Cohen, J. (1988). A Power Primer. *Psychological Bulletin* **112**, 155-159.

Cohen, J. (1992). Statistical Power Analysis. *Current Directions In Psychological Science* **1**, 98-101.

Cohen, J. (1994). The Earth is Round ($p < .05$). *American Psychologist* **12**, 997-1003.

Cohen, J., Nagin, D., Wallstrom, G. and Wasserman, L. (1998). Hierarchical Bayesian Analysis of Arrest Rates. *Journal of the American Statistical Association* **93**, 1260-1270.

Columbo, B. (1952). Preliminary Analysis of Recent Demographic Trends in Italy. *Population Index* **18**, 265-279.

Conell, C. and Cohn, S. (1995). Learning from Other People's Actions: Environmental Variation and Diffusion in French Coal Mining Strikes, 1890-1935. *American Journal of Sociology* **101**, 366-403.

Congdon, P. (2001). *Bayesian Statistical Modeling.* New York: Wiley & Sons.

Congdon, P. (2003). *Applied Bayesian Modelling.* New York: Wiley & Sons.

Congdon, P. (2005). *Bayesian Models for Categorical Data.* New York: Wiley & Sons.

Congdon, P. (2010). *Applied Bayesian Hierarchical Methods.* New York: Wiley & Sons.

Conigliani, C., Castro, J. I. and O'Hagan, A. (2000). Bayesian Assessment of Goodness of Fit Against Nonparametric Alternatives. *Canadian Journal of Statistics* **28**, 327-342.

Conquest, R. (1978). *Kolyma: The Arctic Death Camps.* New York: Viking Adult.

Consonni, G. and Veronese, P. (1992). Conjugate Priors for Exponential Families Having Quadratic Variance Functions. *Journal of the American Statistical Association* **87**, 1123-1127.

Conte, S. D. and de Boor, C. (1980). *Elementary Numerical Analysis: An Algorithmic Approach.* Third Edition. New York: McGraw-Hill.

Cook, D. R. and Weisberg, S. (1982). *Residuals and Influence in Regression.* New York: Chapman & Hall.

Corcoran, J. N. and Tweedie, R. L. (1998). Perfect Sampling From Independent Metropolis-Hastings Chains. *Journal of Statistical Planning and Inference* **104**, 297-314.

Cosmides, L. and Tooby, J. (1996). Are Humans Good Intuitive Statisticians After All? Rethinking Some Conclusions from the Literature on Judgment Under Uncertainty. *Cognition* **58**, 1-73.

Coveyou, R. R. (1960). Serial Correlation in the Generation of Pseudo-Random Numbers. *Journal of the Association for Computing Machinery* **7**, 72-74.

Coveyou, R. R. (1970). Random Numbers Fall Mainly in the Planes (Review). *ACM Computing Reviews*, 225.

Coveyou, R. R. and MacPherson, R. D. (1967). Fourier Analysis of Uniform Random Number Generators. *Journal of the ACM* **14**, 100-119.

Cowles, M. K. (2002). MCMC Sampler Convergence Rates for Hierarchical Normal Linear Models: A Simulation Approach. *Statistics and Computing* **12**, 377-389.

Cowles, M. K. and Carlin, B. P. (1996). Markov Chain Monte Carlo Convergence Diagnostics: A Comparative Review. *Journal of the American Statistical Association* **91**, 883-904

Cowles, M. K., Roberts, G. O. and Rosenthal, J. S. (1999). Possible Biases Induced by MCMC Convergence Diagnostics. *Journal of Statistical Computation and Simulation* **64**, 87-104.

Cowles, M. K. and Rosenthal, J. S. (1998). A Simulation Approach to Convergence Rates for Markov Chain Monte Carlo Algorithms. *Statistics and Computing* **8**, 115-124.

Cox, D. R. (1961). Tests of Separate Families of Hypotheses. *Proceedings of the Fourth Berkeley Symposium on Mathematical Statistics and Probability.* Berkeley: University of California Press, 105-123.

Cox, D. R. (1983). Some Remarks on Overdispersion. *Biometrika* **70**, 269-274.

Crain, B. R. and Morgan, R. L. (1975). Asymptotic Normality of the Posterior Distribution for Exponential Models. *Annals of Statistics* **3**, 223-227.

Craiu, R. V. and Meng, X-L. (2011). Perfection Within Reach: Exact MCMC Sampling. In *Handbook of Markov Chain Monte Carlo,* Steve Brooks, Andrew Gelman, Galin L. Jones and Xiao-Li Meng (eds.). Boca Raton: Chapman & Hall/CRC, pp.199-223.

Cramer, J. S. (1994). *Econometric Applications of Maximum Likelihood Methods.* Cambridge: Cambridge University Press.

Creutz, M. (1979). Confinement and the Critical Dimensionality of Space-Time. *Physical Review Letters* **43**, 553-56.

Creutz, M., Jacobs, L. and Rebbi, C. (1983). Monte Carlo Computations in Lattice Gauge Theories. *Physical Review* **95**, 201.

Cuevas, A. and Sanz, P. (1988). On Differentiability Properties of Bayes Operators. In *Bayesian Statistics 3,* J. M. Bernardo, M. H. DeGroot, D. V. Lindley and A. F. M. Smith (eds.). Oxford: Oxford University Press, pp.569-577.

Cui, L., Tanner, M. A., Sinha, D. and Hall, W. J. (1992). Comment: Monitoring Convergence of the Gibbs Sampler: Further Experience with the Gibbs Stopper. *Statistical Science* **7**, 483-486.

Cyert, R. M. and DeGroot, M. H. (1987). *Bayesian Analysis and Uncertainty in Economic Theory.* Totowa, NJ: Rowman & Littlefield.

Dalal, S. R. and Hall, G. J., Jr. (1980). On Approximating Parametric Bayes Models by Nonparametric Bayes Models. *Annals of Statistics* **8**, 664-772.

Dalal, S. R. and Hall, W. J. (1983). Approximating Priors by Mixtures of Natural Conjugate Priors. *Journal of the Royal Statistical Society, Series B* **45**, 278-286.

Dale, A. I. (1991). *A History of Inverse Probability: From Thomas Bayes to Karl Pearson.* New York: Springer-Verlag.

Damien, P., Wakefield, J. C. and Walker, S. (1999). Gibbs Sampling for Bayesian Non-Conjugate and Hierarchical Models by Using Auxiliary Variables. *Journal of the Royal Statistical Society, Series B* **61**, 331-344.

Dar, R., Serlin, R. C. and Omer, H. (1994). Misuse of Statistical Tests in Three Decades of Psychotherapy Research. *Journal of Consulting and Clinical Psychology* **62**, 75.

Darmois, G. (1935). Sur les Lois de Probabilités à Estimation Exhaustive. *Acadamie de Sciences Paris* **200**, 1265-1266.

Datta, G. S. and Ghosh, M. (1991). Bayesian Prediction in Linear Models: Applications to Small Area Estimation. *Annals of Statistics* **19**, 1748-1770.

Davidian, M. and Carroll, R. J. (1987). Variance Function Estimation. *Journal of the American Statistical Association* **82**, 1079-1091.

Davis, D. H. (1992). *Energy Politics.* New York: St. Martins Press.

Davis, P. J. and Rabinowitz, P. (1984). *Methods of Numerical Integration.* Second Edition. San Diego: Academic Press.

Davis, W. W. (1978). Bayesian Analysis of the Linear Model Subject to Linear Inequality Constraints. *Journal of the American Statistical Association* **73**, 573-579.

Dawid, A. P. (1979). Conditional Independence in Statistical Theory. *Journal of the Royal Statistical Society, Series B* **41**, 1-31.

Dawid, A. P. (1982). The Well-Calibrated Bayesian. *Journal of the American Statistical Association* **77**, 605-613.

Dawid, A. P. (1983). Invariant Prior Distributions. In *Encyclopedia of Statistical Sciences*, S. Kotz and N. L. Johnson (eds.). New York: John Wiley & Sons, pp.228-236.

Dawid, A. P., Stone, M. and Zidek, J. V. (1973). Marginalization Paradoxes in Bayesian and Structural Inference. *Journal of the Royal Statistical Society, Series B* **35**, 189-233.

Dearden, R. and Clancy, D. (2002). Particle Filters for Real-Time Fault Detection in Planetary Rovers. *Proceedings of the Thirteenth International Workshop on Principles of Diagnosis*, 1-6.

De Bruijn, N. G. (1981). *Asymptotic Methods in Analysis.* Dover Edition. New York: Dover Publications.

Deely, J. J. and Lindley, D. V. (1981). Bayes Empirical Bayes. *Journal of the American Statistical Association* **76**, 833-841.

Deely, J. J. and Zimmer, W. J. (1976). Asymptotic Optimality of the Empirical Bayes Procedure. *Annals of Statistics* **4**, 576-580.

de Finetti, B. (1930). Funzione Caratteristica di un Fenomeno Aleatorio. *Mem. Accad. Naz. Lincei* **4**, 86-133.

de Finetti, B. (1937). La Prévision: ses Lois Logiques, ses Sources Subjectives. *Annales de l'Institut Henri Poincaré* **7**, 1-68. Translated by H. E. Kyburg Jr. in H. E. Kyburg Jr. and H. E. Smokler (ed.). *Studies in Subjective Probability* (1964). New York: Wiley, 93-158.

de Finetti, B. (1972). *Probability, Induction, and Statistics.* New York: John Wiley & Sons.

de Finetti, B. (1974). *Theory of Probability, Volume 1.* New York: John Wiley & Sons.

de Finetti, B. (1975). *Theory of Probability, Volume 2.* New York: John Wiley & Sons.

DeGroot, M. H. (1970). *Optimal Statistical Decisions.* New York: John Wiley & Sons.

DeGroot, M. H. (1973). Doing What Comes Naturally: Interpreting a Tail Area as a Posterior Probability or as a Likelihood Ratio. *Journal of the American Statistical Association* **68**, 966-969.

DeGroot, M. H. (1986). *Probability and Statistics.* Second Edition. Reading, MA: Addison-Wesley.

De Oliveira, V. (2013). Hierarchical Poisson Models For Spatial Count Data. *Journal of Multivariate Analysis* **122**, 393-408.

Del Pino, G. (1989). The Unifying Role of Iterative Generalized Least Squares in Statistical Algorithms. *Statistical Science* **4**, 394-408.

Delampady, M. and Dey, D. K. (1994). Bayesian Robustness for Multiparameter Problems. *Journal of Statistical Planning and Inference* **50**, 375-382.

Dellaportas, P. and Smith, A. F. M. (1993). Bayesian Inference for Generalized Linear and Proportional Hazards Models via Gibbs Sampling. *Applied Statistics* **42**, 443-459.

Demir, B., Bovolo, F. and Bruzzone, L. (2012). A Novel System for Classification of Image Time Series with Limited Ground Reference Data. 2012 IEEE International on Geoscience and Remote Sensing Symposium (IGARSS). Munich, Germany, 22-27 July 2012, 158-161.

de Morgan, A. (1837). Review of Laplace's Théorie Analytique des Probabilités. *Dublin Review* **2**, 338-354, **3**, 237-248.

de Morgan, A. (1838). *An Essay on Probabilities and their Application to Life Contingencies and Insurance Offices.* London: Longman, Orme, Brown, Green, & Longmans.

de Morgan, A. (1847). *Formal Logic; or, the Calculus of Inference, Necessary and Probable. Formal Logic.* London: Taylor & Walton.

Dempster, A. P., Laird, N. M. and Rubin, D. B. (1977). Maximum Likelihood from Incomplete Data via the EM Algorithm. *Journal of the Royal Statistical Society, Series B* **39**, 1-38.

Denis, D. J. (2005) The Modern Hypothesis Testing Hybrid: R. A. Fisher's Fading Influence. *Journal de la Société Française de Statistique* **145**, 5-26.

DeSouza, C. M. (1992). An Approximate Bivariate Bayesian Method for Analyzing Small Frequencies. *Biometrics* **48**, 1113-1130.

Devore, J. (1999). *Probability and Statistics for Engineering and the Sciences*. Boston: PWS Publishing.

Devroye, L. (1986). *Non-Uniform Random Variate Generation*. New York: Springer-Verlag.

Dey, D. K. and Birmiwal, L. R. (1994). Robust Bayesian Analysis Using Entropy and Divergence Measures. *Statistical Probability Letters* **20**, 287-294.

Dey, D. K. and Micheas, A. (2000). Ranges of Posterior Expected Losses and ε-Robust Actions. In *Robust Bayesian Analysis*, David Ríos Insua and Fabrizio Ruggeri (eds.). New York: Springer-Verlag, pp.71-88.

Dey, D. K., Ghosh, S. K. and Mallick, B. K. (2000). *Generalized Linear Models: A Bayesian Perspective*. New York: Marcel Dekker.

Dey, D. K., Lou, K. and Bose, S. (1998). A Bayesian Approach to Loss Robustness. *Statistics and Decisions* **16**, 65-87.

Diaconis, P. (1988). *Group Representations in Probability and Statistics*. Hayward, CA: Institute of Mathematical Statistics.

Diaconis, P. and Freedman, D. A. (1980). Finite Exchangeable Sequences. *Annals of Probability* **8**, 745-764.

Diaconis, P. and Freedman, D. A. (1986). On the Consistency of Bayes Estimates. *Annals of Statistics* **14**, 1-67.

Diaconis, P. and Ylvisaker, D. (1979). Conjugate Priors for Exponential Families. *Annals of Statistics* **7**, 269-281.

Diaconis, P. and Ylvisaker, D. (1985). Quantifying Prior Opinion. In *Bayesian Statistics 2*, J. M. Bernardo, M. H. DeGroot, D. V. Lindley and A. F. M. Smith (eds.). Amsterdam: North Holland Press, pp.133-156.

Diaconis, P. and Saloff-Coste, L. (1993). Comparison Theorems for Reversible Markov Chains. *Annals of Applied Probability* **3**, 696-730.

Diaconis, P. and Saloff-Coste, L. (1996). Logarithmic Sobolev Inequalities for Finite Markov Chains. *Annals of Applied Probability* **6**, 695-750.

Diaconis, P. and Stroock, D. W. (1991). Geometric Bounds for Eigenvalues of Markov Chains. *Annals of Applied Probability* **1**, 36-61.

Dias, J. and Wedel, M. (2004). An Empirical Comparison of EM, SEM and MCMC Performance for Problematic Gaussian Mixture Likelihoods. *Statistics and Computing* **14**, 323-332.

DiCiccio, T. J. and Stern, S. E. (1994). Frequentist and Bayesian Bartlett Correction of Test Statistics Based on Adjusted Profile Likelihood. *Journal of the Royal Statistical Society, Series B* **56**, 397-408.

Dickey, J. M. (1974). Bayesian Alternatives to the F-test and Least-Squares Estimator in the Normal Linear Model. In *Studies in Bayesian Econometrics and Statistics*, S. Fienberg and A. Zellner (eds.). Amsterdam: North Holland, pp.515-554.

Diebolt J. and Ip, E. H. S. (1996). Stochastic EM: Method and application. In *Markov Chain Monte Carlo in Practice*, W. R. Gilks, S. Richardson and D. J. Spiegelhalter (eds.). New York: Chapman & Hall, pp.259-273.

Diebolt, J. and Robert, C. P. (1994). Estimation of Finite Mixture Distributions through Bayesian Sampling. *Journal of the Royal Statistical Society, Series B* **56**, 363-375.

Dieter, U. (1975). Statistical Interdependence of Pseudo-Random Numbers Generated by the Linear Congruential Method. *Applications of Number Theory to Numerical Analysis*, S. K. Zaremba (ed.). San Diego: Academic Press, 287-318.

Dieter, U. and Ahrens, J. H. (1971). An Exact Determination of Serial Correlations of Pseudo-Random Numbers. *Numerische Mathematik* **17**, 101-123.

Dieter, U. and Ahrens, J. H. (1973).   A Combinatorial Method for Generation of Normally Distributed Random Variables. *Computing* **11**, 137-146.

Dobson, A. J. (1990). *An Introduction to Generalized Linear Models.* New York: Chapman & Hall.

Doeblin, W. (1940). Éléments d'une Théorie Générale des Chaînes Simples Constantes de Markoff. *Annales Scientifiques de l'Ecole Normale Supérieure* **57**, 61-111.

Doob, J. L. (1990). *Stochastic Processes.* New York: Wiley & Sons.

Doob, J. L. (1996). The Development of Rigor in Mathematical Probability (1900-1950). *American Mathematical Monthly* **103**, 586-595.

Downham, D. Y. (1970). The Runs Up and Test. *Applied Statistics* **19**, 190-192.

Downham, D. Y. and Roberts, F. D. K. (1967).   Multiplicative Congruential Pseudo-Random Number Generators. *Computer Journal* **10**, 74-77.

Draper, D. (1995). Assessment and Propagation of Model Uncertainty. *Journal of the Royal Statistical Society, Series B* **57**, 45-97.

Draper, D., Hodges, J. S., Mallows, C. L. and Pregibon, D. (1993).   Exchangeability and Data Analysis. *Journal of the Royal Statistical Society, Series A* **156**, 9-37.

Duane, S., Kennedy, A. D., Pendleton, B. J. and Roweth, D. (1987).   Hybrid Monte Carlo. *Physics Letters B* **195**, 216-222.

Dudewicz, E. J. (1975). Random Numbers: The Need, the History, the Generators. In *A Modern Course on Statistical Distributions in Scientific Work*, Volume 2, G. P. Patil, S. Kotz and J. K. Ord (eds.). Boston: D. Reidel, pp.25-36.

Dudewicz, E. J. (1976). Speed and Quality of Random Numbers for Simulation. *Journal of Quality Technology* **8**, 171-178.

Dudley, R. M. and Haughton, D. (1997).   Information Criteria for Multiple Data Sets and Restricted Parameters. *Statistica Sinica* **7**, 265-284.

Duflo, M. (1996). *Random Iterative Models.* In Series *Applications of Mathematics*, I. Karatzas and M. Yor (eds.). New York: Springer-Verlag, Volume 34.

DuMouchel, W. and Normand, S. (2000).   Computer-Modeling and Graphical Strategies for Meta-Analysis. In *Meta-Analysis in Medicine and Health Policy*, D. K. Stangl and D. A. Berry (eds.).   New York: Marcel Dekker, pp.127-78.

Earman, J. (1992). *Bayes or Bust: A Critical Examination of Bayesian Confirmation Theory.* Cambridge, MA: MIT Press.

Eaves, D. M. (1985).   On Maximizing Missing Information about a Hypothesis. *Journal of the Royal Statistical Society, Series B* **47**, 263-266.

Ebrahimi, N., Habibullah, M. and Soofi, E. S. (1992).   Testing Exponentiality Based on Kullback-Leibler Information. *Journal of the Royal Statistical Society, Series B* **54**, 739-748.

Edgeworth, F. Y. (1892a). Correlated Averages. *Philosophical Magazine*, 5th Series, **34**, 190-204.

Edgeworth, F. Y. (1892b). The Law of Error and Correlated Averages. *Philosophical Magazine*, 5th Series, **34**, 429-438 & 518-526.

Edgeworth, F. Y. (1893a). Exercises in the Calculation of Errors. *Philosophical Magazine*, 5th Series, **36**, 98-111.

Edgeworth, F. Y. (1893b). Note on the Calculation of Correlation Between Organs. *Philosophical Magazine*, 5th Series, **36**, 350-351.

Edgeworth, F. Y. (1921). Molecular Statistics. *Journal of the Royal Statistical Society* **84**, 71-89.

Edwards, A. W. F. (1992). *Likelihood.* Baltimore: Johns Hopkins University Press.

Edwards, R. G. and Sokal, A. D. (1988).   Generalization of the Fortuin-Kasteleyn-Swendsen-Wang Representation and Monte Carlo Algorithm. *Physical Review Letters* **38**, 2009-2012.

Edwards, W., Lindman, H. and Savage, L. J. (1963). Bayesian Statistical Inference for Psychological Research. *Psychological Research* **70**, 193-242.

Efron, B. (1978). The Geometry of Exponential Families. *Annals of Statistics* **6**, 362-376.

Efron, B. (1979). Bootstrap Methods: Another Look at the Jackknife. *The Annals of Statistics* **7**, 1-26.

Efron, B. (1986). Why Isn't Everyone a Bayesian? *The American Statistician* **40**, 1-11.

Efron, B. (1998). R. A. Fisher in the 21st Century. *Statistical Science* **13**, 95-122.

Efron, B. and Hinkley, D. V. (1978). Assessing the Accuracy of the Maximum Likelihood Estimator: Observed Versus Expected Fisher Information. *Biometrika* **65**, 457-482.

Efron, B. and Morris, C. N. (1971). Limiting the Risk of Bayes and Empirical Bayes Estimators, Part I: The Bayes Case. *Journal of the American Statistical Association* **66**, 807-815.

Efron, B. and Morris, C. N. (1972a). Limiting the Risk of Bayes and Empirical Bayes Estimators, Part II: The Empirical Bayes Case. *Journal of the American Statistical Association* **67**, 130-139.

Efron, B. and Morris, C. N. (1972b). Empirical Bayes on Vector Observations: An Extension of Stein's Method. *Biometrika* **59**, 335-347.

Efron, B. and Morris, C. N. (1973). Stein's Estimation Rule and Its Competitors: An Empirical Bayes Approach. *Journal of the American Statistical Association* **68**, 117-130.

Efron, B. and Morris, C. N. (1975). Data Analysis Using Stein's Estimator and Its Generalizations. *Journal of the American Statistical Association* **70**, 311-319.

Efron, B. and Morris, C. N. (1976). Multivariate Empirical Bayes and Estimation of Covariance Matrices. *Annals of Statistics* **4**, 22-32.

Efron, B. and Morris, C. N. (1977). Stein's Paradox In Statistics. *Scientific American* **May**, 119-127.

Efron, B. and Tibshirani, R. J. (1993). *An Introduction to the Bootstrap.* New York: Chapman & Hall.

Eglese, R. W. (1990). Simulated Annealing: A Tool for Operational Research. *European Journal of Operational Research* **46**, 271-281.

Eichenauer-Herrmann, J. (1996). Equidistribution Properties of Inversive Congruential Pseudorandom Numbers With Power of Two Modulus. *Metrika* **44**, 199-205.

Eliason, S. R. (1983). *Maximum Likelihood Estimation: Logic and Practice.* Thousand Oaks, CA: Sage.

Emerson, J. D. and Hoaglin, D. C. (1983). Resistant Lines for y Versus x. In *Understanding Robust and Exploratory Data Analysis*, D. C. Hoaglin, Frederick Mosteller and John Tukey (eds.). New York: Wiley & Sons, pp.129-165.

Escobar, M. D. (1994). Estimating Normal Means with a Dirichlet Process Prior. *Journal of the American Statistical Association* **89**, 268-277.

Escobar, M. D. and West, M. (1995). Bayesian Density Estimation and Inference Using Mixtures. *Journal of the American Statistical Association* **90**, 577-588.

Espelid, T. O. and Genz, A. (1992). *Numerical Integration: Recent Developments, Software, and Applications.* Dordrecht, NL: Kluwer.

Estlin, T., Volpe, R., Nesnas, I. A. D., Mutz, D., Fisher, F., Englehardt, B. and Chien, S. (2001). Decision-Making in a Robotic Architecture for Autonomy. *The 6th International Symposium on Artificial Intelligence, Robotics, and Automation in Space*, Montreal Canada, 18-21.

Evans, M. J., Fraser, D. A. S. and Monette, G. (1986). On Principles and Arguments to Likelihood. *Canadian Journal of Statistics* **14**, 181-199.

Evans, M. J., Hastings, N. and Peacock, B. (2000). *Statistical Distributions.* New York: John Wiley & Sons.

Evans, M. J. and Swartz, T. (1995). Methods for Approximating Integrals in Statistics with Special Emphasis on Bayesian Integration Problems. *Statistical Science* **10**, 254-272.

Evans, S. J. W. (1994). Discussion of the Paper by Spiegelhalter, Freedman, and Parmar. *Journal of the Royal Statistical Society, Series A* **157**, 395.

Faber, J. (1989). *Annual Data on Nine Economic and Military Characteristics of 78 Nations (SIRE NATDAT), 1948-1983*. Ann Arbor: Inter-University Consortium for Political and Social Research and Amsterdam, and Amsterdam, the Netherlands: Europa Institute, Steinmetz Archive.

Fabius, J. (1964). Asymptotic Behavior of Bayes' Estimates. *Annals of Mathematical Statistics* **35**, 846-856.

Fahrmeir, L. and Kaufmann, H. (1985). Consistency and Asymptotic Normality of the Maximum Likelihood Estimator in Generalized Linear Models. *The Annals of Statistics* **13**, 342-368.

Fahrmeir, L. and Tutz, G. (2001). *Multivariate Statistical Modelling Based on Generalized Linear Models*. Second Edition. New York: Springer.

Falk, M. (1999). A Simple Approach to the Generation of Uniformly Distributed Random Variables With Prescribed Correlations. *Communications in Statistics, Part B, Simulation and Computation* **28**, 785-791.

Falk, R. and C. W. Greenbaum. (1995). Significance Tests Die Hard. *Theory and Psychology* **5**, 396-400.

Falkenrath, R. (2001). Analytical Models and Policy Prescription: Understanding Recent Innovation in U.S. Counterterrorism. *Studies in Conflict and Terrorism* **24**, 159-181.

Fang, K.-T., Hickernell, F. J. and Niederreiter, H. (2002). *Monte Carlo and Quasi-Monte Carlo Methods*. New York: Springer-Verlag.

Fang, K.-T., Kotz, S. and Ng, K. W. (1990). *Symmetric Multivariate and Related Distributions*. New York: Chapman & Hall.

Farnum, N. R. and Stanton, L. W. (1987). Some Results Concerning the Estimation of Beta Distribution Parameters in PERT. *Journal of the Operational Research Society* **38**, 287-90.

Fearnhead, P. (2006). Exact and Efficient Bayesian Inference for Multiple Changepoint Problems. *Statistics and Computing* **16**, 203-213.

Feller, W. (1990). *An Introduction to Probability Theory and its Applications*. Volume 1. New York: John Wiley & Sons.

Feller, W. (1990). *An Introduction to Probability Theory and its Applications*. Volume 2. New York: John Wiley & Sons.

Ferguson, T. S. (1967). *Mathematical Statistics: A Decision-Theoretic Approach*. San Diego: Academic Press.

Ferguson, T. S. (1973). A Bayesian Analysis of Some Nonparametric Problems. *Annals of Statistics* **1**, 209-230.

Ferguson, T. S. (1983). Bayesian Density Estimation by Mixtures of Normal Distributions. In *Recent Advances in Statistics*, H. Rizvi and J. Rustagi (eds.). San Diego: Academic Press, pp.287-302.

Ferguson, T. S. and Phadia, E. G. (1979). Bayesian Nonparametric Estimation Based on Censored Data. *The Annals of Statistics* **7**, 163-186.

Ferrari, P. A., Frigessi, A. and Schonmann, R. J. (1993). Convergence of Some Partially Parallel Gibbs Samplers with Annealing. *Annals of Applied Probability* **3**, 137-153.

Ferreira, M. A. R. and Gamerman, D. (2000). Dynamic Generalized Linear Models. In *Generalized Linear Models: A Bayesian Perspective*, Dipak K. Dey, Sujit K. Ghosh and Bani K. Mallick (eds.). New York: Marcel Dekker, pp.41-53.

Fienberg, S. E. (2006). When Did Bayesian Inference Become Bayesian? *Bayesian Analysis* **1**, 1-40.

Fill, J. A. (1991). Eigenvalue Bounds on Convergence to Stationarity for Nonreversible Markov Chains with an Application to the Exclusion Process. *Annals of Applied Probability* **1**, 62-67.

Fill, J. A. (1998). An Interruptible Algorithm for Perfect Sampling Via Markov Chains. *Annals of Applied Probability* **8**, 131-162.

Fill, J. A., Machida, M., Murdoch, D. J., Rosenthal, J. S. (1999). Extensions of Fill's Perfect Rejection Sampling Algorithm to General Chains. *Random Structures and Algorithms* **17**, 290-316.

Firth, D. (1987). On the Efficiency of Quasi-Likelihood Estimation. *Biometrika* **74**, 233-245.

Fisher, R. A. (1922). On the Mathematical Foundations of Theoretical Statistics. *Philosophical Transactions of the Royal Statistical Society of London A* **222**, 309-360.

Fisher, R. A. (1925a). *Statistical Methods for Research Workers*. Edinburgh: Oliver and Boyd.

Fisher, R. A. (1925b). Theory of Statistical Estimation. *Proceedings of the Cambridge Philosophical Society* **22**. 700-725.

Fisher, R. A. (1930). Inverse Probability. *Proceedings of the Cambridge Philosophical Society* **26**. 528-535.

Fisher, R. A. (1934). *The Design of Experiments*. First Edition. Edinburgh: Oliver and Boyd.

Fisher, R. A. (1935). The Fiducial Argument in Statistical Inference. *Annals of Eugenics* **6**, 391-398.

Fisher, R. A. (1956). *Statistical Methods and Scientific Inference*. Second Edition. New York: Hafner.

Fishman, G. S. and Moore, L. R. (1982). A Statistical Evaluation of Multiplicative Congruential Random Number Generators With Modulus $2^{31} - 1$. *Journal of the American Statistical Association* **77**, 129-136.

Fishman, G. (2003). *Monte Carlo*. New York: Springer-Verlag.

Flegal, J. M. (2012). Applicability of Subsampling Bootstrap Methods In Markov Chain Monte Carlo. In *Monte Carlo and Quasi-Monte Carlo Methods 2010*, Leszek Plaskota and Henryk Wozniakowski (eds.). Springer Berlin Heidelberg, pp.363-372.

Fleischer, M. A. (1995). Simulated Annealing: Past, Present, and Future. *Proceedings of the 27th Winter Simulation Conference* **00**, 155-161.

Flournoy, N. and Tsutakawa, R. K. (eds.). (1991). *Statistical Multiple Integration*. Providence: American Mathematical Society.

Fomby, T. B., Hill, R. C. and Johnson, S. R. (1980). *Advanced Econometric Methods*. New York: Springer-Verlag.

Fort, G., Moulines, E. and Priouret, P. (2011). Convergence of Adaptive and Interacting Markov Chain Monte Carlo Algorithms. *The Annals of Statistics* **39**, 3262-3289.

Fortini, S. and Ruggeri, F. (2000). On the Use of the Concentration Function in Bayesian Robustness. In *Robust Bayesian Analysis*, David Ríos Insua and Fabrizio Ruggeri (eds.). New York: Springer-Verlag, pp.109-126.

Fox, L. (1971). How to Get Meaningless Answers in Scientific Computation and What to Do About It. *IMA Bulletin* **7**, 296-302.

Fraser, D. A. S. (1963). On Sufficiency and the Exponential Family. *Journal of the Royal Statistical Society, Series B* **25**, 115-123.

Fredenhagen, K. and Marcu, M. (1987). A Modified Heat-Bath Method Suitable for Monte Carlo Simulations on Vector and Parallel Processors. *Physical Letters, B* **193**, 486-488.

Freedman, D. A. (1963). On the Asymptotic Behavior of Bayes' Estimates in the Discrete Case. *Annals of Mathematical Statistics* **34**, 1386-1403.

Freedman, D. A. (1965). On the Asymptotic Behavior of Bayes Estimates in the Discrete Case II. *Annals of Mathematical Statistics* **36**, 454-456.

Freedman, D. A. and Diaconis, P. (1982). de Finetti's Theorem for Symmetric Location Families. *Annals of Statistics* **10**, 184-189.

Freedman, L. S. and Spiegelhalter, D. J. (1983). The Assessment of Subjective Opinion and Its Use in Relation to Stopping Rules for Clinical Trials. *Statistician* **32**, 153-160.

Freeman, D. (1983). *Margaret Mead and Samoa: The Making and Unmaking of an Anthropological Myth*. Cambridge, MA: Harvard University Press.

Frieze, A., Kannan, R. and Polson, N. G. (1994). Sampling from Log-Concave Distributions. *Annals of Applied Probability* **4**, 812-837.

Frigessi, A., Hwang, C.-R., Di Stefano, P. and Sheu, S.-J. (1993). Convergence Rates of the Gibbs Sampler, the Metropolis Algorithm, and Other Single-Site Updating Dynamics. *Journal of the Royal Statistical Society, Series B* **55**, 205-220.

Fuller, A. T. (1976). The Period of Pseudo-Random Numbers Generated by Lehmer's Congruential Method. *Computer Journal* **19**, 173-177.

Fulman, J. and Wilmer, E. L. (1999). Comparing Eigenvalue Bounds for Markov Chains: When Does Poincare Beat Cheeger? *Annals of Applied Probability* **9**, 1-13.

Galant, D. (1969). Gauss Quadrature Rules for the Evaluation of $2\pi^{-1/2} \int_0^\infty \exp(-x^2) f(x) dx$. *Mathematics of Computation* **23**, 674.

Galton, F. (1869). *Heredity Genius: An Inquiry into its Laws and Consequences*. Second Edition. London: Macmillan.

Galton, F. (1875). Statistics by Intercomparison, with Remarks on the Law of Frequency of Error. *Philosophical Magazine*, 4th Series (**49**), 33-46.

Galton, F. (1886). Regression Towards Mediocrity in Hereditary Stature. *Journal of the Anthropological Institute* **15**, 246-263.

Galton, F. (1892). *Finger Prints*. London: Macmillan.

Gamerman, D. and Lopes, H. F. (2006). *Markov Chain Monte Carlo*. Second Edition. New York: Chapman & Hall.

Garip, F. and Western, B. (2011). Model Comparison and Simulation for Hierarchical Models: Analyzing Rural-Urban Migration in Thailand. In *Handbook of Markov Chain Monte Carlo*, Steve Brooks, Andrew Gelman, Galin L. Jones and Xiao-Li Meng (eds.), 563-574. Boca Raton: Chapman & Hall/CRC. pp.563-574.

Garthwaite, P. H. and Dickey, J. M. (1988). Quantifying Expert Opinion in Linear Regression Problems. *Journal of the Royal Statistical Society, Series B* **50**, 462-474.

Garthwaite, P. H. and Dickey, J. M. (1992). Elicitation of Prior Distributions for Variable Selection Problems in Regression. *Annals of Statistics* **20**, 1697-1719.

Gates, C. E. (1978). On Generating Random Normal Deviates Using the Butler Algorithm. *Proceedings of the Statistical Computing Section*. Alexandria, VA: American Statistical Association, pp.111-114.

Gauss, C. F. (1809). *Theoria Motus Corporum Coelestium*. Hamburg: Perthes et Besser.

Gauss, C. F. (1823). *Theoria Combinationis Observationum Erroribus Minimis Obnoxiae*. Göttingen: Königlichen Gesellschaft der Wissenschaften.

Gauss, C. F. (1855). *Méthode des Moindres Carrés. Mémoires sur la Combination des Observations*. Translated by J. Bertrand. Paris: Mallet-Bachelier.

Gavasakar, U. (1988). A Comparison of Two Elicitation Methods for a Prior Distribution for a Binomial Parameter. *Management Science* **34**, 784-790.

Gawande, K. (1998). Comparing Theories of Endogenous Protection: Bayesian Comparison of Tobit Models Using Gibbs Sampling Output. *Review of Economics and Statistics* **80**, 128-140.

Gebhardt, F. (1964). On the Risk of Some Strategies for Outlying Observations. *The Annals of Mathematical Statistics* **35**, 1524-1536.

Geisser, S. and Eddy, W. F. (1979). A Predictive Approach to Model Selection. *Journal of the American Statistical Association* **74**, 153-160.

Gelfand, A. E. and Dey, D. K. (1991). On Bayesian Robustness of Contaminated Classes of Priors. *Statistical Decisions* **9**, 63-80.

Gelfand, A. E. and Mitter, S. K. (1993). Metropolis-type Annealing Algorithms for Global Optimization in $\Re^d$. *SIAM Journal of Control and Optimization* **31**, 111-131.

Gelfand, A. E. and Sahu, S. K. (1994). On Markov Chain Monte Carlo Acceleration. *Journal of Computational and Graphical Statistics* **3**, 261-276.

Gelfand, A. E. and Smith, A. F. M. (1990). Sampling Based Approaches to Calculating Marginal Densities. *Journal of the American Statistical Association* **85**, 398-409.

Gelfand, A. E., Sahu, S. K. and Carlin, B. P. (1995). Efficient Parameterizations For Normal Linear Mixed Models. *Biometrika* **82**, 479-488.

Gelman, A. (1992). Iterative and Non-Iterative Simulation Algorithms. *Computing Science and Statistics* **24**, 433-438.

Gelman, A. (1996). Inference and Monitoring Convergence. In *Markov Chain Monte Carlo in Practice*, W. R. Gilks, S. Richardson and D. J. Spiegelhalter (eds.). New York: Chapman & Hall, pp.131-144.

Gelman, A. (2005). Analysis of Variance: Why It Is More Important Than Ever. *Annals of statistics* **33**, 1-53.

Gelman, A. (2006). Prior Distributions for Variance Parameters in Hierarchical Models. *Bayesian Analysis* **1**, 515-533.

Gelman, A., Gilks, W. R. and Roberts, G. O. (1996). Efficient Metropolis Jumping Rules. In *Bayesian Statistics 5*, J. O. Berger, J. M. Bernardo, A. P. Dawid, D. V. Lindley and A. F. M. Smith (eds.). Oxford: Oxford University Press, pp.599-608.

Gelman, A. and Hill, J. (2007). *Data Analysis Using Regression and Multilevel/Hierarchical Models*. Cambridge: Cambridge University Press.

Gelman, A., Meng, X-L. and Stern, H. S. (1996). Posterior Predictive Assessment of Model Fitness Via Realized Discrepancies. *Statistica Sinica* **6**, 733-807.

Gelman, A. and Rubin, D. B. (1992a). Inference from Iterative Simulation Using Multiple Sequences. *Statistical Science* **7**, 457-472.

Gelman, A. and Rubin, D. B. (1992b). A Single Sequence from the Gibbs Sampler Gives a False Sense of Security. In *Bayesian Statistics 5*, J. O. Berger, J. M. Bernardo, A. P. Dawid, D. V. Lindley and A. F. M. Smith (eds.). Oxford: Oxford University Press, pp.223-253.

Gelman, A. and Rubin, D. B. (1995). Avoiding Model Selection in Bayesian Social Research *Sociological Methodology* 25, 165-173.

Gelman, A., Carlin, J. B., Stern, H. S. and Rubin, D. B. (2003). *Bayesian Data Analysis*. Second Edition. New York: Chapman & Hall.

Gelman, A., Huang, Z., van Dyk, D. A. and Boscardin, J. W. (2008). Using Redundant Parameterizations to Fit Hierarchical Models. *Journal of Computational and Graphical Statistics* **17**, 95-122.

Geman, S. and Geman, D. (1984). Stochastic Relaxation, Gibbs Distributions and the Bayesian Restoration of Images. *IEEE Transactions on Pattern Analysis and Machine Intelligence* **6**, 721-741.

Geman, S. and Hwang, C-R. (1986). Diffusions for Global Optimization. *SIAM Journal of Control and Optimization* **24**, 1031-1043.

Gentle, J. E. (1990). Computer Implementation of Random Number Generators. *Journal of Computational and Applied Mathematics* **31**, 119-125.

Gentle, J. E. (1998). *Random Number Generation and Monte Carlo Methods*. New York: Springer-Verlag.

George, E. I. and McCulloch, R. E. (1993). Variable Selection via Gibbs Sampling. *Journal of the American Statistical Association* **85**, 398-409.

George, E. I., Makov, U. E. and Smith, A. F. M. (1993). Conjugate Likelihood Distributions. *Scandinavian Journal of Statistics* **20**, 147-156.

George, R. (1963). Normal Random Variables. *Communications of the Association for Computing Machinery* **6**, 444.

Geweke, J. (1989). Bayesian Inference in Econometric Models Using Monte Carlo Integration. *Econometrica* **57**, 1317-1339.

Geweke, J. (1992). Evaluating the Accuracy of Sampling-Based Approaches to the Calculation of Posterior Moments. In *Bayesian Statistics 4*, J. M. Bernardo, A. F. M. Smith, A. P. Dawid and J. O. Berger (eds.). Oxford: Oxford University Press, pp.169-193.

Geweke, J. (1993). Bayesian Treatment of the Independent Student-t Linear Model. *Journal of Applied Econometrics* **8**, S19-S40.

Geweke, J. (2005). *Contemporary Bayesian Econometrics and Statistics.* New York: Wiley & Sons.

Geweke, J. and Petrella, L. (1998). Prior Density-ratio Class Robustness in Econometrics. *Journal of Business and Economic Statistics* **16**, 469-478.

Geyer, C. J. (1991). Markov Chain Monte Carlo Maximum Likelihood. *Computing Science and Statistics, Proceedings of the 23rd Symposium on the Interface,* 156-63.

Geyer, C. J. (1992). Practical Markov Chain Monte Carlo. *Statistical Science* **7**, 473-511.

Geyer, C. J. (1995). Conditioning in Markov Chain Monte Carlo. *Journal of Computational and Graphical Statistics* **4**, 148-154.

Geyer, C. J. (2011). Importance Sampling, Simulated Tempering, and Umbrella Sampling. In *Handbook of Markov Chain Monte Carlo*, Steve Brooks, Andrew Gelman, Galin L. Jones and Xiao-Li Meng (eds.). Boca Raton: Chapman & Hall/CRC, pp.3-47.

Geyer, C. J. and Thompson, E. A. (1995). Annealing Markov Chain Monte Carlo With Applications to Ancestral Inference. *Journal of the American Statistical Association* **90**, 909-920.

Ghosh, J. K. (1969). Only Linear Transformations Preserve Normality. *Sankhyā, Series A* **31**, 309-312.

Ghosh, M. and Meeden, G. (1997). *Bayesian Methods for Finite Population Sampling.* New York: Chapman & Hall.

Ghosh, M. and Mukerjee, R. (1992). Hierarchical and Empirical Bayes Multivariate Estimation. In *Current Issues in Statistical Inference: Essays in Honor of D. Basu*, M. Ghosh and P. K. Pathak (eds.). Hayward, CA: Institute of Mathematical Statistics Monograph Series 17, pp.1-12.

Giakoumatos, S. G., Delaportas, P. and Politis, D. N. (2005). Bayesian Analysis of the Unobserved ARCH model. *Statistics and Computing* **15**, 103-111.

Gidas, B. (1985). Nonstationary Markov Chains and Convergence of the Annealing Algorithm. *Journal of Statistical Physics* **39**, 73-131.

Gigerenzer, G. (1987). Probabilistic Thinking and the Fight Against Subjectivity. In Krüger, Lorenz, Gerd Gigerenzer and Mary Morgan, eds. *The Probabilistic Revolution.* Volume 2. Cambridge, MA: MIT Press.

Gigerenzer, G. (1998a). We Need Statistical Thinking, Not Statistical Rituals. *Behavioral and Brain Sciences* **21**, 199-200.

Gigerenzer, G. (1998b). The Superego, the Ego, and the Id in Statistical Reasoning. In G. Keren and C. Lewis, eds. *A Handbook for Data Analysis in the Behavioral Sciences: Methodological Issues.* Hillsdale, NJ: Lawrence Erlbaum Associates.

Gigerenzer, G. (2004). Mindless Statistics. *The Journal of Socio-Economics* **33**, 587-606.

Gigerenzer, G. and D. J. Murray. (1987). *Cognition as Intuitive Statistics.* Hillsdale, NJ: Lawrence Erlbaum Associates.

Gilks, W. R. (1992). Derivative-Free Adaptive Rejection Sampling for Gibbs Sampling. In *Bayesian Statistics 4*, J. O. Berger, J. M. Bernardo, A. P. Dawid and A. F. M. Smith (eds.). Oxford: Oxford University Press, pp.641-649.

Gilks, W. R. (1996). Full Conditional Distributions. In *Markov Chain Monte Carlo in Practice*, W. R. Gilks, S. Richardson and D. J. Spiegelhalter (ed.). New York: Chapman & Hall, pp.75-88.

Gilks, W. R., Best, N. G. and Tan, K. K. C. (1995). Adaptive Rejection Metropolis Sampling Within Gibbs Sampling. *Applied Statistics* **44**, 455-472.

Gilks, W. R. and Roberts, G. O. (1996). Strategies for Improving MCMC. In *Markov Chain Monte Carlo in Practice*, W. R. Gilks, S. Richardson and D. J. Spiegelhalter (eds.). New York: Chapman & Hall, pp.89-114.

Gilks, W. R., Roberts, G. O. and Sahu, S. K. (1998). Adaptive Markov Chain Monte Carlo Through Regeneration. *Journal of the American Statistical Association* **93**, 1045-1054.

Gilks, W. R., Thomas, A. and Spiegelhalter, D. J. (1994). A Language and Program for Complex Bayesian Modelling. *The Statistician* **43**, 169-177.

Gilks, W. R. and Wild, P. (1992). Adaptive Rejection Sampling for Gibbs Sampling. *Applied Statistics* **41**, 337-348.

Gill, J. (1999). The Insignificance of Null Hypothesis Significance Testing. *Political Research Quarterly* **52**, 647-674.

Gill, J. (2000). *Generalized Linear Models: A Unified Approach.* Thousand Oaks, CA: Sage.

Gill, J. (2006). *Essential Mathematics for Political and Social Research.* Cambridge, England: Cambridge University Press.

Gill, J. (2010). Critical Differences in Bayesian and Non-Bayesian Inference. In *Current Methodological Developments of Statistics in the Social Sciences.* Stanislav Kolenikov, Lori Thombs and Douglas Steinley (eds.). 135-158. New York: John Wiley & Sons, pp.383-408.

Gill, J. and Casella, G. (2007). Nonparametric Priors For Ordinal Bayesian Social Science Models: Specification and Estimation. *Journal of the American Statistical Association* **104**, 453-464

Gill, J. and Casella, G. (2004). Dynamic Tempered Transitions for Exploring Multimodal Posterior Distributions. *Political Analysis*, **12**, 425-433.

Gill, J. and Freeman, J. (2013). Dynamic Elicited Priors for Updating Covert Networks. *Network Science*, **1**, 68-94.

Gill, J. and King, G. (2004). What to Do When Your Hessian Is Not Invertible: Alternatives to Model Respecification in Nonlinear Estimation. *Sociological Methods and Research* **33**, 54-87.

Gill, J. and Meier, K. J. (2000). Public Administration Research and Practice: A Methodological Manifesto. *Journal of Public Administration Research and Theory*, **10**, 157-200.

Gill, J. and Walker, L. (2005). Elicited Priors for Bayesian Model Specifications in Political Science Research. *Journal of Politics* **67**, 841-872.

Gill, J. and Witko C. (2013). Bayesian Analytical Methods: A Methodological Prescription for Public Administration. *Journal of Public Administration Research and Theory* **23**, 457-494.

Gill, J. and Womack, A. (2013). The Multilevel Model Framework. In *The Sage Handbook of Multilevel Modeling.* Marc A. Scott, Jeffrey S. Simonoff and Brian D. Marx (eds.). Thousand Oaks, CA: Sage Publications, pp.3-20.

Gill, P., Murray, W. and Wright, M. H. (1981). *Practical Optimization.* San Diego: Academic Press.

Gliner, J. A., Leech, N. L. and Morgan, G. A. (2002). Problems With Null Hypothesis Significance Testing (NHST): What Do the Textbooks Say? *The Journal of Experimental Education* **71**, 83-92.

Goel, P. K. (1983). Information Measures and Bayesian Hierarchical Models. *Journal of the American Statistical Association* **78**, 408-410.

Goel, P. K. and DeGroot, M. H. (1979). Comparison of Experiments and Information Measures. *Annals of Statistics* **7**, 1066-1077.

Goel, P. K. and DeGroot, M. H. (1980). Only Normal Distributions Have Linear Posterior Expectations in Linear Regression. *Journal of the American Statistical Association* **75**, 895-900.

Goel, P. K. and DeGroot, M. H. (1981). Information About Hyperparameters in Hierarchical Models. *Journal of the American Statistical Association* **76**, 140-147.

Goldberger, A. S. (1964). *Econometric Theory.* New York: Wiley & Sons.

Golder, E. R. (1976). The Spectral Test for the Evaluation of Congruential Pseudo-Random Generators. *Applied Statistics* **25**, 173-180.

Golder, E. R. and Settle, J. G. (1976). The Box-Müller Method for Generating Pseudo-Random Normal Deviates. *Applied Statistics* **25**, 12-20.

Goldstein, H. (1985). *Multilevel Statistical Models*. New York: Halstead Press.

Goldstein, H. (1986). Multilevel Mixed Linear Model Analysis Using Iterative Generalized Least Squares. *Biometrika* **73**, 43-56.

Goldstein, H. (1987). *Multilevel Models in Education and Social Research*. Oxford: Oxford University Press.

Golstein, H. (2003). *Multilevel Statistical Models*. Third Edition. Oxford: Oxford University Press, Kendall Library of Statistics 3.

Goldstein, H. and Rasbash, J. (1996). Improved Approximations for Multilevel Models with Binary Responses. *Journal of the Royal Statistical Society, Series A* **159**, (1996), 505-513.

Goldstein, M. (1991). Belief Transforms and the Comparison of Hypotheses. *The Annals of Statistics* **19**, 2067-2089.

Golomb, S. W. (1967). *Shift Register Sequences*. San Francisco: Holden-Day.

Golub, G. H. and Van Loan, C. F. (1996). *Matrix Computations*. Third Edition. Baltimore: Johns Hopkins University Press.

Golub, G. H. and Welsch, J. H. (1969). Calculation of Gauss Quadrature Rules. *Mathematics of Computation* **23**, 221-230.

Gönen, M., Johnson, W. O., Yonggang, L. and Westfall, P. H. (2005). The Bayesian Two-Sample t Test. *The American Statistician* **59**, 252-257.

Good, I. J. (1950). *Probability and the Weighting of Evidence*. London: Griffin.

Good, I. J. (1957). On the Serial Test for Random Sequences. *Annals of Mathematical Statistics* **28**, 262-264.

Good, I. J. (1972). Statistics and Today's Problems. *American Statistician* **26**, 11-19.

Good, I. J. (1976). The Bayesian Influence, or How to Sweep Subjectivism Under the Carpet. In *Foundations of Probability Theory, Statistical Inference, and Statistical Theories of Science II*, William L. Harper and Clifford A. Hooker (eds.). Dordrecht: D. Reidel, pp.125-174.

Good, I. J. (1980a). Some History of the Hierarchical Bayes Methodology. In *Bayesian Statistics 2*, J. M. Bernardo, M. H. DeGroot, D. V. Lindley and A. F. M. Smith (ed.). Amsterdam: North Holland, pp.489-515.

Good, I. J. (1980b). The Contributions of Jeffreys to Bayesian Statistics. In *Bayesian Analysis in Econometrics and Statistics: Essays in Honor of Harold Jeffreys*. Arnold Zellner (ed.). Amsterdam: North Holland.

Good, I. J. (1983a). The Bayes/Non-Bayes Compromise or Synthesis and Box's Current Philosophy. *Journal of Statistical Computation and Simulation* **18**, 234-236.

Good, I. J. (1983b). *Good Thinking: The Foundations of Probability and Its Applications*. Minneapolis: University of Minnesota Press.

Good, I. J. (1985). Weight of Evidence: A Brief Survey. In *Bayesian Statistics 2*, J. M. Bernardo, M. H. DeGroot, D. V. Lindley and A. F. M. Smith (eds.). Amsterdam: North Holland Press, pp.249-270.

Good, I. J. (1992). The Bayes/Non-Bayes Compromise: A Brief Review. *Journal of the American Statistical Association* **87**, 597-606.

Good, I. J. and Crook, J. F. (1974). The Bayes/Non-Bayes Compromise and the Multinomial Distribution. *Journal of the American Statistical Association* **69**, 711-720.

Goodman, J. and Sokal, A. D. (1989). Multigrid Monte Carlo Method. Conceptual Foundations. *Physics Review D* **40**, 2035-2071.

Goodman, S. N. (1993). P values, Hypothesis Tests, and Likelihood: Implications for Epidemiology of a Neglected Historical Debate. *American Journal of Epidemiology* **137**, 485-496.

Goodman, S. N. (1999). Toward Evidence-Based Medical Statistics. 1: The P Value Fallacy. *Annals of Internal Medicine* **130**, 995-1004.

Gorenstein, S. (1967). Testing a Random Number Generator. *Communications of the Association for Computing Machinery* **10**, 111-118.

Gossett, W. S. (1908a). The Probable Error of a Mean. *Biometrika* **6**, 1-25.

Gossett, W. S. (1908b). Probable Error of a Correlation Coefficient. *Biometrika* **6**, 302-309.

Goutis, C. and Casella, G. (1999). Explaining the Saddlepoint Approximation. *The American Statistician* **53**, 216-224.

Granger, C. W. J. (1999). Empirical Modeling in Economics: Specification and Evaluation. Cambridge: Cambridge University Press.

Granger, G. W. and Hatanaka, M. (1964). *Spectral Analysis of Economic Time Series.* Princeton: Princeton University Press.

Gray, H. L. (1988). On Unification of Bias Reduction and Numerical Approximation. In *Probability and Statistics Essays in Honor of Franklin A. Graybill*, J. N. Srivastava (ed.). Amsterdam: North Holland, pp.105-116.

Gray, V. and Lowery, D. (1996). *The Population Ecology of Interest Representation: Lobbying Communities in the American States.* Ann Arbor, Michigan: University of Michigan Press.

Gray, V. and Lowery, D. (2001). The institutionalization of state communities of organized interests. *Political Research Quarterly* **54**, 265-284.

Green, P. J. (1984). Iteratively Reweighted Least Squares for Maximum Likelihood Estimation, and Some Robust and Resistant Alternatives. *Journal of the Royal Statistical Society, Series B* **46**, 149-192.

Green, P. J. (1995). Reversible Jump Markov Chain Monte Carlo Computation and Bayesian Model Determination. *Biometrika* **82**, 711-732.

Green, P. J. and Murdoch, D. J. (1998). Exact Sampling for Bayesian Inference: Towards General Purpose Algorithms. *Bayesian Statistics* **6**, 301-321.

Greene, W. (2011). *Econometric Analysis.* Seventh Edition. Upper Saddle River, NJ: Prentice Hall.

Greenwald, A. G. (1975). Consequences of Prejudice against the Null Hypothesis. *Psychological Bulletin* **82**, 1-20.

Greenwald, A. G., Gonzalez, R., Harris, R. and Guthrie, D. (1996). Effect Sizes and p-values: What Should Be Reported and What Should Be Replicated? *Psychophysiology* **33**, 175-183.

Greenwood, J. A. (1974). A Fast Generator for Gamma-Distributed Random Variables. In *Compstat 1974: Proceedings in Computational Statistics*, G. Bruckman, F. Ferschl and L. Schmetterer (eds.). Vienna: Physica Verlag. pp.19-27.

Grenander, U. (1983). Tutorial in Pattern Theory. Unpublished Manuscript, Division of Applied Mathematics. Providence, RI: Brown University.

Grimmett, G. R. and Stirzaker, D. R. (1992). *Probability and Random Processes.* Oxford: Oxford University Press.

Guan, Y., Fleißner, R. and Joyce, P. (2006). Markov Chain Monte Carlo in Small Worlds. *Statistics and Computing* **16**, 193-202.

Guihenneuc-Jouyaux, C. and Robert, C. P. (1998). Valid Discretization via Renewal Theory. In *Discretization and MCMC Convergence Assessment, Lecture Notes in Statistics*, **135**. Christian P. Robert (ed.). New York: Springer-Verlag, pp. 67-98.

Guo, S. W. and Thompson, E. A. (1991). Monte Carlo Estimation of Variance Component Models for Large Complex Pedigrees. *IMA Journal of Mathematics Applied in Medicine and Biology* **8**, 171-189.

Guo, S. W. and Thompson, E. A. (1994). Monte Carlo Estimation of Mixed Models for Large Complex Pedigrees. *Biometrics* **50**, 417-432.

Gupta, S. S. and Berger, J. O. (1982). *Statistical Decision Theory and Related Topics III.* San Diego: Academic Press.

Gustafson, P. (1996). Local Sensitivity of Inferences to Prior Marginals. *Journal of the American Statistical Association* **91**, 774-781.

Gustafson, P. (2000). Local Robustness in Bayesian Analysis. In *Robust Bayesian Analysis*, David Ríos Insua and Fabrizio Ruggeri (eds.). New York: Springer-Verlag, pp.71-88.

Gustafson, P. and Wasserman, L. (1995). Local Sensitivity Diagnostics for Bayesian Inference. *The Annals of Statistics* **23**, 2153-2167.

Gutiérrez-Peña, E. and Smith, A. F. M. (1995). Conjugate Parameterizations for Natural Exponential Families. *Journal of the American Statistical Association* **90**, 1347-1356.

Guttman, I. (1973). Care and Handling of Univariate or Multivariate Outliers in Detecting Spuriosity-A Bayesian Approach. *Technometrics* **15**, 723-738.

Guttman, I., Dutter, R. and Freeman, P. R. (1978). Care and Handling of Univariate Outliers in the General Linear Model to Detect Spuriosity. *Technometrics* **20**, 187-193.

Haag, J. (1924). Sur un Probleme Deneral de Probabilities et Ses Diverses Applications. *Proceedings of the International Congress of Mathematics*. Toronto, 659-674.

Haber, S. (1970). Numerical Integration of Multiple Integrals. *SIAM Review* **12**, 481-526.

Haberman, S. J. (1974a). *The Analysis of Frequency Data*. Chicago: University of Chicago Press.

Haberman, S. J. (1974b). Log-linear Models for Frequency Tables Derived by Indirect Observations: Maximum Likelihood Equations. *Annals of Statistics* **2**, 911-924.

Haberman, S. J. (1977). Product Models for Frequency Tables Involving Indirect Observations. *Annals of Statistics* **5**, 1124-1147.

Hadjicostas, P. and Berry, S. M. (1999). Improper and Proper Posteriors with Improper Priors in a Poisson-Gamma Hierarchical Model. *Test* **8**, 147-166.

Häggström, Olle. (2002). *Finite Markov Chains and Algorithmic Applications*. London: London Mathematical Society (Student Texts **52**).

Hàjek, B. (1985). A Tutorial Survey of Theory and Applications of Simulated Annealing. In *Proceedings of the 24th IEEE Conference on Decision and Control*, New York, IEEE, pp.55-760.

Hàjek, B. (1988). Cooling Schedules for Optimal Annealing. *Mathematical Operations Research* **13**, 311-329.

Haldane, J. B. S. (1938). The Estimation of the Frequency of Recessive Conditions in Man. *Annals of Eugenics* **7**, 255-262.

Haller, H. and Krauss, S. (2002). Misinterpretations of Significance: A Problem Students Share With Their Teachers. *Methods of Psychological Research* **7**, 1-20.

Halpern, E. F. (1973). Polynomial Regression from a Bayesian Approach. *Journal of the American Statistical Association* **68**, 137-143.

Halton, J. H. (1970). A Retrospective and Prospective Survey of the Monte Carlo Method. *SIAM Review* **12**, 1-63.

Hamilton, L. C. (1992). *Regression with Graphics*. Pacific Grove, CA: Brooks/Cole Publishing Company.

Hammersley, J. M. and Morton, J. M. (1956). A New Monte Carlo Techinique: Antithetical Variables. *Proceedings of the Cambridge Philosophical Society* **52**, 449-475.

Hammersley, J. M. and Handscomb, D. C. (1964). *Monte Carlo Methods*. London: Methuen.

Hampel, F. R. (1974). The Influence Curve and its Role in Robust Estimation. *Journal of the American Statistical Association* **69**, 383-393.

Hampel, F. R., Rousseeuw, P. J., Ronchetti, E. M. and Stahel, W. A. (1986). *Robust Statistics: The Approach Based on Influence Functions*. New York: John Wiley & Sons.

Han, C. and Carlin, B. P. (2001). MCMC Methods for Computing Bayes Factors: A Comparative Review. *Journal of the American Statistical Society* **96**, 1122-1132.

Hannan, M. T., Carroll, G. R., Dundon, E. A. and Torres, J. C. (1995). Organizational Evolution in a Multinational Context: Entries of Automobile Manufacturers in Belgium, Britain, France, Germany, and Italy. *American Sociological Review* **60**, 509-528.

Hanushek, E. A. and Jackson, J. E. (1977). *Statistical Methods for Social Scientists.* San Diego: Academic Press.

Haque, M. M., Chin, H. C. and Huang, H. (2010). Applying Bayesian Hierarchical Models To Examine Motorcycle Crashes At Signalized Intersections. *Accident Analysis & Prevention* **42**, 203-212.

Harlow, L. L. E., Mulaik, S. A. and Steiger, J. H. (1997). What If There Were No Significance Tests? Lawrence Erlbaum Associates Publishers.

Haro-López, R., Mallick, B. K. and Smith, A. F. M. (2000). Binary Regression Using Data Adaptive Robust Link Functions. In *Generalized Linear Models: A Bayesian Perspective*, Dipak K. Dey, Sujit K. Ghosh and Bani K. Mallick (eds.). New York: Marcel Dekker, pp.243-253.

Harris, T. E. (1956). The Existence of Stationary Measures for Certain Markov Processes. In *Proceedings of the 3rd Berkeley Symposium on Mathematical Statistics and Probability, Volume II.* Berkeley and Los Angeles: University of California Press, pp.113-124.

Hartigan, J. A. (1964). Invariant Prior Distributions. *Annals of Mathematical Statistics* **35**, 836-845.

Hartigan, J. A. (1969). Linear Bayesian Models. *Journal of the Royal Statistical Society, Series B* **31**, 446-454.

Hartigan, J. A. (1983). *Bayes Theory.* New York: Springer-Verlag.

Harrison, M. (2006) Bombers and Bystanders in Suicide Attacks in Israel, 2000 to 2003. *Studies in Conflict & Terrorism*, **29**, 187-206.

Hartley, H. O. (1958). Maximum Likelihood Estimation From Incomplete Data. *Biometrics* **14**, 174-194.

Hartley, H. O. and Hocking, R. R. (1971). The Analysis of Incomplete Data. *Biometrics* **27**, 783-808.

Harvey, A. C. 1976. Estimating Regression Models with Multiplicative Heteroscedasticity. *Econometrica* **44**, 461-5.

Hastie, D. (2005). Towards Automatic Reversible Jump Markov Chain Monte Carlo. Un-published doctoral thesis, University of Bristol, Bristol, UK.

Hastings, W. K. (1970). Monte Carlo Sampling Methods Using Markov Chains and Their Applications. *Biometrika* **57**, 97-109.

Healy, M. J. R. and Westmacott, M. (1956). Missing Values in Experiments Analyzed on Automatic Computers. *Applied Statistics* **5**, 203-206.

Heath, D. and Sudderth, W. (1976). de Finetti's Theorem on Exchangeable Variables. *The American Statistician* **30**, 188-189.

Heidelberger, P. and Welch, P. D. (1981a). Adaptive Spectral Methods for Simulation Output Analysis. *IBM Journal of Research and Development* **25**, 860-876.

Heidelberger, P. and Welch, P. D. (1981b). A Spectral Method for Confidence Interval Generation and Run Length Control in Simulations. *Communications of the Association for Computing Machinery* **24**, 233-245.

Heidelberger, P. and Welch, P. D. (1983). Simulation Run Length Control in the Presence of an Initial Transient. *Operations Research* **31**, 1109-1144.

Hellekalek, P. (1998). Good Random Number Generators Are (not so) Easy to Find. *Mathematics and Computers in Simulation* **46**, 485-505.

Hernandez, F. and Johnson, R. A. (1980). The Large-Sample Behavior of Transformations to Normality. *Journal of the American Statistical Association* **75**, 855-861.

Heumann, C. (2011). James-Stein Estimator. In International Encyclopedia of Statistical Science. Lovric, Miodrag (ed.). New York: Springer-Verlag, pp.699-701.

Heyde, C. C. and Morton, R. (1996). Quasi-Likelihood and Generalizing the EM Algorithm. *Journal of the Royal Statistical Society, Series B* **58**, 317-327.

Higdon, D. M. (1998). Auxiliary Variable Methods for Markov Chain Monte Carlo With Applications. *Journal of the American Statistical Association* **93**, 585-595.

Higham, N. J. (1996). *Accuracy and Stability of Numerical Algorithms.* Philadelphia: Society for Industrial and Applied Mathematics.

Hill, B. (1974). On Coherence, Inadmissibility, and Inference About Many Parameters in the Theory of Least Squares. In *Studies in Bayesian Econometrics and Statistics: in Honor of Leonard J. Savage*, S. Fienberg and A. Zellner (eds.). Amsterdam: North Holland, pp.555-584.

Hill, J. R. and Tsai, C-L. (1988). Calculating the Efficiency of Maximum Quasilikelihood Estimation. *Applied Statistics* **37**, 219-230.

Hill, J. and Kriesi, H. (2001). Classification by Opinion Changing Behavior: A Mixture Model Approach. *Political Analysis* **9**, 301-324.

Hills, S. E. and Smith, A. F. M. (1992). Parameterization Issues in Bayesian Inference. In *Bayesian Statistics 4*, J. O. Berger, J. M. Bernardo, A. P. Dawid and A. F. M. Smith (eds.). Oxford: Oxford University Press, pp.227-246.

Hinkley, D. V. (1987). Comment. *Journal of the American Statistical Association* **82**, 128-129.

Hipp, C. (1974). Sufficient Statistics and Exponential Families. *Annals of Statistics* **2**, 1283-1292.

Hjort, N. L. (1996). Bayesian Approaches to Non- and Semiparametric Density Estimation. In *Bayesian Statistics 5*, J. O. Berger, J. M. Bernardo, A. P. Dawid, D. V. Lindley and A. F. M. Smith (eds.). Oxford: Oxford University Press, pp.223-253.

Hoaglin, D. C., Mosteller, F. and Tukey, J. W. (1983). *Understanding Robust and Exploratory Data Analysis.* New York: Wiley & Sons.

Hobert, J. P. and Casella, G. (1996). The Effect of Improper Priors on Gibbs Sampling in Hierarchical Linear Mixed Models. *Journal of the American Statistical Association* **91**, 1461-1473.

Hobert, J. P. and Casella, G. (1998). Functional Compatibility, Markov Chains, and Gibbs Sampling with Improper Posteriors. *Journal of Computational and Graphical Statistics* **7**, 42-60.

Hobert, J. P., Robert, C. P. and Titterington, D. M. (1999). On Perfect Simulation for Some Mixtures of Distributions. *Journal of Statistics and Computing* **9**, 287-298.

Hodges, J. S. and Sargent, D. J. (2001). Counting Degrees of Freedom in Hierarchical and Other Richly Parameterized Models. *Biometrika* **88**, 367-379.

Hoel, P. G., Port, S. C. and Stone, C. J. (1987). *An Introduction to Stochastic Processes.* Prospect Heights, IL: Waveland Press.

Hoeting, J. A., Madigan, D., Raftery, A. E. and Volinsky, C. T. (1999). Bayesian Model Averaging: A Tutorial. *Statistical Science* **14**, 382-417.

Hoff, P. D. (2009). Multiplicative Latent Factor Models for Description and Prediction of Social Networks. *Computational and Mathematical Organization Theory* **15**, 261-272.

Hoff, P. D. (2005). Bilinear Mixed-effects Models for Dyadic Data. *Journal of the American Statistical Association* **100**, 286-295.

Hogarth, R. M. (1975). Cognitive Processes and the Assessment of Subjective Probability Distribution. *Journal of the American Statistical Association* **70**, 271-289.

Hogg, R. V. and Craig, A. T. (1978). *Introduction to Mathematical Statistics.* New York: Macmillan Publishing Co.

Holley, R. A., Kusuoka, S. and Stroock, D. W. (1989). Asymptotics of the Spectral Gap with Applications to the Theory of Simulated Annealing. *Journal of Functional Analysis* **83**, 333-347.

Hong, Y. and Lee, Y-J. (2013). A Loss Function Approach To Model Specification Testing and Its Relative Efficiency. *Annals of Statistics* **41**, 1166-1203.

Hora, S. C., Hora, J. A. and Dodd, N. G. (1992). Assessment of Probability Distributions for Continuous

Random Variables: A Comparison of the Bisection and Fixed-value Methods. *Organizational Behavior and Human Decision Processes* **51**, 133-55.

Howson, C. and Urbach, P. (1993). *Scientific Reasoning: The Bayesian Approach.* Second Edition. Chicago: Open Court.

Hsu, J. S. J. (1995). Generalized Laplacian Approximations in Bayesian Inference. *Canadian Journal of Statistics* **23**, 399-410.

Huang, D. S. (1970). *Regression and Econometric Methods.* New York: Wiley & Sons.

Huber, P. J. (1972). Robust Statistics: A Review. *Annals of Mathematical Statistics* **43**, 1041-1067.

Huber, P. J. (1973). Robust Regression: Asymptotics, Conjectures, and Monte Carlo. *Annals of Statistics* **1**, 799-821.

Huber, P. J. (1981). *Robust Statistics.* New York: Wiley & Sons.

Hull, T. E. and Dobell, A. R. (1964). Mixed Congruential Random Number Generators for Binary Machines. *Journal of the Association for Computing Machinery* **11**, 31-40.

Hunter, J. E. (1997). Needed: A Ban on the Significance Test. *Psychological Science* **8**, 3-7.

Hunter, J. E. and Schmidt, F. L. (1990). *Methods of Meta-Analysis: Correcting Error and Bias in Research Findings.* Beverly Hills: Sage.

Hurd, W. J. (1974). Efficient Generation of Statistically Good Pseudonoise by Linearly Interconnected Shift Registers. *IEEE Transactions on Computers* **C-23**, 146-152.

Hurn, M. (1997). Difficulties in the Use of Auxiliary Variables in Markov Chain Monte Carlo Methods. *Statistics and Computing* **7**, 35-44.

Hurst, R. L. and Knop, R. E. (1972). Generation of Random Normal Correlated Variables: Algorithm 425. *Communications of the Association for Computing Machinery* **15**, 355-357.

Hurvich, C. M. and Tsai, C-L. (1991). Bias of the Corrected AIC Criterion for Underfitted Regression and Time Series Models. *Biometrika* **78**, 499-509.

Hurvich, C. M., Shumway, R. and Tsai, C-L. (1990). Improved Estimators of Kullback-Leibler Information for Autoregressive Model Selection in Small Samples. *Biometrika* **77**, 709-719.

Hwang, J. T., Casella, G., Robert, C. P., Wells, M. T. and Farrell, R. H. (1992). Estimation of Accuracy in Testing. *Annals of Statistics* **20**, 490-509.

Hyndman, R. J. (1996). Computing and Graphing Highest Density Regions. *The American Statistician* **50**, 120-126.

Ibrahim, J. G. and Chen, M-H. (2000a). Power Prior Distributions for Regression Models. *Statistical Science* **15**, 46-60.

Ibrahim, J. G. and Chen, M-H. (2000b). Prior Elicitation and Variable Selection for Generalized Linear Mixed Models. In *Generalized Linear Models: A Bayesian Perspective*, Dipak K. Dey, Sujit K. Ghosh and Bani K. Mallick (eds.). New York: Marcel Dekker, pp.41-53.

Ibrahim, J. G. and Laud, P. W. (1991). On Bayesian Analysis of Generalized Linear Models Using Jeffreys Prior. *Journal of the American Statistical Association* **86**, 981-986.

Ibrahim, J. G., Chen, M.-H. and Sinha, D. (2001). *Bayesian Survival Analysis.* New York: Springer-Verlag.

Ibrahim, J. G., Chen, M-H. and Sinha, D. (2003). On Optimality of the Power Prior. *Journal of the American Statistical Association* **98**, 204-213.

Imai, K. and van Dyk, D. (2004). Causal Inference with General Treatment Regimes: Generalizing the Propensity Score. *Journal of the American Statistical Association* **99**, 854-866.

Imai, K. and van Dyk, D. A. (2005). A Bayesian Analysis of the Multinomial Probit Model Using Marginal Data Augmentation. *Journal of Econometrics* **124**, 311-334.

Ingber, L. (1992). Genetic Algorithms and Very Fast Simulated Reannealing: A Comparison. *Mathematical Computation and Modeling* **16**, 87-100.

Ingrassia, S. (1994). On the Rate of Convergence of the Metropolis Algorithm and Gibbs Sampler by Geometric Bounds. *Annals of Applied Probability* **4**, 347-389.

Iosifescu, M. (1980). *Finite Markov Processes and their Applications.* New York: John Wiley & Sons.

Irony, T. Z. and Singpurwalla, N. D. (1996). Noninformative Priors Do Not Exist: A Discussion with José M. Bernard. *Journal of Statistical Planning and Inference* **65**, 159-177.

Jackman, S. (2000). Estimation and Inference via Bayesian Simulation: An Introduction to Markov Chain Monte Carlo. *American Journal of Political Science* **44**, 375-404.

James, W. and Stein, C. (1961). Estimation With Quadratic Loss. *Proceedings of the Fourth Symposium 1* J. Neyman and E. L. Scott (eds.). Berkeley: University of California Press, 361-380.

Jagerman, D. L. (1965). Some Theorems Concerning Pseudo-Random Numbers. *Mathematics of Computation* **19**, 418-426.

Jamshidian, M. and Jennrich, R. I. (1993). Conjugate Gradient Acceleration of the EM Algorithm. *Journal of the American Statistical Association* **88**, 221-228.

Janssen, A. (1986). Asymptotic Properties of Neyman-Pearson Tests for Infinite Kullback-Leibler Information. *Annals of Statistics* **14**, 1068-1079.

Jansson, B. (1966). *Random Number Generators.* Stockholm: Victor Pettersons.

Jarner, S. F. and Roberts, G. O. (2002). Polynomial Convergence Rates of Markov Chains. *Annals of Applied Probability* **12**, 224-247.

Jarvis, Edward. (1858). Distribution of Lunatic Reports. *American Journal of Psychiatry* **14**, 248-253.

Jaynes, E. T. (1968). Prior Probabilities. *IEEE Transactions on Systems Science and Cybernetics* SSC-4, 227-241.

Jaynes, E. T. (1976). Confidence Intervals vs. Bayesian Intervals. In *Foundations of Probability Theory, Statistical Inference, and Statistical Theories of Science II*, William L. Harper and Clifford A. Hooker (eds.). Dordrecht: D. Reidel, pp.175-257.

Jaynes, E. T. (1980). Marginalization and Prior Probabilities. In *Bayesian Analysis in Econometrics and Statistics*, A. Zellner (ed.). North Holland: Amsterdam, pp.43-78.

Jaynes, E. T. (1983). *Papers on Probability, Statistics and Statistical Physics*, R. D. Rosencrantz (ed.). Dordrecht: Reidel.

Jeffreys, H. (1961). *The Theory of Probability.* Third Edition. Oxford, England: Oxford University Press.

Jeng, F-C. and Woods, J. W. (1990). Simulated Annealing in Compound Gaussian Random Fields. *IEEE Transactions on Information Theory* **36**, 94-107.

Jöhnk, M. D. (1964). Erzeugung von Betaverteilter und Gammaverteilter Zufallszahlen. *Metrika* **8**, 5-15.

Johnson, N. L., Kotz, S. and Balakrishnan, N. (1994). *Continuous Univariate Distributions, Volume 1.* New York: John Wiley & Sons.

Johnson, N. L., Kotz, S. and Balakrishnan, N. (1995). *Continuous Univariate Distributions, Volume 2.* New York: John Wiley & Sons.

Johnson, N. L., Kotz, S. and Balakrishnan, N. (1997). *Discrete Multivariate Distributions.* New York: John Wiley & Sons.

Johnson, N. L., Kotz, S. and Balakrishnan, N. (2000). *Continuous Multivariate Distributions, Models and Applications, Volume 1.* New York: John Wiley & Sons.

Johnson, N. L., Kemp, A. W. and Kotz, S. (2005). *Univariate Discrete Distributions.* New York: John Wiley & Sons.

Johnson, R. A. (1984). The Analysis of Transformed Data: Comment. *Journal of the American Statistical Association* **79**, 314-315.

Johnson, V. E. (1998). A Coupling-Regeneration Scheme for Diagnosing Convergence in Markov Chain Monte Carlo Algorithms. *Journal of the American Statistical Association* **93**, 238-248.

Johnson, V. E. and Albert, J. H. (1999). *Ordinal Data Modeling.* New York: Springer-Verlag.

Jones, G. L. and Hobert, J. P. (2001). Honest Exploration of Intractable Probability Distributions via Markov Chain Monte Carlo. *Statistical Science* **16**, 312-34.

Jones, G. L. and Hobert, J. P. (2004). Sufficient Burn-in for Gibbs Samplers for a Hierarchical Random Effects Model. *Annals of Statistics* **32**, 784-817.

Jones, M. C. and Lunn, A. D. (1996). Transformations and Random Variate Generation: Generalized Ratio-of-Uniforms Methods. *Journal of Statistical Computation and Simulation* **55**, 49-55.

Jørgensen, B. (1983). Maximum Likelihood Estimation and Large-Sample Inference for Generalized Linear and Nonlinear Regression Models. *Biometrics* **70**, 19-28.

Kackar, R. N. and Harville, D. A. (1984). Approximations for Standard Errors of Estimators of Fixed and Random Effects in Mixed Linear Models. *Journal of the American Statistical Association* **79**, 853-862.

Kadane, J. B. (1986). Progress Toward a More Ethical Method Clinical Trials. *Journal of Medical Philosophy* **11**, 385-405.

Kadane, J. B. (1980). Predictive and Structural Methods for Eliciting Prior Distributions. In *Bayesian Analysis in Econometrics and Statistics: Essays in Honor of Harold Jeffreys*, Arnold Zellner (ed.). Amsterdam: North Holland, pp.89-109.

Kadane, J. B. and Chuang, D. T. (1978). Stable Decision Problems. *Annals of Statistics* **6**, 1095-1110.

Kadane, J. B. and Srinivasan, C. (1996). Bayesian Robustness and Stability. In *Bayesian Robustness*, J. O. Berger, B. Betró, E. Moreno, L. R. Pericchi, F. Ruggeri, G. Salinetti and L. Wasserman (eds.).Hayward, CA: Institute of Mathematical Statistics Monograph Series 29, pp.139-156.

Kadane, J. B. and Winkler, R. L. (1988). Separating Probability Elicitation From Utilities. *Journal of the American Statistical Association* **83**, 357-363.

Kadane, J. B. and Wolfson, L. J. (1996). Priors for Design and Analysis of Clinical Trials. In *Bayesian Biostatistics*, D. A. Berry and D. K. Stangl (eds.). New York: Chapman & Hall/CRC, pp.157-186.

Kadane, J. B. and Wolfson, L. J. (1998). Experiences in Elicitation. *Journal of The Royal Statistical Society, Series D* **47**, 3-19.

Kadane, J. B., Dickey, J. M., Winkler, R. L., Smith, W. S. and Peters, S. C. (1980). Interactive Elicitation of Opinion for a Normal Linear Model. *Journal of the American Statistical Association* **75**, 845-854.

Kahn, H. (1949). Stochastic (Monte Carlo) Attenuation Analysis. No. RAND-P-88 (rev). Defense Technical Information Center.

Kalyanam, K. (1996). Pricing Decisions under Demand Uncertainty: A Bayesian Mixture Model Approach. *Marketing Science* **15**, 207-221.

Kaplan, A. (1964). *Conduct of Inquiry.* San Francisco: Chandler Publishing Company.

Kaplan, H. B., Johnson, R. J., Bailey, C. A. and Simon, W. (1987). The Sociological Study of AIDS: A Critical Review of the Literature and Suggested Research Agenda. *Journal of Health and Social Behavior* **28**, 140-157.

Karlin, S. and Taylor, H. M. (1981). *A Second Course in Stochastic Processes.* San Diego: Academic Press.

Karlin, S. and Taylor, H. M. (1990). *A First Course in Stochastic Processes.* San Diego: Academic Press.

Kass, R. E. (1989). The Geometry of Asymptotic Inference. *Statistical Science* **4**, 188-219.

Kass, R. E. (1993). Bayes Factors in Practice. *The Statistician* **42**, 551-560.

Kass, R. E. and Greenhouse, J. B. (1989). Comments on the paper by J. H. Ware. *Statistical Science* **4**, 310-317.

Kass, R. E. and Raftery, A. E. (1995). Bayes Factors. *Journal of the American Statistical Association* **90**, 773-795.

Kass, R. E. and Steffey, D. (1989). Approximate Bayesian Inference in Conditionally Independent Hierarchical Models. *Journal of the American Statistical Association* **84**, 717-726.

Kass, R. E., Tierney, L. and Kadane, J. B. (1989). Approximate Methods for Assessing Influence and Sensitivity in Bayesian Analysis. *Biometrika* **76**, 663-674.

Kass, R. E. and Vaidyanathan, S. K. (1992). Approximate Bayes Factors and Orthogonal Parameters, with Application to Testing Equality of Two Binomial Proportions. *Journal of the Royal Statistical Society, Series B* **54**, 129-144.

Kass, R. E. and Wasserman, L. (1995). A Reference Bayesian Test for Nested Hypotheses and Its Relationship to the Schwarz Criterion. *Journal of the American Statistical Association* **90**, 928-934.

Kass, R. E. and Wasserman, L. (1996). The Selection of Prior Distributions by Formal Rules. *Journal of the American Statistical Association* **91**, 1343-1370.

Kass, R. E., Tierney, L. and Kadane, J. B. (1989). Approximate Methods for Assessing Influence and Sensitivity in Bayesian Analysis. *Biometrika* **76**, 663-674.

Kaufmann, H. (1988). On the Existence and Uniqueness of Maximum Likelihood Estimates in Quantal and Ordinal Response Models. *Metrika* **35**, 291-313.

Keating, J. P. and Mason, R. L. (1988). James-Stein Estimation from an Alternative Perspective. *The American Statistician* **42**, 160-164.

Keith, J., K., Kroese, D. P. and Bryant, D. (2004). A Generalized Markov Sampler. *Methodology and Computing in Applied Probability* **6**, 29-53.

Kelleher, C. A. and Yackee, S. W. (2009). A Political Consequence of Contracting: Organized Interests and State Agency Decision-making. *The Journal of Public Administration Research and Theory* **19**, 579-602.

Kempthorne, P. J. (1986). Decision-Theoretic Measures of Influence in Regression. *Journal of the Royal Statistical Society, Series B* **48**, 370-378.

Kendall, M. G. (1949). On Reconciliation of the Theories of Probability. *Biometrika* **36**, 101-116.

Kennedy, W. J. and Gentle, J. E. (1980). *Statistical Computing*. New York: Marcel Dekker.

Kéry, M. (2010). *Introduction to WinBUGS for Ecologists: A Bayesian Approach to Regression, ANOVA and Related Analyses*. Burlington, MA: Academic Press.

Kéry, M. and Schaub, M. (2011). *Bayesian Population Analysis Using WinBUGS: A Hierarchical Perspective*. Burlington, MA: Academic Press.

Keynes, J. M. (1921). *A Treatise on Probability*. London: MacMillan.

Kim, D. (1991). A Bayesian Significance Test of the Stationarity of Regression Parameters. *Biometrika* **78**, 667-675.

Kinderman, A. J. and Monahan, J. F. (1980). New Methods for Generating Student's t and Gamma Variables. *Computing* **25**, 369-377.

Kinderman, A. J. and Ramage, J. G. (1976). Computer Generation of Normal Random Variables. *Journal of the American Statistical Association* **71**, 893-896.

Kinderman, A. J., Monahan, J. F. and Ramage, J. G. (1975). Computer Generation of Random Variables with Normal and Student's Distributions. *Proceedings of the Statistical Computing Section*. Alexandria, VA: American Statistical Association. 128-131.

King, G. (1989). *Unifying Political Methodology: The Likelihood Theory of Statistical Inference*. Cambridge: Cambridge University Press.

Kirk, R. E. (1996). Practical Significance: A Concept Whose Time Has Come. *Educational and Psychological Measurement* **56**, 746-759.

Kirkpatrick, S., Gelatt, C. D. and Vecchi, M. P. (1983). Kirkpatrick, S. Optimization by Simulated Annealing. *Science* **220**, 671-680.

Kitagawa, G. and Akaike, H. (1982). A Quasi Bayesian Approach to Outlier Detection. *Annals of the Institute of Statistical Mathematics* **34B**, 389-398.

Kitagawa, G. and Gersch, W. (1996). *Smoothness Priors Analysis of Time Series.* New York: Springer-Verlag.

Kleibergen, F. and van Dijk, H. K. (1993). Efficient Computer Generation of Matrix-Variate t Drawings with an Application to Bayesian Estimation of Simple Market Models. In *Computer Intensive Methods in Statistics*, W. Hardle and L. Simar (eds.). Heidelberg: Physica-Verlag, pp.30-46.

Kloek, T. and van Dijk, H. K. (1978). Bayesian Estimates of Equation System Parameters; An Application Integration by Monte Carlo. *Econometrica* **46**, 1-19.

Knott, M., Albanese, T. M. and Galbraith, J. (1990). Scoring Attitudes to Abortion. *The Statistician* **40**, 217-223.

Knuth, D. E. (1981). *The Art of Computer Programming, Volume 2: Seminumerical Algorithms.* Second Edition. Menlo Park, CA: Addison-Wesley.

Kollman, K., Miller, J. H. and Page, S. E. (1997). Political Institutions and Sorting in a Tiebout Model. *American Economic Review* **87**, 977-992.

Kolmogorov, A. (1933). *Grundbegriffe der Wahrscheinlichkeitsrechnung.* Berlin: Julius Springer.

Kong, A., Liu, J. S. and Wong, W. H. (1994). Sequential Imputations and Bayesian Missing Data Problems. *Journal of the American Statistical Association* **89**, 278-288.

Koop, G. (1992). "Objective" Bayesian Unit Root Tests. *Journal of Applied Econometrics* **7**, 65-82.

Koopman, L. H. (1936). On Distributions Admitting a Sufficient Statistic. *Transactions of the American Mathematical Society* **39**, 399-409.

Koppel, J. G. S. (1999). The Challenge of Administration by Regulation: Preliminary Findings Regarding the U.S Government's Venture Capital Funds. *Journal of Public Administration Research and Theory* **9**, 641-666.

Korwar, R. M. and Hollander, M. (1976). Empirical Bayes Estimation of a Distribution Function. *The Annals of Statistics* **4**, 581-588.

Kotz, S. and Nadarajah, S. (2000). *Extreme Value Distributions: Theory and Applications.* Singapore: World Scientific Publications.

Kowalski, J., Tu, X. M., Day, R. S. and Mendoza-Blanco, J. R. (1997). On the Rate of Convergence of the ECME Algorithm for Multiple Regression Models With t-distributed Errors. *Biometrika* **84**, 269-281.

Krawczyk, H. (1992). How to Predict Congruential Generators. *Journal of Algorithms* **13**, 527-545.

Kreft, I. G. G. (1993). Using Multilevel Analysis to Assess School Effectiveness: A Study of Dutch Secondary Schools. *Sociology of Education* **66**, 104-129.

Kreft, I. G. G. and De Leeuw, J. (1988). The Seesaw Effect: A Multilevel Problem? *Quality and Quantity* **22**, 127-137.

Kreft, I. G. G. and De Leeuw, J. (1998). *Introducing Multilevel Modeling.* Thousand Oaks, CA: Sage.

Kreuzenkamp, H. A. and McAleer, M. (1995). Simplicity, Scientific Inference, and Econometric Modelling. *The Economic Journal* **105**, 1-21.

Krishnamoorthy, K. (2006). *Handbook of Statistical Distributions with Applications.* New York: Chapman & Hall/CRC.

Krommer, A. R. and Ueberhuber, C. W. (1998). *Computational Integration.* Philadelphia: Society for Industrial and Applied Mathematics.

Kronmal, R. (1964). The Evaluation of a Pseudorandom Normal Number Generator. *Journal of the Association for Computing Machinery* **11**, 357-263.

Kronmal, R. and Peterson, A. V., Jr. (1981). A Variant of the Acceptance-Rejection Method for Computer Generation of Random Variables. *Journal of the American Statistical Association* **76**, 446-451.

Kronrod, A. S. (1965). *Nodes and Weights of Quadrature Formulas.* New York: Consultants Bureau.

Krueger J. (2001). Null Hypothesis Significance Testing: On the Survival of a Flawed Method. *American Psychologist* **56**, 16-26.

Kruschke, J. (2010). *Doing Bayesian Data Analysis: A Tutorial Introduction with R.* Burlington, MA: Academic Press.

Krzanowski, W. J. (1988). *Principles of Multivariate Analysis.* Oxford: Oxford University Press.

Kullback, S. (1968). *Information Theory and Statistics.* New York: Wiley & Sons.

Kurganov, I. A. (1973). Speeches and Writings, 1945-1976. Palo Alto, CA: Online Archive of California, http://www.oac.cdlib.org

Kyung, M., Gill, J. and Casella, G. (2011). New Findings from Terrorism Data: Dirichlet Process Random Effects Models for Latent Groups. *Journal of the Royal Statistical Society, Series C* **60**, 701-721.

Kyung, M., Gill, J. and Casella, G.. (2012). Sampling Schemes for Generalized Linear Dirichlet Process Random Effects Models. With Discussion and Rejoinder. *Statistical Methods and Applications* **20**, 259-290.

Laird, N. M. (1978). Nonparametric Maximum Likelihood Estimation of a Mixing Distribution. *Journal of the American Statistical Association* **73**, 805-811.

Laird, N. M. and Louis, T. A. (1982). Approximate Posterior Distributions for Incomplete Data Problems. *Journal of the Royal Statistical Society, Series B* **44**, 190-200.

Laird, N. M. and Louis, T. A. (1987). Empirical Bayes Confidence Intervals Based on Bootstrap Samples. *Journal of the American Statistical Association* **82**, 739-750.

Laird, N. M., Lange, N. and Stram, D. O. (1987). Maximum Likelihood Computations with Repeated Measures: Applications of the EM Algorithm. *Journal of the American Statistical Association* **82**, 97-105.

Lange, K. L. (2000). *Numerical Analysis for Statisticians.* New York: Springer-Verlag.

Lange, K. L., Little, R. J. A. and Taylor, J. M. G. (1989). Robust Statistical Modeling Using the t Distribution. *Journal of the American Statistical Association* **84**, 881-896.

Laplace, P. S. (1774). Mémoire sur la Probabilité des Causes par le Évènemens. *Mémoires de l'Académie Royale des Sciences Presentés par Divers Savans* **6**, 621-656.

Laplace, P. S. (1781). Mémoire sur la Probabilités. *Mémoires de l'Académie Royale des Sciences de Paris* **1778**, 227-332.

Laplace, P. S. (1811). Mémoire sur les Integrales Définies et leur Application aux Probabilités, et Specialement à Recherche du Milieu Qu'il Faut Chosier Entre les Resultats des Observations. *Mémoires de l'Académie des Sciences de Paris*, 279-347.

Laplace, P. S. (1814). *Essai Philosophique sur les la Probabilités.* Paris: Ve Courcier.

Laud, P. W. and Ibrahim, J. G. (1995). Predictive Model Selection. *Journal of the Royal Statistical Society, Series B* **57**, 247-262.

Lauritzen, S. and Spiegelhalter, D. J. (1988). Local Computations with Probabilities on Graphical Structures and their Application to Expert Systems. *Journal of the Royal Statistical Society, Series B* **50**, 157-194.

Lavine, M. (1991a). Sensitivity in Bayesian Statistics: The Prior and the Likelihood. *Journal of the American Statistical Association* **86**, 396-399.

Lavine, M. (1991b). An Approach to Robust Bayesian Analysis for Multidimensional Parameter Spaces. *Journal of the American Statistical Association* **86**, 400-403.

Lavine, A. M. and Schervish, M. J. (1999). Bayes Factors: What They Are and What They Are Not. *The American Statistician* **53**, 119-122.

Law, A. M. and Kelton, W. D. (1982). *Simulation Modeling and Analysis.* New York: McGraw-Hill.

Lawler, G. F. and Sokal, A. D. (1988). Bounds on the L^2 Spectrum for Markov Chains and Their Applications. *Transactions of the American Mathematical Society* **309**, 557-580.

Lawson, A. B. (2013). Bayesian Disease Mapping: Hierarchical Modeling in Spatial Epidemiology. Boca Raton: Chapman & Hall/CRC.

Le Cam, L. (1986). *Asymptotic Methods in Statistical Decision Theory.* New York: Springer-Verlag.

Leamer, E. E. (1978). *Specification Searches: Ad Hoc Inference with Nonexperimental Data.* New York: John Wiley & Sons.

Leamer, E. E. (1979). Information Criteria for Choice of Regression Models: A Comment. *Econometrica* **47**, 507-510.

Leamer, E. E. (1983). Let's Take the Con Out of Econometrics. *American Economic Review* **73**, 31-43.

Leamer, E. E. (1984). Global Sensitivity Results for Generalized Least Squares Estimates. *Journal of the American Statistical Association* **79**, 867-870.

Leamer, E. E. (1985). Sensitivity Analysis Would Help. *The American Economic Review* **75**, 308-313.

Leamer, E. E. (1992). Bayesian Elicitation Diagnostics. *Econometrica* **80**, 919-942.

Learmonth, G. P. and Lewis, P. A. W. (1973). Some Widely Used and Recently Proposed Uniform Random Number Generators. *Proceedings of Computer Science and Statistics: Seventh Annual Symposium on the Interface*, W. J. Kennedy (ed.). Ames, IA: Iowa State University. 163-171.

L'Ecuyer, P. (1998). Uniform Random Number Generators. *1998 Winter Simulation Conference Proceedings, Society for Computer Simulation.* 97-104.

Lee, P. M. (2004). *Bayesian Statistics: An Introduction.* Second Edition. New York: Hodder Arnold.

Lee, S. (1998). On the Quantile Process Based on the Autoregressive Residuals. *Journal of Statistical Planning and Inference* **67**, 17-28.

Lee, V. E. and Bryk, A. S. (1989). Multilevel Model of the Social Distribution of High School Achievement. *Sociology of Education* **62**, 172-192.

Legendre, A. M. (1805). *Nouvelles Méthodes Pour la Détermination des Orbites des Comètes.* Paris: Courcier.

Lehmann, E. L. (1986). *Testing Statistical Hypotheses.* New York: Springer-Verlag.

Lehmann, E. L. (1993). The Fisher, Neyman-Pearson Theories of Testing Hypotheses: One Theory or Two? *Journal of the American Statistical Association* **88**, 1242-1249.

Lehmann, E. L. (1999). *Elements of Large-Sample Theory.* New York: Springer-Verlag.

Lehmann, E. L. and Casella, G. (1998). *Theory of Point Estimation.* Second Edition. New York: Springer-Verlag.

Lehmer, D. H. (1951). Mathematical Models in Large-scale Computing Units. *Proceedings of the Second Symposium on Large Scale Digital Computing Machinery.* Cambridge: Harvard University Press. 141-146.

Li, L. and Choe, M. K. (1997). A Mixture Model for Duration Data: Analysis of Second Births in China. *Demography.* **34**, 189-197.

Leimkuhler, B. and Reich, S. (2005). *Simulating Hamiltonian Dynamics.* Cambridge: Cambridge University Press.

Lempers, F. B. (1971). *Posterior Probabilities of Alternative Linear Models.* Rotterdam: Rotterdam University Press.

Leonard, T. (1975). A Bayesian Approach to the Linear Model with Unequal Variances. *Technometrics* **17**, 95-102.

Leonard, T. and Hsu, J. S. J. (1999). *Bayesian Methods: An Analysis for Statisticians and Interdisciplinary Researchers.* Cambridge, England: Cambridge University Press.

Leonard, T., Hsu, J. S. J. and Tsui, K-W. (1989). Bayesian Marginal Inference. *Journal of the American Statistical Association* **84**, 1051-1058.

Lesaffre, E. and Kaufmann, H. (1992). Existence and Uniqueness of the Maximum Likelihood Estimator for a Multivariate Probit Model. *Journal of the American Statistical Association* **87**, 805-811.

Levine, R. (1996). Post-Processing Random Variables. Ph.D. Thesis, Biometrics Unit, Cornell University.

Lewis, C. and Thayer, D. T. (2009). Bayesian Decision Theory for Multiple Comparisons. In IMS Lecture Notes Monograph Series, Optimality: the Therd Erich L. Lehmann Symposium **57**, pp.326-332.

Lewis, J. A. (1994). Discussion of the Paper by Spiegelhalter, Freedman, and Parmar. *Journal of the Royal Statistical Society, Series A* **157**, 392.

Lewis, S. M. and Raftery, A. E. (1997). Estimating Bayes Factors Via Posterior Simulation With the Laplace-Metropolis Estimator. *Journal of the American Statistical Association* **92**, 648-655.

Lewis, T. G. and Payne, W. H. (1973). Generalized Feedback Shift Register Pseudorandom Number Algorithm. *Journal of the Association for Computing Machinery* **20**, 456-468.

Li, H-L. (2004). The Sampling/Importance Resampling Algorithm. In *Applied Bayesian Modeling and Causal Inference from Incomplete-Data Perspectives*, Andrew Gelman and Xiao-Li Meng (eds.). New York: John Wiley & Sons, pp.265-276.

Li, L. and Choe, M. K. (1997). A Mixture Model for Duration Data: Analysis of Second Births in China. *Demography* **34**, 189-197.

Lindley, D. V. (1957). A Statistical Paradox. *Biometrika* **44**, 187-192.

Lindley, D. V. (1958). Fiducial Distributions and Bayes' Theory. *Journal of the Royal Statistical Society, Series B* **20**, 102-107.

Lindley, D. V. (1961). The Use of Prior Probability Distributions in Statistical Inference and Decision. *Proceedings of the Fourth Berkeley Symposium on Mathematical Statistics and Probability.* Berkeley: University of California Press. 453-468.

Lindley, D. V. (1965). *Introduction to Probability and Statistics from a Bayesian Viewpoint, Parts 1 and 2.* Cambridge, England: Cambridge University Press.

Lindley, D. V. (1968). The Choice of Variables in Multiple Regression. *Journal of the Royal Statistical Society, Series B* **30**, 31-66.

Lindley, D. V. (1969). Discussion of *Compound Decisions and Empirical Bayes*, J. B. Copas. *Journal of the Royal Statistical Society, Series B* **31**, 397-425.

Lindley, D. V. (1972). *Bayesian Statistics: A Review.* Philadelphia: Society for Industrial and Applied Mathematics.

Lindley, D. V. (1983). Reconciliation of Probability Distributions. *Operations Research, Journal of Operations Research Society of America* **31**, 866-880.

Lindley, D. V. (1985). *Making Decisions.* Second Edition. New York: John Wiley & Sons.

Lindley, D. V. (1986). Comment. *The American Statistician* **40**, 6-7.

Lindley, D. V. (1991). Subjective Probability, Decision Analysis and Their Legal Consequences. *Journal of the Royal Statistical Society, Series A* **154**, 83-92.

Lindley, D. V. and Novick, M. R. (1978). The Use of More Realistic Utility Functions in Educational Applications. *Journal of Educational Measurement* **15**, 181-191.

Lindley, D. V. and Novick, M. R. (1981). The Role of Exchangeability in Inference. *Annals of Statistics* **9**, 45-58.

Lindley, D. V. and Singpurwalla, N. D. (1991). On the Evidence Needed to Reach Agreed Action Between Adversaries, with Application to Acceptance Sampling. *Journal of the American Statistical Association* **86**, 933-937.

Lindley, D. V. and Smith, A. F. M. (1972). Bayes Estimates for the Linear Model. *Journal of the Royal Statistical Society, Series B* **34**, 1-41.

Lindley, D. V., Tversky, A. and Brown, R. V. (1979). On the Reconciliation of Probability Assessments. *Journal of the Royal Statistical Society, Series A* **142**, 146-180.

Lindsay, R. M. (1995). Reconsidering the Status of Tests of Significance: An Alternative Criterion of Adequacy. *Accounting, Organizations and Society* **20**, 35-53.

Lindsey, J. K. (1997). *Applying Generalized Linear Models.* New York: Springer-Verlag.

Linick, T. W., Jull, A. J. T., Toolin, L. J. and Donahue, D. J. (1986). Operation of the NSF-Arizona Accelerator Facility For Radioscope Analysis and Results From Selected Collaborative Research Projects. *Radiocarbon* **28**, 522-533.

Liniger, W. (1961). On a Method by D. H. Lehmer for the Generation of Pseudo-Random Numbers. *Numerische Mathematik* **3**, 265-270.

Liseo, B., Petrella, L. and Salinetti, G. (1996). Bayesian Robustness: an Interactive Approach. In *Bayesian Statistics 5*, J. O. Berger, J. M. Bernardo, A. P. Dawid, D. V. Lindley, (eds.). Oxford: Oxford University Press, pp.223-253.

Little, R. J. A. and Rubin, D. B. (1983). On Jointly Estimating Parameters and Missing Data by Maximizing the Complete-Data Likelihood. *The American Statistician* **37**, 218-220.

Little, R. J. A. and Rubin, D. B. (2002). *Statistical Analysis with Missing Data*. Second Edition. New York: John Wiley & Sons.

Liu, C. and Rubin, D. B. (1994). The ECME Algorithm: A Simple Extension of EM and ECM With Faster Monotone Convergence. *Biometrika* **81**, 633-648.

Liu, C. and Rubin, D. B. (1995). ML Estimation of the t Distribution Using EM and Its Extensions, ECM and ECME. *Statistica Sinica* **5**, 19-39.

Liu, C. and Rubin, D. B. (1998). Maximum Likelihood Estimation of Factor Analysis Using the ECME Algorithm With Complete and Incomplete Data. *Statistica Sinica* **8**, 729-748.

Liu, C., Liu, J. S. and Rubin, D. B. (1992). A Variational Control Variable for Assessing the Convergence of the Gibbs Sampler. *Proceedings of the American Statistical Association, Statistical Computing Section*. 74-78.

Liu, J. S. (1994). The Collapsed Gibbs Sampler in Bayesian Computations with Applications to a Gene Regulation Problem. *Journal of the American Statistical Association* **89**, 958-966.

Liu, J. S. (1996a). Metropolized Independent Sampling with Comparisons to Rejection Sampling and Importance Sampling. *Statistics and Computing* **6**, 113-119.

Liu, J. S. (1996b). Nonparametric Hierarchical Bayes via Sequential Imputations. *Annals of Statistics* **24**, 911-930.

Liu, J. S. (2001). *Monte Carlo Strategies in Scientific Computing*. New York: Springer-Verlag.

Liu, J. S., Liang, F. and Wong, W. H. (2000). The Multiple-Try Method and Local Optimization in Metropolis Sampling *Journal of the American Statistical Association* **95**, 121-134.

Liu, J. S. and Sabatti, C. (1999). Simulated Sintering: Markov Chain Monte Carlo with Spaces Varying Dimension. In *Bayesian Statistics*, J. M. Bernardo, A. F. M. Smith, A. P. Dawid and J. O. Berger (eds.). Oxford: Oxford University Press, pp.389-414.

Liu, J. S., Wong, W. H. and Kong, A. (1994). Covariance Structure of the Gibbs Sampler with Applications to the Comparisons of Estimators and Augmentation Schemes. *Biometrika* **81**, 27-40.

Liu, J. S., Wong, W. H. and Kong, A. (1995). Correlation Structure and Convergence Rates of the Gibbs Sampler with Various Scans. *Journal of the Royal Statistical Society, Series B* **57**, 157-169.

Locatelli, M. (2000). Simulated Annealing Algorithms for Continuous Global Optimization: Convergence Conditions. *Journal of Optimization Theory and Applications* **104**, 121-133.

Loftus, G. R. (1991). On the Tyranny of Hypothesis Testing In the Social Sciences. *Contemporary Psychology* **36**, 102-105.

Loftus, G. R. (1993). A Picture Is Worth a Thousand rho Values: On the Irrelevance of Hypothesis Testing In the Microcomputer Age. *Behavioral Research Methods, Instruments, and Computers* **25**, 250-256.

Louis, T. A. (1982). Finding the Observed Information Matrix when Using the EM Algorithm. *Journal of the Royal Statistical Society, Series B* **44**, 226-233.

Lu, M. (2002). Enhancing Project Evaluation and Review Technique Simulation through Artificial Neural Network-based Input Modeling. *Journal of Construction Engineering and Management* September/October, 438-45.

Lundy, M. (1985). Applications of the Annealing Algorithm to Combinatorial Problems in Statistics. *Biometrika* **72**, 191-198.

Lundy, M. and Mees, A. (1986). Convergence of an Annealing Algorithm. *Mathematical Programming* **34**, 111-124.

Lunn, D. J., Thomas, A., Best, N. G. and Spiegelhalter, D. (2000). WinBUGS–A Bayesian Modelling Framework: Concepts, Structure, and Extensibility. *Statistics and Computing* **10**, 325–337.

Lunn, D. J., Jackson, C., Spiegelhalter, D. J., Best, N. G. and Thomas, A. (2012). *The BUGS book: A Practical Introduction to Bayesian Analysis.* Boca Raton: Chapman & Hall/CRC.

Lwin, T. (1972). Estimation of the Tail Paretian Law. *Scandinavian Actuarial Journal* **55**, 170-178.

Lynn, R. and Vanhanen, T. (2001). National IQ and Economic Development. *Mankind Quarterly* **LXI**, 415-437.

Macdonald, R. R. (1997). On Statistical Testing in Psychology. *British Journal of Psychology* 88, No. 2 (May), 333-49.

MacEachern, S. N. and Berliner, L. M. (1994). Subsampling the Gibbs Sampler. *The American Statistician* **48**, 188-190.

MacLaren, M. D. and Marsaglia, G. (1965). Uniform Random Number Generators. *Journal of the Association for Computing Machinery* **12**, 83-89.

Madras, N. and Sezer, D. (2010). Quantitative Bounds for Markov Chain Convergence: Wasserstein and Total Variation Distances. *Bernoulli* **16**, 882-908.

Makelainen, T., Schmidt, K. and Styan, G. P. H. (1981). On the Existence and Uniqueness of the Maximum Likelihood Estimate of a Vector-Valued Parameter in Fixed-Size Samples. *Annals of Statistics* **9**, 758-767.

Makov, U. E., Smith, A. F. M. and Liu, Y-H. (1996). Bayesian Methods in Actuarial Science. *The Statistician* **45**, 503-515.

Malov, S. V. (1998). Random Variables Generated by Ranks in Dependent Schemes. *Metrika* **48**, 61-67.

Maltzman, F. and Wahlbeck, P. J. (1996). Strategic Policy Considerations and Voting Fluidity on the Burger Court. *American Political Science Review* **90**, 581-92.

Manski, C. F. (1995). *Identification Problems in the Social Sciences.* Cambridge: Harvard University Press.

Marden, J. I. (2000). Hypothesis Testing: From *p* Values to Bayes Factors. *Journal of the American Statistical Association* **95**, 1316-1320.

Marín, J. M. (2000). A Robust Version of the Dynamic Linear Model with an Economic Application. In *Robust Bayesian Analysis*, David Ríos Insua and Fabrizio Ruggeri (eds.). New York: Springer-Verlag, pp.373-383.

Marinari, E. and Parisi, G. (1992). Simulated Tempering: A New Monte Carlo Scheme. *Europhysics Letters* **19**, 451-458.

Maritz, J. S. (1970). *Empirical Bayes Methods.* London: Methuen.

Maritz, J. S. and Lwin, T. (1989). *Empirical Bayes Methods.* Second Edition. New York: Chapman & Hall.

Marsaglia, G. (1961a). Expressing a Random Variable in Terms of Uniform Random Variables. *Annals of Mathematical Statistics* **32**, 894-898.

Marsaglia, G. (1961b). Generating Exponential Random Variables. *Annals of Mathematical Statistics* **32**, 899-900.

Marsaglia, G. (1964). Generating a Variable from the Tail of the Normal Distribution. *Technometrics* **6**, 101-102.

Marsaglia, G. (1968). Random Numbers Fall Mainly in the Planes. *Proceedings of the National Academy of Sciences* **61**. 25-28.

Marsaglia, G. (1972). The Structure of Linear Congruential Sequences. In *Applications of Number Theory to Numerical Analysis*, S. K. Zaremba (ed.). San Diego: Academic Press, pp.249-286.

Marsaglia, G. (1977). The Squeeze Method for Generating Gamma Variates. *Computers and Mathematics with Applications* **3**, 321-326.

Marsaglia, G. (1985). A Current View of Random Number Generators. In *Computer Science and Statistics: 16th Symposium on the Interface*, L. Billard (ed.). Amsterdam: North Holland, pp.3-10.

Marsaglia, G. and Bray, T. A. (1964). A Convenient Method for Generating Normal Variables. *SIAM Review* **6**, 260-264.

Marsaglia, G., MacLaren, M. D. and Bray, T. A. (1964). A Fast Procedure for Generating Normal Random Variables. *Communications of the Association for Computing Machinery* **7**, 4-10.

Marsaglia, G., Tsang, W. W. and Wang, J. (2003). Evaluating Kolmogorov's Distribution. *Journal of Statistical Software* **8**, 1-4.

Marshall, A. W. (1956). The Use of Multi-Stage Sampling Schemes in Monte Carlo Computations. In *Symposium on Monte Carlo Methods*, M. Meyer (ed.). New York: John Wiley & Sons, pp.123-140.

Marshall, A. W. and Olkin, I. (1967). A Multivariate Exponential Distribution. *Journal of the American Statistical Association* **62**, 30-44.

Martikainen, P., Martelin, T., Nihtilä, E., Majamaa, K. and Seppo, K. (2005). Differences in Mortality by Marital Status in Finland from 1976 to 2000: Analyses of Changes in Marital-Status Distributions, Socio-Demographic and Household Composition, and Cause of Death. *Population Studies* **59**, 99-115.

Martin, A. D. and Quinn, K. (2007). Assessing Preference Change on the US Supreme Court. *Journal of Law, Economics, and Organization* **23**, 365-385.

Martín, J., Ríos Insua, D. and Ruggeri, F. (1996). Local Sensitivity Analysis in Bayesian Decision Theory. *Lecture Notes-Monograph Series* **29** (Bayesian Robustness), 119-135.

Martín, J., Ríos Insua, D. and Ruggeri, F. (1998). Issues in Bayesian Loss Robustness. *Sankhyā, Series A* **60**, 405-417.

Martins, A. C. R. (2009). Bayesian Updating Rules In Continuous Opinion Dynamics Models. *Journal of Statistical Mechanics: Theory and Experiment* **2**, P02017.

Mason, R. L. and Lurie, D. (1973). Systematic Simulators of Joint Order Uniform Variates. In *Proceedings of Computer Science and Statistics: Seventh Annual Symposium on the Interface*, W. J. Kennedy (ed.). Ames, IA: Iowa State University, pp.156-162.

Mason, W. M., Wong, G. Y. and Entwistle, B. (1983). Contextual Analysis Through the Multilevel Linear Model. In *Sociological Methodology 1983-1984*, S. Leinhardt (ed.). Oxford: Blackwell, pp.72-103.

Matthews, P. (1993). A Slowly Mixing Markov Chain With Implications for Gibbs Sampling. *Statistics & Probability Letters* **17**, 231-236.

Mayer, M. (1951). Report on a Monte Carlo Calculation Performed with the Eniac. In *Monte Carlo Method*, A. S. Householder, G. E. Forsyth and H. H. Germond (eds.). Applied Mathematics Series 12, Washington: National Bureau of Standards, pp.19-20.

McArdle, J. J. (1976). Empirical Test of Multivariate Generators. In *Proceedings of the Ninth Annual Symposium on the Interface of Computer Science and Statistics*, D. C. Hoaglin and R. Welsch (eds.). Boston: Prindle, Weber, and Schmidt, pp.263-267.

McCloskey, D. N. and Ziliak, S. T. (1996). The Standard Error of Regressions. *Journal of Economic Literature* **34**, 97-114.

McCullagh, P. (1983). Quasi-Likelihood Functions. *Annals of Statistics* **11**, 59-67.

McCullagh, P. and Nelder, J. A. (1989). *Generalized Linear Models*. Second Edition. New York: Chapman & Hall.

McCulloch, C. E. (1994). Maximum Likelihood Variance Components Estimation for Binary Data. *Journal of the American Statistical Association* **89**, 330-335.

McCulloch, C. E. and Searle, S. R. (2001). *Generalized, Linear, and Mixed Models.* New York: John Wiley & Sons.

McCulloch, R. E. and Rossi, P. E. (1992). Bayes Factors for Nonlinear Hypotheses and Likelihood Distributions. *Biometrika* **79**, 663-676.

McCullough, B. D. (1998). Assessing the Reliability of Statistical Software: Part I. *The American Statistician* **52**, 358-366.

McCullough, B. D. (1999). Assessing the Reliability of Statistical Software: Part II. *The American Statistician* **53**, 149-159.

McCullough, B. D. and Wilson, B. (1999). On the Accuracy of Statistical Procedures in Microsoft Excel 97. *Computational Statistics and Data Analysis* **31**, 27-37.

McDonald, G. C. (1999). Letter to the Editor. *The American Statistician* **53**, 393.

McGrath, K. and Waterton, J. (1986). British Social Attitudes, 1983-86 Panel Survey. London, Social and Community Planning Research.

McGuire, J. W. (1996). Strikes in Argentina: Data Sources and Recent Trends. *Latin American Research Review* **31**, 127-150.

McKendrick, A. G. (1926). Applications of Mathematics to Medical Problems. *Proceedings of the Edinburgh Mathematical Society* **44**, 98-130.

McLachlan, G. J. and Basford, K. E. (1988). *Mixture Models: Inference and Application to Clustering.* New York: Marcel Dekker.

McLachlan, G. J. and Krishnan, T. (1997). *The EM Algorithm and Extensions.* New York: John Wiley & Sons.

Mead, M. (1973). *Coming of Age in Samoa.* New York: Morrow.

Mebane, W. R., Jr. (1994). Fiscal Constraints and Electoral Manipulation in American Social Welfare. *American Political Science Review* **88**, 77-94.

Medvedev, R. (1989). *Let History Judge: The Origins and Consequences of Stalinism.* New York: Columbia University Press.

Meehl, P. E. (1978). Theoretical Risks and Tabular Asterisks: Sir Karl, Sir Ronald, and the Slow Progress of Soft Psychology. *Journal of Counseling and Clinical Psychology* **46**, 806-834.

Meehl, Paul E. (1990). Why Summaries of Research on Psychological Theories Are Often Uninterpretable. *Psychological Reports* **66**, 195-244.

Meehl, P. E. (1997). The Problem Is Epistemology, Not Statistics: Replace Significance Tests By Confidence Intervals and Quantify Accuracy of Risky Numerical Predictions. In *What If There Were No Significance Tests*, L. L. E. Harlow, S. A. Mulaik and J. H. Steiger (eds.) Lawrence Erlbaum Associates Publishers, pp.393-425.

Meeus, W., van de Schoot, R., Keijsers, L., Schwartz, S. J. and Branje, S. (2010). On the Progression and Stability of Adolescent Identity Formation: A Five-Wave Longitudinal Study in Early-to-Middle and Middle-to-Late Adolescence. *Child Development* **81**, 1565-1581.

Meier, K. J. and Gill, J. (2000). *What Works: A New Approach to Program and Policy Analysis.* Boulder, CO: Westview Press.

Meier, K. J. and Keiser, L. R. (1996). Public Administration as a Science of the Artificial: A Methodology for Prescription. *Public Administration Review* **56**, 459-466.

Meier, K. J., Polinard, J. L. and Wrinkle, R. (2000). Bureaucracy and Organizational Performance: Causality Arguments about Public Schools. *American Journal of Political Science* **44**, 590-602.

Meier, K. J. and Smith, K. B. (1994). Representative Democracy and Representative Bureaucracy. *Social Science Quarterly* **75**, 798-803.

Meilijson, I. (1989). A Fast Improvement to the EM Algorithm on Its Own Terms. *Journal of the Royal Statistical Society, Series B* **51**, 127-138.

Meng, X-L. (1994a). Posterior Predictive *p*-Values. *Annals of Statistics* **22**, 1142-1160.

Meng, X-L. (1994b). On the Rate of Convergence of the ECM Algorithm. *Annals of Statistics* **22**, 326-339.

Meng, X-L. (2000). Towards a More General Propp-Wilson Algorithm: Multistage Backward Coupling. *Monte Carlo Methods-Fields Institute Communications* **26**, 85-93.

Meng, X-L. and Pedlow, S. (1992). EM: A Bibliographic Review With Missing Articles. *Proceedings of the Statistical Computing Section, American Statistical Association.* Alexandria, VA: American Statistical Association. 24-27.

Meng, X-L. and Rubin, D. B. (1991). Using EM to Obtain Asymptotic Variance Covariance Matrices: the SEM Algorithm. *Journal of the American Statistical Association* **86**, 899-909.

Meng, X-L. and Rubin, D. B. (1993). Maximum Likelihood Estimation via the ECM Algorithm: A General Framework. *Biometrika* **80**, 267-278.

Meng, X-L. and Schilling, S. (1996). Fitting Full-Information Item Factor Models and an Empirical Investigation of Bridge Sampling. *Journal of the American Statistical Association* **91**, 1254-1267.

Meng, X-L. and van Dyk, D. A. (1999). Seeking Efficient Data Augmentation Schemes Via Conditional and Marginal Augmentation. *Biometrika* **86**, 301-320,

Mengersen, K. L. and Robert, C. P. (1996). Testing for Mixtures: A Bayesian Entropic Approach. In *Bayesian Statistics 5*, J. O. Berger, J. M. Bernardo, A. P. Dawid, D. V. Lindley and A. F. M. Smith (eds.). Oxford: Oxford University Press, pp.255-276.

Mengersen, K. L. and Tweedie, R. L. (1996). Rates of Convergence of the Hastings and Metropolis Algorithms. *Annals of Statistics* **24**, 101-121.

Mengersen, K. L., Robert, C. P. and Guihenneuc-Jouyaux, C. (1999). MCMC Convergence Diagnostics: A Reviewww. In *Bayesian Statistics 6*, J. O. Berger, J. M. Bernardo, A. P. Dawid, D. V. Lindley and A. F. M. Smith (eds.). Oxford: Oxford University Press, pp.415-440.

Menzefricke, U. (1981). A Bayesian Analysis of a Change in the Precision of a Sequence of Independent Normal Random Variables at an Unknown Time Point. *Applied Statistics* **30**, 141-146.

Metropolis, N. and Ulam, S. (1949). The Monte Carlo Method. *Journal of the American Statistical Association* **44**, 335-341.

Metropolis, N., Rosenbluth, A. W., Rosenbluth, M. N., Teller, A. H. and Teller E. (1953). Equation of State Calculations by Fast Computing Machines. *Journal of Chemical Physics* **21**, 1087-1091.

Meyer, K. R. and Hall, G. R. Jr. (1992). *Introduction to Hamiltonian Dynamical Systems and the N-Body Problem.* New York: Springer-Verlag.

Meyn, S. P. and Tweedie, R. L. (1993). *Markov Chains and Stochastic Stability.* New York: Springer-Verlag.

Meyn, S. P. and Tweedie, R. L. (1994). Computable Bounds for Convergence Rates of Markov Chains. *Annals of Applied Probability* **4**, 981-1011.

Mihram, G. A. and Mihram, D. (1997). A Review and Update on Pseudo-Random Number Generation, on Seeding, and on a Source of Seeds. *ASA Proceedings of the Statistical Computing Section.* Alexandria, VA: American Statistical Association. 115-119.

Miller, A. (2002). *Subset Selection in Regression.* Second Edition. Boca Raton: Chapman & Hall/CRC.

Mira, A. and Tierney, L. (2001a). On the Use of Auxiliary Variables in Markov Chain Monte Carlo Sampling. *Scandanavian Journal of Statistics* **29**, 1-12.

Mira, A. and Tierney, L. (2001b). Efficiency and Convergence Properties of Slice Samplers. *Scandinavian Journal of Statistics* **29**, 1035-53.

Mitchell, T. J. and Beauchamp, J. J. (1988). Bayesian Variable Selection in Linear Regression. *Journal of the American Statistical Association* **83**, 1023-1032.

Mitra, D., Romeo, F. and Sangiovanni-Vincentelli, A. L. (1986). Convergence and Finite-Time Behavior of Simulated Annealing. *Advances in Applied Probability* **18**, 747-771.

Mkhadri, A. (1998). On the Rate of Convergence of the ECME Algorithm. *Statistics & Probability Letters* **37**, 81-87.

Møller, J. and Schladitz, K. (1999). Extensions of Fill's Algorithm for Perfect Simulation. *Journal of the Royal Statistical Society, Series B* **61**, 955-969.

Monahan, J. F. and Genz, A. (1996). A Comparison of Omnibus Methods for Bayesian Computation. *Computing Science and Statistics* **27**, 471-480.

Monahan, J. F. (2001). *Numerical Methods of Statistics.* Cambridge: Cambridge University Press.

Montgomery, J. M. and Nyhan, B. (2010). Bayesian Model Averaging: Theoretical Developments and Practical Applications. *Political Analysis* **18**, 245-270.

Mooney, C. Z. (1997). *Monte Carlo Simulation.* Thousand Oaks, CA: Sage.

Moore, D. S. and McCabe, G. P. (1989). *Introduction to the Practice of Statistics.* WH Freeman/Times Books/Henry Holt & Company.

Moran, J. L. and Solomon, P. J. (2004). A Farewell To p-values? *Critical Care and Resuscitation Journal* **6**, 130.

Moreno, E. (1997). Bayes Factors for Intrinsic and Fractional Priors in Nested Models. In L_1-*Statistical Procedures and Related Topics.* Y. Dodge (ed.). Hayward, CA: Institute of Mathematical Statistics Monograph Series, pp.257-270.

Moreno, E. (2000). Global Bayesian Robustness for Some Classes of Prior Distributions. In *Robust Bayesian Analysis*, David Ríos Insua and Fabrizio Ruggeri (eds.). New York: Springer-Verlag, pp.45-70.

Moreno, E. (2005). Objective Bayesian Methods for One-Sided Testing. *Test* **14**, 181-198.

Moreno, E. and Cano, J. A. (1991). Robust Bayesian Analysis with ε-Contaminations Partially Known. *Journal of the Royal Statistical Society, Series B* **53**, 143-155.

Moreno, E. and González, A. (1990). Empirical Bayes Analysis of ε-Contaminated Classes of Prior Distributions. *Brazilian Journal of Probability and Statistics* **4**, 177-200.

Moreno, E. and Pericchi, L. R. (1991). Robust Bayesian Analysis for ε- Contaminations with Shape and Quantile Restraints. In *Proceedings of the Fifth International Symposium on Applied Stochastic Models*, R. Gutiéterrez and M. Valderrama (eds.). Singapore: World Scientific, pp.454-470.

Moreno, E. and Pericchi, L. R. (1993). Bayesian Robustness for Hierarchical, ε- Contamination Models. *Journal of Statistical Planning and Inference* **37**, 159-168.

Moreno, E., Martínez, C. and Cano, J. A. (1996). Local Robustness and Influences for Contamination Classes of Prior Distributions. In *Bayesian Robustness*, J. O. Berger, B. Betró, E. Moreno, L. R. Pericchi, F. Ruggeri, G. Salinetti and L. Wasserman (eds.). Hayward, CA: Institute of Mathematical Statistics Monograph Series 29, pp.139-156.

Morey, R. D., Rouder, J. N., Pratte, M. S. and Speckman, P. L. (2011). Using MCMC Chain Outputs to Efficiently Estimate Bayes Factors. *Journal of Mathematical Psychology* **55**, 368-378.

Morgan, B. J. T. (1984). *Elements of Simulation.* London: Chapman & Hall.

Morris, C. N. (1982). Natural Exponential Families with Quadratic Variance Functions. *Annals of Statistics* **10**, 65-80.

Morris, C. N. (1983a). Natural Exponential Families with Quadratic Variance Functions: Statistical Theory. *Annals of Statistics* **11**, 515-529.

Morris, C. N. (1983b). Parametric Empirical Bayes Inference: Theory and Applications. *Journal of the American Statistical Association* **78**, 47-65.

Morrison, D. E. and Henkel, R. E. (1969). Statistical Tests Reconsidered. *The American Sociologist* **4**, 131-140.

Morrison, D. E. and Henkel, R. E. (1970). *The Significance Test Controversy–A Reader.* Chicago: Aldine.

Mouchart, M. and Simar, L. (1984). A Note on Least-Squares Approximation in the Bayesian Analysis of Regression Models. *Journal of the Royal Statistical Society, Series B* **46**, 124-133.

Müller, M. E. (1958). An Inverse Method for the Generation of Random Normal Deviates on Large Scale Computers. *Mathematical Tables and Other Aids to Computation* **12**, 167-174.

Müller, M. E. (1959a). A Comparison of Methods for Generating Normal Deviates on Digital Computers. *Journal of the Association for Computing Machinery* **6**, 376-383.

Müller, M. E. (1959b). A Note on a Method for Generating Points Uniformly on N-dimensional Spheres. *Communications of the Association for Computing Machinery* **2**, 19-20.

Murata, N., Yoshizawa, S. and Amari, S. (1994). Network Information Criterion-Determining the Number of Hidden Units for Artificial Neural Network Models. *IEEE Transactions on Neural Networks* **5**, 865-872.

Murdoch, D. J. and Green, P. J. (1998). Exact Sampling From a Continuous State Space. *Scandinavian Journal of Statistics* **25**, 483-502.

Murdoch, D. J. and Rosenthal, J. S. (1998). An Extension of Fill's Exact Sampling Algorithm to Non-Monotone Chains. Technical Report, `http://www.probability.ca/jeff/ftpdir/fill.pdf`.

Murdoch, D. J. and Takahara, G. (2006). Perfect Sampling for Queues and Network Models. *ACM Transactions on Modeling and Computer Simulation* **16**, 76-92.

Murray, G. D. (1977). Contribution to Discussion of Paper by A. P. Dempster, N. M. Laird and D. B. Rubin. *Journal of the Royal Statistical Society, Series B* **39**, 27-28.

Muthén, B. and Satorra, A. (1995). Complex Sample Data in Structural Equation Modeling. *Sociological Methodology* **25**, 87-99.

Mykland, P., Tierney, L. and Yu, B. (1995). Regeneration in Markov Chain Samplers. *Journal of the American Statistical Association* **90**, 233-241.

Nance, R. E. and Overstreet, C. (1972). A Bibliography on Random Number Generation. *ACM Computing Reviews* **13**, 495-508.

Nandram, B. and Chen, M-H. (1996). Reparameterizing the Generalized Linear Model to Accelerate Gibbs Sampler Convergence. *Journal of Statistical Computing and Simulation* **54**, 129-144.

Natarajan, R. and Kass, R. E. (2000). Reference Bayesian Methods for Generalized Linear Mixed Models. *Journal of the American Statistical Association* **95**, 227-237.

Natarajan, R. and McCulloch, C. E. (1995). A Note on the Existence of the Posterior Distribution for a Class of Mixed Models for Binomial Responses. *Biometrika* **82**, 639-643.

Naylor, J. C. and Smith, A. F. M. (1982). Applications of a Method for the Efficient Computation of Posterior Distributions. *Applied Statistics* **31**, 214-225.

Naylor, J. C. and Smith, A. F. M. (1983). A Contamination Model in Clinical Chemistry: An Illustration of a Method for the Efficient Computation of Posterior Distributions. *The Statistician* **32**, 82-87.

Neal, R. M. (2011). MCMC Using Hamiltonian Dynamics. In *Handbook of Markov Chain Monte Carlo*, Steve Brooks, Andrew Gelman, Galin L. Jones and Xiao-Li Meng (eds.). Boca Raton: Chapman & Hall/CRC, pp.113-162.

Neal, R. M. (1996). Sampling from Multimodal Distributions Using Tempered Transitions. *Statistics and Computing* **4**, 353-66.

Neal, R. M. (2001). Annealed Importance Sampling. *Statistics and Computing* **11**, 125-139.

Neal, R. M. (2003). Slice Sampling. *Annals of Statistics* **31**, 705-767.

Neal, R. M. (1993). "Bayesian Training of Backpropagation Networks by the Hybrid Monte Carlo Method." Technical Report CRG-TR-92-1, Department of Computer Science, University of Toronto.

Neave, H. R. (1973). On Using the Box-Müller Transformation with Multiplicative Congruential Pseudo-Random Number Generators. *Applied Statistics* **22**, 92-97.

Neftçi, S. N. (1982). Specification of Economic Time Series Models Using Akaike's Criterion. *Journal of the American Statistical Association* **77**, 537-540.

Nelder, J. A. (1977). Comments on "Maximum Likelihood From Incomplete Data Via the EM Algorithm." *Journal of the Royal Statistical Society, Series B* **39**, 23-24.

Nelder, J. A. (1985). Quasi-likelihood and GLIM. *Lecture Notes in Statistics* **32**, 120-127.

Nelder, J. A. and Lee, Y. (1992). Likelihood, Quasi-likelihood and Pseudo-likelihood: Some Comparisons. *Journal of the Royal Statistical Society, Series B* **54**, 273-284.

Nelder, J. A. and Pregibon, D. (1987). An Extended Quasi-Likelihood Function. *Biometrika* **74**, 221-232.

Nelder, J. A. and Wedderburn, R. W. M. (1972). Generalized Linear Models. *Journal of the Royal Statistical Society, Series A* **135**, 370-385.

Nevzorov, V. B. and Zhukova, E. E. (1996). Wiener Process and Order Statistics. *Journal of Applied Statistical Science* **3**, 317-323.

Newcomb, S. (1886). A Generalized Theory of the Combination of Observations So As to Obtain the Best Results. *American Journal of Mathematics* **8**, 343-366.

Newton, M. A. and Raftery, A. E. (1994). Approximate Bayesian Inference with the Weighted Likelihood Bootstrap. *Journal of the Royal Statistical Society, Series B* **56**, 3-48.

Neyens, T., Faes, C. and Molenberghs, G. (2012). A Generalized Poisson-Gamma Model For Spatially Overdispersed Data. Spatial and Spatio-Temporal Epidemiology, **3)** 185-194.

Neyman, J. and Pearson, E. S. (1928a). On the Use and Interpretation of Certain Test Criteria for Purposes of Statistical Inference. Part I. *Biometrika* **20A**, 175-240.

Neyman, J. and Pearson, E. S. (1928b). On the Use and Interpretation of Certain Test Criteria for Purposes of Statistical Inference. Part II. *Biometrika* **20A**, 263-294.

Neyman, J. and Pearson, E. S. (1933a). On the Problem of the Most Efficient Test of Statistical Hypotheses. *Philosophical Transactions of the Royal Statistical Society, Series A* **231**, 289-337.

Neyman, J. and Pearson, E. S. (1933b). The Testing of Statistical Hypotheses in Relation to Probabilities. *Proceedings of the Cambridge Philosophical Society* **24**. 492-510.

Neyman, J. and Pearson, E. S. (1936a). Contributions to the Theory of Testing Statistical Hypotheses. *Statistical Research Memorandum* **1**, 1-37.

Neyman, J. and Pearson, E. S. (1936b). Sufficient Statistics and Uniformly Most Powerful Tests of Statistical Hypotheses. *Statistical Research Memorandum* **1**, 113-137.

Ni, S. and Sun, D. (2003). Noninformative Priors and Frequentist Risks of Bayesian Estimators in Vector Autoregressive Models. *Journal of Econometrics* **115**, 159-197.

Nickerson, R.S. (2000). Null Hypothesis Significance Testing: A Review of an Old and Continuing Controversy. *Psychological Methods* **5**, 241-301.

Nicolaou, A. (1993). Bayesian Intervals With Good Frequentist Behaviour in the Presence of Nuisance Parameters. *Journal of the Royal Statistical Society, Series B* **55**, 377-390.

Niederreiter, H. (1972). On the Distribution of Pseudo-Random Numbers Generated by the Linear Congruential Method. *Mathematics of Computation* **26**, 793-795.

Niederreiter, H. (1974). On the Distribution of Pseudo-Random Numbers Generated by the Linear Congruential Method, II. *Mathematics of Computation* **28**, 1117-1132.

Niederreiter, H. (1976). On the Distribution of Pseudo-Random Numbers Generated by the Linear Congruential Method, III. *Mathematics of Computation* **30**, 571-597.

Nigm, A. M. and Handy, H. I. (1987). Bayesian Prediction Bounds for the Pareto Lifetime Model. *Communications in Statistics* **16**, 1761-1722.

Norrander, B. (2000). The Multi-Layered Impact of Public Opinion on Capital Punishment Implementation in the American States. *Political Research Quarterly* **53**, 771-793.

Norris, J. R. (1997). *Markov Chains*. Cambridge: Cambridge University Press.

Novick, M. R. (1969). Multiparameter Bayesian Indifference Procedures. *Journal of the Royal Statistical Society, Series B* **31**, 29-64.

Novick, M. R., Isaacs, G. L. and DeKeyrel, D. F. (1976). CADA User's Manual-1976. Iowa City: Iowa Testing Programs, The University of Iowa.

Novick, M. R. and Hall, W. J. (1965). A Bayesian Indifference Procedure. *Journal of the American Statistical Association* **60**, 1104-1117.

Ntzoufras, I. (2009). *Bayesian Modeling Using WinBUGS.* New York: John Wiley & Sons.

Nummelin, E. (1984). *General Irreducible Markov Chains and Non-negative Operators.* Cambridge: Cambridge University Press.

Oakes, M. (1986). *Statistical Inference: A Commentary for the Social and Behavioral Sciences.* New York: John Wiley & Sons.

O'Hagan, A. (1994). *Kendall's Advanced Theory of Statistics: Volume 2B, Bayesian Inference.* London: Arnold.

O'Hagan, A. (1995). Fractional Bayes Factors for Model Comparison. *Journal of the Royal Statistical Society, Series B* **57**, 99-138.

O'Hagan, A. (1998). Eliciting Expert Beliefs in Substantial Practical Applications. *The Statistician* **47**, 21-35.

O'Hagan, A. and Berger, J. O. (1988). Ranges of Posterior Probabilities for Quasiunimodal Priors With Specified Quantiles. *Journal of the American Statistical Association* **83**, 503-508.

Oh, M.-S. and Berger, J. O. (1992). Adaptive Importance Sampling in Monte Carlo Integration. *Journal of Statistical Computation and Simulation* **41**, 143-168.

Olivera, S. and Gill, J. (2011). Parallel Gibbs Sampling with `snowfall`. *The Political Methodologist* **19**, 4-7.

Orans, M. (1996). *Not Even Wrong: Margaret Mead, Derek Freeman, and the Samoans.* Novato, CA: Chandler & Sharp.

Orchard, T. and Woodbury, M. A. (1972). A Missing Information Principle: Theory and Applications. *Proceedings of the 6th Berkeley Symposium on Mathematical Statistics and Probability* **1**. 697-715.

Orey, S. (1961). Strong Ratio Limit Property. *Bulletin of the American Mathematical Society* **67**, 571-574.

Orton, C. (1997). Testing Significance or Testing Credulity? *Oxford Journal of Archaeology* **16**, 219.

Pagan, A. (1987). Three Econometric Methodologies: A Critical Appraisal. *Journal of Economic Surveys* **1**, 3-24.

Pang, X. (2010). Modeling Heterogeneity and Serial Correlation in Binary Time-Series Cross-sectional Data: A Bayesian Multilevel Model with AR(p) Errors. *Political Analysis* **18**, 470-498.

Pang, X. and Gill, J. (2012). Spike and Slab Prior Distributions for Simultaneous Bayesian Hypothesis Testing, Model Selection, and Prediction, of Nonlinear Outcomes. Technical Report, `http://jgill.wustl.edu/research/current.html`.

Pasta, J. and Ulam, S. (1953). Heuristic Studies in Problems of Mathematical Physics on High Speed Computing Machines. Technical Report, Los Alamos Scientific Lab.

Patterson, R. L. and Richardson, W. (1963). A Decision Theoretic Model for Determining Verification Requirements. *The Journal of Conflict Resolution* **7**, 603-607.

Patterson, T. N. L. (1968). The Optimum Addition of Points to Quadrature Formulae. *Mathematics of Computation* **23**, 847-856.

Payne, W. H. (1977). Normal Random Numbers: Using Machine Analysis to Choose the Best Algorithm. *ACM Transactions on Mathematical Software* **3**, 346-358.

Peach, C. (1997). Postwar Migration to Europe: Reflux, Influx, Refuge. *Social Science Quarterly* **78**, 269-283.

Pearson, K. (1892). *The Grammar of Science.* London: Walter Scott.

Pearson, K. (1900). On the Criterion that a Given System of Deviations from the Probable in the Case of a Correlated System of Variables is Such That It Can Reasonably be Supposed to Have Arisen From Random Sampling. *Philosophical Magazine*, 5th Series, **50**, 157-175.

Pearson, K. (1907). On the Influence of Past Experience on Future Expectation. *Philosophical Magazine*, 6th Series, **13**, 365-378.

Pearson, K. (1914). On the Probability that Two Independent Distributions of Frequency are Really Samples of the Same Population, with Special Reference to Recent Work on the Identity of Trypanosome Strains. *Biometrika* **10**, 85-143.

Pearson, K. (1920). The Fundamental Problem of Practical Statistics. *Biometrika* **13**, 1-16.

Pennington, R. H. (1970). *Introductory Computer Methods and Numerical Analysis*. Second Edition. London: Collier-MacMillan.

Pericchi, L. R. and Nazaret, W. (1988). On Being Imprecise at the Higher Levels of a Hierarchical Linear Model. In *Bayesian Statistics 3*, J. M. Bernardo, M. H. DeGroot, D. V. Lindley and A. F. M. Smith (ed.). Oxford: Oxford University Press, pp.569-577.

Perkins, W. C. and Menzefricke, U. (1975). A Better Additive Congruential Random Number Generator? *Decision Sciences* **6**, 194-198.

Perks, W. (1947). Some Observations on Inverse Probability Including A New Indifference Rule. *Journal of the Institute of Actuaries* **73**, 285-312.

Peskun, P. H. (1973). Optimum Monte Carlo Sampling Using Markov Chains. *Biometrika* **60**, 607-612.

Petrone, S. and Raftery, A. E. (1997). A Note on the Dirichlet Process Prior in Bayesian Nonparametric Inference With Partial Exchangeability. *Statistics & Probability Letters* **36**, 69-83.

Pettit, L. I. (1992). Bayes Factors for Outlier Models Using the Device of Imaginary Observations. *Journal of the American Statistical Association* **87**, 541-545.

Pettit, L. I. and Smith, A. F. M. (1985) Outliers and Influential Observations in Linear Models. In *Bayesian Statistics 2*, J. M. Bernardo, M. H. DeGroot, D. V. Lindley and A. F. M. Smith (eds.). Amsterdam: North Holland Press, pp.473-494.

Pettitt, A. N., Tran, T. T., Haynes, M. A. and Hay, J. L. (2006). A Bayesian Hierarchical Model for Categorical Longitudinal Data From a Social Survey of Immigrants. *Journal of the Royal Statistical Society, Series A* **169**, 97-114.

Phillips, D. B. and Smith, A. F. M. (1996). Bayesian Model Comparison Via Jump Diffusions. In *Markov Chain Monte Carlo in Practice*, W. R. Gilks, S. Richardson and D. J. Spiegelhalter (eds.). New York: Chapman & Hall, pp.214-240.

Phillips, D. T. and Beightler, C. S. (1972). Procedure for Generating Gamma Variates with Non-Integer Parameter Sets. *Journal of Statistical Computation and Simulation* **1**, 197-208.

Phillips, P. C. B. (1991). To Criticize the Critics: An Objective Bayesian Analysis of Stochastic Trends. *Journal of Applied Econometrics* **6**, 333-364.

Phillips, P. C. B. (1995). Bayesian model selection and prediction with empirical applications. *Journal of Econometrics* **69**, 289-331.

Pitman, E. J. G. (1936). Sufficient Statistics and Intrinsic Accuracy. *Proceedings of the Cambridge Philosophical Society* **32**. 567-579.

Pizer, S. M. (1975). *Numerical Computing and Mathematical Analysis*. Chicago: Science Research Associates.

Placket, R. L. (1966). Current Trends in Statistical Inference. *Journal of the Royal Statistical Society, Series A* **129**, 249-267.

Poirer, D. J. (1988). Frequentist and Subjectivist Perspectives on the Problems of Model Building in Economics. *Journal of Economic Perspectives* **2**, 121-144.

Poirer, D. J. (1994). Jeffreys Prior for Logit Models. *Journal of Econometrics* **63**, 327-339.

Polasek, W. (1984). Regression Diagnostics for General Linear Regression Models. *Journal of the American Statistical Association* **79**, 336-340.

Polasek, W. (1987). Bounds on Rounding Errors in Linear Regression Models. *The Statistician* **36**, 221-227.

Pole, A., West, M. and Harrison, J. (1994). *Applied Bayesian Forecasting and Time Series Analysis.* New York: Chapman & Hall.

Pollard, W. E. (1986). *Bayesian Statistics for Evaluation Research.* Thousand Oaks, CA: Sage.

Pollard, P. (1993). How Significant is 'Significance'? In *A Handbook for Data Analysis in the Behavioral Sciences: Methodological Issues*, G. Keren and C. Lewis (eds.). Hillsdale, NJ: Lawrence Erlbaum Associates, pp.448-460.

Pollard, P. and Richardson, J. T. E. (1987). On the Probability of Making Type One Errors. *Psychological Bulletin* **102**, 159-163.

Polson, N. G. (1996). Convergence of Markov Chain Monte Carlo Algorithms. In *Bayesian Statistics 5*, J. O. Berger, J. M. Bernardo, A. P. Dawid, D. V. Lindley and A. F. M. Smith (eds.). Oxford: Oxford University Press.

Polson, N. G. and Scott, J. G. (2012). On the Half-Cauchy Prior For a Global Scale Parameter. *Bayesian Analysis* **7**, 887-902.

Popper, K. (1968). *The Logic of Scientific Discovery.* New York: Harper and Row.

Pratt, J. W. (1965). Bayesian Interpretation of Standard Inference Statements. With Discussion. *Journal of the Royal Statistical Society, Series B* **27**, 169-203.

Prentice, M. J. and Miller, J. C. P. (1968). Additive Congruential Pseudo-Random Number Generators. *Computer Journal* **11**, 341-346.

Press, S. J. (1989). *Bayesian Statistics: Principles, Models, and Applications.* New York: John Wiley & Sons.

Press, S. J. and Tanur, J. M. (2001). *The Subjectivity of Scientists and the Bayesian Approach.* New York: John Wiley & Sons.

Press, W. H., Flannery, B. P., Teukolsky, S. A. and Vetterling, W. T. (1986). *Numerical Recipes: The Art of Scientific Computing.* Cambridge: Cambridge University Press.

Priestley, M. B. (1981). *Spectral Analysis and Time Series: Volumes I and II.* San Diego: Academic Press.

Propp, J. G. and Wilson, D. B. (1996). Exact Sampling with Coupled Markov Chains and Applications to Statistical Mechanics. *Random Structures and Algorithms* **9**, 223-252.

Quinn, K. M., Martin, A. and Whitford, A. B. (1999). Voter Choice in Multi-Party Democracies: A Test of Competing Theories and Models. *American Journal of Political Science* **43**, 1231-1247.

Racine, A., Grieve, A. P., Flühler, H. and Smith, A. F. M. (1986). Bayesian Methods in Practice: Experiences in the Pharmaceutical Industry. *Applied Statistics* **45**, 275-309.

Raftery, A. E. (1995). Bayesian Model Selection in Social Research. *Sociological Methodology* **25**, 111-164.

Raftery, A. E. (1996). Hypothesis Testing and Model Selection. In *Markov Chain Monte Carlo in Practice*, W. R. Gilks, S. Richardson and D. J. Spiegelhalter (eds.). New York: Chapman & Hall, pp.163-188.

Raftery, A. E. (1999). Bayesian Model Selection in Social Research. *Sociological Methodology* **25**, 111-163.

Raftery, A. E. and Adman, V. E. (1986). Bayesian Analysis of a Poisson Process with a Change-Point. *Biometrika* **73**, 85-89.

Raftery, A. E. and Banfield, J. D. (1991). Stopping the Gibbs Sampler, the Use of Morphology, and Other Issues in Spatial Statistics. *Annals of the Institute of Statistical Mathematics* **43**, 32-43.

Raftery, A. E. and Lewis, S. M. (1992). How Many Iterations in the Gibbs Sampler? In *Bayesian Statistics 4*, J. M. Bernardo, A. F. M. Smith, A. P. Dawid and J. O. Berger (eds.). Oxford: Oxford University Press, pp.763-773.

Raftery, A. E. and Lewis, S. M. (1996). Implementing MCMC. In *Markov Chain Monte Carlo in Practice*, W. R. Gilks, S. Richardson and D. J. Spiegelhalter (eds.). New York: Chapman & Hall, pp.115-130.

Raiffa, H. and Schlaifer, R. (1961). *Applied Statistical Decision Theory.* Cambridge: Harvard School ot Business Administration.

Ramsey, J. O. and Novick, M. R. (1980). PLU Robust Bayesian Decision Theory: Point Estimation. *Journal of the American Statistical Association* **75**, 901-907.

Rao, C. R. and Toutenburg, H. (1995). *Linear Models: Least Squares and Alternatives.* New York: Springer-Verlag.

Rathbun, S. L. and Black, B. (2006). Modeling and Spatial Prediction of Pre-Settlement Patterns of Forest Distribution Using Witness Tree Data. *Environmental and Ecological Statistics* **13**, 427-448.

Raudenbush, S. and Bryk, A. S. (1986). A Hierarchical Model for Studying School Effects. *Sociology of Education* **59**, 1-17.

Ravishanker, N. and Dey, D. K. (2002). *A First Course In Linear Model Theory.* New York: Chapman & Hall/CRC.

Regazzini, E. (1992). Concentration Comparisons Between Probability Measures. *Sankhyā, Series B* **54**, 129-149.

Revuz, D. (1975). *Markov Chains.* Amsterdam: North-Holland.

Richardson, S. and Green, P. J. (1997). On Bayesian Analysis of Mixtures with an Unknown Number of Components. *Journal of the Royal Statistical Society, Series B* **59**, 731-732.

Richey, M. (2010). The Evolution of Markov Chain Monte Carlo Methods. *The American Mathematical Monthly* **117**, 383-413.

Ripley, B. D. (1979). Algorithm AS 137: Simulating Spatial Patterns: Dependent Samples from a Multivariate Density. *Applied Statistics* **28**, 109-112.

Ripley, B. D. (1983). Computer Generation of Random Variables-A Tutorial. *International Statistical Review* **51**, 301-319.

Ritter, C. and Tanner, M. A. (1992). Facilitating the Gibbs Sampler: The Gibbs Stopper and the Griddy-Gibbs Sampler. *Journal of the American Statistical Association* **87**, 861-868.

Robbins, H. (1955). An Empirical Bayes Approach to Statistics. In *Proceedings of the 3rd Berkeley Symposium on Mathematical Statistics and Probability* **1**, Berkeley: University of California Press. pp.157-164.

Robbins, H. (1964). The Empirical Bayes Approach to Statistical Decision Problems. *Annals of Mathematical Statistics* **35**, 1-20.

Robbins, H. (1983). Some Thoughts on Empirical Bayes Estimation. *Annals of Statistics* **1**, 713-723.

Robert, C. P. (1995). Convergence Control Methods for Markov Chain Monte Carlo Algorithms. *Statistical Science* **10**, 231-253.

Robert, C. P. (1996). Mixtures of Distributions: Inference and Estimation. In *Markov Chain Monte Carlo in Practice*, W. R. Gilks, S. Richardson and D. J. Spiegelhalter (eds.). New York: Chapman & Hall, pp.441-464.

Robert, C. P. (1997). Discussion of Richardson and Green's Paper. *Journal of the Royal Statistical Society, Series B* **59**, 758-764.

Robert, C. P. (2001). *The Bayesian Choice: A Decision Theoretic Motivation.* Second Edition. New York: Springer-Verlag.

Robert, C. P. and Casella, G. (1999). *Monte Carlo Statistical Methods.* First Edition. New York: Springer-Verlag.

Robert, C. P. and Casella, G. (2004). *Monte Carlo Statistical Methods.* Second Edition. New York: Springer-Verlag.

Robert, C. P. and Casella, G. (2011). A Short History of Markov Chain Monte Carlo: Subjective Recollections from Incomplete Data. *Statistical Science* **26**, 102-115.

Robert, C. P. and Mengersen, K. L. (1999). Reparameterization Issues in Mixture Estimation and Their Bearings on the Gibbs Sampler. *Computational Statistics and Data Analysis* **29**, 325-343.

Robert, C. P. and Cellier, D. (1998). Convergence Control of MCMC Algorithms. In *Discretization and MCMC Convergence Assessment, Lecture Notes in Statistics*, **135**. Christian P. Robert (ed.). New York: Springer-Verlag, pp.27-46.

Robert, C. P. and Richardson, S. (1998). Markov Chain Monte Carlo Methods. In *Discretization and MCMC Convergence Assessment, Lecture Notes in Statistics*, **135**. Christian P. Robert (ed.). New York: Springer-Verlag, pp.1-25.

Roberts, G. O. (1992). Convergence Diagnostics of the Gibbs Sampler. In *Bayesian Statistics 4*, J. M. Bernardo, A. F. M. Smith, A. P. Dawid and J. O. Berger (eds.). Oxford: Oxford University Press, pp.775-782.

Roberts, G. O. (1994). Methods for Estimating L^2 Convergence of Markov Chain Monte Carlo. In *Bayesian Analysis in Statistics and Econometrics: Essays in Honor of Arnold Zellner*, D. Berry, K. Chaloner and J. Geweke (eds.). New York: John Wiley & Sons, pp.373-384.

Roberts, G. O. and Polson, N. G. (1994). On the Geometric Convergence of the Gibbs Sampler. *Journal of the Royal Statistical Society, Series B* **56**, 377-384.

Roberts, G. O. and Rosenthal, J. S. (1998). Markov Chain Monte Carlo: Some Practical Implications of Theoretical Results. *Canadian Journal of Statistics* **26**, 5-32.

Roberts, G. O. and Rosenthal, J. S. (1999). Convergence of the Slice Sampler Markov Chains. *Journal of the Royal Statistical Society, Series B* **61**, 643-60.

Roberts, G. O. and Rosenthal, J. S. (2011). Quantitative Non-Geometric Convergence Bounds for Independence Samplers. *Methodology and Computing in Applied Probability* **13**, 391-403.

Roberts, G. O. and Sahu, S. K. (1997). Updating Schemes, Correlation Structure, Blocking and Parameterization for the Gibbs Sampler. *Journal of the Royal Statistical Society, Series B* **59**, 291-307.

Roberts, G. O. and Smith, A. F. M. (1994). Simple Conditions for the Convergence of the Gibbs Sampler and Metropolis-Hastings Algorithms. *Stochastic Processes and their Applications* **49**, 207-216.

Roberts, G. O. and Tweedie, R. L. (1996). Geometric Convergence and Central Limit Theorems for Multidimensional Hastings and Metropolis Algorithms. *Biometrika* **83**, 95-110.

Roberts, G. O. and Rosenthal, J. S. (1998b). Two Convergence Properties of Hybrid Samplers. *Annals of Applied Probability* **8**, 397-407.

Robinson, D. H. and Levin, J. R. (1997). Research News and Comment: Reflections On Statistical and Substantive Significance, With a Slice of Replication. *Educational Researcher* **26**, 21-26.

Robinson, P. M. (1991). Consistent Nonparametric Entropy-Based Testing. *The Review of Economic Studies* **58**, 437-453.

Roethlisberger, F. and Dickson, W. (1939). *Management and the Worker*. Cambridge: Cambridge University Press.

Romney, A. K. (1999). Culture Consensus as a Statistical Model. *Current Anthropology* **40** (Supplement), S103-S115.

Rosay, A. B. and Herz, C. D. (2000). Differences in the Validity of Self-Reported Drug Use Across Five Factors in Indianapolis, Fort Lauderdale, Phoenix, and Dallas, 1994. ICPSR02706-v1. Ann Arbor, MI: Inter-university Consortium for Political and Social Research [distributor].

Rosenblatt, M. (1971). *Markov Processes. Structure and Asymptotic Behavior*. New York: Springer-Verlag.

Rosenkranz, R. D. (1977). *Inference, Method, and Decision. Towards a Bayesian Philosophy of Science*. Dordrecht: Reidel.

Rosenthal, J. S. (1993). Rates of Convergence for Data Augmentation on Finite Sample Spaces. *Annals of Applied Probability* **3**, 819-839.

Rosenthal, J. S. (1995a). Minorization Conditions and Convergence Rates for Markov Chain Monte Carlo. *Journal of the American Statistical Association* **90**, 558-566.

Rosenthal, J. S. (1995b). Rates of Convergence for Gibbs Sampling for Variance Components Models. *Annals of Statistics* **23**, 740-61.

Rosenthal, J. S. (1995c). Convergence Rates for Markov Chains. *SIAM Review* **37**, 387-405.

Rosenthal, J. S. (1996). Analysis of the Gibbs Sampler for a Model Related to James-Stein Estimators. *Statistics and Computing* **6**, 269-75.

Rosnow, R. L. and Rosenthal, J. S.. (1989). Statistical Procedures and the Justification of Knowledge in Psychological Science. *American Psychologist* 44, 1276-84.

Ross, S. (1996). *Stochastic Processes.* New York: Wiley & Sons.

Rousseeuw, P. J. and Leroy, A. M. (1987). *Robust Regression and Outlier Detection.* New York: John Wiley & Sons.

Rozeboom, W. W. (1960). The Fallacy of the Null Hypothesis Significance Test. *Psychological Bulletin.* **57**, 416-428.

Rubin, D. B. (1980). Using Empirical Bayes Techniques in the Law School Validity Studies. *Journal of the American Statistical Association* **75**, 801-827.

Rubin, D. B. (1981). The Bayesian Bootstrap. *Annals of Statistics* **9**, 130-134.

Rubin, D. B. (1984). Bayesianly Justifiable and Relevant Frequency Calculations for the Applied Statistician. *Annals of Statistics* **12**, 1151-1172.

Rubin, D. B. (1987a). A Noniterative Sampling/Importance Resampling Alternative to the Data Augmentation Algorithm for Creating a Few Imputations When Fractions of Missing Information Are Modest: the SIR Algorithm. Discussion of Tanner & Wong (1987). *Journal of the American Statistical Society* **82**, 543-546.

Rubin, D. B. (1987b). *Multiple Imputation for Nonresponse in Surveys.* New York: John Wiley & Sons.

Rubin, D. B. (1988). Using the SIR Algorithm to Simulate Posterior Distributions. In *Bayesian Statistics 3,* J. M. Bernardo, M. H. DeGroot, D. V. Lindley and A. F. M. Smith (eds.). Oxford: Oxford University Press, pp.395-402.

Rubin, D. B. (1991). EM and Beyond. *Psychometrika* **56**, 241-254.

Rubin, H. (1977). Robust Bayesian Estimation. In *Statistical Decision Theory and Related Topics II,* S. S. Gupta and D. Moore (eds.). San Diego: Academic Press, pp.351-356.

Rubin, H. (1987). A Weak System of Axioms for "Rational" Behavior and the Nonseparability of Utility From Prior. *Statistical Decisions* **5**, 47-58.

Rubinstein, R. Y. (1981). *Simulation and the Monte Carlo Method.* New York: John Wiley & Sons.

Ruggeri, F. (1990). Posterior Ranges of Functions of Parameters under Priors with Specified Quantiles. *Communications in Statistics, Part A, Theory and Methods* **19**, 127-144.

Ruggeri, F. and Wasserman, L. (1993). Infinitesimal Sensitivity of Posterior Distributions. *Canadian Journal of Statistics* **21**, 195-203.

Russell, B. (1929). *Mysticism and Logic and Other Essays.* New York: W.W. Norton & Company.

Sagan, A. (2013). Market Research and Preference Data. In *The Sage Handbook of Multilevel Modeling.* Marc A. Scott, Jeffrey S. Simonoff and Brian D. Marx (eds.). Thousand Oaks, CA: Sage Publications, pp.581-598.

Sakia, R. M. (1992). The Box-Cox Transformation Technique: A Review. *The Statistician* **41**, 169-178.

Sala-I-Martin, X. (1997). I Just Ran Two Million Regressions. *The American Economic Review* **87**, 178-183.

Samaniego, F. J. and Reneau, D. M. (1994). Toward a Reconciliation of the Bayesian and Frequentist Approach to Point Estimation. *Journal of the American Statistical Association* **89**, 947-957.

Sampson, R. J. and Raudenbush, S. W. (1999). Systematic Social Observation of Public Spaces: A New Look at Disorder in Urban Neighborhoods. *American Journal of Sociology* **105**, 603-651.

Sansó, B. and Pericchi, L. R. (1992). Near Ignorance Classes of Log-Concave Priors for the Location Model. *TEST,* **1**, 39-46.

Satten, G. A. and Longini, I. M. (1996). Markov Chains With Measurement Errors: Estimating the True Course of a Marker of the Progression of HIV Disease. *Applied Statistics* **45**, 275-309.

Savage, L. J. (1954). *The Foundations of Statistics.* New York: Wiley.

Savage, L. J. (1962). *The Foundations of Statistical Inference.* London: Methuen.

Savage, L. J. (1971). Elicitation of Personal Probabilities and Expectations. *Journal of the American Statistical Association* **66**, 783-801.

Savage, L. J. (1972). *The Foundations of Statistics.* New York: Dover Publications.

Sawa, T. (1978). Information Criteria for Discriminating among Alternative Regression Models. *Econometrica* **46**, 1273-1291.

Schafer, J. L. (1997). *Analysis of Incomplete Multivariate Data.* London: Chapman & Hall.

Schervish, M. J. (1995). *Theory of Statistics.* New York: Springer-Verlag.

Schervish, M. J. (1996). P values: What They Are and What They Are Not. *The American Statistician* **50**, 203-206.

Schervish, M. J. and Carlin, B. P. (1992). On the Convergence of Successive Substitution Sampling. *Journal of Computational and Graphical Statistics* **1**, 111-127.

Scheuer, E. M. and Stoller, D. S. (1962). On the Generation of Normal Random Vectors. *Technometrics* **4**, 278-281.

Schmidt, F. L. (1996). Statistical Significance Testing and Cumulative Knowledge in Psychology: Implications for the Training of Researchers. *Psychological Methods* **1**, 115-129.

Schmidt, F. L. and Hunter, J. E. (1977). Development of a General Solution to the Problem of Validity Generalization. *Journal of Applied Psychology* **62**, 529-40.

Schruben, L. W. (1982). Detecting Initialization Bias in Simulation Output. *Operations Research* **30**, 569-590.

Schruben, L. W., Singh, H. and Tierney, L. (1983). Optimal Tests for Initialization Bias in Simulation Output. *Operations Research* **31**, 1167-1178.

Schwarz, G. (1978). Estimating the Dimension of a Model. *Annals of Statistics* **6**, 461-464.

Scollnik, D. P. M. (2001). Actuarial Modeling with MCMC and BUGS. *North American Actuarial Journal* **5**, 95-124.

Scott, David W. (1985). Average Shifted Histograms: Effect Nonparametric Density Estimators in Several Dimensions. *Annals of Statistics* **13**, 1024-1040.

Sedlmeier, P. and Gigerenzer, G. (1989). Do Studies of Statistical Power Have an Effect on the Power of Studies? *Psychological Bulletin* **105**, 309-15.

Seewald, W. (1992). Discussion of Hills and Smith (1992). In *Bayesian Statistics 4*, J. M. Bernardo, A. F. M. Smith, A. P. Dawid and J. O. Berger (eds.). Oxford: Oxford University Press, pp.241-243.

Seidenfeld, T. (1985). Calibration, Coherence, and Scoring Rules. *Philosophy of Science* **52**, 274-294.

Seidenfeld, T., Schervish, M. J. and Kadane, J. B. (1995). A Representation of Partially Ordered Preferences. *The Annals of Statistics* **23**, 2168-2217.

Seltzer, M. H., Wong, W. H. and Bryk, A. S. (1996). Bayesian Analysis in Applications of Hierarchical Models: Issues and Methods. *Journal of Educational and Behavioral Statistics* **21**, 131-167.

Selvin, S. (1975). On the Monty Hall problem. (Letter To the Editor.) *The American Statistician* **29**, 134.

Sened, I. and Schofield, N. (2005). Multiparty Competition in Israel, 1988-96 *British Journal of Political Science* **35**, 635-663.

Serlin, R. C. and Lapsley, D. K. (1993). Rational Appraisal of Psychological Research and the Good-Enough Principle. In *A Handbook for Data Analysis in the Behavioral Sciences: Methodological Issues*, G. Keren and C. Lewis (eds.). Hillsdale, NJ: Lawrence Erlbaum Associates, pp.199-228.

Severini, T. A. (1991). On the Relationship Between Bayesian and Non-Bayesian Interval Estimates. *Journal of the Royal Statistical Society, Series B* **53**, 611-618.

Severini, T. A. (1993). Bayesian Interval Estimates Which Are Also Confidence Intervals. *Journal of the Royal Statistical Society, Series B* **55**, 533-540.

Shafer, G. (1982). Lindley's Paradox. *Journal of the American Statistical Association* **77**, 325-351.

Shannon, C. (1948). A Mathematical Theory of Communication. *Bell System Technology Journal* **27**, 379-423, and 623-56.

Shao, J. (1989). Monte Carlo Approximations in Bayesian Decision Theory. *Journal of the American Statistical Association* **84**, 727-732.

Shao, J. (2005). *Mathematical Statistics.* Second Edition. New York: Springer-Verlag.

Shao, J. and Tu, D. (1995). *The Jackknife and Bootstrap.* New York: Springer-Verlag.

Sharp, A. M., Register, C. A. and Grimes, P. W. (1999). *Economics of Social Issues.* Fifteenth Edition. New York: McGraw-Hill.

Shaw, J. E. H. (1988). A Quasirandom Approach To Integration in Bayesian Statistics. *Annals of Statistics* **16**, 895-914.

Sheynin, O. B. (1977). Laplace's Theory of Errors. *Archive for History of Exact Sciences* **17**, 1-61.

Shyamalkumar, N. D. (2000). Likelihood Robustness. In *Robust Bayesian Analysis*, David Ríos Insua and Fabrizio Ruggeri (eds.). New York: Springer-Verlag, pp.127-144.

Sibuya, M. (1961). Exponential and Other Variable Generators. *Annals of the Institute of Statistical Mathematics* **13**, 231-237.

Siegmund, D. (1976). Importance Sampling in the Monte Carlo Study of Sequential Tests. *Annals of Statistics* **4**, 673-684.

Silvapulle, M. J. (1981). On the Existence of Maximum Likelihood Estimates for the Binomial Response Models. *Journal of the Royal Statistical Society, Series B* **43**, 310-313.

Sims, C. A. (1988). Bayesian Skepticism on Unit Root Econometrics. *Journal of Economic Dynamics and Control* **12**, 463-474.

Sims, C. A. and Uhlig, H. (1991). Understanding Unit Rooters: A Helicopter Tour. *Econometrica* **59**, 1591-1599.

Sinclair, A. J. and Jerrum, M. R. (1988). Conductance and the Rapid Mixing Property for Markov Chains: The Approximation of the Permanent Resolved. *Proceedings of the 20th Annual ACM Symposium on the Theory of Computing*, 235-44.

Sinclair, A. J. and Jerrum, M. R. (1989). Approximate Counting, Uniform Generation and Rapidly Mixing Markov Chains. *Information and Computation* **82**, 93-133.

Singleton, R., Jr. and Straight, B. C. (2004). *Approaches to Social Research.* Fourth Edition. New York: Oxford University Press.

Sivaganesan, S. (1993). Robust Bayesian Diagnostics. *Journal of Statistical Planning and Inference* **35**, 171-188.

Sivaganesan, S. (2000). Global and Local Robustness Approaches: Uses and Limitations. In *Robust Bayesian Analysis*, David Ríos Insua and Fabrizio Ruggeri (eds.). New York: Springer-Verlag, pp.89-108.

Sivaganesan, S. and Berger, J. O. (1989). Ranges of Posterior Measures for Priors with Unimodal Contaminations. *Annals of Statistics* **17**, 868-889.

Skates, S. J., Pauler, D. K. and Jacobs, I. J. (2001). Screening Based on the Risk of Cancer Calculation from Bayesian Hierarchical Changepoint and Mixture Models of Longitudinal Markers. *Journal of the American Statistical Association* **96**, 429-439.

Skene, A. M. (1983). Computing Marginal Distributions for the Dispersion Parameters of Analysis of Variance Models. *The Statistician* **32**, 99-108.

Skene, A. M., Shaw, E. H. and Lee, T. D. (1986). Bayesian Modeling and Sensitivity Analysis. *The Statistician* **35**, 281-288.

Skinner, B. F. (1953). *Science and Human Behavior.* Toronto: Macmillan.

Smith, A. F. M. (1973). A General Bayesian Linear Model. *Journal of the Royal Statistical Society, Series B* **35**, 61-75.

Smith, A. F. M. (1975). A Bayesian Approach to Inference About a Change-point in a Sequence of Random Variables. *Biometrika* **62**, 407-416.

Smith, A. F. M. (1984). Present Position and Potential Developments: Some Personal Views: Bayesian Statistics. *Journal of the Royal Statistical Society, Series A* **147**, 245-259.

Smith, A. F. M. (1986). Some Bayesian Thoughts on Modelling and Model Choice. *The Statistician* **35**, 97-101.

Smith, A. F. M. and Roberts, G. O. (1993). Bayesian Computation via the Gibbs Sampler and Related Markov Chain Monte Carlo Methods. *Journal of the Royal Statistical Society, Series B* **55**, 3-24.

Smith, A. F. M. and Spiegelhalter, D. J. (1980). Bayes Factors and Choice Criteria for Linear Models. *Journal of the Royal Statistical Society, Series B* **42**, 213-220.

Smith, C. A. B. (1965). Personal Probability and Statistical Analysis. *Journal of the Royal Statistical Society, Series A* **128**, 469-499.

Smith, Kevin B. and Meier, Kenneth J. (1995). *The Case Against School Choice: Politics, Markets, and Fools.* Armonk, NY: M.E. Sharpe.

Smith, R. L. (1996). The Hit-And-Run Sampler: A Globally Reaching Markov Chain Sampler for Generating Arbitrary Multivariate Distributions. *Proceedings of the 1996 Winter Simulation Conference* J. M. Charnes, D. J. Morrice, D. T. Brunner and J. J. Swain (eds.). 260-264. Coronado, CA.

Smith, W. B. and Hocking, R. R. (1972). Wishart Variate Generator. Algorithm AS53. *Applied Statistics* **21**, 241-245.

Smyth, G. K. (1989). Generalized Linear Models with Varying Dispersion. *Journal of the Royal Statistical Society, Series B* **51**, 47-60.

Sobol, I. M. (1994). *A Primer for the Monte Carlo Method.* New York: Chapman & Hall.

Solzhenitsyn, A. I. (1997). *The Gulag Archipelago. Volume I.* Thomas P. Whitney and H. Willetts, translators. Boulder, CO: Westview Press.

Sowey, E. R. (1972). A Chronological and Classified Bibliography on Random Number Generation and Testing. *International Statistical Review* **40**, 355-371.

Sowey, E. R. (1978). A Second Classified Bibliography on Random Number Generation and Testing. *International Statistical Review* **46**, 89-102.

Spall, J. C. and Hill, S. D. (1990). Least-Informative Bayesian Prior Distributions for Finite Samples Based on Information Theory. *IEEE Transactions on Automatic Control* **35**, 580-583.

Spetzler, C. S. and Staël von Holstein, C. S. (1975). Probability Encoding in Decision Analysis. *Management Science* **22**, 340-358.

Spiegelhalter, D. J., Abrams, K. R. and Myles, J. P. (2004). *Bayesian Approaches to Clinical Trials and Health-Care Evaluation.* New York: John Wiley & Sons.

Spiegelhalter, D. J., Freedman, L. S. and Parmar, M. K. B. (1994). Bayesian Approaches to Randomized Trials. *Journal of the Royal Statistical Society, Series A* **157**, 357-416.

Spiegelhalter, D. J. and Smith, A. F. M. (1982). Bayes Factors for Linear and Log-linear Models with Vague Prior Information. *Journal of the Royal Statistical Society, Series B* **44**, 377-387.

Spiegelhalter, D. J., Thomas, A., Best, N. G. and Gilks, W. R. (1996a). *BUGS 0.5*Examples: Volume 1 (version i).* MRC Biostatistics Unit: http://www.mrc-bsu.cam.ac.uk/wp-content/uploads/WinBUGS_Vol1.pdf.

Spiegelhalter, D. J., Thomas, A., Best, N. G. and Gilks, W. R. (1996b). *BUGS 0.5*Examples: Volume 2 (version i).* MRC Biostatistics Unit: http://www.mrc-bsu.cam.ac.uk/wp-content/uploads/WinBUGS_Vol2.pdf. http://www.mrc-bsu.cam.ac.uk/wp-content/uploads/WinBUGS_Vol3.pdf

Spiegelhalter, D. J., Thomas, A., Best, N. G. and Lunn, D. J. (2000). *WinBUGS User Manual Version 1.4* MRC Biostatistics Unit:
http://www.mrc-bsu.cam.ac.uk/wp-content/uploads/manual14.pdf.

Spiegelhalter, D. J., Thomas, A., Best, N. G. and Lunn, D. J. (2012). *BUGS 0.5*Examples: Volume 3 (version i)*. MRC Biostatistics Unit:
http://www.mrc-bsu.cam.ac.uk/wp-content/uploads/WinBUGS_Vol3.pdf.

Spiegelhalter, D., Best, N. G., Carlin, B. P. and van der Linde, A. (2002). Bayesian Measures of Model Complexity and Fit. *Journal of the Royal Statistical Society, Series B* **64**, 583-640.

Stangl, D. K. (1995). Prediction and Decision Making Using Bayesian Hierarchical Models. *Statistics in Medicine* **14**, 2173-2190.

Steen, N. M., Byrne, G. D. and Gelbard, E. M. (1969). Gaussian Quadrature for the Integrals $\int_0^\infty f(x)dx$ and $\int_0^b \exp(-x^2)f(x)dx$. *Mathematics of Computation* **23**, 661-671.

Steffey, D. (1992). Hierarchical Bayesian Modeling With Elicited Prior Information. *Communications in Statistics* **21**, 799-821.

Stein, C. (1955). Inadmissability of the Usual Estimator for the Mean of a Multivariate Normal Distribution. *In Proceedings of the Third Berkeley Symposium on Mathematical Statistics and Probability*. **1** Berkeley: University of California Press, pp.197-206.

Stein, C. (1965). Approximation of Improper Prior Measures by Prior Probability Measures. In *Bernoulli-Bayes-Laplace Anniversary Volume: Proceedings of an International Research Seminar Statistical Laboratory*, J. Neyman and L. M. Le Cam (eds.). New York: Springer-Verlag, pp.217-240.

Stein, C. (1981). Estimation of the Mean of a Multivariate Normal Distribution. *Annals of Statistics* **9**, 1135-1151.

Stephens, D. A. (1994). Bayesian Retrospective Multiple-Changepoint Identification. *Applied Statistics* **43**, 159-178.

Stephens, M. (2000). Bayesian Analysis of Mixture Models with an Unknown Number of Components-An Alternative to Reversible Jump Methods. *Annals of Statistics* **28**, 40-74.

Stern, S. E. (1997). Simulation-Based Estimation. *Journal of Economic Literature* **XXXV**, 2006-2039.

Stern, H. S. and Cressie, N. (2000). Posterior predictive model checks for disease mapping models *Statistics in Medicine* **19**, 2377-2397.

Stewart, L. (1983). Bayesian Analysis Using Monte Carlo Integration-a Powerful Methodology for Handling Some Difficult Problems. *Statistician* **32**, 195-200.

Stewart, L. and Davis, W. W. (1986). Bayesian Posterior Distributions Over Sets of Possible Models With Inferences Computed by Monte Carlo Integration. *The Statistician* **35**, 175-182.

Stigler, S. M. (1982). Thomas Bayes' Bayesian Inference. *Journal of the Royal Statistical Society, Series A* **145**, 250-258.

Stigler, S. M. (1986). *The History of Statistics: The Measurement of Uncertainty before 1900*. Cambridge, MA: Harvard University Press.

Stigler, S. M. (1999). *Statistics on the Table: The History of Statistical Concepts and Methods*. Cambridge, MA: Harvard University Press.

Stone, M. (1974). Cross-Validatory Choice and Assessment of Statistical Predictions. *Journal of the Royal Statistical Society, Series B* **36**, 111-147.

Stone, M. (1977a). An Asymptotic Equivalence of Choice Model by Cross-Validation and Akaike's Criterion. *Journal of the Royal Statistical Association, Series B* **39**, 44-47.

Stone, M. (1977b). Comments On Model Selection Criteria of Akaike and Schwarz. *Journal of the Royal Statistical Society, Series B*, **41**, 276-278.

Stroud, A. H. (1971). *Approximate Calculation of Multiple Integrals*. Englewood Cliffs, NJ: Prentice-Hall.

Stroud, A. H. and Secrest, D. H. (1966). *Gaussian Quadrature Formulas.* Englewood Cliffs, NJ: Prentice-Hall.

Stuart, A. and Ord, J. K. (1994). *Kendall's Advanced Theory of Statistics: Volume I, Distribution Theory.* Sixth Edition. London: Edward Arnold.

Student, A. (1908a). On the Probable Error of a Mean. *Biometrika* **6**, 1.

Student, A. (1908b). On the Probable Error of a Correlation Coefficient. *Biometrika* **6**, 302.

Suchard, M. A., Weiss, R. E., Dorman, K. S. and Sinsheimer, J. S. (2003). Inferring Spatial Phylogenetic Variation along Nucleotide Sequences: A Multiple Changepoint Model. *Journal of the American Statistical Association* **98**, 427-437.

Sundberg, R. (1974). Maximum Likelihood Theory for Incomplete Data from an Exponential Family. *Scandinavian Journal of Statistics* **1**, 49-58.

Sundberg, R. (1976). An Iterative Method for Solution of the Likelihood Equations for Incomplete Data From Exponential Families. *Communications in Statistics, Part B, Simulation and Computation* **5**, 55-64.

Suppes, P. (1974). The Measurement of Belief. *Journal of the Royal Statistical Society, Series B* **36**, 160-191.

Swendsen, R. H. and Wang, J. S. (1987). Nonuniversal Critical Dynamics in Monte Carlo Simulations. *Physical Review Letters* **58**, 86-88.

Szu, H. and Hartley, R. (1987). Fast Simulated Annealing. *Physics Letters A* **122**, 157-162.

Tadikamalla, P. R. (1978). Computer Generation of Gamma Random Variables. *Communications of the Association for Computing Machinery* **21**, 419-422.

Tadikamalla, P. R. and Ramberg, J. S. (1975). An Approximate Method for Generating Gamma and Other Variates. *Journal of Statistical Computation and Simulation* **3**, 275-282.

Takeuchi, K. (1976). Distribution of Informational Statistics and A Criterion of Model Fitting. *Suri-Kagaku (Mathematical Statistics)* **153**, 12-18.

Tanner, M. A. (1996). *Tools for Statistic Inference: Methods for the Exploration of Posterior Distributions and Likelihood Functions.* New York: Springer.

Tanner, M. A. and Wong, W. H. (1987). The Calculation of Posterior Distributions by Data Augmentation. *Journal of the American Statistical Society* **82**, 528-550.

Taralsden, G. and Lindqvist, B. H. (2010). Improper Priors Are Not Improper. *The American Statistician* **64**, 154-158.

Thatcher, A. R. (1964). Relationships Between Bayesian and Confidence Limits for Predictions. *Journal of the Royal Statistical Society, Series B* **26**, 176-210.

Theil, H. (1963). On the Use of Incomplete Prior Information in Regression Analysis. *Journal of the American Statistical Association* **58**, 401-414.

Theil, H. (1970). On the Estimation of Relationships Involving Qualitative Variables. *American Journal of Sociology* **76**, 103-154.

Theil, H. and Goldberger, A. S. (1961). On Pure and Mixed Statistical Estimation in Economics. *International Economic Review* **2**, 65-78.

Thisted, R. (1988). *Elements of Statistical Computing.* New York: Chapman & Hall.

Thompson, B. (1996). AERA Editorial Policies Regarding Statistical Significance Testing: Three Suggested Reforms. *Educational Researcher* **25**, 26-30.

Thompson, B. (1997). Editorial Policies Regarding Statistical Significance Testing: Further Comments. *Educational Researcher* **26**, 29-32.

Thompson, B. (2002a). Statistical, Practical, and Clinical: How Many Kinds of Significance Do Councilers Need to Consider? *Journal of Counciling and Development* **80**, 64-71.

Thompson, B. (2002b). What Future Quantitative Social Science Research Could Look Like: Confidence Intervals for Effect Sizes. *Educational Researcher* **31**, 24-31.

Thompson, B. (2004). The "Significance" Crisis in Psychology and Education. *Journal of Socio-Economics* **33**, 607-613.

Thompson J. R., Palmer, T. M. and Moreno, S. (2006). Bayesian Analysis In Stata Using WinBUGS. *The Stata Journal* **6**, pp.530-549.

Tiao, G. C. and Zellner, A. (1964a). Bayes's Theorem and the Use of Prior Knowledge in Regression Analysis. *Biometrika* **51**, 219-230.

Tiao, G. C. and Zellner, A. (1964b). On the Bayesian Estimation of Multivariate Regression. *Journal of the Royal Statistical Society, Series B* **26**, 277-285.

Tierney, L. (1991). Exploring Posterior Distributions Using Markov Chains. In *Computing Science and Statistics: Proceedings of the 23rd Symposium on the Interface.* E. M. Keramidas (ed.). Fairfax Station, VA: Interface Foundation, pp.563-570.

Tierney, L. (1994). Markov Chains for Exploring Posterior Distributions. *Annals of Statistics* **22**, 1701-1728.

Tierney, L. (1996). Introduction to General State-Space Markov Chain Theory. In *Markov Chain Monte Carlo in Practice*, W. R. Gilks, S. Richardson and D. J. Spiegelhalter (eds.). New York: Chapman & Hall, pp.59-74.

Tierney, L. and Kadane, J. B. (1986). Accurate Approximations for Posterior Moments and Marginal Densities. *Journal of the American Statistical Association* **81**, 82-86.

Tierney, L., Kass, R. E. and Kadane, J. B. (1989a). Fully Exponential Laplace Approximations to Expectations and Variances of Nonsensitive Functions. *Journal of the American Statistical Association* **84**, 710-716.

Tierney, L., Kass, R. E. and Kadane, J. B. (1989b). Approximate Marginal Densities of Nonlinear Functions. *Biometrika* **76**, 425-433.

Tierney, L. and Mira, A. (1999). Some Adaptive Monte Carlo Methods for Bayesian Inference. *Statistics in Medicine* **18**, 2507-2515.

Titterington, D. M., Smith, A. F. M. and Makov, U. E. (1985). *Statistical analysis of finite mixture distributions.* New York: John Wiley & Sons.

Tobin, James. (1958). Estimation of Relationships for Limited Dependent Variables. *Econometrica* **26**, 24-36.

Tong, Y. L. (1990). *The Multivariate Normal Distribution.* New York: Springer-Verlag.

Toothill, J. P. R., Robinson, W. D. and Adams, A. G. (1971). The Runs Up and Down Performance of Tausworthe Pseudo-Random Number Generators. *Journal of the Association for Computing Machinery* **18**, 381-399.

Torrie, G. M. and Valleau, R. L. (1977). Nonphysical Sampling Distributions in Monte Carlo Free-Energy Estimation: Umbrella Sampling. *Journal of Computational Physics* **23**, 187-199.

Tsai, T-H. and Gill, J. (2012). superdiag: A Comprehensive Test Suite for Markov Chain Non- Convergence. *The Political Methodologist.* **19**, 12-18.

Tsitsiklis, J. N. (1988). A Survey of Large Time Asymptotics of Simulated Annealing Algorithms. In *Stochastic Differential Systems, Stochastic Control Theory and Applications*, W. Fleming and P. L. Lions (eds.). New York: Springer, pp.583-599.

Turkov, P., Krasotkina, O. and Mottl, V. (2012). The Bayesian Logistic Regression In Pattern Recognition Problems Under Concept Drift. 2012 21st International Conference on Pattern Recognition (ICPR). Tsukuba, Japan, 11-15 Nov. 2012, 2976-2979.

Turnbull, B. W. (1976). The Empirical Distribution with Arbitrary Grouped, Censored and Truncated Data. *Journal of the Royal Statistical Society, Series B* **38**, 290-295.

Tversky, A. (1974). Assessing Uncertainty. *Journal of the Royal Statistical Society, Series B* **36**, 148-159.

Tversky, A. and Kahneman, D. (1974). Judgment Under Uncertainty: Heuristics and Biases. *Science* **185**, 1124-1131.

Tweedie, R. L. (1975). Sufficient Conditions for Ergodicity and Recurrence of Markov Chains on a General State Space. *Stochastic Processes Applications* **3**, 385-403.

Ulam, S. M. (1961). On Some Statistical Properties of Dynamical Systems. *Proceedings of the 4th Berkeley Symposium on Mathematical Statistics and Probability* **3**, 315-320.

van Dijk, H. K. and Kloek, T. (1982). Monte Carlo Analysis of Skew Posterior Distributions: An Illustrative Econometric Example. *The Statistician* **32**, 216-223.

van Dyk, D. and Meng, X-L. (2001). The Art of Data Augmentation. *Journal of Computational and Graphical Statistics* **10**, 1-50.

van Houwelingen, H. C. (1977). Monotonizing Empirical Bayes Estimators for a Class of Discrete Distributions with Monotone Likelihood Ratio. *Statistica Neerlandica* **31**, 95-104.

van Houwelingen, H. C. and Stijnen, T. (1993). Monotone Empirical Bayes Estimators Based on More Informative Samples. *Journal of the American Statistical Association* **88**, 1438-1443.

van Laarhoven, P. J. M. and Aarts, E. H. L. (1987). *Simulated Annealing: Theory and Applications.* Reidel: Dordrecht.

van Laarhoven, P. J. M., Boender, P., Aarts, E. H. L. and Rinnooy, K. A. (1989). A Bayesian Approach to Simulated Annealing. *Probability in the Engineering and Informational Sciences* **3**, 453-475.

Vanpaemel, W. (2010). Prior Sensitivity In Theory Testing: An Apologia For the Bayes Factor. *Journal of Mathematical Psychology* **54**, 491-498.

Varian, H. R. (1975). A Bayesian Approach to Real Estate Assessment. In *Studies in Bayesian Econometrics and Statistics in Honor of Leonard J. Savage*, S.E. Fienberg and A. Zellner (eds.). North-Holland, Amsterdam, pp.195-208.

Venables, W. N. and Ripley, B. D. (1999). *Modern Applied Statistics with S-Plus*, Third Edition. New York: Springer-Verlag.

Venn, J. (1866). *The Logic of Chance.* London: Macmillan.

Verma, V., Gordon, G., Simmons, R. and Thrun, S. (2004). Real-time Fault Diagnosis. *Robotics & Automation Magazine, IEEE* **11**, 56-66.

Verma, V., Langford, J. and Simmons, R. (2001). Non-Parametric Fault Identification for Space Rovers. *International Symposium on Artificial Intelligence and Robotics in Space.*

Verma, V., Thrun, S. and Simmons, R. (2003). Variable Resolution Particle Filter. *Proceedings of the International Joint Conference on Artificial Intelligence*, AAAI, August.

Vidakovic, B. (1999). Linear Versus Nonlinear Rules for Mixture Normal Priors. *Annals of the Institute of Statistical Mathematics* **51**, 111-124.

Villegas, C. (1977), On the Representation of Ignorance. *Journal of the American Statistical Association* **72**, 651-654.

Vines, S. K., Gilks, W. R. and Wild, P. (1996). Fitting Bayesian multiple random effects models. *Statistics and Computing* **6**, 337-346.

Volpe, R., Nesnas, I. A. D., Estlin, T., Mutz, D., Petras, R. and Das, H. (2001). The CLARAty Architecture for Robotic Autonomy. *Proceedings of the 2001 IEEE Aerospace Conference*, Big Sky, Montana, March 10-17.

Volpe, R. and Peters, S. (2003). Rover Technology Development and Infusion for the 2009 Mars Science Laboratory Mission. *Proceedings of 7th International Symposium on Artificial Intelligence, Robotics, and Automation in Space*, Nara, Japan, May 19-23.

Von Mises, R. (1957). *Probability, Statistics and Truth.* Second Revised English Edition prepared by H. Geiringer. London: George Allen and Unwin.

von Neumann, J. (1951). Various Techniques Used in Connection with Random Digits, "Monte Carlo Method." *U.S. National Bureau of Standards Applied Mathematics Series* **12**, 36-38.

Vuong, Q. H. (1989). Likelihood Ratio Tests for Model Selection and Non-Nested Hypotheses. *Econometrica* **57**, 307-333.

Waagepetersen, R. and Sorensen, D. (2001). A Tutorial on Reversible Jump MCMC with a View toward Applications in QTL-Mapping. *International Statistical Review/Revue Internationale de Statistique* **69**, 49-61.

Wagner, K. and Gill, J. (2005). Bayesian Inference in Public Administration Research: Substantive Differences from Somewhat Different Assumptions. *International Journal of Public Administration* **28**, 5-35.

Wakefield, J. C., Gelfand, A. E. and Smith, A. F. M. (1991). Efficient Generation of Random Variates via the Ratio-of-Uniforms Method. *Statistics and Computing* **1**, 129-133.

Wakefield, J. C., Smith, A. F. M., Racine-Poon, A. and Gelfand, A. E. (1994). Bayesian Analysis of Linear and Non-linear Population Models by using the Gibbs Sampler. *Applied Statistics* **43**, 201-221.

Walker, A. J. (1974). Fast Generation of Uniformly Distributed Pseudorandom Numbers with Floating-Point Representation. *Electronics Letters* **10**, 533-534.

Wallace, N. D. (1974). Computer Generation of Gamma Random Variables with Non-Integral Shape Parameters. *Communications of the Association for Computing Machinery* **17**, 691-695.

Walley, P., Gurrin, L. and Burton, P. (1996). Analysis of Clinical Data Using Imprecise Prior Probabilities. *Statistician* **45**, 457-485.

Wasserman, L. (1992). Recent Methodological Advances in Robust Bayesian Inference. In *Bayesian Statistics 4*, J. O. Berger, J. M. Bernardo, A. P. Dawid and A. F. M. Smith (eds.). Oxford: Oxford University Press, pp.763-773.

Wedderburn, R. W. M. (1974). Quasi-Likelihood Functions, Generalized Linear Models, and the Gauss-Newton Method. *Biometrika* **61**, 439-447.

Wedderburn, R. W. M. (1976). On the Existence and Uniqueness of the Maximum Likelihood Estimates for Certain Generalized Linear Models. *Biometrika* **63**, 27-32.

Wei, G. C. G. and Tanner, M. A. (1990). A Monte Carlo Implementation of the EM Algorithm and the Poor Man's Data Augmentation Algorithm. *Journal of the American Statistical Association* **85**, 699-704.

Weiss, L. (1961). *Statistical Decision Theory*. New York: McGraw-Hill.

Weiss, R. E. (1996). An Approach to Bayesian Sensitivity Analysis. *Journal of the Royal Statistical Society, Series B* **58**, 739-750.

Wenzel, W. and Hamacher, K. (1999). A Stochastic Tunneling Approach for Global Minimization. *Physical Review Letters* **82**, 3003-3007.

West, M. (1984). Outlier Models and Prior Distributions in Bayesian Linear Regression. *Journal of the Royal Statistical Society, Series B* **46**, 431-439.

West, M. (1992). Modelling With Mixtures. In *Bayesian Statistics 4*, J. O. Berger, J. M. Bernardo, A. P. Dawid and A. F. M. Smith (eds.). Oxford: Oxford University Press, pp.227-246.

West, M. (1993). Approximating Posterior Distributions by Mixtures. *Journal of the Royal Statistical Society, Series B* **55**, 409-422.

West, M. and Harrison, J. (1997). *Bayesian Forecasting and Dynamic Models.* New York: Springer-Verlag.

West, M. and Turner, D. A. (1994). Deconvolution of Mixtures in Analysis of Neural Synaptic Transmission. *Statistician* **43**, 31-43.

Western, B. (1995). A Comparative Study of Working-Class Disorganization: Union Decline in Eighteen Advanced Capitalist Countries. *American Sociological Review* **60**, 179-201.

Western, B. (1998). Causal Heterogeneity in Comparative Research: A Bayesian Hierarchical Modelling Approach. *American Journal of Political Science* **42**, 1233-1259.

Western, B. (1999). Bayesian Methods for Sociologists: An Introduction. *Sociological Methods & Research* **28**, 7-34.

Westlake, W. J. (1967). A Uniform Random Number Generator Based on the Combination of Two Congruential Generators. *Journal of the Association for Computing Machinery* **14**, 337-340.

Wetzels, R., Raaijmakers, J. G. W., Jakab, E. and Wagenmakers, E-J. (2009). How to Quantify Support For and Against the Null Hypothesis: A Flexible WinBUGS Implementation of a Default Bayesian t-Test. *Psychonomic Bulletin & Review* **16**, 752-760.

Wheeler, D. J. (1974). Simulation of Arbitrary Gamma Distributions. *IEEE Transactions* **6**, 167-169.

Wheeler, D. J. (1975). An Approximation for Simulation of Gamma Distributions. *Journal of Statistical Computation and Simulation* **3**, 225-232.

White, H. (1982). Maximum Likelihood Estimation of Misspecified Models. *Econometrica* **50**, 1-26.

White, H. (1996). *Estimation, Inference and Specification Analysis.* Cambridge: Cambridge University Press.

Whittaker, J. (1974). Generating Gamma and Beta Random Variables with Non-Integral Shape Parameters. *Applied Statistics* **23**, 210-214.

Whittaker, J. (1990). *Graphical Models in Applied Mathematical Multivariate Analysis.* New York: John Wiley & Sons.

Whittlesey, J. R. B. (1969). On the Multidimensional Uniformity of Pseudo-Random Generators. *Communications of the Association for Computing Machinery* **12**, 247.

Wiles, P. (1965). Rationalizing the Russians. *New York Review of Books* (October, 28) **5**, 37-38.

Wilkinson, G. N. (1977). On Resolving the Controversy in Statistical Inference. *Journal of the Royal Statistical Society, Series B* **39**, 119-171.

Winkler, R. L. (1967). The Assessment of Prior Distributions in Bayesian Analysis. *Journal of the American Statistical Association* **62**, 776-800.

Wolfson, L. J., Kadane, J. B. and Small, M. J. (1996). Bayesian Environmental Policy Decisions: Two Case Studies. *Ecological Applications* **6**, 1056-1066.

Wong, G. Y. and Mason, W. M. (1985). The Hierarchical Logistic Regression Model for Multilevel Analysis. *Journal of the American Statistical Association* **80**, 513-524.

Wong, G. Y. and Mason, W. M. (1991). Contextually Specific Effects and Other Generalizations of the Hierarchical Linear Model for Comparative Analysis. *Journal of the American Statistical Association* **86**, 487-503.

Wong, W. H. and Li, B. (1992). Laplace Expansion for Posterior Densities of Nonlinear Functions of Parameters. *Biometrika* **79**, 393-398.

Woodward, Phil. (2011). Bayesian Analysis Made Simple: An Excel GUI for WinBUGS. Boca Raton: Chapman & Hall/CRC.

Wright, G. and Ayton, P. (1994). *Subjective Probability.* New York: John Wiley & Sons.

Wu, C. F. J. (1983). On the Convergence Properties of the EM Algorithm. *Annals of Statistics* **11**, 95-103.

Yang, T. Y. and Kuo, L. (2001). Bayesian Binary Segmentation Procedure For a Poisson Process With Multiple Changepoints. *Journal of Computational and Graphical Statistics* **10**, 772-785.

Young, C. (2009). Model Uncertainty in Sociological Research: An Application to Religion and Economic Growth. *American Sociological Review* **74**, 380-397.

Yu, B. and Mykland, P. (1998). Looking at Markov Samplers Through CUSUM Path Plots: A Simple Diagnostic Idea. *Statistics and Computing* **8**, 275-286.

Zabell, S. (1989). R. A. Fisher on the History of Inverse Probability. *Statistical Science* **4**, 247-263.

Zangwill, W. I. (1969). *Nonlinear Programming: A Unified Approach.* Englewood Cliffs, NJ: Prentice-Hall.

Zellner, A. (1971). *An Introduction to Bayesian Inference in Econometrics.* New York: Wiley & Sons.

Zellner, A. (1975). Bayesian Analysis of Regression Error Terms. *Journal of the American Statistical Association* **70**, 138-144.

Zellner, A. (1976). Bayesian and Non-Bayesian Analysis of the Regression Model with Multivariate Student-t Error Terms. *Journal of the American Statistical Association* **71**, 400-405.

Zellner, A. (1985). Bayesian Econometrics. *Econometrica* **53**, 253-269.

Zellner, A. and Chetty, V. K. (1965). Prediction and Decision Problems in Regression Models from the Bayesian Point of View. *Journal of the American Statistical Association* **60**, 608-616.

Zellner, A. and Highfield, R. A. (1988). Calculation of Maximum Entropy Distributions and Approximation of Marginal Posterior Distributions. *Journal of Econometrics* **37**, 195-209.

Zellner, A. and Min, C-K. (1995). Gibbs Sampler Convergence Criteria. *Journal of the American Statistical Association* **90**, 921-927.

Zellner, A. and Moulton, B. R. (1985). Bayesian Regression Diagnostics with Applications to International Consumption and Income Data. *Journal of Econometrics* **29**, 187-211.

Zellner, A. and Rossi, P. E. (1984). Bayesian Analysis of Dichotomous Quantal Response Models. *Journal of Econometrics* **25**, 365-393.

Zellner, A. and Siow, A. (1980). Posterior Odds Ratios for Selected Regression Hypotheses. In *Bayesian Statistics*, J. M. Bernardo, M. H. DeGroot, D. V. Lindley and A. F. M. Smith (eds.). Valencia: Valencia University Press, pp.586-603.

Zellner, A. and Tiao, G. C. (1964). Bayesian Analysis of the Regression Model With Autocorrelated Errors. *Journal of the American Statistical Association* **59**, 763-778.

Zhang, W. and Luck, S. J. (2011). The Number and Quality of Representations in Working Memory *Psychological Science* **22**, 1434-1441.

Zhang, Z., Chan, K. L., Wu, Y. and Chen, C. (2004). Learning a Multivariate Gaussian Mixture Model With the Reversible Jump MCMC Algorithm. *Statistics and Computing* **14**, 343-355.

Zhao, X. and Chu, P. S. (2006). Bayesian Multiple Changepoint Analysis of Hurricane Activity in the Eastern North Pacific: A Markov Chain Monte Carlo Approach. *Journal of Climate.* **19**, 4893-4901.

Zhenting, H. and Qingfeng, G. (1978). *Homogeneous Denumerable Markov Chains.* Berlin: Springer-Verlag Science Press.

Zhou, X. and Reiter, J. P. (2010). A Note on Bayesian Inference After Multiple Imputation. *The American Statistician* **64**, 159-163.

Ziliak, S. T. and McCloskey, D. N. (2007). *The Cult of Statistical Significance: How the Standard Error Costs Us Jobs, Justice, and Lives.* Ann Arbor: University of Michigan Press.

Zuur, G., Garthwaite, P. H. and Fryer, R. J. (2002). Practical Use of MCMC Methods: Lessons from a Case Study. *Biometrical Journal* **44**, 433-455.

Author Index

Subject Index

Printed in the United States
by Baker & Taylor Publisher Services

Printed in the United States
by Baker & Taylor Publisher Services